Arsenic Contamination in the World

Arsenic Contamination in the World
An International Sourcebook 2012

Susan Murcott

Department of Civil and Environmental Engineering,
Massachusetts Institute of Technology

IWA
Publishing
London • New York

Published by

IWA Publishing
Alliance House
12 Caxton Street
London SW1H 0QS, UK
Telephone: +44 (0)20 7654 5500
Fax: +44 (0)20 7654 5555
Email: publications@iwap.co.uk
Web: www.iwapublishing.com

First published 2012

British Library Cataloguing in Publication Data
A CIP catalogue record for this book is available from the British Library

ISBN 9781780400389 (Paperback)
ISBN 9781780400396 (eBook)

Printed by Bell & Bain Ltd., Glasgow

FSC www.fsc.org MIX Paper from responsible sources FSC® C007785

Contents

Types of arsenic contamination and number of occurrences	
A = Anthropogenic-related arsenic	54 countries/territories
C = Coal-related arsenic	28 countries/territories
G = Geogenic-related arsenic	68 countries/territories
M = Mining-related arsenic	74 countries/territories
P = Petroleum-related arsenic	17 countries/territories
V = Volcanogenic-related arsenic	35 countries/territories

List of Figures

Executive Summary

Section 2. Region of the Americas

Section 3. Asian Region

Section 4. European Region

Section 5. Region of Australia and Oceania

List of Tables

Executive Summary

Section 1. African Region

Section 2. Region of the Americas

Section 3. Asian Region

Section 4. European Region

Section 5. Region of Australia and Oceania

List of Acronyms and Abbreviations

AAS	atomic absorption spectrometry
AIDS	acquired immunodeficiency syndrome
AMD	acid mine drainage
ANZECC	Australia and New Zealand Environment Conservation Council
As	arsenic
BP	before present
CI	confidence interval
CSIRO	Commonwealth Scientific and Industrial Research Organisation (Australia)
DACAAR	Danish Committee for Aid to Afghan Refugees
DDT	dichlorodiphenyltrichloroethane
DMA	dimethylarsinic acid
Eh	oxidation–reduction potential (in mV)
ENPHO	Environment and Public Health Organization
EU	European Union
FINNIDA	Finnish International Development Agency
GDP	gross domestic product
GIS	geographic information system
HDI	Human Development Index
HIV	human immunodeficiency virus
ICP-MS	inductively coupled plasma mass spectrometry
MMA	monomethylarsonic acid
OECD	Organisation for Economic Co-operation and Development
PAH	polycyclic aromatic hydrocarbon
ppb	parts per billion ($\times 10^9$) (= μg/kg or μg/l)
ppm	parts per million (= mg/kg or mg/l)
SMR	standardized mortality ratio
TDS	total dissolved solids
UNICEF	United Nations Children's Fund
USA	United States of America

USAID	United States Agency for International Development
USEPA	United States Environmental Protection Agency
USGS	United States Geological Survey
WB	World Bank
WHO	World Health Organization

Acknowledgments

This book is a small attempt to reduce suffering from arsenic contamination in water and food by providing detailed and transparent information. It would not have been possible without the contributions of all the authors cited in this book and without their advancement of our understanding of arsenic contamination in the environment. I would like to first acknowledge and honor inspirational arsenic experts, scientists, engineers and activists whom I have had the privilege to know, and who are raising awareness of arsenic contamination and bringing medical care and/or safe drinking water and arsenic-safe food solutions to those who need it, including: Roshan Shrestha, Arinita Masky, Tommy Na Kit Ngai, Yukie Mato, D.H.Guha Mazumder, Willard Chappell, Dipankar Chakraborti, Allen and Meera Smith, Mickey Sampson, Richard Wilson, Anirban Gupta, Arup Sen Gupta and Rick Johnson, to name a few. Other arsenic experts were generous with their help at times when I most needed it including Jan Rosenboom, Dave Polya, Michael Berg, Alan Welch, Jochen Bundschuh, Paul Westerhoff, Hans Rudolf-Pfeifer and Ravi Naidu.

This work is an expansion of the report prepared in 2006 by the author while a consultant to the Conrad N. Hilton Foundation. The Conrad N. Hilton Foundation deserves praise for their role in support of this original study and for their commitment to global safe water. I would like to extend thanks to the World Health Organization (WHO) for their instrumental role in providing editorial assistance for the first version of the complete monograph including Jamie Bartram, Robert Bos, Bruce Gordon, Jennifer de France, Federico Properzi and Maria Sheffer. Appreciation is also extended to Sergio Gelli for coordinating this effort within WHO, to Giuliana Profeti for her excellent work on updating the maps, and to the regional offices for coordinating feedback in 2009–2010. Special thanks goes to Terrence Thompson, Senior Environmental Health Adviser, WHO, Office of the Representative to Nepal, for his cogent questions, for presentation of some key findings from this work at the International Conference on Water Quality with special reference to Arsenic (Kolkata, India,18-20 February 2011) and for his continuing commitment within the WHO to the arsenic issue.

Also, appreciation goes to my MIT colleagues, Eric Adams, Pete Shanahan, Harry Hemond and to all my MIT students, especially those who have devoted their Masters of Engineering thesis studies to arsenic detection and treatment solutions, including Jessie Hurd, Tommy Na Kit Ngai, Patricia Halsey, Barika Poole, Georges Tabbal, Soon Kyu Hwang, Claudia Espinoza and Maclyn O'Donnell. Many dear colleagues from Environment and Public Health Organization in Nepal aided and taught my students and I a great deal about arsenic, especially Bipin Dangol, Raju Shrestha, Bhushan Tuladhar and Makhan Maharjan. I am grateful to IWA Publishing for their commitment to bringing this book to the public. Particular thanks goes to Maggie Smith and Michelle Jones.

Finally, this work would not have been possible without the love and support of my beloved husband, Ralph J. Coffman Jr., who travelled with me to a number of arsenic-affected countries and communities, who checked innumerable references and who cooked my meals during the long years it has taken to get this book to the public.

All errors are my own. Corrections and/or additional examples of scientific studies of arsenic contamination in drinking-water, the food chain and the environment to update this *Sourcebook* are most welcome and should be sent to murcott@mit.edu.

Executive Summary

A = Anthropogenic-related arsenic: 54 countries/territories

Anthropogenic sources of arsenic, according to the classification scheme of this book, are those created by human intervention, exclusive of mining or petroleum extraction, which are addressed in separate categories below. Arsenic compounds number in the thousands, with arsenopyrite being the most abundant ore mineral of arsenic. Anthropogenic sources of arsenic in ground waters and surface waters are related to the use of arsenic compounds in a wide variety of applications, including wood preservatives, agricultural chemicals, glass, nonferrous alloys, electronics, chemical munitions and more. Anthropogenic arsenic is obtained from copper, gold and lead smelter dust as well as from roasting. Contamination can occur either at the site of manufacture or at the sites of implementation and use. The incineration of materials containing arsenic, including coals, is here categorized as anthropogenic. Burning high-arsenic coals either in industrial facilities, power generation utilities or in enclosed domestic spaces releases toxic fumes and ashes, which can be richer in arsenic than the naturally occuring coals themselves.

C = Coal-related arsenic (geogenic coal, excluding coal-burning): 28 countries/territories

Arsenic in coal can be released naturally into groundwater and surface water through weathering processes, in glacio-fluvial deposits with underlying arsenic-rich coal-bearing units below the aquifers, or in mining wastes and mines. As the economic feasibility for mining low-grade coals and lignite becomes more attractive, these additional arsenical sources will proliferate, since they frequently contain high concentrations of arsenic that are mobilized in even slightly decreased pH environments.

G = Geogenic-related arsenic: 68 countries/territories

Naturally occurring (also known as "geogenic") arsenic in concentrations above 10 μg/l in groundwater is most commonly released from iron oxide. Elevated arsenic concentrations in aquifers are caused by reducing conditions in which arsenic sorbs to metallic oxides such as iron (III). However, in anaerobic conditions, iron (III)-reducing microorganisms can couple the reduction of solid-phase iron (III) with the oxidation of organic matter, releasing arsenic to groundwater. This varies regionally due to geological and climatological conditions through reaction of iron oxide with either naturally occurring carbon or alkaline groundwater in felsic volcanic rocks and alkaline aquifers.

Arsenic may also be released into marine estuarine ecosystems. In estuarine systems, arsenic(V) is detectable for most of the year using the usual laboratory hydride generation technique, which essentially transforms all arsenic species into arsenate. However, in the spring in the northern hemisphere, ultraviolet radiation intensifies, estuarine temperatures rise to about 12°C and biological activity is elevated; under these conditions, forms of arsenic occur that have not been transformed into arsenate (possibly as quaternary arsonium compounds or arsenosugars), and these cannot be detected by the hydride method.

M = Mining-related arsenic (exclusive of coals): 74 countries/territories

All arsenic-containing minerals, especially sulfides, which are frequently associated with gold ores, are a source of arsenic. Mining and smelting of these minerals create environmental hazards of arsenic leaching into groundwater and surface waters from slag pits, waste dumps, extraction basins and derelict mines.

P = Petroleum-related arsenic: 17 countries/territories

Petroleum hydrocarbon compounds are a source of arsenic release. In anaerobic conditions, iron(III)-reducing microorganisms can couple the reduction of solid-phase iron(III) with the oxidation of anthropogenic carbon, such as petroleum hydrocarbons, and release arsenic into the groundwater – a mechanism of arsenical release similar to that induced by naturally occurring carbon.

V = Volcanogenic-related arsenic: 35 countries/territories

Arsenic is a ubiquitous component of active and fossil geothermal systems together with other contaminants including mercury, antimony, selenium, thallium, boron, lithium, fluoride and hydrogen sulphide. Geothermal hot springs can be either naturally occurring or anthropogenically manipulated as sources of thermal and electrical power, and are potential sources of arsenic release.

Introduction

This Executive Summary is an overview of the author's research, undertaken from January 2005 through to March 2012, and refers to cited arsenic incidences in drinking-water and the environment, by country/territory and by category: anthropogenic, coal, geogenic, mining, petroleum and volcanogenic. These arsenic incidences deserve close attention, as remediation projects are being planned and implemented to meet the World Health Organization's (WHO) guideline for arsenic in drinking-water or tolerable intake levels for arsenic contaminated food. By identifying the global extent of arsenic, country-by-country, we hope to bring the full impact of its adverse health effects into international perspective and better enable timely remediation.

Annex 1 is available on the IWA Water Wiki as an ancillary summary database. Please visit http://www.iwawaterwiki.org/xwiki/bin/view/Articles/ExecutiveSummaryofArsenicContaminationintheWorld. For each identified country in Table 1, Annex 1 includes the area affected, the population of that area, the types and levels of arsenic exposure, populations exposed and references that support these data.

Arsenic Exposure

Arsenic is one of the most toxic elements on earth and is listed as the third most toxic substance in the US Toxic Substances and Disease Registry, after lead and mercury. Arsenic has been used therapeutically for centuries and it has also been used as a poison. Today, human exposure to arsenic occurs primarily by drinking of arsenic contaminated water, ingestion of food prepared with or grown in arsenic contaminated water, or eating food crops irrigated with high-arsenic water. To a lesser extent people are exposed to arsenic through inhalation or minimally from dermal contact. There are major negative health effects of arsenic exposure, yet there is considerable uncertainty regarding the exact numbers of people exposed to arsenic worldwide. Moreover, the effect of exposure to arsenic differs in individuals, population groups and geographic environments.

In this book, in identifying arsenic contamination's impact, we have followed the WHO definitions of hazard and risk (WHO, 2011a) and arsenic tolerable intake levels and guideline values (WHO, 2011a, b, FAO/WHO, 2010):

(1) A hazard is "a biological, chemical, physical or radiological agent that has the potential to cause harm."
(2) A risk is "the likelihood of identified hazards causing harm in exposed populations in a specific time frame, including the magnitude of that hard and/or the consequences."

Arsenic is a hazard. People exposed to arsenic above the guideline values over an extended period of time are at significant risk. People at risk from chronic arsenic poisoning (arsenicosis) are a subset of the people exposed. Depending on the level of exposure, diseases can include skin lesions, hyperkeratosis, melanosis, skin cancer, peripheral neuropathy, gastrointestinal disorders, diabetes, renal system dysfunction, cardiovascular disease and cancers of internal organs. Determining the at-risk population involves multiple judgments based on a knowledge of the boundaries of the affected areas, the percentage of water sources affected, the concentration of arsenic in the water source(s), the amount of water consumed, the amount of arsenic-contaminated food ingested, the arsenic content in that food and the duration of exposure, as well as many other factors.

World Health Organization (WHO) arsenic guidelines

Tolerable intake level

In a review of the latest scientific evidence conducted in 2010, the Joint Food and Agriculture Organization of the United Nations (FAO)/WHO Expert Committee on Food Additives (JECFA) determined the lower limit on the benchmark dose for a 0.5% increased incidence of lung cancer ($BMDL_{0.5}$) from epidemiological data to be 3.0 μg/kg body weight per day (2–7 μg/kg body weight per day based on the range of estimated total dietary exposure). The Committee noted that the previously established provisional tolerable weekly intake (PTWI) of 15 μg/kg body weight (equivalent to 2.1 μg/kg body weight per day) for inorganic arsenic was in the region of the $BMDL_{0.5}$ and therefore was no longer appropriate. This PTWI was therefore withdrawn by the Committee (WHO, 2004b). No new tolerable intake level could be established. In areas where levels in water are below the WHO drinking-water guideline value, human health effects are unlikely.

Drinking-water

10 μg/l (provisional guideline value, in view of scientific uncertainties surrounding the risk assessment for arsenic carcinogenicity) (WHO, 2008).

Air

A safe level of arsenic in air cannot be established (WHO, 2000).

Arsenic exposure is monitored and quantified through analysis of blood, urine and/or hair or nail samples. Hair sample of 0.1–0.5 mg/kg imply chronic arsenic poisoning. Chronic arsenic toxicity has non-specific symptoms at the onset. Arsenic exposure over time leads to multi-system disease which varies among individuals, populations and geographies.

Acute arsenic toxicity typically occurs accidentally or in suicide cases. Lethal doses in acute arsenic poisoning vary from 10–300 mg (Schoolmeester & White, 1980).

By **exposed population**, we mean exposure to arsenic from any of the several exposure routes.

For exposure through ingestion of drinking water, **exposed population** means people exposed to drinking water at either 10 or 50 micrograms per liter (μg/l). The 10 μg/l guideline value for arsenic was adopted by the WHO in 1993. The US EPA lowered its arsenic value from 50 to 10 μg/l in 2001. We have sought published sources of exposed populations.

Contamination of food can occur from either organic or inorganic arsenic that enters the food chain through plants or animals. Populations can be exposed to arsenic in the food chain, for example from eating food grown in arsenic contaminated soil or irrigated with arsenic contaminated water. The US EPA has set a limit of 10 mg/kg of arsenic in soil. Arsenic intake can be higher from food than from drinking water (Thomas, 1999). It has been estimated that, based on daily consumption of 1 kg of vegetables, rice and wheat with an arsenic concentration of 0.1 mg/kg food, the intake of arsenic would be 100 μg (1000 g x 0.1 μg/g) (Ratnaike, 2006). Many issues related to arsenic in the food chain still need further research and clarification.

For exposure through ingestion of arsenic contaminated food, **exposed population** means people exposed to food above the WHO tolerable intake level of 3 μg/kg body weight per day (2–7 μg/kg body weight per day, based on the range of estimated total dietary exposure).

In this book, the cited estimates for population exposed to arsenic contamination in a given country are, in most instances, taken from the published sources. In all instances, citations are given. When citations do not provide an estimate of the exposed population in a given country or area, the most current demographic statistics and the area affected (in km^2) have been used to make a calculation. When a given country has multiple regions and citations of populations exposed to arsenic, we have done our best to add separate regions to come up with a total for the country or to make a sound judgment as to the best estimate. We have provided all our citations, so others can improve on this work. The author welcomes more accurate data on populations exposed to arsenic contamination for inclusion in future revisions of this work; please send to: murcott@mit.edu.

Country and Arsenic Source Summaries

Figure 1 is a global map of the 105 countries including 2 territories (Guam and the Virgin Islands of the United States, hereafter considered "countries" for the purposes of this Executive Summary) identified in this volume as having one or more sources of arsenic in drinking-water or the environment. Our research leads to a global estimate of **226 million** people exposed to arsenic contamination from drinking-water or food. This estimate is a substantial increase over most previous global estimates, but it is still only an estimate. Better data will help to periodically update this estimate.

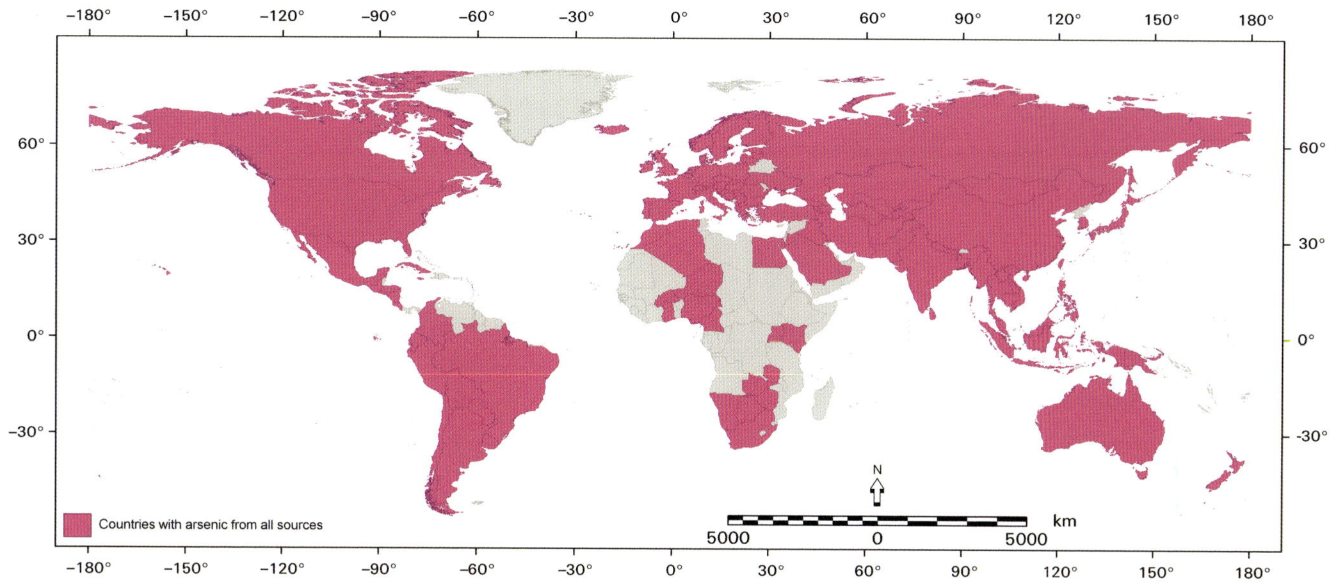

Figure 1. Global map of 105 countries/territories with exposure to arsenic from all sources: anthropogenic, coal, geogenic, mining, petroleum and volcanogenic.

This Executive Summary is a short synopsis of arsenic contamination by source, by region and by major arsenic-affected countries. References and further details are provided in the main body of this work.
Table 1 summarizes the exposed populations, the current arsenic standards and the sources of arsenic for each of the countries considered in this volume, by region.

Table 1. Summary of critical country information, by region.

Chapter/ Section No.	Country/Territory	Population Exposed Estimate (1000s)	Arsenic Standard (ug/l)	A	C	G	M	P	V
1.	**African Region**								
1.1	Algeria	n.a.	50				1		
1.2	Botswana	100.0	50			1	1		
1.3	Burkina Faso	200.0	50			1	1		
1.4	Cameroon	11.9	50				1		
1.5	Egypt	n.a.	50			1			
1.6	Ghana	291.5	10			1	1	1	
1.7	Kenya	n.a.	50				1		
1.8	Morocco	164.6	50				1		
1.9	Namibia	n.a.	50				1		
1.10	Niger	3,375.0	50			1	1		
1.11	Nigeria	11,000.0	50	1	1	1	1	1	1
1.12	South Africa	37.0	10	1	1		1		
1.13	Uganda	n.a.	n.a				1		
1.14	Zambia	2.9	50			1	1		

(*Continued*)

Table 1. Summary of critical country information, by region (*Continued*).

Chapter/ Section No.	Country/Territory	Population Exposed Estimate (1000s)	Arsenic Standard (ug/l)	A	C	G	M	P	V
1.15	Zimbabwe	9,300.0	50			1	1		
	1. Subtotal	**24,482.9**		**2**	**2**	**8**	**14**	**2**	**1**
2.	**Region of the Americas**								
2.1	Argentina	2,000.0	10			1			
2.2	Bolivia (Plurinational State)	1,285.0	10			1	1		
2.3	Brazil	377.7	50	1	1	1	1		
2.4	Canada	n.a.	10	1	1	1	1	1	
2.5	Chile	500.0	10			1	1		
2.6	Colombia	180.0	n.a			1	1		1
2.7	Costa Rica	15.4	n.a						1
2.8	Cuba	110.0	n.a	1			1		
2.9	Dominica	30.0	10						1
2.10	Ecuador	n.a.	n.a.					1	
2.11	El Salvador	n.a.	n.a.						1
2.12	Guatemala	n.a.	n.a	1		1	1		
2.13	Honduras	n.a.	n.a				1		1
2.14	Jamaica	n.a.	n.a.						1
2.15	Mexico	2,000.0	25	1		1	1		1
2.16	Nicaragua	3,000.0	50			1			1
2.17	Paraguay	n.a.	n.a.				1		
2.18	Peru	250.0	n.a.			1	1		1
2.19	United States of America	9,058.6	10	1	1	1	1	1	1
2.20	Uruguay	46.0	n.a	1		1			1
2.21	Virgin Islands of the United States (territory of the United States of America)	n.a.	10			1			
	2. Subtotal	**18,852.7**		**7**	**3**	**13**	**12**	**3**	**11**
3.	**Asian Region**								
3.1	Afghanistan	500.0	10			1	1	1	
3.2	Armenia	n.a.	n.a			1	1	1	
3.3	Azerbaijan	n.a.	n.a				1	1	
3.4	Bangladesh	57,000.0	50	1		1			
3.5	Cambodia	320.0	10			1			
3.6	China	8,237.0	50	1	1	1	1	1	
3.7	Georgia	115.0	50	1			1	1	
3.8	India	78,717.8	50		1	1	1		
3.9	Indonesia	0.3	50			1	1		
3.10	Iran	15.0	50			1	1	1	
3.11	Iraq	n.a.	n.a	1				1	
3.12	Japan	n.a.	n.a	1	1	1	1		1
3.13	Jordan	2,100.0	50			1			
3.14	Kazakhstan	n.a.	n.a	1	1	1	1	1	
3.15	Korea	n.a.	n.a				1		
3.16	Kyrgyzstan	11,342.0	n.a				1		

(*Continued*)

Table 1. Summary of critical country information, by region (*Continued*).

Chapter/ Section No.	Country/Territory	Population Exposed Estimate (1000s)	Arsenic Standard (ug/l)	A	C	G	M	P	V
3.17	Laos	400.0	50			1			
3.18	Malaysia	n.a.	n.a				1		
3.19	Mongolia	200.0	50		1	1	1		
3.20	Myanmar	3,400.0	50			1			
3.21	Nepal	300.0	50			1			
3.22	Pakistan	128.0	50			1			
3.23	Philippines	80.0	10			1	1		1
3.24	Russia	n.a.	n.a.	1	1		1		1
3.25	Saudi Arabia	3,000.0	n.a.			1		1	
3.26	Sri Lanka	n.a.	n.a	1		1			
3.27	Tajikistan	n.a.	n.a		1		1		
3.28	Thailand	268.0	50	1		1	1		
3.29	Turkey	1,000.0	10		1	1	1		1
3.30	Turkmenistan	n.a.	n.a	1					
3.31	Uzbekistan	n.a.	n.a	1		1	1		
3.32	Vietnam	7,000.0	50			1	1		
	3. Subtotal	**174,123.1**		**11**	**8**	**23**	**21**	**9**	**4**
4.	**European Region**								
4.1	Albania	n.a.	n.a	1		1	1		1
4.2	Austria	n.a.	10	1		1	1		1
4.3	Belgium	n.a.	10	1		1	1		
4.4	Bosnia and Herzegovina	n.a.	n.a	1		1	1		1
4.5	Bulgaria	n.a.	n.a	1	1		1		
4.6	Croatia	n.a.	n.a.						1
4.7	Czech Republic	n.a.	n.a.	1	1	1	1		
4.8	Denmark	n.a.	10	1	1				
4.9	Estonia	n.a.	50	1		1			
4.10	Finland	140.0	10	1		1	1		
4.11	France	n.a.	10	1		1	1		1
4.12	Germany	17.3	10	1	1	1	1		
4.13	Greece	40.8	10	1	1	1	1		
4.14	Hungary	1,541.0	10	1		1	1		1
4.15	Iceland	n.a.	10	1					1
4.16	Ireland	n.a.	10	1	1		1		
4.17	Italy	n.a.	10	1	1	1	1		1
4.18	Latvia	n.a.	50	1					
4.19	Lithuania	n.a.	50	1					
4.20	Macedonia (former Yugoslav Republic of)	327.0	50	1		1	1		1
4.21	Netherlands	n.a.	10	1			1		
4.22	Norway	2.1	10	1			1		
4.23	Poland	n.a.	n.a.	1	1	1	1		1
4.24	Portugal	2,800.0	10	1	1	1	1	1	1
4.25	Romania	14.0	50	1		1		1	1

(*Continued*)

Table 1. Summary of critical country information, by region (*Continued*).

Chapter/ Section No.	Country/Territory	Population Exposed Estimate (1000s)	Arsenic Standard (ug/l)	A	C	G	M	P	V
4.26	Serbia and Montenegro	2,034.0	10	1		1	1		1
4.27	Slovakia (Slovak Republic)	n.a.	n.a.	1	1	1	1		1
4.28	Slovenia	n.a.	n.a.	1					1
4.29	Spain	n.a.	50	1	1	1	1		1
4.30	Sweden	n.a.	10	1		1	1		
4.31	Switzerland	9.0	10			1	1		1
4.32	Ukraine	1,560.0	50	1	1				1
4.33	United Kingdom	n.a.	10	1	1	1	1	1	
	4. Subtotal	**8,485.2**		**31**	**13**	**22**	**24**	**3**	**18**
5.	**Region of Australia and Oceania**								
5.1	Australia	12.8	10	1	1	1	1		
5.2	Guam (Territory of the USA)	129.0	10	1					
5.3	New Zealand	n.a.	n.a	1	1	1	1		1
5.4	Papua New Guinea	110.0	n.a				1		
	5. Subtotal	**251.8**		**3**	**2**	**2**	**3**	**0**	**1**
	Total	**226,195.7**		**54**	**28**	**68**	**74**	**17**	**35**

ANTHROPOGENIC-RELATED ARSENIC

Anthropogenically produced chemicals bearing arsenic number in the thousands, and these can become the source of arsenic contamination of groundwater and surface water. Because of the diversity of this category, the following cases are meant to be illustrative, but not exhaustive, of the range of manufactured arsenic compounds that are potential contaminants of groundwater, surface water and food.

Figure 2 is a global map of 54 countries with anthropogenic-related arsenic.

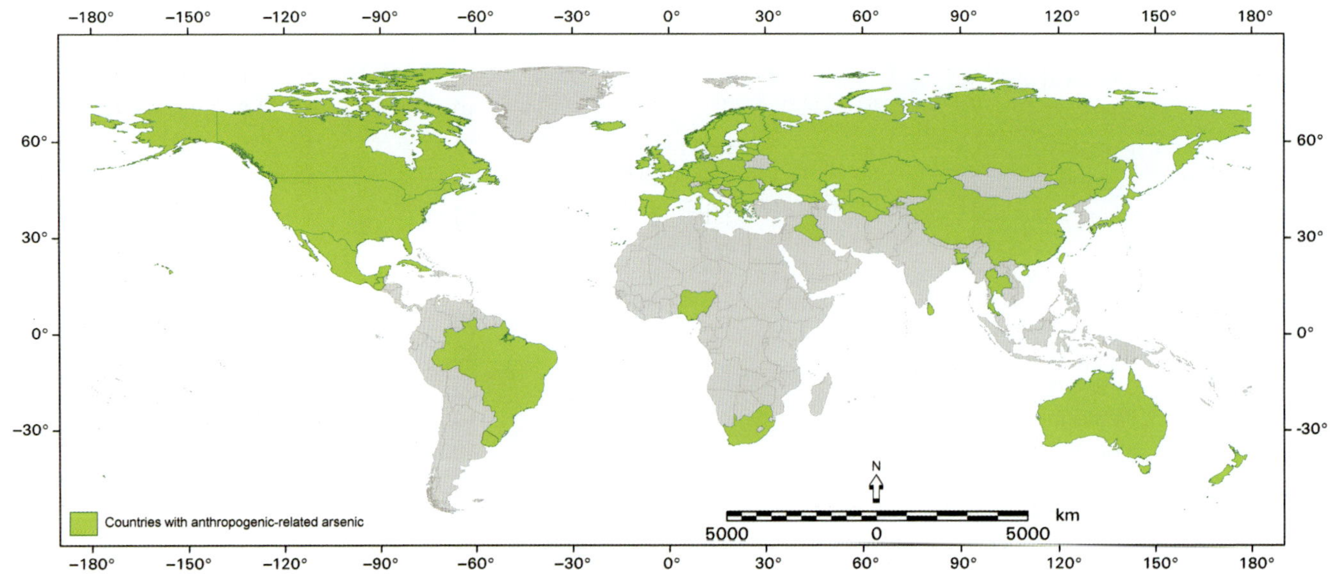

Figure 2. Global map of 54 countries/territories with anthropogenic-related arsenic.

Section 1. African Region

Nigeria. Inland, there is great competition for surface water for drinking and irrigation farming. Water sources in Nigeria are diverse and include springs, dug wells and rain harvesting in the south and mainly borehole wells in the north. The drawdown of the aquifers from overextraction is leading to arsenic entering the food supply from the use of arsenic-rich waters for irrigation.

South Africa. The public outcry regarding some arsenic-contaminated waste being shipped to South Africa reached a crescendo on 14 September 2000 when international and South African environmental organizations voiced their outrage over the Government of South Africa's authorization of the import of 60 tonnes of hazardous waste that had been exported by Australia in contravention of an internationally agreed prohibition on such waste trade.

Section 2. Region of the Americas

Brazil. Exposure to anthropogenic arsenic in Brazil occurs through industrial wastes and atmospheric emissions. No estimates of the exposed population were found.

Canada. An assessment of the top 40 anthropogenic arsenic sources in 2004 revealed power generation, mining (including smelting), wastewater treatment and paper manufacturing as the major sources. These operations affect New Brunswick, Ontario, Manitoba, Alberta and British Columbia, indicating that anthropogenic arsenic contamination is national in extent.

Cuba. Anthropogenic As exposure in Manzanillo Bay is probably contributed by effluent into the Yara River from an alkaline battery factory potentially impacting the city of Manzanillo with a population of about 111000. Cienfuegos Bay in west central Cuba on the south coast with a population of 123600 has been impacted by the manufacture of nitrogen-fertilizers and by a spill of inorganic As oxides in 2001.

Guatemala. An unusual anthropogenic source of As was discovered in 2010, when clay used in ceramic filters for the purification of drinking water in Antigua was found to have As.

Mexico. Industrial wastes containing high arsenic concentrations have leached into irrigation waters. This arsenic has entered the food-chain through produce, through the milk of cows ingesting feed irrigated by arsenic waters and through fish that are contaminated by the release of arsenic from industrial sources into the Rio Grande.

United States of America. Anthropogenic arsenic contamination is found in 30 states, related to power generation plants and industrial facilities burning high-arsenic coal, chemical manufacturing, wood preservation, glass production, agricultural herbicides and pesticides, metal smelting and wastewater treatment. Table 2 indicates the individual states where exposure to arsenic is due to anthropogenic sources as indicated in this report. These data are not complete and only illustrative of the types of anthropogenic exposure for these states.

Table 2. Sources of anthropogenic-related arsenic in 30 states of the USA

State	Anthropogenic source(s)
1. Arizona	Copper smelting
2. Arkansas	Pesticides used formerly in cotton agriculture; manufacture of arsenical wood preservatives; poultry excreta
3. California	Pesticides for agriculture and lawns, bottled water
4. Colorado	Pesticides for agriculture and lawns
5. Connecticut	High-arsenic-bearing landfill materials
6. Florida	Orchards that had been sprayed with arsenic pesticides and fungicides; wooden decks treated with chromated copper arsenate
7. Georgia	Pesticide, herbicide and insecticide manufacture
8. Hawaii	Arsenic chemicals on the site of a former sugar cane plantation
9. Indiana	Arsenic-based chemicals in automotive paint sludge
10. Kansas	Arsenic-based pesticides
11. Kentucky	Arsenic-based pesticides in stream sediments
12. Louisiana	Arsenic-based pesticides for cotton agriculture, now bioaccumulated in rice
13. Maine	Arsenic-based pesticides and herbicides

(*Continued*)

Table 2. Sources of anthropogenic-related arsenic in 30 states of the USA (*Continued*).

State	Anthropogenic source(s)
14. Maryland	Poultry excreta
15. Massachusetts	High-tech manufacturing and testing of arsenic-based chemicals
16. Minnesota	Agricultural arsenic-based pesticides
17. Missouri	Manufacture and use of agricultural arsenic-based pesticides
18. New Hampshire	Lead arsenate sprays, widely used as fungicides and insecticides in apple orchards
19. New Jersey	Lead arsenate sprays, widely used as fungicides and insecticides in fruit orchards
20. New Mexico	Wastewater treatment plant effluents
21. New York	Arsenic-based chemicals used as pesticides and fungicides; landfills
22. North Dakota	Arsenic-based pesticides used to control grasshopper activity
23. Ohio	Industrial arsenic-based chemicals
24. Oregon	Lead arsenate sprays, widely used as fungicides and insecticides in fruit orchards
25. Pennsylvania	High-arsenic fly ash (from a power generating facility?)
26. South Carolina	Elevated arsenic in Coastal Plain and Piedmont
27. South Dakota	Emissions from sewage treatment plants and thermal power cooling plants
28. Texas	Arsenic-based chemicals used as pesticides and fungicides
29. Vermont	Leather tannery arsenic-based chemicals
30. Washington	Arsenic-based pesticides and fungicides for apple orchards; one site was the sole domestic supplier of arsenic for the USA for several decades

Uruguay. Between 1998 and 2002 the World Bank authorized the drilling of over 10000 wells at depths ranging from 30–80 m without any water quality testing, and it has now been discovered that As is present in the Guaraní Aquifer being accessed by these wells, in a situation similar to that of Bangladesh.

Section 3. Asian Region

Bangladesh. The mobilization of arsenic may be connected with the anthropogenic expansion of irrigation in the Brahmaputra delta, in which the arsenic is mobilized in surficial waters and actually seeps down into deeper aquifers.

China. Four million World War II Japanese chemical munitions remain unaccounted for. In 1992, China announced that at least 2 million chemical weapons remained in its territory; most of these were presumably those of Japanese manufacture in the north-eastern provinces. An ambitious plan by the Bonn International Center for Conversion scheduled these to be removed by 2007 under a Chemical Weapons Convention act. The munitions contain poison gas, mainly mustard gas and lewisite, an arsenic-based blistering fluid. Japan has launched six missions since 2000 to recover about 11 200 of these chemical warheads.

Georgia. Anthropogenic sources of arsenic in Georgia come from large industrial processing plants that produce manganese, ammonia, arsenic, copper and gold. Petroleum products have been leached into aquifers by oil refineries and power plants, and arsenic has been released by the action of petroleum on underlying aquifers and has polluted the tributaries of the Black and Caspian sea basins. Heavy oil products and phenols have also heavily polluted the Kuban River.

Iraq. Anthropogenic arsenic in Iraq had its origins in the industrial buildup between 1970 and 1990, during which environmental controls were disregarded. Cement and fertilizer plants, petrochemical factories and heavy industries released tonnes of pollutants, including lead, mercury and arsenic, into surface water and groundwater, especially in the central and southern regions, where general and military manufacturing was concentrated.

Japan. World War II chemical weapons were polluting the groundwater in the area around the town of Kamsu on the island of Honshu near the naval station that was responsible for manufacturing "sneezing gas" or phenyldichlorarsine (called *Sternite* by the French), which causes pulmonary oedema, raised blisters and sneezing, and for the latter effect was thought to be a good battlefield agent to retard the advance of troops. In addition, in Hiroshima Prefecture on Okunoshima Island, from 1929 to 1945, the Japanese Imperial Army produced five types of poison gas, including diphenylchlorarsine. Only recently has the problem begun to be addressed. Another important arsenic contaminant in Japan is high-arsenic dye used in the textile industry. A textile factory's arsenical dyes were emitted into the surface water of the town of Nakajo in Honshu, and its population suffered arsenicosis symptoms. It is not known how many other instances there are where similar dyes are being used but have not been identified.

Kazakhstan. Collective farms promoted the use of arsenical pesticides that have subsequently leached into the groundwater.

Russian Federation. Ecological and medical problems can be caused by Soviet-era chemical weapons, one type of which (lewisite) contains arsenic, which can leach into groundwater if improperly stored. Three marine environments have been exposed, as well as the Republic of Chuvashia and the following oblasts: Bryanskaya, Kurganskaya, Novgorodskaya, Penzenskaya, Samarskaya, Saratovskaya and Volgogradskaya. In 1995, the Moscow-based Union for Chemical Safety began identifying the ecological and medical problems created by the manufacture and storage of Soviet-era mustard gas and arsenic-based gas weapons in the Republic of Udmurtia, which is one area that had a prominent role in this chemical weapons programme.

Sri Lanka. Incidences of arsenic poisoning are related to long-term ingestion of illicit "liquor" (contaminated by either high-arsenic groundwaters or pesticides) and the ubiquitous arsenic-laced opium.

Thailand. There are high concentrations of arsenic in the Mae Moh Reservoir, which may have been caused by the fly ash pond of a local power generation facility in addition to a mining operation nearby.

Turkmenistan. The reduction in volume by almost 75% of the Aral Sea has led to an increasing concentration of geogenic and anthropogenic toxins, including arsenic and various salts, in the water.

Uzbekistan. In the valleys in this region, the rivers supply much-needed irrigation for a variety of crops, and the drawdown of the rivers and aquifers has led to a concentration of arsenic in the groundwater.

Section 4. European Region

Albania. Some of the former Soviet Union's chemical weapons were stored in clandestine bunkers in Albania without adequate safeguards to protect the surrounding environment.

Austria. Nine industrial sites directly discharge 1.70 tonnes of arsenic into surface water each year, and 11 facilities discharge 1.75 tonnes per year into the air.

Belgium. Belgium has 24 industrial sources (as a result of chemical manufacturing and metals smelting and mining) that emit arsenic into air, 21 that emit arsenic directly into surface water and 4 that emit arsenic indirectly into water.

Bosnia and Herzegovina. Anthropogenic sources of arsenic contamination including domestic and industrial waste have polluted the Danube and Sava rivers.

Bulgaria. Anthropogenic arsenic contamination is believed responsible for atmospheric arsenic deposition at Chiprovtzi, a town across the border from Serbia. The smelters in Serbia are thought to be the transboundary source of this atmospheric arsenic.

Czech Republic. Anthropogenic sources of arsenic in the Czech Republic come from 9 industrial facilities that emit 2.23 tonnes of arsenic per year directly into surface water and 33 facilities that emit 4.40 tonnes of arsenic per year into the air.

Denmark. Nine facilities discharge 2.23 tonnes of arsenic per year into surface water, and 33 facilities discharge 4.40 tonnes of arsenic per year into the air. As precipitation from airborne sources is a possibility, these sources are included here, but no further data are available on the concentrations of arsenic in surface water and groundwater adjacent to or downwind from these four sites. In addition, after World War II, 160 tonnes of chemical weapons, including lewisite, were dumped in the Baltic Sea, and these pose a threat of contaminating marine biota with arsenic.

Estonia. The geochemistry of the Tallinn Project, supported by the Scientific and Environmental Affairs Division of the North Atlantic Treaty Organization, has analysed 532 topsoil samples from Tallinn, which is the largest industrial centre of Estonia, with a population of 400 200. In the geochemical anomalies, the levels of arsenic, chromium, manganese and vanadium are more than 3 times higher and levels of lead and zinc are more than 5 times higher than background levels.

Finland. Anthropogenic sources of arsenic in Finland come from 11 industrial, chemical, metal and mining sources that discharge 1.75 tonnes of arsenic per year into the air and 9 that discharge 1.70 tonnes per year directly into surface waters.

France. Anthropogenic arsenic in France comes from 43 pharmaceutical, chemical and metal manufacturers along with mining companies that directly or indirectly annually discharge 5.52 tonnes of arsenic-bearing effluents into air, 33 facilities that emit 4.46 tonnes of arsenic directly into surface waters and 2 that emit 0.0184 tonne of arsenic indirectly into surface waters.

Germany. Anthropogenic arsenic in Germany comes from 21 pharmaceutical, chemical and metal manufacturers and mining companies that directly discharge 1.66 tonnes of arsenic-bearing effluents per year into surface waters, another 14 that indirectly discharge 3.08 tonnes of arsenic-bearing effluents per year into surface waters and another 55 that emit 8.06 tonnes of arsenic effluents per year into the air. Another anthropogenic source of arsenic in Germany was uncovered in groundwaters from Kelheim.

Greece. Anthropogenic sources of arsenic in Greece come from three industrial facilities that discharge 1.32 tonnes of arsenic per year into the air and two facilities that discharge 1.82 tonnes per year into surface waters.

Hungary. Anthropogenic sources of arsenic in Hungary come from three chemical plants that discharge 0.169 tonne of arsenic per year into the surface water and one facility that discharges 0.042 tonne per year into the air.

Iceland. At Nesjavellir's co-generation power plant, hot wastewater is pumped from the plant into shallow drill holes or into the Nesjavellir stream, which disappears into the lava and flows into Lake Thingvallavatn, a rift lake of high conservational value. Concentrations of silicon dioxide, arsenic, aluminium and boron are elevated in geothermal effluents but are diluted before the wastewater reaches the lake.

Ireland. Anthropogenic arsenic in Ireland comes from one alumina production facility and three electrical power generating plants that report 0.16 tonne of arsenic per year released into the air.

Italy. Anthropogenic arsenic in Italy is linked to 22 chemical, coal-burning and thermoelectric utilities that discharge 2.02 tonnes of arsenic per year into the air, 26 facilities that discharge 4.12 tonnes of arsenic per year directly into surface water and 10 facilities that indirectly discharge 0.778 tonne of arsenic per year into surface water.

Latvia. There are approximately 550 landfills in Latvia that had been established under Soviet control in the 1970s and 1980s without any consideration of environmental impacts. The biggest landfill in Liepaja is at Skede, which covers 7.5 ha, of which 6 ha are covered by waste. It is located 10 km north of Liepaja city and is the repository for 79% of all the waste of the entire region. When it was built in 1972, no hydro-isolation barrier was installed, so leaching for 30 years has created serious contamination of groundwater and surface water, which has spread in two plumes, one to Tosmare Lake and the other to the Baltic Sea. Additionally, offshore sediments in the Gotland and Bornholm basins have arsenic contamination from chemical weapons munitions dumps. Elevated arsenic concentrations in sediments near the weapons dumpsites were low in comparison with those at other sites in the Baltic and North seas, but still require monitoring.

Lithuania. Elevated concentrations of anthropogenic arsenic have been detected in mosses around the Akmenės Cementas (cement) factory in Naujoji Akmenė, a city in Akmenė district municipality and one of the newest cities in Lithuania. Akmenės Cementas is the largest cement company in the Baltic region and produces 711 200 tonnes of cement annually, which generates airborne arsenic that is precipitated on soils, bioaccumulates in plants and leaches into surface water.

Macedonia (The former Yugoslav Republic of). Anthropogenic arsenic has been emitted by the ferro-alloy plant at Jegunovce, the lead and zinc smelter at Veles, the zinc and lead mine at Probistip and the thermal power plant at Bitola.

Netherlands. Nineteen chemical, mining and mineral plants discharge 0.968 tonne of arsenic per year directly into surface water, 8 discharge 0.10 tonne of arsenic per year indirectly into surface water and 3 facilities emit 0.226 tonne of arsenic per year into the air.

Norway. Five mineral smelters discharge 0.823 tonne of arsenic per year directly into surface water, and 3 emit 0.242 tonne per year into the air.

Poland. One manufacturing plant and one mining facility discharge 0.351 tonne of arsenic per year directly into surface water, one industrial plant and one sanitation facility discharge 0.023 tonne of arsenic per year indirectly into surface water and seven facilities (including three power plants) discharge 1.29 tonnes of arsenic per year into the air. Other sources of arsenic include tainted groundwater used in beer production, tainted Odra River sediments, post-mining activities and thermal hot springs.

Portugal. Two mining facilities discharge 0.351 tonne of arsenic per year directly into surface water, two waste disposal facilities discharge 0.023 tonne of arsenic per year indirectly into surface water and seven facilities (three power stations and four manufacturing plants) discharge 1.29 tonnes per year arsenic into the air.

Romania. The distribution of arsenic in eastern Romania, measured by bioaccumulation in mosses, is thought to reflect the agricultural use of arsenical pesticides.

Serbia. The industrial release of chemicals into water and soil has subsequently contaminated food supplies. Heavy metals and metalloids (i.e. arsenic) have been detected in a variety of foods harvested near industrial sites, and fish (both marine and freshwater) may be contaminated with arsenic and mercury. The contamination of soils in western Bulgaria at the town of Chiprovtzi is thought to be from emissions from smelters in eastern Serbia.

Slovakia. Power generation stations and steel, cement and other manufacturing plants discharge 31.07 tonnes of arsenic per year into the air, and two facilities discharge 0.099 tonne of arsenic per year directly into surface water. The main producer of hazardous waste is the Novaky Chemical Plant. Fifty per cent of the plant's waste is landfilled; 179 waste dumps have been reported on.

Slovenia. Anthropogenic arsenic is emitted by two thermoelectric plants and one manufacturing plant that discharge 0.109 tonne of arsenic per year into the air and one facility that discharges 0.0143 tonne of arsenic per year directly into surface water.

Spain. Forty-one chemical, mining and mineral plants as well as high-arsenic coal-burning and hydro-thermoelectric utilities discharge arsenic into the air, 12 facilities discharge 2.93 tonnes of arsenic per year directly into surface water and 3 discharge 0.043 tonne of arsenic per year indirectly into surface water. High-arsenic coal burning by electrical utilities discharge arsenic into the air and directly and indirectly into surface water and groundwater.

Sweden. In Sweden, 20 chemical, mining and metal plants and pulp factories discharge 2.27 tonnes of arsenic per year into the air, and 25 facilities directly discharge 1.29 tonnes of arsenic per year into surface water.

Ukraine. Anthropogenic arsenic has contaminated the air and soils of the cities of Donets, Makeevka and Yasinovataya, as well as many smaller towns and villages, because they were subjected to airborne ash from the burning of high-arsenic coals and coke.

United Kingdom of Great Britain and Northern Ireland. Twenty-five power generation plants (including nuclear), metal manufacturing plants and brewing facilities discharge 0.9587 tonne of arsenic per year into the air, 48 facilities discharge 3.24 tonnes of arsenic per year directly into surface water and 11 facilities discharge 0.254 tonne of arsenic per year indirectly into surface water.

Section 5. Region of Australia and Oceania

Australia. Anthropogenic arsenic related to arsenic chemicals used in sheep wool processing has been found at Lake Yangebup, south of Perth, Western Australia, which is part of the Beeliar wetland chain and has high arsenic concentrations in both its water and sediments due to arsenic used in wool scouring. The resuspension of arsenic in these sediments at Lake Yangebup is cause for concern. The question is whether there are other wool processing locations where arsenic chemicals were used. In the Perth suburb of Stirling, arsenic has been found in borehole wells at 1700 μg/l. Households have been instructed to stop using borehole well water. The arsenic is believed to be related to the excavation for new housing in development areas that disturbs underlying peat beds and releases arsenic from its sorbed state. In this context, 3200 households or 12 800 people are exposed to arsenic in groundwater in Stirling. Additionally, anthropogenic arsenic has leached from industrial wastes and from arsenical pesticides and herbicides from agricultural land.

Guam (territory of the USA). Arsenic contamination of the resort area has been a major concern, and it is believed to be caused by the nearby American military base or airport, or both. To date, the cause has not been definitely determined.

New Zealand. One case of anthropogenically introduced arsenic has been reported involving a housing development on a former orchard contaminated with dangerous levels of dichlorodiphenyltrichloroethane (DDT), arsenic, lead and copper from dumped horticultural pesticides.

COAL-RELATED ARSENIC

Elevated arsenic-bearing coals may contaminate surface water and groundwater by natural weathering or by accelerated weathering, when coal seams are exposed due to mining. Cases where contamination is from combusted coal–in unventilated areas or in open air–or by ash contaminating soils, surface water or groundwater are dealt with under anthropogenic arsenic sources.

Arsenic concentrations and distribution in coals vary with the geology of coal basins, with individual beds and even with individual seams. Therefore, it is important, for arsenic-enriched coals, to note the geological formation, the precise location of the particular bed and the position of the seam within the bed. Arsenic-enriched coals often correlate positively with sulfides, such as pyrite (or rarer sulfide minerals), called pyritic arsenic (As_{pyr}). Conversely, low-arsenic coals often correlate positively with organic arsenic (As_{org}).

Although this pertains to anthropogenic arsenic, it should be noted here that the ash arsenic yield is respectively higher for As_{pyr} coals and lower for As_{org} coals. Often the ash arsenic yield is higher than the arsenic concentration of the coal, and this may vary according to the mode of combustion, etc., so it is important to note the measured ash arsenic yield.

Coal geochemistry has made recent advances that necessitate the revision of earlier estimates of arsenic concentrations in coals. As of 1985, world averages of arsenic in coal (called "Clarke values") indicated that bituminous coals have higher arsenic concentrations than lignites (Yudovich *et al.*, 1985):

- Bituminous coals, 20 ± 3 mg arsenic/kg
- Lignites, 14 ± 4 mg arsenic/kg
- Ashes of bituminous coals, 90 ± 74 mg arsenic/kg
- Ashes of lignites, 60 ± 35 mg arsenic/kg

However, in 2002, Ketris & Yudovich revised these preliminary calculations using "successive averaging" (averaging concentrations from one coal bed to several coal beds, from one coal deposit to several coal deposits, and from one coal basin to several coal basins). In this process, the 1985 Clarke values were lowered to:

- Bituminous coals, 9 ± 0.08 mg arsenic/kg
- Lignites, 7.4 ± 1.4 mg arsenic/kg
- Ashes of bituminous coals, 50 ± 5 mg arsenic/kg
- Ashes of lignites, 49 ± 8 mg arsenic/kg

Arsenic-enriched coals occur as two epigenetic types (formed *after* the laying down of the enclosing rock matrix) and two syngenetic types (formed *contemporaneously with* the enclosing rock matrix) (Yudovich & Ketris, 2005):

- *Epigenetic Chinese type*: hydrothermal arsenic enrichment similar to Carlin type of arsenic-bearing gold deposits (found in Bulgaria, Tajikistan, Turkey, Ukraine, the United Kingdom and the Appalachian basin in the USA).
- *Epigenetic Dakota type*: hypergene, low-temperature arsenic enrichment; arsenic (accompanied by molybdenum and uranium) drained from tuffaceous host rocks and sorbed by lignite from groundwaters; found in South Dakota and Wyoming (USA) and British Columbia (Canada).
- *Syngenetic Bulgarian type*: arsenic enrichment from metalliferous surfaces in association with sulfide deposits or from groundwaters in the peat stage.
- *Syngenetic Turkish type*: volcanogenic enrichment from arsenic brines, exhalations and ash in coal-forming peat bogs.

Glacio-fluvial deposits with underlying arsenic-rich coal-bearing units may mobilize arsenic into aquifers. Research on arsenic coals from Arkansas, Ohio and Alabama in the USA is concerned with the leaching problem (Huggins *et al.*, 2002). In addition, indoor burning of arsenic-rich coal has been practised in some countries of the world, such as China and Mongolia. This is a separate anthropogenic vector, mainly through food and inhalation, and some precipitation of arsenic in open liquids can occur. In addition, ash from arsenic-bearing coals and lignites can become arsenic-enriched and pose similar problems of mobilization in soils and groundwaters.

Figure 3 is a global map of the 28 countries/territories with coal-related arsenic.

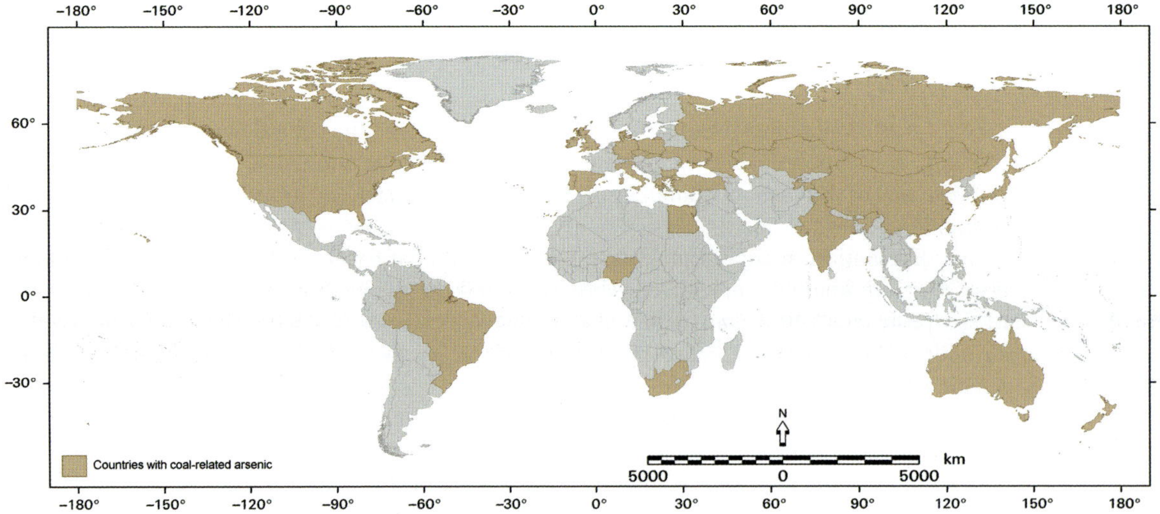

Figure 3. Global map of 28 countries/territories with coal-related arsenic.

Section 1. African Region

Nigeria. Coal mining has recently been found to be a contributor to arsenic and heavy metal pollution. Arsenic and a suite of other heavy metals and metalloids (e.g. manganese, chromium, cadmium, arsenic, nickel and lead) were analysed in coal and sediment samples from the River Ekulu in Enugu, the capital city of Enugu state. The river serves as a source of drinking-water for almost 690 000 people.

South Africa. The arsenic concentration in coal from South Africa's low-sulfur, low-arsenic-bearing coalfields is estimated by Swaine (1990) to average 4 mg/kg. This average concentration is rendered problematic by the age and weathering of older collieries, which date to the 1950s, the increasing numbers of new collieries, the expanding volumes of waste, and the increasing surface areas of worked mines that promote acid mine drainage leaching into soils, surface waters and groundwaters.

Section 2. Region of the Americas

Brazil. Enriched-arsenic coals are found in the Candiota River valley. The emissions of arsenic were scheduled to increase from 9.551 tonnes in 2003 to 17.98 tonnes in 2005. The human health impact of arsenic contamination in the Candiota region has been measured indirectly by measuring the arsenic content in hair; the results have shown that the mean arsenic content varied from 7.4 to 12.8 mg/g within 30 km of the power generation plant, compared with <0.5 mg/g at a distance of 60 km from the plant.

Canada. Enriched-arsenic coals are found in lignite, sub-bituminous and bituminous beds across Canada. The Eocene Hat Creek lignite bed in British Columbia is one of the largest coal beds in the world, with projected resources of about 10 billion ($\times 10^9$) tonnes (580 m thick, 26 km long and 4 km wide); it has elevated arsenic, possibly caused by arsenic leaching from volcanogenic strata above the roof of the bed. Hat Creek A zone coals contain up to 14.0 mg arsenic/kg. The Cretaceous sub-bituminous beds of Vesta Mine with a total thickness of 1.6 m have an anomalous inertinite seam with 3.0 mg arsenic/kg. Vancouver Island has very enriched lignites from Comox coalfield and from Suquash coalfield on Millbank Sound (up to 1400 mg/kg). Nova Scotia coals are also enriched with arsenic, positively correlated with sulfur in Pennsylvanian bituminous coals of the Sydney basin and Cumberland basin (up to 7200 mg/kg coal ash).

United States of America. Eleven states have been identified with arsenic released from coal deposits (see Table 3).

Table 3. Sources of coal-related arsenic in 11 states of the USA.

State	Coal-related occurrence
1. Alabama	Appalachian Basin Pennsylvanian medium- and high-volatile bituminous coals in the Warrior and Plateau coalfields and the Cahaba coalfield in northern Alabama have the highest concentrations of arsenic of all coals in the USA: up to 2500 mg/kg
2. Indiana	Glacio-fluvial deposits with underlying arsenic-rich coal-bearing units in the upper Kankakee River basin
3. Kentucky	High concentration of arsenic in stream sediments along the western edge of the Central Appalachian Basin that follows the exposure of Upper Devonian black shales
4. North Dakota	One-metre-thick Palaeocene Beulah lignite arsenic enriched in 0.4-m-thick bottom bench
5. Ohio	High-volatile bituminous coals cluster in zones parallel to discontinuities (perpendicular to Appalachian folds) that may have facilitated hydrothermal flows (containing arsenic) into the Appalachian Basin during orogenic processes
6. Pennsylvania	High-volatile bituminous coals with elevated arsenic cluster in zones parallel to discontinuities (perpendicular to Appalachian folds) in western Pennsylvania that may have facilitated hydrothermal flows (containing arsenic) into the Appalachian Basin during orogenic processes
7. South Dakota	Lignites with high arsenic from tuffaceous sands in the roof of coal strata with low-temperature groundwaters and high oxidation, a Dakota-type uranium concentration
8. Tennessee	High-volatile bituminous coals with elevated arsenic cluster in zones parallel to discontinuities (perpendicular to Appalachian folds) in central Tennessee's southern Cumberland Plateau capped by Pennsylvanian Warren Point Sandstone and Sewanee Conglomerate (arsenites interbedded with black shales, coals and haematite bands), arsenic-bearing coal beds underlying aquifers may leach arsenic into the sandy loams above these layers that contain seasonal aquifers
9. Texas	Late Eocene Fayette delta system low-density lignites with concentrations up to 30 mg/kg
10. West Virginia	Stream sediments in the north-west of the state have the highest concentrations of arsenic related to the Pennsylvanian Conemaugh Group of coals, which includes the Harlem, Elk Lick and Mahoning coals of Tucker, Mineral and Grant counties; coal fly ash that may be deposited downwind from power plants in the area may also be a contributing factor to elevated arsenic in stream sediments, but natural weathering of arsenic from coal seams is also a possibility
11. Wyoming	The 37-m-thick Wyodak-Anderson coal in the Powder River basin, Wyoming, contains low average concentrations of ash (3.64%), sulfur (0.27%) and trace elements of environmental concern, including arsenic, mercury and selenium; 0.069–1.5 mg/kg; Vermillion Creek also has a Dakota-type uranium infiltration

Section 3. Asian Region

China. Guizhou and Yunnan provinces are the most affected by high-arsenic coal; however, it is not known whether glacio-fluvial deposits have underlying arsenic-rich coal-bearing units that contaminate aquifers in the area. To date, all the attention has been directed to the traditional indoor burning of coal for heating, cooking, smoking and preserving meats, grains and vegetables (corn and chili peppers). Any standing water or soup is likely to be contaminated as well. It has been assumed that arsenicosis in these high-arsenic coal areas is due to ingestion of food exposed to toxic coal fumes or just inhalation of these fumes. By 2004, the overall estimate for those affected by high-arsenic coal fumes in the high-arsenic coal provinces (Guizhou, Inner Mongolia, Ningxia, Shanxi and Yunnan) rose to 1.3 million people. The possibility exists that arsenic leaching from the coalfields into aquifers in these areas is a contributing factor. The drinking-water sources in these areas require further investigation. The current data indicate that in Guizhou, 45 399 people are exposed and 2609 have been diagnosed with arsenicosis in 103 villages tested. However, an earlier assessment indicated that 200 000 were exposed and 3000 were diagnosed, but these values are too low given the actual case rate for the 103 villages. Thus, the actual number of cases is probably much higher than in the official report. In Yunnan, it is indicated that 48 438 are exposed, and those diagnosed would probably be comparable to those of Guizhou. Here, again, these are only the actually assessed villages.

India. Arsenic has been identified in Permian coals of North Karanpura coalfield, Jharkhand State, a western member in the east–west chain of the Damodar Valley Basin, where coal-bearing sediments are distributed over Hazaribagh, Ranchi and Palamau districts.

Japan. Enriched-arsenic coal is found in two Oligocene deposits in the Misassa sedimentary beds on the Island of Hokkaido: the Hiraga deposit and the Igu deposit. In addition, Neogene lignites and sub-bituminous coals have higher average concentrations of arsenic than do Palaeogene bituminous coals. These enriched-arsenic coals are the product of the influx of syngenetic volcanic hydrothermal fluids, which occurred simultaneously with the formation of peat beds. In Tertiary lignites, anomalies are observed with uranium up to 4600 mg/kg, germanium up to 780 mg/kg and arsenic up to 500 mg/kg.

Kazakhstan. Enriched-arsenic coals from mines in Ekibastuz and Shubarkol in Kazakhstan have seriously endangered the environment. Not only have they been incinerated without proper protection domestically, but they have also been incinerated without proper protection by coal-fired power plants. In addition to the local enriched-arsenic coals, Kazakhstan has also incinerated the Russian enriched-arsenic coals from Kuznetsk. For instance, the Alma-Ata power generation plant has a large ash disposal area that has contaminated local soils, surface waters and groundwaters with 1.4 million tonnes of high-arsenic ash annually. Similarly, near to Ekibastuz (town in Pavlodar Province, north-eastern Kazakhstan, with a population of 127 200–1999 census), there are two coal-fired power plants, GRES-1 with a capacity of 4000 MW and GRES-2 with a capacity of 1000 MW. Because of the toxic arsenic burned by the plants, GRES-2 has the tallest chimney in the world, at 419.7 m. The disposition of the ash from these plants is unknown.

Mongolia. High-arsenic coal has been identified from 10 mines in Baganuur (arsenic 183 mg/kg) and Nalaikh (arsenic 121 mg/kg), where it is used in ways similar to those of Guizhou and Yunnan, China. Little work has been done on diagnosis and remediation of this problem in Mongolia. Glacio-fluvial deposits with underlying arsenic-rich coal-bearing units may contaminate aquifers in these areas, but no testing is known to date.

Russian Federation. Arsenic-enriched coals are found in several geological settings: in the Gorlov basin, due to both epigenic hydrothermal fluids and post-Permian syngenic intrusions into Permian strata; in the Altai Mountains Palaeogene lignites of the Taldu-Dyurgun deposit; and in the Chulymo-Enissey basin, western Siberia, where enriched-arsenic coals are in association with grey-coloured alluvial sands rich in uranium.

Tajikistan. Coal-related arsenic is found in the Jurassic Nazar-Ailock anthracite deposit near the tin–mercury ores of the Khshert deposit where hydrothermal fluids are believed to leach from the ores bordering the Tajik basin and penetrate the coal seams.

Turkey. Arsenic-enriched coal beds are found in the Gokler coalfield in western Turkey. They are associated with epigenetic stibnite and iron sulfides that formed during post-Miocene hydrothermal mineralizations and in the lower Eocene coal seam mined in Yeniceltek, Kucukkohne and Ayridam in the Sorgun basin in association with arsenic, zinc, selenium and antimony. Arsenic-enriched lignites occur in the south Marmor Sea, the Aegean Sea, the Mediterranean Sea and the Black Sea.

Section 4. European Region

Bulgaria. Coal-related arsenic is found in Carboniferous Svige anthracites related to epigenetic pyrite, bituminous coals in association with pyrite, marcasite, chalcopyrite and galena, sub-bituminous coals of the Pernik basin with a positive correlation with pyrite and lignite seams in the 20-m-thick Eocene Stanyansky deposit.

Czech Republic. Arsenic-enriched coal is found in five basins: the Cheb sedimentary basin in the north-western Bohemian massif formed between the Late Oligocene and the Pliocene, the Ostrava and Karvina Pennsylvanian bituminous deposits, the

Sokolov brown-coal basin within the Karlovy Vary volcanogenic thermal zone, and the Zitava basin bounded by the Luzice Mountains with the Sluknov Hills in the west and the Jizera Mountains with the Frydlant Hills to the east. The less carbonaceous coals in the Czech Republic are used for power plant consumption. Arsenic in the flue dust from this coal, after having been burned in a power plant, ranged from 43 to 110 mg/kg. This was ingested by 56 10-year-old children living near the power plant. These children showed significant hearing losses in both air and bone conduction at a high frequency range, indicating inner ear damage.

Denmark. Lithogenic coal in Denmark usually contains arsenic within the range of 1–4 mg/kg. However, testing of a domed peat column revealed an arsenic concentration of 28 mg/kg at a depth of 15 cm. Also present in this sample were mercury and lead. For this reason, it has been speculated that the arsenic peak was attributable to coal-fired plants, and the 15-cm depth would place the date of contamination to the 1950s, a time when the United Kingdom was engaged in massive coal combustion.

Germany. Arsenic-enriched coals are found in two Mississippian bituminous deposits. The Ebersdorf deposit has a range of concentration of arsenic from 200 to 400 mg/kg, which is directly correlated with sulfide enrichment, and the Ehrenkammer deposit has a maximal coal ash arsenic concentration of 3000 mg/kg.

Greece. Arsenic-enriched coals are found in north-west Greece in Miocene Serra lignite deposits, in Oligocene Drama lignite deposits and in Neogene Ioannina basin lignite deposits (possibly from syngenetic hydrothermal fluids seeping into coal-forming peat strata). Lignite deposits occur in the Neogene deposits in the Ioannina basin, in the Miocene deposits in the Domeniko region in central Greece, and on the Periphery Island of Crete, also in Middle Miocene lignites. Arsenic-enriched lignites occur in the Periphery of Central Macedonia in Drama and in the Serres Prefecture and may have been hydrothermally present in coal-forming peat bogs from mobilized sulfides (arsenopyrite, sphalerite and galena) or may have been supplied in the Eocene-Oligocene volcanism.

Ireland. Three electrical power generating plants report the release of 0.16 tonne of arsenic per year from the burning of high-arsenic coals.

Italy. Utilities burning high-arsenic coal discharge arsenic directly into the air.

Poland. Arsenic-enriched bituminous coals found in the Pennsylvanian Formation in the Lyublin basin are probably due to pyrite as the carrier because of the positive correlation of arsenic and sulfur and iron(III) oxide.

Portugal. High-arsenic coals in Portugal are a cause of some concern for the contamination of groundwater as well as air. On the basis of estimates in only three coalfields, coal resources of all ranks in Portugal total at least 76 million tonnes. In 2004, 5285 tonnes of bituminous coal of unknown arsenic content were used in power generation facilities. The presence of high-arsenic coal has been definitely established by bioassaying arsenic species in lichen transplants and assaying them in airborne particulate matter (fine and coarse fractions) from samples collected in 1994–1995 near coal-fired power plants; inorganic forms of arsenic (arsenite and arsenate) were present in significant quantities in most of the samples. Studies of bioaccumulation of arsenic by mosses to concentrations above 1.4 mg/kg have confirmed the contamination of soils and surface waters from coal-fired plants in the vicinity of Lisbon, Oporto and Sines.

Slovakia. Soils have been contaminated by arsenic, which has been emitted into the atmosphere by a coal-burning power station in the district of Zemianske Kostolany.

Spain. Arsenic-enriched bituminous coals are found in the Puertollano basin with a positive sulfur correlation, indicating that the source is probably pyrite.

Ukraine. Arsenic-enriched coals are found in the Upper Mississippian Buzhanskaya Formation of the L'vov-Volynsk basin and in Pennsylvanian coal beds in the Donbas region (Lenin mine, central region; Glubokaya mine, Donetsk-Makeyev region; and Dzerzhinski mine, central region). Additionally, the cities of Donetsk, Makeevka and Yasinovataya were subjected to the ash from the burning of coke and from waste pits of mined coals. Many of these pits contained arsenic in concentrations up to 1 kg/tonne and caused an anomalous plume of arsenic concentrations reaching 1.2 mg/l that extended over an area of 15 km by 30 km.

United Kingdom of Great Britain and Northern Ireland. Arsenic-enriched coals are found in several locations in the United Kingdom, including south Yorkshire bituminous coals from the Beeston seam, Newcastle low-ash coals, east Midland coals and south Wales coal beds, associated with hydrothermal fluids, local mineralization and metamorphism that penetrates groundwaters.

Section 5. Region of Australia and Oceania

Australia. Arsenic-enriched coals are found in the Gunnedah basin, New South Wales. The upper strata of Permian coals are dominated by quartz, illite and other epigenetic detrital minerals, whereas kaolinite is abundant in the lower strata. Infills of calcite and dolomite (associated with strontium and barium) occur in upper and lower fractures. Pyrite is abundant in strata where coals formed under marine influence, and these coals are associated with high concentrations of arsenic and minor concentrations of thallium. Chlorine is inversely correlated with pyrite concentrations and appears to be associated with organic matter, as are germanium and gallium.

New Zealand. Enriched-arsenic coals are located in the Upper Eocene brown Buller coal beds on the South Island. The structure of the beds has overlying pegmatites with arsenic mineralization, and this may be the erosional source of arsenic in the underlying brown coals that have anomalies measuring up to 550 mg/kg.

GEOGENIC-RELATED ARSENIC

Naturally occurring arsenic in groundwater is most commonly released from iron oxide. Elevated arsenic concentrations in aquifers are caused by reducing conditions in which arsenic sorbs to metallic oxides such as iron(III). However, in anaerobic conditions, iron(III)-reducing microorganisms can couple the reduction of solid-phase iron(III) with the oxidation of organic matter, releasing arsenic to groundwater.

Geogenic arsenic (and arsenic related to coal and mining in general) may be affected by bacteria, which are generally ubiquitous in arsenic geochemical environments. Bacteria convert arsenic into species with different solubility, mobility, bioavailability and toxicity, by reduction, oxidation and methylation (Stolz & Oremland, 1999; Weeger *et al.*, 1999; Silver & Phung, 2005). Some bacteria may be useful in arsenic bioremediation and detoxification by oxidizing arsenite to arsenate (Xiong *et al.*, 2007). Arsenite is biologically more toxic than arsenate, as it binds to sulfhydryl groups and therefore impairs protein functioning (NRC, 1999). Arsenate, on the other hand, is less biologically toxic and is more readily sorbed than arsenite to surfaces, such as soil and sand particles.

Seawater has an average arsenic concentration of 1.7 μg/l, which can be compared with the USEPAs human health criteria for total dissolved arsenic in seawater from which fishery products are harvested for consumption, which is 0.0175 μg/l. Marine algae bioaccumulate arsenate, metabolize it as arsenite and oxidize it to organoarsenic compounds (including hidden arsenic compounds and arsenobetaine, which is not toxic or carcinogenic to mammals). Arsenite, methylarsonic acid and dimethylarsinic acid are excreted into seawater by algae and are then bioaccumulated by (and therefore potentially toxic to) phytoplankton. Invertebrates, bony fishes and cartilaginous fishes are more limited in bioaccumulating inorganic arsenic, and therefore it is less toxic to them. Most of the remaining organoarsenic compounds that are bioaccumulated by humans from fishes are not metabolized to arsenite and therefore present minimal health risk.

Geogenic hidden arsenic is seasonally excreted into marine estuarine ecosystems by algae and phytoplankton in the spring, when species other than arsenic(V) manifest. The usual laboratory hydride generation technique transforms all arsenic species into arsenate. However, when ultraviolet radiation intensifies, photosynthesis increases, marine temperatures rise to about 12 °C and metabolic activity increases, and arsenic forms that have not been transformed into arsenate are now detected, possibly as quaternary arsonium compounds or arsenosugars (Howard & Apte, 1989; Francesconi & Edmonds, 1998). Therefore, these hidden arsenic forms are absent or less available in seawater during the biologically dormant season of the marine year, which is likely attributable to depressed metabolic levels of marine organisms, such as phytoplankton and algae, that have a concomitantly reduced level of methylarsenicals and other arsenicals that are excreted but not detected by the traditional hydride generation method.

The phenomenon of the manifestation of this hidden arsenic in the spring season in waters bordering on the English Channel was reported for the Scheldt estuary in Belgium (Andreae & Andreae, 1989), the Southampton and Tamar estuaries of southern England (Howard *et al.*, 1988; Howard & Comber, 1989), the Beaulieu and Seine estuaries in France (Howard *et al.*, 1984; Michel, 2001) and further south for the Tagus estuary in Portugal, where hidden arsenic amounted to 19–25% of the total arsenic in the estuarine water (de Bettencourt *et al.*, 1994). It is also reported in the southern Adriatic off the coast of Italy, where arsenical compounds were recovered from three species of bony and cartilaginous fishes (Storelli & Marcotrigiano, 2000). In another instance, hidden arsenic has been identified in Uranouchi Inlet, Tosa Bay, Shikoku Island, Japan and eutrophic Lake Biwa, Shiga Prefecture, Honshu Island, Japan, indicating that hidden arsenicals can occur in both saline and eutrophic fresh waters (Hasegawa, 1996; Sohrin *et al.*, 1997). This latter circumstance has also been identified in Lake Greifen, Switzerland, a eutrophic lake with a seasonally anoxic hypolimnion. In summer, arsenate disappears in the epilimnion and is replaced by arsenite, whereas arsenate remains the dominant species in the hypolimnion (Kuhn & Sigg, 1993). This connection of arsenite manifestation with eutrophic conditions is explicable, because arsenate is dominant in oxic waters, whereas arsenite is dominant in anoxic waters in the presence of iron(II). This speciation of dissolved inorganic arsenic is integral to the biochemistry of redox conditions over the annual lacustrine cycle. As such, many more instances of this phenomenon are expected to be identified. For instance, near Yellowknife in the Northwest Territories, Canada, lakes exhibit a similar hidden arsenic cycle, even though the source of the arsenic was originally from a now-derelict gold mine that produced arsenic trioxide as a by-product (Bright, *et al.* 1996). Figure 4 is a global map of 68 countries with geogenic-related arsenic.

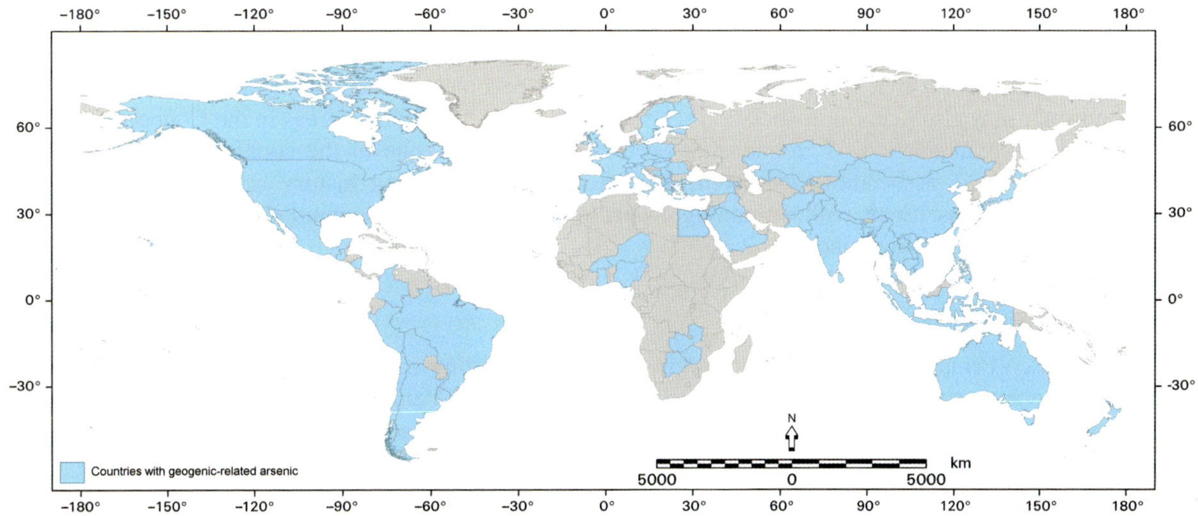

Figure 4. Global map of 68 countries/territories with geogenic-related arsenic.

Section 1. African Region

Botswana. Geogenic arsenic is found in relation to two formations. The first is the natural leaching from amphibolite or Archaean (3 billion years BP) Tati schist in which large reserves of gold are located in crystalline quartz reefs in a 79-km^2 area embracing the area of Francistown. The leaching of arsenic into the groundwater in this highly mineralized area puts Botswana's urban population at risk, especially from historic gold mine waste dumps with both acidic and alkaline tailings. The second source of geogenic arsenic is found in the Sowa and Makgadikgadi salt pans west of Francistown. These salt pans are relicts of an early Holocene lake in the Kalahair Desert. The desiccation has concentrated toxic contaminants.

Burkina Faso. No official data on arsenic in groundwater exist for Botswana (as of January 2006); however, a number of epidemiological accounts of arsenicosis symptoms suggest that arsenic contamination of groundwater does exist in the central zone extending north-east into the Niger. This is probably the result of the leaching of oxidized sulfide minerals (e.g. pyrite, arsenopyrite), which are likely to be concentrated in and around the gold mineralized areas.

Egypt. Arsenic has been found in the Duwi Formation of the Middle Eastern to North African phosphogenic province of Late Cretaceous to Paleogene age in the range of 100 to 2,100 μg/l and an average arsenic concentration of 950 μg/l. Relatively strong positive correlation between arsenic and sulfur is suggestive of the occurrence of arsenic as sulphide.

Ghana. Of 207 rural hand-dug wells and boreholes in a 6500-km^2 area of south-west Ghana, 52 had arsenic concentrations ranging from 50 to 2000 μg/l. High-arsenic waters are associated with Upper Birimian rock, gold–arsenic mineralization, borehole wells and waters with a low redox potential and high total dissolved solids content. Speciation shows significant concentrations of arsenic(III). A strong correlation was found between skin problems and arsenic concentrations in well waters. Hand-dug wells have less arsenic than borehole wells but much more nitrate, and there is a correlation between well depth and arsenic concentration. Column experiments show that Ghana laterite strongly sorbs arsenic. There is no correlation between sulfate and arsenic, although the arsenic source most probably is due to sulfide mineralization. High-arsenic wells are commonly less than 100 m from wells that have trace amounts, indicating that arsenic mobility is limited and there is little transport between water-bearing fractures.

Niger. South-western Niger is at risk for geogenically occurring arsenic in the groundwater because of the similarity of its geology to that of north-eastern Burkina Faso, where keratosis cases have been diagnosed. Although no cases have been reported in the literature for Niger, aquifers in proximity to gold mining operations strongly suggest that a problem exists and should be investigated.

Nigeria. In a 2005 study, arsenic contamination of groundwater aquifers in the south and south-west ranged from 3 to 25 μg/l in the basement rocks and from 6 to 40 μg/l in sedimentary rocks. Sources of arsenic in these regions have been found to be both geogenic and anthropogenic. Limestone and black shale aquifers around Ewekoro and Shagamu contain arsenic concentrations in the range of 360–3020 μg/l, whereas arsenic concentrations in groundwater ranged from 4 to 16 μg/l. The coastal plain's sandy sediments of southern Nigeria facilitate arsenic leaching, similar to that found in the sandy sediments in New South Wales, Australia. Populations exposed potentially range in the tens of thousands.

Zambia. Crystalline basement rocks and slightly acidic groundwater aquifers as well as sedimentary formations yield groundwater with low dissolved arsenic concentrations. This is especially true of carbonate rock types (limestone, dolomite of the Katanga System) and sand and gravel deposits (Kalahari Formation). Arsenic mobilization may occur in the finer-grained clays of the Kalahari aquifer, if the groundwater conditions become anaerobic with dissolved iron and manganese. Arsenic in groundwater may also occur in association with sulfides and in mineralized areas in the Copper Belt.

Zimbabwe. Arsenic contamination of groundwater occurs in the central zone of the country, which contains naturally occurring arsenic minerals in association with gold. Mining activity has a long history here, and so populations have clustered around mining areas. Even without mining, water resources could be potentially contaminated by arsenic by natural weathering of arsenic-bearing minerals.

Section 2. Region of the Americas

Argentina. In the Chaco-Pampean Plain in north-western Argentina, covering around 1 million km^2, about 1.2 million rural people in dispersed settlements use groundwater resources without treatment and are exposed to arsenic through drinking-water from groundwater aquifers with arsenic levels exceeding 50 μg/l. These aquifers comprise Tertiary loess deposits (in the Pampean Plain) and Tertiary and Quaternary fluvial and Aeolian sediments (in the Chaco Plain). Tertiary volcanism is thought to be the primary source of contamination, with groundwater arsenic concentrations in some places ranging from 100 to 2000 μg/l.

Bolivia (Plurinational State of). The Andean highlands of the Plurinational State of Bolivia have elevated arsenic levels in soils. Geogenic sources of arsenic in surface water are seldom reported for the Plurinational State of Bolivia; levels between 210 and 12 600 μg/l from surface water in the south-west have been reported, although it is not clear whether these were used as sources of drinking-water. Similarly, the Poopó basin catchment area (27 700 km^2) with Poopó Lake as its centre serves a population of 14 984 in the Plurinational State of Bolivia's Andean highlands. Surface water from tributary rivers and from Poopó Lake contains arsenic concentrations ranging from 90 to 140 μg/l without contamination from mining sources, so the arsenic must be attributable to geogenic sources.

Brazil. Geogenically occurring arsenic has been identified in an alkaline lake ecosystem in Nhecolandia within the Pantanal of Mato Grosso in central Brazil's Amazon River basin. Encompassing highlands consist of Precambrian and Palaeozoic rocks that contain iron, manganese and limestone, with Quaternary sediments covering the vast floodplain underlain by arsenic-containing clays, covered by an iron cemented coarse-grained sand overlain by fine white sands. When torrential rains accumulate from November to March, the plain floods, forming lakes that then desiccate into a mosaic of grasslands, marshes and forests. Arsenic is usually sorbed by metallic oxides and hydroxides, clay minerals and organic matter, but in Nhecolandia, two other more powerful adsorbents (bicarbonate and fluoride ions) prevent arsenic adsorption from occurring. Consequently, adsorption is not a factor in limiting arsenic concentrations, which leaves evapotranspiration as a mechanism that concentrates arsenic during the dry season. As arsenic(III) is the redox state, it may be proportionally higher in the sediment of the saline lakes of Nhecolandia with a low redox potential and a long residence period.

Canada. Geogenic sources of arsenic are found in glacially transported arsenic minerals in Alberta, Manitoba, Saskatchewan and Nova Scotia. New Brunswick (York and Charlotte counties) has been reported to have arsenic and heavy metals in elevated concentrations in rural domestic water supplies. This appears to be naturally occurring arsenic. Examples of other non-anthropogenic occurrences should be investigated.

Chile. In northern Chile, arsenic in drinking-water is mainly of natural origin due to the hydrological characteristics of the area and the predominance of Quaternary volcanism. Water treatment plants in northern Chile remove arsenic by a precipitation process. The level of exposure of the urban populations of northern Chile to inorganic arsenic fluctuates between 126 and 2363 μg/l per day. Ingestion of water was found to be one of the most important routes of exposure. The numbers exposed are of the order of tens of thousands.

Colombia. Arsenopyrite is found in association with other gold-bearing minerals especially in the northeast.

Guatemala. Geogenic arsenic has only recently become a matter of concern in Guatemala with the evaluation of arsenopyrite-associated ore bodies in the highlands.

Mexico. Contamination of groundwater resources by naturally occurring arsenic was first identified in the late 1990s in the states of Coahuila, Durango, Chihuahua in northern Mexico, Zimapán, Hidalgo, in central Mexico, Baja California Sur, Guanajuato, Sinaloa, Morelos and Sonora, with arsenic levels ranging from 30 to >200 μg/l. Arsenic occurrence appears to be associated with major sandstone aquifers and surface geology and is therefore similar to that of Argentina, where the depth of wells is inversely proportional to the concentration of arsenic. The number of those exposed to elevated levels of arsenic (10–2500 μg/l) is reported as 2 million.

Nicaragua. There is geogenic contamination of groundwater by arsenic in the north-eastern and south-eastern regions, which are close to mineralized areas. This zone is along the active, tectonic outer eastern margin of the Depression of Nicaragua, the Sébaco Valley near Matagalpa, with an area of 52 km^2 and a total population of 3 225 000. The main causes are linked with volcanism in the area.

Peru. One geogenically affected area is in the semi-arid region of southern Peru, where water containing elevated arsenic levels is drawn from rivers deriving from the Andes. Arsenic in the environment is a result of the natural mobilization of natural geological sources, mining and volcanism.

United States of America. The USA has arsenic contamination in groundwater (in both reducing and oxidizing environments) occurring in geological formations. Table 4 is intended to summarize the more important geogenic arsenic occurrences.

Table 4. Sources of geogenic-related arsenic in 36 states of the USA.

State	Geogenic occurrence
1. Alaska	Leaching of naturally occurring arsenopyrite that is often associated with gold deposits
2. Arizona	High-arsenic groundwater present in aquifers composed of felsic volcanic rocks
3. Arkansas	Shallow, Quaternary, alluvial aquifers
4. California	Highly mineralized water due to interaction with various unconsolidated geological deposits and layers of organic material; authigenic pyrite in organic-rich sediments
5. Colorado	Weathering and leaching of arsenic minerals, especially arsenopyrite in gold-bearing rocks
6. Connecticut	Calcareous meta-sedimentary bedrock from the Hebron Formation
7. Florida	The most highly concentrated arsenic occurred in the Leon-Lee belt along the Ocala uplift district extending to the south-western Flatwoods district; coastal barrier sediments from the Jacksonville Ocean Dredged Material Disposal Site; annual atmospheric deposition of Saharan dust rich in arsenic
8. Georgia	Transport mechanism that carries arsenic compounds along the bedrock under a Pleistocene barrier to the sea
9. Idaho	Arsenopyrite weathering in a matrix associated with gold; Miocene fluvial and lacustrine unconsolidated sedimentary aquifers and confining units underlying the rapidly developing Boise Metropolitan Area
10. Illinois	Mahomet aquifer with heavy concentrations of high-arsenic wells south of the Mackinaw River
11. Indiana	Oxidation of black shale fragments within glacial deposits in parts of north-eastern Indiana
12. Kansas	Leaching of arsenical minerals, some of which are associated with zinc deposits
13. Maine	Bedrock interbedded with Ordovician (Penobscot Formation) sulfidic pelite, quartzite, arsenian pyrite and arsenopyrite, intruded by Silurian granitic and dioritic plutons (muscovite, biotite, garnet granites) and inundated by the sea after the glacial retreat; three distinct water types have been identified: calcium bicarbonate, sodium bicarbonate and high arsenic
14. Massachusetts	Natural dissolution of arsenic-bearing minerals; Bald Mountain massive sulfide deposit
15. Michigan	Sulfide-bearing minerals, glacial deposits, shale and sandstone, which contains arsenic-rich pyrite
16. Minnesota	Glacially deposited fine-grain Late Wisconsinan till (10000–15000 years BP) and ubiquitous organic matter
17. Mississippi	The youngest soils (of the Delta and Interior Flatwoods) in the western part of the state have the highest mean arsenic concentration
18. Missouri	Redox conditions in both alluvial aquifers and buried bedrock valley aquifers
19. Nebraska	Varied conditions; highest concentrations were found in the Panhandle and the western Sand Hills, but isolated occurrences are found in the central Platt River valley and the East-central Dissected Plains
20. Nevada	Evaporative concentration of the Carson River (the primary recharge source) and mobilizing redox reactions involving metal oxides and sedimentary-organic matter
21. New Hampshire	Correlates with Late Devonian Concord-type granitic bedrock

(Continued)

Table 4. Sources of geogenic-related arsenic in 36 states of the USA (*Continued*).

State	Geogenic occurrence
22. New Mexico	Weathering of arsenical minerals
23. North Carolina	Geological contact between the stratigraphically underlying metarhyolites and the stratigraphically higher metavolcanics, with very low pH groundwater associated with pyrite along this contact
24. North Dakota	Arsenic trioxide in a glacial drift zone in south-eastern North Dakota associated with glacial drift aquifers in Dakota sandstone
25. Ohio	Weathering of arsenical minerals has been identified in the south-western and north-eastern areas of the state with iron concentrations above 1000 μg/l, suggesting that arsenic was released from iron oxides under reducing conditions
26. Oklahoma	Related to alkaline aquifers in the central Oklahoma aquifer, which encompasses about 8000 km^2, serves 37% of wells >90 m deep (which are public supply wells) with drinking water >10 μg arsenic/l, which together with another 13% of wells with water >10 μg/l from other agencies means that 50% of all drinking-water from public supply wells in central Oklahoma have arsenic >10 μg/l
27. Oregon	Formations of the Oligocene and Upper Eocene in the Willamette Basin
28. Pennsylvania	Mesozoic rift basin and Triassic- through Jurassic-age sedimentary sequences of sandstones and shales intruded by diabase with hydrothermal mineral deposits including arsenic in alteration zones and replacement deposits of copper and zinc in sedimentary rocks
29. South Carolina	Elevated arsenic in Coastal Plain and Piedmont
30. South Dakota	Related partly to felsic volcanic rocks where the groundwater is commonly alkaline; Quaternary Sand and Gravel and the Tertiary Oglalla Group; these units are also responsible for elevated arsenic in groundwater in Nebraska to the south; pH conditions probably control the adsorption of arsenic, which is present in its highest concentration locally in volcanic ash in the Arikaree Formation
31. Tennessee	Southern Cumberland Plateau in the vicinity of Sewanee capped by Pennsylvanian Warren Point Sandstone and Sewanee Conglomerate, in which arsenite is interbedded with shales and coals and haematite bands
32. Texas	Concentrations are highest in the Ogallala aquifer and much lower in the Dockum aquifer; potential sources of arsenic include volcanic ash beds in the Ogallala, black shales in the Cretaceous (Kiamichi Shale) and saline lakes
33. Utah	Millard County is a desert area with groundwater arsenic concentrations of 180–210 μg/l
34. Vermont	Elevated levels of arsenic in groundwater are related to the presence of the St Catherine Formation
35. Washington	Arsenopyrite weathering
36. Wisconsin	Ordovician carbonate and clastic rocks in association with sulfide mineralization in St Peter Sandstone

Uruguay. The source of geogenic arsenic appears to be from volcanic ash in the Raigón Formation that also encompasses proximate areas of Argentina (Santa Fe and Buenos Aires Provinces).

Virgin Islands of the United States. The Virgin Islands of the United States has increased levels of arsenic deposited on its soils by the unusual case of Saharan dust rich in arsenic being transported from Africa by summer westerly winds across the Azores and the North Atlantic to the Caribbean. This African dust has elevated arsenic derived from gold-bearing rocks.

Section 3. Asian Region

Afghanistan. A team (United Nations Children's Fund (UNICEF), the Afghani Water and Sanitation Group and the Danish Committee for Aid to Afghan Refugees) initiated a rapid investigation in the eastern and western regions to be followed by the central and northern regions. In July 2003, of 647 wells in the 15- to 65-m depth range, 565 were negative for arsenic. In Logar Province, 8 wells were found with arsenic concentrations above 10 μg/l, and in the Ghazni metropolitan area, 56 wells contained arsenic above 10 μg/l. In October 2004, more public wells were tested in Logar and Ghazni, with the following results: in Logar, 282 wells were tested, of which 239 were negative and 43 contained arsenic in the range 1–100 μg/l; in Ghazni, 171 wells were tested, of which 97 were negative and 56 contained arsenic in the range 10–50 μg/l. The population exposed to elevated arsenic in drinking-water in the Ghazni and Logar areas is estimated at 500 000, but the other two regions–the central and northern–are not factored into this 21 November 2004 estimate.

Armenia. Armenia has extensive and varied mineral resources. Geogenic sources of arsenic are associated with natural weathering of arsenic in association with sulfide polymetallic ores, including gold, silver, lead, zinc, copper, molybdenum and iron.

Bangladesh. Bangladesh is the classic case of hydrological mismanagement, as deep borehole wells were drilled in the 1970s without adequate testing of the water for arsenic contamination. Today, the most recent estimate of those exposed over 2 decades to water containing arsenic at concentrations ranging from 10 to 2300 μg/l is 57 million over an areal extent of 150 000 km^2. The number of arsenicosis cases is estimated at 5 million.

Cambodia. Elevated arsenic concentrations (>50 μg/l) in groundwater in Cambodia are intimately but not exclusively associated with young Holocene sediments in the Mekong and Bassac river deltas downstream of Phnom Penh. However, high risk levels of arsenic (10–50 μg/l) are also found in Quaternary deposits and even occasionally in older deposits. The 2004 estimate of the population exposed to arsenic levels above 50 μg/l is 320 000. More detailed mapping of arsenic-affected wells is making it possible to indicate correlations between Holocene sediments and potential elevated-arsenic wells, but the caveat remains that each well must be tested, because there are many exceptions to this general pattern. New methods are being explored for correlating arsenic in hair and nails as early warning signals for arsenic ingestion because, until 2005, there had been a lack of positive symptoms of arsenicosis from people ingesting water in arsenic-elevated areas, probably because their duration of exposure had not been sufficiently long.

China. Geogenic zones associated with arsenic contamination are areas subject to seasonal aridity with low flushing and recharge rates of young anaerobic Holocene aquifers. This leads to high arsenic concentrations in wells 20–30 m deep. This situation occurs in areas such as Inner Mongolia. In 2005, the World Bank estimated that 5.6 million people are exposed to arsenic concentrations above 50 μg/l in just three provinces: Inner Mongolia, Shanxi and Xinjiang (the last being related to petroleum extraction, see below) over an areal extent of 68 000 km^2. With a case rate of 5%, this accounts for 283 000 cases of arsenicosis. The total for those in the rest of China exposed to geogenic arsenic contamination is probably about another 2 million; with a case rate of 5%, this would compute to another 100 000 cases of geogenic arsenicosis. The total exposed to arsenic levels above 50 μg/l in China is therefore about 7.6 million, and the projected case total is 383 000. In the Province of Taiwan, arsenicosis is a medical classic, since blackfoot disease with an etiology of cutaneous fibrosis resulting in gangrene was first identified in 1922, but was not seriously addressed until 1958. In the south-west of the province, 100 000 were exposed, and this is where remediation began. Today, most in this region have access to alternative water supplies. However, artesian wells still remain a problem in the north-east whenever they are deep (100–280 m) and, therefore, anaerobic. In the north-east of the province, 10 000 are exposed to arsenic levels above 50 μg/l over an area of 6000 km^2.

India. India is affected along the Ganga-Padmameghna-Brahmaputra plain that extends into Bangladesh. The north-eastern provinces (West Bengal, Assam, etc.) are affected, as well as Bihar, Uttar Pradesh, Madhya Pradesh and Jharkhand. In West Bengal, more than 6 million are exposed to arsenic at concentrations above 50 μg/l, 200 000 arsenicosis cases have been diagnosed and 1 million new cases are projected. Those exposed to arsenic concentrations of 10–50 μg/l would probably add 1 million to the total for West Bengal. Those exposed to concentrations above 10 μg/l in the other north-eastern provinces would conservatively total 3 million. The total number of people exposed to more than 50 μg/l for north-eastern India is 9 million. The Gangetic Plain is another area of arsenic contamination. The total number of people exposed in Bihar in 12 districts on the banks of the Ganges River has been estimated at 540 000. To this should be added the populations exposed to arsenic concentrations above 10 μg/l in West Bengal, the other north-eastern provinces, Madhya Pradesh and Bihar, which total about 1 million. The total for India is therefore 12 million people exposed to arsenic concentrations above 10 μg/l.

The case of West Bengal, covering an areal extent of 23 000 km^2, is very prominentbecause the numbers of those exposed to elevated levels of arsenic (>50 μg/l) has expanded so dramatically and has become recognized as a critical public health calamity. Much remediation has been under way for the last 10 years, For instance, in 1996, only 1 million people in 560 villages were thought to be at risk; the percentages of contaminated water from wells showed that 27% contained arsenic concentrations of ≥50 μg/l, with a range of 50–3700 μg/l. Then in 2000, of the 19 districts of West Bengal, 9 districts (representing 85 blocks) were considered to be arsenic-affected (>50 μg/l). In 2004, there were an estimated 26 000 contaminated wells, and these served in excess of 5 million private and 114 000 public sources. The 2005 estimate is that 6 million people from 9 districts (total population ~50 million) of 18 total districts (total population ~80 million) are exposed to arsenic concentrations above 50 μg/l, and >300 000 cases are probable.

New areas of arsenic contamination are constantly being identified. For instance, areas south of Kolkata have now been added to those other eight districts (to give a total of nine districts) as new housing starts and hotels are being built there on landfill with underlying arsenic-rich sediments.

Bihar in the Middle Gangetic Plain has a population of 450 million, but the number exposed to arsenic levels above 50 μg/l is uncertain. One researcher claims that most are exposed. If so, this would dramatically increase the total exposed for India, but we have chosen not to indicate any numbers for Bihar until the testing is more conclusive.

Indonesia. Geogenic arsenic at a maximum of 65 μg/l has been detected in the lowlands of south Sumatra, Indonesia, by Winkel *et al.* (2008) during hydrogeochemical survey conducted there. The large area covers 100,000 km^2 and is thought to be vulnerable to groundwater arsenic enrichment under reducing aquifer conditions. This may be the first published groundwater data for the highly populated island of Sumatra. (Winkel *et al.* 2008).

Iran (Islamic Republic of). Based on partial field studies, rural villages have water containing geogenic arsenic at concentrations higher than 1000 μg/l, with mean concentrations of 290 μg/l, 6 times greater than the Iranian standard. Primary investigations have shown that the source of this arsenic is mainly geogenic due to dissolution of arsenic-containing compounds in the Earth's crust.

Japan. Geogenic hidden arsenic has been discovered in Uranouchi Inlet, Tosa Bay, Shikoku Island, Lake Biwa, Shiga Prefecture and Honshu Island, indicating that hidden arsenicals can occur in both saline and fresh water. Uranochi Inlet exhibits increases in hidden arsenicals during biologically elevated seasons. Lake Biwa, near Kyoto, exhibits hidden arsenicals in the southern eutrophic basin where arsenic(V) increases 2–4 times during summer and is positively correlated with levels of iron, manganese and phosphorus.

Jordan. Concentrations of arsenic in sediments of the King Talal Reservoir as measured from 1987 to 1989 ranged from 1.53 to 2.80 mg kg^{-1} (Gideon, 1991). This potentially affects a population of 2.1 million.

Kazakhstan. In addition to petroleum and mining contamination, the drawdown of the Aral Sea has concentrated contaminants, including geogenic and anthropogenic arsenic, to high levels.

Lao People's Democratic Republic. A population of 400 000 in 1400 villages is exposed to arsenic at concentrations ranging from 10 to 112 μg/l. The same strictures as for Cambodia about the identification of elevated-arsenic wells apply to Lao People's Democratic Republic: each well must be tested, because Holocene aquifers are no true index of the elevation of arsenic levels, as tainted wells can occur in Quaternary and older sediments.

Malaysia. Peninsular Malaysian surface sediments of the Johor Strait have been found to contain several trace elements including arsenic (27.30 ± 17.11 ug/g). The population of Johor alone is 2.75 million. Elevation of arsenic levels in the vicinity of the Malacca Straits has been reported to originate from terrestrial, marine and atmospheric sources.

Mongolia. A national survey on arsenic was conducted by the Public Health Institute, Ministry of Health, with UNICEF technical and financial support from October 2003 to July 2004. These studies revealed that the country average for arsenic in wells tested is 14 μg/l and that 10% of all wells tested were positive. If we extrapolate that 10% of the wells serve 10% of the rural population of 990 000, the total population exposed to arsenic levels above 10 μg/l comes to 99 000. It was independently estimated that 300 000 are currently at risk from arsenic in drinking-water. As of 1989, 1774 people had been diagnosed with arsenicosis-related ailments. However, this figure is evidently too low, as 82.4% of people in high-risk areas showed symptoms but were not positively diagnosed.

Myanmar. Myanmar has an arsenic-exposed population of 3 million. A rapid survey showed that 35% of 125 tube wells had arsenic concentrations in excess of 50 μg/l.

Nepal. Arsenic concentrations of up to 2620 μg/l have been reported. Although levels are typically much lower than this, there is a wide variation in arsenic concentration. Water quality testing has shown that about 5% of samples (from a total of 14 932) have arsenic concentrations greater than 50 μg/l, whereas 23% exceed 10 μg/l. Based on these figures, the estimated population exposed to levels of arsenic above 50 μg/l is over 500 000.

Pakistan. Pakistan has a reported population of 128 000 exposed to water containing arsenic at concentrations ranging from 10 to 906 μg/l, mainly along the riverbeds of the Indus valley in the south-west Punjab, Muzaffargarh District and the Sindh. The case rate is 140 per 1000, which translates into 17 920 diagnosed cases projected. The geological phenomenon is that the depth of wells appears to follow no overall correlation with arsenic contamination, so each well, even if it is adjacent to a non-contaminated one, must be tested. Also, the situation is not static. In rural areas, arsenic concentrations are generally below 25 μg/l because arsenic in oxic shallow groundwater is sorbed to aquifer sediments. Canal irrigation in semi-arid regions can lead to widespread waterlogging, and evaporative concentration of salts can lead to arsenic concentration in shallow groundwater to levels above 10 μg/l.

Philippines. The Philippines is a major producer of chromite, copper, gold, nickel, coal, gypsum, silver and sulfur, most of the ores that are associated with arsenic occurrences. Therefore, the Philippines are at risk for arsenic contamination of surface water and groundwater from at least two sources: volcanism and mining, as well as perhaps geogenic arsenic sources that naturally leach into the environment.

Saudi Arabia. Jubail petrochemical complex covers an area of about 170 km^2 on the southern Saudi coast of the Arabian Gulf. About 40 km^2 was lanfilled with sand from the nearby desert to construct petrochemical industries. The area has witnessed a considerable growth in the size of the complex since 1992 which is impacting the use of shallow aquifers (up to about 30 meters in depth). These shallow aquifers are inter-connected to the Arabian Gulf so that contamination of them may adversely impact terrestrial vegetation and the Gulf marine ecosystem. Reports indicate that these shallow aquifers are threatened by the leaching of industrial effluent containing arsenic into the soil and the marine ecosystem.

Sri Lanka. The 2004 tsunami brought up black sediments that had resulted from Gangetic–Brahmaputra outwash from West Bengal, India, and Bangladesh over the last 10 000 years. These are the same sediments that are responsible for the arsenic crisis in those areas. These Gangetic black sediments have penetrated from tens to hundreds of metres inland. The effects of toxic exposure to this deep sea sedimentation which had now been brought to the surface were immediate. In the south-west town of Matara, residents complained of itchiness after contact with wave water. In the south-east town of Hambantota, a slimy water texture was reported. The land and the littoral areas have become the repository of arsenic contamination. At risk are agricultural crops and aquatic plankton, fish and other animals in the food-chain that will ingest this toxin. Remediation also poses a threat, as accumulated debris concentrates arsenic, which can then leach over time into groundwater.

Thailand. Approximately 250 000 are at risk from exposure to elevated arsenic, with 2000 skin cancers attributed to arsenicosis from exposure to arsenic levels above 50 µg/l. The first case of arsenicosis was diagnosed in 1987. Testing in other provinces has not been published but because of similar geology is predicted to be positive.

Turkey. Geogenic arsenic occurs in Turkey from arsenic minerals embedded in dense clay matrices that promote the mobilization of arsenic into the groundwater. One instance is in the Hisarcik area that encompasses Igdeköy, Emet and Kütahya villages.

Uzbekistan. Uzbekistan is bounded by two river systems, the Amu Darya and the Syr Darya, which join the Aral Sea. Overuse of the Amu Darya for irrigation has caused serious depletion of the Aral Sea, reducing its volume by almost 75% and leading to an increasing concentration of geogenic and anthropogenic toxins, including arsenic and various salts, in the water. The concentration of arsenic in the Syr Darya has been measured as 1.7 mg/l.

Viet Nam. The two areas most affected by arsenic contamination are the Red River (Song Hong) in the north, including the Hanoi area and Hanam Province, and the Mekong delta in the south, extending into the bordering area with Cambodia. A small area in central Viet Nam near Hai Phong is also affected. In northern Viet Nam, estimates of people exposed to arsenic concentrations above 50 µg/l range from as low as 0.5 million to 10 million. We estimate that the population at risk of exposure to arsenic levels above 10 µg/l (10–430 µg/l) in Viet Nam is potentially 5.8 million people. This range of exposure is dependent on the seasonal monsoon flooding period when arsenic concentrations are diluted and the dry season when they are concentrated, so the season of testing must be recorded to assess the extent of contamination properly. Arsenic in the groundwater of Cuu Long delta of the Mekong River in southern Viet Nam originates in the upper Mekong basin in Cambodia, is present in concentrations ranging from 10 to 30 µg/l and is usually but not necessarily associated with Holocene aquifers.

Section 4. European Region

Albania. Arsenic contamination related to geogenic sources is due to extensive mineral deposits of chromite and copper ores that are associated with secondary or trace amounts of arsenic minerals, such as Hollingworthite, Anduoite and Sperrylite, which can leach into surface water and groundwater through natural weathering.

Austria. Geogenic sources of arsenic are associated with arsenic minerals that are found in many locales in Austria. For instance, most prominent is arsenopyrite, but others include Loellingite (named after the locality in which it was originally identified), which is an iron arsenate, sometimes found in association with cobalt and nickel. If the iron is replaced by more than 50% nickel, the mineral is Rammelsbergite, and if it is replaced by more than 50% cobalt, the mineral is Safflorite. These arsenic minerals, although rare, have an environmental impact.

Belgium. Geogenic hidden arsenic has been identified in the Scheldt estuary in Belgium. The absence of this hidden species of arsenic during the biologically dormant period of the estuarine and lacustrine year may be attributed to depressed metabolic levels of phytoplankton and algae; with lower levels of methylarsenicals or arsenicals that are excreted but not detected by the traditional hydride generation method, they are hidden from traditional detection methods.

Bosnia and Herzegovina. Geogenic sources of arsenic are related to Bosnia and Herzegovina's mineral resources, which include bauxite, coal, industrial minerals, iron ore, lead and zinc. Natural leaching of arsenopyrite and related arsenic minerals are responsible for the presence of arsenic in surface waters and groundwaters.

Czech Republic. Geogenic sources of arsenic have been detected from natural weathering of arsenic-rich mineralized granodiorite and arsenic-rich volcano-sedimentary rocks in the vicinity of the Mokrsko gold deposit. Surface waters and groundwaters were supersaturated with crystalline and hydrous ferric oxides and calcium–iron and potassium(barium)–iron arsenates of arsenic(V), but undersaturated with scorodite, the oldest secondary arsenate mineral in the district. In the stream water, ponds, groundwater and pore water in the district. evidence of high concentrations of arsenite and arsenic(III), the most toxic species, was present.

Estonia. Geogenic arsenic retention in peat bogs occurs in conjunction with extremely high concentrations of iron (43% by weight), lead (2.5% by weight) and other metals and metalloids. The long-term exposure of the peat bog to metal-rich groundwater is a natural analogue to municipal solid waste landfills. Geochemical and transport models have identified different distributions of metals at different depths with iron; arsenic and manganese are in a lower anaerobic–aerobic stratum.

Finland. Finland has many geologically distinct regions containing higher than average concentrations of arsenic. The most prevalent arsenic mineral in Finland is arsenopyrite, not the arsenic-bearing pyrite that is significant in many other parts of the world. Elevated concentrations of arsenic are found especially in the Tampere–Hämeenlinna region, where there are also several gold occurrences. These deposits have been found to contribute to elevated arsenic levels in surface water and groundwater.

France. Geogenic hidden arsenic has been identified in the Beaulieu and Seine estuaries in France. The absence of this hidden species of arsenic during the biologically dormant period of the estuarine and lacustrine year may be attributed to depressed metabolic levels of phytoplankton and algae; with lower levels of methylarsenicals or arsenicals that are excreted but not detected by the traditional hydride generation method, they are hidden from traditional detection methods.

Germany. Geogenic arsenic is also found in Germany. Northern Bavaria, an area of 8000 km^2, has been found to have elevated levels of arsenic released from the Upper Triassic Keuper Sandstone into the groundwater, and no anthropogenic sources have been detected in the area. A survey of approximately 500 deep-water wells revealed that 160 had arsenic at concentrations ranging from 10 to 150 μg/l. The hypothesis for the release of arsenic is that it is related to specific lithofacies of the aquifers and is geogenic in origin.

Greece. Geogenic sources of arsenic have been identified in the drinking-water of the northern cities of Aksios and Kalikratia. Of 21 representative locations, arsenic concentrations were typically 10–70 μg/l. Aksios groundwater has redox potential values of 87–172 mV and a pH of 7.5–8.2, with arsenic(III) predominating. Manganese concentrations were mostly above the European Union standard of 50 μg/l (100–700 μg/l). In Kalikratia, groundwater has redox potential values of 272–352 mV and a pH of 6.7–7.5, and arsenic(V) predominated. In addition, Pleistocene Drama lignites in the Periphery of Central Macedonia are enriched with arsenic in the range of 102–606 mg/kg. In the Serres Prefecture deposit, the arsenic range is 856–1131 mg/kg. In both deposits, arsenic may have been hydrothermally present in the coal-forming peat bogs from mobilized sulfides (arsenopyrite, sphalerite and galena) or been supplied in the Eocene-Oligocene volcanism.

Hungary. Geogenic sources of arsenic in Hungary occur in zones of arsenopyrite, which leach into surface water and groundwater, such as in the Matra Mountains in north-central Hungary, at Recsk and at Gyongyosoroszi. Recsk is characterized by porphyry copper–polymetallic veins and a copper/gold skarn overlain by an argillic alteration in a matrix of Palaeogene volcanic andesitic rocks on a basement of Triassic limestone. Within the argillic alteration, copper–gold ores occur in association with arsenopyrite. The formation has weathered into an upland valley where there are several small villages.

Italy. Geogenic arsenic in Italy is attributable to highly mineralized areas in Tuscany, Sardinia and Sicily, often with supplemental volcanogenic mechanisms that further aid in the dissolution of arsenic from arsenopyrite and other arsenic-bearing minerals. Additionally, geogenic hidden arsenic has also been found in skate (*Raje* spp.), hake (*Merluccius merluccius*) and blue whiting (*Micromesistius poutassou*), which measured elevated organic arsenical compounds in decreasing amounts, respectively.

Macedonia (The former Yugoslav Republic of). Geogenic arsenic contamination can be identified in four areas adjacent to monitored arsenic contamination hot spots in the Periphery of Central Macedonia in Greece. Groundwaters in the municipalities of Novaci and Prilep adjacent to the Pella Prefecture in Greece contain arsenic concentrations of at least 11–30 μg/l. Gevgelija and the area around Dojran Lake in Star Dojran municipality adjacent to the arsenic hot spots in Kilkis Prefecture in Greece also have arsenic contamination of at least 11–30 μg/l.

Poland. Poland has arsenic-contaminated groundwater in southern and central regions, including Krakow, Silesia and Lodz. This was discovered when high concentrations of iron, zinc, tin, copper, lead and arsenic were found in Polish beers that made use of groundwater sources from this area. The residual arsenic in the human body from these sources has also been confirmed by analysis of human bone.

Portugal. Geogenic hidden arsenic has been identified in the Tagus estuary in Portugal, where hidden arsenic amounted to 19–25% of the total arsenic in the estuarine water. The absence of this hidden species of arsenic during the biologically dormant period of the estuarine and lacustrine year may be attributed to depressed metabolic levels of phytoplankton and algae; with lower levels of methylarsenicals or arsenicals that are excreted but not detected by the traditional hydride generation method, they are hidden from traditional detection methods.

Romania. Romania has concentrations of arsenic in drinking-water associated with arsenic mineralization. This is probably related to the area being adjacent to the Great Pannonian Plain.

Serbia. Portions of Serbia bordering the southern part of the Great Pannonian Plain have been exposed to elevated concentrations of arsenic in a geological situation identical to that of Hungary.

Slovakia. Geogenic sources of arsenic are related to three types of arsenic mineralization: quartz-sulfidic ore veins in Palaeozoic rocks with calcium–magnesium–bicarbonate; carbonate-sulfidic ore veins in Palaeozoic rocks with calcium–magnesium–bicarbonate–sulfate; and sedimentary manganese deposits in Palaeogene sediments with calcium–bicarbonate–carbon dioxide.

Spain. Geogenic sources of arsenic are associated with the mineralogy of certain areas. In a series of tests of groundwater, researchers found that the Rio Tinto ore field, which is rich in copper, arsenic and selenium and contains 6000 mg arsenic/kg, is potentially leaching into the groundwater. Another study has identified naturally occurring arsenic leaching in the groundwater of the Madrid basin.

Sweden. Geogenically occurring arsenic in drinking-water has been experimentally determined from a survey completed in 2003 in which zones of high arsenic content in tube wells in the central and northern regions were found. These areas of arsenic concentration were also found to be hydrologically similar to the concentrations of arsenic found in water from dug wells and from surface water.

Switzerland. Geogenic arsenic occurs in four major areas in Switzerland: in the north-east where several mineral and thermal springs are located, in the Jura, where it occurs with iron, limestone and clays, in the Alps, where it is found in ore deposits and crystalline formations and isolated thermal and mineral springs, and in Lake Greifen, a eutrophic lake with a seasonally anoxic hypolimnion. In summer, arsenate disappears in the epilimnion and is replaced by arsenite, whereas arsenate remains the dominant species in the hypolimnion.

United Kingdom of Great Britain and Northern Ireland. Geogenic sources of arsenic are linked to highly mineralized areas such as Cornwall, where arsenopyrite and other arsenic minerals are found in association with copper and tin. In addition, another unusual pathway of geogenically occurring arsenic is through the eating of mutton from sheep that have grazed on kelp from naturally occurring arsenic-tainted littoral sediments. Thirteen different arsenic species (arsenite, arsenate, dimethylarsinic acid, monomethylarsonic acid, trimethylarsine oxide, tetramethylarsonium ion, arsenobetaine, arsenocholine, dimethylarsinoyl ethanol and four common dimethylarsinoylribosides or arsenosugars) were identified in seaweed extracts and in the urine and blood serum of seaweed-eating North Ronaldsay sheep from northern Scotland. Related to this is the occurrence of geogenic hidden arsenic in the Southampton and Tamar estuaries of southern England. The absence of hidden species of arsenic during the biologically dormant period of the estuarine and lacustrine year may be attributed to depressed metabolic levels of phytoplankton and algae; with lower levels of methylarsenicals or arsenicals that are excreted but not detected by the traditional hydride generation method, they are hidden from traditional detection methods.

Section 5. Region of Australia and Oceania

Australia. Surface soil and groundwater in Australia have been found to contain high concentrations of arsenic. At Stuarts Point, New South Wales, elevated arsenic contamination in the groundwater is associated with coastal barrier sand dune environments. The extent of this hydrogeochemical occurrence of arsenic in groundwater in coastal communities needs to be investigated. The number of exposed people is not known.

New Zealand. Two cases are associated with natural leaching of arsenic into the environment: at Waikato River near Hamilton, in which the arsenic level measured 32.1 μg/l, and at Ohakuri Lake, where arsenic concentrations of 38.3–441 μg/l were measured. The number of exposed people is not known.

MINING-RELATED ARSENIC

All arsenic-containing minerals, especially sulfides, which are associated frequently with gold ores, are a source of arsenic. Mining and smelting of these minerals create environmental hazards of arsenic leaching into groundwater and surface waters from slag pits, waste dumps, extraction basins and derelict mines.

Figure 5 is a global map of 74 countries/territories with mining-related arsenic.

Section 1. African Region

Algeria. An old Sb-mine in a semi-arid area of the Northeast Algeria. The mining lodes are localized on the slopes of a mountain at the bottom of which flows out a wadi whose edges are used as pasture. Results show a severe pollution, as values of arsenic are definitely higher than those which are allowed in 70% of the soils (55 to 450 mg/kg^{-1}). It is to be predicted that similar Asenic leachings are to be found in similar mining locations in Algeria.

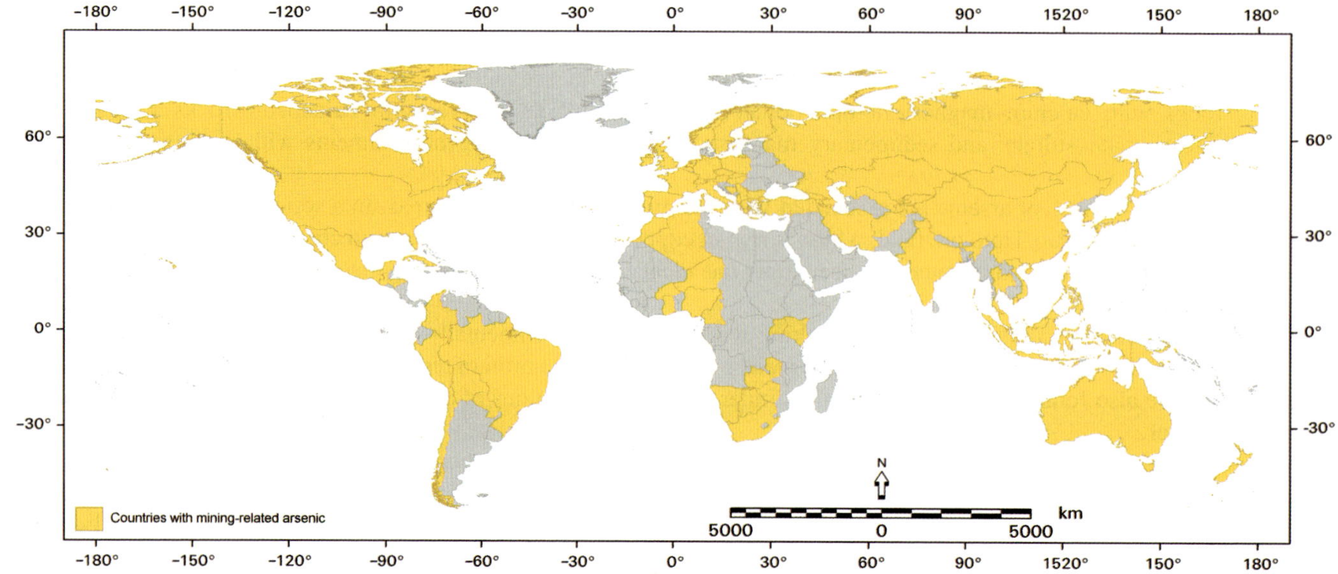

Figure 5. Global map of 74 countries/territories with mining-related arsenic.

Botswana. Botswana has relied on mining, especially for gold, from its earliest days. Historic gold mine waste dumps occur in the Francistown area, where both acidic and alkaline tailings have been discovered. Acidic tailings, which are the result of pyrite weathering, were found at Monarch South and Shashe mines. These surficial deposits with a high concentration of arsenic leach into the surface water each time it rains and provide an ample source of serious groundwater pollution. Furthermore, when the mines are abandoned and the pumps used to keep them dry are removed, they fill with water and leach directly onto the surface, and the acid mine drainage finds a way into regional groundwater or surficial water bodies.

Burkina Faso. Burkina Faso has a rich reserve of gold deposits in the north, occurring along with quartz and occasionally sulfide minerals bearing arsenic in veins from a number of areas in the crystalline basement (mainly Birimian-age rocks, as in Ghana). With the exception of the Essakane and Poura deposits, which have been mined by larger commercial operations, most gold mining activity is artisanal. Arsenic problems occur in some Burkina Faso groundwaters. The high concentrations were taken to be the result of oxidation of sulfide minerals (e.g. pyrite, arsenopyrite) and are likely to be concentrated in and around the gold mineralized areas.

Cameroon. In the Bétaré Oya region, in east Cameroon, with a population of 11,866, gold mineralization was implicated with As, Bi, Ag, Mo, W, Pb and Au. The Bétaré Oya region geology consists of Proterozoic (Birrimian) volcano sediments covered by 25 m of weathered ferruginous mantle. It was found that As is strongly associated with Fe and widely distributed.

Ghana. The economic gain from mining, unrealizable by the ordinary citizen, is achieved at great environmental cost, as slag and tailings as well as dust from smelters have contaminated soils and groundwater and surface water with high levels of arsenic. The most serious problems are in Obuasi and Prestea, where ores are rich in arsenic-bearing pyrite and arsenopyrite.

Kenya. The Migori Gold Belt, Kenya, which up to 20,000 people have invaded from outside, revealed concentrations of heavy metals, mainly Hg, Pb and As above acceptable levels. The Nairobi mine tailings had 76.0 mg kg^{-1} As.

Morocco. The Moulouya Oued basin, an early 20th-century lead mining region, covers 74,000 km^2 with sub-basins of Moulouya Kert, Isly, Kiss, Chott Tigri, and a portion of the Bouarfa-Figuig zone. Over 2.2 million people live in this region. High concentrations of arsenic in tailings have been leaching into the groundwater.

Namibia. As of 23 November 2011, Namibia Custom Smelters (NCS) is still awaiting a final verdict on allegations that its smelting of Bulgarian copper is causing illnesses at Tsumeb. Members of the Tsumeb community (with a population of 12,190) and NCS employees are alleging that emissions from the Bulgarian copper concentrate, which is said to contain arsenic, pose a serious hazard to their health and the environment.

Niger. Niger's geology is similar to that of neighbouring Burkina Faso, in that gold occurs with quartz and occasionally sulfide minerals (including arsenopyrite in veins from a number of areas in the crystalline basement—mainly Birimian-age rocks), as well as deposited in alluvial gravels. It is in this environment that mining, mainly for gold, occurs. Once disturbed, the arsenic-bearing minerals can leach into the surface water and groundwater.

Nigeria. Arsenic-rich effluents from mining (gold and tin) occur in the same geological matrix as in Burkina Faso and Ghana: Birimian-age rock with ores rich in arsenic-bearing pyrite and arsenopyrite.

South Africa. Acid mine drainage represents the most widespread and pervasive mining-related impact in three basins identified. Almost all of the mineralized targets for mining consist of sulfide ore bodies or are contaminated with arsenopyrite. Whenever these ores are exposed to air and moisture, the sulfide minerals present start to oxidize via a well-known set of autocatalytic reactions to give high concentrations of total dissolved salts (particularly sulfates), low pH values and high concentrations of dissolved metal ions (especially iron). The resulting solution is toxic to most forms of aquatic life and can lead to dramatic changes in ecosystem functioning, as well as changes in the structure and chemical composition of soils. The process of sulfide oxidation is very difficult to stop once it has started, and the resulting acid mine drainage can persist for centuries, as proven in some English lead and tin mines. Acid mine drainage in high-sulfur coalfields within the Olifants Basin is exacerbated by both the number of collieries and their size, as well as the decades of operation and the relatively abundant supplies of water available in the area that can disperse the contaminants.

Uganda. The abandoned Kilembe copper mine in western Uganda in the township of Kilembe with a population of 7,914, is a potential source of arsenic and heavy metals from mine tailings that leach into the Rukoki River flowing through wetlands into Lake George. An investigation of lake water and shore and lakebed sediments indicated heavy metals associated with the Kilembe sulphide mineralization.

Zambia. Mining is a major cause of pollution in the Zambezi Basin, causing water acidification and discharging a variety of highly toxic metals, including arsenic. In addition, disposal of solid mine waste in the Zambian Copper Belt, without due attention being paid to the correct siting and operation of tailings dams and proper treatment of effluent, has polluted the Kafue River.

Zimbabwe. International and national mining has been a prominent aspect of the economy of Zimbabwe, but it is now in an uncontrolled spiral that has led to the arsenic contamination of aquifers reaching an all-time high. Examples include the Mutare River and the Odzi River. Groundwater adjacent to the mines is also contaminated with arsenic.

Section 2. Region of the Americas

Bolivia (Plurinational State of). Polymetallic deposits containing arsenic-bearing ores such as the Porco Mine are at risk, since poor oversight is the norm. This mine released 210 000 tonnes of toxic tailing slurries, including arsenic, into the Agua Castilla River in 1996, but no remediation was performed and no public health assessments were undertaken. In addition, arsenic-related mining contamination is rather serious in the Bolivian Amazon.

Brazil. Arsenic contamination in the Iron Quadrangle, Minas Gerais state, at the site of hydrothermal gold deposits, has been a problem since the later 1990s, when 2.8 million tonnes of tailings with a median arsenic concentration of 14 500 mg/kg were deposited along valleys without any environmental concern. From 1983 to 1993, more than 2.1 million tonnes of arsenic were discarded directly into rivers, and from 1962 to 1975, an arsenic trioxide factory produced about 90 tonnes of arsenic trioxide, without environmental monitoring.

Canada. According to a 2003 report, 2838 tonnes of arsenic effluents were discharged from various plants associated with mining operations. In addition, Muggah Creek, Sydney, Nova Scotia, the site of coke ovens and tar pits, relics from previous mining and smelting operations, has been cited (2004) as Canada's worst arsenic-contaminated site.

Chile. Cases of surface water contamination have been caused by the discharge of untreated heavy metals in mining effluents, as in the case of the El Salvador Mine or Planta Osvaldo Martínez Mines, but there is no documentation of the monitoring of these mining operations. The Antofagasta Region is generally at risk of arsenic contamination of drinking-water from mining operations.

Colombia. Colombia has 200 000 artisanal gold miners, many of whom take their ore for processing to six urban centres with a combined population of 150 000 inhabitants (Segovia, Remedios, Zaragoza, El Bagre, Nechí and Bolivar). The number impacted with As is potentially of the order of 180 000.

Cuba. As exposure via the Delita gold mine (which was closed in 1997) in Isla de la Juventud has resulted from oxidation of As(III) in groundwater or from As(V) in surface waters due to leaching of arsenopyite within a mica schist matrix into the El Soldado River and adjacent watersheds of the Majagua, Itabo, and Los Indios Rivers. La Mina Spring and its associated ground waters encompass a 72 km^2 zone, and inhabitants in this zone have historically exhibited cerebrovascular disorders and cancerous tumors. The closed Zinc-lead Santa Lucía Mine has tailings measuring 1724 mg/kg for As and has polluted local streams.

Guatemala. Mining in Guatemala needs further assessment because As concentrations are dependent on rainfall variations from wet months (May to October) and dry months (November to April). The Marlin mine in San Marcos District, western

Guatemala, is a case in point. Leaching from the mine into downslope creeks has impacted villages with elevated concentrations of As in blood (3.20-8.50 μg/l) and urine (0.04-16.71 μg/l).

Honduras. The San Martin gold mine in the Valle de Siria of central Honduras is alleged to have "polluted local rivers, destroyed forests, and dried up the local water supply." Further complicating the environmental impact of this mining venture is that the extraction process utilizes cyanide.

Mexico. Mining is a major cause of high arsenic in groundwater. Although leaching of slag is a concern, airborne particulate arsenic can enter the water supply or food-chain. In northern Mexico, arsenic concentrations in dust are associated with ore smelters that have contaminated several areas across the border in Texas, USA. Furthermore, inactive mines are even more hazardous, because acid mine drainage goes unnoticed and unmonitored.

Paraguay. The Agua Castilla River, which is used by the local population for drinking-water, was contaminated by 210 000 tonnes of toxic tailing slurries containing arsenic on 29 August 1996 by a dyke that ruptured at the Porco Mine upstream in Bolivia.

Peru. The Doe Run Company (based in St Louis, Missouri, USA) mining smelter in La Oroya, which has been in operation since 1922, had gone unmonitored until 1995 and created not only lead poisoning in 98% of the children but arsenic poisoning as well.

United States of America. Mining sulfide-bearing rocks has led to high arsenic concentrations in surface water and groundwater, especially in the western states, including Montana, California, South Dakota and Alaska. Arsenopyrite with gold ore is common at most of these deposits. Arsenic concentrations in groundwater and sediment can be transported hundreds of kilometres downstream from these sources, and an additional threat is from acid mine drainage from many abandoned mines and mine tailings, which can be either oxidized acidic or neutral, as in California's Mother Lode and Nevada's polymetallic deposits. Table 5 summarizes some of the more important sites.

Table 5. Sources of mining-related arsenic in 14 states of the USA.

State	Sources of arsenic from mining
1. Alaska	Gold mining, which exposes arsenopyrite to weathering
2. Arizona	Leaching of arsenic into the groundwater of many Arizona mineralized areas
3. California	Melones Fault Zone in the Mother Lode Gold Belt with rocks that typically range from <100 to 8000 mg arsenic/kg
4. Colorado	Weathering and leaching of arsenic minerals, especially arsenopyrite in gold-bearing rocks
5. Florida	Mining phosphate rock releasing arsenic, cadmium and radium
6. Idaho	Gold mining promotes the release of arsenopyrite
7. Indiana	Iron mining with associated arsenopyrite
8. Kansas	Arsenic mineralization in association with zinc mining
9. Nevada	Mining of gold related to these Carlin-type gold deposits with arsenopyrite in the Roberts Mountains thrust
10. New Mexico	Mining of uranium in association with arsenical minerals in sandstones of the Westwater Canyon and Brushy Basin Members and in the Jackpile Sandstone of the Jurassic Morrison Formation and in the Salt Wash Member of the Morrison Formation in the Shiprock district
11. Oklahoma	Mining of zinc and arsenic in Tar Creek Basin in north-eastern Oklahoma, near Peoria in Ottawa County
12. South Dakota	Mining of gold associated with arsenopyrite deposits
13. Tennessee	Mining of uranium in association with arsenical minerals at Kingston, Poplar Creek and Rogers quarries
14. Washington	Associated with gold mining sites

Section 3. Asian Region

Afghanistan. Mineral resources may be some index of arsenic contamination of groundwater: reserves of copper occur south of Kabul, and petroleum occurs south of Mazar-i-Sharif. For centuries, Afghanistan has been an important source for copper, gold and iron ore but the infrastructure to utilize these resources is crippled or missing.

Armenia. The Debed River, a tributary of the Kura River, was highly polluted with copper and zinc and associated arsenic compounds discharged from the Alaverdi mine in north-east Armenia (which used to yield 50 000 tonnes of cathode copper annually) from the Kanaker aluminium plant and local copper mines.

Azerbaijan. In the area around Nakhichevanskaya, mining for copper, molybdenum, zinc and lead also exposed arsenopyrite and other arsenic-bearing ores that are still present in abandoned mine tailings, where they are leaching into the groundwater.

China. Arsenic has been found leaching into groundwater and surface water from mining operations in zones where arsenopyrite-associated minerals are found, such as in Anhui, Jilin, Ningia, Qinghai and Sichuan. Populations exposed are in the tens of thousands.

Georgia. In Tyrnyauz (Kabardino-Balkaria, the North Caucasus), tungsten and molybdenum had been extracted and enriched since 1945 with arsenic in associated minerals. Environmental safeguards were never implemented. Consequently, 5527 tonnes of suspended solids, 0.1 tonne of molybdenum and 0.07 tonne of arsenic flow annually into the Baksan River.

India. In Madhya Pradesh, Rajnandgaon District, arsenic contamination of drinking-water was first recognized in 1999. Gold mining had been undertaken in Chowki Block. In the worst-affected village, Kondikasa, arsenic in drinking-water had a concentration range between <10 and 880 μg/l. By implication, the gold mining or natural leaching of arsenopyrite contributed to this contamination. In Rajasthan, release of arsenic into the surface water and groundwater has been noted in two regions, Zawar and Khetri, from mine tailings with high concentrations of arsenic.

Indonesia. The mining of heavy metals in Sulawesi has led to the discovery of anthropogenically induced arsenic poisoning in Buyat Bay, the site of the US Newmont Mining Company's gold mine waste site. This issue continues to be controversial, despite an Indonesian court's ruling in favour of the mining company. To date, 300 villagers are exposed to the leaching of ore containing arsenic at a concentration of 666 mg/kg.

Iran (Islamic Republic of). Mining sources of arsenic are pre-eminently found in Kurdistan Province in the western part of the Islamic Republic of Iran. Here arsenic occurs naturally in association with many minerals and metals. Arsenic related to mining is found in the area of the Sari Gunay gold deposit. The Islamic Republic of Iran is the world's 57th largest gold producer. Arsenic may leach into many groundwaters by mining extraction.

Japan. Arsenic-contaminated water supplies were identified in 1971 on Kyushu where the Toroku mine, Matsuo mine and Saganoseki smelter were responsible for leaching arsenic into the river, contributing to arsenicosis cases among the population of two villages, Matsuo Kogai and Sasagadani Kogai, in Miyazaki Prefecture. Air and soil pollution were also discovered. This situation has since been remedied.

Kazakhstan. Kazakhstan has a large mining industry and is a producer of beryllium, tantalum, barite, uranium, cadmium and arsenic. The mining practices prior to 2000 were old-fashioned, and the tailings have been responsible for contaminating drinking-water with metals and metalloids like arsenic for years. One important advance that has been made is the production of a stable form of arsenic from arsenopyrite. This was accomplished at the gold mine in Auezov in north-eastern Kazakhstan in 1994, the first commercial application of the redox process to pre-oxidize sulfide ore concentrates. The Bakyrchik ore is basically quartz, with microscopic gold associated with arsenopyrite and pyrite mineralization.

Korea (Republic of). Three abandoned gold mines, Jingok, Cheonbo and Sino mines, had high concentrations of arsenic(V) under oxic conditions, with total concentrations of arsenic in the mine tailings up to 62 350 mg/kg. This leached into the groundwater, producing drinking-water resulting in an arsenic intake totalling 320 μg/day per person in 1975. The condition of similar mining operations in the Republic of Korea is not known, nor is any remediation effort known.

Kyrgyzstan. Twenty three tailing dumps and thirteen waste rock dumps are scattered throughout Mailuu-Suu (with a population of 26,000, which is the location of the Soviet uranium plant that operated from 1946 to 1968 and processed more than 10,000 metric tons of uranium ore). Unsecured radioactive waste combined with with high seismic activity threatens to contaminate the drinking water supply of the entire Ferghana valley, 22 000 km^2, stretching throughout Kyrgyzstan, Uzbekistan, and Tajikistan, which has a population of 11 342 000. Arsenic is associated with these highly toxic tailings and much work needs to be done in the proper remediation of the tailing dumps.

Malaysia. Malaysia holds the world's greatest reserves of tin and extensive deposits of other heavy metal ores. The uptake of arsenic and heavy metals by aquatic and terrestrial biota around gold and complex sulfide mining operations was investigated at selected sites in Malaysia. Arsenic in filamentous algae from flooded alluvial gold workings showed evidence of acute accumulation (up to 7000 mg/kg). By implication, high arsenic concentrations are present in Malaysia's groundwater near alluvial gold working. This needs to be investigated. In 2005, an environmental paper reported that ex-tin mining pools, contaminated with heavy metals and metalloids such as arsenic, could be contaminating groundwater sources. In Kuala Lumpur, mining has potentially leached arsenic into groundwater. Abandoned mining sites also pose a critical threat and contribute to high arsenic levels in drinking-water. After operations are terminated, little monitoring is done. At the Penjom Mine, the reported concentration range of arsenic in the tailings is 5–3700 mg/kg; at the Bau Mine in Sarawak, Borneo, the range reported is 15–5000 mg/kg. Drinking-water for a nearby village of 300 has an arsenic concentration of 87 μg/l. From all of these sources, the exposed population is potentially in the tens of thousands.

Mongolia. Mongolia has extensive but largely untapped mineral resources, due to a poor infrastructure. Nevertheless, long-term programmes to exploit metallic and non-metallic minerals, such as coal, copper, fluorspar, gold, oil and silver, are under way with foreign investment. One problem is that adequate environmental safeguards are not being instituted, and several areas are experiencing elevated arsenic ranging from 18 to 75 μg/l in groundwater; these include Govi-Sumber, Dornod, Dornod Govi, Khatanbulag soum, Flat Govi-Altai Dornod and Sukhbaatar Dornod.

Philippines. Marinduque Mine on Marinduque Island about 160 km south of Manila was the site of a toxic spill of 1.4 million tonnes of pulverized mining waste and toxic tailings laden with arsenic. This spilled into the local river and surface waters, and 1500 people were exposed.

Russian Federation. The Tuva Republic is a 600 m-high mountain basin surrounded by the Sayan and Tannu-Ola ranges, with Mongun Tayga at 3970 m, the highest mountain in Siberia. The region bordering these mountains has arsenic-bearing ores and many mine tailings that pose a threat to the safety of the groundwater. In 2001, a report was issued that detailed the arsenic leaching of the tailings of Khovu-Aksy Mine into the groundwater.

Tajikistan. Arsenic occurs in association with gold and other minerals in the highly mineralized Zeravshan Valley, in Leninabad Oblast in far north-west Tajikistan, east of Samarkand and south of Tashkent in Uzbekistan.

Thailand. Ron Phibun District in Nakhon Si Thammarat Province in peninsular Thailand includes mining activities that were first recognized in 1987 as a source of arsenic poisoning, with 1000 cases diagnosed with arsenic-related skin disorders. In total, 15 000 people were exposed. The affected area lies within the South-east Asian tin belt. Arsenic in both surface water and shallow groundwater is thought to be leached by oxidation of arsenopyrite, exacerbated by a rise in groundwater levels. Highest arsenic concentrations (≤580 μg/l) were 2–7 km downstream of the mining area.

Turkey. In the Tuztafl gold–silver–lead–zinc–copper mineralized zone in the Kazda dome in the Biga Peninsula in western Turkey, arsenic occurs as the most significant geochemical anomaly in the northern half of the Ayale area. Granitoids intrude in this matrix where arsenic occurs both as a coherent marker for gold deposits and as a potential leachate.

Uzbekistan. In the Amantaytau region of the Kyzylkum Desert, three mining areas are associated with gold mining: Amantaytau Centralny, Amantaytau OP Gold Mine and Uzunbulak. These all have associated arsenopyrite deposits. In addition, arsenic minerals in association with gold and silver–lead ores were intensively mined in the mountain areas of Koksu and Chatkal adjacent to the Brichmulla-Nanai area.

Viet Nam. Extensive data confirm anthropogenically released arsenic from ores in the northern Viet Nam area. Arsenic in mining areas mineralized with sulfide minerals and gold are in Pia Oac, Ngan Son, Pac Lang, Chay River, North Tam Dao, Cho Don-Cho Dien, Dao Vien-Chiem Hoa-Na Hang-Tung and Bac Me. Arsenic also occurs in metamorphic intrusive volcanic formations containing arsenopyrite, pyrite and gold, such as in the Hien River, Binh Gia, east of the Chay River, in Nam Xe-Tam Duong and in the downstream part of the Da River. The arsenic concentration in gold ore in the basalts of the Viet Nam formation in Doi Bu (Hoa Binh) is 50–204 mg/kg. In sandstone, siltstone and the siliceous shale of the Than Sa formation in a mineralized zone of sulfur–gold–quartz in Khau Au-La Hien (Bac Can, Thai Nguyen), the arsenic concentration is 13.2 mg/kg; in the ore, this value is 1292–1442 mg/kg. In the sericite shale and clay shale of the Coc Xo formation in the zinc–lead ore area of Cho Don (Bac Can), the arsenic concentration is 98 mg/kg; in the ore, it is 8–262 mg/kg. The arsenic concentration in groundwater of the ore zone in Lang Vai (Dam Hong) is 730 μg/l.

Section 4. European Region

Albania. Albania has extensive mineral deposits of chromite, copper ore, natural gas, nickeliferous iron ore and petroleum, Several of these are associated with secondary or trace amounts of arsenic, which, when mined, have the potential to leach into the groundwater.

Austria. Austria has been exposed to mining, with arsenic released directly into surface waters.

Belgium. As of 2004, Belgium had 47 facilities engaged in the metal industry and metal ore roasting or sintering installations and installations for the production of ferrous and non-ferrous metals, emitting 1.8 tonnes of arsenic into the air and 0.6575 tonne of arsenic into water.

Bosnia and Herzegovina. Waste from aluminium production operations is discharged into Lake Skadar, which officially has been identified as "too polluted" to provide water for the city of Podgorica. Copper and uranium mining in Bor has led to extensive pollution of the Borska Reka River, a tributary of the Timok River, which is a tributary of the Danube. Solid wastes from the smelter are estimated at 22 million tonnes per year, including 91 tonnes of uranium and many more tonnes of arsenic per year.

Bulgaria. In the Srednogorie-Panagjurishte region, three mines were in operation, working a polymetallic deposit containing arsenic, germanium, gold, iridium, palladium and silver, in addition to copper. Gold is the most important economic element; however, the high arsenic content of the ore presents serious environmental hazards. Owing to the high arsenic content of the

Topolnitz tailing dam, the environmental law of 1991 caused the Government of Bulgaria to prohibit the processing of Chelopech ore. One result of mining such as this in Bulgaria was that groundwater became seriously polluted; in some areas, metal working factories have so polluted the soil and water with lead, arsenic and cadmium that the land is unusable for agriculture and rivers are considered dead or nearly dead.

Czech Republic. Mining has exposed many arsenic minerals in the Czech Republic. Deposits of gold, silver, copper, tin and tungsten in association with arsenic minerals have been mined since the Middle Ages. Arsenic concentrations in surface water near deposits average 0.5 mg/l, and those not in proximity to mine wastes average about 0.02 mg/l.

Finland. In Finland, 21 industrial, chemical, metals and mining sources discharge 1.75 tonnes of arsenic directly into the air, and 9 discharge 1.70 tonnes of arsenic into surface waters.

France. France has arsenic in abundance in association with arsenopyrite and gold. The acid mine tailings of many mines pose an environmental threat. For instance, acid mine drainage of arsenic in Carnoules has posed an environmental threat for some time.

Germany. Mining in Germany for arsenic-related minerals has a long history. The northern Palatinate region in Germany is characterized by elevated levels of arsenic and antimony in the soil due to the presence of ore sources. Former mining activities in a skarn complex east of Schwarzenberg District, Ergebirge, Saxony, has tailings where arsenic-bearing ores were mined for centuries. Since mining is so ancient in Germany, remediation of high-arsenic tailings that leach into the groundwater is an issue of great current interest.

Greece. The former mining site (nowadays used as a technological and cultural park) at the city of Lavrion, Greece, has arsenic pollution mainly from slags, sulfur-containing waste, smelting waste and low-grade lead concentrates. The existence of other arsenic trouble sites in Greece has not been determined.

Hungary. Mining in Hungary has exposed zones of arsenopyrite that are leaching into surface water and groundwater, such as in the Matra Mountains in north-central Hungary, at Recsk and at Gyongyosoroszi.

Ireland. Ireland has major stores of lead and zinc, and these have been mined at Tara Mine (County Meath, active), Glenadalough Mines (County Wicklow, abandoned), Galmoy Mine (County Kilkenny, active) and Lisheen Mine (County Tipperary, active). In addition, there are stores of silver in County Tipperary that was once mined at the Silvermine and stores of copper and sulfur in County Wicklow that have been mined at Avoca Mine. These deposits and their tailings suggest the probable presence of arsenic.

Italy. Arsenic in Italy is known from its occurrence on the island of Sardinia at the Furtei Volcanic Complex, where enargite, a copper arsenic sulfide (Cu_3AsS_4), is found as a potential source of mobilizable arsenic, which leaches into the groundwater. Until now, this has been minor, but is nonetheless of concern. In southern Tuscany, two mines had hazardous arsenic levels in their tailings in 1997 and were leaching into the groundwater. The current status is unknown.

Macedonia (The former Yugoslav Republic of). The Lojane mine has an open dumpsite for flotation waste, which was created by the over 1 million tonnes of tailings containing arsenic, antimony and various hazardous substances. Soils measured at the dump's edge contained 8.093 mg arsenic/kg. Lead and zinc mines are located at Zletovo, Toranica and Sasa. Contamination of water and soil by heavy metals has been observed near the Probistip lead and zinc mine since the area was first mined during the Roman period. Acid mine drainage has polluted the Kiselica and Koritnica rivers, which flow into the Zletovica River, which flows through Bregalnica and Vardar to the Aegean Sea, thus creating transboundary pollution problems.

Netherlands. Twenty-three chemical, mining and mineral plants discharge arsenic into the air, and directly and indirectly into surface water and groundwater.

Norway. Potential arsenic contamination is from the Killingdal area, where arsenic in massive basaltic sulfide deposits measured 1200 mg/kg at Lokken mine and 500 mg/kg at Skorovas mine. Additional testing of groundwaters is recommended in areas where mining of arsenopyrite takes place.

Poland. Mining in the vicinity of the Zelezniak former arsenic ore and smelting site in the Kaczawskie Mountains in the south-west has contributed to high levels of arsenic in the groundwater and in the Odra River sediments.

Portugal. Two metallic ore mines with arsenic minerals are still active, the Neves-Corvo with copper and antimony and the Panasqueira with tungsten and tin (which has been operating continuously since 1896).

Serbia. Mining sources of arsenic in eastern and central Serbia and northern Montenegro are found in porphyry copper and massive sulfide ores in the Bor metallogenic zone where very old mining deposits are still extant.

Slovakia. Arsenic contamination and its potential leaching into the groundwater can come from ancient, derelict mines, such as "Golden" Kremnica, "Silver" Banská Štiavnica and "Copper" Banská Bystrica.

Spain. Acid mine tailings leaching into groundwater have posed a constant threat, and a recent spill from a pyrite mine has confirmed fears about the existence of other such incidents that have gone unnoticed.

Sweden. Three mining smelters in the north of Sweden are rich in arsenic and have been hazardous to the groundwater into which they leached.

Switzerland. Mining has contributed to arsenic contamination, especially of the surface water and groundwater surrounding several former gold mines. In derelict mines and isolated groundwater drill holes with reducing conditions, arsenic concentrations can approach 300 μg/l.

United Kingdom of Great Britain and Northern Ireland. South-west England is at risk, with occurrences of arsenic ranging from 3 to 2700 mg/kg in mine tailings potentially leaching into groundwater and surface water. Arsenic is also at hazardous levels in the Caron and Humber rivers.

Section 5. Region of Australia and Oceania

Australia. In Western Australia, in the region adjacent to the silver, gold and copper tailings of the Nevoria Mine, arsenic levels are 500 mg/kg; at the gold, silver and zinc tailings of the Marvel Loch Mine, the level of arsenic is 2875 mg/kg. These tailings leach directly into the groundwater.

New Zealand. In regions of volcanic hydrothermal veins where gold and mercury are found, gold mining has exposed many veins, and so the possibility of arsenic being released into the groundwater in these regions must be considered.

Papua New Guinea. Elevated arsenic concentrations have been determined in dietary seafood and scalp hair of the population of 110,000 inhabiting the Fly River delta, with an area of 7 100 km^2. Other areas are also at risk due to mining operations.

PETROLEUM-RELATED ARSENIC

Elevated arsenic concentrations in aquifers are usually attributed to reducing conditions in which arsenic sorbs to metallic oxides such as iron(III). However, in anaerobic conditions, iron(III)-reducing microorganisms can couple the reduction of solid-phase iron(III) with the oxidation of organic matter, potentially releasing arsenic to groundwater. While this scenario has been well documented for cases of naturally occurring carbon, little information exists for cases in which anthropogenic sources of carbon, such as petroleum hydrocarbon compounds, are released into aquifers.

This is the key issue when dealing with oilfields of the world, notably in the Middle East oil corridor and surrounding countries, in Nigeria, where water is at such a premium, and in Ecuador, which is particularly vulnerable, as it is located at the headwaters of the Amazon watershed, which can contaminate all of Amazonia.

Apart from the vested oil interests in maintaining silence on environmental issues, there has been a lack of scientific understanding of the release mechanism of arsenic induced by anthropogenically introduced petroleum. Figure 6 is a global map of 17 countries/territories for which petroleum-associated arsenic release into groundwater has been ascertained.

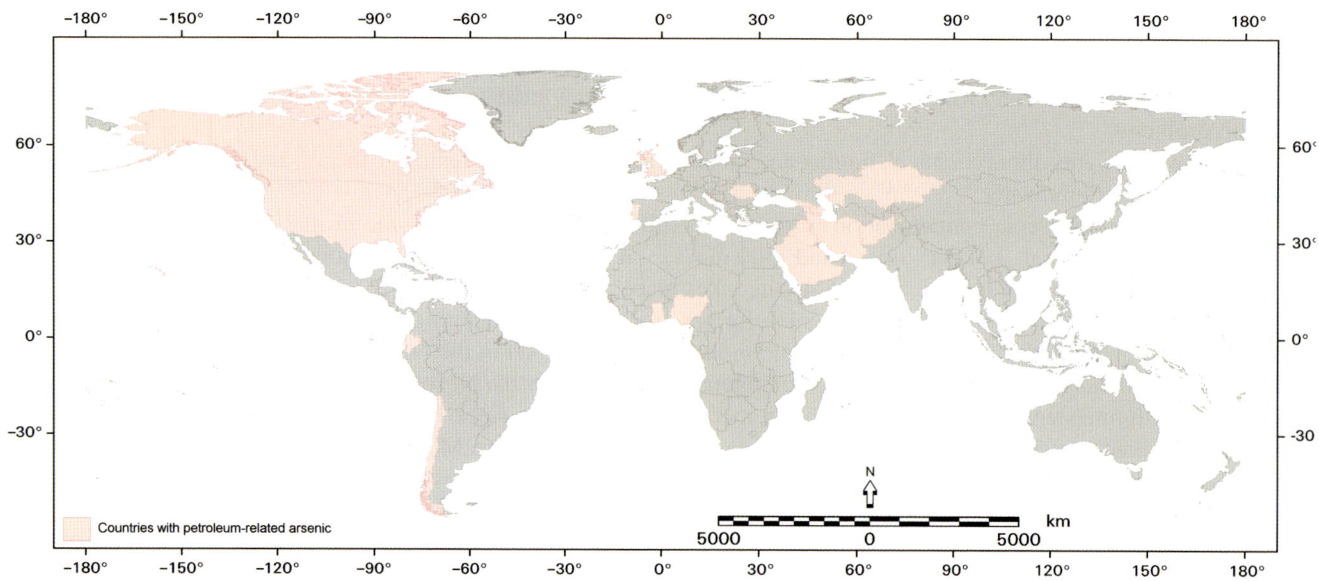

Figure 6. Global map of 17 countries/territories with petroleum-related arsenic.

Section 1. African Region

Ghana. A new member of the club of oil-producing nations, Ghana began oil production in December 2010. Arsenic from petroleum sources has not yet been reported in peer-reviewed literature for Ghana, but it is a potential concern for the future, as production increases from 80 000 to 120 000 barrels per day. Lessons from neighbouring Nigeria provide a cautionary tale.

Nigeria. Arsenic has been found in dug wells in the Niger Delta region and is currently attributed to the petroleum extraction practices in which toxic light crude oil containing a deadly cocktail of hydrocarbons, including benzene (a known human carcinogen), toluene, xylene, butane and propane, has facilitated the release of arsenic from geological sources. Long-term exposure has been linked to a variety of illnesses, including hypertension, diabetes, cardiovascular disease, infertility and cancers of the skin, lungs, urinary bladder and kidney. Other anthropogenic sources of petroleum have released arsenic and are related to industrial dumping by factories located on riverbanks that use the rivers as open sewers for their effluents. Oil spillages occur mainly in the rural areas where there is little or no monitoring, further endangering local sources of already polluted water supply and freshwater biota.

Section 2. Region of the Americas

Canada. The Alberta Athabascan oilfields have come under exploitation for petroleum extraction since the price per barrel reached US$ 60 in 2005. The areal extent of the fields totals 13.8 million hectares, or 21% of the province. Samples from the Quaternary-Tertiary buried channel aquifer contained arsenic concentrations above the analytical detection limit. Arsenic concentrations in the Quaternary-Tertiary buried channel aquifer appear to be controlled by a combination of the following mechanisms: (1) competition between arsenic species and phosphorus or bicarbonate ions for adsorption sites; (2) release of arsenic brought about by changes in redox potential; (3) dissolution of carbonate or evaporite minerals releasing associated arsenic; and (4) proximity to bedrock.

Ecuador. The arsenic pollution due to Texaco's petroleum extraction from the Oriente, which ended in 1992, left environmental destruction in its wake, with 42 million litres of crude spilled into the Amazon River, 12 million litres per day of brine and drilling muds high in arsenic dumped in unlined pits near drinking-water wells.

United States of America. Elevated arsenic concentrations have been detected in the groundwater at many field sites that are either oil extraction sites or petroleum leakage sites. Only three states have been identified in this report–Ohio, Oklahoma and Texas–and they are intended to be merely illustrative of the types of problems encountered with petroleum-associated releases of arsenic. For instance, petroleum-facilitated release of arsenic resulted in the contamination of groundwater along Paddys Run and New Haven roads in Ohio from a petroleum plant, with contamination consisting of arsenic, petroleum and benzene, which moved in three distinct plumes discovered in 1973, with cleanup beginning in 1986. Another example of suspected petroleum-facilitated arsenic release is in the areas of oil well drilling in the central part of Oklahoma and the Panhandle (comparable with Xinjiang Autonomous Region, China). Fort Worth, Texas, Joint Reserve Base, Naval Air Station, is another site. The source of the elevated concentrations of arsenic, iron and manganese in groundwater is thought to be mobilization from naturally occurring minerals as a result of reducing conditions produced by hydrocarbon degradation from petroleum spills. The groundwater at many more such military sites should be tested.

Section 3. Asian Region

Afghanistan. Petroleum reserves may contribute to arsenic release into groundwater. In 2006, the United States Geological Survey and the Afghanistan Ministry of Mines and Industry Joint Oil and Gas Resource Assessment Team found that in northern Afghanistan's Afghan-Tajik and Amu Darya basins and in four total petroleum systems that collectively encompass 87 000 km^2, there are 1596 million barrels of crude oil (plus natural gas) that have yet to be extracted.

Armenia. The petrochemical industry is based in Sumgait, Azerbaijan, Rustavi, Georgia, and Yerevan, Armenia, which collectively supply 70% of Russian petroleum demands. Oil and gas prospecting companies in the Baku-Sumgayit area and the North Caucasus (Krasnodar, Grozny and Maikop) are the epicentres of pollution of groundwater with the release of arsenic in anaerobic conditions.

Azerbaijan. See **Armenia** above.

China. The oil production centre in the Karmay petroleum fields in Xinjiang Autonomous Region has exacerbated arsenic leaching in the vicinity of Wusu (Usu) city, with a population of 100 000 potentially exposed. Kuitun city has s population of 50 760 potentially exposed. Another 463 rural villages are also exposed. Of a total Xinjiang population of 19.2 million (2000), 429 858 have been exposed to >50 mg/l arsenic, but only 3530 confirmed cases of arsenicosis-related diseases were officially recognized as of 2004.

Georgia. Large industrial facilities producing manganese and ammonia, together with arsenic, copper and gold mining and processing plants, oil refineries and power plants, polluted the river bodies of the Black and Caspian seas with heavy metals, oil products, phenols and other toxic substances. Petroleum products have been leached into aquifers by oil refineries and power plants, and arsenic has been released by the action of petroleum on underlying aquifers and has polluted the tributaries of the Black and Caspian sea basins. Heavy oil products and phenols have also heavily polluted the Kuban River.

Iran (Islamic Republic of). Groundwater adjacent to many of the ageing well heads and the inefficiency and wastage of oil in the extraction and storage process indicate that groundwater in adjacent aquifers is at risk of arsenic pollution, as petroleum hydrocarbons enhance the arsenic release mechanism. The National Iranian Oil Company's onshore production is geared only to maintaining output levels from large, ageing fields, where enhanced oil recovery (such as natural gas injection) is being implemented (i.e. at Marun and Karanj). Oil production apparatus in the Islamic Republic of Iran is old and inefficient and needs revamping, with advanced technology and foreign investment.

Iraq. Oil spills from the Basra refinery's damaged and leaking pipes, both above ground and underground, have polluted catchment sands, and spills went undetected during the initial post-conflict inspections in 2003. Some of the damage was apparently caused by military action by the USA and some from a buildup of pressure when the oil system was hastily shut down by the Iraqis. The resultant oil spills have seeped into the Iraqi sands, and the fate of the underlying aquifers is unknown but of concern, as petroleum is known to facilitate the release of arsenic.

Kazakhstan. The environmental situation has steadily worsened with the expansion of oil extraction. Historically, many collective farms created during the Soviet Union period were located very close to the oil and gas reserve fields at a time when extraction of resources was not extensive. During the last decade, oil and gas extraction has expanded significantly into the steppes due to the rising number of commercial projects. Arsenic leaching has been one result.

Saudi Arabia. Jubail petrochemical complex covers an area of about 170 km^2 on the southern Saudi coast of the Arabian Gulf. About 40 km^2 was landfilled with sand from the nearby desert to construct petrochemical industries. The area has witnessed the considerable growth in size of the complex since 1992 which is impacting the use of shallow aquifers (up to about 30 m in depth). These shallow aquifers are inter-connected to the Arabian Gulf so that contamination of them may adversely impact terrestrial vegetation and the Gulf marine ecosystem.

Section 4. European Region

Portugal. Petroleum-related arsenic releases into air and soils have been confirmed by bioaccumulation studies of mosses, with arsenic concentrations above 1.4 mg/kg in the region of Sines, where refineries and petrochemical plants release discharges and emissions.

Romania. Petroleum-related arsenic emissions are associated with the oil refineries in the south, which cause atmospheric pollution across the border in northern Bulgaria. The Romanian National Oil Company runs refineries in the Ploiesti and Campinu regions, which produce 60 000 tonnes of acid tars annually, a major agent in generating arsenic contaminants to the air and water supplies.

United Kingdom of Great Britain and Northern Ireland. Petroleum terminals fractionate petroleum products and emit arsenic.

Section 5. Region of Australia and Oceania

None.

VOLCANOGENIC-RELATED ARSENIC FROM GEOTHERMAL SPRINGS AND WELL EFFLUENTS

Arsenic released from geothermal systems into surficial and ground waters may impact drinking water, aquatic ecosystems and terrestrial biota. However, arsenic is not the sole contaminant in geothermal systems because it occurs in the context of a suite of other elements and chemicals such as antimony, boron, fluoride, hydrogen sulfide, lithium, mercury and thallium. For the purposes of this book we only consider the role of arsenic in geothermal systems.

Geologically, geothermal systems occur in two contexts: tectonic plate action (rifting, sliding and colliding) and magmatic action near crustal surfaces. Tectonic plate geothermal fields occur in the Pacific Rim, in the east African Rift Valley (Ethiopia, Kenya and Tanzania), the Rio Grande Rift Valley (Colorado and New Mexico), and in the Icelandic rift system. Magmatic geothermal systems can occur either as hot spots at shallow depths in the earth's crust (Azore Islands, the Big Island of Hawai'i and Wyoming's Yellowstone National Park). Therefore, the mechanisms for arsenic release in geothermal systems can involve faulting, magmatism and metamorphism.

A brief overview, Table 6, of the actual measured arsenic concentrations relating to geothermal springs and well effluent provides a snapshot of the range of localities affected.

Table 6. Arsenic Concentrations in Selected Geothermal Springs and Geothermal Well Effluents.

Chapter No.	Country	Location	Type	As, mg/kg	Reference
2	Chile	El Tatio	*Well effluent*	45–50	Ellis & Mahon, 1977
2	Mexico	Los Azufres, Mexico	*Geothermal springs*	0.001–4.0	Birkle & Merkel, 2000
2	USA	Hawai'i	*Well effluent*	<0.01–0.07	Webster & Nordstrom, 2003
2	USA	Salton Sea, CA	*Geothermal springs*	0.03–12	White, 1968
2	USA	Valles Caldera, NM	*Geothermal springs*	0.021–2.4	Criaud & Fouillac, 1989
2	USA	Yellowstone Nat. Park, WY	*Geothermal springs*	0.16–10	Stauffer & Thomson, 1984
4	Iceland		*Geothermal springs*	0.001–0.048	Amorsson & Linval, 2001
4	Italy	Phlegraean Fields	*Geothermal springs*	0.012–5.6	Celico *et al.*, 1992
3	Russian Federation	Kamchatka	*Geothermal springs*	2.0–3.6	Karpov & Naboko, 1990
		Kamchatka and Kurile Islands	*Well effluent*	2.0–30	Goleva, 1974
3	Russian Federation	Ebeko volcano, Siberia	*Well effluent*	0.19–28	Khramova, 1974
3	Japan	Tamagawa, Japan	*Geothermal springs*	2.3–2.6	Noguchi & Nakagawa, 1969
5	New Zealand	Wairakei	*Geothermal springs*	0.23–3.0	Ritchie, 1961
			Well effluent	1.0–5.2	Ritchie, 1961
5	New Zealand	Waiotapu	*Geothermal springs*	0.71–6.5	Webster, 1990
			Well effluent	2.1–3.9	Ellis & Mahon, 1977
5	New Zealand	Ohaaki and Broadlands	*Geothermal springs*	5.7–9.0	Ewers & Keays, 1977
5	Philippines	Mt Apo, Philippines	*Geothermal springs*	3.1–6.2	Webster, 1999

Since one major source of volcanogenic arsenic is through the effluent from geothermal energy plants, Table 7 provides an overview of worldwide geothermal energy plants as *potential* sites of arsenic contamination of surficial and groundwater. As fossil fuel prices increase, the economic attractiveness of geothermal energy as a renewable energy source for power generation will also increase. As geothermal energy production increases, heightened awareness and monitoring of hydrothermal effluents containing arsenic (and other potential toxic chemicals) from geothermal energy plants will be needed.

Figure 7 is a global map of the 35 countries/territories with volcanogenic-related arsenic.

Table 7. Seventy-eight Countries with Direct-use Geothermal Systems, 2010 (Lund *et al.*, 2010).

Country	Capacity, MWt	Annual Use, TJ/yr	Annual Use, GWh/yr	Capacity Factor
Albania	11.48	40.46	11.2	0.11
Algeria	55.64	1,723.13	478.7	0.98
Argentina	307.47	3,906.74	1,085.3	0.40
Armenia	1	15	4.2	0.48
Australia	33.33	235.1	65.3	0.22
Austria	662.85	3,727.7	1,035.6	0.18

(*Continued*)

Table 7. Seventy-eight Countries with Direct-use Geothermal Systems, 2010 (Lund *et al.*, 2010) (*Continued*).

Country	Capacity, MWt	Annual Use, TJ/yr	Annual Use, GWh/yr	Capacity Factor
Belarus	3.422	33.79	9.4	0.31
Belgium	117.9	546.97	151.9	0.15
Bosnia & Herzegovina	21.696	255.36	70.9	0.37
Brazil	360.1	6,622.4	1,839.7	0.58
Bulgaria	98.3	1,370.12	380.6	0.44
Canada	1126	8,873	2,464.9	0.25
Caribbean Islands	0.103	2.775	0.8	0.85
Chile	9.11	131.82	36.6	0.46
China	8,898	7,5348.3	20,931.8	0.27
Columbia	14.4	287	79.7	0.63
Costa Rica	1	21	5.8	0.67
Croatia	67.48	468.89	130.3	0.22
Czech Republic	151.5	922	256.1	0.19
Denmark	200	2,500	694.5	0.40
Ecuador	5.157	102.401	28.4	0.63
Egypt	1	15	4.2	0.48
El Salvador	2	40	11.1	0.63
Estonia	63	356	98.9	0.18
Ethiopia	2.2	41.6	11.6	0.60
Finland	857.9	8370	23.25.2	0.31
France	1345	12,929	3591.7	0.30
Georgia	24.51	659.24	183.1	0.85
Germany	2,485.4	12,764.5	3,546.0	0.16
Greece	134.6	937.8	260.5	0.22
Guatemala	2.31	56.46	15.7	0.78
Honduras	1.933	45	12.5	0.74
Hungary	654.6	9767	2,713.3	0.47
Iceland	1,826	24,361	6,767.5	0.42
India	265	2,545	707.0	0.30
Indonesia	2.3	42.6	11.8	0.59
Iran	41.608	1,064.18	295.6	0.81
Ireland	152.88	764.02	212.2	0.16
Israel	82.4	2,193	609.2	0.84
Italy	867	9,941	2,761.6	0.36
Japan	2,099.53	15,697.94	7,138.9	0.39
Jordan	153.3	1,540	427.8	0.32
Kenya	16	126.624	35.2	0.25
Korea (South)	229.3	1,954.65	543.0	0.27
Latvia	1.63	31.81	8.8	0.62
Lithuania	48.1	411.52	114.3	0.27
Macedonia	47.18	601.41	167.1	0.40
Mexico	155.82	4,022.8	1,117.5	0.82
Mongolia	6.8	213.2	59.2	0.99
Morocco	5.02	79.14	22.0	0.50
Nepal	2.717	73.743	20.5	0.86

(*Continued*)

Table 7. Seventy-eight Countries with Direct-use Geothermal Systems, 2010 (Lund *et al.*, 2010) (*Continued*).

Country	Capacity, MWt	Annual Use, TJ/yr	Annual Use, GWh/yr	Capacity Factor
Netherlands	1,410.26	10,699.4	2,972.3	0.24
New Zealand	393.22	9,552	2,653.5	0.77
Norway	3,300	25,200	7,000.6	0.24
Papua New Guinea	0.1	1	0.3	0.32
Peru	2.4	49	13.6	0.65
Philippines	3.3	39.58	11.0	0.38
Poland	281.05	1,501.1	417.0	0.17
Portugal	28.1	386.4	107.3	0.44
Romania	153.24	1,265.43	351.5	0.26
Russia	308.2	6,143.5	1,706.7	0.63
Serbia	100.8	1,410	391.7	0.44
Slovak Republic	132.2	3,067.2	852.1	0.74
Slovenia	104.17	1,136.39	315.7	0.35
South Africa	6.01	114.75	31.9	0.61
Spain	141.04	684.05	190.0	0.15
Sweden	4,460	45,301	12,584.6	0.32
Switzerland	1 0,60 9	7 7,14 6	2 1,43 1	0 23
Tajikistan	2.93	55.4	15.4	0.60
Thailand	2.54	79.1	22.0	0.99
Tunisia	43.8	364	101.1	0.26
Turkey	2,084	36,885.9	10,246.9	0.56
Ukraine	10.9	118.8	33.0	0.35
United Kingdom	186.62	849.74	236.1	0.14
United States	12,611.46	56,551.8	15,710.1	0.14
Venezuela	0.7	14	3.9	0.63
Vietnam	31.2	92.33	25.6	0.09
Yemen	1	15	4.2	0.48
Total	**50,583**	**438,071**	**121,696.0**	**0.27**

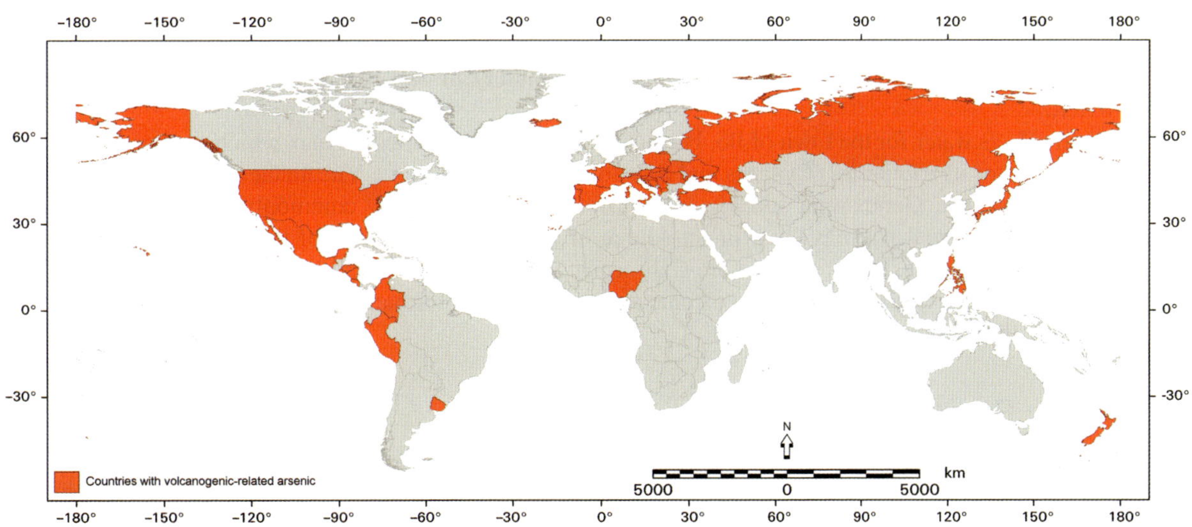

Figure 7. Global map of 35 countries/territories with volcanogenic-related arsenic.

Section 1. African Region

Nigeria. Evidence of volcanism is found in the Biu Plateau in the north-east, which is composed of basaltic lava flows. Relatively recent (perhaps Holocene) cinder cones with uncertain eruption dates with well-defined craters are aligned in a north/north-west–south/south-east direction, and they often contain abundant peridotite xenoliths.

Section 2. Region of the Americas

Colombia. Volcanogenic arsenic in association with antinomy occurs in association with sixty hot springs in Cundinamarca Department, one of which in Yacopi had a concentration of 37 μg/l.

Costa Rica. Volcanogenic sources of arsenic in Costa Rica are just beginning to be investigated at three localities (Rincón de la Vieja, Miravalles and Choco-Suela Platanar). The Rincon de la Vieja Volcano in the province of Guanacaste and Alajuela, 27-km northeast of Liberia City, experienced a phreatic eruption on September 16 and 22, 2011, the first after 13 years of dormancy. Surficial and groundwaters are affected by contaminants, including arsenic; fish were found dead in the River Penjamo, and streams and ditches in the area were infused with sediment which was observed 18 kilometers downstream, with communties potentially affected including Aguas Claras, Canas Dulces and Curubande.

Dominica. This is an island nation that lies on a major volcanogenic zone in the Caribbean Sea. This is evidenced by the volcanic ridge of the island, which is still being formed by volcanogenic activity. Perhaps half of its population is exposed to arsenic levels above 10 μg/l in its thermal waters.

El Salvador. This is the smallest of the seven Central American countries, having an area of 21 040 km^2 and a population of 6.47 million (2003). It is the first and largest producer of geothermal energy in Central America. However, highly mineralized waters containing elevated concentrations of arsenic represent a major problem for the exploitation of the Ahuachapán geothermal field; if only a small proportion of its inhabitants are exposed to these high concentrations of arsenic in its drinking-water, this poses a serious health hazard.

Honduras. Volcanogenic arsenic exposure is related to at least four identified geothermal hot springs at Azacualpa, Pavana, Plantares and Vanes Caldera with concentrations of 70, 110, 1260 and 1160 μg/l, respectively.

Jamaica. Jamaica has a geothermal anomaly in St Elizabeth that was reported in 1999. This is similar to the volcanogenic geology of other islands in the Caribbean and poses a health threat to those drinking untested water from geothermal springs on the island. There is also possible contamination of Jamaican foods by waters tainted by arsenic.

Mexico. Evaporation ponds associated with geothermal production leach arsenic at the Los Azufres geothermal field.

Nicaragua. Arsenic occurs naturally in groundwater in the north-eastern and south-eastern regions along the active, tectonic outer eastern margin of the Depression of Nicaragua, the Sébaco Valley near Matagalpa, with an area of 52 km^2 and a total population of 3 225 000. In the sub-watershed south-east of the valley, high concentrations of arsenic from hydrothermal processes of Tertiary volcanism have been detected. The basement rock produced an intense hydrothermal alteration of faults and fractures close to the flowing groundwater, the conduit of arsenic. Drilling for a drinking-water well above hydrothermal faults or fractures allows arsenic to readily enter wells. Lake Xolotlan, with a surface area of 1000 km^2, has both geogenic and anthropogenic arsenic. A large number of hot springs leach arsenic(III) into the lake. Also, arsenic(III) in the range of 10.2–30.1 μg/l is released from the geothermal plant in the Momotombo Volcano on the north-west coast of the lake. As a result, most of the 10 000 inhabitants of the local community of San Francisco Libre are exposed to arsenic in their drinking-water.

Peru. Volcanism affects water for the town of Ilo, because it is drawn from Aricota Lake, which contains elevated levels of arsenic as a result of two of its tributaries (Callazas and Salado) passing through the Yucamane volcano region.

United States of America. Geothermal springs in Yellowstone National Park commonly have arsenic concentrations exceeding 1000 μg/l. Geothermal water is the primary source of arsenic in other parts of the USA, including some surface water in the eastern Sierra Nevada. States affected are California, Montana, Nevada and Wyoming.

Uruguay. The source of As appears to be from volcanic ash in the Raigón Formation that also encompasses proximate areas of Argentina (Santa Fe and Buenos Aires provinces).

Section 3. Asian Region

Japan. Japan's island of Honshu is most noted for its hot springs in Akita and Aomori prefectures. The effect of the leaching of the arsenic into the groundwater has not been thoroughly examined but is well known; since the use of the hot springs is culturally important, little has been done to understand the potential health effects.

Philippines. High arsenic contamination on Mount Apo has been convincingly attributed to the geothermal plant of the Philippine National Oil Company; 39 cases have been diagnosed. In addition, Marinduque Mine, on Marinduque Island about 160 km south of Manila, was the site of a toxic spill of 1.4 million tonnes of pulverized mining waste and toxic tailings laden with arsenic, in which 1500 people were exposed.

Russian Federation. Fossil fuel power stations in the Kamchatskaya Oblast currently produce 69% of the total electric power generated in the Russian Federation. Hydroelectric and nuclear stations produce 20% and 11%, respectively. Geothermal power stations contribute only 0.06%, despite some growth in the number of power plants. Electric geothermal power is generated at the Kamchatka Pauzhetka power plant (11 MW). This has been joined by the new Verkhne-Mutnovka plant (12 MW) and two other units of 50 MW on the Severo-Mutnovka hydrothermal system. The total power of this plant is planned to be 300 MW. The effluent from these hydrothermal plants is of concern, since the waters being used are very high in arsenic.

The Northern subprovince includes 16 thermal spring clusters, including Tymlat, Palana and Rusakov, which are most attractive for geothermal use. In the Median subprovince, 24 hydrothermal springs include Esso and Anavgai springs and two high-temperature systems, Kireuna and Apapel. About one third of all springs are located in the Eastern subprovince. Thermal springs in the Nalycheva River valley provide water for agriculture and swimming pools. The Karym, Semyachik, Uzon and Geyser valleys in this subprovince are located in Kronotskii Reservation, where any industrial or agricultural activity is forbidden. Finally, 55 hydrothermal springs connected partially with the Pauzhetka, Nizhne-Koshelev, Severo-Mutnovka and Bolshe-Bannoe high-temperature systems are found in the Southern subprovince. The construction of a transmission pipeline approximately 80 km in length to transport waste geothermal fluids from the Severo-Mutnovka field is of critical concern, because of the high concentrations of arsenic. For instance, arsenic in the range 100–5900 µg/l is present in thermal springs in the Viluch River valley, at Nalychevo, the biggest in Kamchatka, which discharges between the Goryachaya and Zholtaya rivers, at Taloviye Thermal Springs, 6 km from Nalychevo, and at Malkinski on the left bank of Kluchoyvka River (left tributary of Bistraya River).

Two small geothermal power plants are in operation in the Kurile Islands (Kunashir and Iturup islands), with a capacity of 2.6 MW and 6 MW, respectively. Dissolved arsenic in geothermal springs on the Kurile Islands is present at concentrations in the range 100–5900 µg/l. A recent study has shown the fate and transport of these high arsenic concentrations in the groundwaters of these regions.

The Tuva Republic is a 600 m-high mountain basin surrounded by the Sayan and Tannu-Ola ranges, with Mongun Tayga at 3970 m, the highest mountain in Siberia. Many of the 8000 rivers are Yenisei River tributaries, and there are numerous hot springs that leach arsenic-bearing minerals into the groundwater.

Turkey. Volcanogenic sources of arsenic have been indirectly identified for Izmir residents dependent on the water from the thermal wells in the district of Manisa. Manisa is located in an area of high thermal activity, with hot springs and thermal baths that go back to antiquity. Indirect proof of the elevated arsenic in the groundwater is the fact that breastfeeding mothers have elevated arsenic in their milk (4.219 ± 0.079 µg/l), and cows have even higher levels (4.932 ± 0.38 µg/l).

Section 4. European Region

Albania. Volcanogenic sources of arsenic are potentially found in numerous thermal mineral springs in Albania with temperatures up to 65.5°C. Three main geothermal areas are Kruje, Ardenica and Peshkopi, with a combined geothermal potential of 39–63 GJ/m^2. In the Kruje geothermal area, the largest geothermal area in Albania, aquifer temperatures range from 145°C to 250°C, surface water temperatures from 30°C to 65.5°C, and flow rates from 3.5 to 15 l/s. Water temperatures from deep wells in the Ardenica geothermal area are 32–38°C; flow rates are 15–18 l/s. Thermal springs in the Peshkopi geothermal area, situated in north-east Albania in the Korabi hydrogeological zone, have a water temperature of 43.5°C and flow rates of 14–17 l/s. In addition to Kruje, Ardenica and Peshkopi, numerous thermal springs are found across Albania, situated primarily in tectonic fractures.

Austria, Bosnia and Herzegovina, Croatia, Hungary, Serbia, Slovakia, Slovenia, Ukraine. One of the areas most prone to geothermal arsenic is associated with countries surrounding Hungary's Great Pannonian Plain: Austria, Croatia, Romania, Serbia, Slovakia, Slovenia and Ukraine. In addition, Bosnia and Herzegovina should be added, with hot springs at Ilidze near Sarajevo, the chalybeate spring at Kiseljak, a spring impregnated with arsenic at Srebrenica, the Ilidža area (the location of the Roman town *Aquae Sulphurae*) and Banja Luka, the second largest city. Arsenic concentrations range from 10 to 50 µg/l, and the affected populations number in the hundreds of thousands.

France. France has volcanogenic sources of arsenic stemming from geothermal springs in the Massif Central, and their drainage into groundwater and rivers poses threats to the environment as well.

Iceland. Iceland's geothermal springs are associated with relatively young basalts, and in theory should pose no serious arsenic contamination threat; however, laboratory tests have shown that concentrations in the range 50–120 μg/l in thermal waters are not unusual, so there is a real health hazard in areas where effluent from geothermal springs has seeped into the groundwater or surface water. In view of this situation, since the use of geothermal energy has increased since 1980, the effluent from geothermal plants demands regular monitoring, especially those discharging into Lake Thingvallavatn.

Italy. Mining and volcanism are often closely interwoven in Italy. The tailings pond of the Campiano mine, which closed in 1995, feeds a spring with a discharge of about 16 l/s of acid mine drainage with elevated concentrations of sulfate, iron, copper, zinc and arsenic that is mixed with geothermal waters (also with arsenic possibly) with a pH of 4. This pond is leaching into the Merse River, which happens to be an area of particular natural and tourist interest.

Macedonia (The former Yugoslav Republic of). The source of arsenic in Macedonia potentially connected with volcanism lies in the south, adjacent to the Periphery of Central Macedonia in Greece. This area is associated with the Alshar metallogenic ore body, a Carlin type of gold deposit with 15 000 tons of arsenic, which exhibits a hydrothermal alteration of host rocks similar to those in Nevada. Total water resources in the country are estimated to be 6.63 billion m^3, which are carried in the Vardar basin (4.77 billion m^3), the Strumica basin (0.18 billion m^3) and the Crni Drim basin (1.68 billion m^3), which is the predominant drinking-water source, encompassing 30 springs with a flow greater than 100 l/s m^3. These springs have not been monitored for arsenic contamination, and they may be a possible source of arsenic contamination, as are similar springs in the neighbouring Periphery of Central Macedonia of Greece. In addition, there is a thermal power plant at Bitola ("REK Bitola").

Poland. Poland's natural geothermal springs located in the Sudetes Mountains at Cieplice and Lądek have associated arsenic concentrations that should be monitored. Little information is available about the arsenic concentrations in these waters.

Portugal. Volcanogenic arsenic may exist in hot springs that may have geothermal energy potential in the Minho district in the north-west. The 1527 m Larouco Mountains of Minho province form a western-facing escarpment of metamorphic crystalline schists, slates and quartzites and intrusive granitics, cut with fault lines from which hot springs emerge. The metamorphic and granitic matrices do contain some arsenic minerals, but the extent of their leaching into the hot springs is not well known. Furthermore, the geothermal potential has not been developed. Geothermal energy resources exist in São Miguel Island, Azores. A programme of evaluation and exploration with technical assistance from the United States Geological Survey was undertaken in the 1980s.

Romania. Portions of Romania bordering the southern part of the Great Pannonian Plain are potentially exposed to elevated concentrations of arsenic in a geological situation identical to that of Hungary in the presence of elevated concentrations of ammonium, iron, manganese, humic acids, dissolved gases and temperature >30°C.

Spain. In Spain, 46 chemical, mining and mineral plants as well as high-arsenic coal-burning and hydro-thermoelectric utilities discharge arsenic into the air and directly and indirectly into surface water and groundwater.

Switzerland. Major areas of volcanogenic arsenic occur in cold mineral springs and hot springs. In the Alps, arsenic occurs in isolated thermal and mineral springs. Granites and schists are the sources of arsenic in the thermal springs in the northeast Jura Mountains at Baden, Zurzach and Schniznach. In the southwest Wallis (Valais) region, thermal springs are found at Saxon Valroc, Combioula and Leukerbad and a cold mineral spring is located at Martigny-Brocard. The lower Engagardine has high arsenic concentrations in the cold mineral springs at Val Sinestra, Andeer and St. Moritz.

Section 5. Region of Australia and Oceania

New Zealand. Volcanogenic release of arsenic can occur in thermal springs related to hydrothermal deposits in areas where there is still active volcanogenic activity.

Section 1
African Region

1.1 ALGERIA (G, M)

Algeria, with a population of 37.1 million (2012 est., Algerian Office of National Statistics), has a Human Development Index of 96 out of 169 and a per capita gross national income (GNI) of US$ 7,658 (US Census Bureau, 2011; UNDP, 2010).

A study showed that As pollution originates both from mining activities and from naturally occurring As in the highly fractured rock in a zone extending from northeast Algeria into northwest Tunisia (Jdid *et al.* 1998). Naturally occurring arsenic in this zone pollutes wadis at the base of affected mountain slopes as well as the Medjerda (Majardah, Wadi Mejerha, Oued Majardah, and Bagradas) River and its principal tributary, the El Melah River. The Medjerda River, the longest river in Tunisia, originates in northeastern Algeria and flows east into Tunisia for 450 km before emptying into the Gulf of Utica in the Mediterranean Sea. Medjerda is Tunisia's crucial waterway providing water to the country's water supply facilities and to villages, towns and farms, and is crucial for the region's agriculture. At the Tunisian/Algerian border, arsenic concentrations in the Medjerda River can exceed 35 mg/l. In leaching tests on samples of ore, flotation tailings and an arseniferous mass indicate that, in the leachate solutions, arsenic levels reach 5,344 and 9,400 mg/l, for each respectively. Arsenic concentrations in the soils range from 5 to 450 mg/kg (Bentellis-Mosbah & Rachel, 2006).

1.2 BOTSWANA (G, M)

Botswana, with a land area of 600 370 km^2 (Central Intelligence Agency, 2009) and a population of 2.1 million, has a Human Development Index rank of 98 out of 169 and a per capita GNI of US$ 13 204, giving it a rank of 58 out of 172 (US Census Bureau, 2011; UNDP, 2010). Botswana has one of the most dynamic economies in Africa, which relies primarily (38%) on mining (of gold and diamonds) and services (44%), including tourism for its extensive and well-managed nature reserves. In 2009, the country suffered from the world's second highest known rate (38.8%) of human immunodeficiency virus (HIV)/ acquired immunodeficiency syndrome (AIDS) infection of adults between the ages of 15 and 49, after Swaziland, although it has a comprehensive programme for dealing with the disease (Avert, 2009).

Francistown, located in an arseniferous geological belt, is a mining city 420 km to the north–north-east of Gabarone, the capital, which more than doubled its population in the decade 1990–2000, having 113 315 residents (WHO & UNICEF, 2006; World Resources Institute, 2006; Botswana National Council on Population and Development (2010)). This rise in the urban population, coupled with a semi-arid climate with hot winters (30–35°C) and warm summers that exacerbate evapotranspiration, having an annual rainfall of 460 mm (which falls mostly in November–February), has severely impacted local water supplies. Furthermore, surface water discharges from impermeable surfaces in built-up areas carry toxic metals (including arsenic), pesticides and hydrocarbons and enter local rivers without treatment. Rural water supplies rely mainly on groundwater, but as the water table is falling, this source is stressed in many areas (Mafa, 2004).

Geogenic arsenic is found in relation to two formations. The first comes from the natural leaching from amphibolite or Archaean (3 billion years BP) Tati schist in which large reserves of gold are located in crystalline quartz reefs, in a 79 km^2 area embracing Francistown itself. The leaching of arsenic into the groundwater in this highly mineralized area puts Botswana's urban population at risk, especially from historic gold mine waste dumps with both acidic and alkaline tailings. At the abandoned Lady Mary mine, for instance, arsenic concentrations in groundwater from two boreholes measured 26 and 244 μg/l (Mafa, 2004).

Except for the Monarch mine, which is unique in the Francistown area, in that its mineralization is not associated with arsenic deposits, all other mine tailings in the Francistown area are characterized by amphibolite and Archaean Tati schist, which contain high levels of arsenic. The highest concentrations of arsenic in a single soil profile of 1 m depth in amphibolite and Archaean Tati schist were 3,215 and 5,288 mg/kg, respectively. Even when mines are active, tailings leach directly into the surface water. Furthermore, when mines are abandoned and pumps that had been used to keep them dry are removed, the shafts fill with water and leach acid mine drainage directly on to the surface, which ultimately drains into groundwater or surface water (Mafa, 2004).

The second source of geogenic arsenic is found in the Sowa and Makgadikgadi salt pans west of Francistown, which are also high in nitrates (Hutton & Lewis, 1980). These salt pans are relics of an early Holocene lake now desiccated owing to the Kalahari Desert, which has concentrated toxic contaminants at an alarming rate (Foster *et al.* 2004). In the Gomoti Valley, Lower Okavango Delta, researchers have been developing a geographic information system (GIS) approach to groundwater modeling of arsenic to help measure the rate of this contamination (Mased *et al.* 2006). Surface deposits have such a high concentration of geogenic arsenic that they leach into the surface water each time it rains and provide an ample source of arsenic for serious groundwater contamination (Mafa, 2004).

Prior to 2002, Botswana had no national drinking-water standard that could be used for monitoring the quality of drinking-water. The result was that drinking-water suppliers (e.g. Water Affairs, Water Utilities Corporation) and councils used company standards, which were sometimes based on the World Health Organization (WHO) guidelines. Then, in 1999, the Botswana Bureau of Standards, through its technical committee on water quality, developed a national standard that specifies three classes of drinking-water: class I which specifies limits of "ideal" drinking-water and is closely comparable to international standards for water quality; class II which is based on the WHO guidelines, and which set the maximum level of arsenic in drinking-water at 10 μg/l; and class III which set limits pertaining to special problems, such as the saline water in Tshabong, which, because of its toxicology, can be consumed continuously only for a maximum on the period of 1 year (Anon, 2002; Botswana Bureau of Standards, 2004).

1.3 BURKINA FASO (G, M)

Burkina Faso, with a population of 16.8 million and a land area of 273 800 km^2, ranks 161 out of 169 in the Human Development Index, with a per capita GDP of US$ 1,215 as of 2010 (US Census Bureau, 2011; UNDP, 2010). It is a landlocked country in the Sahel region of West Africa, bordered to the south by Côte d'Ivoire, Ghana and Togo, to the east by Benin and the Niger, and to the north by Mali. Land use comprises mainly woodland, forest and pastureland, with around 13% being used for arable agriculture. Some 90% of the population is engaged in subsistence farming. Principal crops produced are peanuts, shea nuts, sesame, cotton, sorghum, millet and corn (British Geological Survey, 2002a).

The geology of Burkina Faso is composed of predominantly Precambrian crystalline rocks: metamorphosed sediments, meta-igneous rocks and intrusive granite. The backbone of the country is composed of Archaean (3 billion years BP) crystalline rocks, but intercalations of younger (Birimian) crystalline rocks are also common. These are similar to the formations found in neighbouring Ghana and other parts of West Africa. The margins of a large ancient sedimentary basin (Precambrian to Ordovician age) occupy the western border area. This formation is represented by sandstone, shale and dolomite (limestone). Younger rocks (Mesozoic to Quaternary age) are restricted to sedimentary basins in the extreme north of the country (Gondo Plain), along the border with Mali (British Geological Survey, 2002a).

Most of the gold reserves are located in the northern part of the country, although Poura and Dossi-Kière are located in the south-west. Gold occurs with quartz and occasionally sulfide minerals in veins from a number of areas in the crystalline basement, mainly Birimian-age reefs, which are mostly granitoid deposits, as well as in alluvial gravels. The gold deposits exploited in the north are, according to Gueye (2002):

- the Bouroum reef of Yalogo, Djibo and Dori-Yalogo-Bouroum with Guiro deposits (reserves of 220 000 ounces of gold), discovered by Stremco in 1995;
- the reef of Taporko (reserves of 1 150 000 ounces of gold), discovered by High River Gold in 1997;
- the reef of Bouroum (reserves of 250 000 ounces of gold), discovered by Channel Resources;

- the gold district of Aribinda and the indications of Belahouro (15 million tonnes of ore);
- the gold district of Ouahigouya, Essakane and Essakane deposit (31 million tonnes of ore).

In the west, the reefs of Hounde and Boromo are exploited within the following districts:

- Yako-Kaya gold district, with gold sites in Bouboulou, Bouda, Nagsene and Guibare;
- Dossi-Kière gold district, with indications in Kari, Kwademen, Bagassi and Dama.

Gold mining has become increasingly artisanal since 1986; then, only 10% was artisanal, whilst now it is integral to the economic health of the rural population, with 87% being artisanal. It is estimated that 200 000 people are engaged in artisanal mining at 200 sites; these are mostly locals, especially farmers, who take advantage of the seasonal end of the harvest period (November–May) to mine, but certain sites are exploited throughout the year. Increasingly, artisanal mining has become as important an economic activity as agriculture. In addition to helping reduce the rural exodus, artisanal mining greatly contributes to generating foreign currency for the national economy (Guaye, 2002). Artisanal miners quickly took over operations at Poura when commercial operations were suspended in 1999. Derelict shafts, which are unmonitored, pose a serious health threat due to cave-ins, and living conditions in the area also pose serious public health concerns (Medilinks, 2006).

Given the climatic conditions of Burkina Faso, surface water is in limited supply and groundwater is therefore an important resource. Rural water supply projects rely mainly on groundwater, although this is also scarce in many areas. Dammed river courses typically provide urban supplies. Only around 38% of the total population has improved drinking-water coverage (WHO & UNICEF, 2006). Hand-dug wells are the traditional source of water (as well as ponds used in the rainy season). Today, groundwater is also abstracted from a number of tube wells, typically equipped with hand pumps. Tube wells generally source groundwater from fractures within the basement complex, whereas hand-dug wells are largely sourced from the weathered overburden layer. Yields from tube wells and dug wells in the basement areas are generally low, and many dry up seasonally (UN, 1983). Groen *et al.* (1988) noted groundwater levels in north-western Burkina Faso to be some 10–60 m below the surface, with a seasonal variation of around 1–2 m. In the sedimentary rock formations along the north-west border region, groundwater availability is also limited, although the dolomitic limestones form the best aquifers. The younger sedimentary formations cropping out in the north constitute a small aquifer, which is connected with underlying dolomitic limestone.

As with many arid African countries, groundwater quality in Burkina Faso has been an issue of lower priority than groundwater availability and has therefore received relatively little attention. Hence, few data exist on which to base an adequate assessment of the quality of the nation's groundwater. From the data available, the groundwater from basement rocks is typically fresh, calcium–magnesium–bicarbonate-dominant water. Investigations suggest that pollution from contaminants such as nitrate is a common problem, especially in shallow groundwater sources. Arsenic has also been identified as a problem in some areas. The extent of the occurrence of high-arsenic water is not known; recognized problems appear to be localized to one or two villages. Few other water quality problems have been singled out as major issues.

Arsenic occurrences in groundwater in Burkina Faso were made public on 29 December 2006 (Marshall, 2006):

> With signs that people are developing diseases linked to arsenic poisoning, the government said on Thursday [28 Dec 2006] that it had turned off pumps to 11 deep-water wells in areas in the north of Burkina Faso where water is particularly scarce, adding that it may still have to close down hundreds more wells. "People were coming to our health centres with skin diseases in abnormally high numbers", regional director for health in the north, Moussa Dadjouari, told IRIN [the humanitarian news and analysis service of the United Nations Office for the Coordination of Humanitarian Affairs] on Friday [29 Dec 2006]. "So we conducted investigations and found that it was linked to drinking water." Over the last 2 years officials from the Ministry of Agriculture and Hydraulic Resources found that the level of arsenic was often higher than 10 micrograms per liter, a threshold set by the World Health Organization (WHO). Short-term symptoms of arsenic poisoning include vomiting, stomach aches and bloody diarrhea, according to the Minister of Health. Prolonged exposition can damage kidneys and lungs and cause skin cancer. People in the north complained bitterly about losing their precious water sources, according to the state-owned newspaper. The government said that it had started building more wells to replace those closed. On Friday, UNICEF [United Nations Children's Fund] told IRIN that it has also been testing water in 360 wells it built in the northern provinces of Yatenga and Lorum between 1999 and 2004 and found that 7 wells were contaminated. UNICEF said it too had started building new wells to replace them. Tidiane Zougouri, an environmentalist critical of the policy of closing wells in areas where people are in desperate need of water, said that up to 80 percent of arsenic can be eliminated using a sand filter. "Studies have shown that arsenic is stopped by the iron contained in the sand", Zougouri said. "And it's a cheap solution because sand is everywhere in the area." Zougouri warned that with erosion, arsenic is not only affecting underground water but also water that has risen to the surface. "The danger comes from any water that has not been able to oxygenate", he said. The risk is also not limited to the north but to the whole country, according to the General Director of Water Resources, Francis Bougaire. "Arsenic was not on the agenda when officials were evaluating soil before digging the wells", said Dadjouari. The government has now set up an inter-ministerial committee to try to address the problem.

The occurrence of arsenic-related diseases, such as blackfoot disease, indicates that high concentrations of arsenic (>50 μg/l) do exist in drinking-water in parts of Burkina Faso associated with gold and its associated mineralization. The extent and scale of the problem can only be defined as being coterminous with gold and Birimian-age rocks, which means that those who engage in artisanal mining of gold are at risk. That number is approximately 200 000 at 200 sites; therefore, it is important to monitor the drinking-water at these sites.

This problem has been known since 1975, when an original investigation carried out by IWACO (1975) showed that arsenic contamination occurred in Burkina Faso groundwater as the result of the oxidation of sulfide minerals (pyrite, arsenopyrite) and that the contamination was concentrated in the gold mineralized areas, especially in the north-east towards the Niger. Another early investigation of arsenic contamination of groundwater was carried out by De Jong & Kikietta (1980), who noted the occurrence of locally high arsenic concentrations in groundwater near a village close to Mogtedo in central Burkina Faso. Appelo & Postma (1993) called attention to this growing problem in 1993.

Testing, identification and verification of the locations of arsenicosis cases and their associated groundwater sources are the first steps towards the initiation of a water-testing programme for arsenic in Burkina Faso, especially in the mineralized areas. These measures are essential to define the scale of the problem, so that appropriate remediation regimens can be instituted as soon as possible.

1.4 CAMEROON (M)

Cameroon, with a population of 20 million, has a human development index of 150 out of 169 and a per capita GNI of US$ 2,031 (US Census Bureau, 2011; UNDP, 2010).

In the Bétaré Oya region, in east Cameroon, with a population of 11 866, gold mineralization was implicated with arsenic, bismuth, silver, molybdenum, tungsten, lead and gold. The Bétaré Oya region geology consists of Proterozoic (Birimian) volcano sediments covered by a 25 m thick weathered ferruginous mantle. Mineralization of silicified structures is completely oxidized and weathered at the surface. Arsenic and associated trace element samples were studied from deep drill holes and surface soils, and it was found that arsenic is strongly associated with iron and widely distributed, while silver is completely leached. Mineralized veins are dissolved in the top horizons, and trace elements are dispersed and concentrated in the surrounding ferruginous horizons (Freyssinet, 1989).

1.5 EGYPT (G, M)

Egypt, with a population of 82.5 million, has a human development index of 113 out of 169 and a per capita GNI of US$ 5,269 (US Census Bureau, 2011; UNDP, 2010).

Arsenic has been found in the Duwi Formation of the Middle Eastern to North African phosphogenic province of the late Cretaceous to Paleogene age. The formation is composed of phosphatic, structureless pelloids and bioclasts of francolite composition. Three localities were sampled using inductively coupled plasma (ICP) analyses that indicated an arsenic range from 0.1 to 2.1 ppm with an average of 0.95 ppm, and a cadmium range from 0.37 to 11.55 ppm with an average of 3.2 ppm. The Abu-Tartur phosphorite deposit attains a thickness from 3 to 7 m along a 100 km perimeter of the southern sector of the Abu Tartur plateau. Approximately at the mid-point of this deposit and 50 km west-southwest of the El Kharga Oasis is the Abu Tartur mine.

"The Abu-Tartur phosphorites show higher arsenic than the Red Sea and Nile Valley phosphorites. … The lack of correlation between cadmium (Cd) and arsenic (As) suggests different sources and/or diagenetic behavior. Correlation between both elements and Al_2O_3 is also lacking, suggesting that both Cd and As do not have a detrital origin. Lack of correlation between Cd and As and CaO and P_2O_5 indicates there is no genetic relationship between Cd and As and francolite. Lack of correlation between Cd and TOC suggests that Cd does not have an organic origin, while the relatively strong positive correlation between As and TOC indicates that As contents may be related to the organic fraction. Relatively strong positive correlation between As and sulfur is suggestive of the occurrence of As as sulfide. The geochemistry of Cd and As is completely different and diagenesis seems to be the most important controlling factor of their distribution" (Baioumy, 2005). Any leaching of As from the mining operation or from naturally occurring As into groundwater that might affect the El Kharga Oasis is unknown.

1.6 GHANA (G, M)

Ghana, with a land area of 230 020 km^2 and a population of 25.0 million, has a Human Development Index rank of 130 out of 169 and a per capita GNI of US$ 1385, (US Census Bureau, 2011; UNDP, 2010).

Ghana's relatively low GDP reflects an economic and environmental reality. First, although gold mining accounts for the largest proportion of foreign exchange earned by Ghana, it is a wealth invisible to most Ghanaians: the net in-country wealth is still from cocoa and other crops (see below) and not from gold mining, since most of the wealth flows to the rich who invest outside Ghana and to the gold mining companies who also invest their earnings outside Ghana or pay off debt. Second, the economic gains from mining that accrue to the rich and the foreign companies involved are achieved at great environmental cost. Exploitation of gold puts immense stress on air, water, soil and vegetation and also frequently poses potential and real hazards to human health, including arsenic contamination of surface water and groundwater.

Arsenic contamination in the Ashanti Region in southwestern Ghana has been associated with gold mining in the Obuasi region, with an urban population of 122 600. Arsenopyrite (FeAsS) in the ore has created airborne arsenic pollution from the ore-roasting chimney in the town as well as leaching from tailings into ground and surface waters. Gold ore smelters process arsenopyritic containing about 0.8% arsenic that is not recovered but is dispersed in the surrounding environment. In a 2–11 km perimeter of the smelter, mine tailings leach into streams, flue dusts permeate the atmosphere and surface soils (severely contaminating edible crops). Arsenic in drinking water is 24 times higher than the WHO recommended guideline. The effects of arsenic contamination on the health of local inhabitants are yet to be documented (Manful & Verloo, 1989). Arsenic concentrations in water from streams, shallow wells and boreholes were found to range between <2 and 175 μg L^{-1}. Arsenical sources are mine pollution and natural oxidation of sulphide minerals, predominantly arsenopyrite. Streamwaters contain high As(III) concentrations, resulting from methylation and reduction mediated by bacteria and algae. Concentrations of As in groundwaters reach up to 64 μg L^{-1}, being highest in deeper (40–70 m depth) and more reducing waters. High arsenic concentrations increase as a result of residence times in groundwater and reducing conditions in the aquifer. Interestingly, deep mine exploration boreholes (70–100 m) have exhibited low arsenic concentrations of 5–17 μg L^{-1}, due to arsenic adsorption on hydrous ferric oxides (British Geological Survey, 2011).

The extraction of gold here involves roasting, which releases airborne particles and large quantities of arsenic: 13–17 tonnes daily in Obuasi. In Obuasi, the effects of mining activities and possibly some arsenopyrite oxidation have led to arsenic contamination. Elevated levels of arsenic have been reported in surface water, groundwater and drinking-water around the Obuasi gold mining area (Amasa, 1975; Smedley *et al.* 1996; Thornton & Farago, 1997). The highest drinking-water arsenic concentration reported by Amasa (1975), in Obuasi-Kwabrafoso, was 1400 μg/l. River water arsenic concentrations up to 7900 μg/l have been reported, but many of the ground waters generally contain low levels of arsenic (median 2 μg/l; Smedley *et al.* 1996). Smedley *et al.* (1996) reported arsenic levels in groundwater up to 64 μg/l, but these were not generally in the vicinity of the mines or directly related to mining activity. In another report, arsenic concentrations in drinking-water in Obuasi were 10–38 times higher than levels permitted by the Ghanaian Environmental Protection Agency's general guidelines and over 1800 times higher than the WHO guideline value of 10 μg/l, or 18 000 μg/l (ActionAid, 2007). In a related aspect of arsenic contamination in Obuasi, in areas of historic gold mining activities, the local soils, called Obuasi orange, had arsenic levels 24 and 1,226 times the Ghanaian Environmental Protection Agency and WHO guidelines, respectively (ActionAid, 2007). As a result of elevated arsenic levels in Obuasi soils, the total arsenic content of some food and cash crops from Obuasi farms and markets is elevated. In a comparison of food crops from Kumasi and Obuasi, 266 samples were measured for arsenic; arsenic levels in Kumasi samples ranged from 0.07 to 7.20 mg/kg, whereas those for Obuasi ranged from 0.12 to 70.50 mg/kg, confirming that arsenic levels for Obuasi, in the mining area, are much higher than those for Kumasi (Amonoo-Neizer & Amekor, 1993). As a further consequence, the Ankobra, Birim, Ofin and Owen river systems have been polluted by mining activities, and fish from these rivers have elevated arsenic in their flesh.

Diamond and manganese mining operations have also been contributors to pollution. A study by Mensah (1976) provided a quantitative analysis of the concentrations of pollutants, including arsenic, copper, cyanide, iron, zinc and suspended solids, present within the waste effluent streams discharged from major mines. According to the data, the Ghana Consolidated Diamond Mine at Akwatia and Ghana National Manganese Corporation generated the largest quantities of contaminated liquid effluent, discharging, on a monthly basis, 202.93 million litres and 283.67 million litres, respectively.

Mining pollution continues, despite the efforts of the Ghanaian Environmental Protection Agency, which was established in 1994. It has been unable to regulate mining activities that affect rivers upon which local communities depend. An increasing amount of land is under the control of large-scale mining companies. In Tarkwa alone, 70% of the land area is under their control (Britwum *et al.* 2001). The Ghanaian Environmental Protection Agency has little control over how mineral rights are awarded (Hilson, 2004).

Gold benefits the companies, not the artisanal miners or the farmers of cocoa. In 1998, Africa Recovery, an official report of the United Nations, stated that gold has been Ghana's leading export since 1992, providing over 45% of total foreign earnings by the end of 1995 (Mate, 1998). The United Nations failed to note that this is in gross terms. In net terms, cocoa is still the largest export earner for the people, because mining companies may retain 25–80% of their earnings in foreign accounts for debt servicing and other foreign costs, an important provision of the Minerals and Mining Law.

Good environmental practices are required of mining companies by the Minerals Commission and Ghanaian Environmental Protection Agency, as well as by the Minerals and Mining Law. However, a persistent problem is pollution of surface watercourses, many of which provide drinking-water for communities downstream. In some cases where, as required, companies have supplied wells and pumps for drinking-water, responsibility for maintenance costs has become an issue. Besides the long-running dispute over water between the community and Ashanti Goldfields Co. in the area of its Sansu surface mine and the sporadic clashes between artisanal miners and the semi-private security forces of the mining firms, reports in 1996 of cyanide spillage into watercourses in the Tarkwa area heightened concerns over the weakness of regulators.

Gold is found in locally occurring auriferous arsenopyrite, which is the traditional ore mined in Ghana and which is associated with high levels of arsenic. The gold extraction process often produces airborne particles containing arsenic (Mead, 2005). Some of the health impacts of mining in the Wassa West district are described in a report written by Thomas Akabzaa, a lecturer at the University of Ghana, and sponsored by the Africa Secretariat of the Third World Network. Akabzaa (2000, 2002) noted an increased susceptibility to diseases such as malaria, respiratory tract infections, skin diseases and diarrhoea, the last three being potentially related to arsenic. These health impacts are due partly to the arsenic contamination of groundwater that is attributable to Ghana's gold mining industry.

Activities in Ghana that increase arsenic mobility and their associated environmental and health impacts are listed below (Akabzaa, 2000, 2002):

- Blasting creates dust, increasing particulate matter in air and water, which exacerbates respiratory ailments associated with arsenicosis.
- Unsafe processing methods produce toxic chemicals such as cyanide, arsenic and sulfur dioxide.
- Air pollution is caused by unsafe smelting, resulting in airborne particulate matter. In 1995, the incidence of conjunctivitis in the area was 2.4 per 1,000. [It should be noted, however, that roasting of sulfide gold ores, which produces sulfur and arsenic oxide emissions, was somewhat mitigated in 1995, when recovery circuits were added to two of the three gold-processing plants (Coakley, 1995).]
- High decibel noise and vibration affect hearing.
- Improper industrial effluent treatment results in surface water and groundwater pollution.
- Excessive use of groundwater results in dewatering effects.
- Skin rashes are caused by pollution of air and water by toxic chemicals, including arsenic used in the mining process.
- Mental disorders are related to arsenicosis.

The problem of arsenic in groundwater in Ghana was first identified by Hungarian scientist Emile Sandy in 1966. He found that arsenic contamination of groundwater had been caused by the smelting process associated with a gold mine near Prestea. As Sandy (2005) recounted it:

> The gold-bearing material in this mine was pyrite, which was thinly dispersed in the rocks. The mined rocks were milled, and the pyrite was separated out by flotation. Roasting was then used to get rid of sulfur (as well as of arsenic), after which gold was obtained from the remaining concentrate by a cyanidization process. Cyanide was not a problem for the surrounding population, because it was emptied into the nearby river where it quickly oxidized into harmless substances. On the other hand, the chimney of the roaster emitted 6 tons of arsenic trioxide every day, and this contaminated the nearby village as well as the countryside around it. The population obtained its drinking water from rainwater flowing off rooftops, containing 4,000–6,000 μg/L of arsenic according to our measurements. Dermal and internal symptoms of arsenic poisoning were clearly present in the population of the village. We were able to establish that the arsenic content of well water was below 10 μg/L, and was therefore safe to drink. Since we had no other instruments with us, we collected 10 × 10 cm samples of fresh (third) leaves of the ubiquitous banana tree, in both the villages and the countryside around them, from up to 50 km away.... A few years later, when I was already in Canada, I learned that Ghanaian scientists repeated my study of the area using more modern methods, as well as the area around an even larger gold mine near Kumasi, and found the same levels of arsenic contamination.

In a recent study, the occurrence of arsenic in Ghana's groundwater was explored (Norman *et al.* undated):

> Ghana, the former Gold Coast, is richly mineralized with gold that is accompanied by arsenic mineralization. The Upper Birimian formation has over 10 times the average crustal abundances of Au and As. Our working hypothesis was that groundwater in the 10 belts of Au mineralization and Upper Birimian rocks that cross Ghana have elevated As concentrations.
>
> Arsenic and water chemistry were analyzed in all rural hand dug wells and boreholes in a 6,500 sq. km area of SW Ghana. In addition arsenic was measured in several wells at and near gold mines. Arsenic was analyzed in the field with an Arsenator™ that has a detection limit of 0.5 ppb [μg/l], and the results validated by GFAA [graphite furnace atomic absorption] analyses. Speciation was done on selected wells. Skin problems in villagers similar to those associated with arsenic poisoning were noted.

Arsenic was detected in 52 of 207 wells studied. Concentrations range up to 2,000 ppb [μg/l] and 25 wells have arsenic >5 ppb [μg/l] [see Table 1.1]. In some wells arsenic concentrations vary daily by more than a factor of 2. High arsenic waters are associated with Upper Birimian rock, gold–arsenic mineralization, borehole wells and low-Eh [redox potential], high-TDS [total dissolved solids] waters. Speciation shows significant concentrations of As(+3). A strong correlation was found between skin problems and arsenic concentrations in well waters [see Table 1.2]. Hand dug wells have less arsenic than borehole wells, but much more nitrate, and there is correlation between well depth and arsenic concentration. Column experiments show that Ghana laterite strongly sorbs arsenic. There is no correlation between sulfate and arsenic, although the arsenic source most probably was sulfide mineralization. High-arsenic wells are commonly less than 100 m from wells that have trace amounts indicating that arsenic mobility is limited and there is little communication between water-bearing fractures.

Table 1.1 Summary of Ghana well analyses.

Concentration	No. tube wells	No. hand-dug wells
As >5 μg/l	23	2
As 0.5–5 μg/l	30	6
As <0.5 μg/l	74	68
Fe >0.5 mg/l	52	15

Source: Norman *et al.* undated

Table 1.2 Summary of noted skin problems in south-west Ghana similar to those associated with arsenic poisoning.

As concentration (μg/l)	Wells for which health effects noted in at least one person	Total no. wells	% of wells with noted health effects
2–120	14	27	52
0.5–2	4	22	18
<0.5	6	153	4

Source: Norman *et al.* undated

Therefore, any attempt to address arsenic poisoning in Ghana must address the mining industry and its economics. Some specifics are that gold extraction in Ghana rose from 10.6 tonnes in 1988 to 67.5 tonnes in 1998. This translates to US$ 651.1 million of a total mineral export of US$ 682.2 million. The total of all exports from Ghana in 1998 was US$ 1.5 billion (Akabzaa, 2000). When gold mining operations are initiated in an area, farmers are given a one-time payment for their land without hopes of acquiring a comparable plot for farming (Mate, 1998).

An initial step in addressing the problems associated with arsenic contamination as related to gold mining could be to identify the locations of the major gold mines and analyse the areas around tailings for potential leaching into the groundwater (Akabzaa, 2000, 2002):

- Ashanti Goldfields Co. Ltd, owned by Lonhro, Government of Ghana, Anglo American and others, operating mines in Obuasi, Ayanfuri and Bibiani;
- GoldFields (Gh) Ltd, owned by Goldfields S. Africa, Government of Ghana, operating mines in Tarkwa;
- Teberebie Goldfields Ltd, owned by Pioneer Corporation and Government of Ghana, operating mines in Teberebie (Tarkwa);
- Ghana Australian Goldfields Ltd, owned by Ashanti Goldfields Co. Ltd/Golden Shamrock, International Finance Corporation and Government of Ghana, operating mines in Iduapriem (Tarkwa);
- Abosso GoldFields Ltd, owned by Ranger Minerals and Government of Ghana, operating mines in Damang (Tarkwa);
- Satellite Goldfields Ltd, owned by Government of Ghana and others, operating mines in Subri (Tarkwa);
- Sankofa Gold Ltd, owned by Samax and Government of Ghana, operating mines in Prestea (Tarkwa);
- Bogoso Gold Ltd, owned by Gencor, International Finance Corporation and Government of Ghana, operating mines in Bogoso (Tarkwa);
- Bonte Gold Mines Ltd, owned by Akrokeri Ashanti Gold Mines of Canada and Government of Ghana, operating mines in Akrokeri;

- Dunkwa Continental Goldfields Ltd, owned by Continental Construction Ltd and Government of Ghana, operating mines in Dunkwa;
- Obenamase Gold Resources Ltd, owned by the Resolute Group, operating mines in Obenamase;
- Goldfields Ltd, owned by People's Republic of China, Government of Ghana and others, operating mines in Nkawkaw;
- Amansie Resources, owned by Resolute Group and Government of Ghana, operating mines in Amansie.

It is also important to understand the governmental regulatory infrastructure for the gold mining sector, because there are many overlapping and differing jurisdictions and interpretations of the regulations. The following organizations are chartered to work with the mining industry (Akabzaa, 2000, 2002):

- *Ministry of Mines and Energy*—the Minister is empowered to act on behalf of the government on all matters relating to the industry. The Ministry grants exploration licences and mining leases.
- *Minerals Commission*—recommends mineral policy, advises government on mineral matters, reviews and promotes mining sector activities.
- *Geological Survey Department*—conducts broad-scale geological studies in the country, prepares maps and reports.
- *Environmental Protection Agency*—tries to balance demands of rapid economic growth and the need to protect natural resources and health and welfare of the people. It conducts and promotes studies, investigations, surveys, research and analysis relating to the country's environment.
- *Mines Department*—responsible for establishing mining standards and conducting safety inspections.
- *Lands Commission*—maintains legal records of exploration licences and mining leases, participates in the examination of new licence applications.
- *Chamber of Mines*—represents both public and private sector actions of the mining industry. Seeks to promote the interests of mining companies, and communicates with government and other public and private bodies regarding mining matters.
- *Forestry Department*—responsible for management of the country's forests. The department is chartered to work with the Minerals Commission on the granting of exploration licences and mining leases.

If these governmental agencies could work in a concerted effort to redress the impact of mining on the population, the true value of gold to the country's gross national product may be more fully realized.

Arsenic pollution due to petroleum extraction may occur in Ghana's future. The Ghanaian Jubilee oil field was discovered in 2007 by the US firm Kosmos and is located 60 km offshore at a water depth of 1,250 m. Reserves of between 600 million to 1.8 billion barrels of high-quality oil-bearing reservoir sandstones are estimated. It is one of the largest finds in West Africa. The extraction process involves a plateau oil rate of 120 000 barrels of oil per day (BOPD), a water injection capacity of 230 000 barrels of water per day (BWPD), and a gas export and injection capacity of up to 160 thousand standard cubic feet per day (MSCFD) (OffshoreOil.com, 2010).

The arsenic content of the crude oil is and how this affects the effluent from the water injection process is unknown. Production came on line in December 2010, producing about 80 000 BOPD in 2011 rather than the expected 120 000 BOPD. Observers are worried about the so-called oil curse, whereby large oil revenues lead to an abandonment of other economic sectors to conflict, large-scale corruption and environmental degradation. Environmental pollution from oil spills, such as has already occurred in neighboring Nigeria, and the subsequent release of arsenic, is one of the associated risks. However, to date, releases of arsenic from oil production have not been reported in the peer-reviewed literature.

1.7 KENYA (M)

Kenya, with a population of 41.6 million, has a human development index of 143 out of 169 and a per capita GNI of US$ 1,492 (US Census Bureau, 2011; UNDP, 2010).

The Migori Gold Belt, Kenya, into which district up to 20 000 people have migrated looking for work, reveals heavy metals concentrations, mainly of mercury, lead and arsenic above acceptable levels. In particular, tailings have been shown to have 0.06–76.0 mg/kg^{-1} arsenic and stream sediments in the area have 0.014–1.87 mg/kg^{-1} arsenic. The Nairobi mine tailings had 76.0 $mg\ /kg^{-1}$ arsenic (Ogola *et al.* 2002).

1.8 MOROCCO (M)

Morocco, with a population of 32.3 million, has a human development index of 130 out of 169 and a per capita GNI of US$ 4,196 (US Census Bureau, 2011; UNDP, 2010).

The Moulouya Oued basin, an early 20th-century lead mining region, covers 74,000 km^2, including the Moulouya Kert, Isly, Kiss and Chott Tigri sub basins and a portion of the Bouarfa-Figuig zone. Over 2.2 million people live in this region but the number those exposed to As reside in the High Moulouya Valley, also known as the Upper Moulouya lead district, where high concentrations of As in tailings have been leaching into groundwater. In the upper Moulouya district the As contamination of lake and quarry pools ranges from 53.05 to 114.22 μg/l (the corresponding pH range is 8.14–9.02) and mining residue sediments range from 65.61 to 200.03 ppm (Makhoukh *et al.* 2011). The ZaYda mine and other abandoned mines in this district have been left without any rehabilitation (Bounakhla, 2011). The 2004 census estimated the population of the High Moulouya Valley at 151,605 inhabitants. The rural communities of Zeïda and Mibladen comprise respectively, 9,920 and 3,087 inhabitants; at Aouli, only a few families (a dozen) still stay in the village.

The Hassan II dam has efficiently trapped As, so downstream surface waters are of average to excellent quality but As concentrations in downstream river sediments remain very high, since they were contaminated before the construction of the dam, and these contaminated sediments constitute a potential risk to the entire aquatic ecosystem (Makhoukh *et al.* 2011).

1.9 NAMIBIA (M)

Namibia, with a population of 2.3 million, has a human development index of 120 out of 169 and a per capita GNI of US$ 6,206 (US Census Bureau, 2011; UNDP, 2010).

As of 23 November 2011, "Namibia Custom Smelters (NCS) is still awaiting a final verdict on allegations that its smelting of Bulgarian copper is causing illnesses at Tsumeb. Members of the Tsumeb community with a population of 12,190 and NCS employees are alleging that emissions from the Bulgarian copper concentrate, which is said to contain arsenic, pose a serious hazard to their health and the environment. At a September 29, 2011 meeting at Tsumeb, which involved Mines and Energy Minister Isak Katali, Deputy Minister of Health and Social Services Petrina Haingura, Labour Minister Immanuel Ngatjizeko and the management of Canadian mining company Dundee Precious Metals (DPM) and NCS, it was decided that the Ministry of Health would appoint an investigating team to assess the situation (Anon, 2011). The result of this meeting is that as of 2 March 2012 the Tsumeb smelter will undergo a N$ 560 million upgrade that is aimed at reducing its environmental impact through the reduction of sulphur dioxide emissions, as well as the upgrade of the plant's arsenic processing facility (Allafrica, 2011).

1.10 NIGER (G, M)

Niger is a landlocked country north of Togo and is bordered to the west by Burkina Faso and Mali, to the north by Algeria and the Libyan Arab Jamahiriya, and to the east by Chad. Niger, with a population of 16.5 million, ranks 167 out of 169 in the Human Development Index, with a per capita GNI of US$ 675 (US Census Bureau, 2011; UNDP, 2010).

Ninety-two per cent of the population is engaged in farming subsistence crops (millet, sorghum and cassava) and husbandry, so water availability is a chief concern. Although 46% have access to improved drinking-water coverage, frequent drought cycles, increasing desertification and exposure of arsenic compounds in the Birimian-age formation through gold mining affect surface water and groundwater supplies that are crucial for domestic uses, farming and husbandry (WHO & UNICEF, 2006).

Niger's far-western geology resembles the Birimian-age greenstone deposits in Burkina Faso across the border and has a similar mineralization, with crystalline quartz and occasionally sulfide minerals, including arsenopyrite, as well as alluvial gravels. The first commercial gold mines were opened at Samira Hill near the Burkina Faso border 90 km west of the capital city of Niamey in October 2004, ending a decade of exploration (Harris & Duncan, 2005). The United States Geological Survey (Mobbs, 1997) reported that:

> The three major greenstone gold belts in the Liptako region of western Niger were the focus of the gold rush. The Gorouol (or Kourki) Belt bordered Burkina Faso and Mali in the northwest of the Liptako area, the Sirba Belt was west of Niamey, and the Tera Gassa Belt was between the Gorouol and the Sirba greenstone belts (Bureau of Geological Consultancy, 1993). A number of small-scale miners were producing from operations in the Sirba Belt. The government estimated that production of artisanal gold from the entire Liptako region was about 1 metric ton per year; artisanal miners numbered in the tens of thousands.

The Liptako region covers 26 000 km^2, bounded by the Niger River and Burkina Faso on its north-east in a 110 degree trending terrain of stabilized dunes and inter-dune deflation in a band 10–50 km wide along the south bank of the Niger River (Stratamodel Inc., 2006). In essence, a number of significant gold deposits are known to be in this emerging gold belt (Mining Review Africa, 2003); in 2005, gold was the third most important of the country's exports, accounting for 12.8% of the total (Travel Document Systems, 2007). The population in the Liptako region (3.375 million) accounts for about one quarter of the country's total (International Development Research Center, 1994).

Indirect evidence from the identification of arsenicosis cases associated with arsenic contamination in the Niger River suggests that there is a high probability that arsenic compounds are being liberated by these mining starts and are leaching into surface water and groundwater. Positive epidemiological data on arsenicosis from Burkina Faso (see Section 2) and from Nigeria's Niger Delta (see Section 5), both of which have a geology almost identical to that of north-western Niger, with arsenic mineralization, indicate that Niger is at risk for arsenic contamination of surficial water, riverine water and groundwater (especially from borehole wells). However, although the Niger River flows through Niger and is predictably contaminated with arsenic, no testing has been undertaken Niger itself. As indicated above, the number of artisanal miners is in the tens of thousands, and the population of the Liptako region served by the potentially contaminated groundwater is presumed to be comparable.

It is hoped that the Isotope Hydrology Section of the International Atomic Energy Agency will address this problem when it implements its "Assessment of Water Resources for Sustainable Development in the Tillabery Region of Niger" (IAEA, 2005).

1.11 NIGERIA (A, C, G, M, P, V)

Nigeria, with a population of 165.8 million, has a Human Development Index that ranks it 142 out of 169 and a per capita GNI of US$ 2156 (US Census Bureau, 2011; UNDP, 2010). Its area of 910 770 km^2 is bordered to the north by the Niger and Chad, to the south by the Atlantic Ocean, to the east by Cameroon and to the west by Benin. Nigeria has the eighth largest national population in the world and about a quarter of the total population of all the countries in sub-Saharan Africa. It is ethnically very diverse, which affects the implementation of any water resource programme as a result of competition among 14 ethno-linguistic groups.

Nigeria has abundant water resources, although they are unevenly distributed over the country. River systems divide the country into four major drainage systems. The first is a series of tributary rivers that drain north into Lake Chad: the Kano, Hadejia, Jama'are Misau, Komadougou-Yobe, Yedoseram and Ebeji. The other three river systems drain into the Atlantic: the Niger (with major tributaries being the Benue, Sokoto-Rima, Kaduna, Gongola, Katsina-Ala, Donga, Tarabe, Hawal and Anambara), the Atlantic west of the Niger (with the Ogun, Oshun, Owena and Benin rivers) and the Atlantic east of the Niger.

In addition, Nigeria has eight hydrogeological systems: the Sokoto, Chad, Upper Niger, Benue, Ogun/Osun, Lower Niger and Cross River sedimentary areas, and the Birimian basement complex rock area, which coincide with high mineralization (Gbadebo, 2005).

Evidence of volcanism is found in the Biu Plateau in the north-east, which is composed of basaltic lava flows. Relatively recent (perhaps Holocene) cinder cones with uncertain eruption dates and with well-defined craters are aligned in a north/north-west–south/south-east direction, and they often contain abundant peridotite xenoliths (Smithsonian Institution, undated, b).

Inland, there is great competition for surface water for drinking and farming. As the Atlantic Ocean borders only the lands of the Ijaw and the Ibibio, coastal waters are also much in demand. Water collection is diverse, including springs, dug wells and rain harvesting in the south and mainly borehole wells in the north. The drawdown of irrigation aquifers is concentrating arsenic, which is contaminating agricultural produce and domestic animals used in the food supply.

The climate, which affects the quality and quantity of the country's water resources, results from the influence of two main wind systems: the moist, relatively cool monsoon wind, which blows from the south-west across the Atlantic Ocean towards the country and brings rainfall; and the hot, dry, dust-laden Harmattan wind, which blows from the north-east across the Sahara Desert, with its accompanying dry weather and dust-laden air. The mean temperature is generally between 25 and 30°C, although the mean daily and annual maximum temperatures increase from the coast towards the interior because of the moderating influence of the sea. In the dry season, the temperatures are more extreme, ranging between 20 to 30°C.

The highest annual precipitation of about 3,000 mm occurs in the Niger Delta and mangrove swamp areas of the south-east, where rain falls for more than 8 months of the year. There is a progressive reduction in precipitation northwards, with the most arid north-eastern Sahelian region receiving as little as 500 mm precipitation from about 3–4 months of rainfall. Widespread flooding occurs in the southern parts of the country, whereas the northern parts experience chronic water shortages during the dry season when rain-fed springs, streams and boreholes dry up.

Along the lower reaches of the Niger Delta, there are documented arsenic problems in groundwater and the environment. These arsenic-contaminated areas are in proximity to known occurrences of geothermal and mining-related arsenic problems.

In a 2005 study, it was found that although arsenic contamination of groundwater had not been known with certainty owing to a paucity of data, it was believed to occur naturally in aquifers located in the south and south-west, at concentrations ranging from 30 to 250 µg/l in the Birimian basement rocks and from 60 to 400 µg/l in the sedimentary rocks of Sokoto State in the north-west. The sources of arsenic in these regions have been found to be both geogenic and anthropogenic. Limestone and black shale aquifers around Ewekoro and Shagamu contain arsenic concentrations in the range of 3,600–30 200 µg/l, whereas arsenic concentrations in underground water range from 40 to 160 µg/l. The coastal plain's carbonate-rich sandy sediments of southern Nigeria enhance arsenic leaching in a way similar to that of New South Wales, Australia. It is essential to understand the arsenic cycle in the hydrogeological system of Nigeria in order to identify its impacts on public health (Gbadebo, 2005).

Anthropogenic sources of arsenic contamination are related to industrial dumping by factories located on riverbanks that use the rivers as open sewers for their effluents. The major industry involved is petroleum, and oil spills occur which endanger local sources of water supply and living freshwater resources, especially in the rural areas (UNICEF & WHO, 2002; Achebe & Epstein, 2004).

Effluents from mining (gold, tin and coal) have been responsible for arsenic contamination of stream sediments and groundwater. Other anthropogenic sources of contamination include wood processing, distilleries and industrial production of pulp, pharmaceuticals, textiles, plastics, iron and steel, paint and food (FGN, 1988a,b, 1993; FEPA, 1989, 1991a,b,c).

Coal mining has recently been found to be a contributor to arsenic and heavy metal pollution. Arsenic and a suite of other heavy metals and metalloids (e.g. manganese, chromium, cadmium, arsenic, nickel and lead) were analysed in coal and sediment samples from the River Ekulu in Enugu, Nigeria, using atomic absorption spectrophotometry. The River Ekulu, contaminated with arsenic concentrations above 10 μg/l, is the largest inland water in Enugu Urban, serving the population of the capital city of Enugu State (nicknamed Coal City State), Enugu, of about 688 827 (World Gazeteer, 2012). The Ekulu coal mine is situated on the banks of the River Ekulu, into which the coal mine directly discharges its effluents, which carry toxic heavy metals and arsenic, contaminating riverine sediments and the water. Adaikpoh *et al.* (2005) found that:

> mean concentrations of Mn (0.256–0.389 mg/kg) and Cr (0.214–0.267 mg/kg) are high relative to concentrations of Cd (0.036–0.043 mg/kg), As (0.016–0.018 mg/kg), Ni (0.064–0.067 mg/kg) and Pb (0.013–0.017 mg/kg). Except for Mn, there is a steady decrease in the concentration of all metals from November to March corresponding to decreasing run-off. Metal concentrations are generally higher in the coal samples than in the sediments. Selective leaching of these metals is suspected from the trends of their concentration. Seasonal fluctuations are higher in the sediments than the coal samples. Correlation coefficients show that all the metals are positively correlated with each other implying that they are from the same source or origin. The presence of toxic metals in the area is established, calling for the assessment of their impact on the health of human and aquatic lives around the area.

Oil exploration and extraction are particularly volatile issues in Nigeria. Since the discovery of oil in commercial quantities in 1956 and the oil boom of the 1970s, the oil industry has become the backbone of the Nigerian economy, accounting for over 90% of the total foreign exchange and nearly 80% of government revenues (NigeriaBusinessInfo.com, 2005). This is exacerbated by the fact that most of Nigeria's oil supply is "light crude", which means that it is cheaper to process. In the half century since Nigeria became free from British colonial rule, Nigeria has become the world's eighth largest oil exporter and the fifth largest exporter to the United States of America (USA). Successive dictatorships have empowered the Nigerian government to enact laws that have placed oil resources under the control of multinational oil companies.

Nigeria's oil pipelines in the delta have always been vulnerable to natural leakage, as the pipes traverse a precarious ecological region of swampy streams, along which are situated isolated villages. This has created environmental pollution resulting from oil spillages and the subsequent release of arsenic, which had remained bound to previously undisturbed riverine sediments. For instance, one of the settlements in the delta, Iwhrekan, is home to 1,000 villagers living in traditional mud-brick houses on the banks of deltaic streams. When oil pipelines leak, they pollute both the streams on which the local residents depend for fishing and the groundwater on which they depend for drinking.

In addition, oil has become a significant political issue, as villagers in the delta area, through which the pipelines pass, do not feel that they have received any remuneration from the government in proportion to the large profits resulting from international oil sales. As a result, the delta area is currently under siege from local militias intent on disrupting the government's revenues. On 2 March 2006, Reuters reported that there were fears of more violent attacks on Nigeria's oil industry: "The Movement for the Emancipation of the Niger Delta (MEND) warned … that it was concentrating its resources on 'one huge crippling blow to the Nigerian oil industry' with an aim of completely discontinuing exports of onshore crude oil" (Demongeot, 2006). On 10 May, 2 June, 20 August and 2 October 2006, separate incidents heightened the movement's demands for more resources being made available to the delta's populations, both monetarily and for the responsible oversight of the pipelines (MEND, 2012). Destroying vulnerable pipelines is no solution to Nigeria's endemic political and economic problems, but they demand a solution nonetheless. The sabotage of oil reserves and the concomitant infiltration of oil into underlying sediments could release arsenic and place at least 40 million villagers in the troubled region at risk.

Oil, a potential economic boon to poor villagers, has therefore exacerbated problems associated with the lack of adequate water resources in Nigeria. Studies by the World Bank (1990) suggest that it would cost in excess of US$ 1 billion a year to correct such problems if groundwater and surface water contamination goes unchecked. The people most affected tend to be the landless poor.

Arsenic has been found in dug wells in the Niger River delta and is currently attributed to petroleum extraction practices in which toxic light crude oil containing a deadly cocktail of hydrocarbons, including benzene (a known human carcinogen), toluene, xylene, butane and propane, has facilitated the release of arsenic from geological sources. Long-term exposure has been linked to a variety of illnesses, including hypertension, diabetes, cardiovascular disease, infertility and cancers of the skin, lungs, urinary bladder and kidney.

In the long term, the present level of environmental degradation could create health problems from arsenic release into aquifers and also from waterborne diseases for most of this population. However, many Nigerians are already affected by having to consume unsafe drinking-water. Water contamination also places other resources at risk: fisheries and agrarian resources have already been affected significantly.

Additional environmental pollution problems involving arsenic arise from anthropogenic sources, mainly from domestic and industrial activities. The federal military government has placed great importance on the environment and has established the Federal Environmental Protection Agency (FGN, 1988a).

Efforts are being made to build zonal laboratories in various parts of the country to adequately monitor the domestic, recreational and industrial causes and effects of environmental degradation. Six zonal laboratories were proposed in the following cities: Lagos, Abuja, Benin, Kano, Jos and Port Harcourt. The Lagos laboratory has already been commissioned and is also serving as the national reference laboratory.

The turning point for water resource development and management in Nigeria occurred after the severe drought of the 1960s. The government's response to the catastrophe was the initiation of strategies for coordinated and effective water resource development, culminating in the mid-1970s in the creation of the Federal Ministry of Water Resources and the River Basin Development Authorities. The activities of these institutions were further strengthened in 1981 by the establishment of the National Committee on Water Resources and the Water Boards at the state level. These bodies were charged with taking an inventory and ensuring rational and systematic planned management and conservation of the country's water resources. In the 1970s and early 1980s, water resource management in Nigeria faced many problems, which slowed down the development of the resource. Some of these problems included:

- the deficiency of the resource itself;
- unnecessary duplication and overlap in organizations, structures and functions of the relevant bodies;
- the ill-defined and uncoordinated roles of the federal, state and local government agencies responsible for water resource development;
- failure to recognize the interrelationship between surface water and groundwater and between water resources and land use;
- lack of effective water and environmental protection laws and the means to enforce the already existing laws.

In the late 1980s, Nigeria began to make serious efforts to address these problems: a national body was created to coordinate all environmental protection activities in the country. A comprehensive national environmental policy was formulated, which, among other things, addressed the issue of water resources, and the Hazardous Waste Decree was promulgated with the intention of discouraging reckless and illegal dumping of hazardous and harmful wastes on land and into watercourses (FGN, 1988b). Surface water and groundwater contamination, air pollution, solid waste dumps and general environmental degradation, including the loss of land and aquatic resources, are major environmental problems caused by industrialization in Nigeria. Improper disposal of untreated industrial wastes has resulted in coloured, murky, odorous and unwholesome surface waters, fish kills and a loss of recreational amenities. A significant proportion of the population still relies on surface waters for drinking, washing, fishing and swimming. Industry also needs water of acceptable quality for processing (Helmer & Hespanhol, 1997).

The Directory of Industries in Nigeria, published by the Federal Ministry of Industry (2004), indicates that over 3,000 industrial establishments exist in the country. These industries vary in process technology, size, nature of products, characteristics of the wastes discharged and the receiving environment. At present, there are 10 major industrial categories readily discernible in Nigeria: metals and mining; food, beverages and tobacco; breweries, distilleries and blending of spirits; textiles; tanneries; leather products; wood processing and manufacture, including furniture and fixtures; pulp, paper and paper products; chemical and allied industries; and others. The current standard for arsenic discharge into surface water is 100 μg/l. There is no standard for land discharge. The maximum permitted concentration of arsenic in industrial effluents in Nigeria for the protection of groundwater is 50 μg/l (FGN, 1988a,b, 1993; FEPA, 1989, 1991a,b,c).

1.12 SOUTH AFRICA (A, C, M)

In recent years, South Africa has experienced the development of an increasingly robust infrastructure that has positively impacted water resources. As of 2006, South Africa had a multicultural population of 49.0 million, with a Human Development Index rank of 110 of 169 and a per capita GNI of US$ 9,812 (US Census Bureau, 2011; UNDP, 2010).

With a diverse land area of 1 219 912 km^2 and rich mineralogical reserves, South Africa has been involved in both diamond mining and gold mining for 200 years. Almost all of the mining areas have sulfide minerals, especially iron sulfide pyrites. Whenever these ores are exposed to air and moisture, the sulfide minerals present start to oxidize via a well-known set of autocatalytic reactions to give high concentrations of total dissolved salts (particularly sulfates), low pH values and high concentrations of dissolved metal ions (especially iron). The resulting solution is toxic to most forms of aquatic life and can lead to dramatic changes in ecosystem functioning as well as changes in the structure and chemical composition of soils. The process of sulfide oxidation is very difficult to curtail, and the resulting acid mine drainage can persist for centuries, as illustrated by English lead and tin mines. In South Africa, the concern for the release of arsenic is related to the high sulfur content of the mineralized areas. Sulfate-rich solutions increase the solubility of arsenic, cobalt, iron, magnesium, nickel and uranium (Brink *et al.* 1990); when acid mine drainage evaporates, metal- and metalloid-bearing salts redissolve with the next watering and release acids and metals into solution, leading to sharp seasonal peaks in metal and metalloid concentrations in water systems (King, 1995).

Three river basins, the Olifants, the Vaal and the Limpopo (Shingwedzi subcatchment), have been affected by arsenic contamination from greenstone formations. The Olifants and Vaal basin gangue rock gold mines are rich in arsenopyrite. Arsenic is mobilized by sulfide oxidation, which produces highly toxic arsenic(III). In the Limpopo River basin, Shingwedzi subcatchment, the Giyani greenstone formation is rich in arsenopyrite, which is exposed by artisanal gold mining, and the unregulated exposure of arsenopyrite is a serious health concern for those relying on surface water for drinking in the upper reaches of the Shingwedzi subcatchment (Ashton *et al.* 2001).

As Naicker *et al.* (2003) summarized the situation:

> The Witwatersrand region of South Africa is famous for its gold production and a major conurbation, centred on Johannesburg, has developed as a result of mining activity. A study was undertaken of surface and ground water in a drainage system in this area. Soils were also analysed from a site within the mining district. This study revealed that the ground water within the mining district is heavily contaminated and acidified as a result of oxidation of pyrite (FeS_2) contained within mine tailings dumps, and has elevated concentrations of heavy metals.

As Robinson (2003) remarked:

> On the Witwatersrand gold mining area alone, it has been estimated that approximately 500 megaliters per day have to be pumped from underground and discharged into surface water courses to prevent seepage back into the mine. The underground water is in vast interconnected reservoirs and the hydrology is complex. If a mine ceases operations, the [acid mine drainage] might flow into an adjacent operating mine or can rise to dangerous levels [that] may overflow into the shallower fresh water reservoirs that feed the springs and rivers in the Vaal River catchment basin.

Coal is also a potential source of arsenic release in South Africa. Swaine (1990) estimated that the average arsenic concentration in South African coals is 4 mg/kg; this average concentration is exacerbated by the number of ageing artisanal mines with surface areas exposed to weathering. South Africa's coal industry is the second biggest mining sector after gold, with sales contributing 16% of export revenue in 2003; production for 2003 was estimated at 232.9 million tonnes, with sales estimated at $US 26.86 billion. Coal reserves are located in Karoo-age sediments in the Mpumalanga, Free State, northern Kwazulu, Natal and Northwest provinces and are also mined at Wankie, Zimbabwe, and Maamba, Zambia. Most collieries are concentrated around the towns of Witbank, Ermelo and Secunda (MBendi, 2004). Coals range from low-rank bituminous to anthracite; in commercial sites, bord and pillar mining is usual, with long walling, strip mining and bord and pillar mining with pillar extraction increasing as supplies are depleted (Barnes, 1981). Acid mine drainage from high-sulfur, presumably arsenic-bearing coal is exacerbated by the age of the collieries, which date to 1950, their expanding numbers and size, and their increasing surface areas, which promote acid mine drainage leaching into surface water and groundwater (Ashton *et al.* 2001). In 1998, coal was ranked as providing 75% of the total primary energy supply for South Africa. Dry and wet acid (acid mine drainage) deposition of South African coal ash contains not only arsenic and selenium, but also uranium, thorium, radium and radon, making coal the largest single source of radioactive contamination of the South African environment (South Africa, 2000). However, an intensive study of South Africa's coal mine waters did not include an analysis of the arsenic concentration in acid mine drainage (Azzie, 2007). Coal is also used as a household fuel in South Africa (Center for Entrepreneurship in International Health and Development, 2005).

In sites where the acid mine drainage remains close to its source, the severity of the problem is perceived to be less. However, in situations where the influence of acid mine drainage can be seen several tens of kilometres away from its source, the problem is perceived to be far worse, even if the total quantities of acid involved are similar. This so-called "mobility" of acid mine drainage is the result of increased quantities of water being available to "move" the acid mine drainage away from its starting point. The water responsible for "moving" the acid from its source can be derived either from high local rainfalls or from mining operations that use large volumes of process water that must be discarded after use (Ashton *et al.* 2001).

On 24 October 2004, News24.com (2004) published a report indicating that:

> A drastic picture of the quality of drinking water and waste water released at a number of Free State towns has emerged from a scientific report published in Bloemfontein this week. It reveals that the quality of drinking water poses a threat to public health at 11 Free State towns. In two of them, Dealesville and Jagersfontein, the status of the drinking water was found to be "red", indicating an immediate threat to public health and an urgent need for intervention. The report was compiled by the Council for Scientific and Industrial Research for the Free State department of local government. It covers the six months up to June 2004. The Democratic Alliance said the removal of experienced officials was partly to blame for the poor quality of water at Free State towns. According to the report, the concerns at Dealesville and Jagersfontein were in relation to the amount of total coliform, fecal coliform, nitrate and arsenic in the drinking water.

Following this report, a number of towns – Boshof, Hertzogville, Fauresmith, Luckhoff, Petrusburg, Excelsior, Vrede, Memel and Brandfort – were cited for being in a yellow status, indicating that their drinking-water posed a threat to public health. Five additional towns – Koffiefontein, Senekal, Viljoenskroon, Edenville and Rouxville – were cited for less severe violations of water quality. The nature of these citations was not specified, but at least some towns had high arsenic levels in their drinking-water.

The contamination of groundwater by acid mine drainage was addressed by the Mining, Minerals and Sustainable Development Project, which was designed to elucidate how the mining and minerals sector can best contribute to the global transition towards sustainable development (Danielson, 2006). Acid mine drainage was found to be the most widespread and pervasive mining-related problem for groundwater in the Olifants River basin. However, hampering any attempt to assess the extent of acid mine drainage with regard to specific mine sites is the fact that water quality data from individual mine properties, although collected by the mines and submitted to the Department of Water Affairs and Forestry as part of routine pollution control monitoring operations, are considered to be "commercial in confidence", and public access to them is denied (MMSD, 2001).

Anthropogenic arsenic has been an issue: 60 tonnes of arsenic-contaminated waste were exported to South Africa. Public opposition reached a crescendo on 14 September 2000 when international and South African environmental organizations publicized South Africa's contravention of the 1994 United Nations treaty, the Basel Convention on the Control of Transboundary Movements of Hazardous Wastes and their Disposal. This Convention banned exports of hazardous wastes for any reason from member states of the Organisation for Economic Co-operation and Development (OECD), such as Australia, to non-OECD countries, such as South Africa. At that time, Australia joined in the decision and stated that it would abide by it (Coalition Press, 2000):

> The toxic waste, called "para-goethite", which originates from the Australian mining and smelting giant Pasminco, contains high levels of lead and arsenic. It was sent to the South African firm Mintek, where, according to the South African government, it was used for recycling "research" and then the residues were to be returned to Australia. So far the fate of the wastes and the amount of exposure to African workers and the contamination left on South African soil remains unknown.

The effect of this hazardous waste on the groundwater has yet to be determined, but it poses a potential threat.

1.13 UGANDA (M)

Uganda, with a population of 34.5 million, has a human development index rank of 161 out of 169 and a per capita GNI of US$ 1,124 (US Census Bureau, 2011; UNDP, 2010).

The abandoned Kilembe copper mine in western Uganda in the township of Kilembe, with a population of 7,914, is a potential source of arsenic and heavy metals from mine tailings that leach into the Rukoki River flowing through wetlands into Lake George. An investigation of lake water and shore and lakebed sediments indicates that there are heavy metals associated with the Kilembe sulfide mineralization. Enrichment of metals has been compared between lakebed sediments, both for wet and dry seasons. Sequential extraction of iron for lakebed sediment samples reveals low iron mobility. Relatively higher mobility and biological availability is seen for cobalt, copper and sulfur. Heavy metal contents in lake waters are not an immediate risk to the aquatic environment (Owor *et al.* 2007).

1.14 ZAMBIA (G, M)

Zambia has a population of 13.9 million, of which over 8 million live within the Zambezi River basin. Zambia has a Human Development Index rank of 150 out of 169 and a per capita GNI of US$ 1359, (US Census Bureau, 2011; UNDP, 2010). Although Zambia is a landlocked country of 752 600 km^2 in southern Africa, lying to the south of the Democratic Republic of the Congo and north of Zimbabwe, the Zambezi River basin covers 540 000 km^2 or 71.7% of the country's area, with a dense network of tributaries and ecosystems as well as high plateaus, mountains and hills that are home to the Kafue and Luangwa national parks (Ashton *et al.* 2001). Elevation varies from over 2300 m in the Mafinga Hills on the north-east national border to 329 m in the Zambezi River basin. The western part of the country consists mainly of plateaus, typically at 1,000–1,300 m above sea level. The terrain is more variable in the east. The Muchinga Mountains form a north-east to south-west ridge of 1788 m in the Central and Northern provinces, and deep valleys are cut by the Luangwa and Zambezi rivers in the south-east. Topographic depressions also occur in parts of the Northern and Luapala provinces, where many low-lying areas are swamps, of which the largest is the Bangweulu.

Zambia's geology is dominated by crystalline rocks and four sedimentary sequences that are critical to the quality of surface water and groundwater (UN, 1989):

(1) Precambrian crystalline basement rocks (gneisses and granitics), mainly in the eastern and southern parts of the country;
(2) the Katanga System (Upper Precambrian to Lower Cambrian), comprising metamorphosed sediments, including shales, dolomites and quartzites. The lower part of the sequence has abundant copper deposits, extending from the extreme north-west through the Copper Belt to Southern Province. The sequence is also enriched in other metals, such as cobalt, zinc and lead, and has been extensively mined in some areas, especially the Copper Belt. Shales of the Katanga System occur extensively in the Bangweulu area and west of the Copper Belt;
(3) the Karoo System (Upper Carboniferous to Jurassic), comprising sandstone, shale, limestone and conglomerate with some coal seams, mainly in Southern Province and along the Luangwa Valley. In addition, Karoo volcanic rocks (basalts) underlie the Kalahari sediments in Western Province;
(4) the Kalahari Formation (Cenozoic), comprising loose sands, gravels, clays and marls up to 150 m thick, in the west and south-west (mainly Western Province).

Crystalline basement rocks and slightly acidic groundwater aquifers as well as sedimentary formations yield groundwater with low dissolved arsenic concentrations. This is especially true of carbonate rock types (limestone, dolomite of the Katanga System) and sand and gravel deposits (Kalahari Formation). Arsenic mobilization may occur in the finer-grained clays of the Kalahari aquifer if the groundwater conditions become anaerobic with dissolved iron and manganese. Arsenic in groundwater may also occur in association with sulfides and in mineralized areas in the Copper Belt (MacDonald & Partners, 1990; Aquastat, 1995; British Geological Survey, 2001).

Mining in the Zambezi Basin includes copper mining in the Kafue sub-basin in Zambia and tin, gold and coal mining in Zimbabwe and Mozambique. Coal and copper are, however, the major minerals mined within the Zambezi Basin (Anon, 1986). Both open and underground mining methods are used in the extraction process, wasting vast tracts of land through excavations of the ground and deposition of mine dumps. The Kafue River area has over 90 000 ha under mining, of which 11% or 9,900 ha (99 km^2) are covered by solid waste dumps. As the population density of the Kafue River basin is 29.6/km^2, the total population potentially at risk in the 99 km^2 of the solid waste dumps is about 2,900. In 1996–1997, these waste dumps amounted to approximately 15 million tonnes, comprising overburden, waste rock, tailings and smelting slag. Arsenic and carbon monoxide penetrate farming land tracts, putting miners, processors and farmers at risk of arsenic poisoning and other related health problems (Bailey, 1991; IMERCSA, 2004).

In Zambia, arsenic contamination may also arise when mining operations disturb arsenic-bearing minerals. Mining is a major cause of pollution in the Zambezi Basin, causing water acidification and discharging a variety of highly toxic metals, such as arsenic, mercury, cadmium and lead.

The disposal of solid mine waste also presents serious problems. Large-scale acid mine drainage is associated with Zambian Copper Belt mines, where numerous, very large copper mines have operated for decades without correct siting, protection of runoff from tailings or proper treatment of effluent before being discharged into the Kafue River. Fortunately, the carbonate-rich local lithology provides a large buffering capacity that can neutralize much of the acid mine drainage produced. However, this situation is not sustainable indefinitely, and concerted effort will be needed to redesign, and perhaps even relocate, some of the tailings dams, while improving general housekeeping practices and effluent treatment process efficiencies (Philip, 1979; Norrgren *et al.* 2000; Pettersson *et al.* 2000).

The Zambian Copperbelt province has been studied with regard to the uptake of arsenic by the arsenic hyperaccumulator *Pteris vittata L.* (Chinese brake fern) and in leaves of *Cyperus involucratus*, which grow on the surface of an old flotation tailings pond. Two types of tailings are: (1) reddish-brown, rich in arsenic, iron and other metals; and (2) grey-green, with a lower content of arsenic, iron and other metals, apart from manganese. The recent discovery of an arsenic hyperaccumulator, the Chinese brake fern, has offered the hope that phytoextraction might be developed into an efficient, environmentally friendly, and cost-effective technology for arsenic decontamination (Kribek *et al.* 2011).

Leachates from six metals (chromium, cobalt, copper, zinc, lead, nickel) and one metalloid (arsenic) accumulation were studied in roadside soil and wild rat (*Rattus* sp.) samples from the lead-zinc mine in the town of Kabwe (with a population of 213 800) and the capital city of Lusaka (with a population of 1 225 000). These metals, and the arsenic in the soil samples and lead in the rat tissue samples, were quantified using atomic absorption spectroscopy showing concentrations much higher than benchmark values. Geographic Information System analysis indicated the source was mining and smelting and that the area south of the mine was more highly contaminated even though the prevailing wind flow was westward.

1.15 ZIMBABWE (G, M)

Zimbabwe, with a population of 12.1 million in a land area of 390 759 km^2, has the lowest Human Development Index rank of 169 out of 169 and a per capita GNI of US$ 176 (UNDP, 2006b; WHO & UNICEF, 2006). It borders South Africa to the south, Botswana to the south-west, Zambia to the north-west and Mozambique to the east. Although it is landlocked, it has two major watersheds, the Zambezi and Limpopo river basins. The former accounts for 64.4% of the total land area (or 251 410 km^2) and 72.1% of the total population (or 9.3 million), whereas the latter covers only 12.8% of the land mass (or 49 900 km^2), with only 3% of the total population (or 387 000) (Ashton *et al.* 2001; US Census Bureau, 2011; UNDP, 2010).

Zimbabwe's political history since 1980 has seriously impacted its infrastructure and the quality of its water supplies. In 2000, compulsory land redistribution was enacted, and many infrastructure problems began to manifest. GDP has decreased 40% since 2000, whilst the inflation rate increased from 55.2% in 2000 to 1913.6% in May 2006. The exchange rate depreciated from 55 Zimbabwean dollars to US$ 1 in 2000 to 109 000 Zimbabwean dollars to US$ 1 in May 2006. Foreign direct investment decreased from minus $0.3 million in 2001 to minus $8.7 million in 2004, and agricultural production is down more than 40% since 2002. There is also capacity erosion as professionals and skilled personnel emigrate, with 60% of nurses and 50% of medical doctors, for instance, having left since 1999. Capacity erosion has severely impacted public health, as Zimbabwe has the fourth highest HIV/AIDS rate in the world (UNDP, 2006b; WHO & UNICEF, 2006).

Ironically, Zimbabwe is rich in many metals and minerals, but most gold mines, which contain naturally occurring arsenic, are located in the Zambezi River basin, the central zone of the country, with the highest population density clustered around these mines. Water resources are potentially endangered even without mining activity, as arsenic-bearing minerals weather naturally and leach into the surface water and groundwater.

Exacerbating these geogenic sources of potential arsenic contamination, Zimbabwe has been thrust into uncontrolled mining by its political inability to enforce a number of laws that have been enacted, including the Natural Resources Act, 1996, the Parks and Wildlife Act, 1998, the Mines and Minerals Act, 1996, and the Water Act of 1977 and its Amendment of 1998. For instance, the Mines and Minerals Act is continually under criticism from environmental groups because it appears to override all other legislation. Prospectors and miners have access to almost all land, including national parks, with the restriction that they obtain written consent from the "owner". However, they are allowed access to all water and timber, which they are allowed to clear-cut by up to 50% (Ashton *et al.* 2001).

Part of the reason for the upsurge in mining activity is related to its bolstering of a faltering economy (Mobbs, 1998):

> Zimbabwe's economy was ravaged by problems, including the devalued Zimbabwe dollar, commercial lending rates exceeding 40%, inflation exceeding 30%, hikes in power (a 67% fuel price increase and a 33% electricity price increase) and transportation costs, and a high unemployment rate. The state-owned mining company, Zimbabwe Mining Development Corp. (ZMDC) had an interest in a number of mining operations, and the Government's Industrial Development Corp. of Zimbabwe Ltd had several subsidiary companies operating in the industrial mineral sector. Zimbabwe had a significant local mining industry. More than 500 gold mining operations were registered. Production from the large mines of Ashanti Goldfields Co. Ltd of Ghana, Kinross Gold Corp. of Canada, Lonrho Zimbabwe Ltd, and Rio Tinto Zimbabwe Ltd dominated the gold sector, and the numerous medium- and small-scale producers contributed about 10% to the nation's formal gold production. Alluvial gold panning was Zimbabwe's rural alternative to subsistence farming. The Mining (Alluvial Gold) (Public Streams) Regulations (1991) authorize selective alluvial gold mining, but extensive riverbank erosion and stream siltation resulted from uncontrolled mining activity.

Although international and national mining has been a prominent aspect of the economy of Zimbabwe, it is now in need of governmental controls, because its growth has led to the arsenic contamination of aquifers reaching an all-time high (Biderman, 2003). For instance, Jonnalagadda & Nenzou (1996) studied the Mutare River in 1993 because of its proximity to gold mine tailings and found groundwater contaminated with arsenic at a level of 13–96 µg/l (range of means). They did a similar study 2.2 km downstream from the confluence of the Odzi and Mutare rivers in 1993 and found arsenic concentrations of 10–30 µg/l (range of means) in the groundwater. A subsequent study investigated the levels of arsenic uptake in potentially edible plant foods and found the plant foods to have elevated levels of arsenic (Jonnalagadda & Nenzou, 1997).

As the Zambezi River basin is the location of highly arseniferous mining sites, these sites may have contributed to arsenic contamination of surface water and groundwater, affecting the 9.3 million people who live in the basin. The list below attempts to outline where arsenic risks are potentially located, in lieu of water quality data from individual mine properties: although the data are collected by the mines and submitted to the government as part of routine pollution control monitoring operations, they are considered to be confidential, and public access to them is denied (MMSD, 2001). The total production of gold in tonnes (t) per mine is given as an index of extraction and therefore of potential exposure of arseniferous minerals to weathering and leaching. All the mines that are listed below are gold mines, and all are considered "possible" arsenic contamination sources, determined by the presence of arsenopyrite and other arsenic minerals. All have a known "high" environmental impact on the local watersheds (Ashton *et al.* 2001). If one uses the groundwater data from the Mutare River area as an index of the impact of arseniferous mines on groundwater, then the following catchments may be at or above the WHO guideline for arsenic, and therefore warrant testing:

(1) Shashe-Tuli sub catchment, Gwanda Greenstone Belt: Champion Arsenic (closed 1924) 1022.7 t; Hydra Arsenic (closed 1946) 553.3 t; Vubachikwe (closed 1949) 110.1 t; Blanket Gold 27.0 t; Freda 10.5 t; Horn Gold Dump retreatment 5.0 t; Vubachikwe, Black Jack 25.1 t; Noel Nickel (closed 1963) 4651.9 t (Ashton *et al.* 2001).
(2) Thuli River alluvial gold mining area (Ashton *et al.* 2001).
(3) Gwayi sub catchment open-pit gold mines: Isabella, B & S or Motapa, 9.5 t; Turk 12.6 t (Ashton *et al.* 2001).
(4) Sanyati and Munyati sub catchment gold mines: Athens (closed) 5.9 t; Bell-Riverlea 10.0 t; Brompton (closed?) 7.8 t; Cam and Motor gold dump retreatment 147.2 t; Comrades Extension gold dump retreatment 5.7 t; Connemara 16.7 t; Sherwood Starr (closed) 15.2 t (Ashton *et al.* 2001).
(5) Mupfure subcatchment (Harare and Kadoma mining districts), Greenstone Belt: Dalny 64.0 t (Ashton *et al.* 2001).
(6) Mazowe sub catchment (Harare Mining District): Acturus/Planet 36.0 t; Freda Rebecca 17.4 t; Gladstone 5.7 t; Kimberley Reefs 8.0 t; Prince of Wales Gold (closed) 15.5 t; Shamva 68.7 t, with known, severe arsenic contamination (Ashton *et al.* 2001).
(7) Mzingwane sub catchment of the Limpopo, Filabusi Greenstone Belt: Fred 17.8 t; How 25.5 t; Nsama River, where it crosses the Giyani greenstone formation; alluvial gold mining contributes mercury and arsenic in the river (Ashton *et al.* 2001).
(8) The Midlands Greenstone Belt between the Kwekwe and Kadoma gold mines, many of which are extensive, with six having produced over 15 t of gold and two having produced over 100 t of gold; they are likely to contribute acid mine drainage containing arsenic and antimony sulfides, potentially releasing these highly toxic metals into the local waters (Ashton *et al.* 2001). At the Idarama mine, for instance, located 17 km outside Kwekwe in the Midlands Province, high-arsenic ore leached an average of 725 μg arsenic/l into a local stream (Mutandwa & Mutandwa, 2007).
(9) Mutare River catchment mines, 13–96 μg arsenic/l, range of means (Jonnalagadda & Nenzou, 1996).
(10) Odzi River catchment mines, 10–30 μg arsenic/l, range of means (Jonnalagadda & Nenzou, 1997).

It is obvious from this list that much testing for arsenic contamination of surface water and groundwater in Zimbabwe is needed.

Section 2

Region of the Americas

2.1 ARGENTINA (G)

Argentina, with a population of 41.8 million, has a Human Development Index rank of 46 out of 169 and a per capita GNI of US$ 14 603 (US Census Bureau, 2011; UNDP, 2010).

Geogenic arsenic occurs in the Chaco-Pampean Plain in north-western Argentina, with an area of 1 million m^3 and a population of about 1.2 million (Bejarano Sifuentes & Nordberg, 2003) to 2 million (Smedley & Kinniburgh, 2002). These rural people live in dispersed settlements and use river water and groundwater from shallow aquifers without treatment. The aquifers contain arsenic and are situated in Tertiary loess deposits (in the Pampean Plain) and Tertiary and Quaternary fluvial and aeolian sediments (in the Chaco Plain). Tertiary volcanism is thought to be the primary source of arsenic contamination (Figure 2.1). Two additional areas which are not indicated in Figure 2.1 include: (1) San Antonio de Los Cobres in Salta Province, which lies to the west of the Chaco Plain; and (2) Copahue, which lies to the west of the Pampa Plain (Bundschuh *et al.* 2011).

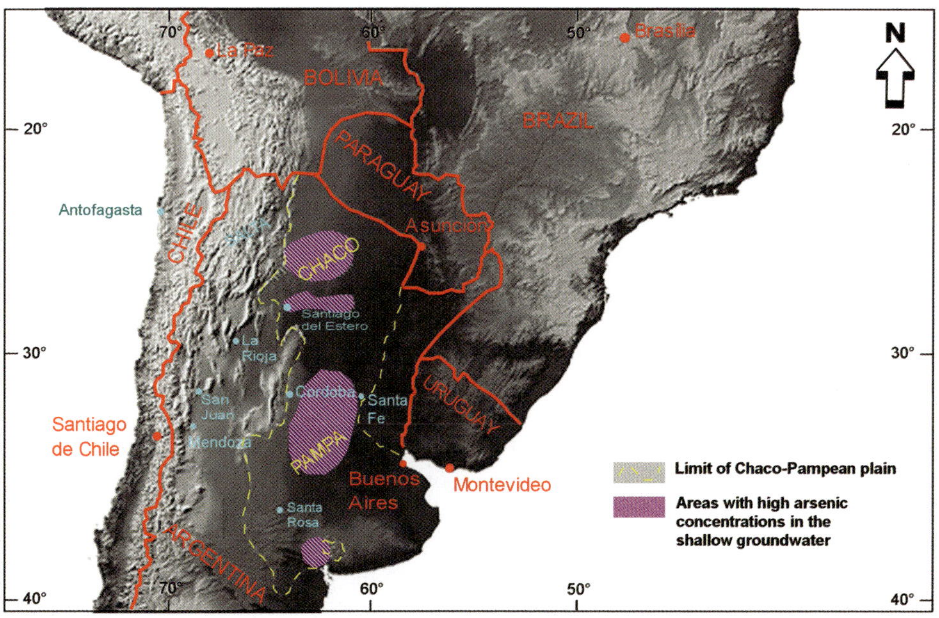

Figure 2.1 Areas with high arsenic concentrations in groundwater of the Chaco-Pampean Plain, Argentina (modified from Bundschuh *et al.* 2004b). (Digital elevation model modified from PIA03388 image; http://photojournal.jpl.nasa.gov).

High concentrations of arsenic in drinking-water (>50 μg/l) have been documented from the provinces of Buenos Aires, Chaco, Córdoba, La Pampa, Salta, San Luis, Santa Fe, Santiago del Estero and Tucumán (Tello, 1986; Esparza, 2000). A wide range of arsenic concentrations in drinking-water sources has been reported. Nicolli *et al.* (1989) reported groundwater arsenic concentrations of 6–1500 μg/l (median 255 μg/l) from Córdoba, whereas Pinedo & Zigarán (1998) recorded levels up to 1900 μg/l from groundwater samples from 40 villages in the same province. Lower levels of between <10 and 720 μg/l (mean 201 μg/l) were reported from ground waters from the Carcarañá River basin (covering the Córdoba and Santa Fe provinces) (Nicolli & Merino, 2002). Smedley *et al.* (2000) reported levels of up to 5300 μg/l (median 145 μg/l) from ground waters sampled in La Pampa province and noted that 73% of the ground waters investigated had arsenic concentrations that exceeded 50 μg/l. The spatial variability of arsenic concentrations tends to be large, and there is no distinct regional trend in the distribution of high-arsenic ground waters. High levels of arsenic are also present in some river waters. Lerda & Prosperi (1996) reported levels of arsenic between 7 and 114 μg/l in river water from Córdoba.

In an effort to characterize these widely varying arsenic concentrations, a shallow, loess aquifer, a river system and a deep aquifer were compared in eastern Tucumán province. The highest arsenic concentrations (>1000 μg/l) were found in a shallow, arsenic-rich loess aquifer, mid-range concentrations were found in fluvial sediments in the Salí River, and lowest concentrations were found in a deep aquifer of alluvial sands/gravels and clays. In the first environment, arsenic was adsorbed onto iron oxyhydroxide in the loess, but more was released into the water with the increase of pH in a west to east flow, as organic wastes and contaminants depleted oxygen and, by reductive dissolution, released the adsorbed arsenic. A similar reductive dissolution of arsenic is thought to occur in shallow aquifers with a rising water table at the end of the rainy season (Garcia *et al.* 2007).

These findings are consistent with arsenic concentrations in the groundwater in Santiago del Estero province. In the counties of La Banda and Robles around the provincial capital of Santiago del Estero, the first symptoms of arsenicosis were detected in 1983. In 1984, the government confirmed arsenic concentrations above 1000 μg/l in shallow aquifers (Martin, 1999) within the alluvial cone of the Río Dulce west of the city; many more aquifers have since been studied in this area to determine the extent and mobilization of arsenic (Bejarano Sifuentes & Nordberg, 2003; Claesson & Fagerberg, 2003; Bundschuh *et al.* 2004b; Bhattacharya *et al.* 2005, 2006a). Arsenic concentrations in groundwater in Santiago del Estero province were mostly in the range from >10 to 430 μg/l, with a highest value of 13 500 μg/l. The arsenic species were mostly arsenic(V) with some arsenic(III). Redox conditions were oxidizing or moderately reducing, consistent with arsenic(V), and elevated pH is consistent with desorption of arsenic, so that highest arsenic concentrations follow the highest pH values (Bhattacharya *et al.* 2006b). The possible mechanisms of arsenic mobilization are the dissolution of volcanic ash, resulting in the release of arsenic (which still remains to be proven), the adsorption of arsenic on iron and aluminium and the release of arsenic under high pH conditions (Bundschuh *et al.* 2004a).

In another attempt to characterize arsenic occurrences in the Chaco-Pampean Plain, Bundschuh *et al.* (2004a) showed that arsenic contamination of wells was inversely proportional to their depth, which is opposite to the situation in Bangladesh, West Bengal, India, Viet Nam and Cambodia:

> In the shallow aquifers, more than 48% of the 63 studied wells show As at toxic levels (maximum 4.8 mg l^{-1}), while in the deep groundwater the concentration is below 0.05 mg l^{-1}. The pH of the shallow ground waters range between 6.5 and 9 and generally have high electrical conductivity with mean values of 2072 and 1693 $\mu S/cm^{-1}$ in the years 1998 and 1999, respectively. Arsenic concentrations are high in the alkaline Na–HCO_3 type ground waters, where As correlates positively with Na^+ and HCO_3^-. Moreover, As correlates positively with Mo, U, and V, while a negative correlation was observed with Ca^{2+} and Mg^{2+}. The potential sources of groundwater As are: (i) layers of volcanic ash with 90% of rhyolitic glass; (ii) volcanic glass dispersed in the sediments; and (iii) clastic sediments of metamorphic and igneous origin. Great lateral variability in the concentration of groundwater As is caused by several hydrogeological and hydrogeochemical factors.

Besides groundwater and river water, another vector of arsenic consumption is through the eating of animal flesh, especially of llamas that drink from contaminated water sources, and further investigation of this vector in rural areas is needed (Faria *et al.* 2004).

Exposure of the population in Argentina to high levels of arsenic in drinking-water has occurred for a considerable length of time (Hopenhayn-Rich *et al.* 1996). Arsenic-related symptoms reported from Argentina include skin lesions (including keratosis and hyperpigmentation), skin cancer and internal cancers, including lung, bladder and kidney cancer (PAHO/ECO, 1993; Hopenhayn-Rich *et al.* 1996; Hopenhayn-Rich *et al.* 1998). Skin lesions were first reported in the early 1900s (Ayerza, 1918; Zaldivar, 1974). When arsenic contamination became known in the late 1970s, aqueducts diverting river water low in arsenic to towns became available, and some people in more remote regions used rainwater harvesting. Since the effects of arsenic exposure manifest only some 20 years after exposure, a study conducted on arsenic exposure and cancer incidence in the period 1986–1991 was necessarily limited to identified populations deemed at risk prior to the installation of the riverine aqueducts. The correlation of bladder, kidney and lung cancers was positive for prolonged exposure to arsenic levels of 178 µg/l (Hopenhayn-Rich *et al.* 1998). Table 2.1 summarizes the results of this study.

Table 2.1 Incidence of cancer in an arsenic-exposed population in Argentina.

Study population	Source and level of arsenic exposure	SMRs for cancer in the high-exposure group
Residents in Córdoba vs rest of Argentina	Up to 533 µg/l; in high-exposure group, over 98% of water samples were ≥40 µg/l	**Males** Bladder: 214 (178–253) Lung: 177 (163–190) Kidney: 157 (117–205) Skin: 149 (71–273) **Females** Bladder: 182 (119–264) Lung: 216 (183–252) Kidney: 181 (119–264) Skin: 278 (161–444)

SMR: standardized mortality ratio
Sources: Hopenhayn-Rich *et al.* 1996; Hopenhayn-Rich *et al.* 1998

Some mitigation measures involving various forms of treatment have been employed, including coagulation at San Antoni de los Cobres in Salta and in Santa Fe province (Figueroa & Montes, undated; Mozziconacci *et al.* 1998). A community trial project using iron(III) chloride for arsenic removal has been conducted in Rufino (Madeic *et al.* 1999), and removal at the household level using mixtures based on alum and iron(III) chloride has been evaluated in Salta (Johnston *et al.* 2001).

2.2 BOLIVIA (PLURINATIONAL STATE OF) (G, M)

The Plurinational State of Bolivia, with a population of 10.1 million, ranks 95 out of 169 in the Human Development Index, with a per capita GNI of $US 4357 (US Census Bureau, 2011; UNDP, 2010).

The principal source of exposure to arsenic is drinking-water contaminated by mining activities. Those affected are located towards the north of El Alto in the Murillo province (La Paz) and in Vinto, in Oruro (Sancha & Castro, 2001). In particular, the city of La Paz, with a population of 1.25 million (2005 estimate), in an area of 470 km^2, receives its public drinking-water from a source that is contaminated with the arsenic-rich effluents from the Milluni mining area (Sancha & Castro, 2001). Although it has been estimated that the total population exposed to arsenic contamination is only 20 000 people (Sancha & Castro, 2001), it is clear from the population of La Paz alone that many more may be exposed to elevated levels of arsenic through drinking-water sources contaminated by mining activities. Furthermore, given the proximity to arseniferous deposits in Argentina and Chile, areas in the Andean highlands have elevated arsenic levels in soils (Esparza, 2000). Geogenic sources of arsenic in surface water are seldom reported, but Quintanilla (1992) reported levels between 210 and 12 600 µg/l from surface water in the south-west of the Plurinational State of Bolivia, although it is not clear whether these were used as sources of drinking-water.

The Poopó basin catchment area (27 700 km^2), with Poopó Lake at its centre, serves a population of 14 984 in the Plurinational State of Bolivia's Andean highlands (Selander & Svan, 2007). Arsenic concentrations in surface water from tributary rivers and from Poopó Lake range from 90–140 µg/l (without contamination from mining sources) to over 200 µg/l with contributory mining activities. Seasonal rainfall has an opposite effect on arsenic concentrations in surface waters of lakes and rivers (Bundschuh *et al.* 2011a): in lakes like Poopó near mining sites, during the dry season, peak arsenic concentrations of 210–220 µg/l are achieved as a result of evapotranspiration whereas, in the rainy season, lake concentrations are diluted by rainfall to 20 µg/l. However, rivers in the dry season have the lowest arsenic concentrations, reaching their maximum arsenic concentrations in the rainy season through increased erosion, dissolution of weathered solids and transport of effluents from mining sites (Garcia, 2006). Two additional areas of arsenic contamination lie to the north and south of Poopó Lake: Oruro and an area to the north of Potosi (Bundschuh *et al.* 2011a).

Soil and dust from arsenic-contaminated mining wastes are also routes of exposure (Esparza, 2000). Arsenic levels in soil from a recreational area close to a mining complex were reported to range between 36 and 953 mg/kg (Esparza, 2000).

Arsenic measurements in water, solids, sediments, vegetables and fish collected from the Pilcomayo River watershed, Chuquisaca Department, southern Bolivia are given in Table 2.2 (Bundschuh *et al.* 2011b).

Table 2.2 Arsenic measured in water, solids, sediments, vegetables and fish collected from the Pilcomayo River watershed, Chuquisaca Department, southern Bolivia.

	Sampling sites (m above sea level)	Water (μg/L) ($n=1$)	Soils (mg/kg) ($n=1$)	Sediments (mg/kg) ($n=1$)	Vegetables (mg/kg·dw) ($n=1$)									Fish (mg/kg·ww) ($n=1$)	
	Detection limit	5	0.05	0.05	0.05									0.05	
					Onion	Carrot	Tomato	Broad bean	Lettuce	Chard	Cabbage	Maize	Beet	Muscle	Head
1	Rio Pilaya (814 m)	7	9.0	8.7	0.4	–	0.5	–	–	–	–	–	–	0.6	0.6
2	Purón Escapana (2431 m)	4	9.1	9.1	2.3	–	–	2.3	–	–	–	–	–	0.6	1.0
3	Monte Sandoval (2311 m)	2	6.6	6.7	0.8	0.9	–	–	–	–	–	–	–	1.3	1.6[d]
4	La Vidriera (2312 m)	30	12	9.7	–	–	–	–	5.5	5.4	–	–	–	0.2	0.2[e]
5	La Quemada (2387 m)	28	11	7.2	5.4	–	–	–	–	5.4	–	–	–	1.2[f]	0.8[g]
6	El Fuerte[a] (1969 m)	2	4.5	4.5	–	–	–	–	2.1	–	2.1	–	–	0.2	0.2
7	El Fuerte[b] (1327 m)	101	–[c]	9.4	–	–	–	–	–	–	–	–	–	18	19
8	Tabasay (1077 m)	62	6.7	10	–	–	–	–	–	–	–	2.1	–	0.1	0.3
9	Taygata (2425 m)	42	9.1	32	2.9	–	–	2.9	–	–	–	–	–	0.6	0.5
10	Sotomayor (2114 m)	91	6.8	26	3.8	3.8	–	–	–	–	–	–	–	0.3	0.3
11	Puente Méndez (2362 m)	47	21	16	–	–	–	–	13	–	–	–	12	1.0	1.0
12	Tasapampa (2297 m)	61	5.1	33	2.3	2.6	–	–	–	–	–	–	–	0.9	0.9

dw: dry weight; ww: wet weight; [a]Santa Elena river before the confluence with Picomayo river; [b]after the confluence of Santa Elena Pikomayo river; [c]Data not available; [d]sample of a 10 cm long fish; [e]not for human consumption (samples correspond to 2 cm long fish); [f,g]not for human consumption (samples correspond to 1–2 cm long fish).
Source: Bundschuh *et al.* 2011b

2.3 BRAZIL (A, C, G, M)

With a population of 203.4 million, Brazil ranks 73 out of 169 in the Human Development Index and has a per capita GNI of $US 10 607 (US Census Bureau, 2011; UNDP, 2010).

Exposure to anthropogenic arsenic in Brazil occurs through industrial wastes and atmospheric emissions. From 1962 to 1975, an arsenic trioxide factory produced about 90 tonnes of arsenic trioxide, without environmental monitoring (Borba *et al.* 2000). No estimates were made of the exposed population.

In Brazil, various forms of arsenic are located mostly in the Iron Quadrangle and Paracatú City in the Minas Gerais state, in the Nova Lima, Santa Barbara, Ouro Preto and Mariana districts (Bundschuh *et al.* 2011), in the Ribeira Valley in the São Paulo and Paraná states, in Fazenda Brasileira in the Bahia state, in the Pantanal in Mato Grosso state, in the Candiota River valley in the Rio Grande do Sul state, and in Santana city in the Amapá state.

Enriched-arsenic coals are found in the Candiota River valley. Because the coals used in these plants have elevated concentrations of heavy metals, monitoring of emissions is important. The emissions of arsenic were scheduled to increase from 9.551 tonnes in 2003 to 17.98 tonnes in 2005, and this was projected to be accompanied by corresponding increases in emissions of chromium, cadmium, nickel and lead (Vaisman & Lacer, 2003). The human health impact of arsenic contamination in the Candiota region has been examined indirectly by measuring the arsenic content in hair using hydride generation atomic absorption spectroscopy. The results have shown that the mean arsenic content varied from 7.4 to 12.8 mg/g within 30 km of the power generation plant, compared with <0.5 mg/g at a distance of 60 km from the plant (De-Moraes-Flores *et al.* 2001).

Geogenically occurring arsenic has been identified at an alkaline lake in Nhecolandia city located in the Pantanal in Mato Grosso, central Brazil. In the saline lakes of Nhecolandia, the redox state of arsenic (As III) appears to be proportionally high due to a low redox potential and a long residence period of sediments in these lakes. However, future research is required to quantify the concentration of arsenic(III) in Nhecolandia's surface water and groundwater (Barbiero *et al.* 2007).

Evidence of arsenic contamination due to mining was found in the 1990s in the Iron Quadrangle, Minas Gerais. The Iron Quadrangle, roughly 60 km × 90 km, rests on the southernmost edge of the Sao Francisco Craton. About 2.8 million tonnes of tailings with a median arsenic concentration of 14 500 mg/kg have deposited along valleys without any environmental constraints, so that any of the 377 652 inhabitants of the watersheds downstream from these tailings are at risk. Additionally, from 1983 to 1993, more than 2.1 million tonnes of arsenic were discarded directly into rivers (Borba *et al.* 2000). These practices have contributed to the release of arsenic into surface water and groundwater, atmospheric emissions into the air and the deposition of particulate matter onto the soil (Deschamps *et al.* 2002).

The highest arsenic concentrations are in two zones: one in the south-east around Ouro Preto, Mariana Carmo River and Passagem de Mariana; and the other in the north-west including Sabara, Raposos and Nova Lima cities; there is also a lesser zone in the north-east including Brumal, Barão de Cocais and Santa Barbara cities. In the north-west, the das Velhas river connects the two main areas and continues northward, so that the arsenic contamination is carried northward and out of the quadrangle. Moreover, in the Iron Quadrangle itself, arsenic concentrations in the mine waters contained up to 2980 μg total arsenic/L and up to 86 μg arsenic(III)/L. These effluents contributed directly to the surface waters that contained up to 350 μg total arsenic/L and to river sediments that contained up to 4000 mg arsenic/kg (Figueiredo *et al.* 2005).

The same investigators also found that arsenic concentrations in groundwater were generally below the Brazilian standard of 10 μg/l but, except in the north-east, the population was less dependent on the groundwater supply, so that for practical purposes the general population was exposed to elevated arsenic in surface waters used for drinking.

Table 2.3 provides the locations of these cities and their populations (totalling 377 652) at risk from exposure to arsenic-contaminated surface water from mineralization in the Iron Quadrangle.

An epidemiological study of the effects of arsenic on human populations in Minas Gerais state was conducted in the last decade. For instance, Minas Gerais was the site of testing of 300 children from three districts (Mariana, Santa Barbara and Nova Lima) of the so-called Iron Quadrangle from 1998 to 2003, and those from Santa Barbara and Nova Lima (near derelict gold mining and smelting sites) were shown to exhibit positive arsenicosis symptoms (Matschullat & Deschamps, 2004).

In each of these districts, the arsenic concentrations in the groundwater were above 50 μg/L. In a survey of 126 children from three towns in the Iron Quadrangle, Matschullat *et al.* (2000) found that the mean arsenic concentration in urine was 25.7 μg/L, and 19.2% of children had urinary arsenic concentrations above 40 μg/L. In addition, a long-term environmental impact of arsenic dispersion in Minas Gerais has been identified in the groundwater (Matschullat & Deschamps, 2004).

The Amazon River has also been subjected to arsenic contamination. Three tributaries were contaminated with arsenic by the Brazilian manganese mining operation, Icomi Mining Company, which started in the early 1970s and operated until the mine was closed in 1997. The Santana city in Amapá state is located in the lower Amazon near to the Atlantic coast in north-eastern Brazil. This area has been mined since 1950 and, although arsenic concentrations in the river sediments were up to 1600 mg/Kg and up to 231 μg/L in their surface water tap water arsenic levels were only up to 0.5 μg/L. The mean urinary arsenic concentration measured in 512 men and women in this area was 0.2 μg/L in their surface water, ranging from 0.063 to 1.936 μg/L (Figueiredo *et al.* 2005, 2007).

Table 2.3 Cities at risk in the Iron Quadrangle, Minas Gerais state, Brazil.

Name of city	Latitude	Longitude	Elevation (m)	Population estimate
Iron Quadrangle, north-east, sites downstream of Conceicão River watershed in proximity to mines with tailings >500 mg arsenic/kg				
Barao de Cocais	−19.93	−43.47	836	7 432
Brumal	−19.98	−43.45	816	4 853
Santa Barbara	−19.93	−43.4	765	4 447
Iron Quadrangle, north-west, sites upstream of Velhas River watershed in proximity to mines with tailings 500–1000 mg arsenic/kg				
Sabara	−19.9	−43.8	922	55 999
Raposos	−19.95	−43.8	934	39 861
Nova Lima	−19.98	−43.85	859	77 632
Outside of north-west of Iron Quadrangle, sites downstream of Velhas River watershed in proximity to mines with tailings 500–1000 mg arsenic/kg				
Santa Lúcia	−19.78	−43.87	813	129 729
Vespasiano	−19.67	−43.92	871	46 220
Iron Quadrangle, south-eastern sites upstream of Velhas River watershed in proximity to mines with tailings 500–1000 mg arsenic/kg				
Ouro Prêto	−20.38	−43.5	1160	6 233
Passagem de Mariana	−20.38	−43.45	901	5 246

Source: Figueiredo *et al.* 2005; Falling Rain Genomics Inc., 2007

A similar situation occurred in the upper Ribeira Valley, in the southernmost part of São Paulo state, south-eastern Brazil (Falling Rain Genomics Inc., 2007). Here, the lead and zinc mine tailings contained 355 mg/Kg and soils contained 2000 mg/Kg of arsenic, but arsenic concentrations in surface water were <9 μg/L (Figueiredo *et al.* 2005, 2007). However, an epidemiological study undertaken in 2002 provided some data suggesting caution, as elevated arsenic levels in urine were demonstrated in a group of 333 subjects who ingested fish as part of their diet. The arsenic concentrations in their urine showed a significant correlation with abnormal urinary excretion of albumin and α-l-microglobulin ($p = 0.019$) (Garlipp *et al.* 2003). In addition, the fish of the Ribeira watershed are regularly consumed, and it is known that bioaccumulation of arsenic (presumably a function of the high arsenic levels in riverine and lacustrine sediments) does occur in many different organisms that fish routinely ingest, including macrophytes, insects, gastropods, macro crustaceans and micro crustaceans (Companhia de Tecnologia de Saneamento Ambiental, 1991).

In 1998, the Brazil Health Ministry, through the Secretary of Health Surveillance and the General Coordination of Environmental Health Surveillance, started the National Program of Water Quality for Human Consumption. The purpose of the program was to develop actions to guarantee water quality along with the Ministry Regulation of standard water quality "Portaria Ministerial nº 518/2004". This regulation specifies that the maximum concentration allowed for arsenic in drinking water consumption is 10 μg/L. The Brazilian drinking-water standard for arsenic was changed from 50 μg/L to 10 μg/L in 2000, and the new standard came into force in December 2002. This reduction is causing many of the sources of arsenic contamination mentioned above to receive closer scrutiny.

Data obtained by The Health Surveillance of Brazil in 2009 shows that the concentration of arsenic in water treatment plant outflows was less than or equal to10 μg/L in almost all the Brazilian states. However, some municipalities in the south-east, south and west-center presented concentrations of As as high as 1000 μg/L (Brazil, 2010).

In addition, an unusual but possible source of geogenic arsenic in Brazil, which has not yet been monitored but which deserves more investigation based on its identification in the Caribbean and south-eastern USA, is the transport of dust from the Sahara Desert in winter dust events called *Harmattan.* This transport process carries 100 million tonnes of dust annually from the Bodélé depression in Chad (and probably elsewhere in West Africa, such as Mali) and is now recognized as providing most of the mineral and organic dust content to the Amazon rainforest (Koren *et al.* 2006).

2.4 CANADA (A, C, G, M, P)

Canada, with a population of 34.0 million, has a Human Development Index rank of 8 out of 169 and a per capita GNI of $US 38 668 (US Census Bureau, 2011; UNDP, 2010).

Enriched-arsenic coals are found in lignite and bituminous beds across Canada. The Eocene Hat Creek lignite bed (580 m thick, 26 km long and 4 km wide) in British Columbia has elevated arsenic, possibly caused by arsenic leaching from volcanogenic strata above the roof of the bed (Goodarzi & Van der Flier-Keller, 1989). It is one of the largest coal beds in the world, with projected resources of about 10 billion tonnes. In another measurement, Zone A of the Hat Creek deposit coals contained between 4.0 and 14.0 mg arsenic/kg. The Cretaceous sub-bituminous beds of the Vesta Mine in the central Alberta Plains are much smaller. They have a total thickness of only 1.6 m. An anomalous inertinite seam has an arsenic concentration of 3.0 mg/kg, whereas the rest of the sub-bituminous coals have concentrations ranging from 0.6 to 1.2 mg/kg (Gentzis & Goodarzi, 1997). In comparison with Hat Creek deposits, lignites from Vancouver Island are much enriched Late Cretaceous deposits (Bickford & Kenyon, 1987). Samples from the Comox coalfield have arsenic concentrations up to 240 mg/kg, and those from the Suquash coalfield on Millbank Sound have concentrations up to 1400 mg/kg (Goodarzi & Swaine, 1993). Nova Scotia coals are also enriched with arsenic. For instance, Pennsylvanian bituminous coals in the Sydney basin exhibit arsenic enrichment in marginal zones, suggesting hydrothermal sources (Hawley, 1955), and the Cumberland basin has elevated arsenic coal seams, which are positively correlated with sulfur concentrations. Coal ash from seam no. 15 measures 7200 mg/kg (Hower *et al.* 2000).

PollutionWatch, a collaborative project of Environmental Defence and the Canadian Environmental Law Association, assessed the nation's vulnerability to pollution and compiled an extensive database on facilities emitting arsenic to water. The top 40 arsenic polluters (Table 2.4) range from the Grand Lake Generating Station in New Brunswick (1452 tonnes) to the South East Oakville Wastewater Treatment Plant in Ontario (82 tonnes) (PollutionWatch, 2004). No facilities from Saskatchewan, Nova Scotia, Newfoundland and Labrador or Prince Edward Island were included in the top 40. Several lesser arsenic polluters are also listed in this source. These statistics may be correlated with areas of arsenic pollution concern in Canadian waters.

Table 2.4 Canada's top 40 arsenic polluters.

Rank	Facility	Company name	Amount of arsenic (tonnes)	Type	Province
1	Grand Lake Generating Station	New Brunswick Power Corporation	1452	A	New Brunswick
2	Red Lake Mine	Goldcorp Inc.	1400	M	Ontario
3	Trail Operations	Teck Cominco Metals Ltd	774	M	British Columbia
4	Lou Romano Water Reclamation Plant	City of Windsor	601	A	Ontario
5	Fonderie Horne	Noranda Inc.	485	M	Quebec
6	Iona Island Wastewater Treatment Plant	Greater Vancouver Regional District	446	A	British Columbia
7	Thompson Operations	Inco Ltd	420	M	Manitoba
8	Leamington Pollution Control Centre	Leamington Pollution Control Centre	413	A	Ontario
9	Gold Bar Wastewater Treatment Plant	City of Edmonton	396	A	Alberta
10	Skyway Wastewater Treatment Plant	Regional Municipality of Halton	326	A	Ontario
11	Inco Ltd Port Colborne Refinery	Inco Ltd	301	M	Ontario

(Continued)

Table 2.4 Canada's top 40 arsenic polluters (*Continued*).

Rank	Facility	Company name	Amount of arsenic (tonnes)	Type	Province
12	Harmac Pulp Operations	Pope & Talbot Ltd	297	A	British Columbia
13	Prince George Pulp and Paper Mills	Canfor	284	A	British Columbia
14	Bonnybrook Wastewater Treatment Plant	City of Calgary Wastewater	248	A	Alberta
15	Ashbridges Bay Treatment Plant	City of Toronto – Works and Emergency Services	247	A	Ontario
16	Kamloops Pulp Division	Weyerhaeuser Company Ltd	235	A	British Columbia
17	Espanola Mill	Domtar Inc.	224	A	Ontario
18	Irving Pulp & Paper Ltd/Irving Tissue Company	Irving Pulp & Paper Ltd/Irving Tissue Company	205	A	New Brunswick
19	Little River Pollution Control Plant	City of Windsor	195	A	Ontario
20	Elk Falls Division	NorskeCanada	189	A	British Columbia
21	Howe Sound Pulp and Paper Mill	Howe Sound Pulp & Paper Ltd Partnership	168	A	British Columbia
22	Papiers Fraser – Pâtes Thurso	Papiers Fraser Inc.	153	A	Quebec
23	Alberta Pacific Forest Industries	Alberta Pacific Forest Industries	152	A	Alberta
24	Cornwall Business Unit	Domtar Inc.	138	A	Ontario
25	Woodward Avenue Wastewater Treatment Plant	City of Hamilton	125	A	Ontario
26	Humber Treatment Plant	City of Toronto – Works and Emergency Services	120	A	Ontario
27	Western Pulp – Squamish Operation	Western Pulp Limited Partnership	119	A	British Columbia
28	Peace River Pulp Division	Daishowa-Marubeni International	116	A	Alberta
29	Mackenzie Pulp Operations	Pope & Talbot Ltd	113	A	British Columbia
30	Annacis Island Wastewater Treatment Plant	Greater Vancouver Regional District	104	A	British Columbia
31	Usine de Windsor	Domtar Inc.	104	A	Quebec
32	Wabamun Generating Station	TransAlta Utilities Corporation	101	A	Alberta
33	South-west Oakville Wastewater Treatment Plant	Regional Municipality of Halton	101	A	Ontario
34	Hinton Division	Weldwood of Canada Ltd	97	A	Alberta
35	Greenway Pollution Control Centre	City of London	97	A	Ontario
36	New Britannia Mine	1126774 Ontario Ltd	96	M	Manitoba
37	Mid-Halton Wastewater Treatment Plant	Regional Municipality of Halton	88	A	Ontario
38	Marathon Pulp Inc.	Marathon Pulp Inc.	87	A	Ontario
39	Kidd Metallurgical Site	Falconbridge Limited	83	M	Ontario
40	South East Oakville Wastewater Treatment Plant	Regional Municipality of Halton	82	A	Ontario

A: anthropogenic; M: mining

Arsenic sources related to smelting in the Sudbury, Ontario, area have been studied as contributors to lacustrine environments, where bioaccumulation by fish is of concern, as well as the use of lake water for drinking (Nriagu, 1983). A mapping of arsenic in lacustrine sediments in Ontario (covering Sudbury, Georgian Bay, Lake Superior, Thunder Bay, Lake of the Woods, Perth, Bancroft, New Liskeard, Kenora and Atikokan) details the extent of other anthropogenic and geogenic arsenic occurrences (Friske *et al.* 1998).

Muggah Creek estuary in Sydney, Nova Scotia, is considered Canada's worst contaminated site. Deposition modelling predicted that the coke ovens and steel plant had a maximum total particulate matter deposition rate of 387 g/m^2 per year and a polycyclic aromatic hydrocarbon (PAH) deposition rate of 10 g/m^2 per year (CBCL Ltd & Conestoga-Rivers & Associates, 1999). It has been argued that the particulate matter contains arsenic, lead, cadmium, chromium and zinc resulting in a 20-cm layer of contaminated material. Deposition modelling predicted that contaminants released from coke and steel production would have deposited in Whitney Pier, Ashby and North End (Table 2.5).

Table 2.5 Levels of arsenic contamination for different sites near the tar ponds in Sydney, Nova Scotia, Canada.

Location/data source	Arsenic concentration (mg/kg)
Whitney Pier (NOCO)	
Government of Canada (a) (GM; $n = 14$)	44
Control locations	
Whitney Pier (outside NOCO)	
Government of Canada (GM, $n = 17$)	94
BH 15	135
BH 16	52.2
BH 17	33.5
Ashby	
BH 5 (under house above grade)	9
Government of Canada (GM; $n = 3$)	60
BH 1	21.7
BH 2	42.7
BH 3	157
BH 4	59.2
BH 6	17.6
BH 7	21.6
North End	
BH 13 (under house beside tar ponds)	36.5
Government of Canada (GM; $n = 6$)	74
BH 8	52.8
BH 9	4.7
BH 10	40.9
BH 11	29.4
BH 12	24.7
BH 14	48.5
20 km outside Sydney	
BH 18	2

BH: borehole; GM: geometric mean; NOCO: north of the coke ovens
Source: Lambert & Lane, 2004

Study results are consistent with the prediction that arsenic, lead and PAHs were deposited in the three communities surrounding the tar ponds. The soil concentrations of lead, arsenic and PAHs in the three communities surrounding the tar ponds are similar but are significantly different from those from the background locations 5–20 km from the coke ovens (Lambert & Lane, 2004).

The Alberta Athabascan oilfields have come under exploitation for petroleum extraction since the price per barrel reached US$ 60 in 2005. In 2007, the production totalled 1 million barrels per day. The areal extent of the fields totals 13.8 million hectares, or 21% of the province, according to the Pembina Institute and the Canadian Parks and Wilderness Society (CBC News, 2007). Alberta oil sands in proximity to arsenic mineralization can promote mobilization of arsenic release. In a study undertaken by the Canadian Government, arsenic concentrations were measured in Quaternary drift and Quaternary-Tertiary buried channel aquifers. The results (Lemay, 2003) showed that:

> The majority of the Quaternary intertill-intratill aquifer water samples have arsenic concentrations below the analytical detection limit. Four of the water samples have arsenic concentrations that exceed the analytical detection limit of 0.01 mg/L. Arsenic conditions appear to be controlled by a combination of the following mechanisms: (1) adsorption to oxide and clay minerals; (2) competition between arsenic species and phosphorus ions for adsorption sites; (3) release of arsenic brought about by changes in pH; and (4) proximity to bedrock.
>
> All three samples from the Quaternary-Tertiary buried channel aquifer contained arsenic concentrations above the analytical detection limit. Arsenic concentrations in the Quaternary-Tertiary buried channel aquifer appear to be controlled by a combination of the following mechanisms: (1) competition between arsenic species and phosphorus or bicarbonate ions for adsorption sites; (2) release of arsenic brought about by changes in Eh [redox potential]; (3) dissolution of carbonate or evaporite minerals releasing associated arsenic; and (4) proximity to bedrock.

Fort Chipewyan, Alberta, has sustained the impact of toxins, including arsenic, in its water supplies from the oil sands development, which has allegedly produced some cancer cases (Anon., 2007). The Athabasca River is used to extract the bituminous tars from the reserves, and consequently river levels have dropped and residents have been warned not to drink river water or eat riverine fish, according to a member of the Fort McKay First Nation, which is located in the midst of the oil development. The Athabasca River flows more than 1500 km from the Columbia Icefield to Lake Athabasca in north-eastern Alberta and then another 2000 km via the Slave and Mackenzie rivers to the Arctic Ocean (Jones, 2006).

Hidden arsenic, probably not from geogenic sources but from mining sources, has been identified in lakes near Yellowknife, Northwest Territories. These lakes have high concentrations of methylated arsenicals contaminated by aqueous or atmospheric arsenic emitted from the Giant Mine, which began processing gold in 1948 and producing arsenic trioxide as a by-product. The arsenic trioxide was placed in subterranean storage chambers and disused stopes (KHS Environmental Management Group Ltd, 2003). Many samples of lake water contained appreciable concentrations of arsenicals, which are hidden to conventional hydride generation analysis and appear to be related directly or indirectly to atmospheric emissions from the gold mine. The occurrence of methylated arsenicals and hidden arsenic at concentrations that comprise a substantial portion of the total arsenic budget in some aquatic systems indicates that biologically mediated methylation–demethylation reactions may influence the local arsenic cycle (Bright *et al.* 1996). The manifestation of hidden arsenic in a eutrophic lacustrine environment is comparable with that which has been encountered in Lake Biwa, Japan (Sohrin *et al.* 1997), and in Greifen Lake, Switzerland (Kuhn & Sigg, 1993).

Geogenic sources of arsenic in groundwater used for commercial sale as bottled water have been cited as a potential health threat, according to the Government of Canada's (1996) Northern River Basins Final Study Report:

> Basin residents should … exercise caution in their choice of bottled waters. [Northern River Basins Study] studies revealed that most bottled waters are of good quality, but a few contain higher than average levels of bacteria, organic contaminants (e.g., chloroform) and minerals (e.g., sodium, arsenic and fluoride).

The occurrence of arsenic in groundwater linked with mining activity occurs mainly in Nova Scotia and New Brunswick. According to Grantham & Jones (1977), the sources of arsenic in wells in Nova Scotia are tailings and waste rock, abandoned mine workings, primary geology and secondary use of waste rock and tailings. A single case of elevated concentrations of arsenic and heavy metals in rural domestic water supplies in York and Charlotte counties in New Brunswick has been reported (Peters, 1977, 1996, 1999; cf. Siddiqi, 2003). These elevated concentrations appear to be from natural sources; examples of other non-anthropogenic occurrences should be investigated.

Ground waters containing elevated levels of arsenic have been identified in south-western New Brunswick, southern Nova Scotia (Bottomley, 1984; Meranger & Subramanian, 1984), British Columbia (Boyle *et al.* 1998) and Saskatchewan (Thompson *et al.* 1999). In some areas, the proportion of affected wells may be as high as 29% (Grantham & Jones, 1977; Peters, 1977; Boyle *et al.* 1998). It is not clear, however, how many people are served by these supplies, so it is not possible to estimate an exposed population.

In Nova Scotia, arsenic levels up to approximately 5000 μg/l in well water have been reported (Grantham & Jones, 1977). In Saskatchewan, the highest arsenic well water concentration recorded was 117 μg/l (Thompson *et al.* 1999).

Arsenic-related skin lesions have been reported from Nova Scotia (Grantham & Jones, 1977). Symptoms were found in almost 30% of people consuming drinking-water with elevated levels of arsenic in the Waverley area of Nova Scotia. Reports from other areas of Canada have not been identified.

Geogenic arsenic in the soils south-west of Manitoba is explained by the fact that arsenic was transported there by the glaciers that carried arsenic-rich schists that composed the Manitoba escarpment (Environment Canada, 2001). Other areas where glacially transported arsenic minerals have been detected that pose a threat to groundwater are Alberta, Saskatchewan and Nova Scotia (Environment Canada, 2001).

2.5 CHILE (G, M)

Chile, with a population of 16.9 million, has a Human Development Index rank of 38 out of 177 and a per capita GNI of $US 13 561 (US Census Bureau, 2011; UNDP, 2010).

The principal source of exposure to arsenic is drinking-water supplied from rivers that originate in the Andes, with high levels of arsenic derived from geological sources. The aquifers in the affected areas of Chile are thought to be Quaternary volcanogenic sediment, although they have not been characterized in detail and the geochemistry is, at present, poorly understood. The groundwater conditions are generally oxidizing, with high salinity and high levels of boron. Conditions are arid, and many river waters also contain elevated arsenic levels (Smedley & Kinniburgh, 2002). Water contamination in Chile occurs without any systematic monitoring or specific studies of acid mine drainage or of the impact of the mining activities on water quality.

High concentrations of arsenic have been recorded in surface waters and ground waters from II Region in northern Chile. This includes the cities of Antofagasta, Calama and Tocopilla (Cáceres *et al.* 1992). Santiago and Rancagua in the central area have also been shown to have elevated levels of arsenic (Sancha & Castro, 2001).

The main period of exposure in urban areas was in the 1960s and early 1970s, when it was estimated that 265 000 people were exposed to high levels (Zaldivar, 1977). This coincided with the inclusion of water from two rivers (which were subsequently found to have high arsenic levels) into the main Antofagasta drinking-water supply. The city installed four water treatment plants that reduced the exposure of the urban population to arsenic to 40 μg/l in their drinking-water, but many rural communities remain exposed to arsenic levels of 600 μg/l (Sancha & Castro, 2001). Current estimates of the number of people exposed to elevated levels of arsenic in drinking-water vary, but estimates for the two main cities (Antofagasta and Calama) are typically around 500 000 (Sancha & Castro, 2001; Smedley & Kinniburgh, 2002).

In northern Chile, arsenic in drinking-water is mainly of natural origin due to the hydrological characteristics of the area and the predominance of Quaternary volcanism. The principal source of exposure is drinking-water supplied from rivers that originate in the Andes, with high levels of arsenic derived from geological sources. Tests have shown that the level of exposure of the urban populations of northern Chile to inorganic arsenic fluctuates between 2363 and 125.63 μg/day (Sancha & Frenz, 2000). Ingestion of water was found to be one of the most important routes of exposure.

In addition, cases of surface water contamination have been caused by the discharge of untreated heavy metals in mining effluents, as in the case of the El Salvador mine or Planta Osvaldo Martínez mines, but there is no documentation of the monitoring of these mining operations (Sancha *et al.* 2000). There are also industrial sources of arsenic in Chile, including copper mining and smelting, which generate atmospheric emissions of arsenic (Wiertz & Rozas, 1996). It is not clear how many people may be exposed to industrial emissions, although air is not thought to contribute greatly to overall exposure (Sancha & Frenz, 2000).

According to CONAMA (2000) and DFID (1997), the main problem of water contamination in the Antofagasta or II Region of Chile is from the presence of minerals, principally arsenic, boron, sulfates, chlorides and total dissolved solids, which generally exceed the water quality standards for both drinking-water and agricultural irrigation water. Positive concentrations of arsenic are found in most surface water systems in the region. In nearly all major tributaries of the Loa River and in the San Pedro and Vilama rivers, which feed into the Salar de Atacama, arsenic concentrations frequently range from 0.2 to 500 μg/l.

The Antofagasta Region is the second largest region in Chile, covering 125 306 km^2 (16.7% of the country), and is located in the north between longitudes 67°00′W and 71°31′W and latitudes 21°00′S and 26°00′S. It is a central valley and Altiplano region bounded by a coastal plain and Cordillera de la Costa to the west and Cordillera Andina to the east. The region may be divided up into six distinct geomorphological provinces that rise step-like from west to east from sea level to over 6000 m, as follows: the Pacific coastal plain, the Cordillera de la Costa, the Atacama Desert (the Depresión Intermedia or Pampa Central), the Cordillera de Domeyko or Cordillera Central, the Precordillera Andina zone (Depresión del Río Loa and the Depresión Prealtiplánica) and the Altiplano (and Cordillera Andina) zone.

Spatial and temporal patterns of rainfall and their variability within the region exert a significant impact on water resources and land use. This poses particular problems for water resource management. The interior climate is arid, while the coast is humid and sometimes cloudy. Seasonal snows fall on the Altiplano and mountain peaks, where clear skies and many hours of high-intensity sunlight are usual. Generally, annual rainfall varies from none or very little in the Atacama Desert to less than 10 mm in the western zone to 50 mm in the upper reaches of the Loa River, 100 mm in the catchments of the Altiplano, and to 200 mm in the eastern zone.

Borregaard *et al.* (2000) summarized the water resources as follows:

> The Region comprises 52 surface water catchments. The majority of these basins are enclosed with internal drainage systems which support few perennial streams and some lagoons. A large number of these enclosed basins are occupied by "Salares" or evaporite basins, in which the principal natural mechanism of discharge is through evaporation. In most of the Central Valley (Pampa Desertica) there is practically no precipitation throughout the year and therefore no infiltration or surface runoff. The Loa River (available flow 870 l/s) and its principal tributaries constitute the largest surface water catchment in the Region and the only system which discharges perennially into the Pacific Ocean. Other important perennial rivers include the San Pedro Grande River and San Pedro Parshall River. Lagoons occur mainly in the highland areas in the east of the Region, where it is likely that precipitation exceeds evaporation and where groundwater recharge is likely to support the water balance of these surface water features. About groundwater resources, the relatively large horst and graben basins filled with thick sequences of Quaternary sediments host the most important and potentially productive aquifers. These include those basins within the Depresión del Río Loa, the Depresión Prealtiplánica and the Cordillera Andina and Altiplano, where there is higher potential for direct and indirect groundwater recharge due to higher precipitation and lower rates of evaporation. Moreover, the Region has an important presence of arid and desertic ecosystems of great interest and importance in the conservation of biodiversity. Twenty areas of greatest regional importance in terms of conservation of terrestrial ecosystem were identified, eight of which are under the protection of the National System of Protected Areas. The areas in general correspond to: coastal desert areas of high flora diversity, pre-cordillera Andina salt pans, Altiplano salt pans, Andean lacustrine systems, Altiplano steppes. The ecosystems lie within four main zones from west to east: coastal, desert, pre-cordillera Andina and Altiplano. [pp. 46–47]

> [T]he regional economic activity depends strongly on mining, a sector which represents about 60% of the regional GDP, and 32% of the country's mining GDP, with a rapid growth tendency in the last few years. The region produces 57% of copper, 71.3% of molybdenum, 43.8% of silver, 100% of sulphur, 100% of lithium carbonate, 83.2% of sodium sulphate, and 100% of saltpetre and iodine in the country. Mining makes up 95% of regional exports, with copper being the most important product; the remaining 5% is made up of fishmeal exports. In total, the zone contributes between 25% and 30% of the country's total exports. [p. 62]

Arsenic concentrations below 100 μg/l in surface water and ground waters are reported to be quite rare, and concentrations up to 21 000 μg/l have been found (Smedley & Kinniburgh, 2002). Karcher *et al.* (1999) reported a range between 100 and 1000 μg/l in raw waters, with an average of 440 μg/l.

The II Region of Chile, with an area of 126 500 km^2, illustrates the problems of aridity and extreme desiccation, as the only source of water for a population of 420 000 is the 440-km-long Loa River, which has arsenic concentrations ranging from 1400 to 21 000 μg/l; its tributary El Salado has arsenic concentrations up to 27 000 μg/l, and its sediments contain arsenic at concentrations of 26–20 000 mg/kg, with a mean of 320 mg/kg (Romero, 2003).

Mining contaminates the rivers San José, Loa (II Region), Limari, Cogotí (IV Region), Aconcagua, Chacabuquito and Rapel (V Region), the Alhué marsh and the Pampa del Tamarugal (I Region) – Quebrada Cahuisa. The Loa River exceeds regulated arsenic levels: from the Lequegna reservoir to before the union with the Salado River, it exceeds it by a factor of 2 (i.e. 100 μg/l) (it is the same for all irrigation water, on average); from the mouth of the Salado River to the Pacific, it reaches a maximum concentration of 1250 μg/l (Universidad de Chile, 2000; Hamamé, 2002).

The aquatic life of the Loa River consists of 80% crustaceans (*Cryphips caementarius* is commercially harvested), 15% molluscs, 3% amphibians and 2% fish (*Basilichthys*, *Salmo gairdneri*, *Salmo trutta fario*), and the Lio's mouth is the spawning area for anchovy (*Engraulis ringens*), sardine (*Strangomera bentinck*) and various littoral species, including corvine (*Cilus gilberti*), striped mullet (*Mugli* spp.) and silver plate (*Basilichthys*) (Servicio Agricola y Ganadero, 2000; Hamamé, 2002). Many of these species experience bioaccumulation of arsenic, so testing should be undertaken.

Pre-1957, Antofagasta's drinking-water contained arsenic at about 90 μg/l. In 1957, a water supply system involving the Tocene and Holajar rivers (which contained arsenic at 800 and 1300 μg/l, respectively) was introduced. This resulted in an average drinking-water arsenic concentration of 870 μg/l (Smith *et al.* 1998). These levels were reduced in 1970 with the introduction of a treatment plant. In Calama, the typical drinking-water arsenic level in the 1950s and 1960s was 120 μg/l. This increased to 240 μg/l in the 1970s until the 1980s, when treatment was introduced (Smith *et al.* 1998).

The first case of arsenicosis in Antofagasta was diagnosed in 1962 (Borgoño *et al.* 1977; Ferreccio *et al.* 2004). Arsenicosis has contributed to high mortality rates from bladder, kidney and lung cancers, and early exposure in children results in the worst effects. Children were found to be at high risk, with a median urinary total arsenic concentration of 50.5 μg/g of creatinine (Cortes *et al.* 2004). In a population study of airborne arsenic, 755 cases were found to have been exposed to 15 times accepted safe levels in Santiago and Temuco, and mining was proposed to be the cause (Flynn *et al.* 2002). Water and food account for the remaining arsenic exposure, with arsenic concentrations in water ranging from 43.2 to 138.8 μg/l (Pino *et al.* 2004). Arsenicosis associated with mining areas accounted for 7% of all deaths in Chile between 1989 and 1993, with levels in the associated groundwater ranging between 800 and 1300 μg/l.

Typical arsenic-related symptoms reported in Chile include skin pigmentation changes, keratosis, skin cancer, cardiovascular problems and respiratory disease (Zaldivar, 1974; Smith *et al.* 2000). Increased infant mortality (Hopenhayn-Rich *et al.* 2000) and lung and bladder cancer have also been attributed to arsenic in drinking-water (Ferreccio *et al.* 1998, 2000; Smith *et al.* 1998). A summary of studies examining arsenic-related cancer incidence is shown in Table 2.6. Smith *et al.* (1998) estimated that, in II Region of Chile, arsenic may account for 7% of all deaths in adults aged 30 years and over.

Table 2.6 Incidence of arsenic-related cancers in Chile.

Study population	Source and level of arsenic exposure	Health effects, metric of exposure and measure of association			Reference
Residents in II Region vs residents in VIII Region	Up to 860 μg/l in drinking-water in II Region; concentrations varied spatially and temporally	Mortality rate ratio and CI			Rivara *et al.* 1997
		All cancer	1.2	1.17–1.21	
		Lung	5.6	5.3–6.3	
		Bladder	6.7	5.9–7.7	
		Kidney	2.7	2.4–3.1	
		Larynx	3.2	2.7–4.0	
Residents in II Region *vs.* rest of Chile	Drinking-water average level ranged from 43 μg/l in the 1990s up to 569 μg/l between 1955 and 1969		Cancer SMR (CI)		Smith *et al.* 1998
			Males	**Females**	
		Bladder	6.0 (4.8–7.0)	8.2 (6.3–10.5)	
		Lung	3.8 (3.0–4.1)	3.1 (2.7–3.7)	
		Kidney	1.6 (1.1–2.1)	2.7 (1.9–3.8)	
		Skin	7.7 (4.7–11.9)	3.2 (1.3–6.6)	
Three regions in northern Chile, case – control study of lung cancer cases	Drinking-water measurements used to derive a mean lifetime exposure to arsenic	Lifetime exposure (mg/l)		OR (CI)	Ferreccio *et al.* 1998, 2000
		0–0.01		1	
		0.01–0.029		1.7 (0.5–5.1)	
		0.03–0.049		3.9 (1.2–13.4)	
		0.05–0.199		5.5 (2.2–13.5)	
		0.2–0.40		9.0 (3.6–22)	

CI: confidence interval; OR: odds ratio; SMR: standardized mortality ratio

Health effects may also be seen as a result of industrial contamination (principally through copper mining). Most studies in the literature have examined effects due to elevated levels of arsenic in drinking-water, although Rivera *et al.* (1997) did measure airborne levels of arsenic.

For 9000 years, people in Chiu Chiu, Atacameño, have drunk Andes river water containing arsenic at concentrations up to 800 μg/l, inducing skin effects despite good nutrition (Pino *et al.* 2004). In northern and central Chile, drinking-water is the most significant route of exposure, while in southern Chile, exposure through diet is thought to be of greater importance (Sancha & Castro, 2001). In Santiago and Temuco, food accounts for 95% of the arsenic exposure, and water 5% (Pino *et al.* 2004); in Temuco, levels of arsenic in water are in the range 11.4–48.6 μg/l (Pino *et al.* 2004).

Arsenic removal was introduced in Chile, in Antofagasta, in the early 1970s. There are now four arsenic removal plants, which produce drinking-water with final arsenic levels of 40 μg/l. This has effectively reduced arsenic exposure in the urban population, but many rural communities are still exposed to levels up to 600 μg/l (Sancha & Castro, 2001). Water treatment plants in northern Chile remove arsenic by a coagulation process (Sancha *et al.* 2000). Current experiments have focused on using iron(III) chloride and iron(III) chloride plus sulfuric acid as adsorption media, and comparing relative efficiencies and costs (Granada, 2002). Some examples of Chile's successful arsenic mitigation measures include the following:

- Remediation of riverine arsenic between 1958 and 1970 in the Toconce River reduced the initial level of 86 µg/l in 1958 to 50 µg/l in 1970 after arsenic removal (Sancha & Frenz, 2000; Sancha *et al.* 2000; Romero *et al.* 2001).
- In Tatal (population 8000), arsenic is removed from groundwater by direct filtration using iron chloride coagulant additive; the resulting arsenic concentration is 30 µg/l (Esparza, 2000).
- In Lasana, arsenic is removed by direct filtration using iron chloride coagulant additive (50 µg/l); use of iron meshes as a source of coagulant has made it possible to reduce the final arsenic concentration to 10 µg/l (Esparza, 2000).
- In Antofagasta (Salar El Carmen complex), arsenic is removed by direct filtration using iron chloride coagulant; the resulting arsenic concentration is 40 µg/l (Esparza, 2000).
- In Calama, the arsenic concentration in drinking-water has been reduced to 45 µg/l (Smith *et al.* 2000).

2.6 COLOMBIA (G, M, V)

Colombia, with a population of 44.7 million, has a Human Development Index rank of 79 out of 169 and a per capita GNI of US$ 8,589 (US Census Bureau, 2011; UNDP, 2010).

Geogenic sources of As have been determined in the estuarine waters of Barbacoas Bay, Sucre department south of Cartagena on the Atlantic Ocean, which contain effluent from the Magdalena river via the early colonial Canal del Dique (Mazo-Gray *et al.* 1997; Bundschuh *et al.* 2011a). The presence of geogenic As in Barbacoas Bay is complicated by added anthropogenic pollution involving offshore currents and tanker discharges.

Colombia's artisanal gold mining sector has 200 000 miners officially producing 30 tonnes Au/a. The gold is in association with arsenopyrite and other gold-bearing minerals in six mining districts. In the northeastern Antioquia department (Bundschuh *et al.* 2011a) there are 17 mining towns with 15–30 000 artisanal gold miners. Not only is the effluent from these artesanal deposits an issue for surficial and ground waters, but also the situation is complicated because the ore is taken to six urban centers for processing (Segovia, Remedios, Zaragoza, El Bagre, and Nechí, Bolivar) with a combined population of 150 000 inhabitants (Cordy *et al.* 2011; Bundschuh *et al.* 2011a). Therefore the total in this district potentially affected by As contamination is 180 000. In addition there are other areas of artisanal mining with an unknown extent of arsenical impact on public health: Valle de Cauca department and Caldas department (Prieto & Gonzalez, 1998; Bustos Rodriguez *et al.* 2005; Bustos Rodriguez *et al.* 2007; Gaeta Tassinari *et al.* 2008; Bundschuh *et al.* 2011a).

Colombia has arsenopyrite in association with gold-bearing minerals in six mining districts located in Nariño (Bundschuh *et al.* 2011), Antioquia (Bundschuh *et al.* 2011a), Bolivar (Bundschuh *et al.* 2011a), Valle de Cauca (Bundschuh *et al.* 2011) and Caldas departments (Bustos Rodriguez *et al.* 2007; Gaeta Tassinari *et al.* 2008; Bundschuh *et al.* 2011a). However, the extent of As leaching from these deposits into groundwater is largely unknown but suspected (Prieto & Gonzalez, 1998; Bustos Rodriguez *et al.* 2005; Gaeta Tassinari *et al.* 2008).

Volcanogenic arsenic in association with antinomy occurs in association with hot springs in sixty hot springs in Cundinamarca department (Bundschuh *et al.* 2011), one of which in Yacopi had a concentration of 37 µg/l (Alfaro-Valero, 2005).

2.7 COSTA RICA (V)

Costa Rica, with a population of 4.6 million, has a Human Development Index rank of 62 of 169 and a per capita GNI of US$ 10,870 (US Census Bureau, 2011; UNDP, 2010).

Volcanogenic sources of As in Costa Rica are just beginning to be investigated at three localities (Hammarlund & Piñones, 2009; Lopez *et al.* 2011): Rincón de la Vieja (Bundschuh *et al.* 2011a); Miravalles (Bundschuh *et al.* 2011a); and in the Choco-Suela Platanar area (Bundschuh *et al.* 2011a). While initial results have shown relatively low As concentrations in the range of 5–6 µg/l, more investigations are projected. In particular, the Rincon de la Vieja volcano located in the 14,083-hectare Rincón de La Vieja National Park is situated in the mountain range of Guanacaste, in the province of Guanacaste & Alajuela, 27-km northeast of Liberia City. Thermal fountains and hot springs called Las Pailas and Hornillas at the foot of the volcano cover 50 hectares. The volcano experienced a phreatic eruption on September 16 and 22, 2011, the first after 13 years of dormancy. Surficial and ground waters have been affected by contaminants such as As; fish were found dead in the river Penjamo, and streams and ditches in the area were infused with sediment which was observed

18 kilometers downstream (Anon., 2011). The communities potentially affected are Aguas Claras (2011 population of 7652) on the north slope and Canas Dulces (population of 2770) and Curubande (population of 4957 on the south slope).

2.8 CUBA (A, M)

Cuba, with a population of 11.1 million, has a Human Development Index rank of 51 out of 187 and a per capita GNI of US$ 5,416 (US Census Bureau, 2011; UNDP, 2011).

Anthropogenic As exposure in Manzanillo Bay is probably contributed by effluent into the Yara River from an alkaline battery factory (Amat, 2000, 2002; Ginarte, 1999). The city of Manzanillo had a population of 110 728 in the census of 2005, but more testing needs to be done to understand the public health impact and the potential number of people affected (Bundschuh *et al.* 2011a).

Cienfuegos Bay in west central Cuba on the south coast, with a population of 123 600, has been impacted for several years by the manufacture of nitrogen-fertilizers by the Empresa de Fertilizantes Nitrogenados de Cienfuegos factory. This situation was exacerbated by a spill of inorganic As oxides in 2001, which was monitored by the testing of *Gracilaria biodgetti*, a sensitive As bioaccumulating benthic algae that registered 11.834 mg/kg As (Castellanos *et al.* 2005; Bundschuh *et al.* 2011a). The monitoring of As levels using this algae continues, as does the monitoring of the potential numbers of people affected.

As exposure via the Delita gold mine (which was closed in 1997) in Isla de la Juventud has resulted from oxidation from As (III) in groundwater to As(V) on the surface through the leaching of arsenopyite within a mica schist matrix into the El Soldado River and adjacent watersheds of the Majagua, Itabo, and Los Indios Rivers (Bundschuh *et al.* 2011a). However, only La Mina spring which is closest to the Delita mine tested with elevated As, i.e. 25 250 μg/l (Toujague *et al.* 2003), which is probably due to the presence of Fe in the surrounding matrix that sorbs As and thus prevents its bioavailability. Those communities dependent on La Mina spring and its ground waters within a 72 km^2 zone have historically exhibited cerebrovascular disorders and cancerous tumors (Robles & Morillo, 2001).

The Fe, Ba, Pb, Zn, As and Cu in the tailings of the closed Zn–Pb Santa Lucía Mine have polluted local streams (Romero *et al.* 2010). The solid waste in the tailings measures 1724 mg/kg for As. The marked fall-off of As in the acidic mine drainage is possibly accounted for by the low solubility of barite and beudantite under acidic conditions (Bundschuh *et al.* 2011a).

2.9 DOMINICA (V)

Dominica, with a population of 73 000, has a Human Development Index rank of 68 out of 177 and a GNI of $US 5643 (US Census Bureau, 2011; UNDP, 2006a) . Its ancient pre-Carib, Taino name is *Wai'tu kubuli*, "Tall is her body". This name reveals the importance of the central ridge of the island, which is still being volcanogenically formed along its 46-km-long, 25-km-wide expanse.

Dominica is situated on the northern end of the Windward chain of the Lesser Antilles between Martinique and Guadeloupe, within the volcanogenic arc in the Caribbean. It has the second largest thermally active "boiling lake" in the world, which is an indicator of the volcanogenic source of arsenic leaching into its groundwater. Perhaps half of its population of 69 029 (2005) is exposed to arsenic at concentrations above 10 μg/l in its thermal waters. Criaud & Fouillac (1989) examined arsenic(III) and arsenic(V) levels in the geothermal waters on the island of Dominica and compared their occurrence with those in other areas of the world, notably the Massif Central of France, the Valles Caldera of New Mexico, USA, and south-west Bulgaria. The results showed that Dominica's waters contain arsenic concentrations that are comparable to some of the higher occurrences in these other areas and therefore demand the scrutiny accorded those areas.

The potential of the island's volcanogenic resources was the object of a thwarted attempt to develop a geothermal plant on the island. In a news story in 2000 (News Dominica.com, 2000), the tale of the failed geothermal plant is recounted:

> In 1969, a [United Nations] report on Dominica's geothermal energy potential stated that "if natural steam can be found on the island of Dominica, it may be the cheapest source of power in the Caribbean region". In 1980, a French geothermal expert, who appraised the country's geothermal energy resources reported that with 10–15 wells spaced 500 meters apart, Soufriere and Wotten Waven sources, once developed, could produce around 50,000 to 100,000 kilowatts each. The Boiling Lake could produce much more the report added. The Labour Party Government of Edward Leblanc, as well as Freedom Party and United Workers Party Governments had all entered into agreements with foreign companies to harness geothermal energy resources for electricity production. The last such agreement was signed in 1996 between Dominica and the Caribbean Power Ltd, an American-based Company. That agreement provided for the formation of the Dominica Geothermal Power Company, a joint venture between Government and the U.S.-based company with Government putting up the land as equity in the venture. The project never got off the ground.

2.10 ECUADOR (P)

Ecuador, with a population of 15 million, has a Human Development Index rank of 77 out of 169 and a per capita GNI of $US 7931 (US Census Bureau, 2011; UNDP, 2010). Ecuador's oilfields lie in the Oriente (the East) or the Ecuadorian Amazon, which is home to one of the most biodiverse rainforest jungles on Earth and also to the Huaorani, an isolated indigenous culture that numbers about 2000 people (various post-2001 estimates), speaks Huao Terero, a language isolate, and, since 1960, lives mostly in permanent forest settlements. At least five Huaorani communities (the Tagaeri, the Huiñatare, the Oñamenane and two groups of the Taromenane) have eschewed contact with non-Huaorani peoples and have increasingly moved into more isolated areas, mostly in the direction of the Peruvian border where the oilfields lie. Because the Huaorani are so endangered and isolated, they require intervention to protect their way of life before it becomes extinct (Brysk, 2000).

Texaco were in the area some 20 years and drilled 339 oil wells in a concession larger than 400 000 ha, extracting roughly 1.4 billion barrels of crude from the Oriente. Texaco's operations generated more than 12 million litres of waste each day, and accidental spills from the pipeline released an estimated 64 million litres of crude into the Amazon River. This exceeds that of the *Exxon Valdez*, the largest oil spill in the history of the USA, which dumped approximately 41 million litres of oil into Prince William Sound. The "drilling muds" produced during Texaco's oil extraction contained arsenic as well as lead and benzene, and these muds were dumped in unlined pits near water wells in the Oriente. Also, Texaco used dirt pits to separate the oil from water, which produced a waste called "produced water", a brine that contains high levels of benzene and arsenic; this was dumped into a nearby stream (Lyons, 2004).

As Texaco was ceasing operations in 1992, another oil company, Maxus Energy, a subsidiary of the international YPF Corporation, was initiating its exploration and extraction programme in a zone 190 km wide and 120–160 km in length from north to south. The zone is located between the Curaray and Napo rivers, about 80 km south of Dureno. Maxus Energy began:

> By bulldozing a 100-mile-long road through the fragile Huaorani Ethnic Reserve and Yasuni National Park, a [United Nations]-recognized World Biosphere Reserve. Environmental organizations based in Ecuador report that the region has suffered numerous oil spills since Maxus began pumping crude two years ago. The Ecuadorian group, Amazon for Life, calculates that Maxus operations have cleared 52,000 acres of pristine, ancient rainforest, and polluted or diverted 540 streams and waterways. In addition, the group reports that Maxus wells are releasing toxic fluids into Yasuni National Park's rivers, that several oil pits containing toxic wastes have overflowed into the environment, and that Maxus has built two landfills inside the park for toxic waste disposal (Enomenga, 1996).

The Ecuadorian government, in response to this situation, formulated an "Intangible Zone" in 1999, comprising 800 000 ha of the Oriente and partially overlapping the Yasuni National Park; it was designed to cover the core territory of the Huaorani and to protect it from all oil extraction and logging. However, the precise boundary of the "Zone" was not to be finalized and made law until 2007. What was needed was the Huaorani to mobilize their political will. As reported by the Yasuní Rainforest Campaign (2005):

> On August 27, 2005, over 250 Huaorani—representing 30 of the 34 Huaorani communities—gathered in the Huaorani community of Tiguino for an extraordinary assembly. In Huaorani culture, the assembly is the most powerful decision making body and the only means to dispose of and elect new leaders. During the assembly, the Huaorani decided to dismiss the seven members of ONHAE—the Huaorani leadership body—and have new elections. This decision included the removal of Huaorani President Juan Enomenga, widely accused of working too closely with oil companies such as Petrobras.
>
> Long-time anti-oil advocate Moi Enomenga became Vice-President. Moi became known to the world in the Joe Kane book *Savages*, which details Moi's fight against the oil companies 10 years ago. For the past several months, Moi has been leading the campaign for a 10 year moratorium on new oil projects within ancestral Huaorani territory. Moi was the mastermind and organizer of the anti-Petrobras march in Quito in July 2005. Now that Moi is Vice-President, the Huaorani campaign against the Petrobras project is likely to escalate.

This prediction took another year and a half to materialize, and, in February 2007, the final boundaries of the "Intangible Zone" were enacted into law. "The Presidential Decree, in addition to prohibiting the extraction of oil within the Intangible Zone, also established a 10-km surrounding buffer zone where the construction of oil access roads is prohibited, but oil extraction is permitted" (Yasuní Rainforest Campaign, 2007).

The net result of this indigenous campaign from the arsenic pollution perspective is that this endangered biological and cultural zone has been reasonably protected by a responsible Ecuadorian government from further oil spillages that could have contributed to future releases of arsenic into the groundwater.

2.11 EL SALVADOR (V)

El Salvador, the smallest of the seven Central American countries, with an area of 21 040 km^2 and a population of 6.1 million, has a Human Development Index rank of 90 out of 169 and a per capita GNI of $US 6498 (US Census Bureau, 2011; UNDP, 2010). It is the first and largest producer of geothermal energy in Central America. The Ahuachapán region, where geothermal energy is tapped, is located approximately 80 km west of San Salvador, 18 km east of the Paz River and 15 km from the Guatemalan border. It is an agricultural region that produces mainly coffee. Ahuachapán was the first geothermal site in El Salvador to be developed for commercial electricity generation. The motivation for this exploration was the "abundance of hot springs, hot wells, fumaroles and other manifestations of hydrothermal activity" (Partida *et al.* 1995) in the south-western volcanic belt of El Salvador, where Ahuachapán lies.

Environmental challenges related to the geothermal energy industry in El Salvador include the impact of waste disposal into the Paz River (Prevost, 2004):

> An advantage of geothermal power generation for the environment is that the generation process emits very little pollutants such as sulfur, carbon dioxide and particulates. Nonetheless, the chemical composition of the waste liquid from the geothermal power plants can pose environmental risks. At Ahuachapán, at one point, the waste liquid was being discharged into the Rio Paz. This was problematic since it reduced the water quality of the river, which was used for irrigation and other purposes. In order to curb the levels of arsenic and boron in the river, this disposal practice had to be discontinued. Today, some waste liquid (about one third) is disposed into the Pacific Ocean. The environmental risks of this are yet to be determined.

The distribution of these geothermal energy resources in El Salvador indicates the potential areas that are susceptible to arsenic contamination. Highly mineralized waters represented a major problem for the exploitation of the Ahuachapán geothermal field (Einarsson *et al.* 1975).

A 2001 study of water consumed in rural households found that of the total number of samples analysed, 43% were contaminated with high levels of arsenic, as well as cadmium, chromium and lead (Alfaro, 2006). Programmes to monitor arsenic in groundwater have been initiated by various volunteer groups. For instance, the Louisiana–El Salvador Partners of the Americas Project of 13–20 February 2005 monitored the Lake Ilopango watershed management programme and compiled a GIS map with the results of their arsenic testing. This was followed by a second monitoring project of 18–25 September 2006 by the same group. Critical areas were identified and will continue to be monitored (Louisiana–El Salvador Partners of the Americas, 2006).

2.12 GUATEMALA (A, G, M)

Guatemala, with a population of 13.8 million, has a Human Development Index rank of 116 out of 169 and a per capita GNI of US$ 4,694 (US Census Bureau, 2011; UNDP, 2010).

An unusual anthropogenic source of As was discovered by Archer & Elmore (2010) when they tested the clay being used in ceramic filters for the purification of water in Antigua. This finding, of course, is consistent with the recent discovery of geogenic As.

Geogenic arsenic has only recently become a matter of concern in Guatemala with the evaluation of arsenopyrite-associated ore bodies and geothermal volcanogenic sources in the highlands. For instance, the leaching of As from volcanic rocks in the municipality of Mixco (Cardoso *et al.* 2010) has been measured at the order of 10 µg/l (Garrido Hoyos *et al.* 2007; Garrido Hoyos & Avilas Flores, 2008; Bundschuh *et al.* 2010, Cardoso *et al.* 2010, Bundschuh *et al.* 2011).

Mining in Guatemala needs further assessment as to the human risk from exposure to As and other heavy metals. One of the issues faced in the interpretation of the data collected by responsible agencies is that the rainfall variations from the wet months (May to October) offset the dry months (November to April). Therefore, it is critical that the actual timing and duration of the testing of As in riverine water and its impact on human blood and urine are accurately identified. The Marlin mine in San Marcos district, western Guatemala, is a case in point. Owned by Canada's Goldcorp company through its Guatemalan subsidiary, Montana Exploradora Inc., it has been the object of a study by environmental health scientists from the University of Michigan, which was coordinated and published by Physicians for Human Rights (Basu & Hu, 2010). The study has shown that leaching from the mine into down slope creeks has impacted the residents of proximate villages with elevated concentrations of As in blood (3.20–8.50 µg/l) and urine (0.04–16.71 µg/l). Seasonal fluctuations are mirrored in the fact that groundwater in the down gradient from the tailings varied between 46 µg/l in the wet season and 261 µg/l in the dry season (ETech International 2010; Bundschuh *et al.* 2011).

2.13 HONDURAS (M, V)

Honduras, with a population of 8.1 million, has a Human Development Index rank of 106 out of 169 and a per capita GNI of US$ 3,750 (US Census Bureau, 2011; UNDP, 2010).

Mining associated with elevated As impacting human health is related to the San Martin gold mine in the Valle de Siria of central Honduras which has been operated by Entremares, a subsidiary of the Canadian-US parent company Goldcorp, since 1999. The Environmental Law Alliance Worldwide (ELAW) has reported that Communities in Siria Valley "have complained for years that the mine has polluted local rivers, destroyed forests, and dried up the local water supply." Further complicating the environmental impact of this mining venture is the fact that the extraction process utilizes cyanide, a practice banned in Montana and restricted in several countries (ELAW).

Volcanogenic As exposure is related to at least four identified geothermal hot springs at Azacualpa, Pavana, Plantares and Vanes Caldera with concentrations of 70, 110, 1260 and 1160 μg/l, respectively (Goff *et al.* 1986b; Bundschuh *et al.* 2011). Plantares geothermal well waters have more complete data showing a range of As concentration from 800–1260 μg/l (Goff *et al.* 1986a; Janik *et al.* 1991).

2.14 JAMAICA (V)

Jamaica, with a population of 2.9 million, has a Human Development Index rank of 80 out of 169 and a per capita GDP of $US 7207 (US Census Bureau, 2011; UNDP, 2010). Heavy metals and metalloids in Jamaican soils are a controversial issue. According to Karanjac (2005), high concentrations of arsenic and cadmium have been detected in soils, and "the elevated distribution of arsenic in soils coincides with the distribution of the White Limestone Formation", which covers about 60% of Jamaica with a topographical pattern of round-top hills with internal drainage. However, "there is no proof that either of these metals has reached ground water in limestone aquifers" (Karanjac, 2005).

The one occurrence of arsenic noted to date on Jamaica is an arsenic anomaly in St Elizabeth that was reported by Lalor *et al.* (1999). It appears to be related to volcanism, which affects some of the Caribbean islands and would likely be found on others. There is a possible contamination of Jamaican foods by waters tainted by arsenic from this source, as reported by Howe *et al.* (2005).

Silva *et al.* (2009) prepared a map of soils in Jamaica that were above the Guideline Limits for Agricultural Soils recommended by Canada. From this map it was calculated that 98.6% of the agricultural land in Jamaica is above the guidelines, which means that only about 48.8 km^2 is below them. However, the levels of arsenic found in agricultural foods are well below the UK recommended guideline of 1 mg/kg. The average levels found are as follows: fruit 0.004 mg/kg, legumes 0.003 mg/kg, leafy vegetables 0.009 mg/kg, root vegetables 0.01 mg/kg and root crops 0.006 mg/kg. Moreover, despite the high soil arsenic content, the arsenic concentration in the surface water is low and there seems to be no immediate health risk to the residents (Laior, GC *et al.* 1999).

2.15 MEXICO (A, G, M, V)

Mexico, with a population of 113.7 million, has a Human Development Index rank of 56 out of 169 and a per capita GDP of $US 13 971 (US Census Bureau, 2011; UNDP, 2010).

Mexico is one of the major arsenic producers in the world, with 1946 tonnes generated in 2002 (COREMI, 2004). Arsenic is abundant in Mexico's bedrock, and there are numerous arsenic-bearing minerals including arsenopyrite, scorodite, mimetite, adamite, tennantite and nickeline (Armienta *et al.* 2007).

Mexican drinking-water arsenic concentrations are regulated at a current standard of 25 μg/l. However, most municipalities have a poor reporting record and data are not centralized. The Official Mexican Standard (NOM 127-SSA1-1994) for arsenic had been 50 μg/l, but it was modified in November 2000 when an annual decrease of 5 μg/l per year was mandated, beginning in 2001 and ending in 2005. In 2005, the 2000 Official Mexican Standard (NOM 018-STPS-2000) of 25 μg/l for arsenic in drinking-water was implemented (Mexican Government, 2000a, b).

Thirteen of the 31 Mexican states have As exposure >50 μg/l in drinking water (Durango, Aguascalientes, Chihuahua, Coahuila, Guanajuato, Jalisco, Morelos, Nuevo León, Oaxaca, Puebla, San Luis Potosí, Sonora, Zacatecas) with a population of 450 000 potentially impacted (Castro de Esparza, 2009; Armienta *et al.* 2008, Sancha & Castro, 2001). In fact, in the Lagunera region alone 400 000 people are exposed to As above 25 μg/l. If the enforceable national regulatory limit of 25 μg/l of As in drinking water is utilized, the total population affected in all states has been estimated between 1 098 887 and 2 million (Exparza, 2000; Smedley & Kinniburgh, 2002; Gleason, 2010) (see Table 2.8). Anthropogenic causes of arsenic contamination of groundwater and surface water include geothermal activities. Evaporation ponds associated with geothermal power plant production can leach arsenic (Birkle & Merkel, 2002). Fish from oxbows (*resacas*) in the Lower Rio Grande Valley in the Texas–Tamaulipas region have been recorded as having elevated arsenic concentrations, suggesting the release of arsenic from industrial sources into the Rio Grande and its subsequent bioaccumulation in fish (Mora *et al.* 2001). Similarly, vegetables grown using irrigation waters with high arsenic concentrations caused by leaching from industrial waste were high in arsenic (Del Razo *et al.* 2002). Cows ingesting alfalfa irrigated with high-arsenic water produce milk that is high in arsenic in Comarca Lagunera (Rosas *et al.* 1999).

Among the major source types, geogenic sources of arsenic are the most prevalent in Mexico's environment. Contamination of groundwater resources by naturally occurring arsenic was first identified in the late 1990s in areas such as the Comarca Lagunera region (located in the states of Coahuila and Durango), Chihuahua in northern Mexico, and Zimapán, Hidalgo, in central Mexico (Cortés Muñoz *et al.* 1999). However, since 2000, the existence of drinking-water resources exceeding the Mexican standard have been discovered in four other states: Baja California Sur, Guanajuato, Sinaloa and Morelos (Rivera & Pena, 2004). Natural volcanic sediment-type aquifers with oxidizing groundwater conditions that have a neutral to high pH are found in north-central Mexico, encompassing an area of 32 000 km^2.

Sonora and Chihuahua currently have the highest concentrations of arsenic in drinking-water, with 30% of the samples tested above 10 µg/l and 4% above 50 µg/l, ranging up to >200 µg/l (Table 2.7). Arsenic occurrence appears to be associated with major sandstone aquifers and surface geology and is therefore similar to that of Argentina, where the depth of wells is inversely proportional to the concentration of arsenic (Westerhoff *et al.* 2005).

Table 2.7 Arsenic concentrations in drinking-water from Sonora and Chihuahua states in Mexico.

Mexican state	City	Average arsenic concentration in samples from city (µg/l)	Maximum arsenic concentration in samples from city (µg/l)
Sonora	Carborca	18 ($n=12$)	31
	Hermosillo	11 ($n=9$)	36
	Lazaro Cardenas	67 ($n=1$)	67
	Oribe de Alba	62 ($n=1$)	62
Chihuahua	Nuevo Delicias	52 ($n=2$)	52
	El Saucillo	43 ($n=4$)	59
	Jimenez	87 ($n=9$)	267

Source: Westerhoff *et al.* 2005

Another study (Meza *et al.* 2004) conducted in the Yaqui Valley in Sonora in 2002 resulted in the following assessment of arsenic contamination of drinking-water:

> The range of total arsenic concentration in the well water from some major towns in the Yaqui Valley was from 3.30 to 49.26 µg/L with an average of 12.70 ± 9.3 µg/L .... Forty-four percent of the sampled towns (Esperanza, Villa Juarez, Tobarito, Campo 5, Cocorit, MR Gomez, Morelos 1 and 2, Ej. Teras, Quetchehueca including Cd. Obregon) had arsenic levels greater than 10 ppb, the current [United States Environmental Protection Agency] drinking water standard .... Esperanza, Villa Juarez, and Tobarito had the highest average total arsenic levels at 43.33, 31.1, and 25.1 µg/L, respectively. Porton, Ej. Independencia, and SJ Bacum had concentrations slightly above those of the [United States] guideline (10.3 ± 0.26 µg/L).

Because of the high mineralization of this region, mining began with the first Spanish colonizers, and it has led to increased arsenic contamination of rivers. Furthermore, mines do not have to be active to be a hazard. Much arsenic contamination is attributable to runoff from abandoned mines; since the mines are derelict, the acid mine drainage is difficult to detect and monitor (Naranjo-Pulido *et al.* 2002).

Water supplies containing levels of arsenic greater than 50 µg/l have been identified in Baja California Sur, the Lagunera region (Chihuahua, Coahuila, Durango, Zacatecas), Guanajuato, Hidalgo, Morelos, Acambro, Delicias, Jiménez and San Antonio (Sancha & Castro, 2001). Wyatt *et al.* (1998) also reported elevated levels in Sonora State.

Zimapán Valley (Hidalgo State)

Lead–zinc–silver mining and ore processing have occurred continuously in Zimapán since at least 1632 and possibly since 1576, resulting in the development of significant slag and tailings. Leaching of ore and tailings has resulted in arsenic contamination of much of the water in the valley, raising the possibility of arsenicosis of the residents (Ongley *et al.* 2003).

High concentrations of arsenic (ranging above 1000 µg/l) were detected in the aquifer system of Zimapán Valley. The main source is natural arsenic-bearing rocks, although anthropogenic sources also contribute to these high concentrations. Variations in groundwater arsenic concentrations correlated with both the precipitation regime and the hydrogeological and geological framework. An induced arsenic transport mechanism related mainly to the rainfall regime has been proposed (Rodríguez *et al.* 2004). Arsenic levels in municipal water supplies ranged from 100 μg/l (Armienta *et al.* 1998) to about 300 μg/l. The Detzani community had an arsenic concentration of 750 μg/l in its municipal water supply in the early 1990s (Armienta *et al.* 2000). Del Razo *et al.* (2003) reported that some inhabitants of the Zimapán area have been exposed to arsenic levels between 150 and 1350 μg/l from at least 1993.

A possible mechanism for increased mobility of arsenic in this region was proposed by Gurdak & Thyne (2005):

> Located in central Mexico, this valley hosts ore deposits that have been mined historically for lead, silver and zinc. Groundwater contamination by arsenic has been previously attributed to the dissolution of arsenic-bearing minerals naturally present in mineralized zones of the highly fractured cretaceous limestone aquifer. Secondary sources include the dissolution and leaching of mine tailings and deposition of arsenic-rich smelter emissions. Although arsenopyrite oxidation and scodorite dissolution/precipitation reactions may be important sources for arsenic in the valley, several geochemical lines of evidence point towards adsorption/desorption as an important process controlling subsequent arsenic mobility. Wells with the highest arsenic concentrations also contained the most reduced water in the valley .... Preliminary results indicate ... sorption processes play a significant role in arsenic mobility of the groundwater system.

Guerrero state is also impacted by leaching from the El Fraile sulfide tailings in Taxco, southern Mexico, which have adversely affected the local populations as Talavera *et al.* (2006) discovered.

Lagunera region (Zacatecas, Durango, Chihuahua and Coahuila states)

In the Lagunera region, the aquifer is composed of volcanic sediments. The groundwater is oxidizing, with a neutral to high pH (Smedley & Kinniburgh, 2002). The arsenic concentration in water sampled from 128 wells ranged between 8 and 624 μg/l, with an average of 100 μg/l and with 50% of the wells exceeding 50 μg/l, exposing more than 400 000 people living in rural areas to high arsenic concentrations. Arsenate was the predominant species in 93% of the samples (Del Razo *et al.* 1990). Del Razo *et al.* (1994) reported the average arsenic level in drinking-water in Santa Ana town (Lagunera) as 404 μg/l. In a review paper, Engel *et al.* (1994) noted that in one village in the Lagunera region, the drinking-water source had an arsenic concentration of 3980 μg/l. A project has been undertaken in the Lagunera region to detect early indicators of health risk from exposure to arsenic (Esparza, 2000).

In the region of Delicias-Meoqui, the arsenic concentration in water varies between 50 and 500 μg/l. (Gleason, 2010). Additional areas with As exposure are: the Santa Barbara mining area, Meoqui, Julimes, Camargo, and Jimenes (Bundschuh *et al.* 2011).

Guariana Valley (Durango State)

Approximately 50% of the monitored wells exceeded 50 μg/l, and nearly all wells exceeded the WHO arsenic guideline value of 10 μg/l (Herrera *et al.* 2001).

San Luis Potosi state

Water from sampled wells had arsenic concentrations ranging from 7 to 740 μg/l, with about 90% of the arsenic present as arsenate (Rosas *et al.* 1999). In San Luis Potosi Valley, 48% of wells had concentrations from 50 to 146 μg/l, with the highest concentrations in the north and northwest of the Valley (Gleason, 2010).

Sonora State

Arsenic concentrations in well water for municipal supplies from the whole state ranged from 6 to 117 μg/l on average for about 20 locations ($n = 173$ wells). Some wells serving Hermosillo (capital of Sonora), which serve 50% of the state's population, contained arsenic at 71–305 μg/l. Fluoride was positively correlated with arsenic occurrence, and many wells had fluoride concentrations above regulatory limits (2000 μg/l) (Wyatt *et al.* 1998). The agricultural valley of Caborca, located within the 160-km border region in the north-west of Sonora, was also intensively sampled. Approximately 30 groundwater wells used for drinking-water and agricultural irrigation were sampled in 2003. Only 15 wells showed arsenic levels between 10 and 245 μg/l, and only 4 wells are used for drinking-water (Espinoza-Ojeda, 2003). Four additional areas have been identified as having As exposure: Magdalena, the Yaqui river watershed, Etchojoa and the Colorado river delta (Bundschuh *et al.* 2011).

Baja California Sur

Most of this state's land is semi-arid. Groundwater is a critical drinking-water source and contamination by mine drainage contributes to increased arsenic levels. Arsenic concentrations range from 10 to 410 μg/l (Carrillo-Chavez *et al.* 2000). Earlier work indicated a potential for arsenic treatment by iron-containing adsorbent material (Carrillo & Drever, 1998). Four areas have been singled out as having As exposure: (1) Cerro Prieto geothermal area; (2) San Ignacio-Mulegé, Sierra

Guadalupe; (3) Sierra El Mechudo, communities of San Juan de la Costa and Punta Coyote; and (4) San Antonio-El Triunfo (Bundschuh *et al.* 2011).

Aguascalientes State

Arsenic concentrations in 39 wells from this state were correlated with fluoride concentrations, which ranged up to 10 000 µg/l (Trejo-Vazquez *et al.* 2002).

Rioverde Valley (San Luis Potosí)

Fifty-three groundwater wells were monitored within a 1200-km^2 area of the basin around the town of Rioverde (21°55′N 99°39′W). Arsenic concentrations varied between 5 and 10 µg/l, and the maximum arsenic concentration was 54 µg/l. Forty-five thousand inhabitants are exposed to this water (Planer-Friedrich *et al.* 2001); however, higher concentrations (700–4900 µg/l) have been determined around three fluoride mines, approximately 80 km west and 45 km south-west of Rioverde (referred to by Planer-Friedrich *et al.* 2001).

Other States

In the region of Jimenez-Camargo, the arsenic concentration is between 50 and 500 µg/l. In the municipality of Zimapán, Hidalgo, high concentrations of arsenic between 32 and 720 µg/l have been detected in wells. In the area of El Mesquite-La Majada the concentrations ranged from 64 to 140 µg/l. Data is sporadic for detection of arsenic in wells in Guanajuato, Zacatecas and Puebla (Gleason, 2010). Puebla state is known for its volcanogenic hot springs in the Acoculco and Los Humeros geothermal areas, which are similar to Michoacan state's Los Azufres geothermal area. Four localities in Tabasco state have been identified as having As exposure but no further details are available: Cactus-Sitio Grande, Luna-Sen, Jujo-Tecominoacán and Pol-Chuc-Abkatún (Bundschuh *et al.* 2011).

Health effects including skin lesions, hyperpigmentation and skin cancer have been attributed to arsenic exposure through drinking-water in Mexico (Cebrián *et al.* 1983, 1994; Diaz-Barriga *et al.* 1992; Del Razo *et al.* 2003).

Gleason (2010) summarized estimates on the average concentration of arsenic in each of 18 municipalities (Table 2.8). She then calculated the individual risk of developing skin cancer for the affected population (Table 2.9).

Table 2.8 Population by municipality and average concentration of arsenic in water for human use and consumption.

State	Municipality	Population	Average arsenic concentration mg/l
Camargo – Jiménez	Camargo	46 386	0.044
	Jiménez	39 746	0.103
	San Francisco de Conchos	2 991	0.261
Delicias – Meoqui	Aldama	19 998	0.085
	Delicias	110 876	0.040
	Julimes	5 335	0.143
	La Cruz	3 844	0.800
	Meoqui	38 152	0.130
	Rosales	14 809	0.080
	Saucillo	31 042	0.470
Coahuila	Francisco I. Madero	47 510	0.200
	Matamoros	88 235	0.510
	San Pedro	91 421	0.140
Durango	Tlahualilo	22 924	0.290
Durango	Valle del Guadiana	74 954	0.064
	Cd. de Durango	389 612	0.051
	Guadalupe Victoria	32 640	0.040
Hidalgo	Zimapán	38 412	0.090
Total		**1 098 887**	

Source: (Gleason, 2010)

Table 2.9 Individual risk of skin cancer by municipality.

State	Municipality	Population	Average arsenic concentration mg/l	Individual risk × 10^{-3} acceptable risk 3.7×10^{-3})
Chihuahua Camargo – Jiménez	Camargo	46,386	0.044	0.82–7.3
	Jiménez	39,746	0.103	0.99–41
	San Francisco de Conchos	2,991	0.261	5.60–36
Delicias – Meoqui	Aldama	19,998	0.085	3.70–18
	Delicias	110,876	0.04	0.65–6.2
	Julimes	5,335	0.143	3.90–44
	La Cruz	3,844	0.8	1.50–24
	Meoqui	38,152	0.13	0.40–45
	Rosales	14,809	0.08	0.90–13
	Saucillo	31,042	0.47	0.40–9.7
Comarca Lagunera Coahuila	Francisco I. Madero	47,510	0.2	1.20–55
	Matamoros	88,235	0.51	0.80–6.4
	San Pedro	91,421	0.14	2.40–20.8
Durango	Tlahualilo	22,924	0.290	15–31.5
Valle del Guadiana	Valle del Guadiana	74,954	0.064	0.35–17
	Cd. de Durango	389,612	0.051	1.28–6.8
	Guadalupe Victoria	32,640	0.040	1.20–5.2
Hidalgo	Zimapán	38,412	0.09	0.04–57
Total		**1,098,887**		

Gleason computed acceptable risk based on the assumption of a less stringent allowed limit of 50 μg/L arsenic. Even so, the individual risk range is consistently above the acceptable risk at the high end of the range, and sometimes even at the low end of the range, for example for San Francisco Conchos, Julimes and Tlahualilo.

The magnitude of the problem of arsenic in Mexico, coupled with the halving of the 1994 standard of 50 μg/l has stimulated the search for low-cost but effective methods of arsenic remediation (Rivera & Pena, 2004). Specific arsenic mitigation measures taken in Mexico, which include both treatment and source replacement, are summarized below:

- Some of the water supplies that exceed the 1994 Mexican government's arsenic limit of 50 μg/l for drinking-water are being partially treated with reverse osmosis systems exclusively for consumption. Although this process is highly effective, the initial investment and maintenance costs are high, so alternative lower-cost methods are being investigated.
- In 1999, the Mexican Institute of Water Technology adopted a methodology based on an oxidation–coagulation–flocculation–adsorption process to remove arsenic from drinking-water. Arsenic levels were reduced to about 0.05 mg/l (Sancha & Castro, 2001).
- In the state of Durango, recommended mitigation includes removal using crushed bone, lime and aluminium sulfate in rural areas and removal using crushed bone, activated aluminium and reverse osmosis and adoption of alternative sources in urban areas (Esparza, 2000).
- A pilot study in Zimapán identified haematite as an alternative sorbent for arsenic removal. Field studies reported levels below 0.05 mg/l in the finished water (Sancha & Castro, 2001).
- In the Lagunera region, a pipeline has been constructed to transport safe groundwater to the most seriously affected towns (Alaerts *et al.* 2001).

2.16 NICARAGUA (G, V)

Nicaragua, with a population of 5.7 million, has a Human Development Index rank of 115 out of 169 and a per capita GNI of $US 2567 (US Census Bureau, 2011; UNDP, 2010).

Geogenic arsenic occurs in Nicaragua as the natural contamination of groundwater in a zone of Tertiary volcanic rocks of the Coyol group. This zone is situated along the active, tectonic eastern margin of the Depression of Nicaragua, the Sébaco Valley near Matagalpa. With an area of 52 km^2 and a total population of 3 225 000 (1990), the inhabitants of the Sébaco Valley are exposed to elevated arsenic levels not only along the east–west fault systems but also along south-west–north-east fault systems. These fault systems intersect both with each other and with streams and rivers within the watersheds of the Sébaco Valley, creating microstructure formations that render specific characteristics of the thickness and depth of the basement of the aquifers (Westerhoff *et al.* 2005). High concentrations of arsenic have been detected in the water supply in the sub-watershed southeast of the valley, the main source of arsenic being fluids from hydrothermal processes of Tertiary volcanism. The basement rock is relatively close to the surface as fault blocks, which have produced an intense hydrothermal alteration of faults and fractures close to the flowing groundwater, the conduit of arsenic. The drilling for a drinking-water well above hydrothermal faults or fractures allows arsenic to readily enter wells (Esparza-Soto *et al.* 2004).

The area of exposure extends from the Sébaco Valley in a 12-km-wide north-west path; therefore, the population exposed is greater than that which has been reported for the valley (Espinoza, 2005):

> Of the 57 water samples taken, 21 show total arsenic concentrations between 10 and 122 μg/L, which are above the [10 μg/l arsenic] guideline established for drinking water [by WHO]. The highest concentrations of arsenic were found in the community of El Zapote. The total arsenic concentration[s] in rocks and soil were 14.98 μg/g and 57.19 μg/g respectively, water concentrations were found to be 122.15 μg/L. Cases of arsenicosis were reported for this community in the year 1996 from people drinking water from a well for a period of 6 months with an arsenic concentration of 1,320 μg/L. People from this community were affected by irreversible health problems, which forced the population to change its water supply from a dug well, which, according to findings from this study, has an arsenic concentration of 122.15 μg/L, which is still above the guideline for drinking water. Two comparative soil samples, located outside our area of study, one at the entrance of La India mine and the other among the surroundings of the community of Agua Fría, showed high arsenic concentrations similar to those of El Zapote community, with values of 95.2 and 59.5 μg/g respectively, giving evidence that the extension of arsenic pollution goes far beyond the area of study, following a north-western route in a 12 km wide path. This means that the population, besides being exposed to arsenic by ingestion, is also being exposed to contaminated soil, which makes this region an environmentally critical zone. The arsenic content in rocks, soil and water are associated to syngenetic (primary) and epigenetic (secondary) hydrothermal processes evidenced along the mainly west-to-east and southwest-to-northeast faults … and fractures characteristic of the eastern Nicaraguan geodynamics dominated by the subduction of the Cocos plate under the Caribbean plate, which is the drive for the formation of the Nicaraguan Depression where the Sébaco Valley is located in its outer eastern margin.

This exposure affects the arsenic in groundwater resources as well as in the adjacent rocks and soils. This was discovered when the situation in El Zapote was analysed in 1996. The highest arsenic concentrations were 1320 μg/l in a borehole well drilled into a heavily fractured hard rock aquifer. This elevated groundwater arsenic concentration correlated with the elevated arsenic measured in nearby rocks at 14.98 mg/kg and soils at 57.19 μg/kg (Bundschuh, 2007). This suggests that agricultural produce is also a vector of human contamination and should be routinely examined for elevated arsenic. A geochemical mechanism involving changing redox conditions promotes the release of arsenic from Nicaragua's matrix of Tertiary volcanic rocks. Because highly fractured Tertiary volcanism extends in zones across Nicaragua and across adjacent countries in Central America, further arsenic contamination from this geochemical mechanism can be expected.

Arsenic exposure in Nicaragua Valley is a factor of both natural and anthropogenic sources relating to soils, volcanic activity, natural hot springs and geothermal extraction plants. Volcanic activity and hot springs are responsible for leaching arsenic(III). Lake Xolotlan, with a surface area of 1000 km^2, leaches arsenic(III) from wastewater from a geothermal plant in the Momotombo Volcano on the north-west coast of the lake. Arsenic(III) in the range of 10.2–30.1 μg/l is released from this geothermal plant. A large number of hot springs potentially leach arsenic(III) as well. As a result, most of the 10 000 inhabitants of the local community of San Francisco Libre are exposed to arsenic in their drinking-water (Lacayo *et al.* 1992).

The natural distribution of arsenic in soils indicates that exposure of the population is permanent, affecting not only drinking-water, but also agriculture (Bigot, 2004). In a study conducted in 2003, 93% of water samples tested contained arsenate, arsenic(V), and 7% contained arsenide, arsenic(III), the most toxic form. Arsenic concentrations in water from public tube wells ranged as high as 1320 μg/l, and from private hand-dug wells ranged between 45 and 66 μg/l (1994–1996). In 2005, national data indicated that 7% of water nationwide had arsenic concentrations above 10 μg/l, with 45% of drilled wells having concentrations above 10 μg/l, 23% of hand-dug wells above 10 μg/l and 20% of springs above 10 μg/l (Gomez, 2004; Espinoza, 2005).

Symptoms manifest in arsenicosis cases in Nicaragua include keratosis, hyperpigmentation, lesions, respiratory fibrosis and portal hypertension (Gomez, 2004).

2.17 PARAGUAY (M)

Paraguay, with a population of 6 million, has a Human Development Index rank of 96 out of 169 and a per capita GNI of $US 4585 (US Census Bureau, 2011; UNDP, 2010).

The Agua Castilla River, which is used by the local population for drinking-water, was contaminated by 213 000 tonnes of toxic tailing slurries containing arsenic on 29 August 1996, by a dyke that ruptured at the Porco Mine upstream in the Plurinational State of Bolivia (Garcia-Guinea & Harffy, 1998). Its remediation is not known. Other areas of concern are with mining operations that continue to go unmonitored.

The Project for Environmental Protection and Sustainable Development of the Guarani Aquifer System conducted sampling of water in the southern and northern parts of the system in 2004 and 2005. A total of 120 samples were collected and analyzed in the laboratory of the Bundesanstalt fur Geowissenschaften und Rohstoffe (BGR). Arsenic at concentrations above the 50 µg/l limit was detected in three wells. It was found that this geogenic source of arsenic is due to the reduction of goethite enriched with arsenic (Fariña, 2009; SAG-PY, 2009).

2.18 PERU (G, M, V)

Peru, with a population of 29.3 million, has a Human Development Index rank of 63 out of 169 and a per capita GNI of $US 8424 (US Census Bureau, 2011; UNDP, 2010).

One geogenically affected area is in the semi-arid region of southern Peru, where water containing elevated arsenic levels is drawn from rivers deriving from the Andes. Arsenic in the environment is a result of the natural mobilization of natural geological sources, mining and volcanism.

For instance, volcanism affects water for the town of Ilo because it is drawn from Aricota Lake, which contains elevated levels of arsenic as a result of two of its tributaries (Callazas and Salado) passing through the Yucamane volcano region (Sancha & Castro, 2001).

It is estimated that approximately 250 000 people are exposed to elevated arsenic concentrations in drinking-water from all sources: geogenic, mining and volcanism (Esparza, 2000; Sancha & Castro, 2001).

Water is drawn from a number of sources containing elevated levels of arsenic. The Locumba River, which serves Puno and Moquegua, is considered to be representative and contains an arsenic concentration of 500 µg/l (Sancha & Castro, 2001). According to Esparza (2000), the average arsenic concentration in the affected area is 140 µg/l.

Groundwater coming from wells located in the Pelipeline community, Puno Department, has been found with geogenic arsenic pollution. Forty-four percent of the wells sampled were found above WHO Drinking Water Guidelines (Esparza, 2005). High levels of arsenic can be found in rivers and water systems, including in Lima, Peru (Esparza, 2005).

The other route of exposure is through industrial emissions and waste products from mining areas and copper foundries, such as those at La Oroya and Cerro de Pasco. Miners were one of the first groups that were identified as having exposure to high levels of arsenic. Arsenic-related disorders have been reported for workers occupationally exposed to arsenic in the mining and metallurgical industries (Esparza, 2000). No estimates of the number of people affected have been provided, however (Sancha & Castro, 2001). An arsenic treatment plant was built at Ilo in 1982. This initially used high doses of lime, but it has been replaced by treatment involving high pH, coagulation and flocculation with natural magnesium hydroxide (Sancha & Castro, 2001).

The general public and especially children have been exposed to high levels of arsenic as a result of mining practices. The Doe Run Company, based in St Louis, Missouri, USA, has been running a mining smelter in the town of La Oroya since 1922. In recent tests of children under 6 years of age, 99.8% were found to suffer from lead poisoning, 1% were suffering from high levels of cadmium and an unspecified percentage were found to be suffering from arsenicosis, with high levels of arsenic in hair and nails. The Doe Run Company, in cooperation with the government of Peru, actively conducted epidemiological tests in late 2005 when these results were made known to the press. Stories were run in the international media on 7 December 2005 (Reuters News Service, 2005) indicating the critical situation, which was being documented by a team of epidemiologists (Canyellas *et al.* 2004). Despite the critical nature of the population's exposure to arsenic and its international media exposure, the Doe Run Company convinced the government to postpone any remediation or curtailment of its mining operations. On 13 May 2006, Carlos Infantas, Peruvian Congressman and member of the Congressional Commission on Energy and Mining, announced that the Doe Run Company "is showing a subversive attitude, in an attempt to use pressure and violence so that the population supports its point of view". The charge of "subversive" was occasioned by the fact that the mayor of La Oroya staged a public demonstration in support of Doe Run (Movement for Health of La Oroya, 2006), even though:

> Doe Run has admitted that its smelter emits daily more than 1000 tons of lead, arsenic, sulphur dioxide, and cadmium over La Oroya's population of 35 000. A study conducted last August [2005] by scientists from Saint Louis University (Missouri) found

that 97.2% of La Oroya's children suffer from "worrisome levels" of lead poisoning and other toxic metals, including arsenic and mercury.

Finally, on 19 April 2007, the Doe Run Company announced to the world (PR Web Press Release Newswire, 2007) that:

"Emissions of particulate matter and heavy metals, including lead, from the main stack of Doe Run Peru's La Oroya metallurgical facility are now within governmental limits…. The progress is attributed to an aggressive modernization campaign and the execution of several projects to address environmental issues including emissions at the main stack," said Juan Carlos Huyhua, president and general manager of Doe Run Peru. "We are gratified to see our environmental work produce this significant reduction in emissions," said Huyhua. "We are continuing our work and are committed to bringing these levels down even further," he said. The company reported a decrease in emissions of particulate matter, lead and arsenic to levels within the Peruvian government's limits. Emissions of particulate matter have fallen more than 50 percent since 1997, when Doe Run Peru purchased the facility. Main stack emissions of … arsenic dropped from 30 mg/Nm^3 in 1994 to 28 mg/Nm^3 in 1997 to 12 mg/Nm^3 in 2007 (through February). The maximum permissible limit is 25 mg/Nm^3…. So far Doe Run Peru has invested more than $116 million on various environmental improvements. Once the [emissions] projects are complete, which is expected in 2009, the company's total investment on environmental issues is expected to reach $254 million, nearly 2.4 times the initial required investment.

2.19 UNITED STATES OF AMERICA (A, C, G, M, P, V)

The USA, with a population of 311 million, has a Human Development Index rank of 4 out of 169 and a per capita GDP of $US 47 094 (US Census Bureau, 2011; UNDP, 2010).

Arsenic contamination in the USA is derived from all six sources treated in this report. Table 2.10 provides an overview of each state.

Table 2.10 Sources of arsenic contamination in 46 states of the USA.

State	Population exposed (1000s)	A	C	G	M	P	V
1. Alabama	n.a.		x	x			
2. Alaska	n.a.			x	x		
3. Arizona	4728.5	x		x	x		
4. Arkansas	n.a.	x		x			
5. California	2400	x		x	x		x
6. Colorado	67.3	x		x	x		
7. Connecticut	6.3	x		x			
8. Florida	n.a.	x		x	x		
9. Georgia	n.a.	x		x			
10. Hawaii	n.a.	x					
11. Idaho	n.a.			x	x		x
12. Illinois	87.1		x	x			
13. Indiana	1.6	x	x	x	x		
14. Kansas	n.a.	x	x	x	x		
15. Kentucky	n.a.		x				
16. Louisiana	n.a.	x					
17. Maine	n.a.	x		x			
18. Maryland	n.a.	x					
19. Massachusetts	n.a.	x		x			
20. Michigan	124.8			x			
21. Minnesota	n.a.	x		x			
22. Mississippi	n.a.			x			
23. Missouri	n.a.	x		x			
24. Montana	n.a.						x

(Continued)

Table 2.10 Sources of arsenic contamination in 46 states of the USA (*Continued*).

State	Population exposed (1000s)	A	C	G	M	P	V
25. Nebraska	100			x			
26. Nevada	n.a.			x	x		x
27. New Hampshire	156	x		x			
28. New Jersey	n.a.	x					
29. New Mexico	1400	x		x	x		x
30. New York	n.a.	x					
31. North Carolina	n.a.			x			
32. North Dakota	3.2	x	x	x			
33. Ohio	72	x	x	x		x	
34. Oklahoma	150			x	x	x	
35. Oregon	n.a.	x		x			
36. Pennsylvania	2.7	x	x	x			
37. South Carolina	n.a.	x		x			
38. South Dakota	n.a.	x	x	x	x		
39. Tennessee	n.a.		x	x	x		
40. Texas	n.a.	x		x		x	
41. Utah	0.3			x			
42. Vermont	n.a.	x		x			
43. Washington	n.a.	x		x	x		
44. West Virginia	n.a.		x				
45. Wisconsin	n.a.			x			
46. Wyoming	n.a.		x	x			x
Totals	**9199.8**	**29**	**12**	**37**	**14**	**3**	**6**

n.a., not available

Arsenic occurs naturally in groundwater under both reducing and oxidizing conditions in different areas (Smedley & Kinniburgh, 2002). Aquifer types and groundwater conditions vary according to the location and are summarized in Table 2.11.

Table 2.11 Aquifer types and groundwater conditions in select regions of the USA.

Region	Aquifer type	Groundwater conditions
Basin and Range, Arizona	Alluvial basins, some evaporites	Oxidizing, high pH
Tulare Basin, California	Holocene and older basin-fill sediments	Internally drained basin; mixed redox conditions
Southern Carson Desert, Nevada	Holocene mixed aeolian, alluvial, lacustrine sediments, some thin volcanic ash bands	Largely reducing, some high pH; some with high salinity due to evaporation
Fairbanks, Alaska	Schist, alluvium, mine tailings	Gold mining, arsenopyrite, possibility scorodite
Coeur d'Alene, Idaho	Valley-fill deposits	River water and groundwater affected by lead–zinc–silver mining
Lake Oahe, South Dakota	Lake sediments	Arsenic in sediment pore waters from gold mining in Black Hills

Source: Smedley & Kinniburgh, 2002

In the USA, arsenic concentrations in public drinking-water supplies are regulated by the United States Environmental Protection Agency (USEPA) because of the associated health risks. In 1999, the National Academy of Sciences concluded that the standard of 50 μg/l for arsenic in drinking water did not sufficiently protect the public from long-term exposure. In response to this conclusion, the USEPA lowered the public drinking-water standard from 50 μg/l to 10 μg/l (USEPA, 2001a). The revised standard of 10 μg/l was fully enforceable for public drinking-water supplies by 2006.

Many areas in the USA have been identified with arsenic problems in groundwater (see Figure 2.2) (Ryker, 2001). Most of the worst-affected and best-documented cases occur in the south-western states (Nevada, California and Arizona). However, aquifers in Maine, Michigan, Minnesota, South Dakota, Oklahoma and Wisconsin have been found with concentrations of arsenic exceeding 10 μg/l (Smedley & Kinniburgh, 2002). Of those reported, however, relatively few have significant numbers of sources with concentrations greater than 50 μg/l. In 1987, it was estimated that more than 350 000 people may be exposed to drinking-water containing arsenic at levels above 50 μg/l (Smith *et al.* 1992). More recent studies suggest that a small percentage of community water supplies have arsenic levels between 10 and 50 μg/l (USEPA, 2000a), although private water supplies may still be affected.

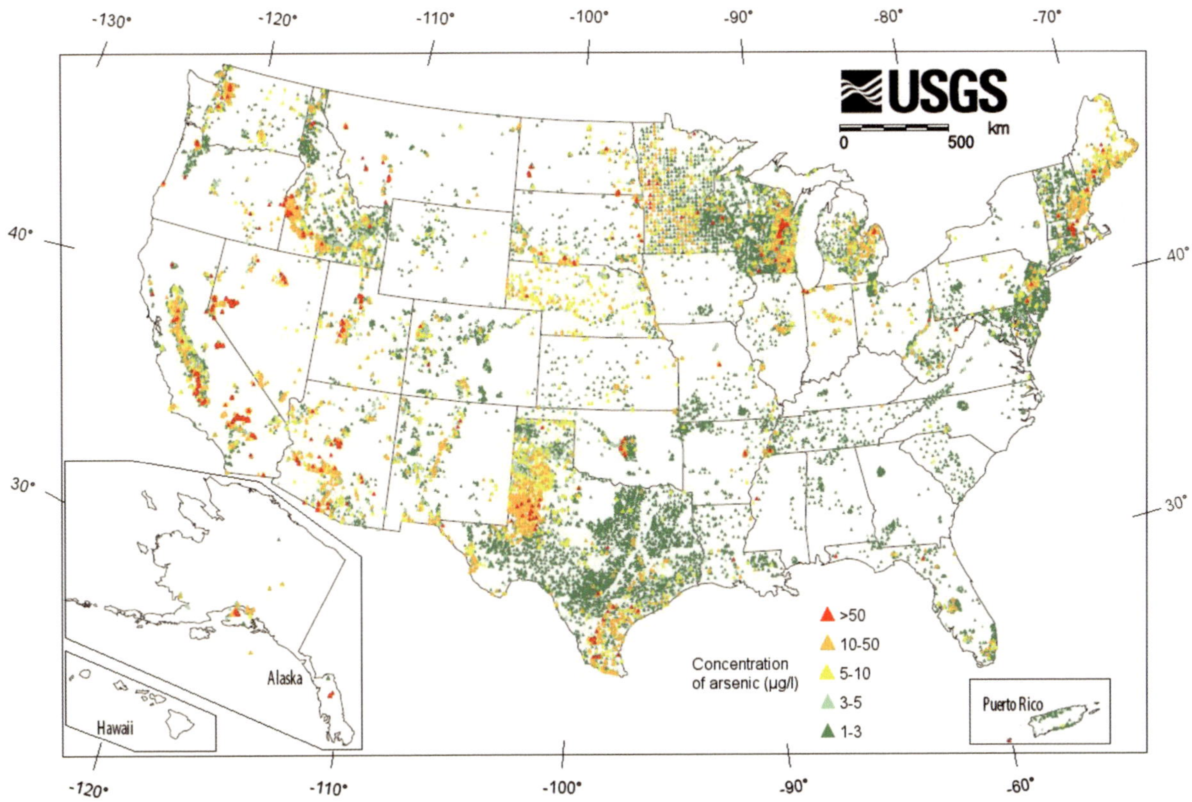

Figure 2.2 Data density-based map showing arsenic concentrations in 31 350 groundwater samples collected in 1973–2001 in the USA (modified from Ryker, 2001).

Geothermal sources are found in the south-western states of California (especially Los Angeles) and Nevada, the area around Yellowstone National Park in Wyoming and mineral-rich areas such as Alaska, Arizona, Colorado, Michigan, New Mexico, South Dakota and Utah, which have arsenic arising from mining activities (Straskaba & Moran, 1990). Geothermal hot springs in Yellowstone National Park commonly have arsenic concentrations exceeding 1000 μg/l (Stauffer & Thompson, 1984; Ball *et al.* 1998). In the Madison River at the park boundary, geothermal water discharge results in arsenic concentrations as high as 360 μg/l; 470 km downstream in the Missouri River, concentrations as high as 19 μg/l have been measured (Nimick, 1994; Nimick *et al.* 1998). Geothermal water is also the primary source of arsenic contamination in other parts of the USA, including some surface water in the eastern Sierra Nevada (Eccles, 1976; Wilkie & Hering, 1998).

Naturally occurring arsenic in groundwater varies regionally due to climate and geology. Examination of 17 000 water analyses suggested that around 40% exceeded 1 μg/l and 5% exceeded 20 μg/l (Welch *et al.* 1999); less than half of 30 000 tests of groundwater found arsenic concentrations of ≤1 μg/l, and about 10% exceeded 10 μg/l (Welch *et al.* 2000). At a broad regional scale, arsenic concentrations exceeding 10 μg/l appear to be more frequently observed in the western states compared with east of the Mississippi River, as a result of metallic mineralization. Arsenic concentrations in the Appalachian Highlands and the Atlantic Plain are very low (≤1 μg/l), with the Interior Plains and the Rocky Mountain System having higher concentrations. In New England, Michigan, Minnesota, South Dakota, Oklahoma and Wisconsin, arsenic concentrations exceeding 10 μg/l are more widespread than previously recognized (Welch *et al.* 2000).

The United States Geological Survey (USGS) estimates that arsenic concentrations in 1% of 54 000 public water supplies in the USA currently exceed 50 μg/l; 3% exceed 20 μg/l; 8% exceed 10 μg/l; and 14% exceed 5 μg/l. For instance, 20% of the over 600 wells tested in Fairbanks, Alaska, contained arsenic at concentrations above 50 μg/l (Weber, 1986).

Anthropogenic sources of arsenic in the USA (Figure 2.3) are related to its use in a wide variety of applications, beginning in about 1930 in wood preservation and glass production (Abernathy, 1983; Lederer & Fensterheim, 1983; Azcue & Nraigu, 1994) and in agriculture, in a host of derivatives including monosodium methanearsonate for cotton and disodium methanearsonate, dimethyl arsinic acid and lead arsenate for fruit, before the introduction of dichlorodiphenyltrichloroethane (DDT) in 1947 (Shepard, 1951; USDA, 1997a).

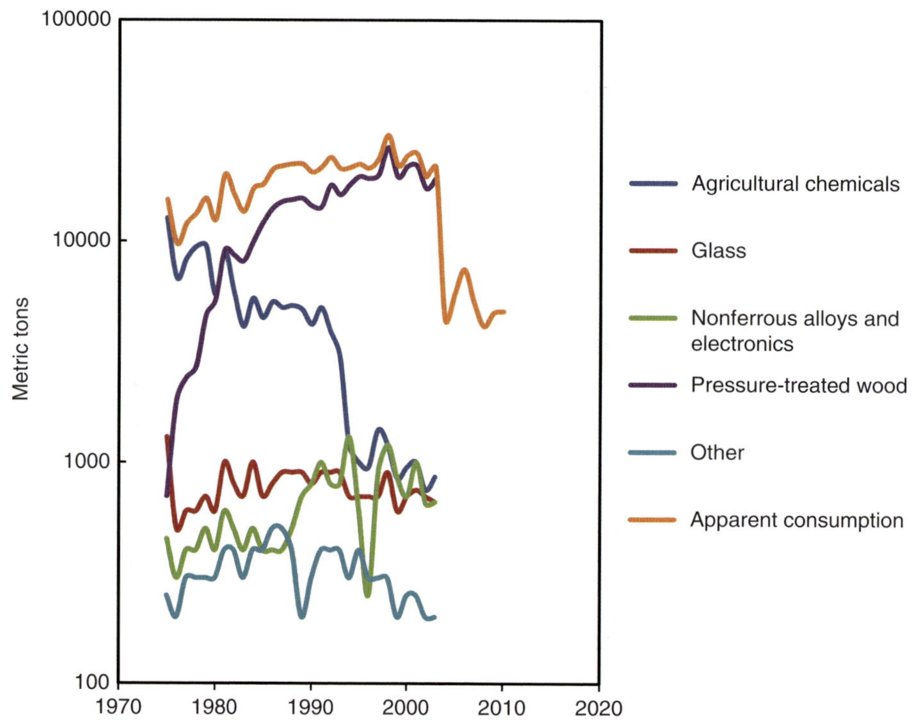

Figure 2.3 The history of anthropogenic arsenic use in the USA (data from USGS, 2012; Chart courtesy of Alan Welch).

Major current industrial sources of arsenic include chromate copper arsenate (CCA) pressure treated wood preservatives, industrial uses, mining and smelting. The production of CCA has accounted for approximately 90% of the arsenic used annually by industry in the USA and one sees in Figure 2.3 that the USGS stopped breaking out arsenic by use category beginning in 2004, which coincided with restrictions and voluntary cutbacks of CCA in the US and Canada dating from that same year. Inorganic arsenic pesticides are no longer used for agricultural purposes (since 1993), although pollution problems remain (Folkes *et al.* 2001; Loiselle *et al.* 2002). The USEPA has a TRI Explorer database that records the emissions of toxic chemicals into air, land and water each year and can be accessed with their TRI search engine in terms of the following reports: Facility, Federal Facility, Trends, Geography, Industry, Maps, Dynamic (USEPA, 2009). Their mapping database provides a per annum snapshot of the total emissions per chemical in pounds (1 pound = 0.454 kg). For 2009, the total emissions of arsenic and arsenic chemicals in the USA were mapped by county, as shown in Figure 2.4.

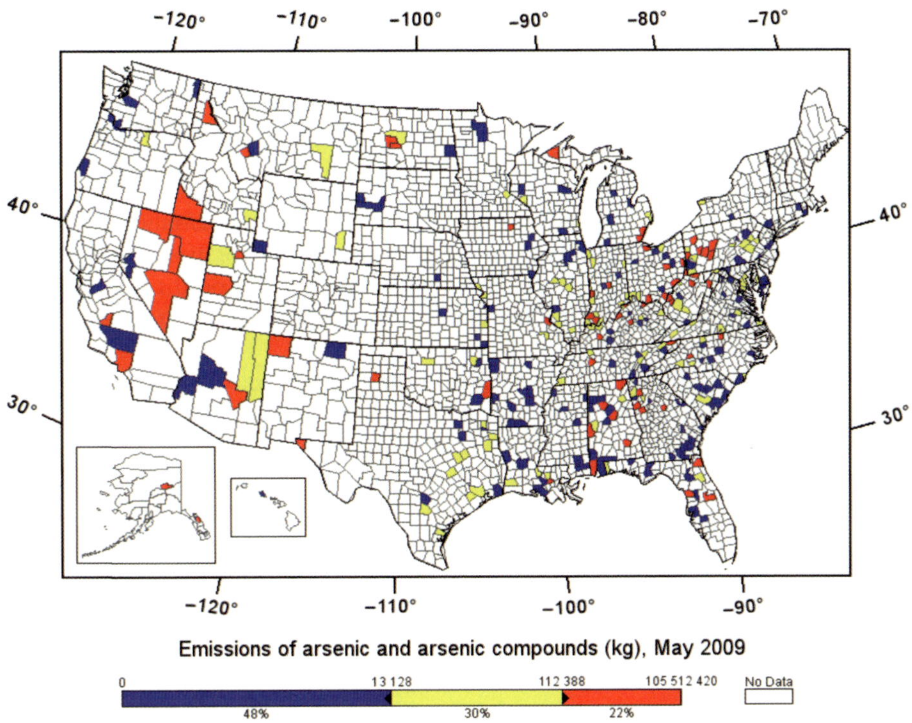

Figure 2.4 Emissions of arsenic and arsenic chemicals from the USEPA TRI Explorer database for 2009 (data from USEPA, 2009).

Mining has been another contributor to the release of arsenic into groundwater and surface water in the USA. Exposure of sulfide-bearing rocks has led to high arsenic releases in surface water and groundwater, especially in the western states, including the upper Clark Fork River in Montana (Brumbaugh *et al.* 1994; Moore, 1994), the Coeur d'Alene mining district of northern Idaho (Mok *et al.* 1988), Bryant Creek drainage of California (Hammermeister & Walmsley, 1985), the Whitewood Creek–Cheyenne River surface water system of western South Dakota (Cherry *et al.* 1986a,b,c; Goddard, 1989) and the Fairbanks mining district of Alaska (Wilson & Hawkins, 1978). Arsenopyrite with gold ore is common at some of these deposits. Arsenic concentrations in groundwater and sediment have been transported hundreds of kilometres downstream from sources (Moore, 1994). In addition, many abandoned mines and mine tailings contain acidic oxidized sulfide minerals (Alpers *et al.* 1994), but neutral to alkaline pH conditions that are normally associated with sulfide minerals can also release arsenic. This latter condition is found in the arsenic-enriched water from California's Mother Lode area and Nevada's polymetallic deposits (Welch *et al.* 2000).

Regional and national solid-phase geochemical surveys conducted by the USGS and the Geological Survey of Canada since 1975 indicate that most of the variability in the distribution of arsenic in stream sediments, lake sediments and soils in Canada and the USA is controlled by the bedrock characteristics (Grosz *et al.* 2004). The USA has arsenic contamination in groundwater (in both reducing and oxidizing environments) occurring from geological formations, geothermal sources and mining-related activities. A comprehensive USGS review (Focazio *et al.* 1999) of a range of existing water quality data indicates that most of the water utilities that are affected are in the western, Midwestern and North-eastern states. At test sites in California, Minnesota, Nevada, New Hampshire and Wisconsin it was found that 60% of total arsenic in source water was removed (Brooks, 2011).

Another source of arsenic in groundwater is related to petroleum contamination (Roller *et al.* 2003):

> Petroleum contamination of shallow aquifers is a serious environmental problem facing the United States today. Benzene, toluene, ethylbenzene, and xylene (BTEX) are the petroleum components that are of greatest concern due to their toxicity and mobility within aquifers. However, BTEX components can be readily degraded coupled to reduction of a variety of terminal electron acceptors by aerobic and anaerobic microorganisms. One of the most important terminal electron acceptors is solid phase Fe(III), in the form of hydrous ferric oxides (HFOs). As solid phase Fe(III) is microbially reduced to soluble Fe(II), metals and nutrients that are sorbed onto the HFO surface are also released to solution. For example, arsenic, which has a strong sorption affinity for HFOs, can be released to water during reductive dissolution of HFOs. Thus, one "unintended consequence" of HFO reduction coupled to BTEX biodegradation is that arsenic that was previously sorbed to HFOs in aquifers can be released to groundwater. This potentially presents a challenge for remediation of petroleum-contaminated aquifers that contain sorbed arsenic. Because arsenic

may not be elevated in background groundwater (due to the fact that initially the arsenic is sorbed to aquifer solids), the elevated arsenic concentrations in the plume may go undetected.

Few health effects associated with arsenic exposure have been reported in the USA. Feinglass (1973) reported 11 cases of acute and subacute arsenic intoxication caused by drinking well water containing extremely high levels of arsenic (between 11 and 21 mg/l). Stohrer (1991) reported a small number of possible cases of arsenic-related skin lesions, although none was considered to be typical. He also reported three cases of skin cancer thought to be arsenic-related.

Engel & Smith (1994) conducted an ecological epidemiological study examining mortality from 30 counties in the USA in relation to arsenic in drinking-water. The results indicated an excess mortality rate for diseases of the arteries and anomalies of the circulatory system. A study by Lewis *et al.* (1999) found a significant excess of deaths from cardiovascular disease. They found no evidence of excess deaths due to lung or bladder cancer, but noted a significant excess of mortality from nephritis and nephrosis.

Arsenic removal technologies employed in the USA include ion exchange, activated alumina, reverse osmosis, modified coagulation/filtration, modified lime softening and electrodialysis reversal (USEPA, 2000b).

A state by state selective survey of arsenic contamination follows.

Alabama (C)

Pennsylvanian bituminous coals from the Appalachian Basin are locally enriched with arsenic in a zone that extends from northern Alabama through central Tennessee, eastern Kentucky, most of West Virginia, eastern Ohio and western Pennsylvania (Britton *et al.* 1989; see Figure 2.5). This zone also appears to conform to the arsenic emissions that are represented in Figure 2.4 above. This implies that emissions are from coal-fired power plants and other sources firing local high-arsenic coals in this zone.

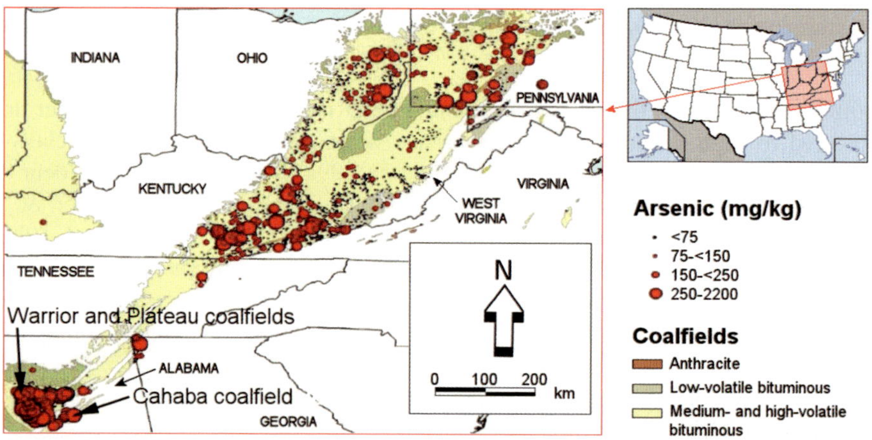

Figure 2.5 Map of regional concentrations of arsenic in the Appalachian coal region in the eastern USA (modified with coalfield names from Goldhaber *et al.* 2001).

An intensive study by the USGS has provided more precise measurements of arsenic concentrations in coals from this region. The Warrior and Plateau coalfields and the Cahaba coalfield in northern Alabama contain the highest concentrations of arsenic of all coals in the USA: up to 2200 mg/kg. Arsenopyrite in the Warrior Basin has up to 4.7 weight per cent arsenic, with elevated mercury, selenium and molybdenum (Goldhaber *et al.* 2001). Pyrite is the dominant carrier of arsenic in these coals, with the highest average arsenic concentration in banded pyrite of 11 210 mg/kg (Diehl, 2002; Diehl *et al.* 2004). Arsenic concentrations in the coal seam production groundwater in the Warrior coalfield have a maximum of 475 μg/l, a mean of 25 μg/l and a median of 6.5 μg/l. Arsenic concentrations in drinking-water from wells in the Warrior coalfield have a maximum of 44 μg/l, a mean of 2 μg/l and a median of 1 μg/l (Goldhaber *et al.* 2001).

Alaska (G, M)

Geogenic sources of arsenic in southern Alaska are related to the leaching of naturally occurring arsenopyrite that is often associated with gold deposits. pH is inversely proportional to the levels of arsenic concentration. The Kenai Peninsula is known for groundwater concentrations of arsenic, which is probably naturally occurring, above 10 μg/l (Glass, 1996).

Gold mining, which exposes arsenopyrite to weathering, exacerbates arsenic release into surface water and groundwater, as Table 2.12 indicates. For instance, elevated arsenic concentrations in wells near Fairbanks (Wilson & Hawkins, 1978) are probably induced by mines in the vicinity (Weber, 1986); well water samples collected from 59 homes had a mean arsenic concentration of 224 μg/l, with a range from 1 to 2450 μg/l (Harrington *et al.* 1978). Kreiss *et al.* (1983) found a maximum concentration of 440 μg/l in the same area.

Table 2.12 Arsenic concentrations in surface water and groundwater at gold mining projects in southern Alaska, USA.

Location	Maximal arsenic concentration in surface water (mg/l)	pH	Maximal arsenic concentration in groundwater (mg/l)	pH
Alaska (southern)	0.011	7.9	0.47	8.2

Source: After Straskraba & Moran, 1990

Arizona (A, G, M)

Geogenic sources of arsenic are ubiquitous in the lower two thirds of the state (Welch *et al.* 2000). This would include the following populations of the following counties (or approximate areas; United States Census Bureau, 2000) as being at risk (rounded off): Mohave (46 500), Yavapai (151 200), Navajo (49 000), Apache (34 500), La Paz (13 300), Yuma (160 000), Maricopa (3 million), Pinal (180 000), Gila (51 000), Graham (34 000), Greenlee (9000), Pima (844 000), Santa Cruz (38 000) and Cochise (118 000), or about 4.7 million total population at risk. High-arsenic groundwater is present in aquifers composed of felsic volcanic rocks. In this environment, groundwater is commonly alkaline (Welch *et al.* 2000). Robertson (1989) noted the occurrence of elevated arsenic concentrations in the Basin and Range Province in Arizona. Of the 467 samples analysed, 7% had arsenic concentrations greater than 50 μg/l. Geogenically, arsenic in groundwater also occurs in basin fill sediments, including volcanic, alluvial and lacustrine deposits, in conjunction with iron oxide deposits (Owen-Joyce & Bell, 1983; Owen-Joyce, 1984). Mining has contributed to leaching of arsenic into the groundwater of many mineralized areas in Arizona. The following (Foust *et al.* 2004) illustrates examples of this situation:

> Forty-one water samples were collected and analyzed from throughout the Verde Valley watershed to identify the source of arsenic in well water used for domestic and agricultural purposes. Each water sample was analyzed for anions, cations and trace chemical constituents by atomic absorption spectroscopy, anion chromatography and traditional wet chemical procedures. Arsenic concentrations ranged from 10 to 210 μg/L, with the highest values observed for water pooled on tailings from an abandoned Cu mine.

Anthropogenic sources of arsenic contamination have been identified from copper smelting. One case involved Gila County (Arizona Department of Health Services, 1999):

> The towns of Hayden and Winkelman, Arizona are located near a smelter operated by ASARCO, Inc. Hayden, Arizona, is located approximately 90 miles southeast of Phoenix on State Highway 177, along the Gila River, below the confluence of the Gila and San Pedro Rivers. The town was founded in 1912 to provide housing for the workers at the Ray open pit mine complex and the copper smelter complex, originally built by the Kennecott Copper Company. The town has a current population of approximately 900, and shares many services, including the local school district, with the town of Winkelman, population 600, located one mile to the south. The towns also share a common history regarding the emissions from the smelter operations.
>
> There have been several smelters on the site of the current Asarco smelter, with emissions being discharged into the air in Hayden and Winkelman since operations began in 1909. Historic emissions … [of] large quantities of lead, arsenic, sulfur dioxide (SO_2) … drifted over the entire region in the air and many of the materials contained in these emissions fell out of the atmosphere and settled on the ground. The first emission controls placed upon the smelters were installed in 1920. These electrostatic precipitators were designed to remove particulate matter from the stack emissions. The smelters within the complex operated with these minimal emission controls prior to 1969, when requirements under the forthcoming Clean Air Act amendments of 1970 required that controls be installed to limit SO_2 emissions. Acid plants were added in 1969 and 1971 to reduce SO_2 and particulate emissions. Additional controls were added to the current smelter in 1984, further reducing the SO_2 emissions. The current smelter stack height was designed to elevate the emissions above the valley air shed and better facilitate dispersion.
>
> The ASARCO smelter is a large complex of approximately 200 acres at the eastern end of the town. Two large emission stacks dominate the horizon. One is about 1,000 feet tall and the other is 250 feet tall. There are also several smaller stacks visible at buildings throughout the complex. A large slag pile, a solid glass-like waste material, is on the eastern end of the facility, adjacent to the town. In 1999, the Arizona Prevention Center, an affiliate of the University of Arizona dealing with environmental and occupational issues, and [the Arizona Department of Health Services] conducted a health survey of the residents of Hayden and Winkelman to determine … child and adult urinary arsenic levels. Adults and children were screened for urinary arsenic. The urinary arsenic analysis measured both organic and inorganic arsenic compounds. Inorganic arsenic compounds [AsV] were the

> focus of this screening since organic arsenic is significantly less toxic than inorganic arsenic [AsIII].... The average urinary total arsenic concentration of individuals tested in Hayden and Winkelman was 13.7 micrograms per liter (μg/L), which is substantially less than the study reference level of 30 μg/L ... established as the marker level by the University of Arizona investigators for persons who consume water with an arsenic concentration of 50 μg/l.

However, the reference level for 10 μg/l of arsenic, the current standard for arsenic in the USA, is in the range of the current urinary average, which should be monitored accordingly. In addition, "Five (2%) of 224 individuals tested in Hayden and Winkelman had inorganic urinary arsenic concentrations exceeding 30 μg/l. The maximum concentration measured was 47 μg/l." These individuals are definitely at risk.

Arkansas (A, G)

Arkansas has both anthropogenic and geogenic sources of arsenic, which have contaminated soils and groundwater.

Anthropogenic arsenic contamination in Arkansas (as well as in most of the south-central states) has been the result of pesticides used formerly in cotton agriculture (*Gossypium* spp.) to control the boll weevil (*Anthonomus grandis*) and as a cotton desiccant. The transfer of land used for cotton agriculture into rice paddies meant that the arsenic in the soils not only continued to leach into the groundwater, but also became part of the food-chain (Loebenstein, 1994; Living on Earth, 2007). Rice grown in arsenical soils exhibits a physiological deformity called "straight head"; in fact, "straight head" rice is preferred for arsenical soils (Marin *et al.* 1992; Dilday *et al.* 2000). It has been reported that 42% of the arsenic found in rice grown in the south-central states is in the more toxic inorganic form (Peplow, 2005). However, even organic arsenic, usually considered less toxic, can be carcinogenic when administered at high doses in the diet or drinking-water (Cohen, 2002), and organic arsenic can be converted into inorganic arsenic in the human body (Akter *et al.* 2005).

Another anthropogenic source of arsenic in Arkansas is the manufacture of arsenical wood preservatives. In 1972, Ouachita-Nevada Wood Treater Co., 0.4 km north of Reader, switched from pentachlorophenol to a water solution of chromate, copper and arsenic. On-site and off-site contamination of soils, sediment and groundwater resulted, with soil contaminated with arsenic concentrations as high as 4500 mg/kg. In 2001, the USEPA began a Remedial Investigation/Feasibility Study to determine the extent of groundwater contamination. The programme is ongoing (Arkansas Pollution Control and Ecology Commission, 2002).

Perhaps the most pernicious source of anthropogenic arsenic in Arkansas is the 4600 tonnes of poultry excreta that is generated in the state daily. The source of arsenic in the excreta is roxarsone (4-hydroxy-3-nitrobenzenearsonic acid), an arsenical compound that is commonly used as an additive in poultry feed for increasing weight and feed efficiency and improving pigmentation. Broiler chickens are fed 21–41 g of arsenic in the form of roxarsone per tonne (Animal Health Institute, 1996). Roxarsone is directly excreted, and poultry litter from commercial broilers can contain 15–35 mg total arsenic/kg (Morrison, 1969; Jackson & Miller, 1999; USGS, 2000). Because of roxarsone's increasing solubility in warmer water, the potential for its mobilization in surface water is greater in the spring and summer months. Furthermore, the disposal of the excreta threatens the Buffalo River, and one section of the river has already been designated as contravening state water quality standards. The river is one of the few free-flowing rivers in Arkansas, with its headwaters in the Ozark National Forest and the rest in the Buffalo National River, established by the 1972 National Wilderness Act. In addition to leaching wastes into the surface waters (and some cases of direct dumping), arsenic-tainted poultry excreta are dumped directly onto agricultural fields as a fertilizer (Hollerman, 1992).

Geogenic sources of arsenic in Arkansas are related to shallow, Quaternary, alluvial aquifers. The Arkansas Water Resources Center (2006) has indicated that:

> water quality in the Bayou Bartholomew watershed ... revealed that 21 out of 118 irrigation water wells completed in shallow Quaternary alluvial deposits (alluvial aquifer) with a depth of 80–100 feet had As concentrations exceeding 10 μg/L. Most municipal water supply wells in eastern Arkansas are advanced predominantly into deeper Tertiary Formations because of the problems associated with the high concentrations of Fe and Mn in the shallow alluvial aquifers. However, domestic wells completed in alluvial deposits may present risks to private well owners not protected by the Safe Drinking Water Act (SDWA).

Acknowledging the mobilization of iron and arsenic in reducing alluvial aquifers, a condition that is found in other Midwestern states such as Illinois and Ohio, the Arkansas Water Resources Center (2006) produced a report stating that:

> The ground water flow system in the alluvial aquifer of the Arkansas River at Dardanelle has iron, manganese and arsenic with trends of increasing concentrations with decreasing Eh [redox potential], especially Fe and As. All the samples with Eh greater than 200 mV have low (<1 mg/l) to non-detect Fe concentrations and low (<5 μg/l) As concentrations. At Eh values <200 mV, Fe concentrations exceed 10 mg/l and As concentrations generally are above 5 μg/l. One sample had an As concentration of 19 μg/l, exceeding the [USEPA

maximum contaminant level] of 10 μg/L. The Fe and the As appear to be from reductive dissolution of Fe-oxides and oxyhydroxides in the alluvial sediment.

More research needs to be done on this important aspect of geogenic arsenic mobilization in Arkansas groundwater.

California (A, G, M, V)

Anthropogenic sources of arsenic contamination in California probably include the use of pesticides for agriculture and lawns, but another unusual source is the bottled water industry. A survey conducted in 1995 found arsenic present in name brands of bottled water (Figure 2.6). Crystal Geyser Company provided the Natural Resources Defense Council with test results indicating that beginning in April 1999, Crystal Geyser substantially reduced the arsenic levels in its spring water below 10 μg/l. Although other bottled waters may have been remediated, it is of interest that such high levels were found initially, and one should be alert to other instances in other states.

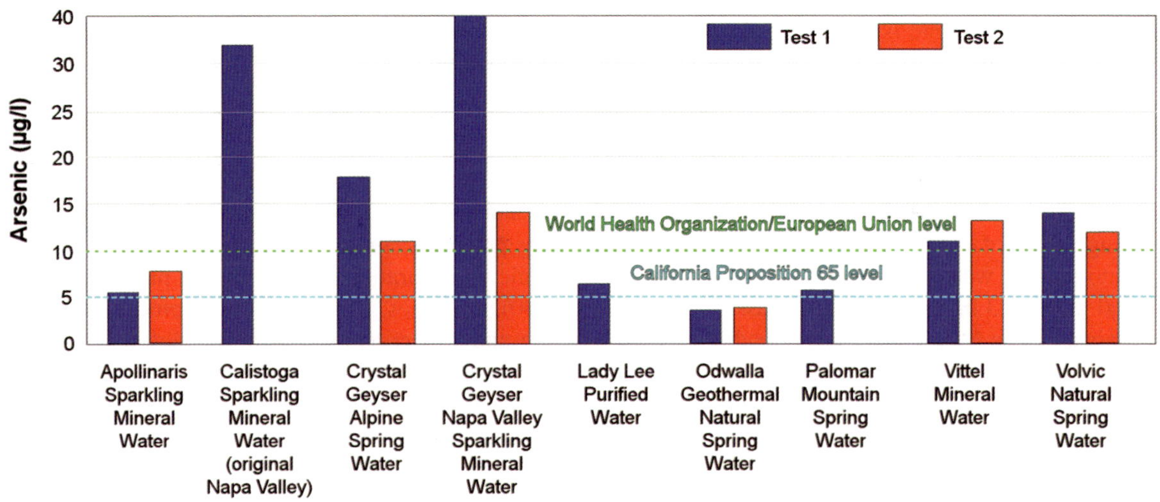

Figure 2.6 Arsenic levels in selected bottled waters as of 1999 (modified from NRDC, 1997–1999).

In ground waters of the Tulare Basin of the San Joaquin Valley, California, concentrations of arsenic up to 2600 μg/l have been reported (Fujii & Swain, 1995). The San Joaquin Valley, 350 km in length to a maximum of approximately 130 km in width, has high-capacity wells ranging in depth from 60 to 600 m. Water is highly mineralized due to geological interactions as it travels westward for great distances, and intermingled with the various unconsolidated geological deposits are layers of organic material. Arsenic is present in many wells.

A key arsenic contamination site in the San Joaquin Valley associated with arsenopyrite reduction is the Kern Water Bank, approximately 25 km south-west of Bakersfield. The inferred decrease of iron oxides with depth and the occurrence of authigenic pyrite in organic-rich sediments suggest that pyrite formation resulted from the reduction of iron oxides and sulfates. Authigenic pyrite served as a sink for arsenic released from iron oxides. Subsequent dissolution of pyrite by oxygenated water results in intermediate redox conditions and elevated groundwater arsenic. The presence of arsenic(III), arsenic(V) and measurable oxygen in pumped water is consistent with intermediate redox conditions resulting from oxygenated water in contact with reduced, pyrite-containing sediments (Boockhoff *et al.* 2005).

The City of Hanford and other municipalities are dependent on this groundwater for a major portion of their water supply. In 1985, the wells in the Hanford area all contained arsenic, frequently at concentrations exceeding 50 μg/l (Cline *et al.* 1985). In 1996, the same municipal water supply had arsenic concentrations ranging from 9 to 75 μg/l, with the percentage of arsenic(III) in most wells averaging 89 ± 6% (Hering & Chiu, 2000).

In addition, aridity can cause evaporative concentration of arsenic in shallow groundwater aquifers, particularly in the hydrologically closed basins of eastern California. Evaporative concentration is a contributing factor leading to high arsenic concentrations in areas such as the Carson Desert and southern San Joaquin Valley (Boockhoff *et al.* 2005).

Mining sources of arsenic release in California are mostly related to Mother Lode Gold Belt mines in central California, which are rich in arsenian pyrite and arsenopyrite. The Mother Lode Gold Belt is situated geologically in Melones Fault Zone rocks, which have arsenic levels typically ranging from <100 to 8000 mg/kg. In addition, residual soil may contain arsenic at concentrations that are at least 10% of the arsenic concentration of the Melones Fault Zone rocks, or from 10 to 800 mg/kg. Groundwater arsenic concentrations in mineralized and mined areas are highly variable and closely associated with bedrock type and Melones Fault Zone mineralization. Although most Mother Lode mines ceased operations by 1960, their tailings have remained in situ, with typical arsenic concentrations above 6000 mg/kg (Motzer & Will, 2007).

Volcanism plays an important role in the mobilization of arsenic in some susceptible zones in California. Thermal springs and wells in Long Valley, for instance, are composed of sodium bicarbonate – chloride waters containing 1000–1420 mg/l of dissolved solids and high concentrations of arsenic (up to 2.2 mg/l). In Long Valley, high-arsenic(III) waters flow into tributaries of Lake Crowley and thus to the Los Angeles Aqueduct, specifically at Hot Creek, a tributary of the Owens River. Water supplied to the City of Los Angeles through the Los Angeles aqueduct contains an average arsenic concentration of 20 μg/l. Arsenic in the 370-m Hot Creek Gorge is magmatic in origin, with high-arsenic(III) hot springs and seeps and other minerals (chlorine and boron) mixing with dilute, oxic surface waters, producing an alkaline pH of 8.3 (Wilkie & Hering, 1998). Crowley Lake provides 75% of the water supply for the 3.2 million residents of Los Angeles: 2.4 million people are exposed to arsenic concentrations above the WHO guideline of 10 μg/l. To meet an ocean discharge standard of 12 μg/l, the arsenic is being removed by the addition of iron(III) chloride to co-precipitate iron hydroxide and arsenic (Mariner & Willey, 1976; Kneebone & Hering, 2000; Kneebone *et al.* 2002).

Colorado (A, G, M)

Geogenic and mining sources account for much of the arsenic in the surface water and groundwater of Colorado. Geogenic sources of arsenic in central Colorado are related to the weathering and leaching of arsenic minerals, especially arsenopyrite in gold-bearing rocks; of course, mining exacerbates the weathering processes. In the presence of acidic conditions, arsenic is released into surface water and groundwater, with maximal arsenic concentrations of 26 mg/l at pH 3.2 in surface water and 19.0 mg/l at pH 4.3 in groundwater (Straskraba & Moran, 1990). At mines, tailings from ore processing were dumped into local drainage. The Yak Tunnel near Leadville is 5.8 km long and was completed as a drainage in 1912. Since then, 1600–3800 m^3/day of acid water (pH 3.5) with a high trace metal content have discharged into the upper reaches of the Arkansas River. In another instance, the Minnesota Mine near Empire discharges about 800 m^3/day of acid water (pH 4.5); however, due to the disposal of tailings into the local drainage, the surface water has been reduced to pH 2.8, with an arsenic concentration of 26 mg/l (Straskraba & Moran, 1990).

Over 17 000 new cases of cancer are registered annually in Colorado, and, on average, approximately one in three Coloradans will develop cancer in their lifetime. As of 2006, the Colorado Department of Public Health and Environment's Water Quality Control Division developed the Colorado Strategy for Arsenic Reduction (CO-STAR) programme to help identify water systems at risk of failing to comply with the new limit for arsenic of 10 μg/l by the 23 January 2006 compliance date announced by the USEPA. Communities that are selected as part of this remediation programme are Agri-health International (Yuma County), Curicanti National Recreation Area (Gunnison County), East Alamosa (Alamosa County), High Valley Mobile Home Park (Alamosa County), Town of Eckley (Yuma County), Town of Telluride (San Miguel County), Rock at Ute Trail Ranch (Gunnison County), Pinebrook Hills (Boulder County) and Town of Akron (Washington County). All of these counties had elevated arsenic levels in their groundwater, and their total population is 67 300.

One arsenic Superfund site indicates the magnitude of contamination from mining and smelting ores. The Vasquez Boulevard and I-70 Superfund site (VB/I-70) covers a 10-km^2 area in north-central Denver. Elevated levels of arsenic and lead have been detected in area soils and are believed to have come from historic smelting activities in the area. Smelters operated at various times from the 1870s through to the 1950s (USEPA, 2007d):

> This was a vast residential soils cleanup project encompassing more than four square miles and approximately 4,500 properties in all of Cole, Clayton, Swansea, Elyria, Globeville and Curtis Park neighborhoods. During the investigation, EPA took 30 soil samples at each of the 4,315 residential yards sampled in the VB/I-70 boundaries. EPA then removed, replaced and re-landscaped yards at 761 properties because of lead and/or arsenic contamination. During the process, EPA removed more than 91 000 cubic yards of soil and planted more than 1.5 million square feet of sod.

Anthropogenic sources of arsenic contamination of soils and groundwater in Colorado are associated with domestic arsenical lawn care products commonly used before they were banned in the 1970s (Folkes, Helgen & Litle, 2001; Folkes, Kuehster & Litle, 2001; Colorado Department of Public Health and Environment, 2003). O'Grady *et al.* (2003) reported a panel discussion on a survey of residential soils of pre-1960s homes in the Denver area that revealed high levels of arsenic in a random pattern associated with these arsenical pesticides. Because of this random dispersion of use, testing of soils is advised whenever such domestic use is suspected.

Connecticut (A, G)

In 1999, 50 wells were sampled in Woodstock, Colchester and East Hampton counties, and 35 of these had detectable levels of arsenic. Because naturally occurring arsenic had been shown to be a problem in other states in the New England region, data from public wells in this area were surveyed based on the hypothesis that the contamination was associated with a particular bedrock type – that is, calcareous meta-sedimentary bedrock from the Hebron Formation that extends from Woodstock (1299 population at risk, assuming that 18% of wells contain arsenic), Colchester (2619 population at risk) to East Hampton (2403 population at risk), with a total population at risk of 6321 (United States Census Bureau, 2000). Over 250 public wells in this region were monitored and results indicate that 18% (45/254) of the wells contained detectable but usually low levels of arsenic, but 3% had >10 μg/l and one had >50 μg/l. Over 20 wells in Windham County, near Woodstock, have been monitored and exhibit elevated arsenic levels due to geogenic sources (ATSDR, 2000). Figure 2.7 illustrates geologically how the contaminated wells fall within the Hebron Formation transect, and Table 2.13 provides the laboratory-determined levels of arsenic contamination from these wells.

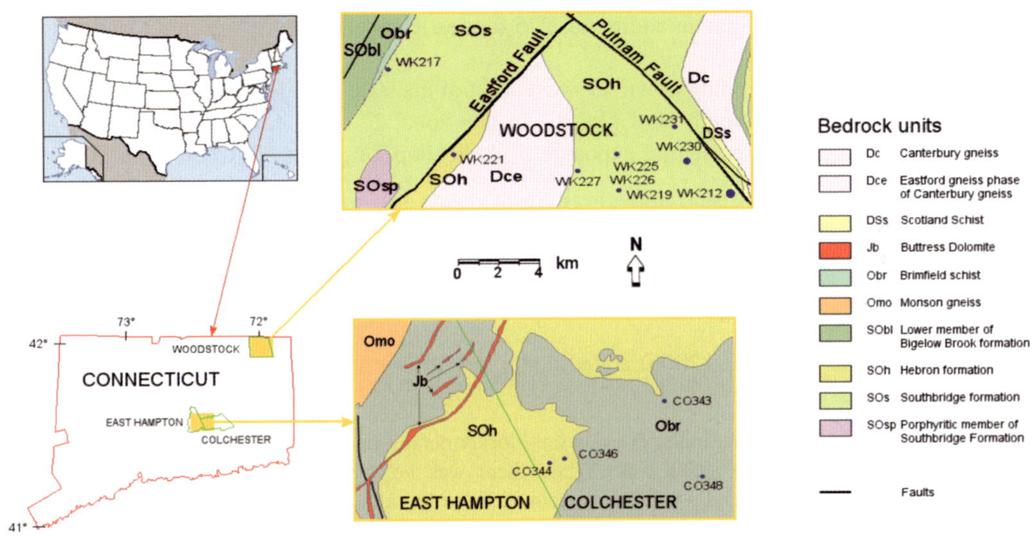

Figure 2.7 Arsenic contamination sites in East Hampton, Colchester and Woodstock counties, Connecticut, USA (modified from ATSDR, 2000). The geological map is an excerpt from the electronic version (http://www.tmsc.org/geology/bedrock/) of the Generalized Bedrock Geological Map of Connecticut, produced by the Connecticut Geological and Natural History Survey in 1990 and reprinted in 1996.

Table 2.13 Wells tested for arsenic in Woodstock (WK) and Colchester/East Hampton counties (CO), Connecticut, USA.

Location code[a]	Arsenic concentration (μg/l)
WK231	0.34
WK221	0.86
WK217	2.9
WK226	4.0
WK227	5.6
WK225	6.0
WK212	14[b]
WK230	24[b]
CO346	0.83
CO348	1.2
CO343	2.7
CO344	4.5

[a]Approximate location of wells indicated on the maps in Figure 2.7.
[b]Exceeds the USEPA limit of 10 μg arsenic/l in drinking-water.
Source: ATSDR, 2000

A press release from the USEPA (2001b) reported that soil samples collected in the Newhall Street neighbourhood in Hamden revealed that some of the residential properties in the neighbourhood are contaminated with high-arsenic-bearing landfill materials.

Florida (A, G, M)

Anthropogenic sources of arsenic contamination have been identified in former farmlands or orchards that had been sprayed with arsenic pesticides and fungicides. In Sanford, an elementary school was scheduled to be built on top of arsenic-contaminated farmland, but it was determined by soil and water tests that arsenic levels were elevated at 27 of 65 test sites, including two soils above the limit of 0.8 mg/kg, the modified Florida default residential exposure Soil Cleanup Target Level, and one groundwater above 10 µg/l, the WHO guideline for water. Consequently, the contaminated soil was removed in the next few months (Weber, 2007). In addition, high levels of arsenic were identified in runoff and in soil under treated decks in 2005. It is estimated that 28 000 tonnes of arsenic had been imported into Florida by 2000, that 4600 tonnes have already leached into surface water and groundwater, and that an additional 275 tonnes leach each year (National Coalition for Alternatives to Pesticides, 2007).

Geogenic arsenic occurs in a zone extending down the west coast of the state from Leon and Madison counties in north-west Florida to Lee and Charlotte counties in south-west Florida. In this zone, 27 of 51 counties had arsenic levels in soil above 0.8 mg/kg, the modified Florida default residential exposure Soil Cleanup Target Level (Chen *et al.* 2001):

> Soil arsenic background concentrations were generally higher in south Florida than in north and central Florida, and associated with wet soils. Individual high arsenic sites were scattered throughout the state, but the most highly concentrated of these occurred in the Leon-Lee belt along the Ocala uplift district extending to the south-western Flatwoods district. Extrapolation of the data using a single arsenic value regardless of the taxonomic and geographical differences in soil arsenic distribution would underestimate potential arsenic contamination in upland soils.

To Chen *et al.*'s (2001) sites, Teaf *et al.* (2006) added a site in Hillsborough County in west-central Florida with 2.8 mg arsenic/kg, with the important caveat that:

> while the 2.8 mg/kg UCL [upper confidence limit] exceeds the Florida default residential cleanup target of 2.1 mg/kg, it does not represent a significantly increased human health risk. While there was no regulatory involvement, preparation of the initial background survey, a site risk evaluation, and a local background evaluation allowed the Site owner and prospective developer to determine that the site was suitable for residential improvement.

Contamination of ground waters in this zone has been measured and is due to the leaching of arsenic from these geogenic sources in the soils. In south-west Florida, arsenic concentrations are elevated with respect to recharge water levels; concentrations exceed 10 µg/l and increase up to 8-fold compared with recharge waters and natural background waters. The mobilization of arsenic is caused by pyrite oxidation by recharge water (Mirecki, 2004).

Mining of phosphates was discovered by the Division of Environmental Health of the Florida Department of Health (2004) to be an underlying cause of arsenic release in Hillsborough County (Weber, 2007), and it is not known on what scale this scenario was repeated elsewhere, as phosphate deposits are prolific:

> Cancer incidence rates were calculated for the population residing adjacent to Coronet Industries, involved in mining phosphate rock for nearly 100 years. Until the 1960s, phosphate rock was extracted using shallow (less than 7.5 m) open pits from areas north of the plant. Documented releases of arsenic-tainted wastewater into local creeks and groundwater sources led the investigators to find "arsenic, cadmium, and radium in a number of nearby residential drinking water wells within a quarter mile radius of the Coronet plant. Approximately 28 of about 146 tested wells showed concentrations exceeding the Maximum Contaminant Levels (MCL) for arsenic, cadmium and/or radium."

Another source of geogenic arsenic was discovered in a situation similar to that in Stuarts Point, NSW, Australia, and in coastal Georgia, USA. In Duval County, 12 samples from placer deposits of heavy metals and metalloids, including arsenic, were found in 1995 in coastal barrier sediments from the Jacksonville Ocean Dredged Material Disposal Site (Parsons, 1999). These findings correlate with the formation of Pleistocene and Holocene beach barriers, which are receiving much attention.

The last source of geogenic arsenic is the unusual case of atmospheric deposition. Saharan dust rich in arsenic is annually transported aloft from Africa by summer westerly winds across the Azores and the North Atlantic to the Caribbean and the tropical Atlantic shore of the south-eastern USA (Perry *et al.* 1997; Prospero, 1999; Holmes *et al.* 2000; Shinn *et al.* 2000). Measurement of African dust has shown that 140 million tonnes are transported aloft in the summer to the Tropical Atlantic shore (Koren *et al.* 2006) and that the dust contains elevated arsenic concentrations. Dust samples from Mali, for instance, which are deemed representative, contained 17 µg arsenic/g (British Geological Survey, 2002b). The dust from other African countries affected by the summer dust events also contributes to this phenomenon, but to date little has been done to characterize the total available arsenic in the dust of these regions.

Holmes & Shinn (2001) summarized the evidence they have collected relating to elevated arsenic found at three sites in Florida: one in Everglades National Park in south Florida, one a Chassahowitzka National Wildlife Refuge along the Gulf Coast in Central Florida and one in the Okefenokee Wildlife Refuge on the Florida/Georgia border:

> In the Everglades National Park in south Florida, the data suggests approximately 50% of the arsenic is associated with other than soil material. At Chassahowitzka Wildlife Refuge in central Florida, only ~25% were linked with the soil fraction. At the Okefenokee Wildlife Refuge, only ~ 2.5% had an association with soil; the remaining fraction had a strong relationship to organic carbon concentrations. Of all the sites, Chassahowitzka samples appeared to have the highest arsenic concentrations ranging from 0.5 to 198 mg/kg with a mean value of 33 mg/kg. Recent studies have demonstrated that there is a high degree of bioaccumulation of arsenic in a fern commonly found in Florida. Time series analysis indicates a strong link between fires in Florida and aerosol arsenic. In addition, ground water data shows that arsenic in central Florida is highest in the region of high aerosol arsenic. It is hypothesized that there is a link between the dust and ground water, with the dust transporting the arsenic to Florida, which is flushed from the troposphere by summer storms, it is accumulated in the plants which subsequently burn during the dry winter season becoming available to leaching into the groundwater systems.

The arsenic bioaccumulating plant in question is *Pteris vittata*, the Chinese brake fern, which extracts arsenic rapidly and efficiently from soils with as much as 1500 mg arsenic/kg. Arsenic concentrations in the fronds increased from 29.4 to 15 861 mg/kg in just 2 weeks. Ferns growing in soil containing only 6 mg arsenic/kg accumulated 755 mg arsenic/kg, a 126-fold enrichment (Ma *et al.* 2001).

Georgia (A, C, G)

The geomorphology of the Georgian coast is similar to that of the New South Wales coast north of Sydney, Australia. Hails & Hoyt (1968) proposed a sequence of barrier island development from examination of elevated Pleistocene barriers in Georgia. These barrier islands begin with a dune or beach ridge forming adjacent to the shoreline. Then they submerge partially, inundating the ridge; and islands subsequently form in the lagoon behind the ridge. In New South Wales at Stuarts Point, there is a transport mechanism that carries arsenic compounds along the bedrock under a Pleistocene barrier to the sea, and this mechanism has facilitated arsenic to invade groundwater wells that were once thought safe from such contamination. As the geomorphology of the Georgian coast is similar to those of New South Wales and Florida, attention should be paid to similar possible transport mechanisms for contaminants.

Grosz *et al.* (1986) identified certain heavy metal placer deposits, including arsenic, in Atlantic coastal shelf sediments. Gillespie *et al.* (1989) obtained sediment samples from the proposed Ocean Dredged Material Disposal Site and from nearby areas and analysed them for high molecular weight hydrocarbons, pesticides, halogenated hydrocarbons, heavy metals and metalloids, including arsenic. Trawl macrofauna were also analysed for bioaccumulation of organics, heavy metals and metalloids. The dredged sediments indicated bioaccumulation of the macrofauna and the need to mitigate the sediments.

Fort Valley, in Peach County, has a 13-ha arsenic contamination site (Woolfolk Chemical Works), which resulted from the manufacture of pesticides, herbicides and insecticides since 1910. The site was put onto the National Priorities List in August 1990. Contamination had spread from the 7-ha facility onto the surrounding residential properties. Seventeen properties were converted to commercial use, and 24 000 tonnes of arsenic-contaminated soils and debris were removed. House dust in a remaining eight homes had excessive arsenic levels, and cleanup was completed in July 1997; the homes were monitored until June 1998 (USEPA, 2007b).

Augusta, in Richmond County, has a 30-ha site that began manufacturing phosphoric acid in 1962. This process results in several waste products, including arsenic trisulfide. Approximately 680 kg of arsenic were placed in two landfills that leached into the surficial aquifer, resulting in concentrations exceeding 50 µg/l. In November 1983, Monsanto excavated the contents of the two on-site landfills and transported the waste to a permitted landfill in Emelle, Alabama. Two small groundwater plumes of arsenic contamination remained on-site, and groundwater was extracted from the surficial aquifer and discharged for treatment. This site was delisted from the National Priorities List on 9 March 1998, and the most recent groundwater monitoring events indicate that only two on-site monitoring wells still exceed the maximum limit for arsenic; the concentrations in these two wells have decreased significantly over time (USEPA, 2007a).

A 1998 programme monitoring concentrations of arsenic in river waters (especially the Chattahoochie) found that levels of dissolved arsenic were below 50 µg/l but that arsenic concentrations in all species of riverine fish were high enough to warrant concern for the general public, which was warned not to consume more than six fish per year (Rosi-Marshall *et al.* 2001).

Hawaii (A)

Anthropogenic contamination by arsenic chemicals on the site of a former sugar cane plantation has been identified by Peard (2006):

> Environmental investigations at a planned hotel site and former sugar plantation camp in Kea'au, Hawai'i, led to the discovery of significant arsenic contamination in soil (>500 mg/kg total arsenic). The arsenic is related to the use of sodium arsenite in sugar cane fields during the early 1900s. During this time, over 100 000 acres of land were under cultivation for sugar cane. The arsenic is sorbed to iron hydroxyoxide, allophane and imogolite particles in thin Andesols. The soil investigation was expanded to include multi-incremental sampling of eighteen other areas in Kea'au. The highest levels of total arsenic were reported for community gardens (mean 331 mg/kg), followed by undeveloped land adjacent to residential subdivisions (mean 278 mg/kg), parks (mean 121 mg/kg) and schools (mean 37.0 mg/kg). Background total arsenic is typically <20 mg/kg.
>
> The elevated arsenic posed several initial concerns, including leaching and impacts to groundwater, uptake in produce and direct exposure of residents. Elevated levels of arsenic have not been reported in local water supplies. Lab-based leaching studies are underway. Total arsenic in produce from community gardens is similar to marketplace data compiled by the USFDA [United States Food and Drug Administration]. Bioaccessibility tests were conducted on soil samples to further evaluate direct-exposure concerns. Reported bioaccessibility ranges from 1.5% to 22%, with the highest levels of bioaccessible arsenic reported for the hotel site (186 mg/kg) and the community gardens (111 mg/kg). Bioaccessible arsenic in other areas ranges from <1 mg/kg to 27 mg/kg. An in-vivo study carried out on soils from the hotel site yielded a bioavailability of only 5–6%, however, or approximately 35 mg/kg bioavailable arsenic. A urine investigation carried out on residents appears to rule out unusual arsenic exposure from living in the area. Measures are being taken to minimize long-term exposure and educate the community on chronic risks posed by high arsenic in soils.

The concern is that there may be many more sites where sugar cane and pineapple plantations have used arsenic-based pesticides for their crops.

Idaho (G, M, V)

Geogenic sources of arsenic in central and north-east Idaho are related to arsenopyrite weathering in a matrix associated with gold. Consequently, gold mining promotes the release of arsenopyrite by accelerating its exposure to weathering processes, which, in the presence of acidic conditions, releases the arsenic into surface water and groundwater, as the data in Table 2.14 indicate.

Table 2.14 Arsenic concentrations in surface water and groundwater at gold mining projects in central Idaho, USA.

Location	Maximal arsenic concentration in surface water (mg/l)	pH	Maximal arsenic concentration in groundwater (mg/l)	pH
Idaho (central)	0.031	5.6	1.4	6.8
Idaho (north-east)	0.029	6.9	2.7 (contaminated by cement)	>11.0
			1.0	6.5

Source: Straskraba & Moran, 1990

Geogenic arsenic at concentrations up to 100 μg/l also occurs in Miocene fluvial and lacustrine unconsolidated sedimentary aquifers and confining units underlying the rapidly developing Boise Metropolitan Area. Because of the demographics of Boise, it has become urgent to understand the mechanisms of the fate and transport of arsenic in the underlying aquifers and groundwater. Arsenic in unconsolidated sediments of the Boise Treasure Valley aquifer rarely exceed 6 mg/l, but redox reactions could increase this concentration of naturally occurring arsenic. This could be an issue, because there is a reduced "blue clay" lens that overrides an oxidized stratum of sand and gravel (Domato *et al.* 2005).

Volcanism also promotes the release of arsenic by increasing the temperature of the ambient fluid in which it is dissolved and also by changing the redox environment. Examples of this phenomenon are found in the south-western part of the state (Maurer *et al.* 1996).

Illinois (C, G)

Geogenic sources of arsenic in drinking-water in Illinois are associated with the Mahomet aquifer, which covers an area of 10 000 km^2 and serves a population of 670 000. The sand and gravel unit averages 30 m thick but is 60 m thick in places, producing a yield of approximately 1700 million litres per day. A series of surveys have been carried out on the aquifer since 1988, which were summarized by Wilson *et al.* (2001). These surveys and their interpretation shed light on the processes of arsenic mobilization and transport in Illinois ground waters. Since the Mahomet aquifer is so large and serves such a large population, it has been deemed important to share the available mapping of the arsenic concentrations throughout the aquifer (see Tables 2.15 and 2.16).

Table 2.15 Arsenic concentrations in the Mahomet aquifer, Illinois, USA.

Study or database	No. wells surveyed	Arsenic concentrations (μg/l)		
		Minimum	Maximum	Median
IEPA/ISWS data bases	460	<1	135	2
Tazewell County Health Department	494	<1	259	26
Holm & Curtiss (1988) (public)	19	<1	80	46
Holm & Curtiss (1988) (private)	20	9	226	42
Holm (1995)	33	<3	99	9
Warner (2001)	30	<1	84	17
Kelly & Wilson (2000)	21	1	46	13

IEPA = Illinois Environmental Protection Agency; ISWS = Illinois State Water Survey
Source: Wilson *et al.* 2001

Table 2.16 Percentage of samples with various arsenic concentrations in the Mahomet aquifer, Illinois, USA.

Study or database	Arsenic concentrations (μg/l)			
	≤10	10–20	20–50	>50
IEPA/ISWS data bases	68.8	11.9	13.7	5.6
Tazewell County Health Department	39.3	6.9	18.6	35.2
Holm & Curtiss (1988) (public)	42.1	5.3	21.1	31.6
Holm & Curtiss (1988) (private)	10.0	5.0	45.0	40.0
Holm (1995)	54.6	21.2	18.2	6.1
Warner (2001)	56.7	3.3	33.3	6.7
Kelly & Wilson (2000)	38.1	38.1	23.8	0.0

Source: Wilson *et al.* 2001

In 30 domestic wells, arsenic(V) was the dominant species (Holm & Curtiss, 1988; Warner, 2001), but in 43 municipal wells, the percentages of arsenic(V) and arsenic(III) were equal (Holm & Curtiss, 1988). In the western part of the aquifer, shallow wells had relatively elevated arsenic concentrations (Wilson *et al.* 2001).

Coals may also be another source of arsenic contamination in Illinois, as underlying coal-bearing units containing arsenic in a redox environment have been identified (Yarling, 1992).

Indiana (A, C, G, M)

Anthropogenic and possibly mining sources of arsenic have been found in the town of Pines (population 800) on Lake Michigan, which was the site of a Ford Motor Company plant where arsenic paint sludge was dumped (Sarver, 2004). The site was also the location of an iron mine that had arsenopyrite associated with it, so there may have been additional arsenic accumulated in the soil from mine tailings as well (Thomas, 2003). The extent of similar iron mining sites with arsenopyrite mineralization is not known.

Coal-related sources of arsenic contamination of Indiana's groundwater may be attributable to glacio-fluvial deposits with underlying arsenic-rich coal-bearing units. For instance, the upper Kankakee River basin in Indiana has high arsenic levels in its surface water, sediment and biota; this suggests that there may be leaching from arsenic-rich underlying coal beds

(Schmidt & Blanchard, 1997; Fitzpatrick *et al.* 1998). In the ash of high-sulfur coal from Springfield, the weighted average arsenic concentration was 92 mg/kg, with a peak in the fourth bench of 130 mg/kg (Mastalerz *et al.* 2004). In the ash of the Danville coal bed, a weighted average arsenic concentration of 8.1 mg/kg was measured, and two upper benches measured 53.2 and 13.7 mg arsenic/kg, which are not correlated with the sulfur content (Yudovich & Ketris, 2005).

Geogenic arsenic has been identified as being released by oxidation of black shale fragments within glacial deposits in parts of north-eastern Indiana. Arsenic concentrations above 50 µg/l in the upper Illinois River basin, Elkhart County, Wakarusa, have been identified in groundwater from the deep lower Mahomet aquifer, which is composed of unconsolidated Quaternary deposits from glacial outwash and tills. The geochemical release of arsenic is related to the redox weathering of Antrim, New Albany and Ellsworth black shales (Yarling, 1992), which are also present in the Kankakee River basin in east-central Illinois (Schmidt & Blanchard, 1997; Fitzpatrick, Arnold & Colman, 1998). Deep aquifers in the St Joseph River basin near South Bend, near the Kankakee River headwaters, had arsenic concentrations significantly higher than those in shallow aquifers due to redox weathering (Fenelon *et al.* 1995).

The state has enacted an Arsenic Remediation Grant Award through the Indiana Finance Authority that is designed to remediate some of these elevated arsenic levels in public drinking-water supplies. A case in point is the award given to the Fayette Country School Corporation for $US 51 000 after the Department of Environmental Management's Drinking Water Branch discovered arsenic concentrations exceeding 10 µg/l in the school's drinking-water (McGoff, 2007). Another case in point is the groundwater of Cedar Lakes, which had also been identified as having arsenic concentrations above 10 µg/l (Weitz & Luxenberg, 2007).

A list of those communities assessed for arsenic contamination as of 2000 is included in Table 2.17.

Table 2.17 List of public water systems in the state of Indiana in which arsenic was found and reported to the USEPA from 1980 to 1998.

County	Water system name	Population served	Average arsenic concentration (mg/l)
Adams	Oakhurst Mobile Home Park	48	3.4
Allen	Fort Wayne, 3 Rivers Filtration Plant	180 000	0.1
Bartholomew	Meadows Metal Products	N/A	1
De Kalb	Nucor Fastener	19 525	1.8
De Kalb	Vulcraft	284	9.1
Delaware	DeSoto Elementary School	175	3
Elkhart	World Missionary Press	40	2.2
Grant	Three Acres Mobile Home Park	32	2
Henry	Indiana Sailors & Soldiers Home	800	4.1
Howard	Indiana American Water	48 827	1
Jasper	Demotte Christian School	350	23
Kosciusko	Atwood Elementary School	25	1.8
	H & H Mobile Village	125	4.6
	Hamelin Industries	55	4.7
	Hide Away Hills Mobile Homes	477	5
	Hoosier Water Co. – Warsaw	8 006	1.7
	Mentone Water Co.	912	2.1
	North Webster Day Care	61	4.6
	Pleasant Acres Mobile Homes	65	3.7
	Shamrock Mobile Homes	350	3.1
LaGrange	Brock's Mobile Homes	55	4.2
	United Technologies Auto Division	110	4
Madison	BPA-JLS Companies	100	6.8

(*Continued*)

Table 2.17 List of public water systems in the state of Indiana in which arsenic was found and reported to the USEPA from 1980 to 1998 (*Continued*).

County	Water system name	Population served	Average arsenic concentration (mg/l)
Owen	Valley View	45	3
Putnam	Fillmore Elementary School	195	12
Steuben	West Side Village	195	12
Tippecanoe	Perdue Univ. Water Works	16 291	4.4
Whitley	Columbia City Community Care	30	6.1
	Chain O'Lakes	25	7.1
	Hessen Utilities	360	3.9
	Plymouth Sub-district	30	7.7
	Southcrest Mobile Homes	150	1.6

Source: NRDC, 2000

Kansas (A, C, G, M)

Anthropogenic sources in the form of arsenical pesticides may have contributed to elevated soil arsenic levels in parts of the Marais des Cygnes National Wildlife Refuge, where sediment samples from six locations had elevated levels of arsenic compared with US norms. Fortunately, arsenic was not detected in any fish composite. Geogenic sources related to leaching of arsenical minerals may also have contributed to these elevated arsenic levels (Allen *et al.* 1995).

The arsenic concentration in coals in the south-eastern part of the state is 22 mg/kg, which is higher than concentrations in any surrounding states. These coals also have elevated levels of silver, lead, germanium, zinc and cadmium (Kansas Geological Survey, 1997).

Arsenic occurs naturally in the south-western part of the state and is associated with the zinc mines that extend across the state line into Oklahoma, where the problem is more pronounced. Tar Creek Basin in south-western Kansas and north-eastern Oklahoma was mined from 1891 to 1970, during which period 1.5 million tonnes of zinc and 7.3 million tonnes of lead with an unknown quantity of associated arsenic minerals were extracted using 480 km of underground tunnels. As a result, 59 million tonnes of tailings have seriously impacted the Tar Creek Basin, with derelict mine runoff from the tailings into surface water and groundwater (USGS, 2001).

Kentucky (A, C)

In some stream sediments, arsenic enrichment from coal is exacerbated by anthropogenic arsenical pesticides used in agriculture (Porter *et al.* 1995).

Arsenic-bearing coal occurs in Pennsylvanian coal seams that leach into riverine sediments (Tuttle, 2001). The Springfield no. 9 seam in western Kentucky has arsenic increasing in concentration from the Illinois basin periphery to the centre, which may have been a peat bog or was alternatively washed by thermal fluids (Coleman & Bragg, 1990). The low-ash, low-sulfur Blue Gem seam in south-west Kentucky indicates that arsenic is correlated positively with ash (Hower *et al.* 1994). In the south-eastern county of Clay, the Manchester coal bed has an average arsenic concentration of 47 mg/kg (with a range from 5.6 to 292 mg/kg), with the highest arsenic concentration correlated with the highest sulfur content in the roof of the bench (Sakulpitakphon, 2004). Also in the south-east of the state, the upper Hance seam displays arsenic anomalies in the ash, with concentrations that range from >200 to 1750 mg/kg, but these are not strongly correlated with high sulfur (Hubbard *et al.* 2002).

High-arsenic coals are located in over 50% of eastern Kentucky coals ($n = 780$), and they exceed the geometric mean concentration for coals in the USA of 7.6 mg/kg (Bragg *et al.* 1997), with maximum concentrations of 680 mg/kg. These high-arsenic coals cluster in zones parallel to discontinuities (perpendicular to Appalachian folds) that may have facilitated hydrothermal flows (containing arsenic) into the Kentucky Appalachian Basin during orogenic processes (Pohn, 2001). As can be seen in Figure 2.8, the most striking correlation is the high concentration of arsenic in stream sediments along the western edge of the Central Appalachian Basin that follows the exposure of Upper Devonian black shales (Tuttle *et al.* 2001).

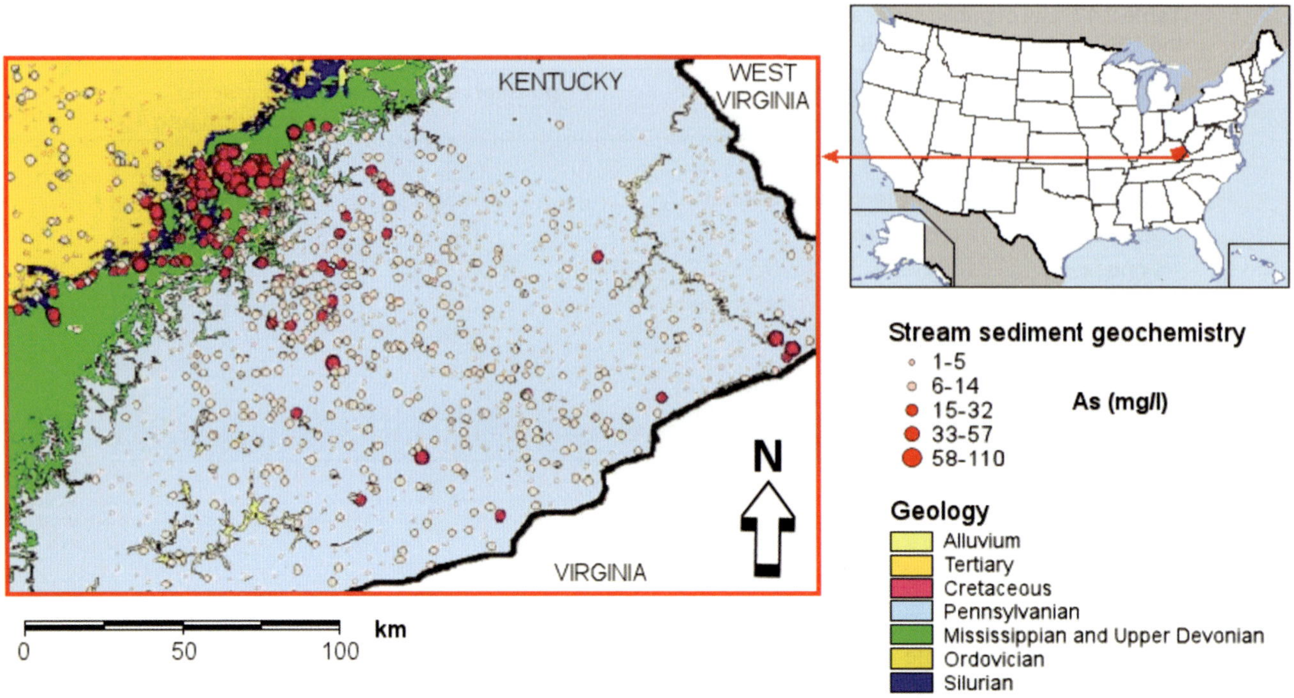

Figure 2.8 Arsenic concentrations in the Appalachian Basin, Kentucky, USA (modified from Tuttle *et al.* 2001).

Louisiana (A)

Anthropogenic sources of arsenic contamination in Louisiana come from agricultural fields that once had grown cotton (*Gossypium* spp.) and had been treated with arsenical pesticides. Many of these fields were then transformed into rice paddies, and the arsenical pesticides that had been used to control the boll weevil (*Anthonomus grandis*) and as a cotton desiccant continued to leach into the groundwater from the soil and have now entered the food-chain (Ori *et al.* 1993). In Louisiana, the highest concentration of arsenic in rice (0.66 µg/g) was found in a sample of rice from Louisiana Mills (Meharg, 2007). In fact, the rice from Louisiana Mills is affected physiologically by being grown in arsenical soils, which produces a "straight head" condition; indeed, this deformity has been selected as the best type of rice to grow in high-arsenic soils (Marin *et al.* 1992; Dilday *et al.* 2000).

Agricultural land once used for growing cotton and now used for growing rice is common to many south-central states, which average 0.27 µg arsenic/g in rice. In contrast, arsenic concentrations in rice from California, where cotton was not an earlier crop and hence arsenical pesticides were not used, average 0.16 µg/g. Organically grown rice in California averages 0.11 µg/g (Meharg, 2007).

Other unrelated sources of arsenic in groundwater have been identified in Calcasieu River and Calcasieu Lake surface sediments, which have a background arsenic concentration of 0.60 mg/kg (together with concentrations of cadmium, chromium, copper, lead, mercury, silver and zinc). Although levels were higher at sewage and industrial outfalls, elevated arsenic was not found in the river and lake complex (Beck *et al.* 1990).

Maine (A, G)

The geogenic occurrence of arsenic in a small watershed in coastal Maine near Northport is illustrative of the complexity of weathering materials in fracture surfaces and the difficulties of interpreting bedrock groundwater hydrology. The maximum level of arsenic in the groundwater of Northport was 230 µg/l, the highest in the state, in a series of wells that are clustered in the centre of a discharge area near Kelly's Cove. Northport bedrock is interbedded with Ordovician (Penobscot Formation) sulfidic pelite, quartzite, arsenian pyrites and arsenopyrite, later intruded by Silurian granitic and dioritic plutons (muscovite, biotite, garnet granites) and then inundated by the sea after the glacial retreat. Three distinct water types have been identified: calcium bicarbonate, sodium bicarbonate and high arsenic. The calcium bicarbonate water type is recharge water and is dominated by Ca^{2+} and HCO_3^- from weathered carbonates. The sodium bicarbonate water type is dominated by

Na^+ and HCO_3^- from plagioclase dissolution and cation exchange of calcium for sodium. The high-arsenic water type (a less evolved sodium bicarbonate water than the sodium bicarbonate water type with which it has no hydrological communication) has the highest arsenic levels. The delineation of these three distinct types of waters and their lack of hydrological communication provide a basis for groundwater flow modelling and the ascertaining of the flow pattern of arsenic-rich waters. This technique of distinguishing water types and their flow patterns may prove valuable in applications elsewhere in Maine where a similar geochemistry exists (Lipfert & Reeve, 2004).

Other geogenic sources of arsenic in Maine groundwater have been linked to a release mechanism caused by microbial action of *Sulfurospirillum barnesii* and *S. arsenophilum* found in soils and aquifers. These arsenic-reducing bacteria were identified in a test site in Maine in a study by MacRae (2005):

> In 2001, an isolate named NP4 was obtained from a contaminated well in Northport Maine. The well is among a cluster of wells with very high arsenic concentrations, and with no known anthropogenic sources of arsenic. At the time of sampling, the total arsenic concentration in the water was 1400 ppb [parts per billion]. The isolate was identified by sequencing the 16S rRNA [ribosomal ribonucleic acid] gene and found to be closely related to two other arsenate reducing bacteria, *Sulfurospirillum barnesii*, and *S. arsenophilum*, both of which were isolated from surface water sources. Its presence in groundwater, and its ability to reduce arsenic as well as a variety of other electron acceptors, including Fe(III) and Mn(IV), prompted a fluorescence in situ hybridization (FISH) study to determine its prevalence in the environment. Well water was taken from wells in the Northport area and in the Branch Lake area of Ellsworth, Maine, where the groundwater has much lower concentrations of arsenic, but with some readings still higher than the proposed drinking water standard of 10 ppb. While NP4 as a percentage of total bacterial numbers does not correlate with total As concentrations in groundwater, it does correlate with As(III). A positive correlation was also found between *Geobacter*, a genus that includes many iron-reducing bacteria, and total arsenic. These results indicate that microorganisms may be important in arsenic mobilization and speciation in groundwater.

This phenomenon of arsenic-reducing bacteria facilitating the release of sorbed arsenic is similar to the situation in Bangladesh. This mechanism begins with the erosion of mountain belts aided by glaciers, which facilitate the transport of arsenic by surface waters, the adsorption of arsenic by stream sediments and the deposition of stream sediments and organic matter in alluvial deposits. The subsequent mechanism of iron(III)-reducing bacteria present in alluvial aquifers causes the release of sorbed arsenic into groundwater under moderately reducing conditions (Saunders *et al.* 2005).

In addition to the geogenic sources of arsenic in Maine groundwater, anthropogenic sources also play a role. Ayuso & Robinson (2001) documented the persistence of arsenic contamination on sites that had been used for agriculture and treated with a variety of arsenic compounds; their report documents the variety of these arsenic compounds:

> Arsenical pesticides and herbicides were extensively used on apple, blueberry, and potato crops in New England. Lead arsenate was the most heavily used arsenical pesticide until it was officially banned .... Arsenic and lead concentrations in stream sediments are higher in agricultural areas that used arsenical pesticides than in other areas; arsenic and lead concentrations are positively correlated ($r = 0.68$). Historical pesticide residues remain in soils and stream sediments when agricultural land previously contaminated with arsenical pesticides becomes residential. The common arsenical pesticides have similar Pb isotope compositions: $^{208}Pb/^{207}Pb =$ 2.3839–2.4721, and $^{206}Pb/^{207}Pb = 1.1035–1.2010$. For other arsenical pesticides such as copper acetoarsenite (or Paris green), methyl arsonic acid and methane arsonic acid, as well as arsanilic acid (used as feed additives to promote swine and poultry growth) range widely. The pesticides partially overlap the composition of the stream sediments from areas with the most extensive agricultural use. Soil profiles from a watershed containing arsenic-enriched groundwater contain labile Pb showing a moderate range in $^{206}Pb/^{207}Pb = 1.1870–1.2069$, and $^{208}Pb/^{207}Pb = 2.4519–2.4876$. The isotope values vary as a function of depth: the lowest Pb isotope ratios (e.g., $^{208}Pb/^{206}Pb$) representing labile lead (acid leach) are in the uppermost soil horizons. Lead contents decrease with depth in the soil profiles. Arsenic contents show no clear trend with depth. Labile lead in the profiles represents lead that is held in soluble minerals (Fe- and Mn-hydroxides, carbonate, etc.). Lead isotope compositions of stream sediments and soils from areas with heavy use of pesticides include contributions from rock, silicates, sulfides and their weathering products, in addition to industrial lead (mostly from atmospheric deposition). In agricultural regions, the extensive use of arsenical pesticides and herbicides can also be a significant anthropogenic source of arsenic and lead in soils and sediments.

Maryland (A)

Anthropogenic sources are suspected for arsenic and other trace elements that are within or above the high range of what could be expected for natural waters of the Chesapeake Bay basin and sediments. For instance, concentrations of dissolved arsenic collected in the Pocomoke River basin in February 1999 ranged from 0.014 to 0.84 µg/l (USGS, 2000). The source of this arsenic may be the excreta of the 600 million broiler chickens that are produced on the Delmarva Peninsula each year (USDA, 1997b). Roxarsone (4-hydroxy-3-nitrobenzenearsonic acid) is commonly used in poultry feed as an intestinal palliative for to increase weight, improve feed efficiency and enhance skin pigmentation. Roxarsone has the advantage that it is not retained in the flesh or eggs of chickens. Broiler chickens are fed 21–41 g of arsenic in the form of roxarsone per tonne (Animal Health Institute, 1996). Roxarsone is directly excreted, and poultry litter from commercial broilers can contain 15–35 mg total arsenic/kg (Morrison, 1969; Jackson & Miller, 1999; USGS, 2000). It has been estimated that between 20 and 50 tonnes of roxarsone are excreted per year on the eastern seaboard (Christen, 2001). Because of roxarsone's increasing solubility in warmer water, the potential for its mobilization in marine and surface water environments is greater in the spring and summer months. Additionally, when roxarsone is biologically degraded, arsenic(V) can be mobilized into local surface water and groundwater (Gabarino *et al.* 2003; Rutherford *et al.* 2003; Brown *et al.* 2005). Public health data suggest that roxarsone is at least partially responsible for an elevated cancer incidence on Maryland's shores, with cancer rates on the Lower Shore exceeding the national average (1993–1999) of 206 deaths/100 000: Somerset County, with 267 deaths/100 000; Wicomico County, with 233 deaths/100 000; and Worcester County, with 229 deaths/100 000 (Vandiver, 2004).

Massachusetts (A, G)

Anthropogenic sources of arsenic in groundwater have been identified in Massachusetts in groundwater and an adjacent wetland near the Industri-Plex Site in Woburn on Route 128, a technology belt of high-tech manufacturing and testing of arsenical chemicals (Gawel *et al.* 2004). Research initiated in 2004 studied the feasibility of attenuating such a site (Ford, 2004) and one such project (Ford, 2006) is designed:

(1) To assess the geochemical processes that control arsenic mobility in the Aberjona watershed;
(2) To develop methods for characterizing solid phase arsenic speciation; and
(3) To develop guidance for the assessment of natural attenuation on remediation of arsenic in soils and sediments.

Geogenic sources of arsenic in Massachusetts are related to bedrock aquifers. As groundwater arsenic concentrations exceed the existing USEPA standard of 10 µg/l in many parts of New England, it has become increasingly important to identify the sources of this contamination (Blute *et al.* 2009). For instance, one central Massachusetts site adjacent to a tributary to a public water supply reservoir was tested, and concentrations in overburden and bedrock groundwater exceeded 10 µg/l. Multiple lines of evidence indicated that elevated concentrations of arsenic were due to natural dissolution of arsenic-bearing minerals. These lines of evidence included: historical data that did not reveal any anthropogenic source; geological mapping of overburden and bedrock from microprobe analyses and samples taken during well drilling that did identify geogenic arsenic sources; and geochemical indicators (pH, dissolved oxygen, oxidation/reduction potential) of groundwater samples that provided positive conditions for the release of arsenic (Nelson *et al.* 2006). Another geogenic source of arsenic in groundwater in Massachusetts is the Bald Mountain massive sulfide deposit, where arsenic has been measured in the groundwater at concentrations ranging from 2 to 430 µg/l in a pH range of 5.9–8.0 (Welch *et al.* 2000; Seal *et al.* in press).

Michigan (G)

Geogenic sources of arsenic in Michigan are associated with sulfide-bearing minerals. Glacial deposits, shale and sandstone are the basic geological units associated with arsenic mobilization, the last of which contains arsenic-rich pyrite, a possible source of arsenic in the overlying glacial aquifer in the far eastern "thumb" region (Kolker *et al.* 1998; Westjohn *et al.* 1998).

An 11-county area in south-eastern Michigan has high levels of arsenic in water from domestic and municipal wells; concentrations range from 1 to 1310 µg/l, with most levels being in the range 5–50 µg/l.

The area is heavily populated, and nearly all water supplies come from local groundwater. Many wells exceed the USEPA limit for arsenic in drinking-water of 10 µg/l. The source of arsenic in groundwater is Marshall Sandstone, a major bedrock aquifer (Figure 2.9). The USGS, in a study to pinpoint the geological source of the arsenic, drilled two wells, one in Huron County and one in Lapeer County, into the Marshall Sandstone aquifer. Marshall Sandstone has black shale beds associated with arsenopyrite, with arsenic concentrations ranging from 140 to 310 mg/kg pyrite. Arsenic contents of water from the new wells can be as high as 125 µg/l. The arsenic levels in water samples collected from isolated parts of the wells show no direct relation to the arsenic content of corresponding core samples. This is consistent with preliminary geochemical modelling, which suggests that pyrite in the bedrock aquifer is not reacting with the groundwater, so there may be an organic source of the mobilized arsenic (Slotnick *et al.* 2003).

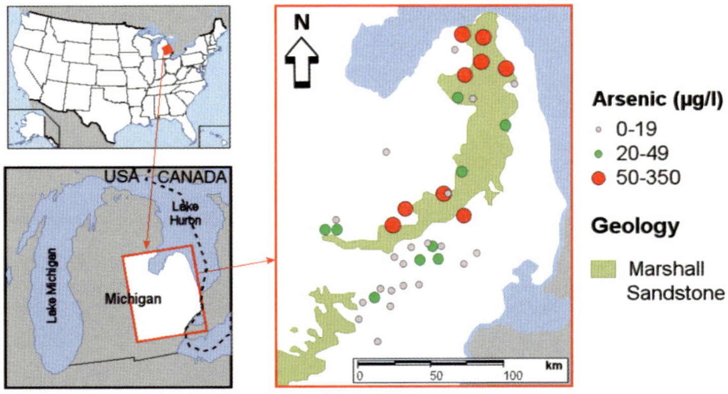

Figure 2.9 Arsenic concentrations measured in well water in the south-eastern region of Michigan, USA. The Marshall Sandstone is thought to be the natural source of arsenic (modified from USGS, 1999).

Minnesota (A, G)

The Minnesota Arsenic Study, conducted from April 1998 to February 1999 by the Minnesota Department of Health, examined nine counties (Big Stone, Clay, Douglas, Grant, Otter Tail, Stevens, Swift, Traverse and Wilkin) and had four goals:

- To measure arsenic in private wells in areas where concentrations reach up to 150 µg/l;
- To develop geochemical, geographic, and geological indicators to identify wells at risk;
- To determine if people using high arsenic wells have elevated levels of urine and hair arsenic;
- To determine if these people show evidence of arsenicosis.

Results showed that arsenic in Minnesota occurs in glacially deposited fine-grain Late Wisconsin till (10 000–15 000 years BP), and that ubiquitous organic matter fosters biological activity, causing arsenic(V) to be reduced to arsenic(III) and then to be mobilized into the groundwater (Erickson, 2005; Erickson & Barnes, 2005a,b). Also, arsenic(III) is highly mobile due to groundwater flow and anaerobic conditions, so the hydrology is very dynamic. Of 869 wells sampled, 65% had arsenic concentrations below 20 µg/l, 25% had 20–50 µg/l and 10% had more than 50 µg/l (Minnesota Department of Health, 2001a), with a maximum arsenic concentration of 91 µg/l (Minnesota Pollution Control Agency, 1998). To put it another way, over 50% of the 869 sampled wells in western Minnesota had arsenic concentrations above 10 µg/l (Minnesota Department of Health, 2001b). Based on data for the entire state, 15% of all wells in Minnesota may have arsenic levels of at least 10 µg/l (Erickson & Barnes, 2005b). The problem is therefore dynamic and state-wide and requires testing of any wells in or near glacial drift areas.

In addition to geogenically occurring arsenic in groundwater, Minnesota has also had anthropogenic arsenic leaching into surface water from agricultural arsenical pesticides: 61 properties were identified as having arsenic concentrations above 95 mg/l in the soil. Of these 61 properties, 30 were remediated in 2004, and the remainder were to have been remediated by the spring of 2006. However, the widespread use of agricultural arsenical pesticides provides a note of caution, as other properties that had been exposed to similar chemicals and not yet identified may pose threats of arsenic contamination (USEPA, 2006).

Mississippi (G)

Arsenic occurs geogenically in Mississippi, with soil concentrations in the range of 1–15 mg/kg. However, arsenic levels in the Delta and the Blacklands are significantly higher, but health problems associated with these levels have not been documented (Mississippi Department of Environmental Quality, 2002).

There are eight major soil resource areas in Mississippi. The Coastal Plain (Upper and Lower) comprises 41.8% of the land mass and tends to be acidic with highly weathered loams and quartz sands. The Upper and Lower Loess is mainly quartz, feldspar, mica and carbonate silt covering 28.9% of the land mass. The rich, clayey sediments of the Delta cover 17% of the land mass. The Blackland Prairie accounts for 6.1% of the land mass and is underlain by Selma chalk (calcium carbonate). The Interior Flatwoods has soapstone in its silty clay, with lesser amounts of kaolinite and illite, and constitutes 2.6% of the state (Pettry & Switzer, 2001).

Trace metal content (including arsenic) of soils is strongly influenced by the parent materials in which they form, with ages of parent materials of soils decreasing from east to west. The oldest materials, including Tombigbee sand and Selma chalk, are in the north-eastern part of the state. The youngest major soil parent material is the Mississippi River alluvium (Pettry & Switzer, 2001). Table 2.18 indicates that the youngest parent material is in the west, and Table 2.19 shows that the youngest soils have the highest mean arsenic concentration.

Table 2.18 Mean elemental arsenic concentrations in selected parent materials in Mississippi, USA.

Parent material	Area	Locations	Arsenic levels (mg/kg of soil parent materials)		
			Mean ± SD	Minimum	Maximum
Citronelle formation	LCP	2	10.28 ± 4.92	6.80	13.76
Neshoba sand	UCP	2	5.58 ± 3.72	2.95	8.22
Porters Creek	IF	2	8.98 ± 6.90	4.10	13.87
Peoria loess	L	5	6.90 ± 2.00	5.02	9.82
Clayey alluvium	D	5	21.05 ± 5.81	13.87	28.69
Selma chalk	BP	3	6.68 ± 3.44	3.88	10.53
Basic city shale	UCP	2	4.32 ± 0.54	3.94	4.71
Tombigbee sand	UCP	2	2.55 ± 0.21	2.40	2.71
Wilcox group	UCP	4	1.77 ± 1.02	0.32	2.69
Winona formation	UCP	8	33.76 ± 14.35	19.70	54.76
Yazoo clay	BP	2	16.57 ± 0.14	16.56	16.58
All parent materials		37	14.42 ± 13.71	0.32	54.76

BP = Blackland Prairie; D = Delta; IF = Interior Flatwoods; L = Loess; LCP = Lower Coastal Plain; UCP = Upper Coastal Plain; SD = Standard Deviation
Source: Pettry & Switzer, 2001

Table 2.19 Mean arsenic concentrations across all soils in Mississippi, USA.

Region	No. samples	Mean arsenic concentration (mg/kg of soil parent materials)[a]
Delta	31	15.02 a
Interior Flatwoods	7	12.73 ab
Blackland Prairie	24	11.11 b
Loess	42	9.31 b
Upper Coastal Plain	61	6.58 c
Lower Coastal Plain	81	5.79 c
Coastal Flatwoods	14	4.42 c

[a]Means with same letter are not significantly different ($P = 0.05$).
Source: Pettry & Switzer, 2001

Soil arsenic concentrations, averaged across all soils analysed in the state, had the following relationships: Delta > Interior Flatwoods > Blackland Prairie > Loess > Upper Coastal Plain > Lower Coastal Plain > Coastal Flatwoods. Clayey soils of the Delta, Interior Flatwoods and Blackland Prairie had the highest arsenic levels, and lowest concentrations occurred in the coarser-textured Coastal Plain and Coastal Flatwoods soils. Soil arsenic concentrations increased with increasing distance from the Gulf of Mexico, reflecting effects of parent materials, weathering and pedogenesis (Pettry & Switzer, 2001).

Background arsenic concentrations provide essential data to indicate the extent of anthropogenic arsenic accumulation from cultural activities, but total arsenic concentration may not be the most reliable indicator of toxicity to plants or potential leaching and movement in the soil. Therefore, the toxicity of arsenic is influenced more by its water-soluble form than by the total amount present in soils. If arsenic is tightly bound in soils, it is unlikely to be bioavailable or mobile. Therefore, the amount of water-soluble arsenic (Table 2.20) is an important indicator to determine critical levels of concern (Pettry & Switzer, 2001).

Table 2.20 Water-soluble (bioavailable) arsenic in selected parent materials and soils in Mississippi, USA.

Material	Water-soluble arsenic concentration (ng/g of soil and parent materials)
Sharkey horizon	13.02
Byram horizon	3.27
Selma chalk parent material	10.36
Yazoo clay parent material	2.18
Winona formation parent material	4.24
Winona formation parent material	2.02
Winona formation parent material	1.83
Citronelle formation parent material	0.68
Wilcox group parent material	0.66

Source: Pettry & Switzer, 2001

Table 2.20 indicates that the two highest water-soluble arsenic samples of nine samples analysed are from the Selma chalk and Sharkey horizon. Measurements from the soils indicate a similar scenario of arsenic contamination. Selma chalk in the Blackland Prairie had arsenic concentrations ranging from 27.4 mg/kg in the surface in Monroe County, to 3.8 mg/kg in Rankin County (Pettry & Switzer, 2001; Table 6). The mean arsenic concentration was 11.11 mg/kg, and there were no significant correlations with organic matter, pH or clay. Sharkey horizon soils in the Delta in the Mississippi River alluvium are the youngest soils and have the highest mean arsenic concentrations, with arsenic levels ranging from 2.86 mg/kg in the coarse-textured, fluvial Bruno soil to 26.85 mg/kg in the cultivated surface (Ap) horizon of Sharkey soil in Sharkey County (Pettry & Switzer, 2001; Table 8).

Missouri (A, G)

Anthropogenic sources of arsenic contamination in Missouri are mainly related to the manufacture of herbicides for agriculture and the use of those herbicides on agricultural land.

One example of contamination of groundwater and soil by the manufacture of arsenical herbicides is the Armour Road site, which operated from 1948 to 1986 in North Kansas City, Clay County. Arsenic contamination at high levels was found in on-site soil and on-site and off-site groundwater. Although the contaminated groundwater is not being utilized for potable water, a number of industries and the city of North Kansas City use the aquifer for a water source. Soil contamination extends to the water table (4–5 m) over much of the site. Arsenic levels were measured at 121 000 mg/kg near the processing area and varied from 3.8 to 27 000 mg/kg elsewhere on-site. Arsenic concentrations decreased with depth, with a maximum subsurface level of 6800 mg/kg at 5 m. Groundwater flows to the south and east towards the Missouri River. A well 0.4 km down gradient from the site had an arsenic concentration of 1.67 mg/l. An additional seven monitoring wells had arsenic concentrations ranging from 0.050 to 0.171 mg/l (ATSDR, 1998).

Additionally, the use of arsenical pesticides in Missouri was prevalent in the growing of cotton (*Gossypium* spp.) to prevent boll weevil (*Anthonomus grandis*) infestations and as a cotton desiccant (Loebenstein, 1994). In the 20th century, as cotton-growing land was turned into rice-growing land, the earlier arsenical pesticides deposited into the soil continued to leach into the groundwater and entered the food-chain (Wells & Gilmour, 1977). The Marais des Cygnes National Wildlife Refuge Area is an example of the contamination of soil and possibly surface water and groundwater by arsenical pesticides. Six locations had elevated arsenic levels, but arsenic was not detected in any fish samples taken from the wetlands. The historic use of arsenical pesticides for cotton and then rice agriculture suggests that there exists a legacy of elevated soil arsenic levels in other areas (Allen *et al.* 1995). In fact, "straight head" rice (a physiological deformity caused by arsenical pesticides) was specifically chosen because it grew well in soils treated with arsenical pesticides (Marin *et al.* 1992; Dilday *et al.* 2000).

Other types of anthropogenic sources of arsenic contamination in Missouri include those associated with industry. Soils containing high levels of arsenic (as well as petroleum, asbestos and lead-based paint) were discovered on the 9-ha abandoned Kansas City Railway Terminal when the property was chosen for use as a community centre by the Hispanic

Redevelopment Corporation and its Westside partners. The land was quickly remediated, and an insurance-based risk transfer strategy facilitated its implementation, resulting in the USEPA awarding the project the Region 7 Phoenix Award as an excellent example of public/private partnership and community participation (Forrester Group, 2003).

In addition to anthropogenic sources of arsenic in groundwater in Missouri, the presence of high arsenic concentrations in groundwater is caused geogenically by redox conditions in both alluvial aquifers and buried bedrock valley aquifers in which arsenic minerals leach because of a change in redox conditions. As these aquifers are relatively high in clay content and have a relatively low yield, they are tapped mainly by private wells, which may not have been adequately tested for arsenic. In alluvial aquifers, iron oxides and associated arsenic are deposited with the alluvium during stream flow. Once the alluvium is buried, slow groundwater movement promotes reducing conditions: iron(III) oxides are reduced to iron(II) oxides, which are soluble; they mix with adsorbed arsenic, which also becomes mobilized in these reducing conditions; and they enter the groundwater (Gulen *et al.* 1979; Korte, 1991). This is a Midwestern geogenic arsenic phenomenon that affects Ohio and Illinois as well as Missouri.

Montana (A, V)

Anthropogenic sources of arsenic in soils of south-western Montana have been attributed to copper and other metal smelter emissions (Galbraith *et al.* 1995).

Volcanogenic sources of arsenic in groundwater within the Madison and upper Missouri river valleys appear to be directly or indirectly related to the Yellowstone geothermal system (Nimick, 1994, 1998), and this volcanism is responsible for the presence of arsenic in the adjacent aquifers (Sonderegger & Ohguchi, 1988):

> The arsenic concentration in 13 water samples from wells in the thin, alluvial aquifer of the Madison River Valley, Montana, USA ranged from 26 to 150 μg/l. The Madison River, which originates in Yellowstone National Park, has a mean arsenic concentration of 51 μg/l ($\sigma = 26$ μg/l), based upon very limited sampling in the study area during the main irrigation period. Groundwater arsenic concentration increases down the valley can be best correlated with the intensity of ditch irrigation in this semiarid area. No other sources of dissolved arsenic as concentrated as that of the river water have been identified. Evaporative concentration of river-derived irrigation water is believed to have been the overwhelming factor in the contamination of this shallow aquifer. In a follow-up study, elevated arsenic was found at Three Forks on the Missouri River near Gallatin, caused by irrigation with diluted geothermal water in the lower Madison valley (Sonderegger *et al.* 1989).

Nebraska (G)

Geogenic sources account for most of the arsenic contamination of groundwater in Nebraska. In two sets of data, the USGS tested 534 sites, and the National Uranium Resource Evaluation programme tested 3396 sites and documented spatial and geological variability in the geogenic occurrence of groundwater arsenic (Figure 2.10). According to its Nebraska Health and Human Services System, 81 public water systems serving 100 000 people have arsenic concentrations above 10 μg/l (El-Hajjar, 2004). Highest concentrations were found in the Panhandle and the western Sand Hills, but isolated occurrences were found in the central Platt River valley and the East-central Dissected Plains ($>$15 μg/l), whereas arsenic concentrations in the eastern groundwater regions were generally higher, between 5 and 15 μg/l. These results reflect differences in geological, groundwater and geochemical environments for sites that are close to each other; although they make predictions of arsenic concentrations at specific sites infeasible, they can help in screening areas suspected of containing high groundwater arsenic. A survey has been ongoing since 2004 to assess arsenic occurrences during an annual cycle at specific test sites (Gosselin *et al.* 2004; Gosselin *et al.* 2004).

Nevada (G, M, V)

Volcanogenic arsenic is responsible for the cluster of moderate and high arsenic concentrations in north-central Nevada's geothermal groundwater, which is associated with dewatering of sulfide-bearing gold deposits (Maurer *et al.* 1996). The USEPA has called for Fallon, Nevada (population 8300), to reduce the arsenic concentration in its drinking-water from 100 μg/l to the current standard (10 μg/l). The city of Fallon's drinking-water was, for many years, treated only to reduce microbial contamination (Smedley & Kinniburgh, 2002). The state epidemiologist is investigating whether six cases of leukaemia in children, diagnosed between March and July 2000, are related to arsenic ingestion, since no other county in Nevada has ever reported more than one case in a year. In fact, Churchill County, where Fallon is located, leads the nation with a maximum measured arsenic concentration of 1400 μg/l.

Geogenic arsenic in groundwater is found throughout the southern Carson Desert in concentrations ranging from 100 to 2600 μg/l. Shallow groundwater aquifers tapping into this water are used for domestic supply. Arsenic appears to be affected by evaporative concentration of the Carson River (the primary recharge source) and mobilizing redox reactions involving metal oxides and sedimentary organic matter. Arsenic is also found sorbed in iron oxides. This situation, as described by Welch & Lico (1998), may actually be a precursor of other, more mature hydrogeological systems:

> Hydrologic and geochemical conditions in the Carson Desert are similar to other areas with high As and U concentrations in ground water, including the Salton Sea basin and southern San Joaquin Valley of California. Hydrologic and geochemical conditions that produced some sandstone-type U-ore deposits, including those in the non-marine, closed-basin sediments of the Morrison Formation near Grants, New Mexico, suggest that the Carson Desert may be a modern analogue for those systems.

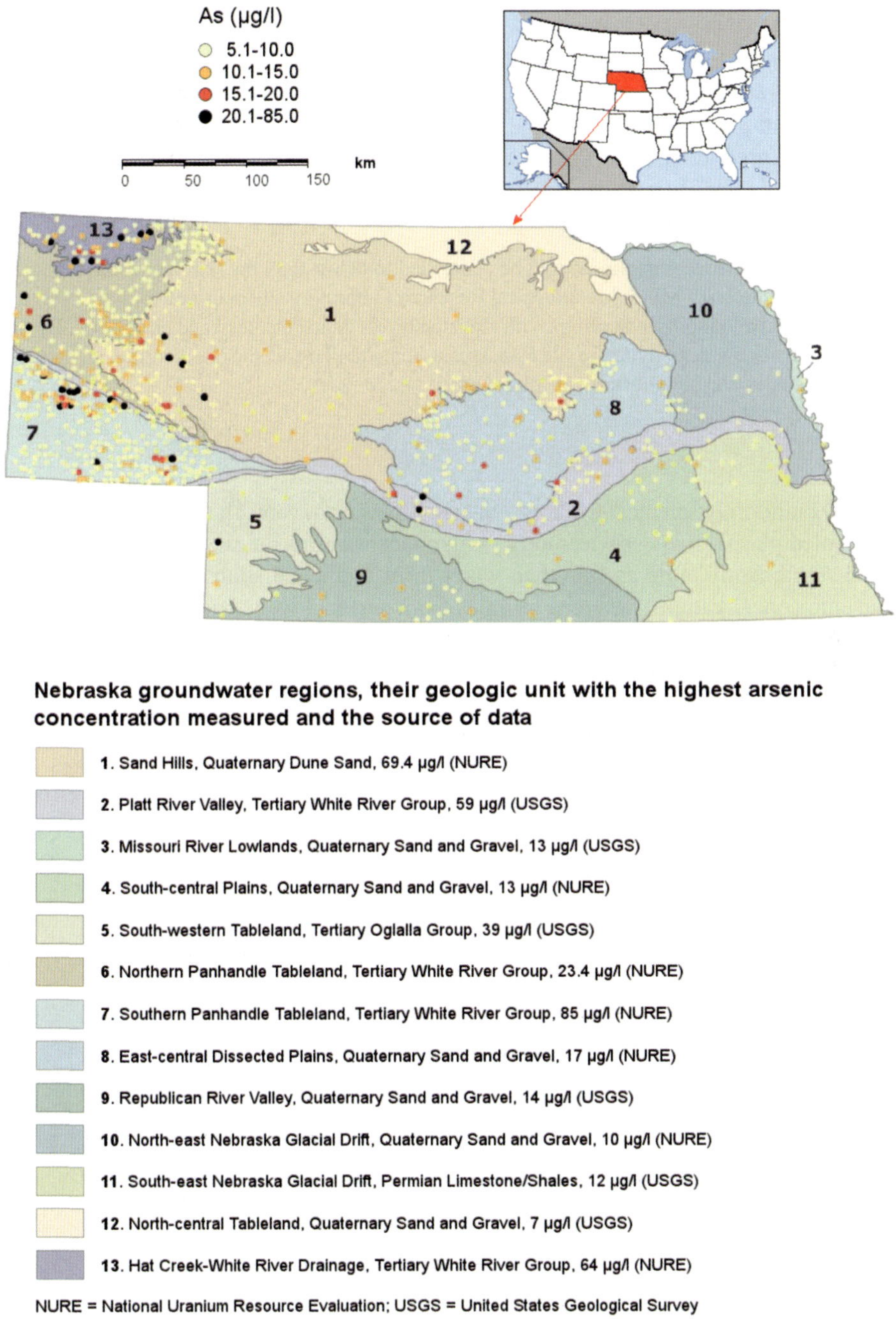

Figure 2.10 Concentrations of arsenic above 5 µg/l in groundwater in Nebraska, USA (modified from Gosselin *et al.* 2004).

In addition, there is a striking alignment of sedimentary rock-hosted Carlin-type gold deposits (with associated arsenopyrite) along trends of the Roberts Mountains thrust. This alignment suggests igneous activity at depth coupled with ore genesis and host rock permeability that served as conduits of the gold-related fluids, including arsenic compounds (Cunningham, 1988; Christensen, 1993; Li & Peters, 1998). These deposits have a similarity to those deposits in Guizhou Province, China, along the Yangtze craton.

Mining of gold in Nevada is related to these Carlin-type gold deposits, which are associated with arsenopyrite in the Roberts Mountains thrust.

Arsenic exhibits enrichment in hypogene gold deposits (Boyle, 1979). Gold-bearing polymetallic deposits have elevated arsenic, as do the typical quartz veins at Fortitude Mine, where the arsenic concentrations are from tens of parts per million to percentage amounts in the ore. Because gold is associated with arsenopyrite and arsenic-bearing sulfosalts in some deposits, it constitutes a concomitant of these minerals. Disseminated gold deposits in north-central Nevada all have elevated arsenic levels (Berger & Bethke, 1984). As mining of gold subjects deposits of arsenopyrite to weathering processes and as surface waters are generally acidic in gold mining areas, this set of conditions contributes to arsenic mobilization. In a seminal study of arsenic occurrences in surface water and groundwater associated with gold mines in the western USA, the maximal arsenic concentration in groundwater in central Nevada was measured to be 0.85 mg/l; the pH was alkaline (8.7). No perennial streams were measured in Nevada (Straskraba & Moran, 1990).

Stinchfield *et al.* (2005) found naturally occurring arsenic in area soils at concentrations ranging from about 3 to 9 mg/kg in the small residential community of Boulder City, Nevada:

> Yet current health-based cleanup goals recommend arsenic be controlled to below 0.39 mg/kg in a residential, one in a million, cancer endpoint model. The former US Bureau of Mines Metallurgical Research Laboratory site operated for half a century in the center of this community and resulted in uncontrolled soils and tailings with elevated levels of arsenic, lead and chromium. By involving specialists in regulatory decision making, risk assessment, dynamic site assessment, remediation and site engineering, an eighteen year stalemate was ended and a successful remediation and site closure was obtained.

New Hampshire (A, G)

Arsenic was measured in 992 randomly selected drinking-water samples from New Hampshire households using online hydride generation inductively coupled plasma mass spectrometry, with concentrations ranging from <0.0003 to 180 μg/l; water from domestic wells contained significantly more arsenic than water from municipal sources. Water samples from drilled bedrock wells had the highest arsenic concentrations, whereas samples from surficial wells had the lowest arsenic concentrations. It is probable that much of the groundwater arsenic in New Hampshire is derived from weathering of bedrock materials and not from anthropogenic contamination. Water samples voluntarily submitted to the New Hampshire Department of Environmental Services for arsenic testing were significantly elevated, and spatial distribution of elevated arsenic concentrations (>50 μg/l) correlates with Late-Devonian Concord-type granitic bedrock (Peters *et al.* 1999).

South-eastern New Hampshire (Hillsborough, Rockingham and Strafford counties) has a 2006 estimated population of 819 046, or 62.3% of the state population (United States Census Bureau, 2006a), of which 285 048 use private bedrock wells for drinking-water. Of these wells, 19% have concentrations of arsenic above 10 μg/l (Peters *et al.* 1999; USGS, 2003). High-arsenic drinking-water (>10 μg/l) from these wells affects at least 41 000 people (see Table 2.21; Figure 2.11).

Table 2.21 Arsenic concentration in wells in four counties in New Hampshire, USA.

County	Estimated population, 2006	No. samples	Arsenic concentration (mg/l)			% of wells with arsenic concentration (μg/l):		
			Minimum	Median	Maximum	>5	>10	>50
Hillsborough	402 789	158	<0.001	0.002	0.075	32	21	3
Rockingham	296 267	125	<0.001	0.001	0.215	26	14	2
Strafford	112 233	70	<0.001	0.003	0.090	37	21	1
Overall	819 046	353	<0.001	0.002	0.215	31	19	2

Source: USGS, 2003; United States Census Bureau, 2006a

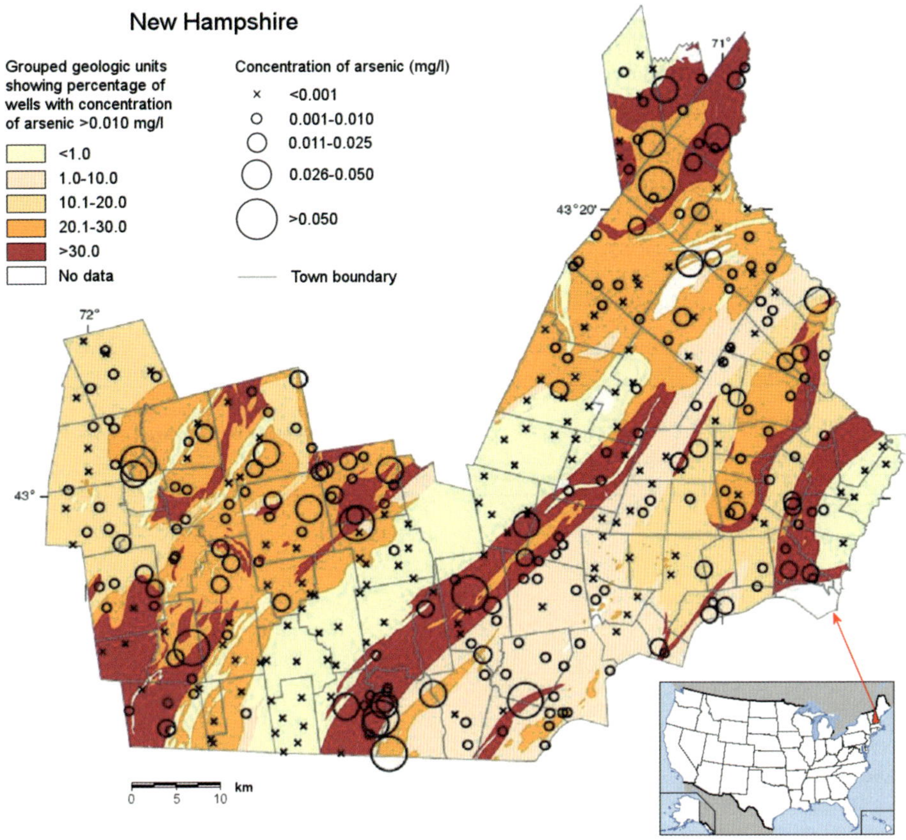

Figure 2.11 High-arsenic occurrences in south-western New Hampshire, USA (modified from USGS, 2003).

These high arsenic concentrations in south-western New Hampshire are mostly geogenic from grandiorite and diorite deposits and are more prevalent in south-western New Hampshire than in other regions of the state (Ayotte *et al.* 1999, 2003; Peters *et al.* 1999). However, it should be noted that the grandiorite and diorite deposits extend to the north-east and may pose threats farther north.

Some occurrences of high arsenic concentrations in groundwater may be related to present or past land use practices, including apple orchards that had used arsenic-based pesticides and fungicides. In fact, apple orchards are extremely susceptible to arsenic contamination, since earlier chemicals used on orchard trees have loaded the soil with high concentrations.

One study has sought to better understand the arsenic cycle for orchards as a possible source of arsenic leaching into surface water and groundwater. Lead arsenate sprays, widely used as fungicides and insecticides in apple orchards for more than a century, represent the largest single anthropogenic input of arsenic into the environment. Wong *et al.* (2002) developed the following scenarios:

> The applied As may (1) have remained in the surface soil, (2) have moved downward in the soil column and become stored in deeper soil horizons and/or regional groundwater system, or (3) have been transported as a result of overland surface runoff and/or erosion to surface reservoirs.

They propose to

> examine these pathways using two types of samples collected from a Southern New Hampshire apple orchard: soil profiles from apple orchards having different pesticide application (sprayed or not sprayed with lead arsenate) and land use (tilled or untilled) history, and stream sediment cores that may have accumulated sediments transported from nearby apple orchards. Preliminary analyses provide the following observations. First, apple orchards which used lead arsenate pesticides contain significantly elevated As and Pb concentrations (up to 80 µg/g and 600 µg/g, and about 1 and 2 orders of magnitude above the background levels, respectively) in the surface soils. Second, As and Pb are generally limited to the upper 10–15 cm of soil, showing little evidence of downward transport. This suggests that As is largely chemically immobile in the soil environment and that the main mechanism for As removal from its source may be physical erosion. We hypothesize that, if left undisturbed, lead arsenate remains immobile in the soil column.

> However, any disturbances that increase physical erosion of the soil may mobilize the arsenic and lead and concentrate these metals in nearby stream and lake sediments. We test this hypothesis by comparing apple orchard sites with different land management history (tilled versus untilled) and by examining stream sediment samples potentially originating from As and Pb contaminated soils.

Note that in Table 2.22, the highest percentage of samples with arsenic concentrations greater than 10 μg/l are grandiorites and granites of the Ayer, Eliot, Kittery, Rangeley and Berwick formations, whereas the next highest are those granites and diorites associated with the Spaulding, Exeter Littleton Concord Two-mica and Perry Mountain formations. Although Table 2.22 provides a broad view of the likelihood of elevated arsenic occurrences in groundwater, one should keep in mind that groundwater from wells in different subdivisions of a geological member can vary markedly.

Table 2.22 Geological units associated with elevated arsenic in groundwater in south-western New Hampshire.

Groups of geologic units Keyed to arsenic map above	Geologic unit	No. of samples	Percent of samples with concentrations of arsenic greater than 0.01 mg/L	Percent of study area underlain by geologic unit
	Greater than 30 percent of samples			
	Ayer Granodiorite	2	50	<1
	Eliot Formation, Calef Member	2	50	<1
	Kittery Formation	11	46	3
	Rangeley Formation, lower part	16	31	4
	Rangeley Formation, upper part	16	31	5
	Berwick Formation, unnamed member	32	31	6
	20.1 to 30 percent of samples			
	Spaulding Tonalite	40	28	10
	Exeter Diorite	11	27	3
	Littleton Formation	4	25	2
	Concord Granite	28	25	7
	Two-mica granite of northern and southeastern New Hampshire	4	24	2
	Perry Mountain Formation	21	24	6
	10.1 to 20 percent of samples			
	Eliot Formation	20	20	8
	Kinsman Granodiorite	28	11	8
	1 to 10 percent of samples			
	Berwick Formation	54	7	16
	Less than 1 percent of samples			
	Smalls Falls Formation, undivided	3	0	1
	Massabesic Gneiss Complex	32	0	10
	Rangeley Formation, upper part, pink to green calc-silicate and purple biotite granofels	1	0	<1
	Madrid Formation, undivided	1	0	<1
	Rangeley Formation, undivided	3	0	<1
	Berwick Formation, Gove Member	3	0	<1
	Rye Complex	4	0	2
	Breakfast Hill Granite of Novotny (1964)	3	0	<1
	Mesoperthitic granite	3	0	1
	Gray biotite granite	11	0	3

Source: from USGS, 2003

New Jersey (A)

Anthropogenic sources of arsenic in New Jersey come in many forms. In a highly visible arsenic study, it has been claimed by the New Jersey Environmental Federation that as many as 700 000 residents are drinking water that is contaminated with dangerously high, unacceptable levels of arsenic (Jacobs, 2000). In fact, the USGS and the New Jersey Geological Survey found significant sources of arsenic contamination in soils at and near a former industrial site in Monmouth County and residual amounts of arsenic from lead arsenate pesticides used in fruit tree orchards before the 1950s. The problem is state-wide, as agricultural land accounts for about 20% of the state, and pesticides, including those containing arsenic, have been detected in groundwater and surface water in these areas (USGS, 1997).

A highly contaminated site in Vineland has been the subject of a Superfund remediation programme. Dendrochemical measurement of arsenic in annular tree rings was tested to determine whether the past history of arsenic concentrations in groundwater can be reconstructed (Cheng *et al.* 2005). The rhizosphere (woody plant tissue) has not normally been used for such a measure due to various constraints, such as lack of adequate sampling methods. Research conducted on several tree species at the Vineland Chemical Superfund site was made possible by its known history of arsenic contamination by arsenical pesticide production from 1950 to 1994. Annular rings of several species were tested (Cheng *et al.* 2005), including

> post oak (*Quercus stellata*), white oak (*Quercus alba*), pitch pine (*Pinus rigida*), Eastern red cedar (*Juniperus virginiana*), and pignut hickory (*Carya glabra*) growing on soil containing 20–1000 mg/kg As.... Results [were] compared to trees of the same species growing in two nearby uncontaminated control areas. High-resolution inductively coupled plasma mass spectrometry (HR ICPMS) provided the analytical sensitivity required to measure the As content of single tree rings....

Given the precise measurements applied in this study covering both deciduous and coniferous species, it is notable that generalized arsenic transport across annual rings was observed and may render it impossible to accurately reconstruct the past history of arsenical contamination from dendrochemical evidence. While this was an important experiment, its results were inconclusive.

New Mexico (A, G, M)

New Mexico has geogenic arsenic in many parts of the state, and groundwater arsenic in these areas is found commonly in the range of 10–50 μg/l. The population in New Mexico exposed to elevated arsenic is large: over 80% (1.56 million) of the total population of 1.95 million (2005 estimate) is served by public water supplies, of which 90%, or 1.4 million people, are potentially at risk for using arsenic-contaminated groundwater (United States Census Bureau, 2006b). Densely populated urban areas are affected, such as Albuquerque, with a 2003 estimated population of 471 856, or 25.2% of the state's population (United States Census Bureau, 2006d), which has average arsenic levels of 10–30 μg/l and maximal levels above 50 μg/l in drinking-water. Tiny communities may be even more severely impacted, such as San Ysidro, with a 100% Hispanic population of 238 (United States Census Bureau, 2006c), where drinking-water arsenic levels are as high as 200 μg/l. The New Mexico Environmental and Health Effects Tracking Program was instituted in 2004 to monitor these and other health risks, and a Drinking Water Advisory Group has scheduled meetings around the state to involve the public and water systems officials (Flowers, 2004).

The hazards of ingesting arsenic-contaminated fish were investigated for the Rio Grande from 1994 to 1996, and the extent of arsenic in fish tissue was reviewed, especially for the Middle Rio Grande (Wilcox, 1997):

> Eighteen surface-water sampling sites on a reach of the Rio Grande from the Pueblo of San Felipe to Los Lunas, New Mexico, were sampled quarterly from October 1994 to August 1996. The sites include eight Rio Grande sites, one Jemez River site, five riverside drain sites, and four wastewater-treatment plant outfalls. Trace-element protocol was used to collect and process the samples. Field and laboratory quality-control samples were analyzed, and the results are included in this report. Fish-tissue samples were collected from four of the Rio Grande sites and the Albuquerque Riverside Drain, the Atrisco Riverside Drain, and three lakes at a recreational fishing area on the Isleta Indian Reservation.
>
> Arsenic in the Rio Grande is nearly all in the dissolved phase. There was little temporal change in arsenic concentration at the Rio Grande sites [from previous testing]. The mean dissolved-arsenic concentration in the Rio Grande increased downstream from 1.8 micrograms per liter at the Pueblo of San Felipe to 3.6 micrograms per liter at Los Lunas. Mean dissolved-arsenic concentrations in the riverside drains were slightly higher (2.8 to 4.5 micrograms per liter) than those in the Rio Grande and were higher still in the wastewater-treatment plant outfalls (7.9 to 16.2 micrograms per liter) and the Jemez River (18.2 micrograms per liter). The mean total-arsenic concentration in fish-tissue samples from the Rio Grande and Albuquerque Riverside Drain was 14.53 micrograms per kilogram.

The contamination of agricultural soils either by pesticides or by arsenic-rich irrigation water also poses a threat to consumers of vegetables grown in arsenic-rich soils. Nemmers *et al.* (2006) found that arsenic accumulated within the soil profile in New Mexico soils irrigated with water containing arsenic at an amount greater than 80% of the arsenic in the irrigation water.

Mining of uranium in two major districts has created arsenic and uranium contamination hazards in New Mexico, with more than 200 mines in 18 counties in operation from 1948 to 2002. Grants uranium district between Gallup and Laguna in the San Juan Basin of north-western New Mexico produced 157 900 tonnes of uranium oxide between 1948 and 1978, more than one third of all production in the USA and more than any other district in the world. The uranium ores mined in this zone are in association with mineralized arsenic. These deposits are located mainly in sandstones of the Westwater Canyon and Brushy Basin Members and in the Jackpile Sandstone of the Jurassic Morrison Formation. A smaller uranium deposit is found in the Salt Wash Member of the Morrison Formation in the Shiprock district, the site of the Shiprock Navajo Reservation, where the host rocks were laid down about 150 million years ago in ancient stream channels and playa formations (Bland & Scholle, 2007). Navajo residents were employed as shallow-pit miners from 1948 to the 1990s, when 40 mines were abandoned. Current testing of drinking-water by a team from Northwestern University has determined levels of arsenic in 63 reservation wells; 4 wells had water containing arsenic at concentrations above 10 µg/l: 19, 22, 82 and 104 µg/l (Hoover *et al.* 2004). Because of the potential increase in the utilization of uranium for power generation, the monitoring of uranium extraction will remain an important component of environmental concern.

Another illustrative instance of arsenic contamination is the historic mining of metals and metalloids. Lead–arsenic carbonate tailings from the derelict Stephenson-Bennett silver mine, 17 km north-east of Las Cruces, had leached down an arroyo into a residential trailer home area, where elevated levels of lead (32 200 mg/l) and arsenic (860 mg/l) in the soil and elevated levels of lead (7647 mg/l) and arsenic (310 mg/l) in indoor dust were measured by the USEPA in 1997. These tailings had also leached into the groundwater, but the concentrations were not measured, apparently. Soil remediation began in 1998, and acceptable levels of 500 mg/l and 20 mg/l were established for lead and arsenic, respectively. Several vacant lots in the same residential area as the trailer homes were not remediated, because the USEPA was denied access to the lots by the property owners. The homes where the indoor dust was found to be contaminated with lead and arsenic were remediated in November–December 1998 after soil removal activities had been completed. Contaminated soil under the homes was not removed because it was claimed that this would have compromised the structural integrity of the homes. Although the soil under the mobile homes and on the vacant lots continues to be unremediated, the final assessments by the United States Department of Health and Human Services (1998) are (respectively) “No apparent health hazard” and “Indeterminate health hazard”.

Volcanogenic sources of arsenic in New Mexico are mainly related to the Valles Caldera in the southern Jemez Mountains, in which waters can be classified into five general types: (1) surficial meteoric waters; (2) acid sulfate waters at Sulphur Springs, Valles Caldera; (3) thermal meteoric waters in the ring fracture zone in Valles Caldera; (4) deep geothermal waters of the baca geothermal field; and (5) mineralized waters near San Ysidro. The Valles Caldera thermal and acid waters carry high arsenic levels, with a maximum concentration of 1100 µg/l (Goff & Shevenell, 1987).

Where the Jemez River joins the Rio Grande near Bernalillo, water contains 30–80 µg arsenic/l, resulting from volcanogenic sources in the Valles Caldera. Arsenic in the Rio Grande is compounded anthropogenically by the discharge of the Bernalillo, Rio Rancho and Albuquerque wastewater treatment plants, the effluents of which contain between 7 and 38 µg/l. When the Rio Grande flows past the Mogollon-Datil volcanic field from Bernardo to below Truth or Consequences, arsenic in the groundwater is elevated to about 47 µg/l (Dunbar *et al.* 2002).

New York (A)

Mortefolio & Ridenour (2006) investigated a site in upstate New York that had been previously used for agriculture, where a number of arsenic-based chemicals had been used as pesticides and fungicides from the 1930s to the 1970s:

> During that period arsenic released through air emissions was deposited on soil in areas surrounding the site, and arsenic was also discharged into a nearby stream which contaminated sediment and floodplain soils. The land surrounding the site has historically been used for residential and agricultural purposes (including orchards).

They provide a model of how historic arsenic pesticide usage in area orchards can be determined and discuss how land use categories were established based on suspected degrees of historic arsenic usage and the development of a soil sampling strategy based on the relative size of each land use type in the area surrounding the site. The value in the study is that it can be replicated at many other orchards and agricultural sites in the state.

Parisio *et al.* (2005) investigated the effects of redox conditions created by landfills on the release of naturally occurring arsenic:

> Iron flocculate or “floc” deposits are commonly observed in groundwater discharge zones such as seeps, springs or stream margins down gradient of unlined solid waste landfills. Typically, these deposits are bright orange and may manifest as a surface coating on soil or sediment particles or as suspended particles loosely adhering to submerged rocks, plant stems or vegetative debris. Composed predominantly of amorphous iron oxy-hydroxides, these deposits generally have been regarded as aesthetically unpleasant, but environmentally benign. In recent years, there has been increased awareness of the widespread occurrence of elevated arsenic in sediments and groundwater. Research carried out at a municipal landfill in Maine indicates that naturally occurring arsenic exhibits

redox-sensitive mobility and may be associated with iron as a dissolved constituent in leachate impacted (reduced) groundwater. If iron precipitates in discharge zones where reduced ground waters are exposed to atmospheric oxygen, it follows that arsenic may co-precipitate with iron in these areas. To assess the prevalence of this problem within the lower Hudson Valley of south-eastern NY, staff from the regional office of the New York State Department of Environmental Conservation (NYSDEC) and researchers from Columbia University's Lamont-Doherty Earth Observatory sampled iron floc deposits at seven inactive solid waste landfills, and at one control site where iron floc deposition is derived from a natural spring. At six of seven landfill sites, arsenic concentrations exceeded the NYSDEC sediment guidance value of 30 mg/kg (severe effects level for aquatic life) at one or more sampling points. These results indicate that arsenic contamination is a potential concern down gradient of landfills wherever iron-stained leachate discharges are observed. Sampling and analysis of iron flocs associated with such leachates could provide a means of identifying landfills which may present risks to down gradient water supply wells, especially in cases where groundwater monitoring wells are not available for sampling.

Hydraulic Fracturing and Arsenic Release

Marcellus Shale is an extensive formation of shale (a type of sedimentary rock that is high in carbon) in New York, Pennsylvania, Ohio, West Virginia, and other states in the region (Catskill Mountainkeeper, 2009). It underlies about 18,700 square miles in southern New York, including New York City's entire 1,585-square-mile watershed west of the Hudson River (Esch, 2009). This shale has received renewed attention because of new estimates of the quantity of methane gas believed to be sequestered by the formation (Esch, 2008).

The method of extracting methane from Marcellus shale is called hydraulic fracturing or hydrofracking, a process first used experimentally in 1947 at the Hugoton gas field, Kansas, and first used commercially in 1949. Chemicals used in hydrofracking including hydrochloric acid, diesel fuel (benzene, tuolene, and xylene) as well as formaldehyde, polyacrylimides, arsenic and chromates, which are pumped into a wellbore (Anon, 2009). These fluids then seep into the surrounding rock. No comprehensive list of hydraulic fracturing chemicals exists because the information is propietary. In a letter to Congress, an Environmental Protection Agency employee describes how, during the G.W. Bush Administration, the EPA produced a scientifically insupportable conclusion that hydrofracking should not be regulated under the Safe Water Drinking Act because some of these chemicals include known carcinogens and other hazardous substances (Wilson, 2004).

After pumping hydraulic fracturing fluid into the wellbore, a proppant is injected to maintain open conduits so that more fluids can be injected into the shale to expand and extend the fractures. Proppants are sized particles including sand grains, resin-coated sand and high-strength ceramics such as sintered bauxite, which, because of their size and sphericity, hold fractures open after hydraulic fracturing in order to drain hydrocarbons from the gas field. In the US as of 2009, 90% of 493 000 active natural-gas wells across 31 states have been hydraulically fractured (Britt & Jones, 2009; Urbina, 2011).

Nationwide, residents living near hydrofracked gas wells have filed over 1,000 complaints regarding tainted water, severe illnesses, livestock deaths and fish kills (Guynup, 2012). We present only a selected example of the problems faced when this method of gas extraction is applied to a geological watershed that supplies drinking water to 17 million people who could be at risk if no scientifically verifiable measures are put in place to insure there is little or no risk of contaminating their water source from arsenic and other toxic chemicals.

The case of the Upper Delaware River watershed between New York and Pennsylvania

A regional issue has been raised in the Upper Delaware River watershed between New York and Pennsylvania which provides drinking water to over 17 million people. Two companies, Chesapeake Appalachia and Statoil, have a stated objective of developing 13 500 to 17 000 gas wells in the region over the next twenty years (American Rivers, 2010). However, there appears to be little oversight of hydraulic fracturing in the region. "H.R. 1084: Fracturing Responsibility and Awareness of Chemicals Act of 2011" was sponsored by Rep. Diana DeGette [D-CO] to the 112th Congress, 2011–2012 "To repeal the exemption for hydraulic fracturing in the Safe Drinking Water Act, and for other purposes," but it has been referred to Committee as of 1 April 2012 and is unlikely to pass (see: www.govtrack.us/congress/bills/112/hr1084).

However, some data suggesting arsenic is a potential contaminant of this watershed has already been released. On November 9, 2009, Reuters reported that the owner of 480 acres of land in southwest Pennsylvania claimed Atlas Energy Inc. ruined his land with toxic chemicals used in or released there by hydraulic fracturing, and he also claimed to find seven potentially carcinogenic chemicals above permissible levels set by the US Environmental Protection Agency. He performed tests on his well water a year before drilling began and said the water conditions were "perfect." After the drilling began, water tests found arsenic at 2,600 times acceptable levels, benzene 44 times above limits and naphthalene five times the federal standard. He has decided to sue Atlas Energy Inc. for negligence and is seeking an injunction against further drilling, and unspecified financial damages. Jay Hammond, general counsel for Atlas, said Zimmermann's claims are "completely erroneous" and said Atlas will "vigorously" defend itself in court and declined further comment (Huddle, 2009).

The role of arsenic, among other toxic compounds, in hydrofracking is not limited only to the example provided above for the Upper Deleware River watershed in New York and Pennsylvania. Comparable battles are occurring in a number of other US states and in other countries that are engaging in hydrofracking as a means to procure natural gas. Arsenic contamination of groundwater and surface water supplies could potentially be occurring in many other states and countries; however, we don't have the data to support these claims. Therefore, what can be said at this time is that more research and more transparent data sharing is necessary for this issue to be publicly addressed and to avoid preventable public health consequences of contamination of aquifers with toxic chemicals, which possibly includes arsenic.

North Carolina (G)

The Rock in the Lake Tillery area contains small amounts of gold; pyrite is ubiquitous, and arsenopyrite, a chief host of arsenic, was reported in an interval where pyrite is concentrated along a fracture in quartz veins from 72 to 93 m. The water associated with stream sediments shows a pronounced low pH that follows the geological contact between the stratigraphically underlying metarhyolites and the stratigraphically higher metavolcanics. The very low pH groundwater is associated with pyrite along this contact (Reid & Wooton, 2002).

A study of arsenic concentrations in streams of the Piedmont (Figure 2.12) provided additional support for the leaching of arsenic into surficial waters as well as ground waters. The importance of these findings is that arsenic does exist in streams of the Piedmont region and that more needs to be understood about the bioaccumulation of arsenic in species, including birds and fish that are taken as game food (Reid & Carpenter, 1993). Reid *et al.* (2007) found that arsenic-contaminated groundwater was a potential health risk to users of domestic water supply wells in the eastern slate belt of North Carolina:

> Integrated geological and geochemical investigations have determined that ambient arsenic is present in groundwater in two discrete geologic settings: molybdenum porphyry and volcanogenic-gold deposits. The presence of arsenic was determined by analyzing precipitates from first, second and third order streams under base flow conditions. When groundwater discharges into surface water, arsenic and other metals are precipitated out of solution, because of redox changes between the subsurface and surface environments. These oxides are then deposited as coatings on streambed materials.

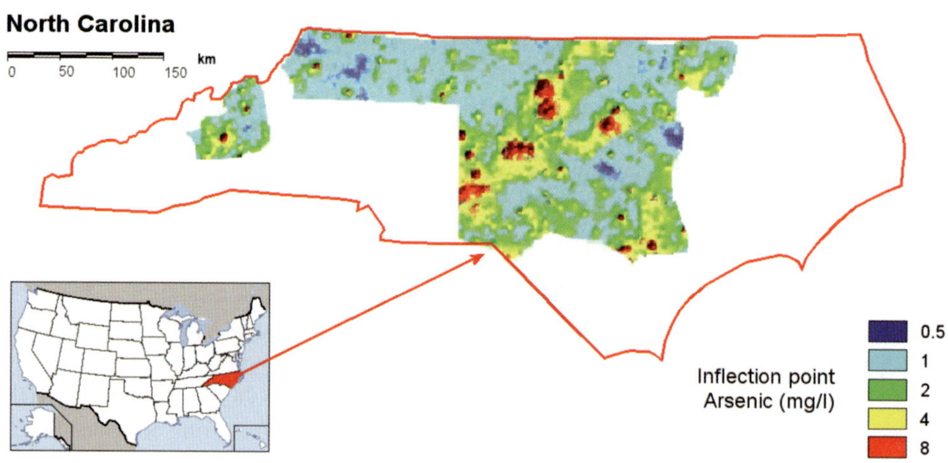

Figure 2.12 Piedmont streams with measured arsenic concentrations in North Carolina, USA (modified from Reid & Carpenter, 1993).

North Dakota (A, C, G)

Anthropogenic arsenic has been found in a south-eastern North Dakota agricultural area where an arsenical pesticide was used to control grasshopper activity (Roberts *et al.* 1985). As grasshopper damage to crops covers a wide area of the state, other similar locations are suspected to exist.

Coal contributes arsenic in the 1-m-thick Palaeocene Beulah lignite beds in the 0.4-m-thick bottom bench at a concentration of 374 mg/kg (compared with 2–22 mg/kg higher in the beds); this stratum also has high sulfur content, which is not linearly proportional to arsenic concentrations (Karner *et al.* 1986).

Geogenic sources of arsenic in the form of arsenic(III) are associated with glacial drift aquifers in Dakota sandstone. The USEPA (2007c) identified arsenic trioxide in a glacial drift zone in south-eastern North Dakota that covers approximately 20 townships (about 1470 km^2), encompassing portions of Richland, Ransom and Sargent counties:

> The site area is sparsely populated farmland with a few small towns, including Lidgerwood, Wyndmere, Milnor and Hankinson. The site topography is primarily low rolling hills and flat plains. Ground water aquifer systems include shallow glacial drift aquifers approximately three to 150 feet deep and the Dakota Sandstone aquifer approximately 200 to 1,000 feet deep. Ground water with elevated arsenic levels appears to be limited to the upper, unconfined glacial drift aquifers and does not extend into the deeper sandstone unit. The upper aquifer is commonly used as a drinking water source in the region.

Ohio (A, C, G, P)

High-volatile bituminous coals with elevated arsenic cluster in zones parallel to discontinuities (perpendicular to Appalachian folds) that may have facilitated hydrothermal flows (containing arsenic) into the Appalachian Basin during orogenic processes (Pohn, 2001).

Geogenically released arsenic from weathering of arsenical minerals has been identified in the south-western and north-eastern areas of the state.

In the Miami River basin of south-western Ohio (see Figure 2.13), mobilization of arsenic into groundwater has been found in 57 domestic wells in Preble, Miami and Shelby counties (see Table 2.23) by Thomas *et al.* (2005). Ground waters there had a median arsenic concentration of 7.1 μg/l and a maximum of 67.6 μg/l; 37% of samples contained arsenic levels above 10 μg/l.

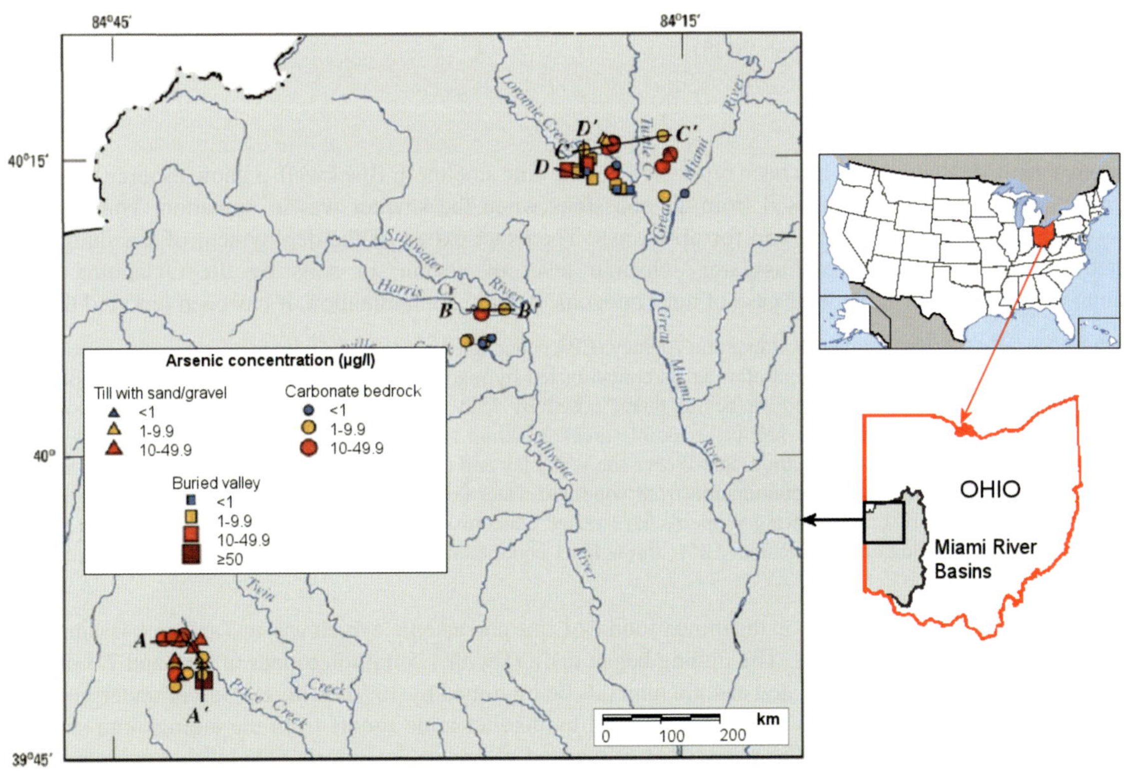

Figure 2.13 Wells with elevated arsenic concentrations in south-western Ohio, USA (modified from Thomas *et al.* 2005).

Depth of wells was no index of the level of contamination, but in all elevated arsenic groundwater samples (see Table 2.23), iron concentrations were above 1000 μg/l, suggesting that arsenic was released from iron oxides under reducing conditions. The study was unable to resolve if the observed differences among aquifer types are related to variations in: (1) arsenic content of the aquifer material, (2) organic carbon content of the aquifer material, (3) mechanisms of arsenic mobilization (or uptake) or (4) rates of arsenic mobilization (or uptake).

Table 2.23 Arsenic concentration in 57 wells tested in South-Western Ohio, USA.

Study site	No. wells sampled	Arsenic concentrations (μg/l)		Percentage of samples exceeding 10 μg/l
		Median	Maximum	
Preble County	21	9.8	67.6	48
Miami County	8	7.7	13.7	12
Shelby County	28	5.4	49.1	36

Source: Thomas *et al.* 2005

Groundwater in north-eastern Ohio in the Sharon conglomerate of Pennsylvanian age is of the calcium–magnesium bicarbonate type, but could become calcium–magnesium chloride/sulfate type in some of the shallow wells where elevated levels of arsenic were identified (Eshete & Chyl, 2003).

There is also a case of petroleum- and anthropogenic-facilitated release of arsenic that has resulted in the contamination of groundwater along Paddys Run and New Haven roads from an Exxon/Mobil petroleum plant and the adjacent Rutgers-Nease Chemical Co. The contamination, consisting of arsenic, petroleum, benzene and various other metals, is moving in three distinct plume vectors. The contamination was first discovered in 1973, but the USEPA began negotiations on a cleanup only in 1986. One problem has been the proper understanding of the release mechanism of arsenic into the groundwater by anthropogenically introduced petroleum-based carbon (Klepal, 2005).

Oklahoma (A, G, M, P)

Anthropogenic arsenic contamination has been reported from the zinc smelter in Blackwell, a known source of lead, arsenic and cadmium. Area soils became contaminated from air emissions when the smelter was in operation. The sludge from these operations was also used as fill material and for driveways. The exact extent of the deployment of the sludge material is of concern, but there is no established regulatory cleanup level of arsenic in soils by the Oklahoma Department of Environmental Quality (2007). In another case of anthropogenic arsenic contamination, it has been reported that:

> From 1996 to 1997, the US Environmental Protection Agency (EPA) conducted a cleanup action on a former zinc smelter operated by Eagle-Picher Mining and Smelting, Inc. and other areas contaminated by materials from this site near Henryetta, Oklahoma. EPA's cleanup focused on removing the immediate health threat posed by lead- and arsenic-contaminated soil transported from the Eagle-Picher site to residential and other highly accessible areas. Cleanup actions on the Eagle-Picher site involved establishing proper drainage on the site, encapsulating the Eagle-Picher site with clay and cover soil, and establishing vegetative cover at the site to protect nearby residents from the recontamination from wind and water erosion of hazardous materials. Since completion of the cleanup, private landowners of a neighboring property have raised concerns about contamination of their property resulting from EPA's cleanup actions. Landowners allege that EPA, through its contractors, transported and negligently disposed of hazardous substances on their property.

The Blackwell smelter was dependent on the mined source of zinc and arsenic, which was in Tar Creek Basin in north-eastern Oklahoma, near Peoria in Ottawa County. The mining began in 1891 with 1.5 million tonnes of zinc and 7.3 million tonnes of lead, with an unknown quantity of associated arsenic minerals being mined by 1970. With 480 km of underground tunnels and 59 million tonnes of tailings, this area is seriously impacted by derelict mine runoff from the tailings into surface water and groundwater (USGS, 2001). In view of a failed 1993 Superfund effort, a Governor's Council has initiated the process of addressing these contamination issues, which impact Native American populations, among others (Oklahoma Office of the Secretary of the Environment, 2000).

Geogenic arsenic concentrations above 50 μg/l are found in the central Oklahoma aquifer, which encompasses about 8000 km^2. The aquifer is composed of Quaternary alluvium and terrace deposits and underlying Permian Hennessey sandstone and mudstone. Although the waters in the alluvium and terrace deposits generally do not contain elevated arsenic levels, the Permian units (Garber Sandstone, Wellington Formation, Chase, Council Grove and Admire Groups), which are coloured yellow-brown by iron oxides (i.e. goethite), are strongly associated with arsenic up to 62 μg/g. The desorption of arsenic(V) is facilitated by increasing pH, which explains the strong correlation between dissolved arsenic and pH and the fact that levels of arsenic above 50 μg/l are associated with water at or above pH 8.5 (Schlottman & Breit, 1992; Norvell, 1995; Welch *et al.* 2000).

The central Oklahoma aquifer serves 37% of mostly public supply wells more than 90 m deep, and these wells contain water with arsenic levels above 10 μg/l. Additionally, another 13% of historical public supply wells have drinking-water arsenic concentrations above 10 μg/l, so the two combined data sets indicate that 50% of all drinking-water from public supply wells in central Oklahoma have arsenic concentrations above 10 μg/l (Schlottman, 2005). Oklahoma City and Norman lie within the boundary of this aquifer and have a combined population of over 600 000 (United States Census Bureau, 2006e). If only 25% have access to public supply well water, 150 000 are at risk.

In addition, petroleum-facilitated arsenic release is suspected to occur in the areas of oil well drilling in the central part of Oklahoma and the Panhandle (comparable with Xingjiang Autonomous Region, China).

Oregon (A, G)

Lingering anthropogenic sources of arsenic may stem from the policy of the state board of horticulture that "encouraged orchardists to use a variety of arsenic and copper solutions, lead arsenate, and lime sulphur to prevent damage from coddling moths, aphids, pear fruit worms, red spider mites, and fungal canker diseases" in the early 20th century, even though it was recognized by the 1920s that arsenic-based chemicals were injurious to human health (Tucker, 2002). Oregon is very proactive in remediating these situations.

Geogenic sources of arsenic are related to formations of the Oligocene and Upper Eocene in the Willamette Basin in the upper north-west of the state. Data on arsenic in Willamette Basin were collected from 728 discrete sites by Hinkle & Polette (1999), who provided the following preliminary assessment:

> Regionally, the distribution of arsenic concentrations in ground water of the Willamette Basin appears to be primarily related to aquifer geology. High arsenic concentrations (concentrations exceeding the USEPA current MCL [maximum contaminant level]) are widespread in bedrock areas in south-central and eastern Lane County, and Linn County. High concentrations of arsenic also are present in some ground water in the Tualatin Basin (a sub-basin in the north-western part of the Willamette Basin). High arsenic concentrations in Lane and Linn Counties appear to be associated with two regionally extensive associations of rocks, (1) the Fisher and Eugene Formations and correlative rocks, and (2) the undifferentiated tuffaceous sedimentary rocks, tuffs, and basalt .... The undifferentiated tuffaceous sedimentary rocks, tuffs, and basalt are approximately equivalent to the Little Butte Volcanic Series.... At land surface, these two rock associations cover 24 percent of the Willamette Basin. These associations of rocks include extensive volumes of silicic (rhyolitic) volcanic rocks, which are commonly associated with high concentrations of arsenic. High concentrations in the Tualatin Basin are associated with alluvial deposits. At a regional scale, well depth does not appear to be a useful predictor of arsenic concentration in the Willamette Basin. However, depth may be an important parameter on a local scale, particularly where wells of different depth tap aquifers in different geologic units.
>
> For alluvial ground water of the Tualatin Basin, presence of competing anions and occurrence of reducing conditions may be important controlling factors in arsenic adsorption and desorption reactions. These two factors might be more important than pH controls over arsenic adsorption and desorption. Reducing conditions and high concentrations of dissolved iron also suggest that dissolution of iron oxides, with subsequent release of adsorbed and (or) coprecipitated arsenic, may play a role in arsenic mobility in the Tualatin Basin.

More testing is ongoing as other areas are reporting elevated levels of arsenic in drinking-water, such as Tualatin Basin near Portland; the Ontario region in Eastern Oregon; areas around Creswell and Cottage Grove south of Eugene; and much of Puget Sound (Andrews, 2007).

Pennsylvania (A, C, G)

Anthropogenic sources of arsenic contamination are associated with fly ash, presumably from a power generation facility. On 2 February 2005, the following press release (Environmental Integrity Project, 2005) indicated that thousands of tonnes from a fly ash hill overlooking a section of Forward Township tumbled down, sweeping through the neighbourhood and then into the Perry Mills Run, which feeds directly into the Monongahela River:

> A sample of ... fly ash spilled in Forward Township from an adjacent hillside reveals elevated levels of arsenic, as well as the presence of lead, cadmium, mercury and other contaminants. The sample was analyzed by GLA Laboratories in King of Prussia, Pennsylvania; the laboratory is state-certified. Forward Township is twenty miles south of Pittsburgh. The sample measured arsenic at 170 parts per million, or more than fourteen times the level of cleanup recommended for arsenic contaminated soils in Pennsylvania.... The high levels of arsenic in the sample are only part of the concern. Because the site has not been stabilized or contained, the fly ash and leachate continues to flow into a creek that leads to the Monongahela River. Residents report that in portions of the creek, fly ash is

> nearly five feet deep, and that the neighborhood is threatened by an additional landslide from fly ash remaining on the nearby hillside. Meanwhile, lacking any guidance from the Pennsylvania Department of Environmental Protection, the township has apparently removed some of the fly ash and deposited it on a nearby playground.

The disposition of this problem is unknown.

High-volatile bituminous coals with elevated arsenic cluster in zones parallel to discontinuities (perpendicular to Appalachian folds) in western Pennsylvania may have facilitated hydrothermal flows (containing arsenic) into the Appalachian Basin during orogenic processes (Pohn, 2001). In Pennsylvania, arsenic in pyrites and marcasites occur in nodules in coal of the Upper Freeport seam (up to 19 000 mg arsenic/kg) as wood pseudomorphs, in the Lower Kittanning seam (15–3420 mg arsenic/kg) and in the Pittsburgh seam no. 8 (227 mg arsenic/kg) (Ruppert *et al.* 1992; Hickmott & Baldridge, 1995; Palmer & Lyons, 1996). In the case of the lower Kittanning seam, a positive correlation of arsenic and salinity was observed using a modified particle-induced X-ray emission analysis.

Geogenic sources of arsenic contamination of groundwater are a problem in Bucks County, with drinking-water containing arsenic at concentrations from >10 to 40 μg/l in Hilltown, Perkasie, Quakertown, Richland, Sellersville, Solebury and Telford and at >40 μg/l in Palisades and Sellersville. Schools also have contamination problems, with arsenic concentrations of 20 μg/l in drinking-water at Palisades High School, 27 μg/l at Durham-Nockamixon Elementary School and 40 μg/l at Palisades Middle School (Associated Press, 2006).

In 2000, the USEPA found south-eastern Pennsylvania groundwater concentrations of arsenic and fluoride above 10 μg/l and 4 mg/l, respectively, and of boron above 600 μg/l along the north-western edge of the Newark Basin. In 2003, the USGS sought to identify sources of arsenic, boron and fluoride in the groundwater in the Congo Road area, which included possible anthropogenic releases and naturally occurring mineralization in the local bedrock aquifer. The Newark Basin is the largest of 13 exposed Mesozoic rift basins that stretch from Nova Scotia to South Carolina and includes Triassic- through Jurassic-age sedimentary sequences of sandstones and shales intruded by diabase with hydrothermal mineral deposits that include arsenic in alteration zones and also occur as replacement deposits of copper and zinc in sedimentary rocks. Groundwater from about 10% of wells throughout the Newark Basin of south-eastern Pennsylvania contained arsenic concentrations above 10 μg/l, with a maximum concentration of 70 μg/l (Senior & Sloto, 2006).

A project at Lehigh University's Earth Observatory has sought to address these and other problems (Peters, 2007):

> Arsenic exposure and health effects for the 31.2 million people that live in Pennsylvania are currently unknown. Recent graduate student research has located some existing datasets of arsenic samples and produced a GIS suggesting a connection between geology and arsenic occurrence. Some samples with arsenic concentrations above the EPA regulatory limit located in south-eastern Pennsylvania suggest that arsenic may be a problem in this area, and merits further investigation.

South Carolina (A, G)

Elevated arsenic is found in the coastal plain area with an average concentration of 2 mg/kg; in the Piedmont region, the average concentration of arsenic in soils is 11 mg/kg. It is thought that the lower values on the coast reflect weathering processes (Canova, 1999; South Carolina Department of Health and Environmental Control, 2003). It is unknown whether the occurrence of arsenic on the coastal plain is attributable to the geomorphology of Pleistocene barrier island development in which a transport mechanism carries arsenic along the bedrock, facilitating invasion of groundwater wells (cf. Hails & Hoyt, 1968).

Areas with elevated arsenic levels that exceed the South Carolina residential arsenic soil limits and the USEPA 503 residential regulations were identified in a study of 50 sites in all land resource and use categories. Elevated arsenic concentrations were found in soils, exceeding 30% of the USEPA 503 regulations that stipulate arsenic concentrations of ~0.4 to ~40 mg/kg, corresponding to a 1 in 1 million to a 1 in 10 000 incremental cancer risk range via ingestion of or direct contact with contaminated residential soils (Franklin *et al.* 2003).

Cancer is not the only risk from exposure to elevated levels of arsenic. An epidemiological study found significant clusters of developmental delay and mental retardation in children born in South Carolina that appeared to parallel elevated arsenic in soils. Although it is difficult to correlate one factor with developmental delay and mental retardation, exposure of fetuses to heavy metals and metalloids can induce developmental delay and mental retardation, because maternal blood bioaccumulates heavy metals and metalloids (Caldéron *et al.* 2003). This is also true of mammals in general. Elsewhere, a strong association has been found between lower intelligence quotient scores in children and arsenic concentrations in drinking-water after adjusting for social and economic variables (Wasserman *et al.* 2004).

Soil samples were taken from the top 5 cm in an elevated developmental delay and mental retardation incidence area and an unaffected control area. In order to prevent the location of the collected soil from being associated with the specific location of a developmental delay and mental retardation incident, 50 randomized samples were taken within each of the two areas. A subset of soil samples was then analysed by inductively coupled plasma atomic emission spectroscopy for arsenic, lead and chromium (USEPA method 6010). Results indicated that the percentage of moisture was significantly correlated with relative toxicity units, because high moisture soils have higher sorptive capacities for heavy metals and metalloids (Phillips, 1999). The highest arsenic concentration measured in the soils was 30 times higher than the South Carolina residential soil limit. The measured range was 0.64–12 mg arsenic/kg, compared with the residential soil preliminary remediation goal of 0.39 mg arsenic/kg and the industrial preliminary remediation goal of 1.6 mg arsenic/kg (Aelion & Davis, 2006). The source of the arsenic is not known, so it may be inferred that it is either anthropogenic or geogenic or both.

In another unnamed site in South Carolina, remediation of arsenic contamination was achieved by injecting a 50% slurry of a commercial blend of treatment materials into an aquifer with low permeability at rates up to 95 litres/min; visual impact was observed more than 6 m from the injection point. The pH increased from highly acidic to near equilibrium, and arsenic concentrations dropped to regulatory levels (Haselow *et al.* 2003). The source of the arsenic is not known, so it may be inferred that it is either anthropogenic or geogenic or both.

South Dakota (A, C, G, M)

Coal-related sources of arsenic are found in uranium and molybdenum lignites measuring 3.2% arsenic from ash (King, 1956), with an average arsenic concentration of 1600 mg/kg (Piperingos, 1966). Molybdenum, uranium and arsenic were mobilized in a low-temperature, highly oxidized groundwater flow, similar to the Vermillion Creek bed in Wyoming, from tuffaceous sands of the roof of the coal strata (Hatch, 1987).

Geogenic sources of high-arsenic groundwater in South Dakota are related partly to felsic volcanic rocks, where the groundwater is commonly alkaline (Welch *et al.* 2000). In the extreme south-central region of the state, two geological units account for elevated arsenic in groundwater: Quaternary Sand and Gravel and the Tertiary Oglalla Group; these units are also responsible for elevated arsenic in groundwater in Nebraska to the south (Gosselin *et al.* 2004).

Another geogenic source of arsenic has been reported in groundwater in the Grass Mountain area of the Rosebud Indian Reservation (Carter *et al.* 1997):

> Dissolved arsenic in groundwater was significantly positively correlated with dissolved vanadium and molybdenum, two of the elements commonly associated with naturally occurring arsenic in rocks. Additionally, elevated uranium concentrations were detected in two of the groundwater samples that had high arsenic concentrations. Because the only probable local source of vanadium, molybdenum, and uranium is natural and because of these elements' association with naturally occurring arsenic in rocks, this supports the conclusion that the source of arsenic is natural.
>
> Both dissolved arsenic (As^{+3} plus As^{+5}) and arsenite (As^{+3}) were significantly negatively correlated with calculated oxidation–reduction potentials (Eh) and dissolved oxygen. The decrease in dissolved oxygen concentrations, increase in dissolved iron and manganese concentrations, and relations between concentrations of dissolved solids, soluble ions, and calculated oxidation–reduction potentials in the ground-water samples provide evidence that ground-water conditions become less oxidizing, accompanied by increased dissolution of rock constituents in the more down gradient wells. Arsenic concentrations were found to increase with pH, which indicates greater solubility at high pH and possibly indicates desorption from solid adsorbent surfaces as pH increases. The pH conditions probably control the adsorption of arsenic, which is present in its highest concentration locally in volcanic ash in the Arikaree Formation.
>
> High pH (above 8.0) appears to allow the release of arsenic, uranium, vanadium, selenium and molybdenum from the volcanic ash and/or aquifer sediments. These trace elements from the ash then are transported in solution down gradient, and the redox conditions control the valence state of dissolved arsenic. Farther down gradient where the Arikaree Formation is in contact with alluvium, the oxidation of organic carbonaceous materials may react with dissolved oxygen, resulting in a more reduced groundwater. Decreases in pH and Eh probably reach a point where trace elements are converted to a more reduced species and ultimately may be precipitated or adsorbed from solution….
>
> Elevated arsenic concentrations presently do not appear to be a problem in local surface water because arsenic was not elevated in the dissolved phase, suspended phase, or bed-sediment samples collected from the Little White River and its tributaries nearby. The elevated groundwater arsenic concentrations appear to occur in less than a 1- to 2-mile radius around Grass Mountain Community. However, it is possible that elevated arsenic concentrations could be present in the groundwater in other parts of Todd County.

Mining of gold, associated with arsenopyrite deposits, is also responsible for high arsenic levels in groundwater. In gold mining areas, surface waters are generally acidic, which, in turn, facilitates the mobilization of arsenic. In a seminal study of arsenic occurrences in surface waters and ground waters associated with gold mines in the western USA, the results shown in Table 2.24 were obtained for South Dakota.

Table 2.24 Arsenic concentrations in surface water and groundwater at gold mining projects in South Dakota, USA.

Location	Maximal arsenic concentration in surface water (mg/l)	pH	Maximal arsenic concentration in groundwater (mg/l)	pH
South Dakota	0.446	4.6	0.263	7.5

Source: Straskraba & Moran, 1990

In addition, Whitewood Creek in the Black Hills has been subject to both anthropogenic and mining wastes containing arsenic. The creek is fed by perennial melt water into which arsenic has leached from gold mine tailings, and anthropogenic arsenic has been emitted by sewage treatment plants and power generation cooling plants. Although gold mining began in 1876 and lasted to 1977, the tailings are still leaching along a 57-km length of the creek banks. Arsenic concentrations range between <0.1 μmol/l upstream to 1.5 μmol/l downstream (Goddard, 1988; Kuwabara *et al.* 2003).

Additional anthropogenic arsenic has been found in the Grass Mountain area aquifers of the Rosebud Sioux lands. Arsenic concentrations were substantially higher than 50 μg/l. Also, pesticides were found in some samples. Knowledge of the extent and thickness of the aquifers and their relationships to each other is needed to determine the extent, sources and direction of movement of the plumes of contamination (Heakin, 1997).

Tennessee (C, G, M)

High-volatile bituminous coals with elevated arsenic cluster in zones parallel to discontinuities (perpendicular to Appalachian folds) in central Tennessee that may have facilitated hydrothermal flows (containing arsenic) into the Appalachian Basin during orogenic processes (Pohn, 2001). The presence of coal beds underlying aquifers in this area may contribute to the elevated levels of arsenic along with dark shales in stratigraphically higher and highly eroded Whitwell shale. Sandy loams above these layers contain seasonal aquifers, which were used as drinking-water sources for the region until 1960. Arsenic concentrations in these seasonal aquifers ranged from 20 to 95 μg/l, which resulted in the local population's exposure to high levels of arsenic until 1960 (11.1–35.1 μg/l) (Knoll *et al.* 2005).

Mining sources of elevated arsenic were found in acid mine drainage from Kingston, Poplar Creek and Rogers uranium quarries, with arsenic concentrations above 50 μg/l; additionally, arsenic is associated with the releases of hazardous waste from radioactive ore contaminants into the Clinch River/Watts Bar Reservoir system (Ford *et al.* 1995; Knoll *et al.* 2005).

Naturally occurring sources of arsenic contamination of groundwater are associated with the southern Cumberland Plateau in the vicinity of Sewanee, Tennessee. This plateau is capped by Pennsylvanian Warren Point Sandstone and Sewanee Conglomerate, in which arsenite is interbedded with shales and coals and haematite bands.

Texas (A, C, G, P)

Arsenic-enriched coals in the San Miguel lignite deposit in McMullen County are thought to have been deposited in a non-deltaic freshwater swamp behind a barrier island (Gennett & Raymond, 1986) and are associated with low-density lignites with arsenic concentrations up to 30 mg/kg. There appears to be no sulfide correlation, as arsenic concentrations in sulfide-rich heavier fractions are only 2–4 mg/kg. In this case, organic arsenic dominates (Clark *et al.* 1980).

A regional approach to the massive elevated arsenic concentrations in groundwater of the Southern High Plains in Texas is being undertaken by Scanlon (2006). Initial results indicate:

> Contamination is focused in the Southern High Plains (32% of wells exceed the MCL [maximum contaminant level]) and the southwestern Gulf Coast (29% of wells exceed the MCL). The Southern High Plains (SHP) region was subdivided into two areas: a northern area (SHP-N) characterized by low total dissolved solids (TDS <500 mg/L) and a southern area (SHP-S) characterized by high TDS (>500 mg/L). Arsenic contamination is much greater in the SHP-S region (51% of wells >10 μg/L; 2% >50 μg/L) than in the SHP-N region (7% of wells >10 μg/L). Regional analyses of groundwater arsenic concentrations do not support a surficial source of arsenic contamination. Arsenic concentrations are not correlated with land use, with cotton production, with distance from cotton gins, or with nitrate concentrations. Results of drilling and sampling 18 boreholes in the Southern High Plains indicate that the distribution of arsenic is not related to the distribution of cotton production. High arsenic concentrations in a rangeland profile (peak 77 μg/kg) indicate that background levels of water-soluble arsenic are high in soils. Arsenic levels in cultivated areas are variable. Some profiles have highest arsenic levels near the surface, which are correlated with nitrate and phosphate that may suggest a fertilizer or arsenical pesticide source. These data indicate that arsenic related to arsenical pesticides is probably restricted to the near-surface zone. Other profiles have peak concentrations in the middle of the profile or at depth. It is unlikely that arsenical pesticides associated with cotton production would have reached the water table. The unsaturated zone data indicate a widespread

source of water-soluble arsenic in soils in the Southern High Plains that may contribute to groundwater arsenic contamination. Groundwater arsenic contamination occurs in generally oxidizing conditions in the High Plains. Correlations between arsenic and other constituents (vanadium, r^2 0.65; fluoride r^2 0.30; molybdenum r^2 0.18; boron r^2 0.17; selenium r^2 0.14) suggest a geologic rather than an anthropogenic source. Arsenic concentrations are highest in the Ogallala aquifer and much lower in the Dockum aquifer. Potential sources of arsenic include volcanic ash beds in the Ogallala, black shales in the Cretaceous (Kiamichi Shale), and saline lakes. Additional studies will be required to assess geologic sources, including geophysical logging and stratified sampling.

Petroleum-related arsenic contamination has been detected in the groundwater at the Joint Reserve Base, Naval Air Station, Fort Worth, Texas. The source of the elevated concentrations of arsenic, iron and manganese in groundwater is thought to be mobilization from naturally occurring minerals as a result of reducing conditions produced by petroleum hydrocarbon degradation presumably from leakage spills (Vroblesky, 2001). The groundwater at many more such military sites should be tested.

Anthropogenic sources of arsenic pollution in Texas are related to the production and use of arsenical herbicides. One such site will be highlighted to serve as an example (ATSDR, 1988): the 2-ha Crystal Chemical Company, located on Westpark Drive in Houston, Texas. The company manufactured arsenic-based herbicides from 1968 to 1981, including monosodium and disodium methanearsenic acid and cacodylic acid. The site was contaminated with arsenic, endangering a domestic population within a 1.6-km radius of about 20 000 and a commercial population of 65 000. When the site was abandoned, 2.3 million litres of process wastewater remained in ponds. The flood control channel was contaminated by surface water runoff from the site, and sediment contributes to contamination downstream by being carried in the current. Groundwater is heavily contaminated with arsenic in the shallower zones (well no. 5 exceeding 400 mg/l at a depth of 11 m). Wind erosion of arsenic-contaminated dusts is of concern, since there are high levels of arsenic in the surficial soils. The site has since been remediated.

Utah (G)

Arsenic has been identified in drinking-water and assessed for its role in promoting bladder cancer in a case–control study in Utah (Bates *et al.* 1995). Millard County is a desert area with about 250 people drinking well water with arsenic concentrations of 180–210 µg/l (Southwick *et al.* 1983). Lewis *et al.* (1999) investigated the association of drinking-water arsenic and mortality in a cohort of residents from Millard County. Median drinking-water arsenic concentrations in public and private samples collected and analysed under the auspices of the State of Utah Department of Environmental Quality, Division of Drinking Water, ranged from 14 to 166 µg/l. Standardized mortality ratios (SMRs) were calculated:

> Using residence history and median drinking water arsenic concentration, a matrix for cumulative arsenic exposure was created. Without regard to specific exposure levels, statistically significant findings include increased mortality from hypertensive heart disease [SMR = 2.20; 95% confidence interval (CI), 1.36–3.36], nephritis and nephrosis (SMR = 1.72; CI, 1.13–2.50), and prostate cancer (SMR = 1.45; CI, 1.07–1.91) among cohort males. Among cohort females, statistically significant increased mortality was found for hypertensive heart disease (SMR = 1.73; CI, 1.11–2.58) and for the category of all other heart disease, which includes pulmonary heart disease, pericarditis, and other diseases of the pericardium (SMR = 1.43; CI, 1.11–1.80). SMR analysis by low, medium and high arsenic exposure groups hinted at a dose relationship for prostate cancer. Although the SMRs by exposure category were elevated for hypertensive heart disease for both males and females, the increases were not sequential from low to high groups. Because the relationship between health effects and exposure to drinking water arsenic is not well established in US populations, further evaluation of effects in low-exposure populations is warranted.

Vermont (A, G)

The elementary-age schoolchildren of Pawlet and Rupert in south-western Vermont are served by the Mettawee Community School (Mango, 2005):

> When the school was built in 1998, groundwater from two newly-drilled wells was in compliance with all state drinking water regulations. However, when the State of Vermont lowered its drinking water standard for arsenic from 50 ppb [parts per billion] to 10 ppb, the school discovered it had an arsenic problem, with concentrations of the order of 30–40 ppb As. Forty wells within a ten-mile radius of the school were sampled for this study, in an effort to determine the extent and source of the arsenic contamination. Three other wells were found to have arsenic concentrations above 10 ppb, with one as high as 91 ppb. Temperature, pH, total dissolved solids, conductivity and oxidation-reduction potential [ORP] were also measured at each well. pH ranged from 6.3 to 9.1; higher arsenic concentrations are associated with higher pH values. There is also a correlation between arsenic concentration and oxidation–reduction potential, with higher levels of arsenic associated with lower ORP values.
>
> Because there is no obvious source of anthropogenic arsenic pollution (the region's main sources of income are dairy farming, slate quarrying and tourism), bedrock is being targeted as the source of arsenic. The study area is in the Taconic Slate Belt. The dominant lithologies are Lower Cambrian and Lower-Middle Ordovician slate, phyllite and schist. The Cambrian Saint Catherine Formation

contains green phyllite, quartzite, purple and green slate, and chlorite schist. There are abundant quartz veins, sometimes including small, as-yet-unidentified masses of black, fine-grained, granular material. This unit is highly folded and faulted, and holds up the high ridges of the region. The Ordovician Pawlet Formation is a mixture of black slate and metagraywacke. The Ordovician Poultney Formation is dominantly gray and green slate, and is the leading quarried rock of the local area; a narrow north-south belt of this formation runs along the Vermont–New York border less than five miles west of the school. Pyrite is found is variable concentrations in all formations. Preliminary results suggest that elevated levels of arsenic in groundwater are related to the presence of the Saint Catherine Formation.

Vermont also has at least one site of anthropogenic arsenic contamination. This is the Pownal Tannery Superfund site in Pownal. The Pownal Tannery on the Hoosic River tributary of the Hudson River, between the Green and Taconic mountains, is located on the narrow lowlands of the Vermont Valley and has twice been covered by Pleistocene glacial lakes to more than 300 m above mean sea level. The topography is characterized by low-gradient fluvial terraces associated with Pleistocene glaciation and modern floodplain sediment, underlain by bedrock associated with the Taconic Range, specifically the Cambrian Saint Catherine formation of purple, grey-green and variegated slate and phyllite with interbeds of white to green quartzite. In addition to elevated levels of chromium (10 000–70 000 mg/kg) and lead (1000–2000 mg/kg) in the sludge, arsenic was detected at levels above 10 μg/l in the Hoosic River and associated wetlands, in on- and off-site private wells and in the former tannery area (McVickar, 2002).

Washington (A, M, G)

The major source of anthropogenic arsenic contamination was the Tacoma Smelter Plume Site, which was the sole domestic supplier of arsenic for the USA for several decades. Smelter emissions were also a source for other metals, such as lead, cadmium, antimony and mercury. The site was declared a Superfund site and remediated in the late 1990s (Glass, 2003). Owing to its longevity and the nature of airborne particulate matter, the continued testing of soils for contamination within the plume area is advised.

The sources of arsenic in tidal waterways near Tacoma may be due to geogenic sources as well (Carpenter *et al.* 1978; Davis *et al.* 1997), and perhaps this phenomenon of tidal/estuarine arsenic should be compared with possibly Quaternary arsonium compounds or arsenosugars that are released seasonally into marine estuarine ecosystems in Belgium and Portugal.

Other anthropogenic sources of arsenic are found in eastern Washington and are attributed to the apple industry's use of arsenical pesticides. These have contributed arsenic loading rates up to approximately 490 kg/ha in apple orchards (Benson, 1976), leading to arsenic concentrations in soil in excess of 100 mg/kg (Davenport & Peryea, 1991).

Geogenic sources of arsenic in central Washington State are related to arsenopyrite weathering. These deposits are also in areas of the mining of gold, which subjects the deposits of arsenopyrite to accelerated weathering processes. The lower values of arsenic in surficial waters and ground waters in north-central Washington appear to be related to higher pH levels. Additionally, elevated arsenic concentrations have been reported in alkaline aquifers in north-western Washington, which are attributed to geogenic sources (Welch *et al.* 2000).

Mining sources of arsenic release into surficial water and ground waters in Washington are associated with gold mining sites, as indicated in Table 2.25.

Table 2.25 Arsenic concentrations in surface water and groundwater at gold mining projects in north-central Washington, USA

Location	Maximal arsenic concentration in surface water (mg/l)	pH	Maximal arsenic concentration in groundwater (mg/l)	pH
Washington (north-central)	0.02	8.1	0.012	8.1

Source: Straskraba & Moran, 1990

West Virginia (C)

Arsenic contamination in West Virginia is related to its occurrence in coal, primarily with pyrite and arsenopyrite. Since pyrite is more abundant than arsenopyrite in coal, most arsenic is in solid solution within pyrite in small quantities. Arsenic is highest in the Pennsylvanian Conemaugh Group of coals, which includes the Harlem, Elk Lick and Mahoning coals of Tucker, Mineral and Grant counties, but these coals have not been mined because they are thin and not extensive. As the affected area lies south of a battery of coal-fired power plants, a test of whether the arsenic was anthropogenically introduced from the plants was devised by taking a core from the sediment in a reservoir in the area. Very preliminary results indicate that arsenic was from fly ash from the coal-fired plants, because the ash peaks in the mid-1970s and then declines, which conforms to the power plants' activity profile (Goldhaber *et al.* 2001). However, some natural leaching of arsenic from coal seams is also a possibility.

Elimination of the Conemaugh Group from coal-burning processes has become critical economically, because arsenic can damage expensive catalysts used in the "scrubbing process" that are used to reduce or eliminate nitrogen oxide emissions into the atmosphere. Therefore, high-arsenic coals are not cost-effective in power generation plants where catalyst scrubbers are used (Hatt, 2007).

It is also possible that some arsenic contained in Conemaugh Group coals can leach into groundwater. For instance, the groundwater of the Appalachian Plateau in the Kanawha River basin (underlain by Pennsylvanian-age sedimentary rocks) flows by gravity to the river basin through bedding-plane separations that contain arsenopyrite. Analysis of water from 4 of the 30 wells in this region revealed that 13% contained arsenic at levels up to 16 μg/l (Sheets & Kozar, 2000). Although these arsenic concentrations are minimal, they do indicate a potential for arsenic to leach into ground waters in the Kanawha River basin, suggesting the need to conduct further testing.

Wisconsin (G)

Arsenic is ubiquitous in the ground waters of Wisconsin, especially in the east-central and the south-eastern regions of the state, often characterized by Ordovician carbonate and clastic rocks in association with sulfide mineralization in Saint Peter Sandstone (Simo *et al.* 1996), but sufficiently reducing conditions may cause arsenic mobilization in Quaternary formations as well in the Fox River valley and in Lake Geneva (Gotkowitz, 2005):

> In the Fox River valley of east-central Wisconsin, the primary source of arsenic to groundwater is sulfide mineralization in the Saint Peter sandstone. Up to 30 percent of wells in some parts of this region have arsenic concentrations greater than 5 mg/L. In areas south and west of the Fox River valley, arsenic concentrations in groundwater from the Saint Peter sandstone exceed 5 mg/L in 3 to 10 percent of the wells tested. In southeast Wisconsin, groundwater in the sand and gravel aquifer and some shallow bedrock formations is susceptible to arsenic contamination. Research at a site near Lake Geneva demonstrated that where groundwater is under sufficiently reducing conditions, low solid-phase concentrations of arsenic in Quaternary sediments are sufficient to cause arsenic in groundwater at concentrations up to 100 μg/L. Data from water-sampling programs in northern Wisconsin showed that similar geologic and geochemical conditions likely exist in shallow aquifers in other glaciated areas of the state, resulting in similar levels of arsenic in groundwater.

Wyoming (C, G, V)

Coal as a source of arsenic release in Wyoming occurs in the 37-m-thick Wyodak-Anderson coal bed in the Powder River basin, which contains low average concentrations of ash (3.64%), sulfur (0.27%) and trace elements of environmental concern, including arsenic, mercury and selenium, at concentrations of 0.069–1.5 mg/kg (Flores *et al.* 2006). The arithmetic mean concentration of arsenic based on whole coal, remnant moisture measurements is 2.6 mg/kg, with a minimum of 0.2 mg/kg and a maximum of 19 mg/kg (Stricker & Ellis, 1999). For the areal distribution of the arsenic concentrations, see Figure 2.14.

The high arsenic levels in groundwater within the Madison and upper Missouri river valleys appear to be at least partly related, directly or indirectly, to the Yellowstone geothermal system (Sonderegger & Ohguchi, 1988; Nimick, 1994). A study by Nimick *et al.* (1998) found that:

> Geothermal As from Yellowstone National Park causes high As concentrations (10–370 μg/L) in the Madison and Missouri Rivers in Montana and Wyoming. Arsenic transport is largely conservative in the upper basin as demonstrated by the near equivalence of dissolved and total-recoverable As concentrations, the constancy of As loads, and consistent ratios of concentrations of As to conservative geothermal tracers. Diurnal cycling of As between aqueous and solid phases in response to pH-induced changes in sorption equilibria causes small variations of about 10–20% in dissolved As concentrations. HCl-extractable As concentrations in river and lake sediment in the upper basin are variable depending on position relative to the As-rich headwaters and geochemical and physical processes associated with lakes. In the lower Missouri River, large quantities of suspended sediment from tributaries provide sufficient sorption sites for substantial conversion of As from the aqueous phase to the solid phase.

Geogenic arsenic may be associated with the Tertiary White River Geologic Unit in the extreme south-eastern region of the state, which may contain arsenic mineralization. The highest level of arsenic in groundwater of this geological unit, just over the state line in Nebraska, is 85 μg/l (Gosselin *et al.* 2004).

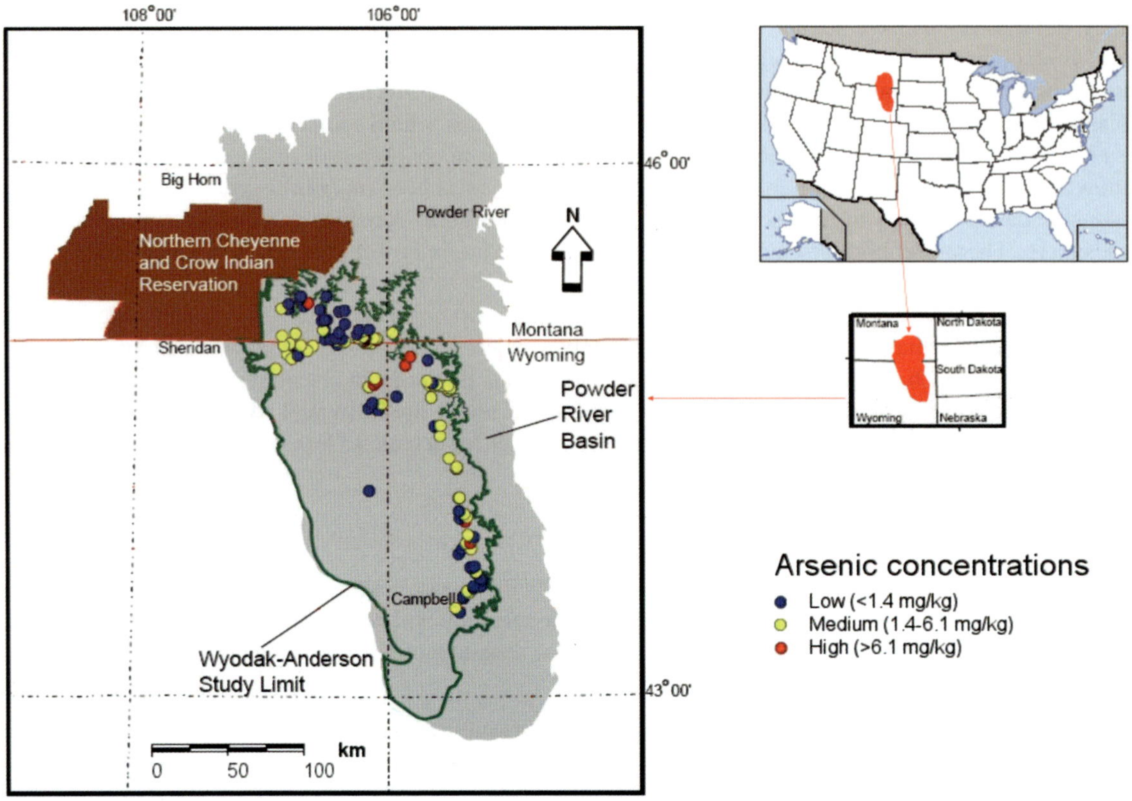

Figure 2.14 Arsenic concentrations in the Wyodak-Anderson coal zone, Powder River basin, Wyoming, USA (modified from Stricker & Ellis, 1999).

2.20 URUGUAY (A, G, V)

Uruguay, with a population of 3.3 million, has a Human Development Index rank of 52 out of 169 and a per capita GNI of US$13,808 (US Census Bureau, 2011; UNDP, 2010).

Between 1998 and 2002 the World Bank (WB) authorized the Proyecto Sistema Acuífero Guaraní (2011) drilling of over 10,000 wells at depths ranging 30–80 m without any water quality testing for As in the Guaraní Aquifer being accessed (Bundschuh *et al.* 2011). (Bangladesh was also funded by the WB without testing for As in water from bore-hole wells with dire consequences). In San José department in the southwest, researchers found As concentrations in groundwater between 25 and 50 μg/l (Guérèquiz *et al.* 2006, 2008, 2009). The source of this As appears to be from volcanic ash in the Raigón Formation that encompasses proximate areas of Argentina (Santa Fe and Buenos Aires provinces). While the potential impacted population affected by As exposure is about 46,000, public interest has only been stimulated recently (Diario El País, 2007).

2.21 VIRGIN ISLANDS OF THE UNITED STATES (G)[1]

The Virgin Islands of the United States, with a population of 106 400 (US Census Bureau, 2010), has increased levels of arsenic deposited on its soils by the unusual case of Saharan dust rich in arsenic being transported from Africa by summer westerly winds across the Azores and the North Atlantic to the Caribbean (Prospero, 1999; Holmes *et al.* 2000; Shinn *et al.* 2000). African dust has elevated arsenic derived from gold-bearing rocks. Samples from Mali, for instance, contained 17 μg/g (British Geological Survey, 2002b).

At the eastern end of St John Island, arsenic in cistern material measured 38 μg/g. Of 679 samples collected at this site from 1990 to 1994, 35% had detectable arsenic, with concentrations ranging from 0.5 to 44 μg/g, with a mean of 17 μg/g (Holmes *et al.* 2001).

[1]Note that, unlike most of the other entries in this sourcebook, the Virgin Islands of the United States is not a country, but a territory of the USA.

Section 3
Asian Region

3.1 AFGHANISTAN (G, M, P)

Afghanistan, with a population of 29.8 million (US Census Bureau, 2011), encompasses approximately 647 500 km^2 of rugged terrain and a harsh seasonal climate. The country is very poor. Afghanistan is close to the bottom of both the GDP list (having a per capita GNI of US$ 1419 – exacerbated by the fact that the poorest 30% of the population receive only 9% of the national income, whereas the upper 30% receive 55%) and the Human Development Index (in which it ranks 155 out of 169) UNDP, 2010). These facts were reported by the first ever Afghanistan National Human Development Report, "Security with a Human Face: Challenges and Responsibilities" (UNDP, 2004), initiated in 2003 by the government and the United Nations Development Programme, with financial support from the Canadian International Development Agency and the World Bank.

Since the 1970s, war, extreme political upheaval and civil strife have damaged or crippled the infrastructure of water supplies as well as social services, education, transportation, industry and mining, and 4–5 million people have been displaced. In recent years, however, there has been a flood of returning refugees into the cities of Kabul, Mazar-i-Sharif, Kandahar and Herut. This, in turn, has placed demands on water supplies, creating a drawdown of aquifers and increasing the potential for arsenic contamination, especially in areas where groundwater flow is from old mining areas. As over 80% of Afghani girls never attend school and as they are the ones who provide water for families, their lack of education has severe implications for understanding the parameters of what constitutes safe drinking-water. In fact, only 23% of the population (mostly in urban areas) has access to safe drinking-water.

Geogenic sources of arsenic are indicated by Afghanistan's mineral resources. Reserves of copper, gold and iron (with associated arsenopyrite) are found in mineralized belts, one of which is south of Kabul. These have been mined for centuries, and mine tailings have leached into groundwater. In the period since 1980, however, an infrastructure to extract these resources and to monitor groundwater has been handicapped by war and political instability (Lyday, 2003).

Petroleum reserves may also contribute to arsenic release into groundwater. In 2006, the USGS and the Afghanistan Ministry of Mines and Industry Joint Oil and Gas Resource Assessment Team found that in northern Afghanistan's Afghan-Tajik and Amu Darya basins and in four total petroleum systems (the Amu Darya Jurassic-Cretaceous, the Kalaimor-Kaisar Jurassic, the Afghan Tajik Jurassic and the Afghan-Tajik Palaeogene), which collectively encompass 87 000 km^2, there are 1596 million barrels of crude oil, 444 billion m^3 of natural gas and 562 million barrels of natural gas liquids, estimates that are far larger than hitherto known (USGS & Afghanistan Ministry of Mines and Industry, 2006).

In July 2003, a rapid investigation of the quality of water sources (including testing for arsenic using field kits exclusively) was begun as a multiphase project, sponsored jointly by UNICEF, the Afghani Water and Sanitation Group and the Danish Committee for Aid to Afghan Refugees (DACAAR), a nongovernmental organization. Testing was limited in Phase 1 to the eastern and western regions and was completed in November 2004 (Figure 3.1). The southern region was omitted due to security problems. Arsenic testing was restricted to wells equipped with hand pumps; of those, only one well per 100 km^2 was tested, with priority given to those in highly populated areas. In July 2003, of 647 wells in the 15- to 65-m depth range, 565 were negative. In Logar Province, 8 wells were found with arsenic concentrations above 10 μg/l, and in the Ghazni metropolitan area, 56 wells contained arsenic above 10 μg/l. In October 2004, more public wells were tested in Logar and Ghazni, with the following results: in Logar, 282 wells were tested, of which 239 were negative and 43 contained arsenic in the range 1–100 μg/l; in Ghazni, 171 wells were tested, of which 97 were negative, and 56 contained arsenic in the range 10–50 μg/l (Figure 3.2).

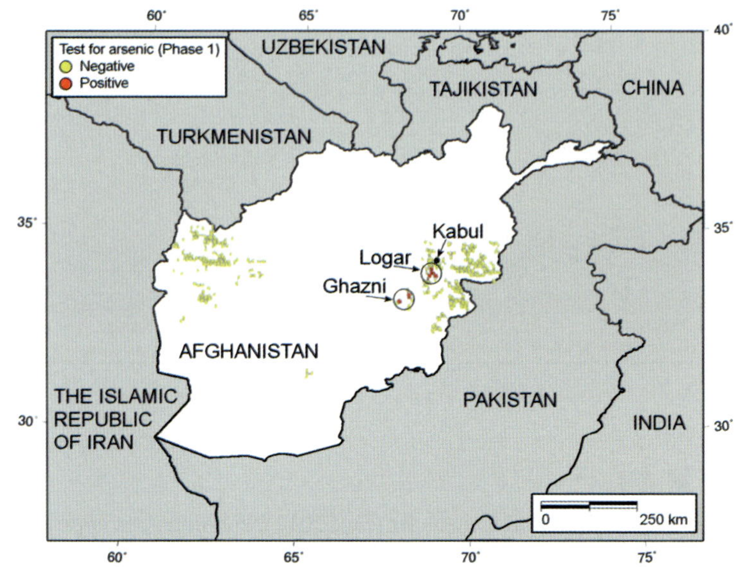

Figure 3.1 Results of Phase 1 testing for arsenic contamination in wells in Afghanistan in 2004 (modified from Saltori, 2004).

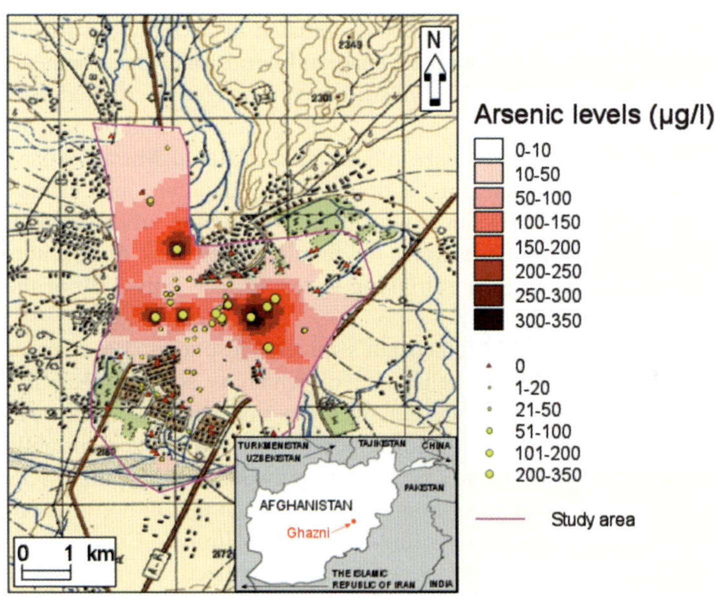

Figure 3.2 Arsenic contamination of wells in metropolitan Ghazni, Afghanistan: Phase 1 (modified from Saltori, 2004).

Phase 2 began in October 2004 and was geared to test every public well in the Logar and Ghazni areas. Field kits were used that have an accuracy of 1 μg/l. The arsenic contamination of the Logar area south of Kabul is clearly the result of old copper mining, which has leached arsenic from the arsenopyrite in tailings and waste dumps into the groundwater that has flowed towards the Kabul area. Forty-three wells, or 15%, contained arsenic in Logar. However, if the flow of groundwater is on a trajectory to Kabul, then there is a critical arsenic problem in the offing, since the population of Kabul is 700 000. The 2004 figures for those exposed to arsenic concentrations above 10 μg/l are for those in the Logar area only, where three mountain villages or 500 residents are exposed (Saltori, 2004). The arsenic contamination of the Ghazni urban water supply is apparently not from mining but from geogenic processes. There are more than 3000 wells, which are either public or private. Of these, 171 wells were tested (~6% of the total); 97 wells were negative, 18 wells were in the 1–10 µg/l range and 56 wells were in the 10–500 µg/l range (33% of the tested wells). This extrapolates to about 2% of the total wells being contaminated, or 600 that are at risk of arsenic contamination above 10 µg/l. It is estimated that these wells serve a population of up to 500 000 residents (Lyday, 2003). Therefore, the population exposed to elevated arsenic in drinking-water in the Ghazni and Logar areas is estimated at 500 000, but the other two regions – the central and northern – are not factored into this 21 November 2004 estimate. The flood of in-migration, which has significantly overexploited the water resources by causing a drawdown of aquifers in Ghazni and Kabul, is critical to the ongoing arsenic threat.

During 2005, DACAAR, on behalf of UNICEF, completed surveys of arsenic in more than 600 wells in Ghazni City and its surrounding areas and additional wells in Logar Province. Those affected wells in Logar Province were found to be associated with copper mines. Since the population in Logar Province is relatively small, the remediation effort is thought to be relatively affordable. On the other hand, Ghazni City's 500 000 inhabitants are seriously affected and require serious attention. According to DACAAR, taking into account the high child mortality rate in Afghanistan (257 per 1000 live births), which is at least partly due to the lack of safe drinking-water, alerts on arsenic should be given particular attention. Only about 40% of the population in Afghanistan has access to safe drinking-water (WHO & UNICEF, 2006), and a sensitive evaluation and action plan to address the arsenic contamination are highly recommended (DACAAR, 2005).

DACAAR has established a water monitoring database and, during 2005, monitored and recorded data, including global positioning system coordinates and arsenic levels, from 2041 previously established water points (DACAAR, 2005). DACAAR's water quality laboratory, the only fully equipped facility in Afghanistan with a quality control policy, has a technical staff that trains Afghani engineers from the Ministry of Rural Rehabilitation and Development. DACAAR is instrumental in training in the proper use of water quality test kits made available by UNICEF to all provinces (DACAAR, 2005).

3.2 ARMENIA (G, M, P)

Armenia, with a population of 3.0 million, has a Human Development Index rank of 76 out of 169 and a per capita GNI of $US 5495 (US Census Bureau, 2011; UNDP, 2010).

Geogenic and mining sources of arsenic in Armenia are related to gold, silver and copper deposits. Mining of them contributes to Armenia's GDP and can be traced back to around 6000 BC, with recent archaeological investigations revealing copper utensils and ornaments from previously mined deposits. Around 3000 BC, Transcaucasus was supplying copper, gold and silver to Mediterranean countries. At the beginning of the 10th century, mine production from a massive sulfide polymetallic deposit was recorded, which contained high concentrations of arsenic-related minerals. In the former Soviet Union, the mining industries of Armenia supplied copper, molybdenum and zinc, but after 1989 these mining operations were left in poor condition (Nishikawa, 2000).

Armenia has extensive and varied mineral resources, including gold, silver, lead, zinc, copper, molybdenum, iron and rare earths and elements such as bismuth, tellurium, selenium, gallium, indium, thallium, rhenium and germanium. Thirty-three mining and metallurgical enterprises are mainly involved in non-ferrous metals (gold, copper and molybdenum). The largest are Ararat Gold Recovery Co. and Zangezur Copper-Molybdenum Combine. Ararat Gold Recovery Co., a subsidiary of Sterlite Gold Ltd, is developing the Zod and Meghradzor gold deposits and is operating a gold processing plant in the vicinity of Mount Ararat. These gold mines have arsenopyrite associated with the matrix in which gold is found and so require regulation. The mining code and the concession law of 2003 clarify the rights and responsibilities of the companies and the Department of Mineral Resources under the Ministry of Nature Protection, which issues licences and monitors both exploration and environmental controls. In 2002, Armenia's mining and metallurgical sector reported a 21.8% growth, and the sector exported more than 80% of its production (MBendi, 2007).

The Debed River, a tributary of the Kura River, was highly polluted with copper and zinc and associated arsenic compounds discharged from the Alaverdi mine in north-east Armenia, which used to yield 50 000 tonnes of cathode copper annually. Prospecting for additional copper reserves is currently under way. A new copper integrated enterprise with expected annual production of 36 000–45 000 tonnes of cathode copper has been constructed (Senza, 1994).

Armenia has the following oil reserves, which deserve monitoring because petroleum leaks promote the mobilization of arsenic from sediments (Business Wire, 2004):

(1) Armavir sub-basin of the Ararat Intermontane Depression, 30 km west of Yerevan: the Karmir prospect, 283.2 million m^3 of reserves; South Artamet prospect, 84.95 million m^3; and North Artamet prospect, 141.6 million m^3;

(2) Yerevan basin, Central Depression, 10 km east of Yerevan: Shorakhpur reserve in the Palaeocene and Nubarashen reserve in the Palaeocene, 100 869 m^3.

During the period from 1999 to 2001, the World Bank funded the development of an Integrated Water Resources Management Plan for Armenia. The United States Agency for International Development (USAID) is one of the active donors, financing integrated water resources management activities within the Caucasus region, including the national project for Sustainable Water Resources Management in Armenia. USAID has implemented the South Caucasus Water Project to strengthen water resources management in the Kura-Araks basin, based on integrated river basin water resources management principles (CEO, 2002b).

Arsenic is also associated with petroleum production in Armenia. The petrochemical industry is based in Sumgait, Azerbaijan, Rustavi, Georgia and Yerevan, which collectively supply 70% of Russian petroleum demands. Oil and gas prospecting companies in the Baku-Sumgayit area and the North Caucasus (Krasnodar, Grozny and Maikop) are the epicentres of pollution of groundwater with the release of arsenic.

3.3 AZERBAIJAN (M, P)

Azerbaijan, with a population of 9.4 million, has a Human Development Index rank of 67 out of 169 and a per capita GDP of $US 8747 (US Census Bureau, 2011; UNDP, 2010).

Azerbaijan's Kura-Araz lowlands and the Absheron peninsula, where 70% of the population lives, are completely dependent on irrigation for their agricultural production but have a significant water deficit due to low precipitation and high evapotranspiration. The main river, the Kura, originates in Turkey and flows through Georgia, Armenia and the Islamic Republic of Iran to Azerbaijan before discharging into the Caspian Sea. The Kura River basin covers 80% of Azerbaijan, has no wastewater treatment for a population of 11 million and yet supplies more than 70% of Azerbaijan's drinking-water supply. Coastal areas north and south of the Kura basin drain to the Caspian Sea by smaller rivers like the Samur, on the border with Dagestan of the Russian Federation. The Samur River supplies Baku with drinking-water and Absheron with irrigation water. Elsewhere, groundwater is important for irrigation and water supply to provincial towns and rural areas, especially in the foothills of the mountain ranges, although it provides only about 5% of the total water requirements of the country (UNDP, 2007).

Azerbaijan and the Russian Federation are parties to the 1972 London Convention on the Prevention of Marine Pollution by Dumping of Wastes and Other Matter. Azerbaijan participates in the Caspian Sea Environmental Program, operated since the late 1990s. The United Nations Development Programme, the Tacis Programme, the World Bank and USAID fund the programme. The significant output of this programme is the establishment of the Caspian Centre for Pollution Control and the development of a GIS-friendly database. Along with this, some large-scale investment projects for infrastructure rehabilitation, which were identified as priorities in National Environmental Action Programmes, are being implemented within the region. For example, the World Bank finances long-term irrigation drainage system rehabilitation projects in Georgia and Azerbaijan, together with investment components, including capacity building for sound management of the systems – specifically, the establishment of local community-based water use associations.

An 18-month United Nations Development Programme – Global Environment Facility Project aimed at "Reducing Trans-boundary Degradation of the Kura-Aras/Araks River Basin" began in April 2005 (UNDP, 2005b):

> The Kura-Aras/Araks river system is the principal source of water for industry, agriculture, residential uses and energy in Armenia, Azerbaijan and Georgia, as well as for sizeable populations in Iran. Integrated, inter-country efforts are urgently required to reverse the ongoing degradation of these river ecosystems. The river systems of the Caucasus flow within and between countries and the health of these rivers systems depends on human activities. The current challenges facing the Kura and Aras/Araks river system include political, social and economic barriers to collaborative transboundary cooperation among the people of Armenia, Azerbaijan, Georgia and Iran.
>
> A lack of coordinated management of these waters reduces the over all quality and quantity of this vital resource. Further, there are implications for exacerbated tensions in the region between countries if these issues are not addressed. The conflict between Armenia and Azerbaijan regarding territorial claims creates an especially tenuous set of circumstances for collective water management, but also emphasizes the importance of addressing these issues through coordination with creativity on multiple levels. There is a need to increase the capacity for water management through wider public involvement, relying on multiple stakeholder inputs, enhanced NGO [nongovernmental organization] coordination and pilot projects that serve as key public educational components. The environmental resource management will be strengthened through multiple channels of input, rather than focusing solely on the government entities, and long term changes must occur at the grassroots level.

Mining in Azerbaijan is located at 441 registered fields, including 18 ore deposits, where arsenic is in association with iron, alunite, copper, lead, gold, cobalt, mercury and molybdenum. In addition, 155 mineral deposits exist in the occupied territories (five of gold, six of mercury and two of copper). Increased extraction techniques have negatively impacted the environment. For instance, mining in Dashkasan has encroached on forests and degraded other natural habitats. In Absheron and Gobustan, winter pasturage has been invaded by unplanned, unsystematic extraction from quarries, accelerating erosion, destroying natural habitats and generating thousands of hectares of tailings, which pollute the land, water and air. Acid mine drainage flows onto soils, making them unsuitable for agriculture, and into rivers, which pollutes riverine biota and downstream ecosystems. The pollution of land and water by arsenic from mines in the Caucasus is a major problem in Azerbaijan. Under 2006 regulations, when a mine is decommissioned, the area must be reclaimed and plant cover restored. However, these regulations do not address the situation of already derelict and abandoned mining operations (Republic of Azerbaijan, 2004).

The territory of Nakhichevan Autonomous Republic is rich in minerals, including deposits of molybdenum, cobalt, copper, zinc, silver, lead, gold, mercury and chromite. Nakhichevan is a landlocked exclave of Azerbaijan of 5500 km^2 that borders Armenia (221 km), Turkey (9 km) and the Islamic Republic of Iran (179 km). In the area around Nakhichevan, copper, gold, zinc and lead deposits are in association with arsenic minerals, including arsenopyrite and other arsenic-bearing ores. Many of these arsenic minerals are still present in abandoned mine tailings, from which they continue to leach into the groundwater (Republic of Azerbaijan, 2004).

Because of the long history of petroleum extraction in the region, Azerbaijan is at risk for arsenic release to the groundwater (CEO, 2002a):

> Oil has been of greatest importance for Azerbaijan all through its history. Evidences of oil extraction on the Absheron peninsula have existed since the 5th century AD. In 1897–1907, the 833 km long Baku–Batumi oil pipeline was built, which was the largest in the world at that time (one of the first oil pipelines in the world was built in Azerbaijan in 1878). In 1901, Absheron was the largest oil producer in the world (it accounted for over 50% of the world's extraction). Before World War II, Azerbaijan was the greatest oil extracting and supplying region of the USSR, with a share of total oil extraction. However, this was not followed by important rise in socio-economic conditions in the republic. Moreover, the state of the environment has become worse. Current levels of oil production in Azerbaijan are far below the 70s and 80s levels explained by reduced oil reserves, out-of-date technologies, inadequate investments, etc. It is noteworthy to mention that Caspian oil has made the Caucasus a strategic region.

Currently, the energy industry in Azerbaijan is dominated by oil production (Republic of Azerbaijan, 2004):

> In 2001, 14.8 million tonnes of crude oil was produced (89% from off-shore deposits, along with 5.5 billion m^3 of gas (96% from off-shore deposits). Of the hydrocarbons produced in 2001, 28% were used for the production of electricity. In that year oil- and gas-powered power stations produced 92.5% of electricity (with hydroelectric power accounting for a further 5.9% and generators for 0.6%). However, the oil- and gas-based power stations are major contributors to air pollution in the country. The oil industry has underpinned the economy of Azerbaijan for many years, and as well as extraction it also supports a number of related industries such oil refining and chemical processing. The importance of oil extraction to the economy of Azerbaijan is increasing. After independence, companies such as the State Oil Company (SOCAR) were created in order to extract and refine oil. The companies were given the jurisdiction explore, extract and refine oil and gas deposits from the Azeri sector of the Caspian Sea and from on-shore sources. At present SOCAR operates within 22,000 ha of land and 10,000 km^2 of sea, and employs 60,400 people, of which the majority (46,000) work in extraction.

The economic impact of the oil industry in Azerbaijan should not overwhelm the regulatory agencies that are concerned with protecting the groundwater from the reductive dissolution of arsenic compounds in the geological matrices surrounding the groundwater aquifers in the oil extraction areas.

3.4 BANGLADESH (A, G)

Bangladesh, with a population of 160 million, has a Human Development Index rank of 129 out of 169 and a per capita GNI of $US 1587 (US Census Bureau, 2011; UNDP, 2010). Bangladesh borders the Bay of Bengal and is at the southernmost confluence of both the Ganges and the Brahmaputra rivers, which flow from high mountain ranges rich in arsenic-bearing minerals. Dhaka is the capital and the largest city, with a population of 11 million.

The principal source of exposure is drinking-water supplied from groundwater containing high levels of arsenic derived from geological sources. The affected aquifers are generally shallow (less than 100–150 m deep) and of Holocene age and comprise a mixed sequence of micaceous sands, silts and clays (Smedley & Kinniburgh, 2002; Rosenboom, 2004). The groundwater conditions are strongly reducing, with slow flow rates.

Tube wells have been in use in Bangladesh since the 1940s, but their numbers increased markedly from the 1970s onward. It was not until 1993, however, that the elevated levels of arsenic in many wells were identified (Smith *et al.* 2000; Quamruzzaman *et al.* 2002). The worst-affected areas are to the south and east of Dhaka, where more than 90% of the wells in some villages have arsenic concentrations greater than 50 μg/l. The reported level of arsenic in drinking-water supplies ranges from <0.5 to 3200 μg/l (DPHE/BGS/MML, 1999; Smedley & Kinniburgh, 2002; Rosenboom, 2004).

The four maps of Bangladesh given below show the regional pattern of arsenic contamination of the groundwater. The worst-affected area is in southern Bangladesh. However, some water from the southernmost coastal region has low arsenic concentrations, because the water is drawn from wells greater than 200 m in depth in order to avoid salinity problems in the shallower horizons. Most of the other wells are from a depth of 15–80 m, with an arsenic concentration peak occurring at 20–70 meters (Ahmed *et al.* 2006). While some wells in central and north-west Bangladesh have arsenic concentrations below <5 μg/l, two zones along the Brahmaputra and Megha rivers in central Bangladesh and east of Bangladesh have been identified as critical, where arsenic concentrations in groundwater can exceed 2000 μg/l.

The first map (Figure 3.3) shows the percentage of positive arsenic occurrences identified using field test kits. This generalized map shows the zones that are of concern. The second map (Figure 3.4) presents arsenic content of groundwater in Bangladesh by district. The most severely contaminated districts are Chandpur and Munshiganj; severely contaminated districts include the greater parts of Comilla, Noakhali, Faridpur, Madaripur, Gopalganj, Shariatpur and Satkhira. The third map (Figure 3.5) provides a spatial distribution of arsenicosis in Bangladesh. A total of 13 458 patients were identified in 62 *upazilas* at prevalence rates varying from 0 to 0.32%, with an average of 0.078%. Yet a closer look at the data shows that there is no correlation between the percentage of contaminated wells and the prevalence of arsenicosis (Figure 3.6).

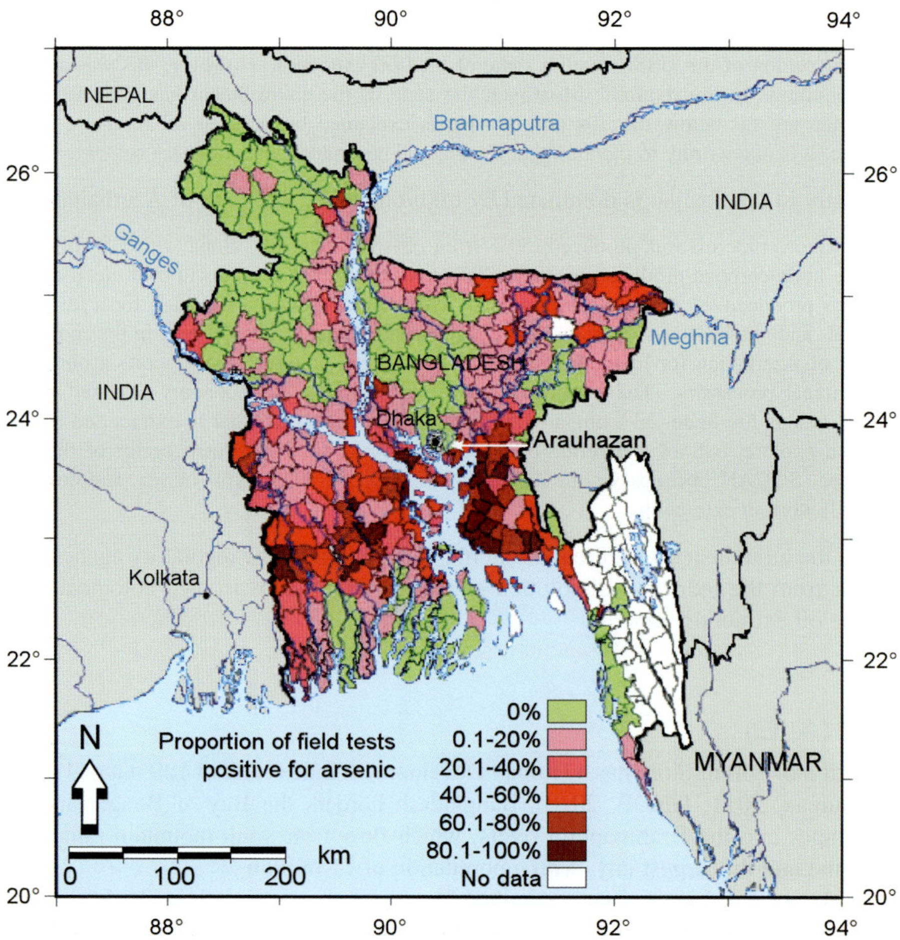

Figure 3.3 Arsenic occurrences in shallow aquifer wells ($n = 8$–12 million) in Bangladesh [modified from a map from UNICEF-Dhaka; http://international.usgs.gov/images/projects/big_pics/bg-map-lrg.png; accessed December 19, 2011)].

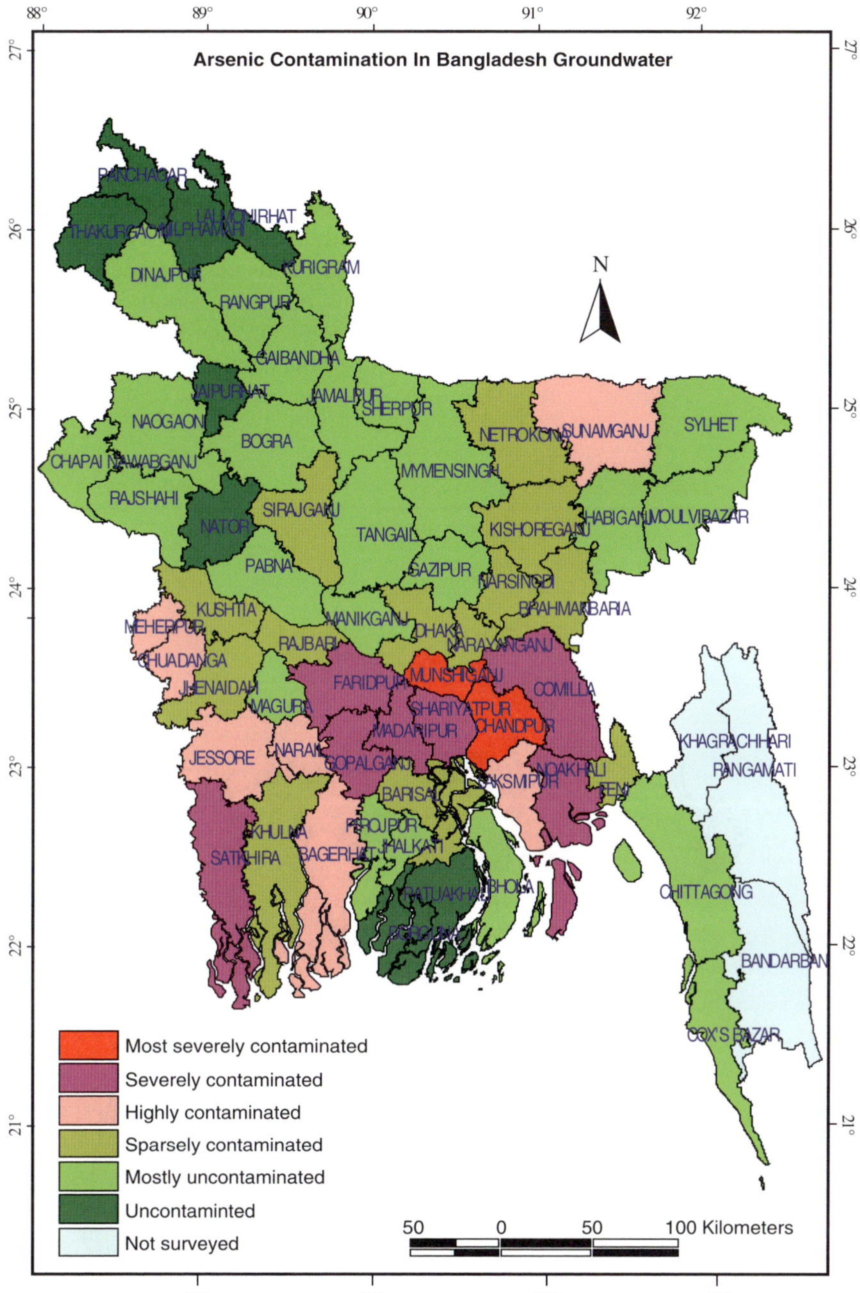

Figure 3.4 Arsenic content of groundwater in Bangladesh (Ahmed *et al.* 2006).

Smith *et al.* (2000) indicated that there would be a rapid rise in the rate of cancer from drinking contaminated drinking water in the years to come. The risk also comes from ingestion of food prepared with arsenic contaminated water, crops irrigated with high concentrations of arsenic water, foods prepared with tainted water or consumption of animal products exposed to arsenic though water or soil. Arsenic contamination of agricultural products is, therefore, a concomitant of arsenic contamination of groundwater in Bangladesh, India and elsewhere.

One organization that has been very active in pursuing food security is the International Food Policy Research Institute (IFPRI), which was established in 1975 as one of 15 agricultural research centers that receive principal funding from governments, private foundations, and international and regional organizations, most of which are members of the

Consultative Group on International Agricultural Research (CGIAR). In a recent report of the Institute by Mahabub Hossain (2009) prepared for the project on "Millions Fed: Proven Successes in Agricultural Development" (www.ifpri.org/millionsfed) it is reported that "The fast diffusion of Boro rice in the rain-fed, low-yielding "*aus*" rice area contributed to accelerated growth in rice productivity, a reduction in the unit cost of production, and a decline in real rice prices. Together, these factors have been a major factor behind attainment of food security and moderately reduced poverty in Bangladesh over the last two decades." While it is admitted that "It is now established that the arsenic contamination of groundwater is a serious problem for Bangladesh," the position of the paper is that the link of arsenic contamination "with the exploitation of groundwater for *boro* cultivation is yet to be firmly established." The report does end with a caveat: "However, it has recently become apparent that arsenic-contaminated groundwater that is used for irrigation is seeping into soils and rice in several countries of South and Southeast Asia, thus posing a serious threat to sustainable agricultural production and to the health and livelihoods of the affected people in these countries (Brammer, 2009). Since groundwater-dependent *boro* rice now accounts for nearly 60 percent of the annual rice production, the risk is very high for Bangladesh,"

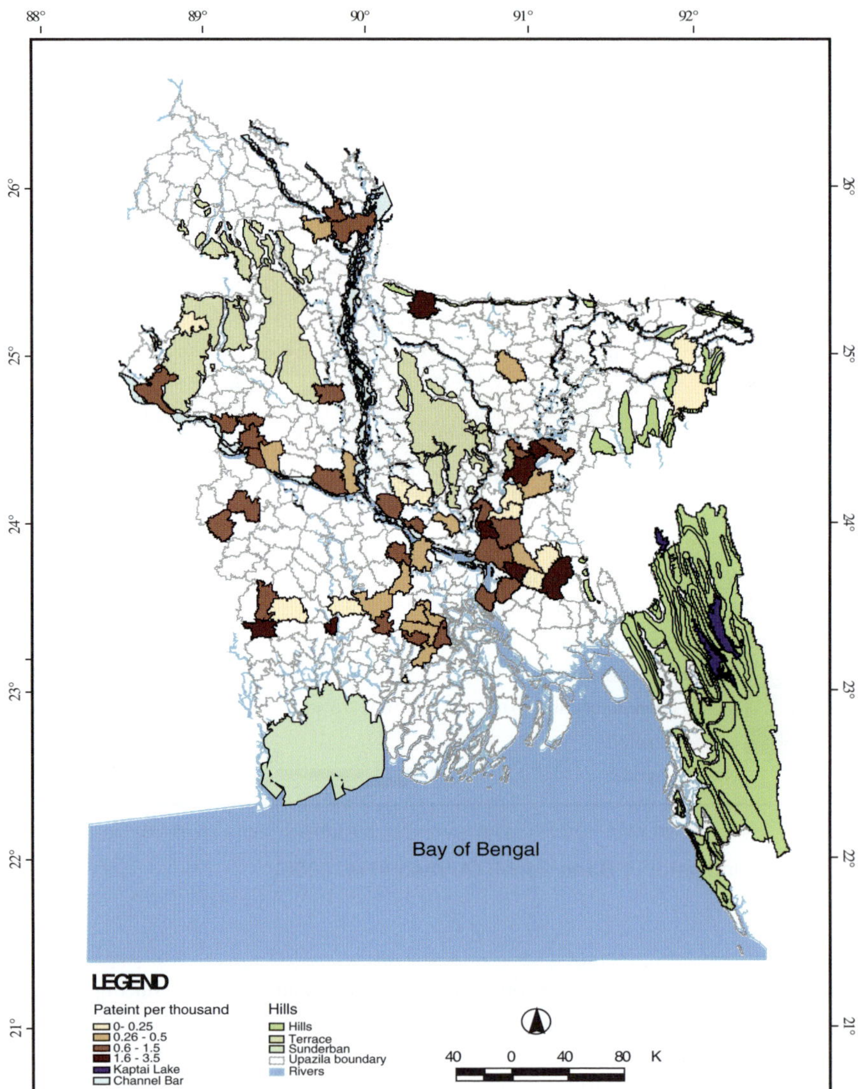

Figure 3.5 Spatial distribution of arsenicosis in Bangladesh (Ahmed *et al.* 2006).

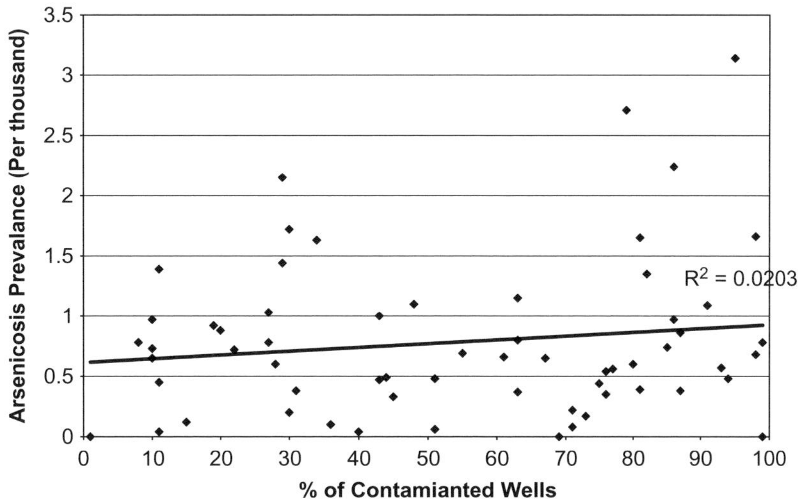

Figure 3.6 Relationship between arsenic contaminated wells and number of arsenicosis patients (Reprinted with permission from Ahmed *et al.* 2006 with data from UNICEF).

One estimate of those exposed to >50 µg/l is 57 million as of 2005 (Mazumder, 2005), with 5 million projected to develop arsenicosis and 200 000 with symptoms of arsenicosis. However, the number of those exposed to concentrations of arsenic above 10 µg/l would be much larger, perhaps approaching 63 million, based on Caussy (2004). Some had claimed back in 2000 that only 30 million would be exposed (Mudur, 2000) yet other estimates were much higher at 79.9 million (Chowdhury *et al.* 2000). Other estimates of the number of people exposed to elevated levels of arsenic in their drinking-water above 50 µg/l vary between 28.1 to 35 million (Kinniburgh & Smedley, 2001; Rahman, 2002, World Bank, 2005a).

The most common arsenic-related health effects seen in Bangladesh are skin lesions and skin cancer, with the first cases being detected as early as 1984 (Dhar *et al.* 1997; Tondel *et al.* 1999; Karim, 2000; Smith *et al.* 2000; Anawar *et al.* 2002; Saha, 2002). Internal cancers, as well as increased rates of hypertension and diabetes mellitus, have also been reported (Rahman *et al.* 1998, 1999a,b, 2001). In some areas, high numbers of people may be affected. Biswas *et al.* (1998), for example, reported that in a survey of 52 villages, 57% of people examined showed skin lesions. Rahman (2002) reported that in a two-phase study, almost 11 000 patients suffering from arsenic-related health effects were identified. In one study, a strong association was found between lower intelligence quotient scores in Bangladeshi children and arsenic concentrations in drinking-water after adjustments were made for social and economic variables (Wasserman *et al.* 2004).

Field research continues in the effort to more precisely identify the components of arsenic contamination in Bangladesh. The USGS is currently conducting research on the levels of arsenic contamination of deeper aquifer systems in Bangladesh as well as in areas associated with high levels of arsenic in the shallow groundwater. This research integrates geological, hydrological, microbiological and geochemical data to characterize the deep aquifer system and to develop GIS coordinates that will identify locations of safe groundwater. This characterization of the hydrogeological framework is critical to identifying potential safe groundwater sources in the area.

Important questions to be addressed are: To what extent can deep aquifers of the Brahmaputra–Ganges delta be reliably used to provide an alternative drinking-water source? What is the quantity of water that can be produced from these deep aquifers? What is the expected longevity of these aquifers before they begin to draw down the water table or before arsenic begins to invade them?

The mobilization of arsenic may be connected with the anthropogenic expansion of irrigation in the Brahmaputra delta, in which the arsenic is mobilized in surficial waters and actually seeps down into deeper aquifers. Saunders *et al.* (2005) provided a model of how erosion of mountain belts aided by glaciers facilitates the transport of arsenic by surface waters, the adsorption of arsenic by stream sediments and the deposition of stream sediments and organic matter in alluvial deposits. The subsequent mechanism involving iron(III)-reducing bacteria present in alluvial aquifers is shown to cause the release of sorbed arsenic into groundwater under moderately reducing conditions.

Since the cost of very deep wells is prohibitive for the population as a whole, multiple options need to be considered based on risk factors and local conditions. There is no single best solution. A number of mitigation measures have been put forward. These include marking wells as safe or unsafe and using a known safe supply, identifying alternative water supply options to borehole wells, such as rainwater harvesting, treatment of surface waters, protected shallow well and/or use of arsenic treatment technologies, which can be either household-scale or community-scale arsenic removal technologies. A cost-effectiveness comparison of source and household systems to treat either arsenic or microbially contaminated water is shown in Figure 3.7.

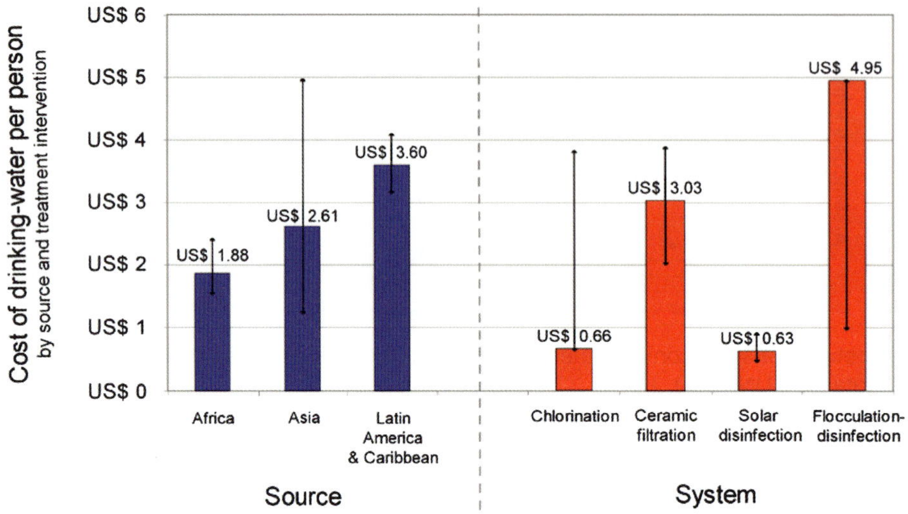

Figure 3.7 Mean annual cost of drinking-water per person in US$ by source (average costs of standposts, boreholes, dug wells) and by household drinking-water treatment intervention (error bars represent range of costs) (modified from Clasen, 2006).

3.5 CAMBODIA (G)

Cambodia, with a population of 14.7 million, has a Human Development Index rank of 124 out of 169 and a per capita GNI of $US 1,868 (US Census Bureau, 2011; UNDP, 2010).

Some 100 000 family-based wells are used for drinking water in Cambodia (Buschmann, 2007) and over 60 000 of these wells are identified in the WellMap database. Elevated arsenic concentrations (>50 μg/l) in groundwater in Cambodia are intimately but not exclusively associated with young Holocene sediments in the Mekong and Bassac River deltas downstream of Phnom Penh. However, high-risk levels of arsenic (10–50 μg/l) are also found in Quaternary deposits and even occasionally in older deposits (Figure 3.8). Six provinces have known levels of elevated arsenic risk: Kampong Cham, Kampong Chnang, Kampong Thom, Kandal, Kracheh and Prey Veng (RDIC, 2012), as shown in Table 3.1 below. UNICEF identifies peri urban Phnom Penh as a seventh area (UNICEF, 2009).

The total population of these 6 provinces and 1 per-urban area is 2.25 million of which 25% use tube wells. Thirty-five percent of tube wells within this area were found to be contaminated with arsenic above the current Cambodian National Standard of 50 μg/l (UNICEF, 2009). The 2005 estimate of the maximum population at risk from arsenic exposure is 320 000 (World Bank, 2005a). In contrast, Ravenscroft, in a report prepared for UNICEF, estimates the population at risk as 2.42 million (Ravenscroft, 2007). By "at-risk" Ravenscroft is referring to "the number of people who develop arsenicosis, who are only a small proportion of the people who consume more than the maxium recommended levels of arsenic."

The Mekong River extends 400 km within Cambodia before entering the Viet Nam delta; except for the northernmost 40 km, annual flooding (May–October) inundates the river course (Polizzotto *et al.* 2005; Stanger *et al.* 2005). Tube wells provide a significant source of drinking-water, and many of these are found within the flood zone. However, in the dry season, less flushing occurs, and higher concentrations of contaminants are found in the groundwater. So far, this regime has not been properly documented.

Arsenic in Cambodia originates in the geology of the upper Mekong basin and then manifests as alluvial arsenic in the lower Mekong basin. Three different scenarios affect the presence of arsenic: higher concentrations of arsenic in Holocene aquifers, lower concentrations of arsenic in Pleistocene aquifers and the ubiquitous presence of arsenic in soils that are regularly inundated during the monsoonal period. For instance, of 599 sampled tube wells in Holocene aquifers in the lower Mekong basin, 50% have arsenic concentrations below 10 μg/l, whereas 50% have arsenic levels above 10 μg/l. Of the latter, 113 tube wells (or 19% of the total) had arsenic concentrations above 50 μg/l (World Bank, 2005a).

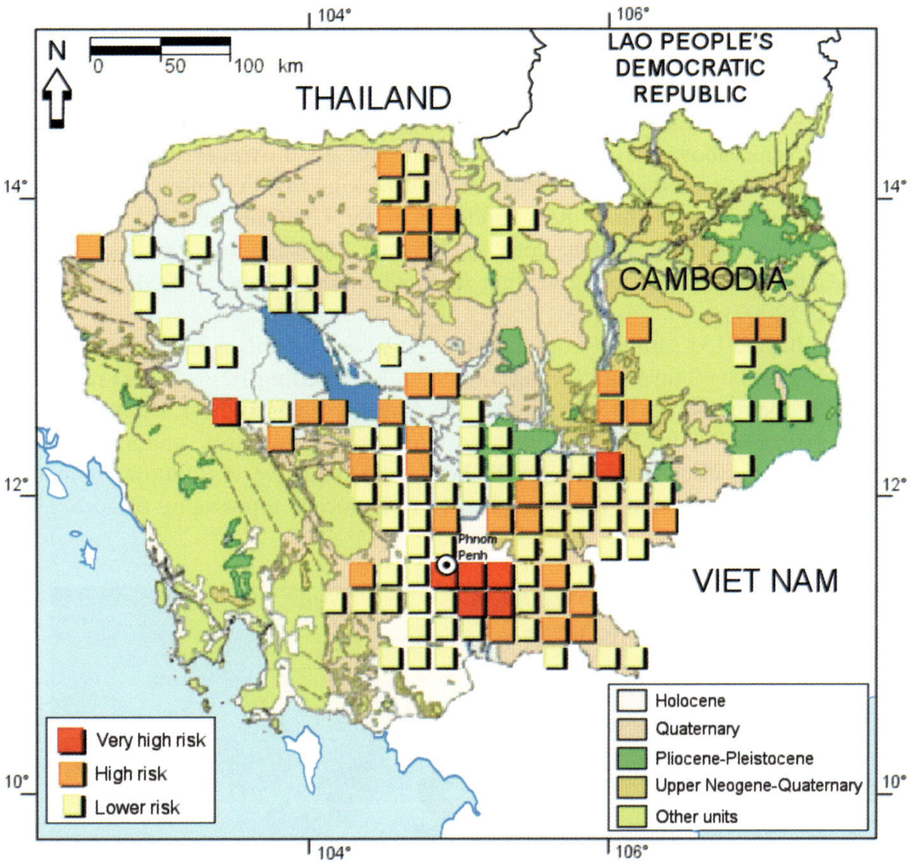

Figure 3.8 Geological map of Cambodia (modified after ESCAP, 1993; and Diebe, 2004), with 20-km grid squares showing areas of very high arsenic hazard (>50 μg/l) and high arsenic hazard (10–50 μg/l) in well water in Holocene sediments, especially near the Mekong and Bassac rivers downstream of Phnom Penh. Note, however, that high groundwater arsenic hazard is also found in Quaternary deposits and even occasionally in older deposits. (Reproduced with the kind permission of the Mineralogical Society of Great Britain and Ireland from a paper by Polya *et al.* 2005; for a higher resolution groundwater hazard map of Cambodia, see Figure 3.13).

Table 3.1 Provincial Summary of Arsenic in Cambodia.

Province	Tested Wells	Average As (ppb)	% Wells ≥ 50ppb	% Wells ≥ 250ppb	Estimated Impacted Pop.
Kampong Cham	7775	52.5	36.5	6.8	25552
Kampong Chhnang	1083	11.0	8.1	0.7	3932
Kampong Thom	1955	10.9	6.6	0.1	2358
Kandal	17 709	94.7	35.9	17.4	69593
Kracheh	1010	16.3	13.8	1.3	1144
Prey Veng	10 648	32.0	16.8	5.2	40922

Source: RDIC, 2012.

Elevated arsenic levels in these Holocene aquifers are believed to be related to the release of arsenic in a reducing and slightly alkaline environment, which occurs in both shallow aquifers (25–70 m, with a peak arsenic concentration at 60 m) and deeper aquifers (100 m or more). This chemical environment is not present in Pleistocene aquifers, and so the levels of arsenic contamination are lower.

Arsenic Contamination Testing Results

The history of testing for arsenic contamination of drinking-water sources is relatively recent. One hundred tube wells from 13 provinces were monitored in a project supported by WHO Cambodia in 1999, and this included analysis of arsenic, fluoride, trace metals and pesticides. Nine per cent of the samples contained arsenic levels above 10 µg/l, with a range of 10–500 µg/l. Concentrations of arsenic above 10 µg/l were found in 5 of 13 provinces, with Kandal Province, near Phnom Penh, having concentrations greater than 50 µg/l (Feldman & Rosenboom, 2001). Following the investigations of Feldman and Rosenboom, several thorough surveys have been undertaken by UNICEF and the Cambodian Ministry of Rural Development (MRD). Significant work has been done on arsenic measurement by Mickey Sampson of Resources Development International-Cambodia, as well as by visiting researchers Dave Polya (2005), Benjamin Bostick *et al.* (2009) and Johanna Buschmann *et al.* (2007).

Buschmann *et al.* (2007) collected 131 samples from the Mekong River floodplain in 3 provinces (Kandal, Takeo and Prey Veng). Arsenic ranged from 1–1 340 µg/l (average 163 µg/l) with 48% exceeding 10 µg/l. This research team also determined that elevated manganese (57% above 0.4 mg/L) was an additional health threat to the 1.2 million people in these provinces. Arsenic in Kandal Province averaged 233 µg/l while to the west and east of the Bassac and Mekong Rivers it was <10 µg/l. Release of arsenic from Holocene sediments is thought to be due to reductive dissolution of metal oxides. Low arsenic regions appear to coincide with low topographic relief, a relic of pre-Holocene topography (Figure 3.9).

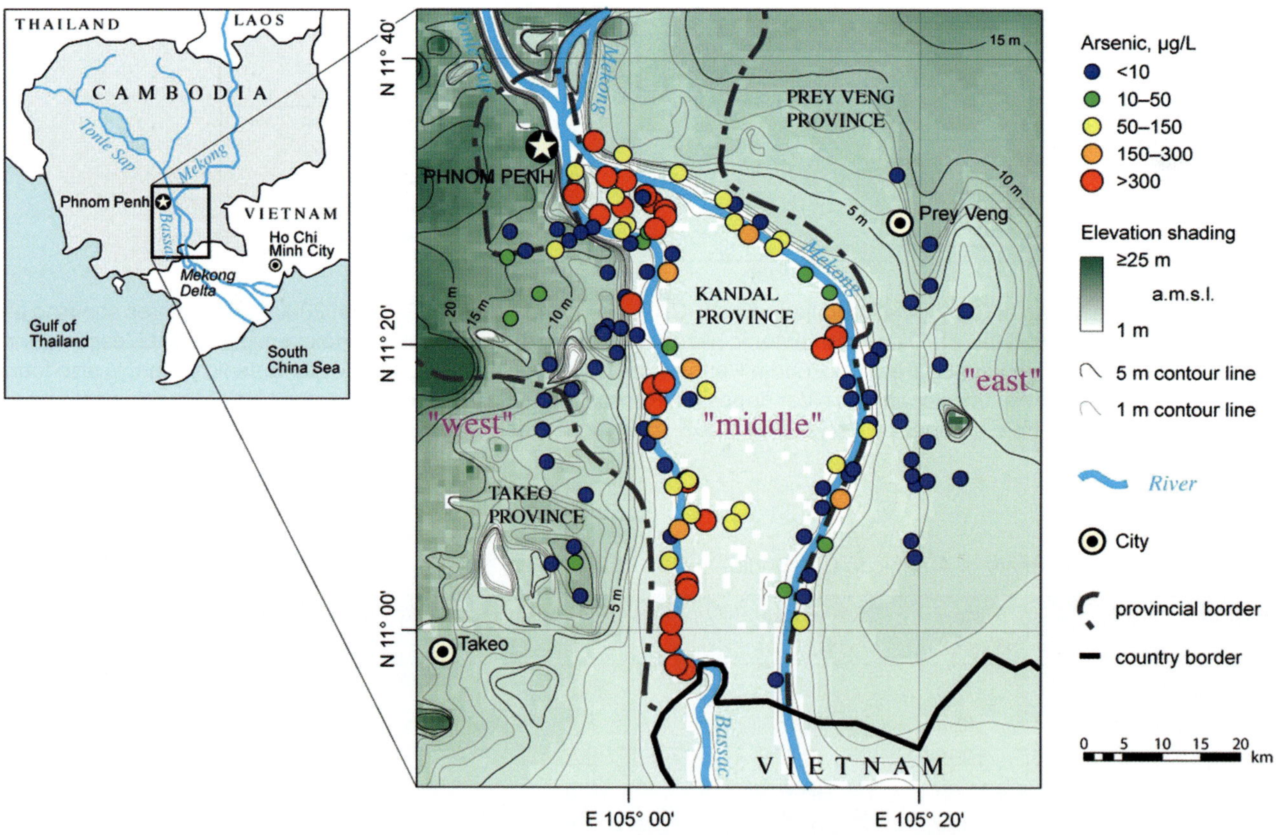

Figure 3.9 Arsenic concentrations measured in Cambodian groundwater collected in November/December 2004 and March 2006. Low relief topography is visualized by green shading as well as 1 and 5 contour lines. Note the highest elevation on this map is only 25 meters above sea level; only a few samples could be collected in the inner floodplain between the Mekong and Bassac Rivers because this area is swamp, scarcely populated and almost inaccessible (*Source*: Buschmann *et al.* 2007)

RDIC has conducted province-by-province groundwater surveying activities from 2005–2012, collecting samples and analysing them in their laboratory. Over 10,000 wells have been tested in five priority provinces. Data from 4 provinces (Kandal,

Prey Veng, Kampong Cham and Kampong Chnang) (Figure 3.10), showed mean and maximum arsenic concentrations of 28 µg/l and 1000 µg/l, respectively. Other parameters investigated included fluoride, nitrate, manganese and iron (RDIC, 2012) (Table 3.2).

Table 3.2 Groundwater quality data from tube wells in Kandal, Prey Veng, Kampong Cham and Kampong Chnang Provinces.

Parameter	Units	Samples	Mean	Max	Min	CDWQS[1]	# Exceeding CDWQS[1]	% Exceeding CDWQS[1]
Health-impacting								
Arsenic	µg/L	8488	28.28	1000	0	50	1060	12.49
Fluoride	mg/L	8484	0.68	110	0	1.5	357	4.21
Nitrate	mg/L as NO_3	8486	7.96	1361	0	50	262	3.09
Manganese	mg/L	8486	0.37	28	0	0.4[2]	1653	19.48
Aesthetic-impacting								
Iron	mg/L	8485	3.43	100	0	0.3	6492	76.51
Manganese	mg/L	8486	0.37	28	0	0.1	3433	40.45
Turbidity	NTU	8488	26.18	2960	0	5	3951	46.55
Chloride	mg/L	8487	107.63	2303	0	250	1733	20.42
pH		8482	6.72	9.95	0	6.5–8.5	2502	29.50
Hardness	mg/L $CaCO_3$	8488	202.92	3186	0	300	1787	21.05
Salinity	ppt	8473	0.32	19	0	N/A	N/A	N/A

[1] Cambodian Drinking Water Quality Standard (CDWQS)
[2] Manganese health guideline value of 0.4 mg/L from WHO

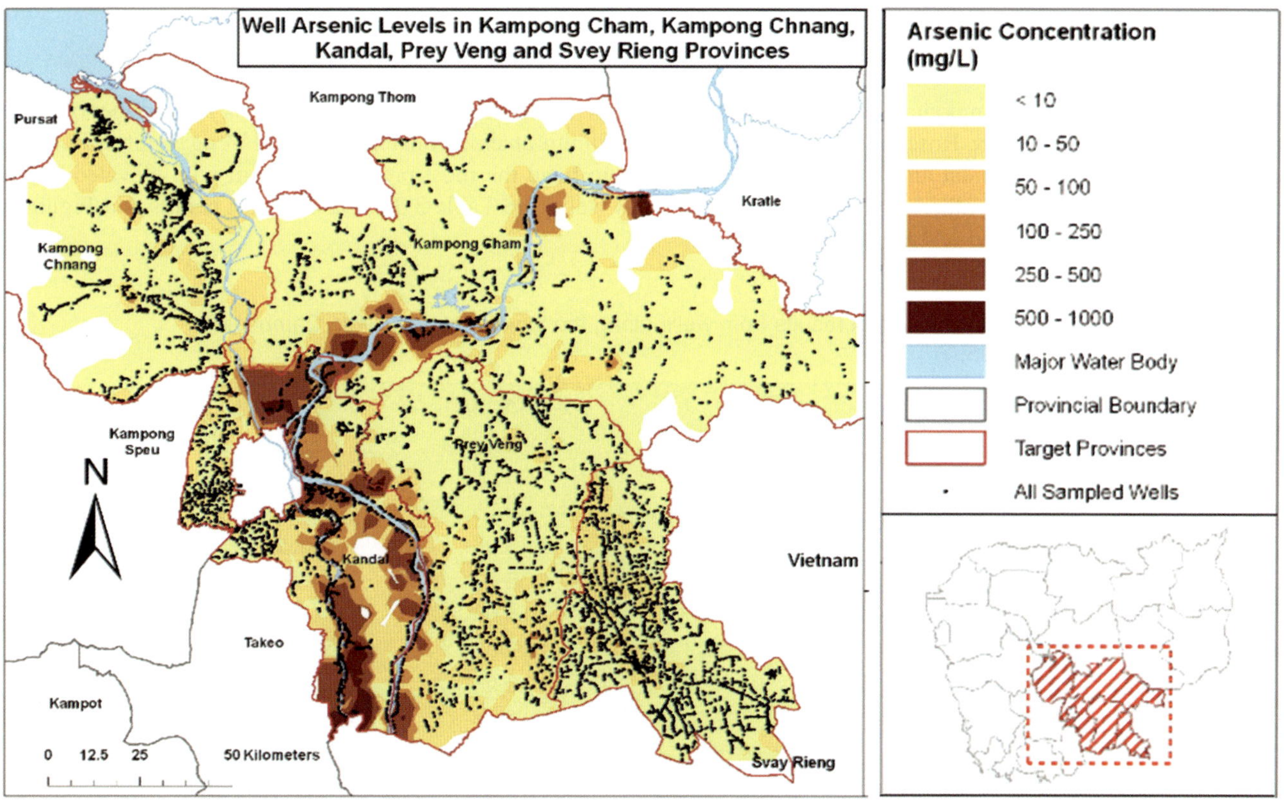

Figure 3.10 Arsenic Levels in Kampong Cham, Kampong Chnang, Kandal, Prey Veng, Svey Rieng Provinces (RDIC, 2012).

The Director of Research and Development at the Institute of Technology, Cambodia, reports on the overall national program of well testing as of 2009, indicating that the total number of tube wells tested to date is 47 950, of which 31 320 have been testing for arsenic (Uy, 2009).

There were 9,047 (= 29%) tube wells in risk areas (AS > 50 ppb), out of the 31 320 tube wells tested for arsenic.

The break down by province is as follows:

–Kandal	35%
–Kampong Chamampong	33%
–Prey Veng	19%
–Kracheh	15%
–Kampong Chhang	6%
–Kampong Thom	3%

Arsenic testing kits have been used to determine wells with arsenic concentrations above 10 μg/l in two additional provinces. Positive correlations between arsenic and iron and manganese concentrations as well as anaerobic conditions are common features of the groundwater throughout the lowland areas of Cambodia, which suggests that reducing conditions release arsenic into the water.

The results of groundwater arsenic screening in selected Mekong aquifers by UNICEF and the Japan International Cooperation Agency (World Bank, 2005a) are given in Table 3.3.

Table 3.3 Summary of arsenic concentrations in groundwater from tube wells in the Mekong Valley, Cambodia.

	No. samples (%)			Total no. samples
	As < 10 μg/l	As 10 –50 μg/l	As > 50 μg/l	
Holocene aquifer	301 (50)	185 (31)	113 (19)	599
Pleistocene aquifer	1184 (95)	59 (5)	3 (0.2)	1246

Source: World Bank, 2005a

Most tube well drilling in Cambodia was initiated in the late 1990s. In an early study designed to look at a number of possible water contaminants, Feldman & Rosenboom (2001) found a maximum arsenic concentration of 504 μg/l. Of the provinces investigated, Phnom Penh, Kandal and Kampong Cham seemed to have the most pressing problem with elevated levels of arsenic (>50 μg/l), but several other provinces are also at risk as well. (see Table 3.3 for arsenic test results and Figure 3.11 for a map of sites and associated arsenic levels).

Table 3.4 Arsenic test results from wells along the lower Mekong River, Cambodia, 2001–2005.

Province	No. samples	% >50 μg/l	% >500 μg/l
Phnom Penh	1	100	0
Kandal	1263	40.78	8.87
Kampong Cham	1454	15.06	0.34
Prey Veng	1188	6.06	2.27
Kracheh	1343	5.88	0.15
Kampong Chhnang	442	3.85	0
Krong Pailin	92	3.26	0
Kampong Thom	375	2.13	0
Svay Rieng	1607	1.43	0
Stung Treng	426	0.47	0

Source: Cambodia National Arsenic Database, 2005

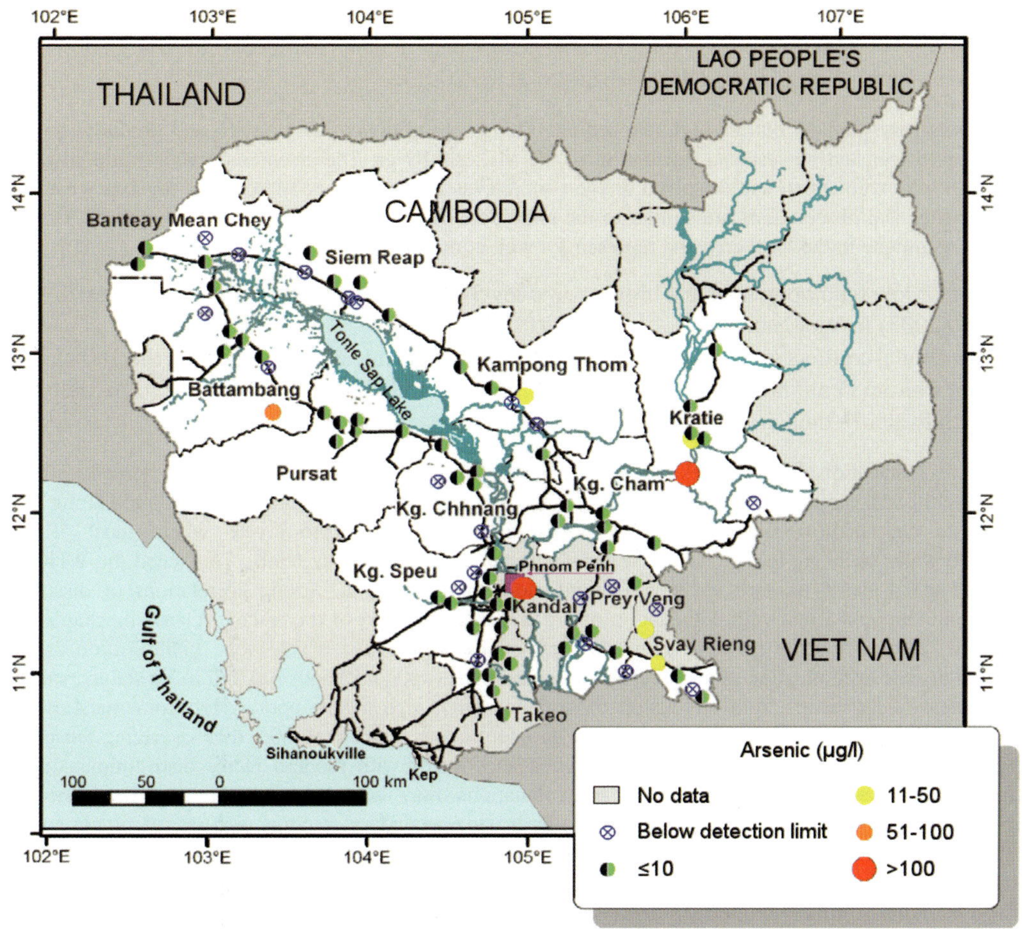

Figure 3.11 Arsenic levels of wells tested in Cambodia in 2005 (modified from Feldman *et al.* 2006).

The highest arsenic concentrations were found in Kien Svay and Takhmau districts, Kandal province, with 11 samples (nearly 70%) containing arsenic levels above 10 μg/l and 10 samples containing arsenic above 50 μg/l. Concentrations ranged up to 504 μg/l and averaged >130 μg/l. Here, household and community wells were approximately 40 m deep, while urban supply wells ranged in depth from 52 to 209 m (Feldman *et al.* 2006); hence, in this in research, well depth is not an indication of potential contamination.

However, according to Polya *et al.* (2005), there appears to be some correlation between arsenic concentrations and shallow wells in Holocene aquifers ranging from 25 to 70 m deep, with peak concentrations of arsenic at about 60 m (Figure 3.12; Polya *et al.* 2005).

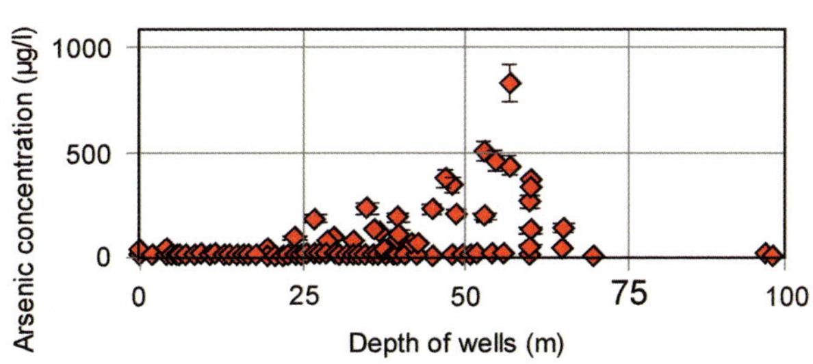

Figure 3.12 Correlation between depth of wells and arsenic concentrations in Cambodia (modified from Polya *et al.* 2005).

All arsenic-bearing wells in the lower Mekong River have high iron and manganese levels and a low oxidation–reduction potential, a characteristic of Quaternary-age alluvial and Holocene deltaic sediments in Bangladesh, north-east India's Gangetic Plain and Viet Nam's Song Hong delta (Polizzotto *et al.* 2005):

> Arsenic concentrations vary spatially but are high (exceeding 100 µg/L) throughout the aquifer, and groundwater levels along with recharge gradients are dominated by seasonal fluctuation of the Mekong River. The arsenic-containing groundwater is reduced, as evidenced by low Eh [redox potential] and dissolved oxygen and elevated ferrous iron; additionally, in some cases dissolved sulfide is detected. Our geochemical observations are similar to the conditions reported for the Ganges-Brahmaputra Delta in Bangladesh, while the hydrology is unique to the local area and represents a well constrained system.

This co-occurrence of low redox potential and high arsenic levels has been confirmed by Feldman *et al.* (2006) in Cambodia and by Treng & Viet (2005) in Viet Nam.

The age of shallow fluvial aquifers is another parameter determining arsenic concentration. In the 180 km^2 study area of Kien Svay district, south of Phnom Penh, the composition and age of aquifer sediments are believed to partially control the arsenic concentration of groundwater (Yount *et al.* 2005):

> The sedimentary facies association along the Mekong differs from the larger, sandier fluvio-deltaic environments of Bangladesh in that distinct channel-levee–flood basin sequences are closely associated across the floodplains of the sub parallel Mekong and Bassac Rivers. Organic-rich marshes and shallow lakes within the flood basins lie adjacent to levees where most of the villages and arsenic-contaminated water wells are located. Of 27 water wells sampled (20 to 65 m depth), 16 exceed the WHO arsenic drinking water guideline (10 µg/L) and 5 produce water containing more than 500 µg/L. Strong correlations of dissolved arsenic with alkalinity and ammonia are consistent with degradation of organic matter as part of the arsenic release mechanism. Excavations as deep as 5m exposed sediment with as much as 11 wt.% Corg interbedded with red paleosols. Juxtaposition of contrasting redox conditions results in arsenic accumulation in peaty layers and gray muds (up to 450 ppm As) and in Fe-stained sands (230 ppm As). High dissolved arsenic likely results from interaction of organic carbon-rich sulfate-poor water from marshes with ferric oxide coated sand. Drillers logs indicate aquifers are more tabular and continuous from 40 to 60m than overlying lensoidal levee-channel complexes of present rivers. Yet the co-occurrence of high arsenic wells with modern facies distribution suggests the modern environment influences arsenic concentration. Possible explanations may be: a) juxtaposition of organic-rich facies against channelized or tabular sand bodies containing significant Fe-hydroxide secondary coatings or b) a hydrologic connection of facies produced by the common practice of lining well casings with coarse sand to the ground surface. An area of consistently low arsenic water draws from erosional remnants of Pleistocene sediment based on surface exposures. The likely irregularity of the buried Holocene-Pleistocene contact further complicates interpretation of aquifer distribution. Although wells in the Pleistocene sediment are low in arsenic, high manganese may limit these aquifers as drinking water supplies.

These results show that micro testing of each well is necessary, because nearby wells can differ in orders of magnitude (by as much as 250 µg/l) in the arsenic concentrations they carry due to the irregularity of the distribution of aquifers. Since testing of well water requires field sampling and laboratory testing within time and sampling parameters, this effort has been hindered by physical, financial and laboratory constraints. One possible solution to this problem is the sampling of human biomarkers that contain stable arsenic: nails, hair and urine. These human biomarkers are more stable than arsenic in water, and the samples are not subject to the vagaries of in-field assessment. Another possible solution is the application of computer modelling.

Figure 3.13 shows results obtained by a regression-kriging computer modelling approach. This appears to provide a good indicator of the potential areas at risk for arsenic exposure since it includes most of the rich organic sediments within the most critical areas. In this model (Lado *et al.* 2008):

> Arsenic concentration in shallow groundwater in Cambodia was estimated using 1329 georeferenced water samples collected during the period 1999–2004 from wells between 16–100 m depth. Arsenic concentrations were estimated using block regression-kriging on the log transformed arsenic measurements. Auxiliary raster maps (DEM-parameters, remote sensing images and geology) were converted to 16 principal components that were used to explain the distribution of arsenic over the study area. The regression-kriging model was validated using an external set of 276 samples, and the results were compared to those obtained by ordinary block kriging. The regression analysis revealed that there is a good correlation between topographic environmental variables and the content of arsenic in groundwater. This result is broadly consistent with the findings of previous studies and is not unexpected given models of microbial mediated arsenic mobilization in recent low lying sediments. Kândal, Prey Vêng and Kâmpóng Cham are the provinces with the highest potential arsenic hazard, indicating the requirement for development and implementation of policy control measures. The regression-kriging model explained 48% of the variability in the validation set. However, the model does not show good results for the prediction of high arsenic concentration. This points to the existence of local environmental factors, not captured by this model, that highly influence the mobilization of arsenic in groundwater… The regression-kriging approach provides a realistic description of the distribution of arsenic since it captures the large-scale variation of arsenic in the study area.

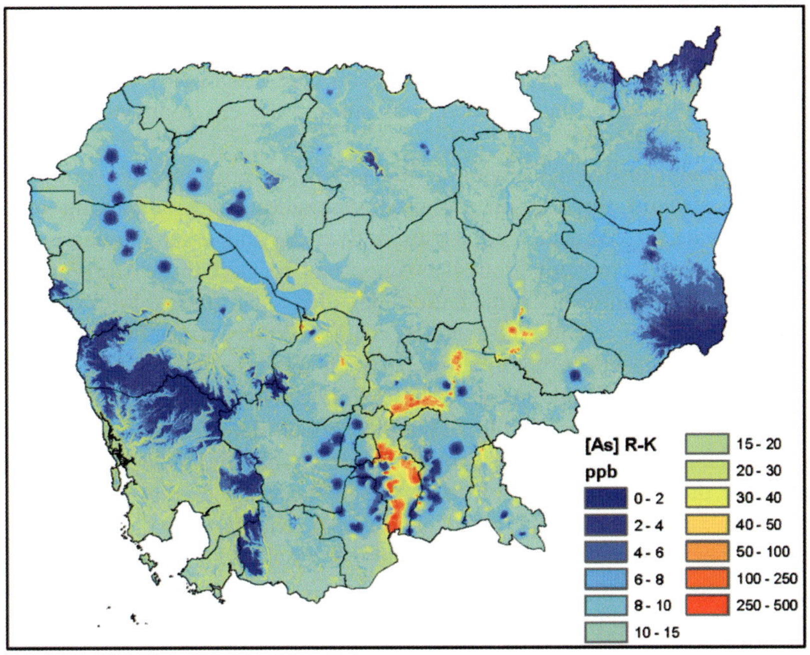

Figure 3.13 Estimation of arsenic content in 16–100 m depth groundwater (lg/L) by regression mapping. Lado *et al.* (2008). (Reprinted with permission from Elsevier, 2012)

Policy and Remediation

External constraints, such as the interaction of government agencies and nongovernmental organizations involved in the testing programme, the levels of their equipment and/or expertise and the funding and personnel levels involved, are vitally connected with the success of any arsenic remediation programme. For this reason, it is important to review the current status of the arsenic testing programme in Cambodia.

Dr Mao Saray, Chairman of the Arsenic Task Force, outlined the following scenario of current and future ministerial and nongovernmental organization coordinated activities to expand the current level of testing of groundwater in Cambodia (M. Saray, personal communication, 2005). As of December 2005, 7 provinces and 1600 villages had been identified as being at high risk (arsenic levels above 50 μg/l), but this represents only 20% of the total number of villages in these provinces. Therefore, a rapid assessment of water from rural borehole wells and surface-dug wells in 13 provinces that are not now tested should be tested to assess the wider scope of the arsenic contamination problem. The Provincial Department of Rural Development is responsible for testing and for education assisted by UNICEF, and only one commercial test kit has been used extensively. The current practice is to paint the wells red if water contains arsenic levels above 50 μg/l. This is to indicate that people are not supposed to use the water for drinking or cooking but can use it for gardening and animals. However, in a study by UNICEF in 2009, it was found that only 12% of 4,448 respondents (52.5% women) were aware of arsenic (UNICEF, 2009). This indicates that education on the dangers of arsenic is not reaching the people responsible for water management in the home. Private wells are even more difficult to test and monitor, because testing requires the approval of the owners, so these wells fall outside of the jurisdiction of testing and education generally. However, in those areas identified at high arsenic risk, all wells are being tested.

The governmental responsibilities for water quality in Cambodia are divided among the following ministries: Ministry of Industry, Mines and Energy, Ministry of Environment (urban water quality), Ministry of Water Resources and Meteorology (water policy) and Ministry of Rural Development (safe water supply and sanitation at the rural level). These ministries are in theory coordinated by an Inter-Ministerial Coordinating Committee, vice-chaired by the Under-Secretary of State of Rural Development. The Inter-Ministerial Committee on Arsenic is vice-chaired by directors from the Ministry of Rural Development, Department of Rural Health Care and Department of Rural Water Supply. Each Ministry has 5–7 Secretaries of State as well as an equal or greater number of Undersecretaries of State, creating jobs but not necessarily carrying out important duties in the service of peoples' health.

Laboratory facilities in Cambodia are spread among different governmental agencies that are in competition for limited funding. The Ministry of the Environment has a large central laboratory. As of 2010, there is a new, well-equipped central laboratory of the Ministry of Industry, Mines and Energy. Also, the laboratory at the Phnom Penh Water Supply Authority (PPWSA) can be used for specific tests, although it is not available on a commercial basis. Additional laboratories include small facilities, such as the Dr Pasteur Laboratory, which is a health care laboratory, not an environmental laboratory, plus two university laboratories which are grossly underfunded, and one private environmental lab run by the NGO Resources Development International-Cambodia (RDIC),

There is a 2-fold need in water quality testing in Cambodia: first, to develop university, governmental and provincial staffs for testing water quality (arsenic, magnesium, fluoride, microbial contamination) and for training and education in hygiene and sanitation; and second, to coordinate government laboratories, university and other laboratories. These needs are basic to expanding in-country testing for arsenic contamination. It is important to alert those working in the field to this in-depth water quality information, since responsibilities for remediation may cross institutional lines of command and control.

A positive step in this direction of coordinated action is WellMap, an online well database created in 2010 (www.cambodiawellmap.com). This is a cooperative effort between the government of Cambodia, the Water Sanitation Programme (WSP), the Asian Development Bank (ADB), UNICEF, and two the non-profit organizations, World Vision and Resources Development International-Cambodia (RDIC). This online well databases include over 62 000 wells. The system allows users to query and show data available for each known well in Cambodia, including construction details and physical characteristics, plus water quality data. Arsenic is included as one of the water quality parameters; however, entries for this parameter appear to be scant compared to other parameters such as pH, chloride, iron, manganese, salinity, nitrate, turbidity and hardness. Nevertheless, in terms of inter-organizational cooperation and transparent data-sharing, it is an excellent resource and is a model for other countries to follow.

Government agencies, water experts and public health professionals must take vigorous action to mitigate the problem of chronic arsenic poisoning in order to protect 1–2 million people in the Mekong Delta who are drinking untreated groundwater (Buschmann *et al.* (2008). Even though the National Arsenic Strategic Plan 2007–2012 was developed by the Ministry of Rural Development, with assistance from International agencies and donors, in recent years the political environment has not been favourable to addressing arsenic. Action has been slow and the time is long overdue to implement this Strategic Plan. The following are pressing recommendations for Cambodia as regards arsenic remediation:

(1) Take action on arsenic mitigation with attention to the following issues: areas affected, timeline for implementation and system costs.
(2) Hold politicians accountable. Engage government in strategic discussions and lobby for the United Nations-recognized human right to safe drinking water.
(3) Update and maintain the database of arsenic contaminated wells, and ascertain how many people are actually affected. This includes annually updating WellMap, which appears not to have been updated since its creation in 2010.
(4) Remap arsenic-contaminated water areas, especially in mining sites on the borders with neighbouring countries (Viet Nam, Thailand and the Lao People's Democratic Republic).
(5) Test and remediate school wells, because many have high arsenic concentrations and are generally left out of government planning.
(6) Widespread evidence of arsenicosis is lacking for Cambodia due to the nature of the erratic occurrence of highly contaminated wells. More extensive public education programmes are needed to convince affected populations of the risks of exposure to arsenic contamination.

Mitigation has been the focus of the research of Samnang Chea *et al.* (2008) and Davin Uy (2009) together with partner organizations. Uy (2009) describes the efforts in arsenic mitigation currently taking place in Cambodia. He urges government, international agencies, NGOs and stakeholders to:

- Encourage the use of alternative water (surface, rain and dug well) by the Ministry of Rural Development and UNICEF;
- Initiate the Environmental Technology Verification process under the support of the Water Sanitation Programme;
- Apply piped water systems provided under private and/or public partnerships only in the capital;
- Experiment with household and village scale arsenic removal systems.

3.6 CHINA (A, C, G, M, P)

China, with a population of 1.337 billion, has a Human Development Index rank of 89 of 169 and a per capita GNI of $US 7258 (US Census Bureau, 2011; UNDP, 2010).

Elevated levels of arsenic have been reported from the south-west coastal zone of the Province of Taiwan and also the north-eastern parts of the island (Hsu *et al.* 1997). In mainland China, elevated arsenic concentrations have been reported from

Inner Mongolia Autonomous Region (Ba Meng, Tumet Plain, Huhhot Basin and Biyingao), Xinjiang Uygur Autonomous Region and Shanxi Province (Luo *et al.* 1995, 1997; Niu *et al.* 1997; Guo *et al.* 2001; Sun *et al.* 2001b; Tian *et al.* 2001), affecting an area of approximately 68 000 km^2 (Smedley & Kinniburgh, 2002). In 2001, Jinlin, Ningxia Hui Autonomous Region and Qinghai were also found to be affected (Sun *et al.* 2002). Over 20 000 cases of chronic arsenicism have been reported from China (Sun *et al.* 2001b). Reported effects include skin lesions, including skin cancer, lung, liver and bladder cancer (Luo *et al.* 1995; Luo *et al.* 1997; Tsai *et al.* 1999; Guo *et al.* 2001; Sun *et al.* 2002).

The initiation of studies concerned with the presence of arsenic in drinking-water began in 1975 (1983 according to official sources) in Xinjiang Uygur Autonomous Region, an area composed mainly of Uygurs, a Turkic-Mongolian people who had the first diagnosed arsenic-induced cutaneous cases of keratosis (Jin, 2006). This diagnosis was followed by epidemiological studies in Inner Mongolia Autonomous Region in 1989, where positive results were found. In 1992, the National Ministry of Health in China initiated a national survey of arsenic contamination and surveyed Jilin, Liaoning, Ningxia Hui Autonomous Region, Qinghai and Shanxi provinces in the 1990s (Luong *et al.* 2004). By 2003, 50 000 wells had been tested, and those containing arsenic concentrations above 50 μg/l were mapped and physically marked by the Ministry of Water Resources, revealing that 8 continental provinces/regions (excluding Taiwan) and 40 counties had 2 343 238 people at risk (Figure 3.14; Table 3.5), 522 566 people exposed to drinking-water with arsenic levels of ≥50 μg/l and 7821 confirmed cases of arsenicosis (Jin *et al.* 2003). It was estimated in 1997 that over 5.6 million people were affected by drinking-water arsenic concentrations greater than 50 μg/l (Zhang & Chen, 1997).

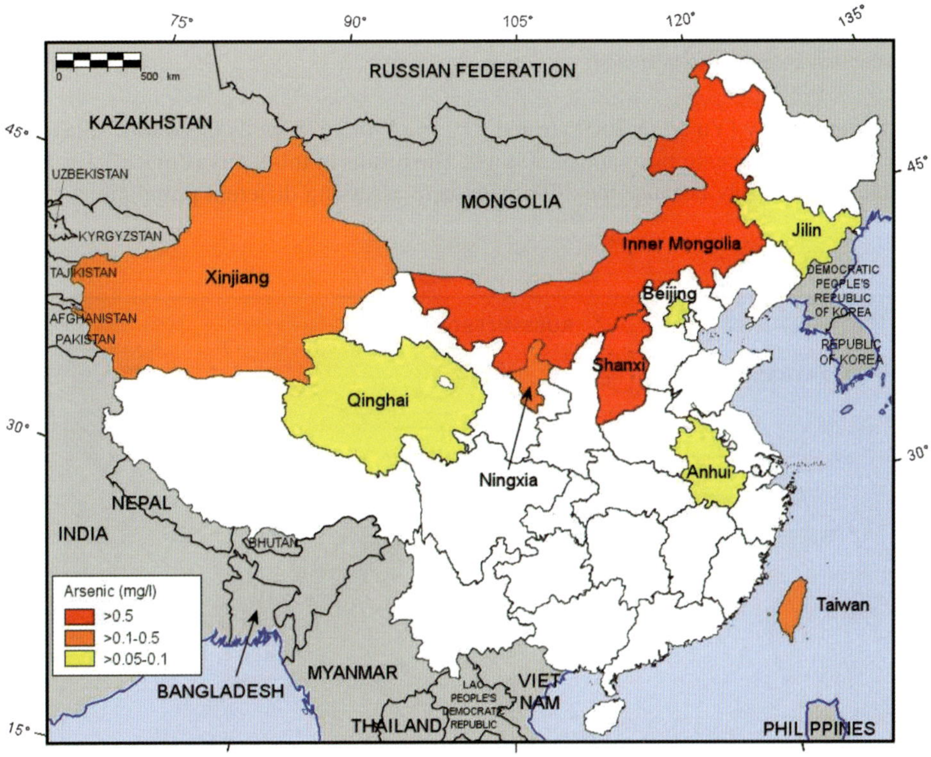

Figure 3.14 Elevated arsenic levels in drinking-water in several provinces/regions in China (modified from Xia & Liu, 2004).

By 2004, 21 of the 30 provinces of China were known to have drinking-water containing arsenic at concentrations above 50 μg/l (Luong *et al.* 2004). It was estimated in 2004 that 4.5 million were exposed to arsenic concentrations ranging from 200 to 4440 μg/l in Liaoning, Jilin, Ningxia Hui Autonomous Region and Qinghai alone (Luong *et al.* 2004). By 2006, from these and additional data sources, the population exposed to arsenic levels above 10 μg/l in China as a whole was estimated at greater than 4.8 million people, or more than double the 2003 estimate. The World Bank's 2005 estimate is 5.6 million (World Bank, 2005a). While the data on arsenic contamination of drinking-water in China are often fragmentary, they are consistently incremental, indicating that the population exposed to arsenic-contaminated drinking-water at levels above 50 μg/l is of the order of 5–6 million. To this figure should be added the 1.3 million exposed to high-arsenic coal fumes as well as the little-investigated leaching of arsenic from high-arsenic coal beds into aquifers in Guizhou, Inner Mongolia Autonomous Region, Ningxia Hui Autonomous Region, Shanxi and Yunnan.

Table 3.5 Populations exposed to high arsenic concentrations in drinking-water in several provinces/regions in China.

Province/region	No. counties	No. towns	No. villages	High-arsenic exposed population	Maximum arsenic level (mg/l)
Inner Mongolia Autonomous Region	13	83	776	1 025 175	1.86
Shanxi	18	44	116	932 086	0.783
Xinjiang Uygur Autonomous Region	1	10	50	143 164	0.09
Taiwan	n.a.	119[a]	n.a.	110 000[a]	1.8 (mean 0.5)[a]
Jilin	2	5	71	58 859	0.36
Ningxia Hui Autonomous Region	2	3	22	24 550	0.10
Qinghai	1	1	2	12 219	0.318
Anhui	2	5	5	86 879	0.15
Beijing	1	3	5	60 306	0.143
Total	**40**	**154**	**1047**	**2 343 238**	

n.a.: not available
[a]Lo *et al.* (1977)
Source: Xia & Liu, 2004 (except where otherwise indicated)

From 2001 to 2005, another study (Yu *et al.* 2007) tested 21 155 of 445 638 wells in 20 517 villages in 292 counties in 16 provinces from China with arsenic concentrations above 50 µg/l. The results of this study (Figure 3.15; Table 3.6) are at variance with those shown in Table 3.5, possibly because the 2004 data were assessing different populations.

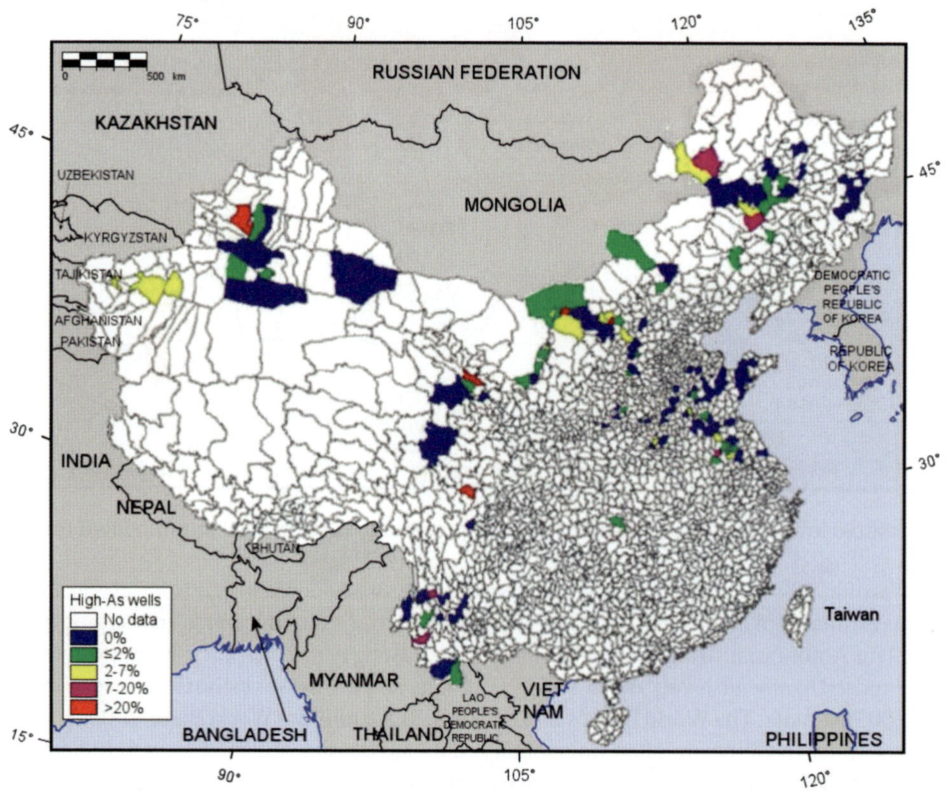

Figure 3.15 Percentage of wells containing high concentrations of arsenic (>50 µg/l) at the county level in China from a 2005 survey. Percentages are based on field tests of 445 638 wells (modified from Yu *et al.* 2007).

Table 3.6 Distribution of wells containing high levels of arsenic in 16 provinces/regions in China.

Province/region	No. wells tested	No. villages tested	No. counties tested	Population tested	No. wells >50 µg/l	No. villages >50 µg/l	No. counties >50 µg/l	Population exposed >50 µg/l	Maximum arsenic (µg/l)	Years tested
Shanxi	48 583	2 120	30	1 455 294	6 077	144	18	115 865	≥500	2001–2003
Qinghai	10 003	102	18	106 344	1 081	30	7	21 745	≥500	2002–2003
Sichuan	3 870	53	3	36 000	359	8	1	860	≥600	2003
Inner Mongolia Autonomous Region	130 104	2 691	23	1 429 564	8 142	324	18	109 161	≥600	2001–2003
Jilin	42 900	2 063	9	1 027 198	2 598	283	6	127 920	≥500	2002–2003
Xinjiang Uygur Autonomous Region	29 747	1 086	12	1 226 090	903	193	12	100 000	≥500	2002–2003
Gansu	5 016	538	13	495 059	133	33	2	22 954	≥250	2005
Anhui	24 033	739	5	115 427	630	45	3	2 833	≥1000	2003
Jiangsu	9 604	410	15	773 609	220	n/a	3	No data	≥300	2005
Ningxia Hui Autonomous Region	27 432	2 517	22	285 144	510	112	6	34 590	≥100	2002–2003
Henan	28 068	1 635	61	1 470 112	192	4	3	7 855	≥500	2003
Heilongjiang	43 344	420	15	184 542	220	73	5	No data	≥100	2003
Yunnan	9 535	2 978	16	1 029 755	33	10	5	6 839	≥200	2005
Shandong	19 899	2 419	46	n.a.	49	30	8	31 799	≥50	2003
Hunan	10 000	721	1	614 877	8	5	1	348	≥50	2005
Liaoning	3 500	25	3	13 000	0	0	0	0	<50	2002
Total	445 638	20 517	292	10 262 015	21 155	1 294	98	582 769		2001–2005

n.a.: not available
Source: Yu *et al*. 2007

Based on this 2005 study, the percentage of wells in each of the 15 of 16 provinces tested that contained levels of arsenic above 50 μg/l are given in bar graph form in Figure 3.16.

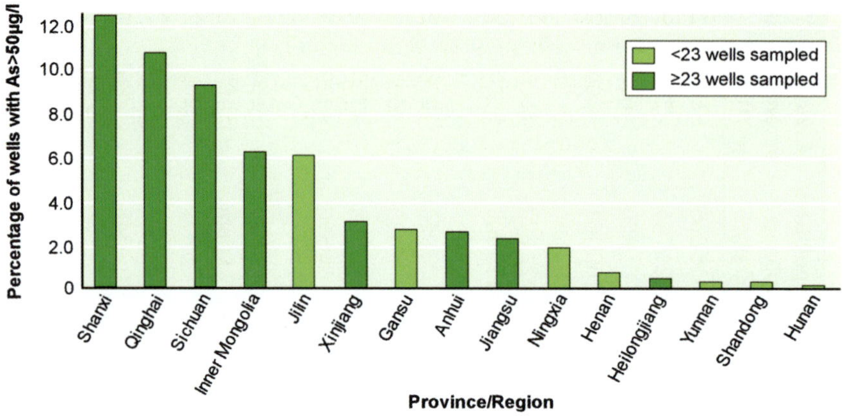

Figure 3.16 Wells containing arsenic levels above 50 μg/l in 15 provinces/regions in China. Lighter shaded bars indicate that less than the median value of 23 wells were tested in each village from these provinces/regions ($n = 6$) and that the province/region is therefore under surveyed (modified from Yu *et al.* 2007).

Based on Yu *et al.* (2007), the rates of diagnosed arsenicosis patients are given in Figure 3.17. The figure indicates a greater percentage of the total number of cases being moderate in Shanxi, Qinghai, Jilin, Xinjiang Uygur Autonomous Region and Ningxia Hui Autonomous Region.

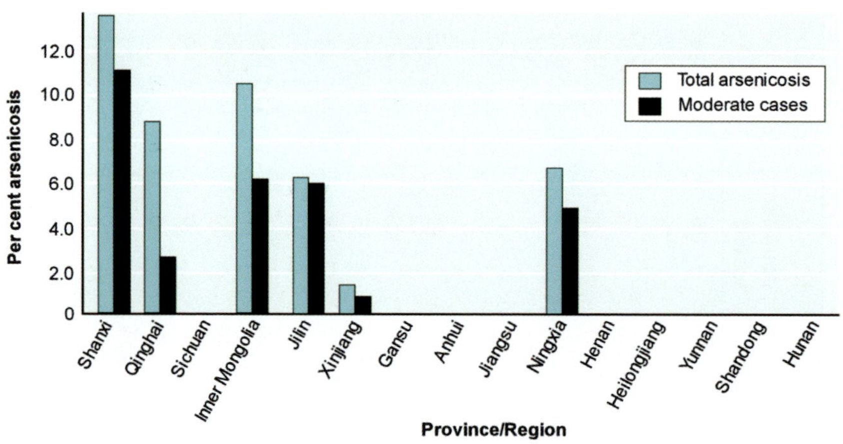

Figure 3.17 Rate of arsenicosis in six provinces/regions in China (modified from Yu *et al.* 2007).

Exposure to arsenic-contaminated water has increased in rural China since the initiation of the economic reforms in 1978 that enabled peasants to afford household wells 20–30 m deep fitted with hand pumps that tap water mostly from arsenic-rich aquifers. Previously, this sector had used highly saline, bitter and astringent or fluoride-rich water from traditional shallow wells and/or surface water from afar (Luong *et al.* 2004).

Mitigation measures include source replacement (e.g. in Xinjiang Uygur Autonomous Region and Province of Taiwan), with the provision of a tap water supply and treatment. Treatment options include chlorine addition to oxidize arsenite, followed by the addition of iron(III) chloride, sedimentation and filtration, and use of a reverse osmosis system on tube wells where a safe alternative supply cannot be provided (Sun *et al.* 2001b).

Four different geological zones are associated with arsenic contamination in China. The first is associated with areas subject to seasonal aridity and with low flushing and recharge rates of young anaerobic Holocene aquifers, which are most susceptible to high arsenic concentrations in wells 20–30 m deep. This situation occurs in areas such as Inner Mongolia Autonomous Region. A second zone is associated with mining or natural leaching where arsenopyrite-associated minerals are found, such as in Anhui, Jilin, Ningia, Qinghai and Sichuan. Often this zone overlaps with the next zone, being associated with Carlin-type gold deposits that may overlay coal seams. This third zone occurs in areas where high-arsenic coal is mined, such as in Guizhou, Yunnan, Shanxi and Ningia provinces and Inner Mongolia Autonomous Region. A fourth geological zone is associated with petroleum extraction, such as in Xinjiang Uygur Autonomous Region and Heilongjiang.

In the first geological zone indicated above, anaerobic Holocene aquifers are found in the Province of Taiwan, Inner Mongolia Autonomous Region, Xinjiang Uygur Autonomous Region and, to a lesser extent, Shanxi. Aquifers associated with reducing conditions and their accompanying groundwater conditions are summarized in Table 3.7.

Table 3.7 Aquifer type and groundwater conditions in several provinces/regions in China.

Province/region	Aquifer type	Groundwater conditions
Taiwan	Sediments, including black shale	Strongly reducing, artesian conditions, some ground waters contain humic acid
Inner Mongolia Autonomous Region	Holocene alluvial and lacustrine sediments	Strongly reducing conditions, neutral pH, high alkalinity; deep ground waters often artesian, some have high concentrations of humic acid
Xinjiang Uygur Autonomous Region	Holocene alluvial plain	Reducing, deep wells are artesian
Shanxi	Alluvial plain	Possibly reducing

Source: Smedley & Kinniburgh, 2002

The second and third geological zones mentioned above are associated with the mining of minerals associated with arsenopyrite (especially in association with gold) or the mining of high-arsenic coal. In the first two instances, the mining tailings are at risk of leaching into the groundwater and surface water. In the third instance, the end recipients are those who burn coal for domestic use, as a fuel for heating and/or for drying and smoking foods or companies that burn coal as a fuel to generate electricity.

Relatively high concentrations of arsenic are found in four types of ore associated with Carlin-type gold deposits: siliceous (1525.2 mg/kg), clay (1415.4 mg/kg), pyrite (1262 mg/kg) and arsenopyrite (4914.9 mg/kg) (Li & Peters, 1998). Arsenic-rich pyrite and arsenopyrite are the most important host minerals for gold in sedimentary rock. Deposits of arsenopyrite in argillite and siltstone are located in Hunan Province, Guizhou Province and Sichuan Province. Of 56 sampled deposits, arsenopyrite was present in 41, or 73.2% (Li & Peters, 1998). Because of the positive correlation between arsenic and gold in pyrites in sedimentary hosted gold deposits in these mining areas, they are more prone to arsenic leaching, since they are being exposed to weathering (Wells & Mullens, 1973; Bakken *et al.* 1989; Arehart *et al.* 1993; Fleet & Mumin, 1997). Therefore, environmental mitigation methods are needed in these high-arsenic mineralized areas.

Coals are found in six zones in China: I, Permian-Carboniferous in the north-east, corresponding to the North China Plate; II, Early-Middle Jurassic in the north-west, corresponding to the Talimu Plate and Aertai Fold Zone; III, Late Jurassic in the east-central, corresponding to the North China Plate; IV, Mesozoic-Cenozoic in the Tibetan Plateau, Qinhai, western Sichuan and western Yunnan, corresponding to the Himalayan Fold Zone and the Gangdisi-Nianqing Tanggula Fold Zone; V, Late Permian in the south, corresponding to the Yangtze Plate and the Southeast China Fold Area; and VI, Tertiary, corresponding to the Province of Taiwan (Figure 3.18).

Geological occurrences of arsenic in Chinese coals often follow geotectonic boundaries. The North China Plate (Late Jurassic age, III) and the Yangtze Plate (Late Permian age, V) encompass the largest coal basins and arsenic occurrences. Coals situated in or near igneous rocks associated with magma intrusions (and also where Carlin-type gold deposits are likely to be found) have the highest arsenic concentrations, >30 mg/kg, such as on the edge of the North China Plate of Permian-Carboniferous age, with an arsenic concentration range of 16–94 mg/kg; this includes the Mentougou deposit of Beijing (Kunli, 2005). Other high-arsenic coal deposits are found in the Talimu Plate in Xinjiang Uygur Autonomous Region's Hatu coal reserves (arsenic range of 18–94 mg/kg), in the Aertai Fold Zone and in the following areas of Late Permian age (V): western Hubei, northern Shaanxi, south-western Guizhou and north-eastern Yunnan.

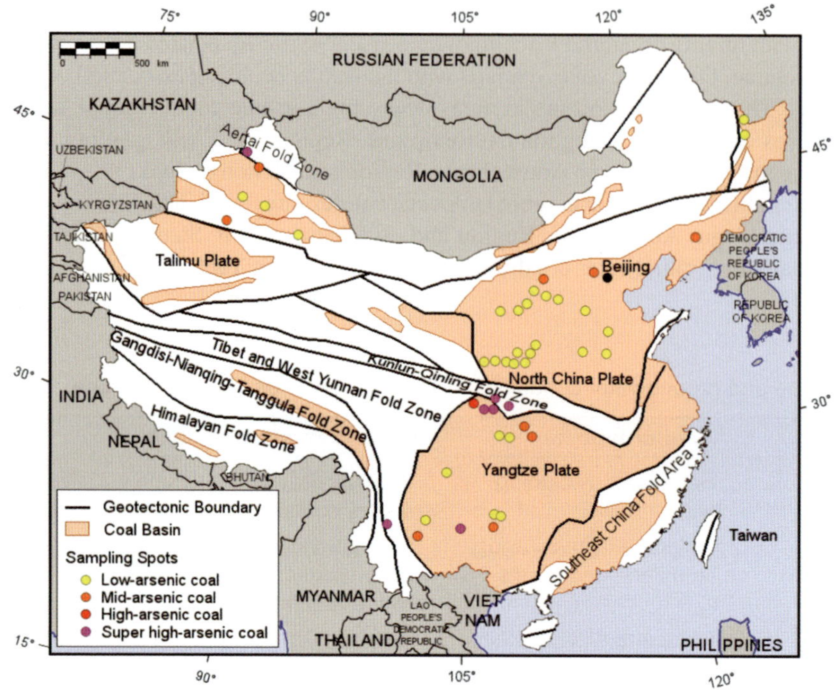

Figure 3.18 Map showing geological occurrences, distribution patterns and sampling of arsenic content in Chinese coals (modified from Kunli, 2005).

Table 3.8 indicates arsenic concentrations (determined by the molybdenum blue coloration method) in 297 coal samples of mainland China, excluding the Province of Taiwan (Wang *et al.* 1992).

Table 3.8 Arsenic concentrations in coal deposits from 26 provinces/regions in China, arranged alphabetically.

Province/region	Low concentration (mg/kg)	High concentration (mg/kg)	Mean concentration (mg/kg)	No. samples
Anhui	1.8	20	5.4	10
Beijing	–	–	2.9	1
Chongqing	0.66	8.6	4.3	8
Fujian	6.7	19	12	3
Gansu	1.3	6.5	3.6	5
Guangdong	4.6	5.5	5.1	2
Guangxi Zhuang Autonomous Region	7.7	50	24	5
Guizhou	0.52	45	7.5	17
Hebei	1.3	9.6	4.3	15
Heilongjiang	1	5.9	3.2	10
Henan	0.36	71	6.2	23
Hubei	1.3	7.9	5.2	3
Hunan	0.52	71	6.1	10
Inner Mongolia Autonomous Region	1	53	10	17
Jiangsu	1.8	9	4.6	6
Jiangxi	2.8	24	9.3	7

(*Continued*)

Table 3.8 Arsenic concentrations in coal deposits from 26 provinces/regions in China, arranged alphabetically (*Continued*).

Province/region	Low concentration (mg/kg)	High concentration (mg/kg)	Mean concentration (mg/kg)	No. samples
Jilin	4.1	26	10	4
Liaoning	0.61	29	8.1	9
Ningxia Hui Autonomous Region	3.1	9.9	5.7	4
Qinghai			3.3	1
Shaanxi	0.24	15	3.8	12
Shandong	1.4	27	5.7	20
Shanxi	0.32	35	5.7	84
Sichuan	0.66	21	6.7	8
Xinjiang Uygur Autonomous Region	0.68	7.8	4.1	5
Yunnan	2.1	27	10	7
Average	1.94	22.13	6.8	–

Source: After Wang *et al.* 2006

Table 3.9 lists the 26 provinces/regions in descending order of arithmetic mean concentrations of arsenic in coal, showing eight provinces/regions most at risk, with concentrations above 7 mg/kg (Guangxi Zhuang Autonomous Region, Fujian, Inner Mongolia Autonomous Region, Yunnan, Jilin, Jiangxi, Liaoning and Guizhou), which makes them candidates for serious study of arsenic pollution. However, if these same provinces/regions are listed in descending order of the high arsenic concentrations, 21 provinces are potentially at risk of arsenic concentrations above 7 mg/kg. In the treatment of individual provinces/regions below, those provinces/regions with high coal arsenic concentrations above 7 mg/kg are specified as "C" or coal arsenic regions.

Table 3.9 Arsenic concentrations in coal from 26 provinces/regions in China, arranged by arithmetic mean concentration.

Province/region	Zone	Population (millions, 2004)	Low concentration (mg/kg)	High concentration (mg/kg)	Mean concentration (mg/kg)	No. samples
Guangxi Zhuang Autonomous Region	V	48.9	7.7	50	24	5
Fujian	V	35.1	6.7	19	12	3
Inner Mongolia Autonomous Region	III	23.8	1	53	10	17
Yunnan	V	44.1	2.1	27	10	7
Jilin	I	27.1	4.1	26	10	4
Jiangxi	V	42.8	2.8	24	9.3	7
Liaoning	III	42.2	0.61	29	8.1	9
Guizhou	V	39	0.52	45	7.5	17
Sichuan	V	87.3	0.66	21	6.7	8
Henan	III	97.2	0.36	71	6.2	23
Hunan	V	67	0.52	71	6.1	10
Shanxi	III	33.4	0.32	35	5.7	84
Shandong	III	91.8	1.4	27	5.7	20
Ningxia Hui Autonomous Region	III	5.9	3.1	9.9	5.7	4

(*Continued*)

Table 3.9 Arsenic concentrations in coal from 26 provinces/regions in China, arranged by arithmetic mean concentration (*Continued*).

Province/region	Zone	Population (millions, 2004)	Low concentration (mg/kg)	High concentration (mg/kg)	Mean concentration (mg/kg)	No. samples
Anhui	V	64.6	1.8	20	5.4	10
Hubei	V	60.2	1.3	7.9	5.2	3
Guangdong	V	83	4.6	5.5	5.1	2
Jiangsu	III	74.3	1.8	9	4.6	6
Hebei	III	68.1	1.3	9.6	4.3	15
Chongqing	V	31.4	0.66	8.6	4.3	8
Xinjiang Uygur Autonomous Region	III	19.6	0.68	7.8	4.1	5
Shaanxi	V	37.1	0.24	15	3.8	12
Gansu	III	26.2	1.3	6.5	3.6	5
Qinghai	III	5.4		3.3	3.3	1
Heilongjiang	I	38.2	1	5.9	3.2	10
Beijing	III	14.9		3.9	2.9	1

Source: Data from Wang *et al.* 2006

In Zone V alone, for instance (encompassing Anhui, Chongqing, Fujian, Guangdong, Guangxi Zhuang Autonomous Region, Guizhou, Hubei, Hunan, Jiangxi, Shaanxi and Yunnan provinces, with a total population of 640.5 million), it has been estimated that 100 000 people, or about 0.02% of the population, are at risk (Adriano, 2001).

The level of arsenic in coal that is safe to be used in unventilated domestic cooking is an issue of linking a specific level of arsenic in coal that is to be combusted to an experimentally measured level of arsenic in food that is subjected to the coal smoke. This linkage between arsenic concentrations in high-arsenic coal and those in food prepared using high-arsenic coal has been measured experimentally (Figure 3.19).

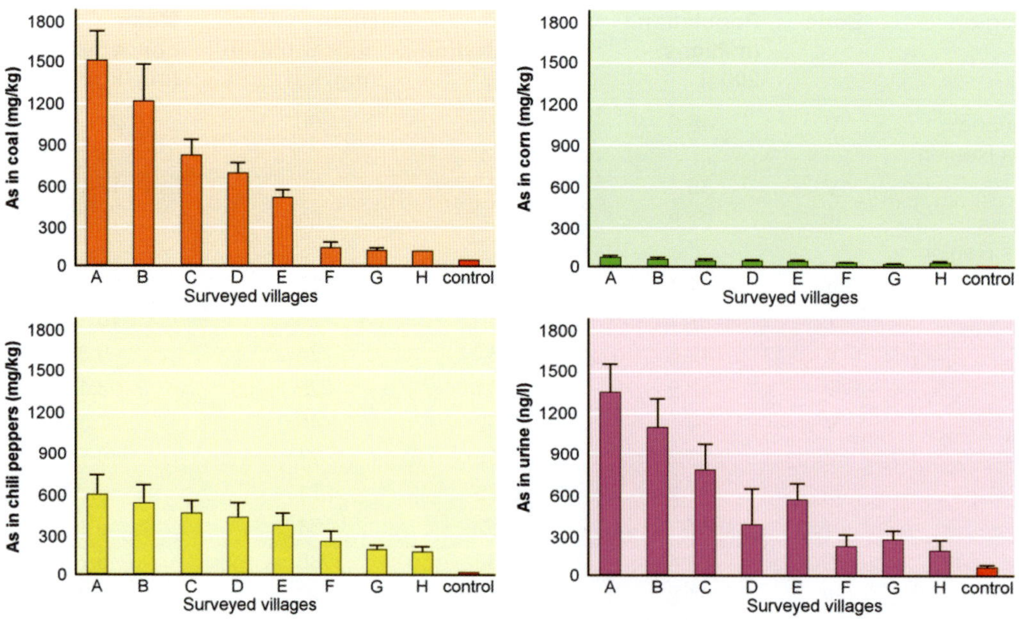

Figure 3.19 The relationship between arsenic concentrations in coal, chilli and corn, and urine among eight selected villages in China using high-arsenic coal and a control village using normal coal (modified from Liu *et al.* 2002; adapted from D.X. Zhou *et al.* 1994 and Y.S. Zhou *et al.* 1994).

These experimental findings provide a relative index of the critical range of arsenic concentrations in coal when used in domestic unventilated heating and cooking. They show that the relative concentrations of arsenic transferred from high-arsenic coal samples to chilli peppers range approximately as shown in Table 3.10.

Table 3.10 Percentage of arsenic transferred from high-arsenic coal samples to chilli peppers (approximate data extrapolated from Figure 3.19).

Villages surveyed	Arsenic concentration in coal sample (mg/kg)	Arsenic concentration in chilli (mg/kg)	% Arsenic infused into chilli
A	1500	600	40
B	1200	550	46
C	800	450	56
D	700	425	61
E	500	350	70
F	100	250	250
G	100	200	200
H	100	180	180

Three different guidelines are available for an acceptable maximum level of arsenic in a medium transferrable to food or inhaled as combusted vapour. One is the USEPA (2001a) maximum contaminant level for soil of 10 mg/kg, but this does not readily translate into the effect of arsenic in coal on smoked foods, because this standard was developed for the leaching potential of soil into surface water and its uptake by crops. A second standard is the United States Occupational Safety and Health Administration's permissible exposure to airborne arsenic of 10 μg/m^3 (Nadu *et al.* 2006). This standard is more applicable, as the combustion of high-arsenic coal in an unventilated area can be measured experimentally. The third is the USEPA's proposed limit of 0.7 mg/kg for arsenic in ingestible solid foods (Adriano, 2001). In Figure 3.19, only the control samples of coal would be acceptable.

For the purposes of the following analysis, a concentration of arsenic in coal of 7.5 mg/kg or higher is considered to produce risk. This assessment is in line with the mean concentration of arsenic in coal in Guizhou Province that resulted in elevated urinary arsenic.

In Table 3.11, Zones I, III and V have average high arsenic concentrations over 20 mg/kg, based on data from Wang *et al.* (2006), but the two super high samples in Zone II from Kunli (2005) have a range of 18–94 mg/kg, so this would place all zones in the critical category. Note that in Table 3.11 there are seven provinces/regions that overlap other zones (indicated by parentheses) that are not included in the calculations. In addition, Tibet Autonomous Region, Sichuan, Qinghai and Xinjian overlap or are in Zone IV and are not represented. One mean arsenic concentration in coal for Yunnan in Zone IV is 3.4 mg/kg and is given in Table 3.12. Provinces overlapping multiple zones indicate regions for further research.

Table 3.11 Listing by region and zone type of each of the 26 provinces/regions in China in which arsenic coal occurs, with indications of provinces/regions that overlap additional zones in parentheses. Zone IV is not represented.

Province/region	Zone	Population (millions, 2004)	Low concentration (mg/kg)	High concentration (mg/kg)	Mean concentration (mg/kg)	No. samples
Jilin	I	27.1	4.1	26	10	4
Heilongjiang	I	38.2	1	5.9	3.2	10
Liaoning	I (III)	42.2	0.61	29	8.1	9
Inner Mongolia Autonomous Region	I (II–III)	23.8	1	53	10	17
Ningxia Hui Autonomous Region	II (III)	5.9	3.1	9.9	5.7	4

(*Continued*)

Table 3.11 Listing by region and zone type of each of the 26 provinces/regions in China in which arsenic coal occurs, with indications of provinces/regions that overlap additional zones in parentheses. Zone IV is not represented (*Continued*).

Province/region	Zone	Population (millions, 2004)	Low concentration (mg/kg)	High concentration (mg/kg)	Mean concentration (mg/kg)	No. samples
Xinjiang Uygur Autonomous Region	II	19.6	0.68	7.8	4.1	5
Gansu	II	26.2	1.3	6.5	3.6	5
Henan	III (V)	97.2	0.36	71	6.2	23
Shanxi	III	33.4	0.32	35	5.7	84
Jiangsu	III (V)	74.3	1.8	9	4.6	6
Hebei	III	68.1	1.3	9.6	4.3	15
Qinghai	III	5.4	0	3.3	3.3	1
Beijing	III	14.9	0/16	3.9/4.1	2.9/21	1/8
Guangxi Zhuang Autonomous Region	V	48.9	7.7	50	24	5
Fujian	V	35.1	6.7	19	12	3
Yunnan	V (IV)	44.1	2.1	27	10	7
Jiangxi	V	42.8	2.8	24	9.3	7
Guizhou	V	39	0.52	45	7.5	17
Sichuan	V	87.3	0.66	21	6.7	8
Hunan	V	67	0.52	71	6.1	10
Anhui	V (III)	64.6	1.8	20	5.4	10
Hubei	V	60.2	1.3	7.9	5.2	3
Guangdong	V	83	4.6	5.5	5.1	2
Chongqing	V	31.4	0.66/0.10	8.6/4.72	4.3/2.3	8/41
Shaanxi	V	37.1	0.24	15	3.8	12
Average	I		1.68	28.48	7.83	
	II		1.7/18	8.07/94	4.47	
	III		0.74	22.7	4.67	
	V		2.47	26.2	8.28	

Source: Data are from Wang *et al.* (2006), except where two sets of data are given; in such cases, the second set is from Kunli (2005).

Table 3.12 Ranges of arsenic concentrations for five of eight coal areas in China (excluding the Province of Taiwan).

Zone	Range of arsenic concentrations (mg/kg)	Mean arsenic concentration (mg/kg)	No. samples
I. North	0.24–71	5.4	184
II. North-west	0.68–7.8	4.2	11
III. North-east	0.61–53	8.8	31
IV. Yunnan	–	3.4	1
V. South	0.52–50	8.5	70

Source: Data from Wang *et al.* 2006

The burning of high-arsenic coal for cooking, heating and drying corn and peppers contaminates not only the food (50–80%), but also the air (10–20%) and any exposed water (1–5%). Coals that have undergone epigenetic mineralization can contain extremely high arsenic concentrations. Typical levels range between 100 and 900 mg/kg but can reach 35 000 mg/kg. Coal is burned inside the home in open pits and non-vented ovens. In areas that traditionally rely on foods dried and cooked by coal fires or that heat using indoor coal stoves, arsenic contamination has affected millions and has resulted in tens of thousands of confirmed arsenicosis cases. In 2002, it was estimated that 333 905 people were at risk from burning high-arsenic coal in the high-arsenic coal provinces/regions: Guizhou, Inner Mongolia Autonomous Region, Ningxia Hui Autonomous Region, Shanxi and Yunnan (Jin *et al.* 2003). By 2004, the estimate had risen to 1.3 million (Luong *et al.* 2004); of 48 438 actual confirmed household residents who have been exposed to high-arsenic coal fumes (Jin *et al.* 2003), more than 10 000 arsenicosis cases had been confirmed (Luong *et al.* 2004). The government's approach has been early detection coupled with provision of alternative arsenic-safe water supplies and improved stoves; however, households do not always wish to adopt better ventilation, because this results in colder temperatures in the winter. Health effects relating to exposure to arsenic from coal burning include skin lesions (including skin cancer), lung dysfunction, neuropathy, nephrotoxicity and liver damage, including cancer (Liu *et al.* 2002).

In a report published in 2005, the following scenario was presented for the remediation of the domestic use of high-arsenic coal. According to the 2003 national endemic disease report, 5 332 632 households were at risk of coal arsenic poisoning before the current remediation plan was put into place, and 21% of these affected households have been fitted with vented stoves, benefitting 2 026 400 households or 7 094 600 people. By the end of 2010, the plan is that vented stoves to lower arsenic content will have been installed in 3 million households, or 75% of the affected households in the government-specified disease areas. However, it was noted that the maintenance rate was low, as many people may revert to prior practices of using unvented stoves, especially when heat is needed in winter, and this is expected to be an ongoing issue of public health (Yu, 2005).

The use of high-arsenic coal in power plants has been studied by Luo *et al.* (2004):

> By combustion of 1 t of Permo-Carboniferous coal (containing roughly 5 mg/kg arsenic), high-temperature power stations emit roughly 0.40 g arsenic into the atmosphere and the arsenic emission rate is about 7.70%; mid-low power stations emit roughly 0.15 g arsenic into the atmosphere and the arsenic emission rate is about 2.97%. A total of 600 million tons coal is burnt annually in Chinese power stations, and the coal comes mainly from Permo-Carboniferous depositing in the North China Plate and northwest China coal mines. Taking the average arsenic content of the coal used at the value of 5 mg/kg, the total annual arsenic emission from steam coal combustion into the atmosphere is about 195.0 t.

Their recommendations to remediate this discharge of arsenic into the atmosphere are the following (Luo *et al.* 2004):

> Most of the arsenic in coal can be released in the process of coal combustion, and most of the released arsenic can be seized by the fly ash and then both of them are seized by the dust catcher of power station, so the arsenic emission ratio to the atmosphere is declined; in addition, research on the arsenic emission amount and emission rules from the coal power stations in China should go on the coal power stations with the dry-process dust catchers by the experiments results. In the wet process of dust catcher, 20% of the arsenic in the fly ash is dissolved in the water of sedimentation tank in high-temperature power station; in the mid-low temperature power station there are 70% of the arsenic in the fly ash dissolved in the water of sedimentation tank, this is an important source of arsenic pollution in environment and should not be overlooked.

The fourth geological zone is associated with petroleum extraction, such as in Xinjiang Uygur Autonomous Region and Heilongjiang. These areas have been developed only within the last 20 years; therefore, the impact of the escalation of the Chinese economy, especially the automotive industry, is a relatively recent phenomenon. Perspectives on health hazards associated with petroleum extraction and use are basically two-fold. First, extraction and the associated leaching of effluents create anoxic conditions, which favour the release of arsenic(III). Second, automotive exhaust produces its own health hazards from air pollution. This study concerns itself only with the areas of petroleum extraction, Xinjiang Uygur Autonomous Region and Heilongjiang, but it should also be remembered that oil pipelines can create the same anoxic conditions that are found in the extraction zone should leakages and spillages occur.

Most of the anthropogenic arsenic poisoning in China has been caused by the lack of attention paid to the location of chemical plants. The most vulnerable plant sites are located on rivers. According to Pan Yue, the administration's deputy chief of the environmental agency (Staff Writers, 2006):

> the frequent incidents were related to the "irrational" distribution of industry, with over 20,000 chemical plants built on rivers. Up to 10,000 plants sat along the banks of China's longest river, the Yangtze, while 4,000 chemical plants were on the Yellow River, he said. Many plants were located near major drinking water sources or large population centers, he added.

Another source of anthropogenic arsenic poisoning is attributable to 4 million unaccounted for Japanese-manufactured World War II chemical weapons. In 1992, China announced that at least 2 million chemical weapons remained in its territory, buried in Heilongjiang, Jiangsu, Jilin, Liaoning and Inner Mongolia Autonomous Region. An ambitious plan by the Bonn International Center for Conversion scheduled these to be removed by 2007 under a Chemical Weapons Convention Act (Evans, 1997). The munitions contain poison gas, mainly mustard gas and lewisite, an arsenic-based blistering fluid. Japan has launched six missions since 2000 to recover about 11 200 chemical weapons. A list of these is given in Table 3.13. Arsenic contamination of groundwater in any of these locations could be evidence of lewisite or arsenic trichloride, the most pernicious arsenic-based chemical weapons that can leach into groundwater from any known or unknown stockpiles.

Table 3.13 Chemical weapons in China.

Location	Type	Quantity	Agent	Status
Heibei Province				
Gaocheng	75 mm	50	Phosgene	Unknown
Shijazhuang	Gas shells	6 of 50 Japanese	Mustard gas, lewisite?	Unknown
Heilonjiang Province				
Acheng	Munitions	>300	Mustard gas, lewisite?	Unknown
	Bulk	10 tonnes	Mustard gas, lewisite?	Unknown
Bayan	Munitions	>100	Mustard gas, lewisite?	Unknown
Fujin	75, 90, 105, 150 mm	>100 000	Mustard gas, lewisite?	Unknown
Shangzhi	Bulk	1.1 tonnes	Mustard gas, lewisite?	Unknown
Sunwu	105, 150 mm	513	Mustard gas, lewisite?	Unknown
	Smoke cylinders	4 boxes	Mustard gas, lewisite?	Unknown
	Bulk	2 barrels	Mustard gas, lewisite?	Unknown
Qiqihaier	75, 150 mm	248	Mustard gas, lewisite?	Partially buried
Shuangcheng	Munitions	11	Mustard gas, lewisite?	Unknown
Sunwu	Poison gas containers	>500	Mustard gas, lewisite?	Unknown
	75, 105, 150 mm	513	Mustard gas, lewisite?	Buried
	Toxic cylinders	4 boxes	Mustard gas, lewisite?	Buried
Inner Mongolia Autonomous Region				
Hohot	Bulk	3 barrels	Mustard gas, lewisite?	Unknown
	Bulk	4 barrels	Mustard gas, lewisite?	Unknown
Jiangsu Province				
Cnuzhou	Bulk	3 containers	Mustard gas, lewisite	Sealed
Nanjing	Bulk	3 containers	Diphenylchlorarsine	Sealed
	Bulk	4 barrels	Mustard	Unknown
	Cans	3000	Mustard gas, lewisite?	Unknown
	Mortar shell	1	Mustard gas, lewisite?	Unknown
Jiangxi Province				
Shanggao	Iron sheet cans, 11 cm diameter × 22 cm length	>200	Hydrogen cyanide	Unknown
Jilin Province				
50 km east of Dunhua			Mustard gas, lewisite	Buried
	75, 90, 105, 150 mm		Mustard gas, lewisite?	Unknown
	Aerial bombs	1 800 000		Unknown
	75, 90 mm	700 000	Mustard	Unknown
	75, 90, 105, 150 mm	700 000	Mustard gas, lewisite	Unknown
		700 000	Mustard gas, lewisite	Possibly leaking
Haerbaling	Munitions	365	Mustard gas, lewisite, pepper	Excavated
Jilin suburbs	75 mm	>40		

(*Continued*)

Table 3.13 Chemical weapons in China (*Continued*).

Location	Type	Quantity	Agent	Status
Meihekou	75, 105, 150 mm			Buried beneath railroad tracks
	Bulk	74 tonnes	Mustard gas, lewisite	Two mounds
	Bulk			Two mounds
Weijin	Bulk	74 tonnes	Mustard gas, lewisite	Two mounds
Laioning Province				
Shenyang	Barrels	6	Mustard gas	Sealed, stored
	75 mm	24		Stored
Zhejiang Province				
Hangzhou	75 mm	33		

Source: Evans, 1997

A 2007 Chemical Weapons Convention deadline is mandated to locate and remove 7000 tonnes of poison gas (including sneezing gas, mustard gas and lewisite) delivered by chemical warheads by Japan to the Chinese mainland (Yamaguchi, 2003). Lewisite (Japanese nomenclature: "Yellow no. 2" or "A3") is an often fatal skin-blistering chemical weapon, dichloro-2-chlorovinyl arsine ($ClCHCHAsCl_2$), which penetrates clothing and even rubber. It was discontinued by the USA before World War II because it was considered ineffective, but Japan continued its manufacture for its war on China, and China, ironically, manufactured it until 1989 for the former Soviet Union, where Chinese-made stockpiles have been uncovered.

Beginning with its inception in 1993, the Convention on the Prohibition of the Development, Production, Stockpiling and Use of Chemical Weapons located stockpiles of chemical weapons in the former Soviet Union, Japan and China. From 1997 to 2005, many of Japan's stockpiles of chemical weapons for use before and during World War II have been located in China. Owing to their age and their clandestine burial, their containers have become corroded and are at risk of leaching into the environment.

A province by province assessment of arsenic pollution in China follows.

Anhui Province (A, G)

The geological occurrence of arsenic contamination in Anhui Province is imperfectly known. In 2003, the Anhui government reported two counties and five villages with 100 000 at risk of arsenic exposure and a case rate of 0.3% (Jin *et al.* 2003).

A study that concluded in 2003 (Xia & Liu, 2004) measured the arsenic concentrations in wells in Anhui Province, as reported in Table 3.5. In 2005, a 5-year study concluded (Yu *et al.* 2007), with the findings as summarized in Table 3.6.

Additionally, in October of 2008, the Gimhae Metallurgy Chemical Company in Ma'anshan City, a prefecture-level city in the east of Anhui Province in eastern China, released arsenic-contaminated water into the local river: 450 people were diagnosed with high arsenic in their urine in the neighboring town of Hech, Guangxi Province, and 4 have been clinically diagnosed to be arsenic-poisoned (Business news agency, 2009. http://china.toocle.com).

Beijing (C, G)

Luong *et al.* (2004) reported that Tianzhu village of Shunyi district is at risk for arsenic contamination of groundwater sources. An additional four villages have since been added (Jin *et al.* 2003; Sun & Yang, 2004), making the total population at risk about 370 000. A study that concluded in 2003 (Xia & Liu, 2004) measured arsenic concentrations in well water in Beijing, with the findings as recorded in Table 3.5 above. Additionally, In 5 villages and 3 towns in the Beijing suburbs Xia & Liu (2004, p. 25) reported that a population of 60 306 was exposed to high-As (>0.05 mg/L) groundwater. The occurrence of arsenic (As) in deep wells (depth, 100 ~ 120 m) in fourteen villages in Houshayu town, Shunyi District, Beijing which has an area of 42.6 km^2 and a registered population of 21000 was investigated three times from February 2004 to January 2005. Additionally, 12 villages contain As higher than 10 g/L, two of which were over 50 g/L. Only two of the 14 samples had an As concentration below 10 g/L. As(III) species (16–64%) and As(V) species (23–71%) coexist in all of the groundwater samples (Dou *et al.* 2005).

Beijing is also in the region of the Late Jurassic-age Mengtougou coal beds of Zone III, in the North China Plate (see Table 3.11). With a population of 14.9 million, the use of the Mengtougou coal is of environmental concern, as one of its main applications is in the Beijing city power plants.

Chongqing (C)

Chongqing (Chungking), the largest and most populous of China's four provincial-level municipalities, is a major trading and industrial port, with more than 40 kinds of minerals and 4.8 billion tonnes of coal reserves. While there is no evidence of arsenic contamination of its groundwater, the high range of arsenic in its coal reserves is of some concern, as Chongqing lies at the head of the reservoir behind the Three Gorges Dam (Watts, 2006). In a series of multi-province investigations carried out in 2003–2004, 186 villages in Chongqing were studied out of a total of 381. Most of the 381 villages studied exceeded the national standard of 1.5 mg arsenic/kg, with 89% of the villages having smoked food that contained more than 10 mg arsenic/kg and 37% having smoked food that contained more than 100 mg arsenic/kg (Yu, 2005).

Fujian Province (A, C, G)

Fujian Province on the south-east coast of China has a population of 35 million, many of whom have been in the process of de facto urbanization since at least 1995, as industrialization has become intensive and as the agricultural groups that once accounted for 80% of the population gravitate to urbanized centres (Zhu, 1998). This movement to suburban areas of near-industrialization has had health consequences, since arsenic in the soils that are both local and from rice paddy fields has been transferred to plants that make up domestic diets. The sources of this arsenic are presumed to be both naturally occurring and potentially anthropogenic from increased industrialization. As Huang *et al.* (2006) reported:

> The transfer factor based on the soil available arsenic (TF[avail]) was chosen to compare the accumulation ability of the various crops. The TF[avail] values of rice grains (air-dried weight basis) ranged between 0.068 and 0.44 and were higher than those of the vegetables, ranging from 0.001 to 0.12. The accumulation ability of the crops decreased in the order of rice > radish > water spinach > celery > onion > taro > leaf mustard > fragrant-flowered garlic > pakchoi > Chinese cabbage > lettuce > garlic > cowpea > cauliflower > bottle gourd > towel gourd > eggplant. Daily consumption of rice and other As-rich vegetables could result in an excessive intake of arsenic, based on the provisional tolerable intake for adults for arsenic recommended by WHO.[1]

In addition, local coals of Late Permian geologic age in Fujian Province have been recently cited as being another source of arsenic contamination (see Table 3.11; Wang *et al.* 2006). Coal is a domestic source of home cooking energy in Fujian Province, and a recent study has shown high arsenic content in some of the deposits. As in Guizhou Province, there is a tradition of smoking vegetables and meat using coal in unventilated stoves. In addition, some of the arsenic in the coal seams may be contributing to the contamination of the groundwater.

Gansu Province (G)

Gansu (Kansu) derives its name from "Long Mountain in the west". Gansu (454 000 km^2) is more than 1 km above sea level, with its southern mountains sloping towards the Huang He. Most of its economy is derived from mining and significant oilfields at Yumen and Changqing (Bian, 2005).

A 5-year study that concluded in 2005 (Yu *et al.* 2007) assessed the arsenic concentrations of wells in Gansu Province (see Table 3.6). The soil profiles of the area suggest that arsenic mineralization is geogenic and not associated with anthropogenic sources (Xu & Xu, 1996).

Since Gansu supplies most of the water for the upper Huang He, any arsenic contamination of its surface waters and ground waters is of regional concern.

Guangdong Province

Guangdong Province is located in south China and has a population of 83 million. It is located in a zone of high-arsenic coal, with concentrations ranging from 4.6 to 5.5 mg/kg and an average concentration of 5.1 mg/kg (Wang *et al.* 2006). The arsenic found in the coal seams in mining centres like Daxing is possibly related to the general mineralization of arsenic in the general rock matrix and soils of the area (Xu & Xu, 1996). For instance, in a study of the spatial distribution and vertical variation of 260 soil profiles, the arsenic concentration followed an approximately lognormal distribution. The geometric mean concentration of 10.4 mg/kg was decidedly elevated, and an upper baseline concentration of 23.4 mg/kg was estimated for surface soils. High arsenic concentrations were mainly in limestone, sand and shale areas, indicating a correlation with bedrock geochemistry rather than anthropogenic sources. Geometric mean vertical concentrations of arsenic increased from 10.4 to 10.7 to 11.3 mg/kg (A to C horizon) and may be related to the lower soil organic matter and soil degradation and erosion. Consequently, the leaching of arsenic from the soil into surface water and groundwater is a potential threat to the water quality in Guangdong (Zhang, 2006).

[1] Note that WHO withdrew the provisional tolerable weekly intake (PTWI) in 2010. See http://apps.who.int/ipsc/database/evaluations/chemical.aspx?chemID=1863

Guangxi Zhuang Autonomous Region (A, M, C)

With an ageing population of 48.9 million, Guangxi Zhuang Autonomous Region is located in southern China, between 20°54′ and 26°20 N, on the south-eastern corner of the Yunnan-Guizhou Plateau. It borders Viet Nam to the south-west and is surrounded by Guangdong, Guizhou, Yunnan and Hunan provinces.

Guangxi Zhuang Autonomous Region leads the country in the extraction of non-ferrous metals. Mining of arsenic has been a source of economic power: white arsenic, a highly toxic industrial chemical widely used in leather making and the production of pesticides, is produced in Guangxi Zhuang Autonomous Region, and the slag, which is not properly processed and disposed of properly, has contaminated the soil and the groundwater (China Daily, 2004). In December 2002, approximately 20 tonnes of arsenic were spilled into local rivers, causing severe contamination (Planet Ark, 2002).

In Guangxi Zhuang Autonomous Region, local coals of Late Permian geological age have been cited as being a source of arsenic contamination. Zeng *et al.* (2005) studied arsenic and other trace elements in sulfur-rich Late Permian-age coals in the Heshan coalfield. A further study by Wang *et al.* (2006) has provided a precise analysis of five samples of coal from Guangxi Zhuang Autonomous Region (see Table 3.11).

Coal is a domestic source of home cooking energy in Guangxi Zhuang Autonomous Region, and a recent study has shown a high arsenic content in some of the deposits (Wang *et al.* 2006). Just as in Guizhou Province, there is a tradition of smoking vegetables and meat using coal in unventilated stoves. In addition, some of the arsenic in the coal seams may be contributing to the contamination of the groundwater.

Guizhou Province (C, G, M)

Guizhou Province has vast coal and mineral reserves in an expanse of 176 000 km^2 and with a population of 39 million. The total output of coal in 2005 was 100 million tonnes, a 12% growth from the previous year (Guizhou People's Government, 2007).

Since the 1960s, the government of Guizhou has attempted to implement high-arsenic coal mitigation (restricting mining and treating patients with chelating drugs), but the Cultural Revolution from 1966 to 1976, the small, isolated geographic area and the needs of the economically restricted population prevented progress. Beginning in 1978, high-arsenic coal was linked to arsenicosis through cooking foods in unventilated domestic settings, inhaling the smoke and eating tainted foods (Zhou, 1993; Finkelman *et al.* 1999; Zheng *et al.* 1999). The rates of arsenic contamination of corn and chilli when roasted with high-arsenic coal are shown in Table 3.14.

Table 3.14 Arsenic concentrations in corn and chilli roasted with high-arsenic coal in Qianxi'nan Autonomous Prefecture and Bijie Prefecture, Guizhou Province, China (Source: Jin, 2006).

County	As range of coal (mg/kg)	As range in chilli roasted with high As coal (mg/kg)	As range in corn roasted with high As coal (mg/kg)
Xingyi	265–352	545–693	7.0–12.5
Anlong	418–530	586–688	6.7–11.9
Xingren	624–852	512–300	2.8–4.1
Zhijin	58–2167	77–610	16.3–43.6

By the 1990s, after more than 20 years of exposure, arsenicosis was being expressed in much of Xingren county, and the government began closing mines, treating symptomatic individuals and installing vented stoves. However, most residents did not understand the consequences of their heating and cooking with high-arsenic coal, and new cases continued to be recorded by the 21st century (Li & An, 2005; An *et al.* 2007). Because the climate is subtropical with an average rainfall of 1000–1300 mm and with an average annual temperature of 10–20°C, the need to use coal for home heating is mainly restricted to the cool autumn climate (Liu *et al.* 2002). Estimated sources of arsenic exposure in south-western Guizhou are arsenic-contaminated food (50–80%), air (10–20%), water (1–5%) and direct contact in coal-mining workers (1%) (Liu *et al.* 2002). In a 2004 baseline survey of 45 364 residents, Dong *et al.* (2007) noted more than 2,800 individuals diagnosed with arsenicosis. In a similar baseline survey, Jin (2006) noted that of 2609 had been diagnosed with arsenicosis from a total population of 45 399. A remediation program was initiated in 2004 at a cost of 4 million Yuan RMB (US$ 500 000) financed mostly by the government. The program involved a health education campaign that promoted lifestyle changes coupled with the shutting down of local coal pits and the installation of 10 000 new stoves with chimneys for ventilation. Results show that urinary As concentrations in a control population of individuals with arsenicosis decreased from 0.198 ± 0.300 mg/L ($n = 144$) in 2004 to 0.049 ± 0.009 mg/L ($n = 50$) in 2005 ($p < 0.01$).

High-arsenic coal seams in south-western Guizhou Province, some of which contain up to 35 000 mg/kg (Finkelman, 2000), are generally distributed parallel to the long axis of the anticline fault on both sides of the Permian Longtan Formation in association with Carlin-type gold deposits (Figure 3.20).

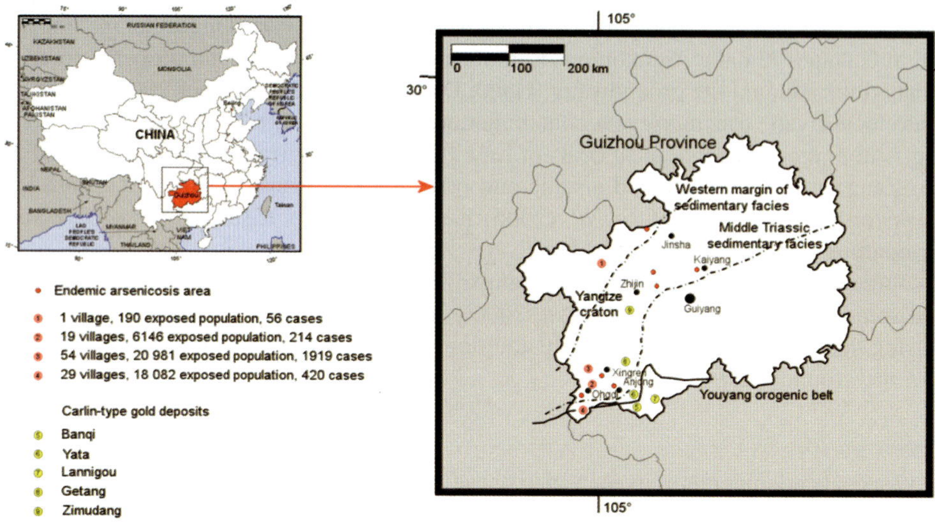

Figure 3.20 Map of Guizhou Province showing areas of endemic arsenicosis as they occur on the Middle Triassic sedimentary facies between the Yangtze craton and the boundary with the Youyang orogenic belt (after Tao, 1990; Yao, 1990; Tan, 1994; Li & Peters, 1998; Liu *et al.* 2002; Jin, 2006).

The nearer the coal seams are to a fault or a gold-bearing ore body, the greater is the concentration of arsenic, a correlation that appears to correspond to most Carlin-type gold deposits (Zhang *et al.* 2002). The genesis of Carlin-type gold deposits is a tantalizing, unresolved issue. Based upon research in the Nevada (USA) Carlin-type deposits, Berger & Theodore (2005) suggested the following explanation: "original sedimentary-exhalative (SEDEX) gold concentrating processes might have been initiated by hot spot activity – i.e., the stratabound Carlin-type deposits with related synsedimentary local uplifts suggest[ing] a specific Paleozoic hot spot…." In Guizhou, this hot spot would be associated with the Permian-age Longtan Formation when a massive eruption of basaltic magma concentrated trace elements, including arsenic, in south-western Guizhou (Huang & Xie, 2006).

The possibility of arsenic leaching from these coal seams and from the Carlin-type gold deposits into groundwater sources should not be ruled out, and much more testing is needed in this area. Since Guizhou Province adjoins Sichuan (and the municipality of Chongqing) to the north, Yunnan to the west, Guangxi Zhuang Autonomous Region to the south and Hunan to the east, those areas bordering the anticline fault of the Permian Longtan Formation may also be at risk.

In Guizhou Province, examples linking arsenicosis to high-arsenic coal mined from seams adjacent to Carlin-type gold deposits are found throughout Qianxi'nan Autonomous Prefecture. Since many of the villages near these areas that use the coal for smoking food have yet to be specifically studied epidemiologically, proven and potential arsenic contamination sites that deserve to be investigated based on the proximity to arsenic-associated minerals and coal mined are listed below (Mindat.org, 2007c):

(1) Ceheng County deposits: Baidi, gold; Bannian, gold; Banqi, gold–mercury–antimony; Ceyang, gold; Qingping, gold; Weihuai, gold–arsenic; Yata, gold;
(2) Pu'an County deposits: Miaolong, gold–arsenic–antimony; Sixiangchang, mercury–gold;
(3) Qinglong County deposits: Dachang, gold–antimony–arsenic–mercury; Qinglong, antimony; Shaziling, gold–arsenic; Tangxinzhai, gold; Yanshang, gold;
(4) Tongren Prefecture mines: Tongren, Wanshan (Wan-Shan-Ch'ang), Yanwuping, Yunchangping
(5) Wangmo County deposits: Daguan, gold–arsenic–antimony; Wangmo (Wangme), gold;
(6) Xingren (Xinren) County deposits: Dayakou, gold–antimony; Getang, gold; Lanmuchang, thallium-(mercury); Nipu, gold; Zimudang, gold–mercury–thallium;
(7) Xingyi County deposits: Lubuge, gold–arsenic; Xuongwu, gold–antimony–arsenic–mercury;

(8) Zhenfeng County deposits: Beiyinpe, gold–mercury; Jinfeng (Lannigou), gold–arsenic–mercury; Loudong, gold; Pegao, gold; Sanchahe, gold.

In Guizhou, gross anatomical symptoms of arsenic exposure recorded include keratosis of the hands and feet, pigmentation on the trunk, skin ulceration and cancers, lung dysfunction, neuropathy, nephrotoxicity, hepatomegaly, cirrhosis, ascites and liver cancer. Apart from these gross symptoms, a study undertaken in Jiaole in Xingren County, Qianxi'nan Autonomous Prefecture, indicated that long-term arsenic exposure may be associated with chromosome and deoxyribonucleic acid damage and gene mutations that are long-term consequences and hence may not be expressed until many years after exposure (Zhang *et al.* 2007). Current estimates of those affected must, therefore, be used cautiously.

Additionally, in January 2008 a mining company in Dushan County in Guizhou Province of southwest China released 1900 tonnes of arsenic contaminated water directly into Duliu River, resulting in risk to people obtaining drinking water downstream. 17 people were found to have various levels of arsenic poisoning (Business news agency., 2009. http://china.toocle.com).

Hebei Province (C)

Hebei Province ("North of the Yellow River") lies mostly within the North China Plain with Taihengshan to the west. Its continental monsoonal climate has cold winters, averaging temperatures of −16°C to −3°C in January, which make its population vulnerable to the high range of the arsenic-bearing coals (see Table 3.11) from the Kailuan mines, which have an annual production of 20 million tonnes (Hebei People's Government, 2007).

Heilongjiang Province (A, C, G, P)

Heilongjiang derives its name from its proximity to the Amur River or "Black Dragon River" that separates China from the Russian Federation. Its topography is varied, with mountain ranges on the perimeter and a low-lying interior dissected by three tributaries of the Amur River. Its climate is sub-Arctic, with long winters ranging in temperature from −31°C to −15°C. Its economy is based on petroleum, coal and production of cold-weather crops (Heilongjiang People's Government, 2007). Because of the need for heating fuel, its high-arsenic coal reserves are of special concern, although wind power is ideally suited to its topography.

High levels of arsenic in drinking-water (>50 μg/l) have been reported (Donglou, 2004). Groundwater monitoring in Heilongjiang Province in 2003 showed that arsenic concentrations in water exceeded the Chinese standard of 50 μg/l (see Table 3.6).

These elevated levels of arsenic in groundwater may be partially related to both the petroleum and the coal deposits. Daqing oilfield, which has been in operation since 1960, is China's largest, with 60 000 well heads, and it realized an output of more than 50 million tonnes of crude oil per year from 1976 until 2003, with an annual oil sewage emission of 400 million m^3. The hydrocarbons released into the groundwater from the Daqing oilfield may have mobilized some arsenic in a mechanism that is similar to that which pertains to the oilfields in Xinjiang Uygur Autonomous Region. In an effort to remediate this situation, a "Green Plan" was put into effect in 2006. It was reported on 10 June 2007 (China Daily, 2007) that:

> Daqing oilfield has built the largest domestic industrial sewage treatment system and the largest industrial sewage treatment sample unit in the world. The harmless treatment of industry wastes reaches 100%, and the green ratio stands at 19.8%, which helped Daqing City win the honour of Environmental Protection Model City, the only reward for inland cities.

In addition, local coals in Heilongjiang have heightened levels of arsenic, being in the Permian-Carboniferous Zone I of the North China Plate. At the Wuyun coal beds, arsenic concentrations were measured as ranging from 0.2 to 12.6 mg/kg, with a mean of 6.2 mg/kg (Kunli, 2005).

Henan Province (C, G, M)

Henan province in north-central China, with a population of 97.2 million, has some of the highest-arsenic coal in China, with concentrations reaching 71 mg/kg (Table 3.11); the province also has the highest rates of oesophageal cancer in the world. The co-occurrence of high-arsenic coal and oesophageal cancer has been noted in many other provinces in China where the practice of smoking vegetables (especially chilli peppers) with unvented indoor coal-fired stoves is followed. In such environments, the total residual arsenic in chilli peppers can exceed 500 mg/kg (USGS, 1998), and arsenic from the breathing of high-arsenic coal smoke comes in contact with the oesophageal lining and the lungs.

Groundwaters in Henan are at risk of arsenic leaching from geogenic sources, a situation that needs monitoring. High-arsenic wells have been identified in Henan by Donglou (2004), and a study that concluded in 2005 (Table 3.6) of 28 068 wells in 1635 villages with a population of 1.47 million revealed 192 wells with arsenic concentrations above 50 μg/l (Yu *et al.* 2007).

Additionally, in 2008 the Seijo Chemical Co., Ltd in Minquan, Henan Province, not having received legal approval to expand their factory operations, went ahead and purchased ore of high arsenic content anyway and disposed of arsenic tainted wastewater into the Daliul River, leading to trans-provincial large-scale pollution. The amount of water polluted is over 6 million m^3, at a level 3 times higher than the safety limits (Business news agency., 2009. http://china.toocle.com).

Hubei Province (C)

Hubei Province, with a population of 60.2 million, has arsenic coal deposits with a high concentration of 7.9 mg/kg (Wang *et al.* 2006; Table 3.11). Practices in Hubei Province are similar to those in Guizhou: coal is burned in unventilated stoves to smoke and dry chilli peppers. The accumulation of arsenic on the smoked food as well as the inhaled smoke are health issues. In a national campaign in 2003–2004, 12 villages from Hubei Province were investigated for arsenicosis. It was found that smoked food in most of the 381 villages studied in eight provinces exceeded the national standard of 1.5 mg arsenic/kg, with 89% of the villages having smoked food that contained more than 10 mg arsenic/kg and 37% having smoked food that contained more than 100 mg arsenic/kg (Yu, 2005).

Hunan Province (A, C, M)

Anthropogenic arsenic contamination in Hunan Province has been contributed by chemical plants involved in smelting metals, such as zinc, that have as their by-products cadmium or arsenic compounds. For instance, a series of environmental accidents from November to December 2005 in Hunan Province was capped by a spill of cadmium and arsenic from the Zhuzhou Smelter on 4 January 2006 after Zhuzhou Water Conservancy Investment Company, which started a silt cleaning project without official approval, dammed a waste drainage pipe from the zinc smelter, and the dammed water spilled into two holding lakes containing high levels of cadmium and arsenic from nearby plants before overflowing into the river. The affected population totalled more than 80 000 (Li, 2006; Yuan, 2006). Another incident involving arsenic pollution occurred in the first week of September 2006 (Staff Writers, 2006):

> waste water from the holding pond of a chemical plant in central Hunan province ... began seeping into Xinqiang river in Hunan's Yueyang county, Xinhua news agency reported. The chemical in question is arsenic trioxide leaked into the [river] prompting the government to urge locals not to drink the water. "Arsenic trioxide could cause vomiting, stomach pain and even cancer and death," Xinhua said.

Morgan (2006) provided more detail for the Shanghai-based *Agence France-Presse*:

> Haoyuan Chemical, a sulfuric acid manufacturer, and Taolin Lead Zinc Ore Chemical, [were identified] as the key perpetrators ... [that] discharged waste water with various types of arsenic content more than 1,000 times the permitted level into the Xinqiang river.... Their unbridled polluting activities had been going on for a "long time", the [Xinghua] report said, underscoring the serious challenge the world's most populous nation faces in cleaning up its environment. According to Pan Yue, the deputy director of the State Environmental Protection Administration, Haoyuan had been dumping an astounding 50,000 tonnes of waste [including arsenic trioxide] a month into the [Xinqiang] river, while Taolin contributed 280 tonnes a month.

Western Hunan's mining area has strata that are rich in arsenic and geochemically abnormal, such as in the Wuxi deposit of gold–antimony–tungsten (Hu *et al.* 1996). Arsenic speciation here is associated with the Neoproterozoic-Cambrian strata that are generally carbonate-bound and sulfide-bound, approaching greater than the background arsenic concentration. According to a study by Lu & Zhang (2005), "The distribution of rock with high arsenic content or high easily transferred arsenic speciation is consistent with the distribution of high arsenic content soil."

High-arsenic minerals have also leached into the groundwater. A 5-year study that concluded in 2005 (Yu *et al.* 2007) assessed the arsenic concentrations of wells in Hunan Province (see Table 3.6).

In the mining region, soils and river/brook waters were polluted by mine tailings and smelting wastewater. The arsenic concentrations of polluted paddy soils and riverine water were 46.26–496.19 μg/g and 0.3–16.5 mg/l, respectively. Although it is known that arsenic in the soil and water is bioaccumulated by agricultural crops and the health of those ingesting the crops and the water is affected, more epidemiological work needs to be done on tracking this region after the danger is remediated (Lu & Zhang, 2005).

A similar situation pertains to the Shimen Mine in Shimen County, Changde Prefecture, which has been extracting ore from the Jiepaiyu arsenic deposit for decades, from which some of the world's finest realgar (As_4S_4) and orpiment (As_2S_3) museum specimens have been mined (Wang *et al.* 1999; Wilson, 2007). Because of the disposal of the tailings and waste, this mining area has also become a zone of severe arsenic pollution and has been the subject of a study by Wang *et al.* (1999), who discovered arsenic concentrations:

> in three villages of the polluted areas, in soil were 84.17 to 296.19 mg/kg, in river water 0.5 to 14.5 mg/L, in well water and tap water lower than 0.05 mg/L, in rice 0.196 to 0.620 mg/kg. Arsenic intake amounted to 195 to 1129 μg/day. In the hair, the arsenic contents were from 0.972 to 2.459 μg/g, increasing with aging.

Their conclusions were that "In the areas near Shimen arsenic mine, arsenic pollution was serious. The levels of arsenic exposure of residents were equal to or even over that of the severe cases of chronic arsenic poisoning reported domestically and abroad." In addition, soils near the Shimen mine have elevated levels of arsenic, and a study of three fern series has shown that ferns have adapted to the elevated levels of arsenic by bioaccumulating the metalloid (Yan *et al.* 2003).

In another mining site at Binzhou, which was highly mineralized with silver and gold and associated arsenic deposits, the mine waters had leached into the local surface water, and there were 220 individual cases recorded of arsenicosis at a 5% case rate, which indicated that 4400 had been exposed (Fan, 2000).

In addition, the antimony deposit at Xikuangshan Mine, Lengshuijiang County, Loudi Prefecture, the largest in the world, has been associated with elevated arsenic in the groundwater (Peng *et al.* 2002).

In January 2008 in Chenxi County, Hunan Province, the groundwater was illegally contaminated by arsenic-containing wastewater release from a coal mine. On January 26, official figures from the city affected showed 73 people had various levels of arsenic poisoning. However, non-official figures from investigations in local communities found that the actual number of people affected may have been much higher (Business news agency. 2009. http://china.toocle.com).

Inner Mongolia (Nei Mongol) Autonomous Region (A, C, G)

This area (4800 km^2) is home to 49 nationalities and lies south of the Gobi Desert and north of the Great Wall of China. The Huhhot region lies in a triangular, segmented and sunken palaeo lake basin south of the Da Qing (Great Green) Mountains and along the left bank of the Yellow River.

Groundwaters occur in a Holocene earth stratum and have a naturally occurring high concentration of arsenic-bearing rock. High levels of arsenic [mainly arsenic(III)], fluoride, oxygen, sulfates, nitrates and selenium occur in the aquifers, providing a reducing environment, rich in organic substances, in which arsenic is easily transported and in which it accumulates.

The western Huhhot alluvial basin is an agricultural area producing wheat, millet, corn, green beets, potatoes and sunflowers. It is home to 1000 species of plants, 93 species of fish, 362 species of birds and 114 species of animals. Huhhot alluvial basin groundwater is contaminated with arsenic along its seasonally arid zone that spans the Great Bend of Huang Ho (the Yellow River) and Hu Bao Plain. High arsenic levels are in shallow and deep wells, but not in surface waters. The deep aquifers spanning much of the central portion of the Huhhot alluvial basin cover an areal extent of 2500 km^2, while the shallow wells dug into limited, discontinuous and often discrete aquifers with elevated arsenic levels are on the periphery of the basin, between the mountains and the plain. The total area affected is about 150 000 km^2, with a population of 2.6 million exposed to arsenic concentrations above 10 μg/l (Xie *et al.* 2006). A 5-year study that concluded in 2005 (Yu *et al.* 2007) assessed the arsenic concentrations in wells in the region (see Table 3.6).

Figure 3.21 provides a map of those tube wells in Huhhot Basin with arsenic concentrations greater than 50 μg/l.

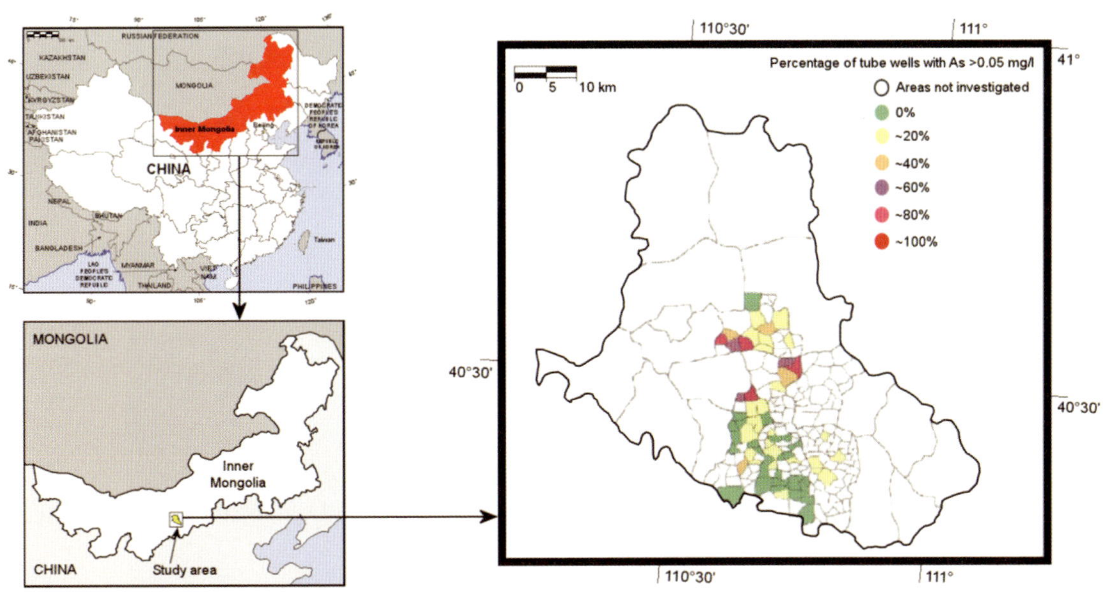

Figure 3.21 Map of Huhhot Basin, Inner Mongolia Autonomous Region, China, showing arsenic contamination based on the percentage of tube wells with concentrations of arsenic in water above 50 μg/l (modified from Sun, 2004a).

Arsenic in the Huhhot alluvial basin is generally mobilized in anaerobic groundwater from Holocene lacustrine and alluvial aquifers covering 2500 km^2 that formed a palaeo lake 8000 years BP (Smedley & Kinniburgh, 2002; Sun, 2004a,b). Most levels of arsenic range between 10 and 200 μg/l from waters flowing downhill from the Da Qing and Man Han mountains, contaminating 665 (11.3%) of 5885 tube wells. In the Hetao plain between the Huang Ho and the Inshan mountains, 96% of wells tested had arsenic concentrations above 50 μg/l (Jin *et al.* 2003).

Arsenic contamination of groundwater in the Huhhot Basin is irregular for shallow to medium deep dug wells and absent from wells 70 m deep or deeper. Shi (2004) found that wells in the 4–50 m range contained arsenic concentrations of 10–4000 μg/l, but two wells of 70 m and 75 m did not exhibit arsenic contamination. Of 29 wells in Tie Men Jing village, 28 exceeded 10 μg/l and 17 exceeded 50 μg/l. Furthermore, arsenic(III) was the dominant species in 22 of 29 wells.

The irregularity of arsenic contamination for shallow to medium deep wells is illustrated by Table 3.15, which gives the level of contamination, and Figure 3.22, which shows the relative location of each well, indicating that nearby wells can vary considerably in their arsenic concentration.

Table 3.15 Dug wells in Tie Men Jing Village, Huhhot Basin, China, showing arsenic concentration, depth, pH and redox potential.

Tie Men Jing well	Arsenic concentration (μg/l)	Depth (m)	pH	Redox potential (mV)
TMJ-1	82.1	7	7	233
TMJ-2	76.5	7	7.12	237
TMJ-3	90.3	8	7.44	239
TMJ-4	124.6	10	7.5	243
TMJ-6	19.8	12	7.45	387
TMJ-7	39.5	3	7.18	458
TMJ-8	5.1	4	7.13	470
TMJ-9	18.6	8	7.59	196
TMJ-10	109	8	7.3	162
TMJ-11	115.1	8	7.37	229
TMJ-12	119.4	50	7.56	192
TMJ-13	140.5	9	7.12	200
TMJ-14	132.6	12	7.12	171
TMJ-15	61.8	4	7.5	669
TMJ-16	84.3	5	7	429
TMJ-18	28.3	45	7.42	135
TMJ-19	52.5	30	7.4	238
TMJ-20	16.2	5	7.4	260
TMJ-21	55.9	5	7.5	212
TMJ-22	127.9	4	7.5	–
TMJ-23	29.3	43	7.5	–
TMJ-24	13.2	46	7.4	–
TMJ-25	137.6	7	7.5	–
TMJ-26	17.8	45	7.5	–
TMJ-27	78.6	8	7.5	–
TMJ-28	51	24	7.5	–
TMJ-29	33.2	5	7.5	–
TMJ-30	23.6	40	7.5	–
TMJ-31	41.8	50	7.5	–
Average	**66.42**	**17.66**	**7.38**	**282.1**

Source: Shi, 2004

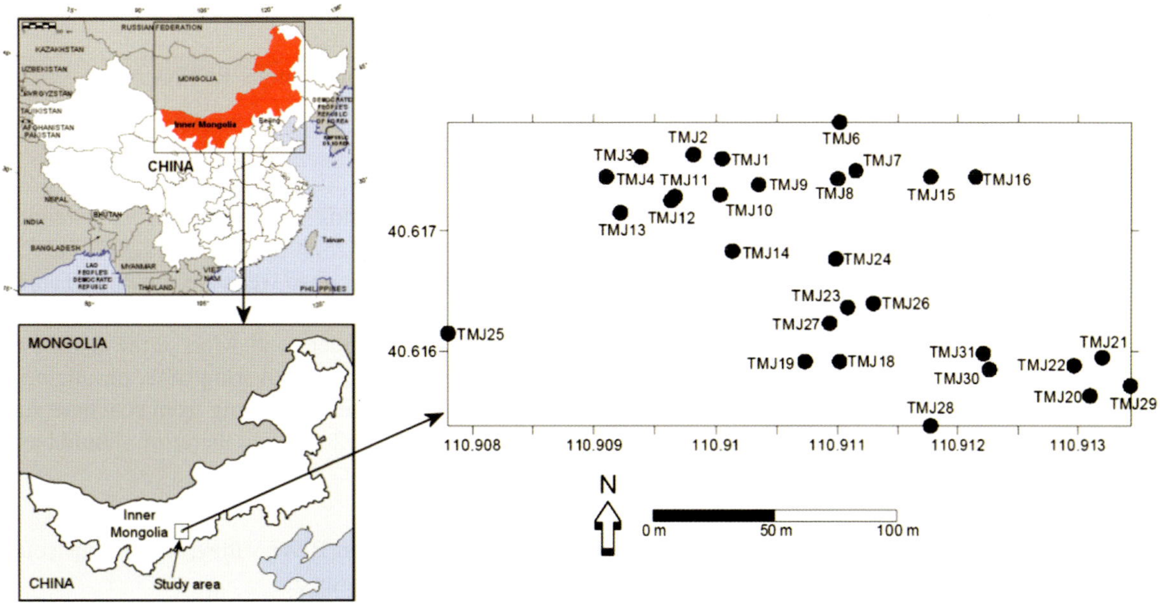

Figure 3.22 Location of 29 hand-dug wells in Tie Men Jing Village, Huhhot Basin, China (modified from Shi, 2004).

A study that concluded in 2003 (Xia & Liu, 2004) found arsenic in wells in Inner Mongolia Autonomous Region (see Table 3.5).

In the western region of Huhhot Basin, a study showed that the depth of wells is directly proportional to arsenic concentrations, with over 50% of deep wells (>100 m) having concentrations above 50 μg/l (see Table 3.16).

Table 3.16 Arsenic concentrations in the western part of the Huhhot Basin, Inner Mongolia Autonomous Region, China, by well type.

Well type	No. wells tested	Concentration range (μg/l)	% above 50 μg/l
Shallow	305	<10–1860	20.7
Deep	33	<10–360	54.6
Surface	17	<10–20	0
Total	**355**	**< 10–1860**	**22.3**

Source: Luo *et al.* undated

Coal is also a major economic resource in Inner Mongolia Autonomous Region. A recent study has shown high arsenic content in some of the deposits (see Table 3.11; Wang *et al.* 2006). Just as in Guizhou Province, there is a tradition of smoking vegetables and meat using coal in unventilated stoves. In addition, some of the arsenic in the coal seams may be contributing to the contamination of the groundwater. This is an issue that demands further study.

Chronic arsenicosis cases were first recognized clinically at Zhi Ji Liang village in the Huhhot area, when the Anti-Epidemic Station of Huhhot investigated requests for dermatological medications in May 1990 (Luo *et al.* 1993, 1997). In a three-village area of Huhhot in 1991, the arsenic content in local wells ($n = 184$) was correlated with local residents ($n = 3228$) and their well-use histories ($n = 3179$). Arsenic exposures ranged from <10 to 2000 μg/l. Thirty-five per cent (64) had a peak arsenic concentration below 50 μg/l, 69% were below 100 μg/l and 86% were below 150 μg/l. Manifestations of arsenicosis were as follows: skin cancer, 8 (with both hyperkeratoses and dyspigmentation and with peak arsenic concentrations of 150 μg/l or greater); hyperkeratosis, 172; dyspigmentation, 121; and both hyperkeratoses and dyspigmentation, 94. Although cases of hyperkeratoses and of dyspigmentation occurred with peak arsenic concentrations under 50 μg/l, they did not reach expected prevalences until higher peak arsenic concentrations, revealing a statistically significant deficit of skin cancer among those with peak arsenic concentrations between 50 and 150 μg/l. In a controlled survey of 832 individuals exposed to water containing arsenic levels above 50 μg/l, 168 (20.1%) were diagnosed with arsenicosis (Luo *et al.* undated). Another study

(Sun, 2004a; PRC National Cooperation Group, undated) reported that in 13 counties, 776 villages had a potential population at risk of 411 200 and a case rate of 15.54%.

In addition, this contamination of groundwater by arsenic in Inner Mongolia Autonomous Region may be attributable to chemical weapons leaching from corroding warheads (Evans, 1997).

Jiangsu Province (C)

Jiangsu Province has an extensive irrigation system for its low-lying agricultural fields, which account for a major part of its economy (Provincial People's Government of Jiangsu, 2007).

Jiangsu Province's groundwater was tested for arsenic in 2005, and the concentrations in wells are shown in Table 3.6 above (Yu *et al.* 2007).

The mechanisms of the liberation of arsenic into the groundwater are not clearly understood. Most of the answer might lie in the fact that Jiangsu is low lying, with water covering 18% of its territory and low plains covering 68%, producing in effect a scenario similar to that of Viet Nam. Another minor part of the answer might lie in arsenic leaching from coal reserves that have in their high ranges very moderate levels of the metalloid (see Table 3.11; Wang *et al.* 2006) and therefore should be monitored.

Jiangxi Province (C, M)

Jiangxi, with a population of 42.8 million (2004), has mountains surrounding the Gan River valley, which runs south to north through its centre. Agriculture is important, with major crops of rice, cotton and rapeseed.

Jiangxi is rich in mineral resources, leading the provinces of China in deposits of copper, tungsten, gold, silver, uranium, thorium, tantalum and niobium, among others (Provincial People's Government of Jiangxi, 2007). Dayu County is associated with Jurassic biotite magmatic to hydrothermal granites and is rich in tungsten (Ishirara *et al.* 2003). In a study of arsenic speciation, it was reported that soil is polluted with arsenic in Dayu County (Akter *et al.* 2005). Dexing County, which is rich in copper, gold and arsenic, was the site of about 200 small-scale gold mining operations from 1990 to 1995 that used the mercury extraction process in private homes. This resulted in not only mercury but also arsenic contamination of the air, wastewater (which then penetrated groundwater) and solid tailings (Chen & Li, 1990). Analyses of the arsenic content in gold ores indicate a range from 7.7 to 13.8 mg/kg (Gunson, 2004). Since 1996, national legislation has prohibited small mining activities, but follow-up data are not available, nor have data been found for other small-scale gold mining areas with similar arsenic concentrations (Lin *et al.* 1997).

Coal is also an economic resource in Jiangxi, and recent investigations have identified high-arsenic content in some of the deposits. Just as in Guizhou Province, there is a tradition of smoking vegetables and meat using coal in unventilated stoves. In addition, some of the arsenic in the coal seams may be contributing to the contamination of the groundwater. A study done by Wang *et al.* (2006) identified an arsenic content range of local coals from 2.8 to 24 mg/kg. This is an issue that demands further study (see Table 3.11). In fact, a study has sought to identify arsenic-resistant bacteria in the 15 mg/kg arsenic of Shangrao's coal mine wastes. Such a process may be able to remediate and detoxify China's coal mines by oxidizing arsenite to arsenate (Xiong *et al.* 2007).

Jilin Province (A, C, G)

Jilin, with a population of 27.1 million (2004), has a continental monsoonal climate, with long, cold winters and short, warm summers. Average January temperatures range from −20°C to −14°C (Provincial People's Government of Jilin, 2007). These winter temperatures may exacerbate the use of high-range arsenic coal to heat homes.

Groundwater in Jilin is known to be contaminated with arsenic. Reports indicate that two counties with 71 villages and a population of 3800 are at risk, with an arsenicosis case rate of 0.95%. It is also reported that 101 (12.21%) of 8200 tube wells are contaminated with arsenic at concentrations above 50 μg/l (Sun, 2004a; PRC National Cooperation Group, undated). A study that concluded in 2003 (Xia & Liu, 2004) found arsenic in wells in Jilin Province at concentrations shown in Table 3.5 above. A 5-year study that concluded in 2005 (Yu *et al.* 2007) also assessed the arsenic concentrations of wells in Jilin Province (see Table 3.6).

In addition, it is possible that some arsenic contamination of groundwater may be attributable to leaching from high-arsenic Permian-Carboniferous coal seams, which have the fifth highest arsenic concentrations in China of 26 mg/kg (Wang *et al.* 2006; see Table 3.9). Jilin has five rivers flowing through it (Yalu, Tumen, Liao, Songhua and Nen), and they may be agents in the leaching of arsenic minerals and coalfields. Just as in Guizhou Province, there is a tradition of smoking vegetables and meat using coal in unventilated stoves. The social practice of drying and smoking foods in poorly ventilated or unventilated areas with high-arsenic coal is an issue that demands further monitoring in Jilin.

Liaoning Province (A, C, G)

Liaoning Province, with a population of 29 million and area of 145 900 km^2, is located in the southern part of China's north-east, and its watersheds drain the low-lying area into the Yellow and Bohai seas. It enjoys a strong industrial base and abundant natural resources, including coal. It is one of the comparatively well-developed coastal areas in northern China that is open to the world (Provincial People's Government of Liaoning.2007). A continental monsoonal climate of below 0°C temperatures in winter promotes the use of high-arsenic coals for home cooking and heating.

Anthropogenic arsenic contamination is a result of this industrialization (Xu *et al.* 2000):

> Shenyang is the capital of Liaoning Province, located in north-eastern China. It is the largest, heavily industrialized city in the country. The urban area of Shenyang covers approximately 230 km^2 and has a population of 3.1 million. The major industries in Shenyang include ferrous and nonferrous smelters, machinery plants, chemical plants, coke plants, and power plants. The particles emitted from these industries contain large amounts of heavy metals, including lead, arsenic, and cadmium.

With Liaoning's industrial outlook, coal is an important economic resource: a 2006 study has identified a maximum arsenic content of 29 mg/kg in some of the Late Jurassic-age deposits (see Table 3.11; Wang *et al.* 2006). Just as in Guizhou Province, there is a tradition of smoking vegetables and meat using coal in unventilated stoves.

Arsenic contamination of the groundwater in Liaoning Province has been lowered below 50 μg/l (Jin *et al.* 2003). A 5-year study that concluded in 2005 reported that none of the 3500 wells in 25 villages in Liaoning Province (see Table 3.6) had arsenic concentrations above 50 μg/l (Yu *et al.* 2007).

Ningxia Hui Autonomous Region (C, G)

Ningxia Hui Autonomous Region, home of the Hui people, is rural, with the second lowest GDP in China. It is located on the north-west loess highland with the Huang He flowing through it. It has a continental climate, with temperatures of −7°C and −10°C in January (Provincial People's Government of Ningxia, 2007).

The case of groundwater being contaminated with arsenic has been documented in several studies. A study that concluded in 2003 (Xia & Liu, 2004) measured arsenic concentrations in wells in this region (see Table 3.5). This same study found that of 4609 patients in the three towns surveyed, 9.46% or 436 were diagnosed with arsenicism; of these, 232 had moderate cases and 204 cases were suspected (Xia & Liu, 2004). Another 2004 report indicates that two rural counties with 22 villages and urban populations of 530 000 were at risk from arsenic-contaminated groundwater, with a case rate of 5.03%. Of 8276 tube wells, 88 (1.06%) being contaminated with arsenic above 50 μg/l (Sun, 2004a; PRC National Cooperation Group, undated). A 5-year study that concluded in 2005 (Yu *et al.* 2007) assessed the arsenic concentrations of wells in Ningxia Hui Autonomous Region (see Table 3.6).

It appears that in Ningxia Hui Autonomous Region, the depth of tube wells is directly proportional to arsenic levels in the groundwater (Sun, 2004a; Sun & Yang, 2004; PRC National Cooperation Group, undated). This groundwater contamination appears similar to that of Inner Mongolia Autonomous Region to the north.

Qinghai Province (G)

Qinghai, at an average elevation of about 3000 m above sea level on the Tibetan Plateau, is the source of three major rivers: the Huange He, the Yangtze and the Mekong. Being rural, it has the lowest GDP in China, with Tibet Autonomous Region.

A 2004 report indicates that six villages with a population of 290 000 are at risk from arsenic contamination of groundwater, with a case rate of 2.49%. The arsenic concentrations in the drinking-water ranged from >50 to 320 μg/l. Of 22 tube wells, 8.33% were contaminated; the depth of these tube wells was directly proportional to the arsenic concentration (Sun, 2004a; PRC National Cooperation Group, undated). Another study that concluded in 2003 (Xia & Liu, 2004) measured arsenic concentrations in wells in Qinghai Province (see Table 3.5). A study that concluded in 2005 (Yu *et al.* 2007) also assessed the arsenic concentrations of wells in Qinghai Province (see Table 3.6).

Shaanxi Province (C)

Shaanxi, with Xi'an as its ancient capital on the Silk Road, spans the central loess to the Qinling Mountains and has a continental climate, with January temperatures ranging from −11°C to 3.5°C (Provincial People's Government of Shaanxi, 2007). Tourism is a major industry, although coal deposits are also important for home heating and cooking. While most of these deposits are benign, there are a few of concern, with high ranges of arsenic up to 15 mg/kg (see Table 3.11; Wang *et al.* 2006).

Shandong Province (A, C, G)

Shandong is located on the hilly, eastern perimeter of the North China Plate, with extensive deposits of coal, oil and minerals, including gold and associated arsenopyrite (Provincial People's Government of Shandong, 2007).

Initial reports had indicated a "high level" of arsenic contamination, presumably above 50 μg/l, in Shandong Province (Donglou, 2004). Then, a study that concluded in 2005 (Yu *et al.* 2007) measured arsenic concentrations above 50 μg/l in 19 899 wells in 30 villages, with a total population at risk of 31 799 (see Table 3.6).

In addition to arsenic contamination of the groundwater, Late Jurassic coal beds also contain high arsenic levels. The findings of a 2006 study are summarized in Table 3.11 above (Wang *et al.* 2006). The leaching potential of arsenic from these coal beds is unknown.

Additionally, in January 2009, because the Linyi Red Allcorn Chemical Co., Ltd in Shandong Province illegally released arsenic tainted wastewater, the downstream region was seriously polluted. The water channel at the Hong Daodong provincial boundary of Shandong and Jiangsu had an arsenic concentration 39 times higher than the safety limit. The input to water in the canal is 5 times higher than safety limits (Business news agency., 2009. http://china.toocle.com).

Shanxi Province (G)

Shanxi borders Hebei to the east, Henan to the south, Shaanxi to the west and Inner Mongolia Autonomous Region to the north. A central plateau dissected by valleys rises in the east to the Taihang Mountains and in the west to the Lüliang Mountains. The Huang He (Yellow River) forms the western border with Shaanxi, with two major tributaries (Fen and Qin) flowing north to south. The continental monsoonal climate has 350–700 mm of precipitation (60% falling from June to August). Temperatures vary between 0°C in January to 21–26°C in July (Provincial People's Government of Shanxi, 2007). This temperature fluctuation plays a role in the use of local high-arsenic coal for heating.

In the 1990s, problems related to arsenic in drinking-water were recognized in Shanxi Province (Luo *et al.* 1997). Initial reports indicated 18 counties, 116 villages and a population of 60 300 at risk. The arsenic-caused cancer death rate was reported at 217 per 100 000, with a case rate of 10.76% (Sun & Yang, 2004; PRC National Cooperation Group, undated). However, by 2006, these initial figures were revised upward. The area in Shanxi Province with contamination of arsenic in well water above 50 μg/l covers about 2000 km^2, where 100 000 people are living in 79 villages with high arsenic contamination. In this area, 33% of 2364 wells tested had arsenic levels above 50 μg/l (Jin, 2006). About 5000 patients with typical symptoms of arsenic poisoning were diagnosed, showing a trend of family accumulation. Morbidity increased with length of exposure and age, but did not differ between men and women. Skin and peripheral nerve lesions were different among different endemic districts. Cancer mortality in the disease areas was several times higher than in other areas (Feng, 2006).

In the Qinba Mountains, a "high level of arsenic in the drinking water" was reported (Donglou, 2004). Of 3079 tube wells, 1612 (52.35%) contained arsenic concentrations above 50 μg/l. The highest measured arsenic concentration was 780 μg/l. A list of contaminated wells in Shanxi counties (number of wells/number of wells containing arsenic levels above 5 μg/l) follows: Shanyin, 2062/715; Yingxian, 74/19; Shuocheng, 86/44; Tianzhen, 28/13; Fenyang, 64/35; Xiaoyi, 27/5; Pingyao, 32/2; total, 2373/833 (Sun & Yang, 2004; PRC National Cooperation Group, undated). A study that concluded in 2003 (Xia & Liu, 2004) measured the arsenic concentrations in groundwater in Shanxi Province, with the results as shown in Table 3.5 above.

From 2001 to 2003, Yu *et al.* (2007) tested 48 583 wells in Shanxi Province. The summary results for this survey are tabulated in Table 3.6 above.

There are also significant Late Jurassic-age coal reserves in Shanxi. Many of these have high arsenic concentrations. A summary of these coal reserves is given in Table 3.11 above (Wang *et al.* 2006). Because of the freezing temperatures in winter, coal has been used in unventilated domestic stoves for both heating and cooking. Additional research is needed to determine the levels of exposure and case rates of arsenicosis among the coal-using population. A further issue is whether any arsenic in these high-arsenic coal deposits is leaching into the groundwater.

Sichuan Province (G)

Sichuan Province, as the "province of abundance", is one of the major agricultural areas and also has 132 mineral resources, of which 11 are the largest in China (Provincial People's Government of Sichuan, 2007). This province is in a zone of high mineralization that has contributed to its having a "high level" of arsenic contamination in the drinking-water (Donglou, 2004). A study that concluded in 2005 (Yu *et al.* 2007) assessed the arsenic concentrations of wells in Sichuan Province, as shown in Table 3.6 above.

In addition, Sichuan has several Late Permian-age coal seams that contain high levels of arsenic. A study that concluded in 2006 measured the arsenic concentrations in these mined coal seams, producing the results summarized in Table 3.11 above (Wang *et al.* 2006). An issue of concern is whether any arsenic in these high-arsenic coal deposits is leaching into the groundwater.

Taiwan, Province of (A, G)

The Province of Taiwan, with an area of 35 801 km^2 and a population of 22.9 million (2005), is densely populated and heavily industrialized. Most natural resources have been depleted, except coal deposits and some insignificant oil and gas deposits (Central Intelligence Agency, 2012).

Arsenicosis was first reported in the 1920s as blackfoot disease, but it was only seriously addressed in conjunction with internal cancers of the bladder, kidney, lungs and skin beginning in 1958. In the 1960s, the south-west was reported as having arsenic concentrations in the drinking-water ranging from 10 to 1800 µg/l (Kuo, 1968), with a mean of 500 µg/l. In 1977, 119 townships were reported as having concentrations in their drinking-water above 50 µg/l, of which 58 had concentrations above 350 µg/l (Lo *et al.* 1977). Other studies have corroborated this wide spectrum of concentrations (Tseng *et al.* 1968; Tseng, 1977; Chen *et al.* 1988; Wu *et al.* 1989; Chen *et al.* 1992; Chiou *et al.* 1995).

In the north-east Province of Taiwan, tested ground waters contain average arsenic concentrations of 135 µg/l, with some concentrations above 600 µg/l (Hsu *et al.* 1997). A case–control study of 582 adults between the ages of 40 and 60 had odds ratio exposure rates of 2.77 (0.84–9.14) and 4.28 (1.26–14.54) for arsenic exposure of 10–1990 µg/l-years and 20.0+ mg/l-years, respectively, compared with those who were not exposed (Tseng *et al.* 1996). In the south-west, Chen *et al.* (1986) conducted a matched case–control study of 204 deceased cancer cases and 368 matched alive community control cancer patients (bladder, lung and liver). A positive dose–response relationship was observed, and age–sex adjusted odds ratios were 3.90, 3.39 and 2.67 for bladder, lung and liver cancers, respectively. An early study (Lo *et al.* 1977) measured arsenic concentrations in well water in the Province of Taiwan (see Table 3.5).

In 2005, an organic process involving bacteria as the releasing mechanism was suggested for this form of arsenic mobilization, in which arsenic cycling involves active erosion of mountain belts, transport of arsenic by surface waters, adsorption of arsenic by stream sediments and deposition of stream sediments and organic matter in alluvial deposits. Subsequently, iron-reducing bacteria present in alluvial aquifers cause the release of sorbed arsenic to groundwater under moderately reducing conditions (Turner *et al.* 2005). Groundwater mineralogy is poorly understood in the Province of Taiwan, but it is suspected to be strongly reducing, in both the south-west, where arsenic(III) is present in deep artesian wells at depths ranging from 100 to 280 m (Tseng *et al.* 1968) and the north-east, where arsenic is present in wells at depths ranging from 16 to 40 m (Hsu *et al.* 1997).

Anthropogenically caused arsenic contamination of ground waters has occurred recently due to aquaculture. Although remediation programmes have been carried out in the south-west, waters laced with arsenic have continued to be used to raise fish, such as trout and milkfish, as a means to sanitize and preserve the product. Studies have shown that the flesh of these aquacultured fish contains high levels of arsenic (>10 µg/l), so the south-west is still considered an arsenic-contaminated area (Lin & Lin, 2005).

Contamination of crops irrigated with high arsenic water is a known danger in Southeast Asia (Achin, 2011) which is now also being recognized in Taiwan. During the lean summer months in Southeast Asia from January to May a large amount of groundwater is used for irrigation which is contaminated with arsenic resulting from the oxygenated decomposition of arsenopyrite forming Fe^{++} and Fe^{+++} sulfates and H_2SO_4 acid which mobilize arsenic. The mobilized arsenic is then accumulated in soils and taken up by crops like rice. While the exact mechanism of arsenic release is controversial (following oxidation of arsenopyrite, reductive dissolution of Fe hydroxides or anion exchange of arsenic with fertilizer phosphates) its effect is dramatic, for it provides another pathway – through the food chain. Arsenic contamination of agricultural soils and products in Taiwan is a potential looming concomitant of arsenic contamination of groundwater. Several studies in Taiwan have revealed high levels of arsenic in Taiwanese paddy rice. Schoof *et al.* (1998) reported 0.76 µg g^{-1} d. wt. of arsenic in Taiwanese paddy rice collected directly from farms. It was also reported that about 0.20 µg g^{-1} d. wt. of arsenic (range 0.19–0.22 µg g^{-1} d. wt.) in Taiwanese firm rice. Lin *et al.* (2004) revealed <0.10–0.63 µg g^{-1} d. wt. of arsenic in Taiwanese paddy rice, a value comparable to that reported by Williams *et al.* (2005).

Xinjiang Uygur Autonomous Region (G, M, P)

Xinjiang Uygur Autonomous Region is a vast, 1.66 million m^3 area in north-western China inhabited mainly by Uygurs who developed their own system for water management and irrigation. The region is rich in minerals, coal and oil, and borders

Mongolia, the Russian Federation, Kazakhstan, Kyrgyzstan, Tajikistan, Afghanistan, Pakistan, India, Tibet Autonomous Region, Qinghai and Gansu.

One area of arsenic contamination is the 38 000 km^2 area of the Dzungar Basin, north of the Tien Shan Mountains. Another is south of the Tien Shan in an area that is very desiccated today and is the region through which the northern route of the Silk Road traversed. A third area that is not given much attention at all is the area south of the Taklimakan Desert, an inhospitable region along which many of the ancient abandoned towns of the southern route of the Silk Road are found. In this survey, owing to a lack of information about the southern route, only the area north through the Tien Shan is considered.

The indigenous population of Muslim Turkic-Mongolian Uygurs had a reverence for the ancient Persian Karwell system of water channelling. This system was a hallmark of the pre-Muslim cultures in Persia and Baluchistan and depended on voluntary contributions to finance the excavation of subterranean water channels from high mountain aquifers to carry water sometimes hundreds of kilometres to villages that were founded at artificial oases where the water surfaced, being used for drinking, cooking, washing, bathing and irrigation. The water was safe because it was from the high mountain aquifers, and it was protected from contamination by flowing in 6- to 12-m-deep excavated conduits. Without this system, these towns on the Silk Road would not have been able to support life. With it, they became an integral connection between Europe and China.

In the 1960s, artesian groundwater in the Dzungar Basin was first extensively tapped into as waves of new immigrants from other provinces settled as farmers in Xinjiang Uygur Autonomous Region and needed more water for irrigation (Sun & Yang, 2004). In the late 1970s, Kuitun city, at 455 m above sea level and with a population of 50 760, was discovered to have arsenic-contaminated groundwater (>50 μg/l) (Wang *et al.* 1995; Luong *et al.* 2004). Of 14 050 tube wells tested, 670 (4.77%) had arsenic concentrations above 50 μg/l (Sun & Yang, 2004). Kuitun's population of 50 760 is now exposed to water that also contains concomitantly high levels of fluoride (Table 3.17).

Table 3.17 Arsenic and fluoride content of 102 wells in Dzungar Basin, Xinjiang Uygur Autonomous Region, China.

Arsenic concentration (mg/l)	No. wells containing following fluoride concentrations:			Total no. wells
	1.0–1.9 mg/l	2.0–3.9 mg/l	>4.0 mg/l[a]	
0.05–0.09	26	11	3	40
0.10–0.19	5	13	6	24
0.20–0.29	1	8	7	16
0.30–0.39	0	5	8	13
0.40–0.49	0	1	4	5
0.50–0.59	0	0	2	2
>0.60[b]	0	1	1	2
Total	**32**	**39**	**31**	**102**

[a]Highest level of fluoride was 21.5 mg/l.
[b]Highest level of arsenic was 0.88 mg/l.
Source: Wang *et al.* 1997

Studies that concluded their surveys of Xinjiang Uygur Autonomous Region's groundwater in 2003 (Xia & Liu, 2004) and 2005 (Yu *et al.* 2007) assessed the arsenic concentrations of its wells, with the results shown in Tables 3.5 and 3.6 above, respectively.

In general, arsenic concentrations range from 40 to 750 μg/l in artesian ground waters and up to 1200 μg/l in borehole wells at a depth of 660 m. It is known that concentrations of arsenic are directly proportional to depth, similar to the situation in the Huhhot Basin of Inner Mongolia Autonomous Region (Wang & Huang, 1994). However, the public health system remains unresponsive, primarily now that oil has been discovered in the Dzungar Basin and the government's policy is to fill Beijing and other cities with fossil fuel–powered vehicles. Therefore, since 1990, the pace of drilling tube wells has accelerated even more, with the recent influx of foreigners from Henan and elsewhere, and, like Bangladesh, the wells have been drilled deep (up to 660 m deep) into the strongly reducing Holocene alluvial sediments.

The impact of the discovery of oil in the Karamay fields in the Dzungar Basin has exacerbated the arsenic problem in Xinjiang Uygur Autonomous Region, not only by increased exploitation of groundwater for use by immigrant workers, but also by the very process of oil extraction. Karamay was touted by the government as the second biggest oilfield in China, producing 21.9 million tonnes of crude oil in 2005 (next to Heilongjiang's 45 million tonnes); the goal was to have Xinjiang Uygur Autonomous Region equal its rival and reach 45 million tonnes by 2008.

Geological surveys show that Karamay's oil and natural gas deposits stand at an equivalent of 27 billion tonnes of oil, accounting for a quarter of the nation's total reserves. China's top oil company aims to build a $14.5 billion, 4170-km pipeline to transport natural gas from Xinjiang Uygur Autonomous Region to China's eastern coastal provinces of Shanghai and Jiangsu. Rival China Petrochemical Group (Sinopec) plans to spend more than 1 billion yuan (US$ 120.8 million) a year to explore for oil and gas in Xinjiang Uygur Autonomous Region (Yang *et al.* 1998).

The oil production centre in the Karamay petroleum fields promotes arsenic leaching into the groundwater in the vicinity of Wusu (Usu) city, with a potential population at risk of 100 000.

Surface waters are also contaminated with high levels of arsenic. For instance, in saline Lake Aibi, a natural catchment, arsenic concentrations in the groundwater are above 170 μg/l, and arsenic levels in shallow groundwater measure 10–68 μg/l. Rivers tend to have lower arsenic concentrations, but concentrations in the Kuitun, for instance, range from 10 to 30 μg/l (Wang & Huang, 1994).

Early-Middle Jurassic-age coal deposits in Xinjiang Uygur Autonomous Region are also of concern because they have high arsenic concentrations. The results of a study concluded in 2006 are summarized in Table 3.11 above (Wang *et al.* 2006). In addition to the domestic use of high-arsenic coal to smoke and cook foods in unventilated areas, thereby releasing arsenic into the foods and into indoor air, another issue of concern is whether these arsenic-rich coal deposits leach into the groundwater.

In summary, of the total population of 19.6 million in Xinjiang Uygur Autonomous Region, 429 858 are exposed to arsenic concentrations above 50 μg/l in drinking-water, and another group of unknown size is exposed to arsenic contamination in high-arsenic coal used in unventilated domestic cooking and heating. Populations exposed include the urban centres of 100 000 (in Wusu), 50 760 in Kuitun city and those in 463 rural villages and towns. The first case of chronic arsenic poisoning was recognized in Xinjiang Uygur Autonomous Region in 1980 (Wang, 1984; Huang *et al.* 1985). The actual death rate associated with arsenicosis in Xinjiang Uygur Autonomous Region is 463 per 100 000, but this is not to be confused with the case rate, which is certainly much higher. As of November 2004, 3530 confirmed cases of arsenicosis-related diseases had been officially recognized (Sun, 2004a; PRC National Cooperation Group, undated) (Figure 3.23).

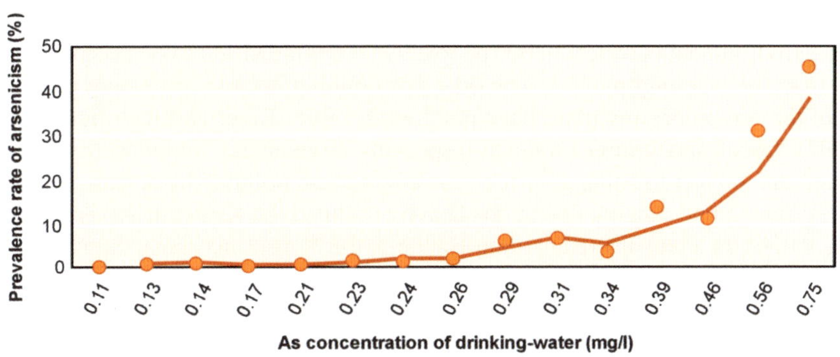

Figure 3.23 The relationship between arsenic concentrations in drinking-water and prevalence rates of arsenicism in Xinjiang Uygur Autonomous Region, China (modified from Feng, 2006).

Yunnan Province (C, G, M)

Reports indicate that this province has a population of 48 438 at risk from arsenic poisoning due to contamination of groundwater (Jin *et al.* 2003). Tube wells are one vector, with many arsenic levels above 100 μg/l. Yu *et al.* (2007) mapped tube wells in south-western Yunnan Province with arsenic concentrations above 50 μg/l in a study completed in 2005.

The second vector of arsenic poisoning is high-arsenic coal. The results of a study concluded in 2006 on arsenic levels in coal in Yunnan Province are summarized in Table 3.11 above (Wang *et al.* 2007). In addition to the domestic use of high-arsenic coal to smoke and cook foods in unventilated areas, thereby releasing arsenic into the foods and into indoor air, another issue of concern is whether these arsenic-rich coal deposits leach into the groundwater (Jin *et al.* 2003).

The third vector for arsenic poisoning is the geological occurrence of high arsenic-bearing ore, the basis for the local Wenshan Arsenic Factory of Yunnan, which produces 2700 tonnes of white arsenic (As_2O_3) and 1100 tonnes of metallic arsenic every year. Wenshan township (population of 100 000) is exposed to arsenic emitted from the smelter into the air at a concentration of 3 μg/m^3. However, there is also a significant contamination of the soil, and therefore agricultural crops, mainly rice and corn, were responsible for more than 90% of the total arsenic intake (300–1100 μg/day) (Sun *et al.* 2001a).

Additionally, in June 2008, water quality in Yangzong Lake in Yunnan Province in south China consistently showed high arsenic levels. In late July, the average arsenic concentration in the lake were 0.116 mg/L. Nearly 30 km^2 of open water was polluted by arsenic. The provincial environmental protection department said the main pollution source was The Chengjiang JinYe Trade Limited Liability Company (Business news agency., 2009. http://china.toocle.com).

Zhejiang Province (G, M)

In Tongxing city, the drinking-water has reportedly "high concentrations of arsenic", which presumably means levels above 50 μg/l (Jin *et al.* 2003). The source of this contamination is presumed to be geogenic. The geological occurrence of high arsenic-bearing ore is the basis for the local Wenzhou Foreign Trade Industrial Product Co. Ltd, which produces arsenic trioxide and arsenic pentoxide (As_2O_5), and its mine tailings are an issue in groundwater contamination.

3.7 GEORGIA (A, M, P)

Georgia, with a population of 4.6 million, has a Human Development Index rank of 74 out of 169 and a per capita GNI of $US 4902 (US Census Bureau, 2011; UNDP, 2010).

Anthropogenic sources of arsenic in Georgia come from large industrial processing plants that produce manganese, ammonia, arsenic, copper and gold. However, the North Caucasus region of the Russian Federation is far more of an untreated wastewater polluter than Armenia, Azerbaijan and Georgia combined. The effluents of similar sources have also been identified flowing into the Black Sea (Academy of Science of Georgia, 1998). Even so, in Armenia and Azerbaijan, different industries also discharged high loads of pollutants into the Kura and Araks rivers and their tributaries, including anthropogenic arsenic. In the North Caucasus, one of the major concerns was the contamination of the Terek River and its tributaries from non-ferrous industries, and the contamination of the Terek-Kuma artesian aquifer with arsenic was and still is a problem (CEO, 2002b). Diffused sources of pollution, including agricultural runoff, drainage waters from legal landfills and illegal dumpsites and open-pit mining operations, as well as urban runoff, also posed high threats to surface water and groundwater.

Georgia has a long history of mining of non-ferrous minerals bearing arsenic compounds, including copper, lead, zinc and gold (Dzhaoshvili, 1968). For instance, the Madneuli mine is one of the largest in the Caucasus and has been in operation since the 1970s. The surrounding environment in which 30 000 people live is degraded and toxic because of the deposition of airborne arsenic and other metals from the smelting process. The dust from mining arsenic and heavy metals annually amounts to 28 tonnes. Open pit waters of the ore processing plant have polluted the Kura River and its tributaries with arsenic, copper and zinc (Tvarlchrelidze, 1997). Local agricultural products (especially vegetables), supplied to the inhabitants of the city of Tbilisi and its surroundings, are subject to being grown in this atmosphere polluted by arsenic and heavy metals.

In Georgia, wastewaters from copper mining operations heavily polluted the Kazretula River (Kura River basin) with heavy metals and arsenic (CEO, 2002b). In Tyrnyauz (Kabardino-Balkaria, the North Caucasus), tungsten and molybdenum had been extracted and enriched since 1945 with arsenic in associated minerals, and environmental safeguards were never implemented. Since 1998, about 125 million tonnes of toxic wastes containing arsenic have accumulated in the Tyrnyauz molybdenum and wolfram quarry and processing plant in Kabardino-Balkaria. Annually, 5527 tonnes of suspended substances, 0.1 tonne of molybdenum and 0.07 tonne of arsenic flow into the Baksan River (Dzhaoshvili, 1968; Bakuradze *et al.* 1995; Tvarlchrelidze, 1997; State Committee on Ecology and Control of Natural Resources Utilization, 1998; Price, 2000; Ministry of Environment of Georgia and UNDP, 2001–2002).

Petroleum-related arsenicals from oil refineries and power plants have been leached into aquifers, and arsenic released by the action of petroleum on underlying aquifers has polluted the tributaries of the Black and Caspian sea basins. Heavy metals, oil products and phenols also heavily polluted the Kuban River. For example, in the late 1980s in the Kuban River, ambient concentrations of oil products and copper were 5–7 times as high as existing surface water quality standards, and arsenic was elevated as well (CEO, 2002b).

3.8 INDIA (C, G, M)

India, with a population of 1.2 billion, has a Human Development Index rank of 119 out of 169 and a per capita GNI of $US 3337 (US Census Bureau, 2011; UNDP, 2010).

India lacks an umbrella framework to regulate all dimensions of freshwater. A number of different principles, rules and acts have been adopted over many decades and in terms of statutory development, irrigation laws are the most developed part of the law. Water law is largely state based and states have the power to regulate water supplies, irrigation and canals, drainage and embankments, water storage, hydropower and fisheries. However, India has no national laws which define "clean" and "potable" water. Local administrations can supply water that is neither potable nor drinkable, and citizens have little recourse. Without this legal framework in place and regulatory agencies, no government authority or institution can be ultimately held responsible for the quality of the water. The Central Public Health and Environmental Engineering Organisation, which comes under the Ministry of Urban Development and Poverty Alleviation, do set guidelines for drinking-water quality, but they are merely guidelines (Jishnu, 2004):

> In 1996, a parliamentary Committee on Subordinate Legislation had suggested that water be treated and supplied by local authorities and should be treated as "food" as "the agency responsible for supplying drinking water to the public has to ensure purity and the statute should bind it to do so". However, the ministry of urban development quashed the idea by pointing out that such agencies were [in] no position to meet legally enforceable standards because they lacked the necessary financial resources.

Arsenic has been identified in Permian coals of the North Karanpura coalfield, Jharkhand State, a western member in the east–west chain of the Damodar Valley Basin, where coal-bearing sediments are distributed over the Hazaribagh, Ranchi and Palamau districts. The North Karanpura coalfield covers 1230 km^2. Coals from Badam, Kerendari, KDH, Rohini, Dakra and Karkatta were analysed using molybdenum blue colorimetry. Arsenic concentrations in coal samples ranged from $<$0.01 to 0.49 mg/kg, with an arithmetic mean of 0.15 mg/kg. Ash measured up to 32.51% arsenic. In sediments, arsenic concentrations varied from 0.2 to 2.0 mg/kg. In Damodar Valley Basin, arsenic is confined mainly to the surface water and sediments closer to the mining area (Priyadarshi, 2004).

Geogenic arsenic is found in the Upper, Middle and Lower Gangetic Plains, which are among the most densely populated areas of India, where much food production depends on the surface water and groundwater supplies. The Gangetic delta region is prone to arsenic contamination of groundwater. In the Bengal Basin 10 000–7500 years BP, a rapid sea level rise and sediment discharge from Himalaya erosion formed deltaic sediments, in which arsenic-bearing aquifers are mainly found (Acharyya *et al.* 2000; Goodbred & Kuehl, 2000). Reduction and ion exchange naturally contribute to the mobilization and release of arsenic into groundwater in this region. In addition, the drilling of deep tube wells in this region in the 1980s has channelled groundwater flow, which has anthropogenically facilitated the further release of arsenic (Bhattacharya *et al.* 1997; Nickson *et al.* 1998; Acharyya *et al.* 1999; McArthur *et al.* 2001; Acharyya, 2002; Appelo *et al.* 2002; Harvey *et al.* 2002).

With the exception of contamination from industrial waste, elevated arsenic levels in India derive from geological sources. The aquifers are of Holocene age and comprise a mixed sequence of micaceous sands, silts and clays deposited by the Ganges, Brahmaputra and Meghna river systems and their precursors. The groundwater conditions are strongly reducing, with slow flow rates (Smedley & Kinniburgh, 2002).

The length of exposure of segments of the Indian population to elevated arsenic levels is not known, but exposure is likely to have started in the late 1960s with the construction of tube wells for irrigation. High levels of arsenic in drinking-water were discovered in 1978, and the first cases of arsenic-related health effects were reported in the early 1980s (Mazumder *et al.* 1998).

Groundwater arsenic contamination in the Lower Gangetic Plain of West Bengal was identified in July 1983 (K.C. Saha, unpublished data). Garai *et al.* (1984) reported 16 patients in three families from one village in 24 Parganas District. Saha (1984) reported 127 cases with lesions from 25 families from five villages in three districts. By July 2002, more than 125 000 water samples and more than 30 000 urine, hair, nail or skin bioassays of 100 000 people identified 8500 cases from 255 affected villages out of 306 screened. More than 3000 villages have tube wells with arsenic concentrations above 50 μg/l. The 2005 estimate is that 6 million people from 9 districts (total population ~50 million) of 18 total districts (total population ~80 million) are exposed to arsenic concentrations above 50 μg/l, and more than 300 000 cases are probable (Chakraborti *et al.* 2003; Mazumder, 2005).

In seven districts adjoining the river Bhagirathi, a large number of people are exposed to arsenic at concentrations above 50 μg/l (Mukhopadhyay & Sanyal, 2005):

> The problem seems to be triggered by large-scale withdrawal of groundwater for agricultural irrigation during the lean period of January to April when the groundwater recharge is at its minimum. However, the surface water bodies in the affected belt have remained effectively free of the toxin. This seems to suggest that the soil, which is receiving the arsenic contaminated irrigation water, is acting as a sink for arsenic.

In experiments designed to examine the decoupling of arsenic and water as the soils were permeated with aqueous arsenic, the arsenic breakthrough curves indicated hydrodynamic dispersion as the aqueous arsenic passed vertically through the iron-rich soils, which sorbed the arsenic and allowed surface water and groundwater recharges to remain relatively free of arsenic.

AS-Soil contamination

The contamination of crops irrigated with high arsenic water is a serious arsenic contamination risk highlighted by Dipankar Chakraborti (Achin, 2011). "Use of contaminated water in agriculture could dramatically expand the scope of the problem." Apart from setting up mobile laboratories to monitor the arsenic affected bore hole wells drilled by the World Bank and other international aid organizations in the 1970s, Chakraborti advocates (for a longer term solution) the better management of surface water: "Where you have rainwater [then] harvest rainwater; where you have lake water, use lake water. … All the arsenic affected villages in West Bengal, Bihar, Uttar Pradesh, Jharkand, all are surrounded by water bodies" (Achin, 2011).

The mechanism is well known. For instance, a main agricultural crop in West Bengal, Bihar, Assam and parts of Uttar Pradesh and Orissa is Boro rice. In Assam for instance, boro rice is farmed in the Brahmaputra and Barak river valleys. Boro rice can be grown in low-lying areas or river deltas in three contexts: (1) rainfed swampy or flood prone areas; (2) irrigated flood prone areas; and (3) irrigated flood-free areas (http://sourcing.indiamart.com/agriculture/food-beverages/rice/boro-rice/).

During the lean summer months from January to May a large amount of groundwater is used for irrigation which is contaminated with arsenic resulting from the oxygenated decomposition of arsenopyrite forming Fe^{++} and Fe^{+++} sulfates and H_2SO_4 acid which mobilize arsenic. The mobilized arsenic is then accumulated in soils and taken up by crops like Boro rice. While the exact mechanism of arsenic release is controversial (following oxidation of arsenopyrite, reductive dissolution of Fe hydroxides or anion exchange of arsenic with fertilizer phosphates) its effect is dramatic, for it provides another pathway – through the food chain – for arsenic poisoning of millions.

Dilip Kumar Das (2007) has proposed four approaches to the problem. He reviewed the geo-chemical mechanism of arsenic adsorption behavior in relation to West Bengali soils and the subsequent uptake of arsenic by rice (*Oryza sativa* L.). The project was to apply $ZnSO_4$ along with farm-yard manure in order to reduce the amount of arsenic-tainted irrigation water needed to nurture a rice crop. The result was the lowering of arsenic in the rice biota assay. His recommendations included: (1) determine the exact course of arsenic release and its mitigation; (2) determine the behavior of arsenic in particular soils with regard to mobility, retention, etc.; (3) determine the upper limit of arsenic toxicity in soils for different crops; and (4) determine the most suitable crops in a particular area that require less arsenic-tainted irrigation water.

Arsenic contamination of agricultural soils as it relates to food security is, therefore, a major concomitant of arsenic contamination of groundwater in India, Bangladesh and elsewhere. One organization that has been very active in pursuing food security is the International Food Policy Research Institute (IFPRI), which was established in 1975 as one of 15 agricultural research centers that receive principal funding from governments, private foundations, and international and regional organizations, most of which are members of the Consultative Group on International Agricultural Research (CGIAR). In a recent report of the Institute by Mahabub Hossain (2009) prepared for the project on "Millions Fed: Proven Successes in Agricultural Development" (www.ifpri.org/millionsfed) it is reported that "The fast diffusion of Boro rice in the rain-fed, low-yielding "*aus*" rice area contributed to accelerated growth in rice productivity, a reduction in the unit cost of production, and a decline in real rice prices. Together, these factors have been a major factor behind attainment of food security and moderately reduced poverty in Bangladesh over the last two decades." While it is admitted that it "is now established that the arsenic contamination of groundwater is a serious problem for Bangladesh," the position of the paper is that the link of arsenic contamination "with the exploitation of groundwater for *boro* cultivation is yet to be firmly established." The report does end with a caveat: "However, it has recently become apparent that arsenic-contaminated groundwater that is used for irrigation is seeping into soils and rice in several countries of South and Southeast Asia, thus posing a serious threat to sustainable agricultural production and to the health and livelihoods of the affected people in these countries (Brammer 2009). Since groundwater-dependent *boro* rice now accounts for nearly 60 percent of the annual rice production, the risk is very high for Bangladesh," and we can add the same risk pertains to northwest India including West Bengal as well.

Beyond West Bengal, the main areas of concern are the north-eastern provinces of Assam, Arunachal Pradesh, Manipur, Nagaland and Tripura, where 5 million people are estimated to be exposed to arsenic concentrations above 50 μg/l. In hilly and mountainous areas, most of the drinking-water comes from rivers, ponds and natural springs, but the latter are becoming seasonal. In valleys, drinking-water comes from shallow tube wells and dug wells, but these have restricted supplies in summer, and the overall quality is questionable (Singh, 2004). To these should be added Bihar and Madhya Pradesh in the central Gangetic Plain. Summing all state estimates gives a staggering overall estimate of 78.8 million people exposed to arsenic contamination in India. A description of the situation in the individual states follows.

Arunachal Pradesh

Villages in the districts of the midland Dibang Valley, Papum Pare, East Kameng, West Kameng, Lower Subansiri and Tirap, situated near the border area of Assam, all have high levels of arsenic in groundwater (Table 3.18).

Table 3.18 Six arsenic-contaminated areas of Arunachal Pradesh, India.

District	Number of samples analysed (% contaminated by arsenic)	Location	pH	Iron concentration (mg/l)	Arsenic concentration (μg/l)
Papum Pare	24 (33.4)	Doimukh	6.2	1.955	74
West Kameng	20 (2.0)	Dirang	7.8	4.931	127
East Kameng	20 (2.0)	Pakke	8.0	1.051	58
Lower Subansiri	28 (42.85)	Yazali, Yada Vill Dokum	6.8–7.5	0.29–1.14	63–159
Dibang Valley	19 (42.10)	Hunli	7.5	0.132	75
		Midland	7.5	0.251	618
Tirap	28 (14.20)	Khonsa	7.4	0.259	90

Source: Singh, 2004

Assam

In Assam, high arsenic content in groundwater was observed in Jorhat (Titabor, Dhakgorah, Selenghat and Moriani Block), Dhemaji (Sissiborgoan and Dhemaji Block), Golaghat District (Podumoni Block) and Lakhimpur (Boginadi, Lakhimpur Block). To these incidences of arsenic contamination of groundwater should be added the presence of fluoride in the highly ferruginous groundwater of Assam valleys (Aowal, 1981), especially in Karbi Anglong and Nagaon districts (Akoijam, 1997; Sengupta, 1999; Sushella, 2001).

Assam's maximum arsenic concentrations are in Jorhat, Lakhimpur, Nalbari and Nagaon districts (Singh, 2004). In Nagaon District, 76 samples were collected, and 13% contained arsenic at concentrations ranging between 112 and 601 μg/l. In Jorhat District, the contamination of arsenic was highest, in the range of 194–657 μg/l. Altogether, 80 samples were collected, and 21% of the samples were found to be contaminated with arsenic. In Lakhimpur District, of 76 samples, 21% had arsenic concentrations between 50 and 550 μg/l. In Nalbari District, 19% of the samples ($n = 72$) had concentrations between 106 and 422 μg/l (see Table 3.19). In most of the samples, iron was above the permissible level, but it did not show a significant correlation with arsenic. In Assam's floodplain districts of Barpeta, Dhemaji, Dhubari, Darrang and Golaghat, arsenic ranged in concentration between 100 and 200 μg/l (Table 3.19). The remaining 11 districts of Assam where arsenic was detected contained arsenic at concentrations between 50 and 100 μg/l. Only three districts – Karbi Anglong, NC Hills and Morigaon – were free from arsenic contamination.

Table 3.19 Arsenic concentration in groundwater of Assam, India.

District	Number of samples analysed (% contaminated by arsenic)	Location	pH	Iron concentration (mg/l)	Arsenic concentration (μg/l)
Nagaon	76 (13)	Lowkhowa, Pachim Koliabor, Doboka, Binakandi, Raha, Borhompur, Pakhimora, Kathiatali	6.9–7.9	0.76–11.02	112–481
Jorhat	80 (21)	Moriani, Titabor, Selenghat	6.6–7.6	0.22–30.59	194–657
Lakhimpur	76 (21)	Lakhimpur, Telahi, Ghaimora, Gossai, Nowboisha, Karunabari, Boginadi, Narayanpur, Bihpuria	6.5–7.0	1.22–49.39	50–550
Nalbari	72 (19)	Barama, Pachim Nalbari, Pub Nalbari, Tamulpur	6.3–6.9	4.95–22.30	100–422
Golaghat	60 (13)	South Golaghat, Podumoni	7.6–8.5	5.17–25.29	100–200
Dhubari	52 (19)	Mankachar, Jangal, Fekamari, Mahamaya	7.6–8.4	6.98–19.39	100–200
Darrang	52 (7.6)	Bahmolla (Sipajhar)	6.8	20.90	200
Barpeta	66 (18)	Mandia, Kolkussia, Chak chaka, Bajali	7.6–8.2	1.70–16.99	100–200
Dhemaji	44 (22)	Sissiborgoan, Dhemaji, Jonai	5.6–6.8	11.45–26.84	100–200

Source: Singh, 2004

Bihar

The province of Bihar has an areal extent of 173 877 km^2 and a population of 83 million. High levels of arsenic have been found in Semria Ojha Patti village (Chakraborti *et al.* 2003):

> where tube wells replaced dug wells about 20 years ago. Analyses of the arsenic content of 206 tube wells (95% of the total) showed that 56.8% exceeded arsenic concentrations of 50 μg/L with 19.9% > 300 μg/L, the concentration predicting overt arsenical skin lesions. On medical examination of a self-selected sample of 550 (390 adults and 160 children), 13% of the adults and 6.3% of the children had typical skin lesions, an unusually high involvement for children, except in extreme exposures combined with malnutrition. The urine, hair and nail concentrations of arsenic correlated significantly ($r = 0.72$–0.77) with drinking water arsenic concentrations up to 1,654 μg/L.

To counter this concern over the release of arsenic in the Gangetic Plain, it has been noted that most riverine floodplains are narrow and entrenched within the broad valleys south of the confluence of the Himalayan Rivers and the Ganges (Acharyya & Shah, 2004). Sedimentation in these floodplains, deposited by Holocene sea level fluctuations, formed large fluvial lakes and swamps (Singh, 2001). Semria Ojha Patti village is located in a floodplain about 8 km south of the Ganges, a setting in which arsenic was released into tube well water. Similar arsenic-affected areas may also occur in the Terai region of Nepal. Although Acharyya & Shah (2004) believe "there is no evidence that As pollution is a health hazard over major parts of the Ganges Alluvial Plain", the recent discovery of arsenic release in floodplains in Bangladesh suggests that mechanisms that could enhance the release of arsenic in similar conditions should be closely investigated.

Chakraborti *et al.* (2003) presented his evidence on arsenic contamination in Bihar based on epidemiological studies and geological data:

> Arsenic-rich sediments derived from the Himalayan Mountains and the foothills of the Shillong Plateau are deposited in the Gangetic Plain, Padma-Meghna-Bramhaputra delta of Bangladesh, Terai region of Nepal, Chandigarh area, and now Bihar. Most of the arsenic-contaminated tube wells are in the depth range of 20–55 m, similar to that of West Bengal and Bangladesh tube wells. The deposition is expected to be in the Holocene-type deposits. The meandering pattern of the river is responsible for the localized depositions of arsenic-rich sediment in selected areas along the course of the river Ganges.... As reported in our analyses of approximately 125,000 tube wells ..., some portions of Bangladesh and West Bengal are geologically free of arsenic. Similarly, the entire Ganges Plain, home of 449 million people, may not be uniformly affected despite our expectations that groundwater will be arsenic contaminated over a wide region.

Since 2003, Chakraborti and his colleagues have tested drinking-water and performed bioassays in Bihar, with important results that support their previous contention that arsenic contamination is quite grave (Chakraborti *et al.* 2004). Arsenic contamination has occurred within a conscribed geological zone, with Bihar, discovered with elevated arsenic in 2002, between Chandrigarh, discovered with elevated arsenic in 1970, and West Bengal, discovered with elevated arsenic in 1983. In due course, other areas followed: Uttar Pradesh in October 2003, Jharkand in December 2003 and Assam in January 2004.

Table 3.20 Arsenic exposure in Bihar District, India.

Bihar District	Potential population exposed to arsenic concentrations >50 μg/l, in millions (1991 census)	Arsenicosis cases
Bhagalpur	3.2	yes
Khagaria	0.99	
Munger including Lakhisarai	3.1	
Begusarai	1.8	yes
Samastipur	2.7	
Patna	3.6	yes
Baishali (Vaisahali)	2.1	
Saran	2.6	yes
Bhojpur including Buxar	2.9	yes
Katihar	1.8	
Total	**24.79**	

Source: Chakraborti, 2006

The severity of the impending health crisis in Bihar is known because of Chakraborti's (2006) research. Table 3.20, based on his data for Bihar, suggests that 24.79 million people are exposed to arsenic concentrations above 50 μg/l, even though Chakraborti (personal communication, 2006) indicates that a far greater number are exposed.

An extract from Chakraborti's (2006) report is included to illustrate the severity of the problem:

> From the last 19 years' survey analyzing around 221 000 water samples, around 50 000 biological samples, screening around 200 000 villagers with our medical team we have found a good portion of the states and countries [Uttar Pradesh, Bihar, Jharkhand, West Bengal, Assam, Manipur and Bangladesh (jointly with Dhaka Community Hospital)] in the Ganges-Meghna-Brahmaputra (GMB) plain–an area of 569 749 km^2, with a population of over 500 million are at risk from groundwater arsenic contamination and its health effects (http://www.soesju.org/arsenic/misc_crip.htm). In June 2002 when we first reported arsenic contamination in Bihar in middle Ganges plain it was vehemently refuted. Today on basis of analyzing 15 000 samples from Bihar we have found arsenic concentrations above 50 μg/l in 12 districts, 32 blocks and 201 villages. The districts, Bhagalpur, Khagaria, Munger, Begusarai, Lakhisarai, Samastipur, Patna, Baishali, Saran, Bhojpur, Buxar and Katihar all lie by the side of Ganges and in 6 of them (Bhojpur, Saran, Begusarai, Bhagalpur, Patna and Buxar) we have already identified hundreds of subjects with arsenical skin lesions. More are coming to fore with the continuing surveys. Our medical team confirmed the presence of arsenical skin lesions, consistent peripheral motor and sensory neuropathy as well other neurological abnormalities in the arsenicosis patients of Bihar. We also observed an apparent increase in fetal loss and premature delivery in the women with the elevated level of arsenic in their drinking water.
>
> Hair, nail and urine are proved biomarkers for arsenic. Around 1500 biological samples from arsenic affected villages have been analyzed; we measured inorganic arsenic and its metabolites in urine and total arsenic in hair, nail and skin scale. High arsenic concentrations in urine is an indicator of recent arsenic exposure. In arsenic affected villages of Bihar, high concentrations in urine of both patients and non patients indicate they were drinking contaminated water. It appears from the analyses that many villagers may be sub-clinically affected and they are continuing drinking the contaminated water since they were mostly unaware of the problem before we surveyed them. Infants and children are considered to be more susceptible to the adverse effect of toxic substances than adults. A note of caution:
>
> Though first case of arsenicosis was revealed in West Bengal in early 1980s, the widespread contamination was not recognized until 1995. Similar pattern followed in the late recognition of groundwater arsenic contamination of Bangladesh. In Bangladesh and West Bengal, at present less people are drinking arsenic contaminated water due to growing awareness and access to arsenic safe water. But no doubt the problem would not have attained such gravity, if it were not ignored for quite a long time. Unfortunately today similar mistakes are being repeated in Bihar, Uttar Pradesh, Jharkhand and Assam where still the villagers are drinking contaminated water. Non recognition of truth continues. The administration in Bihar is yet to learn from the mistakes committed in a neighboring state of West Bengal. A thorough survey is required in these areas to assess the magnitude of contamination.

Madhya Pradesh, Rajnandgaon District

(Location when testing was done in 1999, now located in Chattisgarh State). Water-related arsenic problems were first recognized in 1999. The worst-affected village, Koudikasa, in Chowki Block, had an arsenic concentration range between <10 and 880 μg/l (Haque *et al.* 1999). Of 146 samples analysed, 8% were found to contain arsenic at >50 μg/l and 17% between 10 and 50 μg/l. Three of these samples were from dug wells, with one concentration of 520 μg/l. Most high-concentration samples were from tube wells, which had well depths in the range 10–75 m. Arsenic-related skin disorders have been diagnosed in several villages. Natural springs and dug wells have been considered to be the only cost-effective and viable sources of drinking-water, but these are dug into highly contaminated strata bearing arsenic in solution.

Gold mining has been undertaken in Chowki, but the extent of mining and its impact on groundwater arsenic contamination are not known. To date, no maps of arsenic-affected wells in Chowki Block or of diagnosed cases have been produced to indicate the distribution and scale of the problem.

Manipur

Thoubal District (Kakching Block) has been identified as having high concentrations of arsenic in groundwater (Table 3.21; Singh, 2004).

Nagaland

Nagaland, near Jorhat District of Assam, has seven locations in Mokokchung District and five locations in Mon District with high arsenic levels in groundwater. Low levels of arsenic were also found on the plains in the districts of Wokha and Zunheboto (Table 3.22).

Table 3.21 Arsenic-contaminated area of Manipur, India.

District	No. samples analysed (% contaminated by arsenic)	Location	pH	Iron concentration (mg/l)	Arsenic concentration (μg/l)
Thoubal	12 (50)	Khunyai	7.3	1.95	986
		Kakching Market	7.5	0.74	858
		Ningthou Lekei	7.6	4.32	798

Source: Singh, 2004

Table 3.22 Arsenic-contaminated areas of Nagaland, India.

District	No. samples analysed (% contaminated by arsenic)	Location	pH	Iron concentration (mg/l)	Arsenic concentration (μg/l)
Mokokchung	21 (33)	Tzudikong, Tuli, Merapani	6.9–7.5	0.91–1.92	50–278
Mon	20 (25)	Longling, Borjan, Naginimara, Namtula	7.1–7.4	0.33–1.32	67–159

Source: Singh, 2004

Rajasthan

Anthropogenic releases of arsenic into surface water and groundwater have been noted in north-west India in the state of Rajasthan. In two regions, Zawar and Khetri, arsenic was present in high concentrations in mine tailings. Concentrations of arsenic in sulfide ores in the two regions were 7683 $\pm$ 4414 mg/kg and 145 $\pm$ 72 mg/kg, respectively. Arsenic concentrations in copper tailings in the two regions were 1519 $\pm$ 1493 mg/kg and 1179 $\pm$ 1382 mg/kg, respectively, whereas zinc concentrates contained arsenic at an average concentration of 10 727 $\pm$ 6322 mg/kg (Madhavan & Subramanian, 2000).

Tripura

In Tripura, high arsenic concentrations in groundwater have been found in West Tripura District (Triania Block), Dhalai District (Salema Block) and North Tripura District (Dharmanagar Block) (Singh, 2004) (Table 3.23).

Table 3.23 Arsenic-contaminated areas of Tripura, India.

District	No. samples analysed (% contaminated by arsenic)	Location	pH	Iron concentration (mg/l)	Arsenic concentration (μg/l)
West Tripura	38 (22.2)	Jirania	7.1	10.98	191
Dhalai	19 (42.1)	Salema, Halhali, Halhooli, Kamalpur, Joyanagar	7.3–8.3	0.45–8.91	65–444
North Tripura	21 (57.14)	Sanitala, Rajbari, Dharma Nagar	6.8–8.1	0.48–1.04	122–283

Source: Singh, 2004

Uttar Pradesh

Groundwater in Uttar Pradesh's Ballia District was found to be contaminated with arsenic in late 2003 (Kowalczyk, 2004). Unpublished evidence was presented by Dr Dipankar Chakraborti, Director of the School of Environmental Studies, at Jadavpur University in Kolkata in January 2004, after 2404 test samples of water were drawn from hand pumps in 55 villages in Ballia District. More than half were found to have arsenic levels above the Indian guideline of 10 µg/l, and 8% were above 500 µg/l (D. Chakraborti, personal communication, 2006).

West Bengal

Arsenic contamination of groundwater in West Bengal is a huge public health calamity (Chatterjee *et al.* 1995; Das *et al.* 1995; Mazumder, 2005). In 1996, 1 million people in 560 villages were thought to be at risk (Murdur, 1996). The percentages of wells with contaminated water had shown that 27% contained arsenic at concentrations at or above 50 µg/l, with a range of 50–3700 µg/l (Chadha & Sinha Ray, 1999). Although drinking-water concentrations of arsenic up to 3700 µg/l have been reported in West Bengal (Chowdhury *et al.* 1997), Mazumder *et al.* (1998) suggested that levels between 50 and 500 µg/l are more common. Mazumder (2005) estimated that 6 million people would be drinking water containing arsenic at concentrations above 50 µg/l in 2006. Elevated levels of arsenic in drinking-water have been found in eight districts in West Bengal (Malda, Murshidabad, Bardhaman, Nadia, North 24 Parganas, South 24 Parganas, Haora and Hugli), with the percentage of the population exposed to levels greater than 50 µg/l ranging from less than 1% to almost 25% (CPCB, 2001). Estimates of the total population at risk from high levels of arsenic in West Bengal range between 4.5 million and 6 million (CPCB, 2001; Smedley & Kinniburgh, 2002, World Bank 2005a). Chowdury (2000) has given a staggering estimate of 47.2 million people exposed in West Bengal. The extent of the massive infrastructure problem is highlighted by its enormity. There are an estimated 26 000 contaminated wells, and these serve in excess of 5 million private wells and 114 000 public sources from 9 districts and 85 blocks; this revises upward the earlier 2004 UNICEF data (Figure 3.24) (UNICEF, Kolkata, 2004; Hossain *et al.* 2005). New areas are constantly being identified.

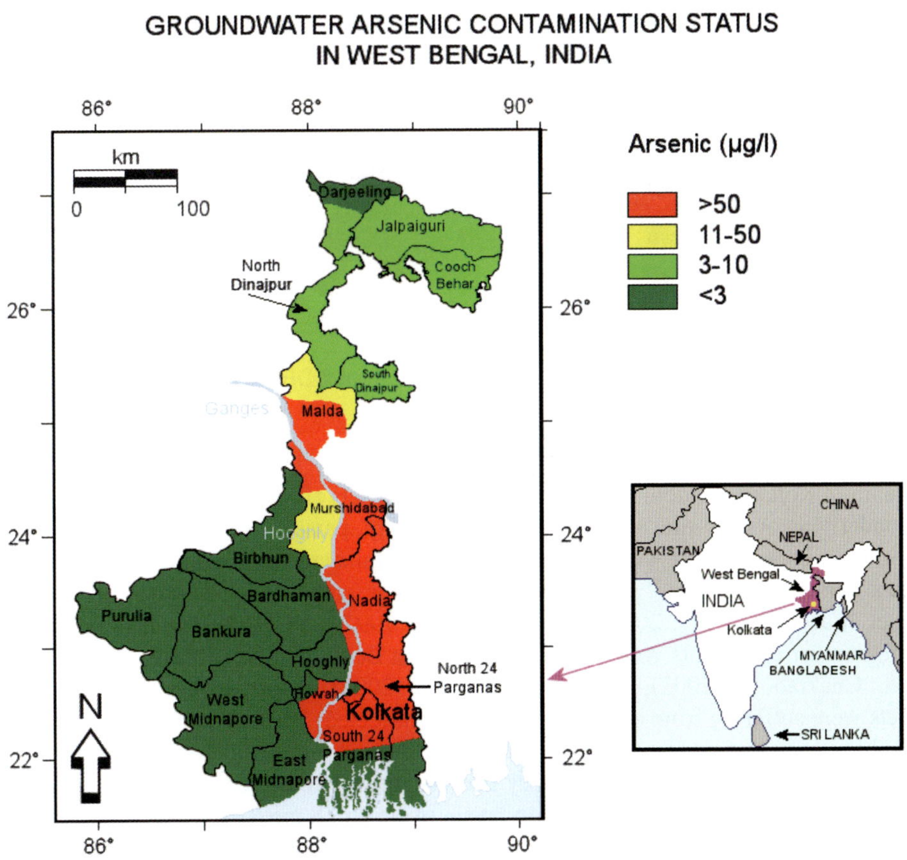

Figure 3.24 Groundwater arsenic contamination in West Bengal, India (modified from India Institute of Hygiene and Public Health, 2006).

Kolkata (Calcutta), is the largest eastern city in India, divided into 141 wards with a population of 4.6 million. The groundwater of Kolkata District is accessed by the population through hand pump tube wells. Most water samples have been shown to exceed the WHO arsenic guideline of 10 μg/l, with the highest concentration measured at 825 μg/l in the Lake Gardens area of ward No. 93. In a 19-year study, a detailed district map of arsenic concentrations (Figure 3.25) was developed for the Kolkata region.

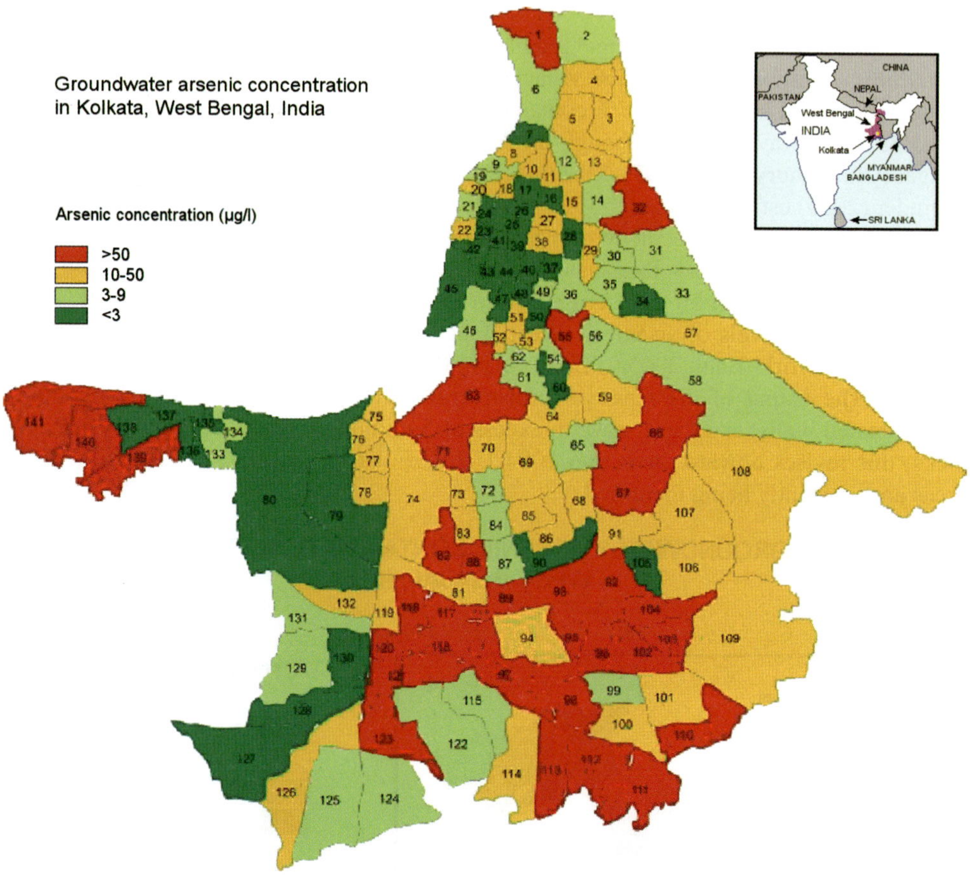

Figure 3.25 Groundwater arsenic situation in Kolkata, India (modified from SOES, 2006).

In addition, areas south of Kolkata are now being added to those districts affected, as new housing starts and hotels are being built on landfill with underlying arsenic-rich sediments (S. Chakraverty, personal communication, 2005).

Health effects reported to be associated with drinking water contaminated with arsenic in India include skin lesions, skin cancer, Bowen disease, internal cancers, gastrointestinal problems, respiratory disease, liver damage and peripheral neuropathy (Mazumder *et al.* 1988, 1997, 2000; Chowdhury *et al.* 1997; CPCB, 2001; Mazumder, 2001; Chakraborti *et al.* 2002). Estimates of the number of patients with skin lesions in West Bengal range between 200 000 to 250 000 (Chowdhury *et al.* 1997; Saha & Chakraborti, 2001). Chakraborti *et al.* (2002) reported that 10% of people examined from arsenic-affected districts were suffering from skin lesions; of those affected, 1% had some form of cancer.

A number of people have been exposed to high levels of arsenic through industrial pollution from a factory manufacturing a copper acetoarsenite insecticide. Effluent from the factory had contaminated the local environment, leading to high levels of arsenic in the soil and also local water sources. Arsenic levels of up to 58 mg/l have been reported in shallow tube wells close to an insecticide factory in South Kolkata (Mazumder *et al.* 1992). Individuals exposed to very high levels of arsenic in drinking-water as a result of industrial pollution have been found to exhibit skin lesions, gastrointestinal problems, anaemia, signs of liver disease and peripheral neuropathy (Mazumder *et al.* 1992).

Mitigation measures for arsenic in drinking-water include (CPCB, 2001):

- Tapping of arsenic-free deep aquifers;
- Water supply schemes based on surface water and groundwater after treatment in arsenic-affected areas;
- Development and application of arsenic removal filters with hand pumps on tube wells.

In some areas, allowing the water to settle (which, in iron-rich waters, results in the co-precipitation of iron and arsenic) overnight may reduce the arsenic concentration.

3.9 INDONESIA (G, M)

Indonesia, with a population of 245.6 million, has a Human Development Index rank of 108 out of 169 and a per capita GNI of $US 3957 (US Census Bureau, 2011; UNDP, 2010).

Geogenic arsenic has recently been identified in Sumatra. The lowlands of Sumatra encompass 100 000 km^2 along the Calik, Banuyasin and Musi River valleys (the last of which flows through the city of Palembang). This area is characterized by peat bogs with a relatively thin organic-rich sediment that distinguishes it from the deep, rich sediments of Southeast Asia. These lowlands are prone to arsenic mobilization in groundwater under reducing conditions. In these swamps, Holocene formations are characteristically the youngest and have produced water from tube wells with concentrations of arsenic ranging from $>$10 μg/l to 65 μg/l along with concentrations of boron, manganese and selenium that exceed WHO guidelines. Middle Miocene to Pliocene formations are the oldest formations in these swamps and are subject to redox conditions with an alkaline pH and high concentrations of sodium, boron, selenium and strontium (Winkel *et al.* 2008). This first published assessment of groundwater data for the highly populated island of Sumatra indicates that much needs to be done to monitor and continue to assess the human impact of these initial findings. (Winkel *et al.* 2008).

The mining of heavy metals and their associated metalloids on Sulawesi has resulted in the discovery of anthropogenically induced arsenic poisoning in Buyat Bay, the site of the USA's Newmont Mining Company's gold mine waste site (Perlez, 2004). The Newmont Mining Company is just one of at least eighty mining companies in Indonesia (Infomine.com, 2012).

The Buyat Baysite continues to be controversial despite an Indonesian court's ruling in favour of the mining company. In a report compiled by WALHI (Indonesian Forum for Environment)–Friends of the Earth Indonesia, Indonesian Mining Advocacy Network and Indonesian Center for Environmental Law, based on the Joint Investigation Technical Team report dated 9 November 2004 (WALHI, 2004a), the top 10 independent findings were that:

(1) The Buyat Bay seabed is polluted with arsenic and mercury.
(2) Arsenic and mercury found in Buyat Bay are not natural.
(3) Newmont dumping breached toxic waste law.
(4) Arsenic and mercury in fish pose an unacceptable risk for the Buyat Bay community.
(5) Mercury is accumulating in seabed creatures of Buyat and Ratatotok bays.
(6) Biodiversity in Buyat Bay has been hit by arsenic pollution.
(7) There is no protective thermocline, despite Newmont claims.
(8) The Buyat Bay human health hazard requires a fish intake reduction and possibly relocation of its residents.
(9) The Buyat Bay human health hazard requires arsenic poisoning investigation and up to 30 years of monitoring by Newmont.
(10) Legal action should be taken over breaches of environmental law, and ocean dumping of mine waste should not be permitted in future.

More specifically, the sampling of Buyat Bay by Australia's Commonwealth Scientific and Industrial Research Organisation (CSIRO) found extremely high arsenic levels (678 mg/l, 491 mg/l and 466 mg/l) in the area most affected by mine tailings. The CSIRO report notes that Newmont's own monitoring has discovered arsenic levels as high as 1190 mg/l. CSIRO findings show that the seabed at Buyat Bay is contaminated by tailings, with arsenic at levels 10–20 times higher than the Australian/New Zealand guidelines and the USA/Canadian probable (toxic) effects level (WALHI, 2004b). A comparison of the pre-mining environmental conditions and measurements taken at control sites irrefutably proves that Newmont's mine waste dump is the source of arsenic and mercury pollution in Buyat Bay.

The epidemiological effects of this pollution were reported by The New York Times on 8 September 2004, indicating that the 300-person village witnessed fish dying in the lake and a malformed newborn dying a few months after birth (Perlez & Rusli, 2004).

Other mining companies with mine properties located in Indonesia include Freeport-McMoran Copper & Gold, 3D Resources Limited, Adavale Resources Limited, and Allied Indo Coal Pt. There is also Batu Hijau and Grasberg, both of which are open pit copper mines, as well as Aceh/Samadua, AGAM Iron Sands, AHC KP. These are just a few of over 80 mining companies in Indonesia.

Sustainable development has been a relatively recent concern of mining companies. One sustainability issue is the presence of toxic chemicals, such as arsenic, in feedstocks. Several popular mining websites have addressed this issue as it pertains to arsenic, as follows:

- as high grade feedstocks become depleted, arsenic bearing concentrates will become a more predominant feedstock. For existing copper producers, arsenic management is a key aspect of business management as arsenic impurity levels rise. (International Mining; im-mining.com , 2010);
- Outotec has developed a two-stage partial roasting process to remove impurities, such as arsenic… More than 90% of the arsenic contained in the concentrate can be removed to produce clean copper calcine, says Outotec CEO Perti Korhonen (Mining.com, 2011).

Arsenic in mining waste is a major issue not only for local mine waste, but also becomes an issue when mined rock materials or cement are exported. For example, Indonesia exports large quantities of rock and cement materials to Singapore in order to create new land and to supply its construction industry. In 2011, exports from Indonesia to Singapore made Singapore Indonesia's fourth largest export trading partner, after Japan, China and the US (CIA, 2012). Reportedly, Singapore has high levels of arsenic in rainwater runoff and surface waters, which could partly be on account of leaching from building materials derived from Indonesia. A scientist from Singapore reported arsenic at the level of 40 µg/l (personal communication). However, because the government of Singapore does not wish to make this information public, there is no peer reviewed literature on this topic (hence no separate section on arsenic in Singapore in this volume). This exported arsenic contamination would also pertain to Indonesia's other export country partners, such as China, Japan and US. For the benefit of public health and transparency in government, this topic deserves wider attention and publicly available research and information.

3.10 IRAN (ISLAMIC REPUBLIC OF) (G, M, P)

The Islamic Republic of Iran, with a population of 77.9 million, has a Human Development Index rank of 70 out of 169 and a per capita GNI of $US 11 764 (US Census Bureau, 2011; UNDP, 2010).

The Islamic Republic of Iran's management of water resources is almost as old as Islam. From the very beginnings of the Islamization of Persia, the area was conscious of the sacred management of water. All water sources (lakes, seas, rivers) were considered common property. With the development of the Karwell system (as it was called in Uyghur-occupied Xinjiang), or *qanats* in Arabic, the series of wells that were linked by subterranean channels running down slope from the highest mountains to ensure purity were both an investment of the donor and the property of all. Such a view is common throughout the Islamic world (Sadr, 2001). In this perspective, the pollution of water sources can never be regarded as a secular matter, and the management of these *qanats* is often in the care of a religious community leader.

The total volume of flowing water in the Islamic Republic of Iran is estimated to be about 7786 million m^3, of which about 14% is used in the province of Kurdistan. Bijaar County, a region affected by naturally occurring arsenic, is 580 km^2 in area, is located in the north-east of Kurdistan and has an average altitude of about 1750 m above sea level (NGO, 2001). There are 17 deep wells, 616 dug wells, 42 springs and 9 subterranean canals in this area (KMPO, 2000). Total annual water consumption of this county for municipal, industrial and agricultural uses is about 30 million m^3, which is mainly supplied from underground resources.

The Iranian arsenic standard is 50 µg/l (ISIR, 1996). Identification and remediation of arsenic-contaminated water are both national health priorities. Field studies have shown elevated levels of arsenic in some rural areas of Bijaar County. Table 3.24 indicates that some of the 18 villages have water containing arsenic at concentrations higher than 1000 µg/l, with a mean concentration of 290 µg/l, six times higher than the Iranian standard. Primary investigations have shown that the source of this arsenic is mainly geogenic due to the dissolution of arsenic-containing compounds in the Earth's crust (Mosaferi *et al.* 2005a).

The first cases of chronic arsenic poisoning due to drinking-water were diagnosed in 1981, but no scientific and systematic study has been conducted in the region pertaining to the prevalence or incidence of health consequences such as skin lesions, keratosis, hyperpigmentation, blackfoot disease and gangrene (Mosaferi *et al.* 2005a).

Table 3.24 Arsenic concentrations in well-water in Kurdistan, Islamic Republic of Iran.

Village number	Type of water source	Type of use	Arsenic concentration[a] (μg/l)
1	Spring	Piped for drinking	32
2	Spring	Piped for drinking	10.7
3	Well	Piped for drinking	21
4	Spring	Piped for drinking	187
5	Well	Piped for drinking	205
	Underground canal	Non-drinking	206
6	Spring	Piped for drinking	11.6
	Spring	Non-drinking	11
7	Spring	Piped for drinking	15.9
8	Spring	Non-drinking	28
9	Spring	Non-drinking	15
10	Spring	Non-drinking	28
11	Spring	Non-drinking	45
12	Spring	Piped for drinking	90
	Spring	Piped for drinking	30
13	Spring	Non-drinking	1040
	Spring	Piped for drinking	210
	Spring	Piped for drinking	801
	Spring	Non-drinking	440
	Spring	Non-drinking	470
14	Spring	Ordinarily used for drinking	43
15	Spring	Piped for drinking	23.9
16	Spring	Piped for drinking	1480
	Spring	Ordinarily used for drinking	20.8
	Spring	Non-drinking	118
17	Spring	Piped for drinking	422
18	Spring	Non-drinking	408

[a]Analysis method: neutron activation. Measurements were conducted in 2001.
Source: Mosaferi *et al.* 2005a

Remediation measures of the Iranian government include recognizing and labelling polluted wells and springs, supplying fresh water to polluted areas with portable reservoirs and identifying alternative water resources. Indeed, there is a large study for assessing arsenic concentrations in total water supplies of the Islamic Republic of Iran, which is in its early phase (Mosaferi *et al.* 2005a). In the special case of Bijaar County, because of a lack of awareness of the full extent of the problem in the area, the following further actions are being undertaken with the goal of reducing exposure to arsenic via drinking-water (Mosaferi *et al.* 2005a):

(1) Development of an arsenic mitigation strategy.
(2) Formation of expert committees for situation analysis.
(3) Analysis of total water supplies regarding presence of arsenic and other heavy metals.
(4) Defining sources and mechanisms of arsenic presence in drinking water.
(5) Immediate detection and treatment of arsenicosis patients.
(6) Conducting a pilot study for removal of arsenic from drinking-water, with special regard to POU [point of use] technology suitable for rural areas.
(7) Education and involvement of local residents in problem solving.

Arsenic occurs naturally in association with many minerals and metals in Kurdistan Province, in the western part of the Islamic Republic of Iran. Mining sources of arsenic are also predominant in this area. The reason why Kurdistan is so mineralogically rich and diverse is that it encompasses the subduction zone between the colliding Eurasian and African tectonic plates: the Arabian micro-plate is being subducted under the Iranian and Anatolian micro-plates. The Persian Gulf once extended to the Kurdish foothills and joined the Indian Ocean to the Mediterranean Sea and the Atlantic, separating Eurasia from Africa and Arabia. About 15 million years ago, Kurdistan was thrust up from the bottom of the Tethys Sea, which covered south-west Asia. The seabed of the Persian Gulf is the remnant Tethys seabed.

This subduction zone is where lead, copper, zinc and gold mineralizations occur, where associated arsenic minerals are to be found and where vast petroleum and natural gas deposits lie. As arsenic mineralization is widespread, arsenic may leach into many ground waters by natural weathering, by mining extraction or when petroleum spills change the redox chemistry of aquifers (Geological Survey of Iran, 1976; Sengör, 1984).

Mineral deposits associated with igneous intrusions folded in north-west to south-east faults have been exposed by many massive volcanisms. This subduction zone has elevated Lesser and Greater Ararat mountains and Numrud Dagh, and lava flows created the Vân and Urmiâ lakes by damming ancient river channels. It has also created zones of accessible mineralizations and oil and gas reserves.

Arsenic related to mining is found in the area of the Sari Gunay gold deposit. This is a sub-volcanic epithermal gold deposit, located 60 km north-west of Hamedan, and contains a group of arsenic association minerals. Gold occurs as invisible mineralizations in fine-grained arsenical pyrite and arsenopyrite deposited by quartz, pyrite and stibnite intrusions in association with realgar (AsS), orpiment (As_2S_3), cinnabar (HgS) and wire gold. The Sari Gunay gold deposit and its host volcanic system are within the Takab belt of middle Miocene, mildly alkaline volcanic plutons, although it is less alkaline than classic deposits of this group, such as Cripple Creek, Colorado, USA. The Sari Gunay is within the larger Zarshuran gold deposit, which extends on a 190-km north–north-west zone of middle Miocene volcanism and is displaced on the south-west by a younger Eocene–Miocene volcanic arc (probably the zone between the Afro-Arabian and Eurasian plates) that hosts several large porphyry copper deposits at Sar Cheshmeh and Meiduk, which are other potential areas of arsenic mineralization (Richards *et al.* 2006). In fact, another gold mine was opened in this zone, the Zarshuran gold mine, in the north-western city of Takab, on 26 April 2007. It is the Islamic Republic of Iran's largest, with an anticipated 3 tonnes annually, and caps a 20-year exploration programme by domestic and international companies. The deposits of Zarshuran were estimated at 4 million tonnes of 5.81 mg/kg gold ores and 7 million tonnes of 4 mg/kg gold ores, with a total 50 tonnes of gold from this mine. The Islamic Republic of Iran is the world's 57th largest gold producer, with about 250 tonnes of proven gold reserves identified and another 750 tonnes yet to be exploited, according to the Geology and Mineral Exploration Organization's director for exploration affairs, Behrouz Borna (IranMania, 2006).

The need to monitor such mining areas is related to the acid mine drainage that can potentially form as runoff, and contaminate local groundwater. A comprehensive statement for the management of Sari Gunay mine was formulated with the aid of the Australian government to guide a systems approach to minimizing the effects of acid mine drainage on the environment in Kurdistan (Australian Government, 2007):

> In relation to water management, a number of environmental values have been recognised. They include: aquatic ecosystems, primary industries (irrigation and general water uses, livestock drinking water, aquaculture and human consumption of aquatic foods), recreation and aesthetics, drinking water, industrial water and cultural and spiritual values. Water quality criteria guidelines for aquatic ecosystems, primary industries and recreation and aesthetics have been given by ANZECC/ARMCANZ (2000). Criteria for drinking water are given by NHMRC (2004). No water quality guidelines are provided for industrial water or cultural and spiritual values. In general, guidelines for the protection of aquatic ecosystems are more stringent than those for drinking water, which are more stringent than those for primary industries and recreation. Background water quality should be the basis for determining environmental values. Mining activities should not lead to water quality degradation such that the most conservative guideline of environmental values defined for a water body is compromised. This does not mean that there must be no measurable impacts, but rather that impacts be minimised so that the water quality is not degraded to the point where any existing environmental value is lost. Strategies that mining companies should employ to demonstrate the preservation of environmental values include: ensuring that pertinent maximum concentration or trigger value guidelines for different water environmental values (or uses) are not exceeded in receiving water bodies ensuring that discharge does not result in a statistically significant change in key water quality parameters (no change occurs that is outside the seasonally relevant background concentration plus (or minus) two standard deviations demonstrating that the discharge will not have ecological impacts on the basis of site-specific ecotoxicological studies.

Geogenic arsenic also occurs in the drinking-water of Kurdistan Province, and this is related to the mineralization geology indicated above, with arsenic being released through various geochemical mechanisms. Arsenic contamination of drinking-water has been studied in an effort to establish a correlation between human bioaccumulation (in hair) and concentrations in drinking-water (Pazirandeh *et al.* 1998; Mosaferi *et al.* 2005b). Two villages (total population ~10 000) in the Bihar city area (World Gazetteer, 2005) were found to have arsenic in drinking-water: Gheslaghnoruz, with 85 μg arsenic/l, and Gavandak, with 455 μg arsenic/l. A test of the arsenic in the hair of 39 women, which included a control village without arsenic contamination, showed a positive correlation between arsenic in water and arsenic in hair: 85 μg/l to 0.5 mg/kg and 455 μg/l to 1.8 mg/kg (Mosaferi *et al.* 2005b). These correlations of arsenic in drinking-water (in μg/l) to arsenic in hair (mg/kg) are comparable with the results found in other studies worldwide:

- Nova Scotia, Canada: <10 μg/l to <1 mg/kg (Hindmarsh *et al.* 1977)
- Egypt: ~10 μg/l to 0.04–1.04 mg/kg (Saad *et al.* 2001)
- Hungary: 50–100 μg/l to 3 mg/kg average (Börzsönyi *et al.* 1992)
- Glasgow, Scotland: [inferred elevated arsenic in water], 0.02–8.17 mg/kg (Liebscher & Smith, 1968)
- Alaska, USA: 400 μg/l to 3.3 mg/kg (Harrington *et al.* 1978)
- Illinois, Chicago, USA: [downwind from power plant burning arsenic-contaminated coal], ~10 mg/kg (Sky-Peck, 1990)
- California, USA: 400 μg/l to 1.2 mg/kg (Valentine *et al.* 1979)
- Nevada, USA: 100 μg/l to 0.5 mg/kg (Valentine *et al.* 1979)

Petroleum extraction may also contribute to arsenic leaching into groundwater, because when petroleum is spilled into the area of an aquifer, it can change the redox geochemistry of the minerals in the rocks of the aquifer and help mobilize any adsorbed arsenic. The Islamic Republic of Iran exports around 2.5 million barrels of oil per day, with major customers including Japan, China, the Republic of Korea and Europe. The Islamic Republic of Iran's main export blends include Iranian Light Heavy, Lavan Blend and Foroozan Blend/Sirri. The Islamic Republic of Iran's domestic oil consumption, 1.5 million barrels per day in 2004, is increasing rapidly as the economy and population grow. The country heavily subsidizes the price of oil products, resulting in a large amount of waste and inefficiency in oil consumption (Ezzati, 2005).

The National Iranian Oil Company's onshore production is geared only to maintaining output levels from large, ageing fields, where enhanced oil recovery (such as natural gas injection) is being implemented (e.g. at Marun and Karanj). The Islamic Republic of Iran's oil production apparatus is old and inefficient and needs revamping, advanced technology and foreign investment (Ezzati, 2005). The location of the Islamic Republic of Iran's groundwater adjacent to many of the ageing well heads and the inefficiency and wastage of oil in the extraction and storage process indicate that groundwater in adjacent aquifers is at risk of arsenic pollution, as petroleum hydrocarbons enhance the arsenic release mechanism.

In 2004, the Japanese consortium INPEX agreed to develop the US$ 2 billion Azadegan oilfield. Azadegan, discovered in 1999, represents the Islamic Republic of Iran's largest oil discovery in 30 years and is located onshore in the south-western province of Khuzestan, a few kilometres east of the border with Iraq. Azadegan contains proven crude oil reserves of 26 billion barrels, but the field is also considered to be geologically complex, making the oil more challenging and more expensive to extract (Ezzati, 2005). This complex geology of the Azadegan oilfield with fractures close to aquifers indicates that Iran's adjacent groundwater aquifers are at risk of arsenic contamination.

3.11 IRAQ (A, P)

Iraq, with a population of 30.4 million (US Census Bureau, 2011)), has undergone severe economic and social upheaval since the War in Iraq began in 2003, and its infrastructures are seriously damaged. Iraq is not listed in the 2010 Human Development Index (UNDP, 2010).

Anthropogenic arsenic in Iraq had its origins in the industrial build-up between 1970 and 1990, during which environmental controls were disregarded. Cement and fertilizer plants, petrochemical factories and heavy industries released tonnes of pollutants, including lead, mercury and arsenic, into surface water and groundwater, especially in the central and southern regions, where general and military manufacturing was concentrated (Brown, 2004).

Poor water quality has been a serious problem since the bombing of sewage treatment plants as a result of the 1991 conflict. Since the renewed conflict in 2003, a million m^3 of untreated sewage enter the Tigris River in Baghdad every day. Arsenic pollution of groundwater and surface water is well above the WHO guideline (WHO, 2004a). Military hardware and factories destroyed during these conflicts have released heavy metals and metalloids such as arsenic from chemical agents and other dangerous substances into the air, soil and water, as well as low-level radioactive depleted uranium munitions fired by American and British forces (Verrengia, 2003).

In addition, oil spills from the Basra refinery's damaged and leaking pipes, both above ground and underground, have polluted catchment sands. This condition was exacerbated because the spills went undetected during the initial post-conflict inspections in 2003. Some of the damage was apparently caused by military action by the USA and some from a build-up of pressure when the oil system was hastily shut down by Iraqis (Verrengia, 2003). The resultant oil spills have seeped into the Iraqi sands. The fate of the underlying aquifers is unknown, but it is feared that the petroleum will facilitate the release of arsenic.

3.12 JAPAN (A, C, G, M, V)

Japan, with a population of 127.5 million, has a Human Development Index rank of 11 out of 169 and a per capita GNI of $US 34 692 (US Census Bureau, 2011; UNDP, 2010). Japan is located between the north Pacific Ocean and the Sea of Japan and east of the Korean Peninsula. The country's land area is 374 744 km^2. Japan is one of the most industrialized, technologically advanced countries in the world, despite its limited indigenous natural resources. It relies heavily on imports of mineral fuels and non-fuel minerals and mineral products for its large manufacturing and utility (electricity and gas) sectors.

Japan's substantial indigenous resources of industrial minerals (dolomite, iodine, limestone, pyrophyllite and silica) are not generally associated with arsenic occurrences.

Anthropogenic arsenic poisoning has occurred from leaching of World War II stores of chemical weapons. These sites are located throughout the countryside in clandestine places. One such occurrence persisted from 1960 to 1978 in Kamisu town, Ibaraki Prefecture, Ibaraki-ken Eisi. Residents had fallen ill after ingesting food, including marine fish, and drinking well water tainted with arsenic. In the late 1970s, it was established that the daily rate of ingestion of arsenic by 20 residents from the town was in the range of 70–170 µg (Nakao, 1960; Horiguchi *et al.* 1978; Ishizaki, 1979). The Imperial Japanese Navy had a chemical weapons factory in the area. The well water now registers an arsenic level of 4500 µg/l (Kyodo World News Service, 2003). The chemical weapon in question is diphenylchlorarsine, or "sneezing gas", which was intended to impede the enemy by causing sneezing and vomiting. In June 2003, the government began paying the medical costs of sick residents under a hastily created aid programme (Japan Times, 2003). Although epidemiological studies of arsenic in bioassays spanned the period from 1960 to the late 1970s, the government failed to recognize an arsenic contamination problem. Only in 2003 did arsenic contamination become an unavoidable fact when a mother and her child expressed full-blown arsenicosis symptoms of numbing sensations in their limbs.

From 1929 to 1945, the Japanese Imperial Army produced five types of poison gas on Okunoshima Island, including diphenylchlorarsine. The island was so secret that it was erased from all official and public maps. Japanese historians have concluded that just over 5400 tonnes of gas were produced on the island during this period. Although the chemical weapons factory buildings are rapidly disappearing on Okunoshima Island, arsenic contamination of the groundwater due to diphenylchlorarsine has just begun. In 1996, the local water supply was discovered to be contaminated with arsenic. In June, the Environment Ministry found arsenic levels 49 times the Japanese environmental standard (i.e. 490 µg/l) at one location on the island. Beginning in 1998, whenever arsenic-tainted soil was discovered, it was sent to Akita Prefecture for decontamination. However, the central government continues to decline to actively search for and remove the hidden stores of poison gas that certainly remain, because it is still a taboo subject (Johnson, 2004).

In 1955, powdered milk treated with sodium phosphate and arsenic pentoxide as a preservative was responsible for creating several cases of arsenic poisoning, but these were of short duration. Another case of arsenic-tainted food occurred in 1956 in a batch of soy sauce, with arsenic again being used as a preservative (Mizuta *et al.* 1956).

In 1959, anthropogenic arsenic poisoning was identified in Nakajo, Nagano Prefecture, with a population of 2000. The cause was well water that had been polluted by wastewater that contained inorganic arsenic (400 µg/l) dumped from a dye factory. Chronic arsenicism was found in 93 of 454 inhabitants in a polluted area, and high arsenic concentrations in well water were still present 36 years later (Nakadaira *et al.* 2000).

Enriched arsenic in Japanese coal is found in two Oligocene deposits in the Misassa sedimentary beds on the Island of Hokkaido. The Hiraga deposit has arsenic concentrations up to 83 mg/kg, and the Igu deposit has arsenic concentrations up to 325 mg/kg. In addition, Neogene lignites and sub-bituminous coals have higher average concentrations of arsenic of 80 mg/kg than do Palaeogene bituminous coals, with an average of 10 mg/kg (Imai *et al.* 1984). These enriched-arsenic coals are the product of the influx of syngenetic volcanic hydrothermal fluids, which occurred simultaneously with the formation of peat beds. In Tertiary lignites, anomalies are observed, with uranium up to 4600 mg/kg, germanium up to 780 mg/kg and arsenic up to 500 mg/kg (Sakanoue, 1960a,b).

Geogenic hidden arsenic has been discovered in Uranouchi Inlet, Tosa Bay, Shikoku Island, and Lake Biwa, Shiga Prefecture, Honshu Island, indicating that hidden arsenicals can occur in both saline and fresh water. In Uranouchi Inlet, the estuarine environment is analogous to that of the sites along the English Channel that have exhibited similar increases in hidden arsenicals during biologically elevated seasons (Hasegawa, 1996). In Lake Biwa, near Kyoto, the northern basin is mesotrophic and the southern basin is eutrophic. In the euphotic zone of both basins, arsenite or arsenic(III) increased in spring and autumn and dimethylarsinic acid or arsenic(V) dominated in summer. In the northern mesotrophic basin, arsenic concentrations remained constant throughout the year, but in the southern eutrophic basin, arsenic(V) increased 2–4 times during summer, which was positively correlated with levels of iron, manganese and phosphorus. In lacustrine environments, it appears that arsenic species depend on metabolism by bacteria, phytoplankton and algae and microbial reduction of iron and manganese oxides in sediments. Eutrophication is positively correlated with the concentration and speciation of lacustrine arsenic (Sohrin *et al.* 1997).

A survey of Japanese rice ($n = 26$) found a total arsenic concentration range from 0.07 to 0.42 (mg/kg) (Meharg *et al.* 2009). This data must be correlated with the per capita rice consumption rate for Japan at 254 g/d (husked rice) and 168 (polished rice) as compared to the even more rice intensive economies like Bangladesh at 445 g/d (polished rice) to assess its oncological implications. It is notable that developed countries like Japan have highly skewed high arsenic grain concentrations, while developing countries have normally distributed total grain arsenic concentrations (Meharg *et al.* 2009). This factor may be due to the local application of arsenical pesticides and herbicides.

Mining has contributed to arsenic contamination. Arsenic-contaminated water leached from the tailings of the Matsuo and Toroku mines in Miyazaki Prefecture, Kyushu Island. The Matsuo mine is a defunct arsenic–gold–tin–antimony–bismuth mine whose main veins contained a polymetallic deposit, with a mean concentration of 4.4 g gold/tonne in association with arsenopyrite and with a maximum concentration of 27 g gold/tonne and correspondingly high arsenic (Mindat.org, 2007a). The Toroku mine, also now derelict, was rich in arsenic as Legrandite ($Zn_2(AsO_4)OH \cdot H_2O$), which occurred in hydrothermal vein cavities related to a Tertiary granitic intrusion into a hard Palaeozoic limestone matrix that leached arsenic (with antimony and lead) into the surficial water and groundwater (Palache *et al.* 1951; Roberts *et al.* 1990). The populations of two villages, Matsuo Kogai and Sasagadani Kogai, in Miyazaki Prefecture were affected by aerosol arsenic dust from a smelter as well as water and soil pollution. This situation has since been remediated.

However, Japan is on the "Ring of Fire" and has active and inactive volcanoes. Consequently, the largest source of arsenic contamination of groundwater in Japan is thermal springs that occur throughout the islands in association with volcanogenic dynamics. The occurrences of arsenic in such instances, due to the tourist nature of many of the thermal springs, is probably under-reported. However, one case of identified arsenic and arsenic–lead sulfide sediments is from the Tamagawa hot springs, Akita Prefecture (Noguchi & Nakagawa, 1969), a heavily visited tourist attraction. One study that takes a look at the historical use of the site and its changing chemistry is that by Yoshike (1993). Other such potential arsenic–lead sulfide sites are Kuroishi springs, Kuroishi Town in Aomori Prefecture, Akinomiya hot springs resort in Akita Prefecture and Goshogake hot springs.

3.13 JORDAN (G)

Jordan, with a population of 6.3 million, has a human development index of 95 out of 169 and a per capita GNI rank of US$ 5300 (US Census Bureau, 2011; UNDP, 2010).

It was found that average concentrations of arsenic in sediments of the King Talal Reservoir as measured from 1987 to 1989 ranged from 1.53 to 4.36 mg/kg^{-1} (Gideon, 1991). This potentially affected a population of 2.1 million. A water quality policy statement (Preul, 1997) acknowledged that wastewaters in Jordan are municipal and industrial, with the largest plants in the Amman-Zarqa region. The As-Samra waste pond system and over 100 wet-type industries in this region constitute by far the largest portion of the total available wastewater flows that require water conservation and quality management. The most immediate priority recommendations are :

- An improved Ain-Ghazal/As-Samra treatment system.
- Implementation of an industrial waste discharge fee system.
- Implementation of an industrial waste minimisation programme.
- Training programmes in water pollution control and wastewater treatment plant operation and maintenance.
- Investigation into a small power station using the As-Samra effluent.

Longer-term water conservation and water quality improvements will result from the following actions:

- Basin-wide water quantity and quality management through river basin authorities or an environmental protection agency.
- Effective water quality monitoring and compliance.
- Technical assistance to industrial waste dischargers.

3.14 KAZAKHSTAN (A, C, G, M, P)

Kazakhstan, with a population of 17.3 million, has a Human Development Index rank of 66 out of 177 and a per capita GNI of $US 10 234 (US Census Bureau, 2011; UNDP, 2010).

Enriched-arsenic coals from mines in Ekibastuz and Shubarkol in Kazakhstan have seriously impacted the environment. Not only have coals been incinerated without proper protection domestically, but they have also been incinerated without proper protection by coal-fired power plants. In addition to the local enriched-arsenic coals, Kazakhstan has also incinerated Russian enriched-arsenic coals, such as those from Kuznetsk. For instance, the Alma-Ata power generation plant has a large ash disposal area that has contaminated local soils, surface water and groundwater with 1.4 million tonnes of high-arsenic ash annually (Baraboshkina & Ogorodnikova, 1999). Similarly, Ekibastuz is a town in Pavlodar Province, north-eastern Kazakhstan, with a population of 127 200 (1999 census). Near Ekibastuz are two coal-fired power plants, GRES-1 with a capacity of 4000 MW and GRES-2 with a capacity of 1000 MW. Because of the toxic arsenic burned by the plants, GRES-2 has the tallest chimney in the world, at 419.7 m. The disposition of the ash from these plants is unknown. Shubarkol and Vostochny pits produce 17.7 million tonnes of coal per year. As Vostochny coal mine produced the 300 millionth tonne of coal in 2004 and has 1284 million tonnes in resources and 790 million tonnes in proven reserves, with a mine life period beyond 2045, it is imperative that monitoring be undertaken from a variety of perspectives (Eurasian Natural Resources Corporation, 2004).

Kazakhstan possesses the largest oil deposits in the Caspian Sea area and currently produces about 1.3 million barrels a day. By 2015, its oil output is expected to reach 3 million barrels a day. In 2005, Kazakhstan opened the 1800-km pipeline that runs from the Azerbaijani port capital of Baku via Georgia to the Turkish Mediterranean port of Ceyhan. In January 2006, Kazakhstan opened a major 1000-km oil pipeline from Atasu to China, a key achievement in China's aggressive efforts to secure foreign energy supplies and a 50/50 joint venture between state companies China National Petroleum Corporation and KazMunaiGaz. By these pipelines, Kazakhstan fulfils its geopolitical strategy of a multivector policy of selling its oil to markets of its own choice, independent of political pressure (Bukharbayeva, 2005).

However, the environmental and human health situation has steadily worsened with the expansion of oil extraction, due to the location of many collective farms (created during the former Soviet Union period) very close to the oil and gas reserve fields. Collective farms use pesticides with arsenic compounds, which leach into the groundwater (Kim, 2004):

> Berezovka, one of the settlements located among the steppe lands of western Kazakhstan, is a vivid example [of adverse effects of the oilfields on health]. Only several kilometres away from Karachaganak, one of the nation's largest petroleum fields, Berezovka's population has been suffering from a number of illnesses ..., and the mortality rate has steadily increased.
>
> The results of a small study undertaken by Crude Accountability, a non-governmental organization which focuses on environmental protection and justice issues in the Caspian basin and which has implemented an environmental health programme in the area, are disturbing. Among nearly 200 students surveyed in 2003, unusual symptoms and illnesses included memory loss, hair loss and muscular-skeletal problems.
>
> The lack of environmental education among the native Kazakh people and the insufficient health care in remote areas make the situation even worse. Berezovka is not the only village in the Caspian region that feels the negative impact of petroleum extraction on the health of its citizens.
>
> There is no special environmental regulation to deal with situations similar to that in Berezovka. According to the norms of Kazakhstan's national law, the village must be relocated as a settlement in a "sanitary-protection zone", and people should be compensated for damages.

In addition to petroleum, Kazakhstan has a large mining industry and is a producer of beryllium, tantalum, barite, uranium, cadmium and arsenic. The mining practices prior to 2000 were old-fashioned, and the tailings were responsible for contaminating drinking-water with metals and metalloids like arsenic for years. One important advance that has been made is to produce a stable form of arsenic from arsenopyrite. This was accomplished at the gold mine in Auezov in north-eastern Kazakhstan in 1994, the first commercial application of the redox process to preoxidize sulfide ore concentrates. The Bakyrchik ore is basically quartz, with microscopic gold associated with arsenopyrite and pyrite mineralization (Kim, 2004).

In addition to the petroleum and mining contaminations, the drawdown of the Aral Sea has concentrated contaminants, including geogenic and anthropogenic arsenic, to high levels. In the climatic conditions of the Aral Sea basin, 60 km^3 of water per year are required for maintenance of the surface area of the Aral Sea. However, this scenario is no longer possible; hence, the sea is disappearing. It is speculated that it would take 10 years for the Amu Darya and Syr Darya to fill the Aral Sea if the rivers were not used for any irrigation projects during this period.

3.15 KOREA (REPUBLIC OF) (M)

The Republic of Korea, with a population of 48.8 million (2004), has a Human Development Index rank of 12 out of 169 and a per capita GNI of $US 29 518 (US Census Bureau, 2011; UNDP, 2010). It occupies the southern half of the Korean Peninsula below the 38th parallel and has a population spread unevenly over an area of 98 480 km^2. The country is bordered by the Yellow Sea, the Korea Strait and the Sea of Japan.

The Republic of Korea's mineral resources are limited to small reserves of antimony, coal, copper, gold, iron ore, lead, molybdenum, silver, tin, tungsten and zinc and relatively larger reserves of kaolin, limestone, pyrophyllite, silica stone (quartzite) and talc. It also has small reserves of natural gas but no petroleum. It is a significant producer of cement, copper, pyrophyllite and talc, but the mining of anthracite coal, ferrous and non-ferrous metals, and industrial minerals was the smallest sector of its economy in 2004, accounting for only 0.33% of the country's GDP (Wu, 2005).

The major source of arsenic contamination in the Republic of Korea is from mine tailings containing arsenopyrite in association with gold and silver deposits in the vicinity of six abandoned mining areas and two functioning mining districts (see map in Kyoung-Woong & Ilwon, 2003).

The first mining area, near Seoul, in Gyeonggi-do Province, on steep Mount Cheonbo, produced mainly gold. The three mines associated with this area (Jingok, Cheonbo and Sino) have high concentrations of arsenic(V) under oxic conditions, up to 62 350 mg/kg in the mine tailings. The main arsenic-associated minerals were pyrite and arsenopyrite (Kim *et al.* 2002, 2003).

The second mining area, Myungbong, about 100 km from Gwangju (Ko *et al.* 2003):

> is located in a south-western district near the southern Korean coast. The mine was active in the early 1900s, but closed in the 1970s. The ore deposit is a metasedimentary-hosted hydrothermal gold-bearing quartz vein containing small amounts of pyrite (FeS), arsenopyrite (FeAsS), galena (PbS), sphalerite (ZnS), and chalcopyrite ($CuFeS_2$); of these the arsenopyrite is the most likely source mineral of As in the tailings.... Mine exploitation has resulted in the deposition of 5,500 m^3 of tailings. Unlike other metal mines, the Myungbong mine tailings had a high As content (up to 1,400 mg/kg), but low concentrations of 1 mg Cd/kg, 150 mg Pb/kg, 48 mg Zn/kg and 8 mg Cu/kg.... The tailings have been abandoned without any remediation measure, and the arsenic-bearing tailings are an obvious candidate as a source of As contamination in the study area.
>
> Two ponds adjacent to the mine site are under the direct effects of tailings.... The larger pond, pond A, always contains water of a maximum depth of ca 1 m, and is apparently under anaerobic conditions. The smaller pond, pond B, undergoes seasonal drying and flooding according to the water levels, and the sediment is subject to a reduction/oxidation cycle. A small stream passing through pond A is used for irrigation of agricultural fields located beneath the mine site.

In this case, the arsenic concentrations in the tailings ranged from 2500 to 6420 mg/kg. The leaching was determined to have occurred from crystalline iron oxides produced by the oxidation of sulfide minerals. The leaching of arsenic was identified in adjacent streams and two nearby ponds with reducing conditions, which probably facilitated the release of arsenic into the groundwater. Downslope of the tailings, agricultural fields have been contaminated by arsenic-tainted groundwater used for irrigation.

The third mining area, Hyungjo, is located about 120 km inland from Busan (Pusan). The next three mining areas are within 120 km south to north of the 36th parallel, with Dongil in the south, Ockdong almost exactly on the parallel and Suyeo to the north. Dongil mining area had arsenic concentrations in its tailings as high as 8720 mg/kg, and these tailing were coupled with arsenic-tainted rice produced from associated soils (Lee & Chon, 2003). Indeed, the transport of arsenic into the Korean food-chain (mainly rice) has been well known and corroborated from measurements of daily ingested levels of arsenic averaging 320 μg/day per person since 1975 in these mining areas (Lee, 1976).

The first of the two functioning mining districts is the Dongjeong gold–silver–copper mine. In a 2005 study, the levels of arsenic and metal contamination of soils, crop plants and waters around the Dongjeong mine were used to assess the bioaccessibility of metals in soil and plants using the EHS (extraction of heavy metals in stomach and small intestine) test. The mean concentration of arsenic in tailings was 4070 mg/kg; cadmium (6 mg/kg), copper (380 mg/kg), manganese (7760 mg/kg), lead (19 150 mg/kg), zinc (2590 mg/kg) and mercury (2.8 mg/kg) levels were also measured (Chung *et al.* 2005):

> Mountain soils adjacent to the tailings and waste rock piles were significantly contaminated by these elements. Farmland soils contained higher concentrations of As, Cd, Hg, Mn, Pb and Zn than soils at the control area, especially As and Pb which were over permissible levels. Metals were accumulated in crop plants grown on farmland soil near the mine area. The concentrations of metals were higher in leafy plants than in grain plants such as rice grain and red pepper. The concentrations of SO_4^{2-}, Mn and Zn were elevated in stream water and were influenced by effluents from the tailings and waste rock piles. The concentrations of Cd and Pb in groundwater were higher than the permissible levels for drinking water in Korea. The results of the EHS test showed *c.* 60% of Cd, 10% of As and 30–40% of Cu, Pb, Mn and Zn were present as a bioaccessible fraction in farmland soils. In crop plants, high percentages of total concentrations of metals except Pb were extracted as bioaccessible concentrations. Bioaccessible concentrations of metals may be valuable for providing the input data for risk assessment at sites subject to anthropogenic soil contamination.

The other functioning mining area is the Gubong mining district in Chungcheong Province (White, 2006):

> The Gubong deposit consists of five massive, gold-bearing mesothermal quartz veins that fill fractures oriented northeast and northwest along fault shear zones over an area of 14 km^2 in Precambrian metasedimentary rocks of the Gyeonggi massif. The veins are divided into three groups, based on their orientation and location. They have a ribbon texture that is interpreted to result from repeated hydraulic fracturing events.

Arsenic was identified in Gubong groundwater with levels exceeding 10 μg/l, and its mobilization from arsenopyrite depended on pH. Dissolved arsenic was controlled by sorption with amorphous iron ($Fe(OH)_3$) and carbonate minerals, processes that are pH dependent, so that a relatively small decrease in pH could result in a drastic increase in arsenic concentrations (Woo & Choi, 2001).

Clearly, the health risks associated with the geological occurrences of arsenopyrite on the Korean Peninsula indicate the need for close monitoring of known areas of arsenic mineralization listed above and areas where arsenic has been known to leach into the groundwater. It is also very real possibility, given the peninsula's geology, that other sites remain to be identified.

3.16 KYRGYZSTAN (M)

Kyrgyzstan, with a population of 5.4 million, has a human development index of 126 out of 169 and a per capita GNI of US $2,036 (US Census Bureau, 2011; UNDP, 2010).

Twenty three tailing dumps and thirteen waste rock dumps are scattered throughout Mailuu-Suu. With a population of 26 000, this is the location of a Soviet uranium plant that operated from 1946 to 1968 and processed more than 10 000 metric tons of uranium ore, leaving 1.96 million m^3 of radioactive mining waste. Unsecured radioactive waste combined with high seismic activity threatens to contaminate the drinking water supply of the entire Ferghana valley, which is 22000 km^2 (8494 square miles), stretching throughout Kyrgyzstan, Uzbekistan and Tajikistan, that has a population of 11 342 000. In May 2002 a huge mudslide blocked the course of the Mailuu-Suu River and threatened to submerge another toxic waste site. In April 2005 the Obschestvenny Reiting newspaper reported that after another earthquake and landslide, about 300 000 m^3 of material fell into the Mailuu-Suu River near the uranium mine tailings (Blacksmith Institute, 2011). Arsenic is associated with these highly toxic tailings and much work needs to be done to properly remediate the tailing dumps.

A series of mines and tailings were evaluated and the results confirmed arsenic contamination. According to Vogt, "soil from the floor of the flood basin leading from the Kadamaji tailing deposits was heavily contaminated with sulfur, sodium, barium, zinc, and especially arsenic and antimony. In the streambank, clear white salt crystals were abundant. A sample including these crystals contained large amounts of sodium and sulfur. The heavy metal content was low and similar to the reference sample. Most likely this salt is a sodium sulfate used as flux in the electrolysis. The liquid sample collected in a small ditch was also highly enriched in sulfate and sodium, as well as the two metalloids arsenic (217 ppm) and antimony. Significant amounts of other heavy metals were also present (i.e. molybdenum, selenium, vanadium, zinc and cobalt). The sediment of the Hidarkan refinery was strongly contaminated with barium, lead, arsenic, zinc, mercury, tin, and especially antimony (1,386 ppm). These tailings represent a potential treat to the water quality in the entire Fargana valley with an estimated population of 11 342 000 as any leakages drain directly down into the basin. Although the Kadamji refinery was shut down after the Tulip revolution in spring 2005, the fate of the contaminated tailing site is critical to the arsenic pollution of the area (Vogt *et al.* 2011).

3.17 LAO PEOPLE'S DEMOCRATIC REPUBLIC (G)

The Lao People's Democratic Republic, with a population of 6.5 million, has a Human Development Index rank of 122 out of 169 and a per capita GNI of $US 2321 (US Census Bureau, 2011; UNDP, 2010).

The present estimate is that 400 000 people are exposed to arsenic levels above 10 μg/l in 1400 villages. Of these, 40% use tube wells, 20% of which are contaminated with arsenic (Anon., 2004b).

UNICEF has preliminarily tested groundwater from wells in the Attapeu, Savannakhet, Champasak, Saravane, Sekong, Khammuane and Bolikamxay areas of the Lao People's Democratic Republic. Results from 200 samples reported by Fengthong *et al.* (2002) suggest that arsenic concentrations ranged from 10 to 50 μg/l. The highest concentration of arsenic observed in the region was 112 μg/l in Attapeu Province. In Attapeu and Champasak provinces, 38% of wells contained arsenic concentrations above 10 μg/l. In Saravane Province, Ba Na Pong, Kongxedone District, wells have tested for arsenic at 42 μg/l. In Champasak Province, Ban Thoang Baeng, Pathoomphone District, wells have tested for arsenic at 54 μg/l. To date, UNICEF, in collaboration with the government and the Adventist Development and Relief Agency, has tested some 680 samples from drinking-water sources and found 1% of sources having arsenic concentrations above 50 μg/l (Table 3.25).

Table 3.25 Summary of arsenic concentrations in groundwater from tube wells in the Mekong Valley of the Lao People's Democratic Republic.

	Number of wells (%)			Total number of wells
	As $<$10 μg/l	As 10–50 μg/l	As $>$50 μg/l	
Holocene aquifer	531 (78)	143 (21)	6 (1)	680

Source: World Bank, 2005a

3.18 MALAYSIA (M)

Malaysia, with a population of 28.7 million, has a Human Development Index rank of 57 out of 169 and a per capita GNI of $US 13 927 (US Census Bureau, 2011; UNDP, 2010). Malaysia is located next to Brunei, Indonesia, Singapore and Thailand in South-east Asia and is a member of the Association of Southeast Asian Nations. It has a total land area of about 328 600 km^2 and a population of 25.6 million.

Malaysia is a federation of 14 states and 3 federal territories. Mineral resources belong to each of the states, which are responsible for approval and issuance of exploration licences and mining rights. In the late 1980s, a decline in metal prices shut down many tin mines; in the new millennium, however, as a result of increased metal prices, many abandoned mines have been reopened, and the federal government has urged the states to issue more mining licences (Malaysian Tin Bulletin, 2004; Tse, 2005a). The net result from an arsenic contamination point of view is that the abandoned mines being reopened and new mines being dug could lead to remediation of acid mine drainage problems, in which arsenic leaches into the surface water and groundwater. However, this requires governmental oversight, which is currently not present, so acid mine drainage is actually increasing.

Malaysia holds the world's greatest reserves of tin and high concentrations of other heavy metal ores in association with arsenic as a metalloid. For instance, gold mining is at about 4 tonnes per year. The Penjom Mine, the largest, is located at Penjom in the State of Penang and is owned by Avocet Mining PLC and operated by Specific Resources Malaysia. There are several small-scale gold mines located in the states of Kelantan, Penang and Terengganu (Tse, 2005a).

In each of these areas, arsenopyrite and pyrites are found in association with gold, and leaching of arsenic into the groundwater is suspected. The same is true for the tin mining areas. Malaysia's major tin deposits are located along the western coast of peninsular Malaysia in the states of Perak and Selangor. However, because of depleted tin resources and lower-ore grades, Malaysia's tin mine production continues to decline, even though Malaysia Smelting Corporation is the world's leading tin producer, with 23 tin mining areas in 2004, reduced from 26 in 2003 (Malaysian Tin Bulletin, 2005; Tse, 2005a).

While the confirmed presence of arsenic in drinking-water in Malaysia is only by implication from its gold and tin mining practices, it needs to be investigated. In 2005, an Indonesian-based environmental paper reported that "Malaysia has the world's largest production of tin, the country has now widespread areas with ex-tin mining pools, contaminated with heavy metals and metalloids such as arsenic. Humans may be exposed to arsenic through different routes in the environment" (Putrohari, 2005).In another report issued by Friends of the Earth International (2004), potential arsenic poisoning was connected with current disposal practices for hazardous wastes, especially heavy metals such as arsenic, copper, lead, mercury, zinc and cadmium, which are monitored in Malaysia. The report goes on to indicate:

> The DOE [Department of Environment] reports the most serious offenders in terms of hazardous waste generation in the total load. Based on this criterion, the metal finishing sub-sector in the machinery, engineering and electronics subsectors are the major offenders. Improper handling of hazardous waste can result in air, water as well as soil pollution. Existing databases suggest that the nine major industrial polluters are metal finishing, electrical and electronics, textiles, food processing, chemicals, palm oil, rubber, wood-based and iron and steel manufacturing units. Most of these are concentrated in Selangor, Penang, Johor, Perak and Kuala Lumpur. Although SMIs [small- and medium-sized industries] account for 60 percent of the total firms in Malaysia and often use "dirty" technologies, sophisticated industries generate larger amounts of hazardous and toxic waste per unit of output. The electronics industry, dominated by Japanese and US [United States] multinationals, uses hazardous chemicals intensively. Information on toxic and hazardous waste is sketchy especially about how much of it there is and what kind. Some experts project that the total industrial waste will rise relative to water and air pollution industry. The estimated generation rate of this type of waste in 1992 was 1.08 million cubic meters per year. In 1994, the DOE reported that 420,000 metric tons [tonnes] of scheduled wastes were generated by industries. Acids, heavy metals, dross, slag, clinker and mineral sludge accounted for over 70 percent of the quantity, with asbestos, heavy metal sludge, oil and hydrocarbons making the balance. In the states of Penang, Perak and Johor, on the west coast of Peninsular Malaysia, where industries are concentrated, heavy metal concentrations [including metalloids such as arsenic] exceeded the standards set. Due to the absence of scheduled waste treatment facilities in Malaysia, industries have to store, pre-treat or export their wastes, and storage capacity is now exhausted. Current hazardous waste management practices in Malaysia threaten public health. Exposure to hazardous waste has had public health consequences in Bukit Merah, Perak, with leukaemia and infant death cases.

Peninsular Malaysian surface sediments of the Johor Strait have been found to contain several trace elements including arsenic (27.30 ± 17.11 ug/g). Elevation of arsenic levels in the vicinity of the Malacca Straits has been reported to originate from terrestrial, marine and atmospheric sources (Zulkifli *et al.* 2010a,b). Arsenic and other pollutants are eventually deposited on the sediment affecting the aquatic environment biota. Geological Survey of Malaysia's Provincial Mineral

Maps (1982–1988) pinpoint where such arsenic contaminations could occur and need to be investigated. Preliminary studies by Williams & Breward (1995) for the British Geological Survey report that:

> The potential and efficiency of induced ferric oxide precipitation for scavenging heavy metals and arsenic from mine-waters at a Cu-Au mining operation in Sabah, East Malaysia, was investigated using a combination of empirical (sequential extractive speciation) and simulative (saturation-speciation and sorption modelling) methods.… The uptake of arsenic and heavy metals by aquatic and terrestrial biota around gold and complex sulfide mining operations was investigated at selected sites in Malaysia and Zimbabwe. Preliminary results for arsenic in filamentous algae from flooded alluvial gold workings showed evidence of acute accumulation (up to 7000 μg/g). The concentrations reported for fish muscle samples from the same localities were, however, surprisingly low (<0.2 μg/g).

The acute accumulation of arsenic indicated here is of special concern, as such levels are being found in home garden produce in local villages near mining operations. Moreover, abandoned mining sites pose an ongoing critical threat and contribute to high arsenic levels in drinking-water and health hazards, because after operations are terminated, little or no monitoring is done.

The Bau Mine area in Sarawak, East Malaysia, is of special concern in this regard. About 300 residents, especially children, in Bau and Kampung Sapur near Lake Tasik Biru are exposed to high arsenic levels via drinking-water, vegetables and other food and bathing using the water from the lake, which was previously used as a leach pit for the gold mining area. Water in Tasik Biru has an arsenic concentration of 87 μg/l, and local residents have been drinking it since 1997, when the clean, piped water supply to the village was cut off by the government. The Sarawak Natural and Environment Board, in a 4-month study, confirmed that arsenic in the lake increases at the rate of 51 μg/l per year (Anon., 1999).

Another example is the Penjom Mine area, where the reported arsenic concentration in the tailings is 5–3700 mg/kg. This compares with the concentrations in the tailings of the Bau Mine; the range reported is 15–5000 mg/kg (Bowell & Parshley, 2004). The expected level of arsenic in the surface water from these tailings would be comparable to the arsenic level in Lake Tasik Biru, about 90 μg/l, but this needs corroboration in the field.

3.19 MONGOLIA (C, G, M)

Mongolia, with an area of 1.565 million km^2 and a population of 3.1 million, has a Human Development Index rank of 100 out of 169 and a per capita GNI of US$ 3619 (US Census Bureau, 2011; UNDP, 2010). It borders the Russian Federation in the north and China in the south. It has an average altitude of 1580 m above sea level and is divided into three distinct ecosystems: the north-western mountain forest steppe, the eastern grasslands and the southern Gobi Desert. The population is 62% (1.92 million) urban, of whom 949,000 (31%) live in the capital Ulaan Baatar (CIA, 2012b). In general, most of the urban sector has access to a municipal water supply and nationally 51% have access to improved sanitation facilities (UNICEF/WHO, 2012). The rural sector is dispersed, with 30% in the central region, 18% in the west and 32% in the south and east.

A national survey on arsenic was conducted by the Public Health Institute, Ministry of Health, with UNICEF technical and financial support, from October 2003 to July 2004. The survey revealed that 10% of all wells tested were positive for arsenic and that the average arsenic concentration in wells was 14 μg/l. If we assume that 10% of the wells serve 10% of the rural population of 990 000, the total population exposed to arsenic levels above 10 μg/l comes to 99 000, a probable estimate. In the country as a whole, the population that has access to safe water is 41.5% (urban 69%, rural 27.5%) (Enkhtsetseg, 2004). As 82.4% of people in high-risk areas showed symptoms of arsenicosis but were not positively diagnosed (WHO, 2005), it is safe to say that 82.4% of 99 000, or 81 576, are symptomatic.

Concentrations of arsenic found in various Mongolian provinces are shown in Table 3.26.

Table 3.26 Arsenic contamination in drinking-water in selected provinces of Mongolia.

Province	Arsenic concentration in drinking-water (μg/l)
Govi-Sumber	30
Dornod	18
Dornod Govi	24
Dornod Govi, Khatanbulag soum	75
Flat Govi-Altai Dornod	18
Sukhbaatar Dornod	18

Source: Enkhtsetseg, 2004; WHO, 2005

Mongolia has extensive but largely untapped mineral resources, due to a poor infrastructure. For instance, only 15% of its total area has been geologically mapped, but infrastructure development began in 2005, with the mining sector expected to play an important role. Long-term programmes to exploit metallic and non-metallic minerals, such as coal, copper, fluorspar, gold, oil and silver, are under way with foreign investment. Even with this underdevelopment, mining accounted for 30% of the country's industrial output and 60% of its export revenue. Erdenet Mining Corporation sells more than 130 000 tonnes of copper and 1500 tonnes of molybdenum to China. Ivanhoe Mines Ltd's Oyu Tolgoi ("Turquoise Hill") mines in Umnogovi Aimag in the southern Gobi are projected to have a sustained ore output of 40 million tonnes per year for 25 years. Mongolia also has four significant copper, gold, arsenic and molybdenum deposits located about 140 km north-west of Oyu Tolgoi, in a highly mineralized 14-km-long corridor (Tse, 2005b). Cases of acid mine drainage that could contribute to arsenic leaching into the groundwater are unknown, but the mining sources indicated above have arsenic-related minerals associated with them. This is a potential hazard, especially in Umnogovi Aimag and the highly mineralized corridor north-west of Oyu Tolgoi.

Mongolian coal output is mainly from mines at Aduun Chuluun, Baganuur, Sharyn Gol and Shivee-Ovoo, which account for 90% of total production (Tse, 2005b). High-arsenic coal from 10 mines was analysed, and arsenic concentrations in Baganuur (183 mg/kg) and Nalaikh (121 mg/kg) were the highest. The leaching of these coal beds into overlying aquifers is hypothesized but has not been tested, so the contamination of the adjacent aquifers is potential but not proven. The population at risk from coal burning is not known, but if it follows the Chinese demographics in high-arsenic coal areas, it would be of the order of tens if not hundreds of thousands.

The geological formations of the eastern Gobi and the Tamsag basins in eastern Mongolia have many similarities with the Erlian and Khailaar basins in China. Current estimates of oil resources in Mongolia range between 4 billion and 6 billion barrels (Tse, 2005b). These reserves are being developed by foreign mining companies. The effects and controls on the petroleum extraction process are unknown, but testing of adjacent aquifers is necessary to ensure safe drinking-water in these regions.

3.20 MYANMAR (G)

Myanmar, with a population of 54 million, has a Human Development Index rank of 132 out of 169 and a per capita GNI of $US 1596 (US Census Bureau, 2011; UNDP, 2010). Shallow tube well construction started in Myanmar in the 1990s. No arsenic-related health effects have been reported to date, but the potential for elevated arsenic in drinking-water is suspected based on the similarities between Myanmar's delta region and Bangladesh.

The Water Resources Utilization Department (WRUD, 1999) in the Ministry of Agriculture and Irrigation of Myanmar conducted arsenic testing in mid-1999. Samples drawn from Lower Myanmar showed traces of arsenic. A survey conducted by Save the Children Fund UK (International Rice Research Institute, 2006) in Thabaung, Laymyethan and Hethada townships during March–May 2000 found that 35% of 125 tube wells tested showed arsenic concentrations in excess of 50 μg/l. An extended survey of 1912 wells in the Ayeyarwady division found that 21% had arsenic levels exceeding 50 μg/l. The percentage of affected wells varied by area, ranging from 3% to 37%. Based on these figures, it has been estimated that 13 099 households are exposed to elevated levels of arsenic (SCUK, 2001). Other data suggest that the current estimate of the population in Myanmar exposed to arsenic at concentrations above 10 μg/l is about 3.4 million (World Bank, 2005a).

The Water Resources Utilization Department has conducted a thorough water quality monitoring programme in five selected townships, namely, Magway, Sittway, Kawmu, Kyaungkone and Henthada. In these townships, localized random distribution of arsenic concentrations of varying magnitudes was observed. The evidence so far reveals that tube wells in delta and coastal areas are generally prone to arsenic contamination. The townships of Thabaung, Laymyethan and Hethada have 42 high-risk wells, of which 12 have arsenic levels in the groundwater above 50 μg/l (Htut *et al.* 2004). In these towns, 6976 villagers are at risk, and 21 cases of dermal change have been reported; of those, 8 are suspected arsenicosis cases. In the Ayerawady, Bago and Mandalay divisions and Rakhine state, approximately 15% of shallow tube wells have arsenic concentrations above 50 μg/l (Htut *et al.* 2004).

The possible mechanism of arsenic contamination of the Salween delta is related to a common origin in the eastern Palaeo-Tethys ocean that was characterized by ferromanganoan arsenic-bearing sediments. This was later subject to headwater erosion by the Himalayan uplift, which released arsenic that was redistributed in Neogene sandstones, silts and iron-rich clays. These secondary and tertiary reservoirs of adsorbed and authigenic arsenic are the source of arsenic in groundwater basins of south and south-eastern Asia, including the Brahmaputra, Salween, Irrawaddy, Mekong, Ganges and Red river drainages (Stanger, 2005).

3.21 NEPAL (G)

Nepal, with a population of 29.4 million, has a Human Development Index rank of 138 out of 169 and a per capita GNI of $US 1201 (US Census Bureau, 2011; UNDP, 2010).

Authorities were alerted to the possibility of elevated arsenic levels in drinking-water by reports that the arsenic problems in Bangladesh and West Bengal, India, were due to the alluvial sediments deposited by rivers draining from the Himalayas. The shallow aquifer of the lowland Terai region of southern Nepal is situated in a Quaternary alluvial sediment (Smedley, 2002).

Arsenic concentrations up to 2620 μg/l have been reported, although levels are typically much lower than this, and there is a wide variation in arsenic concentration (Shrestha *et al.* 2002). Water quality testing has shown that about 5% of samples (from a total of 14 932) have arsenic concentrations greater than 50 μg/l, whereas 23% exceed 10 μg/l. Based on these figures, the estimated population exposed to levels of arsenic above 50 μg/l is over 500 000 (Shrestha *et al.* 2002).

The following Executive Summary is from a report by Shrestha *et al.* (2004):

> The discovery of dangerous levels of arsenic in large numbers of ground water tube wells in West Bengal, India during the early 1990s, followed closely by even higher levels of arsenic in even greater numbers of wells in Bangladesh in the mid to late 1990s, has raised the spectre of a potential environmental disaster in South Asia. The World Health Organization estimates that 40–60 million people are at risk from drinking arsenic-contaminated water in Bangladesh and West Bengal, and that this health risk also may extend into southern Nepal.
>
> The first Nepalese studies on arsenic in groundwater were carried out in 1999 by the Department of Water Supply and Sewerage, and the World Health Organization, and in 2000 by the Nepal Red Cross Society and the Japanese Red Cross Society. Both studies provided evidence of arsenic contamination in Terai region of southern Nepal. Furthermore, an initial health survey in 2001–2003 found evidence of arsenic-related dermatosis and elevated amounts of arsenic in human hair and nail samples in four districts where tube well drinking water contained arsenic above 50 ppb [parts per billion]. This initial evidence of arsenic contamination and associated health effects led to the creation of the National Arsenic Steering Committee (NASC) to help coordinate efforts by government and non-government agencies to address the potential problems of arsenic contamination in the rapidly growing region of southern Nepal.
>
> Approximately 500 000 shallow tube wells provide water to about 11 million people, nearly half of the national population, in the agriculturally important Terai region. Ninety percent of the people in the Terai use groundwater as their primary source for drinking water. Providing clean drinking water and developing groundwater resources for agriculture are significant challenges for government and non-government organizations, and the issue of arsenic contamination must now be evaluated by all agencies and organizations that provide drinking water. Nearly all water organizations have been testing wells in their project areas since 2000 by various methods. Nearly 6% of these tube wells in the Terai have been recently tested for arsenic. This report integrates all qualified arsenic test results from these organizations, collected through the summer of 2003, into a national arsenic database.
>
> The national database in 2003 consists of 18 635 arsenic-tested tube wells. Arsenic distribution and summary statistics for each district and VDC [village development committee] are shown on (Figure 3.26). The majority of tube wells, 76.3%, are below the WHO Guideline of 10 ppb. Approximately 7.4% of all tube wells tested higher than the Nepal Interim Standard of 50 ppb of arsenic concentration and 16.3% of tube wells contained 11–50 ppb of arsenic concentration. Thus, 23.7% of tube wells in the Terai are above the WHO Guideline. The percentage of wells contaminated above 50 ppb varies at district level from 0% in Dang, Chitwan and Sunsari to 25.7% in Nawalparasi. Over half of the arsenic-tested tube wells in Rautahat and nearly half of those tested in Nawalparasi contain more than 10 ppb of arsenic. The highest concentrations of arsenic (up to 2,620 ppb) were measured in Devedaha VDC of Rupaudehi district, near the border with Nawalparasi.
>
> Most of the tube wells with high arsenic values are less than 50 m deep, and the majority of tube wells deeper than 50 m have arsenic values below 50 ppb. The highest percentage of contaminated wells is found from 11 to 30 m deep. In that depth range 11% of arsenic-tested tube wells were above Nepal Interim Standard; 32% of tube wells were above the WHO Guideline; and 68% of the wells tested were below the WHO Guideline. A very low correlation was found between the level of arsenic of a well and the age of the well.
>
> Classes of arsenic vulnerability are defined and presented on maps at the district level based on the percentage of arsenic tested tube wells exceeding the WHO Guideline [Figure 3.26]. Four classes of vulnerability are defined. **Low vulnerability to arsenic:** VDCs where the percentage of tube wells above WHO guideline is zero (0). In these VDCs, all arsenic tested tube wells contain 0–10 ppb of arsenic. **Moderate vulnerability to arsenic:** VDCs where 1–25% of tested tube wells exceed the WHO Guideline for arsenic. **Moderately high vulnerability to arsenic:** VDCs where 26–50% of tested tube wells exceed the WHO Guideline. **High vulnerability to arsenic:** VCDs where over 50% of arsenic-tested tube wells exceed the WHO Guideline.
>
> Among the 1,050 arsenic-tested VDCs, only 20 VDCs had more than 100 tube wells that were tested for arsenic by 2003. Moreover, there were only five VDCs, in Nawalparasi district, that contained more than 200 arsenic-tested tube wells. Although 60% of VDCs have

low vulnerability, nearly 74% of those VDCs have a high uncertainty for that assigned vulnerability level, because of a very low number of wells that were tested in those VDCs. In order to account for VDCs with low levels of arsenic testing, a level of uncertainty was characterized to help describe the confidence that accompanies the level of vulnerability assigned to each VDC. The uncertainty is based on the proportion of household in a VDC that is using a tested tube well.

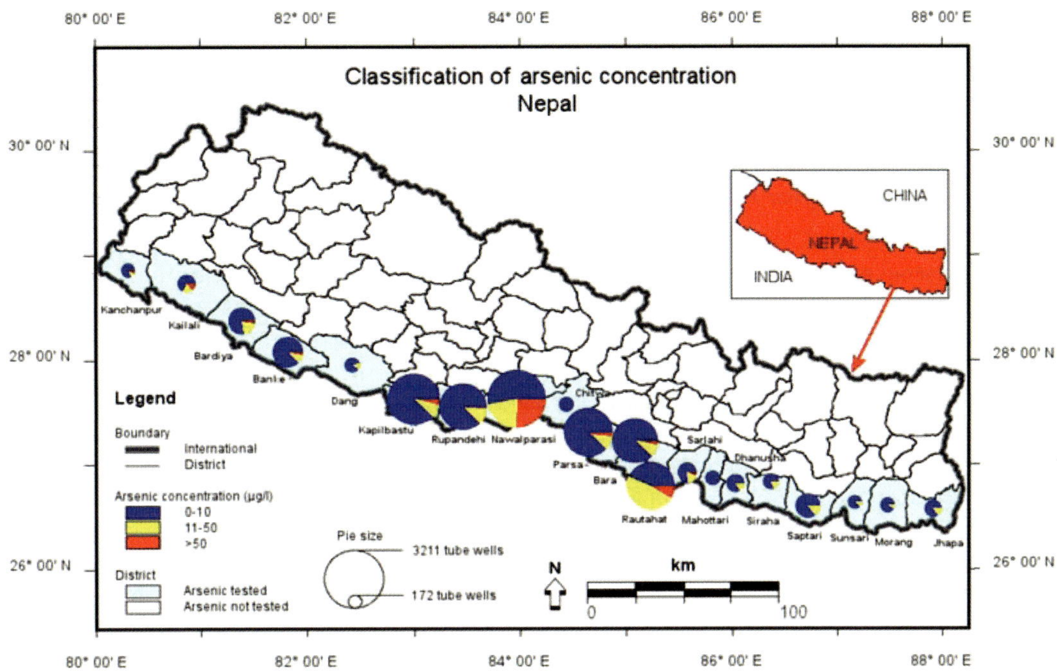

Figure 3.26 Classification of arsenic concentration in the Terai, Nepal (modified from Shrestha *et al.* 2004).

Among the 20 Terai districts only one district, Rautahat, is characterized with high vulnerability to arsenic. The uncertainty level is also low for this district. This indicates that more than 50% of arsenic tested tube wells were above the WHO Guideline, and that the percentage of households using arsenic tested tube wells is 26–50%. Two districts, Nawalparasi and Kailali, have been classified as having moderately high vulnerability to arsenic; however, their uncertainty levels are different. Nawalparasi has medium uncertainty, and Kailali has high uncertainty. Therefore, although both districts have moderately high vulnerability to arsenic, Nawalparasi district has more arsenic-exposed households than Kailali.

The majority of districts in the Terai have been classified as moderate vulnerability to arsenic. However, the level of uncertainty is different in these districts. Among 16 districts classified with moderate vulnerability to arsenic, 9 districts have high uncertainty, 4 districts have medium uncertainty and 3 districts have low uncertainty. There is only one district, Chitwan, with low vulnerability to arsenic. However, this classification has high uncertainty, because only 1% of total households of the district were using arsenic-tested tube wells.

In terms of population exposed to potential arsenic-related health risks, the 2003 arsenic database indicates that 69 126 people (5.4%) were drinking water above the Nepal Interim Standard at the time of testing; 44% of those people live in Nawalparasi district, 21% in Rautahat, and the rest, or 34%, are distributed among the other Terai districts. The population in the Terai estimated to be drinking water that exceeds the WHO Guideline is 272 626 (21.3%). These statistics will likely change in the future, because arsenic testing has not yet taken place in 345 VDCs–25% of the total VDCs–in the Terai.

Recommendations to improve the national arsenic database

The sources of arsenic in the ground water are believed to be of geologic origin. However, the processes that concentrate arsenic to environmentally dangerous levels in the shallow aquifers are poorly understood. It is not yet clear whether or not contamination is entirely natural or is accelerated by human activities. Because the mechanisms of arsenic contamination are not yet defined, it is important to continue arsenic testing of wells in the Terai and, because arsenic levels may increase over time, wells with arsenic levels that exceed the WHO Guideline should be monitored at some regular interval. Areas of anomalously high arsenic concentrations found in the database should be tested further to define the areal extent of the "arsenic hotspot", and mitigation efforts should be introduced to these areas of concern. Health organizations should initially focus on these arsenic hotspots to identify individuals in need of medical attention.

> Standard procedures need to be created for continued field kit testing, laboratory analysis, and data maintenance. Most important is the inclusion of GPS [global positioning system] well location data with arsenic tests. As more new field test kits become available, they should be tested and verified with laboratory results. A national center should be established to maintain the database of arsenic-tested tube wells. VDC and district maps of arsenic distribution should be updated on a regular basis. This center, under the guidance of NASC [National Arsenic Steering Committee], should also evaluate and recommend appropriate arsenic mitigation strategies and methods. More health surveys are needed in the Terai to identify patients that need treatment, and to provide data that will help to clarify the health effects of drinking water that contains different concentrations of arsenic. Basic health related data should be included in the national arsenic database. The database should be used to guide arsenic mitigation and testing strategies, as well as further water resources development, in the 20 districts of the Terai. The testing of VDCs with high data uncertainties, or very few tested wells, should be given priority. The NASC and appropriate water agencies and NGOs [nongovernmental organizations] should communicate to the public on a regular basis with accurate and useful information about arsenic issues at local, district, and national levels.

The possibility that arsenic may be present in some Nepalese ground waters was recognized in 1999. As a consequence, arsenic-related health effects have not been extensively studied, but classic skin lesions have been observed (Shrestha *et al.* 2002). In the better-studied Terai districts, the overall prevalence of arsenic-related dermatitis was found to be 2.6%, ranging between 1.4% and 5%. In the Nawalparasi District, however, prevalences of up to 20% have been seen (Shrestha *et al.* 2002). It is estimated that people have been exposed to elevated levels of arsenic in drinking-water for up to 13–14 years (depending upon when the tube well was constructed; R.R.Shrestha, personal communication).

There are four approaches to arsenic mitigation promoted by the Nepali government, United Nations Children's Fund (UNICEF)-Nepal, Environment and Public Health Organization (ENPHO), the Nepal Red Cross and various other key players in the water sector. These include three alternative safe water source methods: (1) sharing safe tube wells; (2) protected dug wells; and (3) rainwater harvesting; one further treatment approach uses a modified biosand filter with 5 kg of non-galvanized iron nails as the absorption media. The arsenic filter was invented by a joint Massachusetts Institute of Technology, ENPHO, Finnish International Development Agency (FINNIDA) partnership in 2002 (T. Ngai, personal communication, 2007). The dissemination of the filter began in 2002, with 5000 filters in use by the end of 2006. UNICEF-Nepal and the Nepal Red Cross took the lead in disseminating an additional 4000–5000 filters in 2007–2008. The final dissemination of filters occurred between 2008 and 2010, when another 12 000 filters were funded by UNICEF and the United Nations Human Settlements Programme (UN-HABITAT). In all, 24 000 filters have been disseminated, reaching an estimated 200 000 people (Espinoza & O'Donnell, 2011).

3.21 PAKISTAN (G)

Pakistan, with a population of 187.3 million, has a Human Development Index rank of 125 out of 169 and a per capita GNI of $US 2678 (US Census Bureau, 2011; UNDP, 2010).

The presence of arsenic in groundwater supplies in Pakistan is similar to that in arsenic-affected aquifers of Bangladesh and West Bengal, India (Smedley, 2002). Because arsenic contamination of groundwater in Pakistan has undergone a significant re-evaluation since 2002 as a result of laboratory testing using atomic absorption spectrometric (AAS) techniques, the history of the earlier testing procedures and results is given below so that the earlier literature can be interpreted in terms of these constraints.

The Government of Pakistan began undertaking initiatives with UNICEF in 1999 because it was feared that the country could be at risk of contamination similar to Bangladesh. From November 1999 to January 2001, the Pakistan Council of Scientific and Industrial Research and the Pakistan Council of Research in Water Resources of the Ministry of Science and Technology tested for arsenic in drinking-water in six districts in northern Punjab: Jhelum, Chakwal, Attock, Rawalpindi, Sargodha and Gujarat. These districts were selected for the occurrence of coal or iron mining, hydrothermal waters, reducing groundwater and granites/basalts. The tests were carried out solely using field test kits.

It was thought in 1998 that a major difference between the Indus Plain and the Gangetic Plain was that the Indus Plain had predominantly aerobic aquifer conditions, because nitrate and dissolved oxygen were detected (Mahmood *et al.* 1998). Aerobic conditions tend to inhibit arsenic from being released into the groundwater. A study carried out in 2000 by the Provincial Government of Punjab and UNICEF tended to confirm this hypothesis by testing samples solely using field test kits; of a total of 364 water samples, 90% contained arsenic concentrations below 10 μg/l and only 6 samples (2%) contained arsenic concentrations above 50 μg/l (Iqbal, 2001) (Table 3.27). This apparent minimal arsenic contamination was thought to be confirmed by an epidemiological study that diagnosed no confirmed cases of arsenic-related diseases. As of 2005, based solely on field testing of arsenic, fluoride and salinity were thought to be more pressing health issues (World Bank & Pakistan Ministry of Environment, 2005).

Table 3.27 Field data on frequency distribution of arsenic concentrations in groundwater samples from northern Punjab, Pakistan.

District	No. samples (%)			Total no. samples
	As <10 μg/l	As 10–50 μg/l	As >50 μg/l	
Gujarat	33 (87)	3 (8)	2 (5)	38
Jhelum	32 (86)	4 (11)	1 (3)	37
Chakwal	63 (88)	9 (12)	0 (0)	72
Sargodha	49 (83)	7 (12)	3 (5)	59
Attock	68 (92)	6 (8)	0 (0)	74
Rawalpindi	81 (96)	3 (4)	0 (0)	84
Total	**326 (90)**	**30 (8)**	**6 (2)**	**364**

Source: Iqbal, 2001

However, subsequent laboratory testing using AAS analysis showed the degree of arsenic contamination in Pakistan to be much worse than the field data indicated. From samples analysed using AAS, 28% and 7%, respectively, had contamination over 10 μg/l and 50 μg/l, compared with 10% and 1.8%, respectively, from field data. The large deviation between laboratory and field results appears to be attributed to the improved accuracy and sampling and testing methods of AAS compared with field test kits. The conclusion is that six areas of Punjab and Sindh provinces pose a serious arsenic threat to human health, with levels of arsenic in drinking-water above 10 and 50 μg/l, as shown in Table 3.28.

Table 3.28 Arsenic contamination levels in Punjab and Sindh provinces, Pakistan.

Area	% of samples >10 μg/l	% of samples >50 μg/l
Multan	35	1.0
Rahim Yar Khan	14	2.3
Bahawalpur	9	0.6
Sindh	2 of 9 districts >10 μg/l	4 of 9 districts >50 μg/l
Dadu	18	5
Khaipur	21	3

Source: Pakistan Council of Research in Water Resources, 2003a,b, 2004

Punjab Province means "land of five rivers", referring to the Beas, Ravi, Sutlej, Chenab and Jhelum rivers, but not the Indus River, which also flows through Punjab. Punjab, with an area of 205 344 km^2, has a population of 79 429 701 and a population density of 386.8/km^2. In Punjab, over 20% of the population (15.8 million) is exposed to arsenic contamination above 10 μg/l in drinking-water, whereas nearly 3% of the population is exposed to over 50 μg/l (BGS/Water Aid, 2001; Institute of Public Health, 2003; Pakistan Council of Research in Water Resources, 2003a,b; Nickson *et al.* 2004).

Sindh Province has always been intimately connected with water, particularly the Indus River. Sindh, with an area of 140 914 km^2, has a population of 33 038 773 (2003) and a population density of 234.5/km^2. There are 23 districts and 160 towns. In Sindh, the arsenic contamination situation is serious, with 36% and 16% of the population exposed to water contaminated with arsenic at concentrations above 10 μg/l and 50 μg/l, respectively. Both shallow and deep sources have arsenic contamination. A recent epidemiological study of arsenicosis in a controlled population indicates a case rate of 140/1000. The estimated population exposed to arsenic concentrations above 50 μg/l is in the hundreds of thousands (Ahmad *et al.* 2004). Based on a study of 39 000 wells, the population at risk in Pakistan is estimated to be 128 000. This appears to be a conservative estimate. The highest levels of contamination appear to be in Muzaffargarh District, south-west Punjab, central Pakistan, where the maximum arsenic level reached 906 μg/l (Nickson *et al.* 2004).

The depth of wells appears to follow no overall correlation with arsenic contamination. In Sindh Province, in Dadu, wells less than 50 m are more contaminated than those over 50 m. In some areas in Punjab Province, high contamination can be found in shallow and deep wells alike, and in Muzaffargarh in Punjab, a deep well can be more contaminated than a nearby shallow well. The results of arsenic monitoring by Nickson *et al.* (2004) are provided in Table 3.29. The excellent data from Nickson *et al.* (2004) show that proximity of wells is no indication of a common level of arsenic contamination. As a result, every well must be tested because of different underlying geological dynamics, in the same way that proximate wells must be tested in Viet Nam (Ahmad *et al.* 2004).

Table 3.29 Arsenic data collected from Muzaffargarh locations in Pakistan, illustrating the range of concentrations in proximate locations.

			Arsenic concentration (μg/l)[a]	
Locality	**Latitude**	**Longitude**	**Laboratory**	**Field**
Multan	712542	301224	274	200
Multan	712614	301019	905	1000
Multan	712543	301209	88	60
Muzaffargarh	712924	302708	27	20
Muzaffargarh	712642	303225	14	10
Muzaffargarh	712955	303148	4	10
Muzaffargarh	711416	302142	6	0
Muzaffargarh	711341	301602	1	0
Muzaffargarh	713349	303058	2	10
Muzaffargarh	712624	302221	8	0
Muzaffargarh	712642	303421	11	0
Muzaffargarh	711407	302706	6	0
Muzaffargarh	712403	302237	1	10
Muzaffargarh	711434	303658	1	0
Muzaffargarh	711437	303656	6	0
Muzaffargarh	711406	303432	5	0
Muzaffargarh	711356	303502	2	0
Muzaffargarh	711359	303456	96	200
Muzaffargarh	711356	303502	169	100
Muzaffargarh	710441	300358	2	0
Muzaffargarh	711341	301400	3	0
Muzaffargarh	711112	300400	146	100
Muzaffargarh	711051	300510	1	0
Muzaffargarh	710929	300302	15	20
Muzaffargarh	711335	300922	89	100
Muzaffargarh	711010	300320	99	100
Muzaffargarh	711017	300338	147	400
Muzaffargarh	710931	300409	50	25
Muzaffargarh	710857	300249	9	0
Muzaffargarh	711010	300430	6	25
Muzaffargarh	711035	301502	25	0
Muzaffargarh	711029	300412	91	100
Muzaffargarh	711132	300408	4	0
Muzaffargarh	711100	300345	8	0

(*Continued*)

Table 3.29 Arsenic data collected from Muzaffargarh locations in Pakistan, illustrating the range of concentrations in proximate locations (*Continued*).

			Arsenic concentration (μg/l)[a]	
Locality	**Latitude**	**Longitude**	**Laboratory**	**Field**
Muzaffargarh	711808	301303	8	10
Muzaffargarh	711834	301505	27	40
Muzaffargarh	711049	300421	125	300
Muzaffargarh	711351	300917	184	400
Muzaffargarh	711109	300451	1	0
Muzaffargarh	711229	300611	85	100
Muzaffargarh	711326	300838	22	0
Muzaffargarh	711301	300411	4	10
Muzaffargarh	711044	300513	87	200
Muzaffargarh	711807	301304	60	50
Muzaffargarh	711357	300919	80	35
Muzaffargarh	711406	300920	85	20
Muzaffargarh	711032	300444	110	80
Muzaffargarh	710957	300321	90	200
Muzaffargarh	711132	300407	166	50
Muzaffargarh	712948	302659	7	

[a]Laboratory/field indicates whether measurements of arsenic were taken in the laboratory or the field.
Source: Nickson *et al.* 2004

In addition, canal irrigation in semi-arid regions can lead to widespread water logging, and evaporative concentration of salts can lead to arsenic concentration in shallow groundwater to levels above 10 μg/l. In rural areas, arsenic is generally below 25 μg/l, because arsenic in oxic shallow groundwater is sorbed to aquifer sediments (Nickson *et al.* 2004).

A study of bioaccumulation of heavy metals in 19 species of fish in the Indus River from 11 sites indicates that *Labeo calbasu* from Guddu barrage had the highest concentration of arsenic in its flesh, at 3.072 mg/kg (Tariq *et al.* 1996). *Labeo calabasu* is a bottom-dwelling, illophagic fish feeding on decayed organic matter, molluscs, diatoms, plant matter, green algae, blue-green algae and zooplankton, with juveniles selecting zooplanktonic organisms and adults selecting decayed organic matter and molluscs (Kumar & Siddiqui, 1989). Measurement of arsenic in the sediments that sustain the molluscs on which adults feed showed an arsenic concentration of 7.452 mg/kg at Jinnah barrage.

3.23 PHILIPPINES (G, M, V)

The Republic of the Philippines, with a population of 101.8 million, has a Human Development Index rank of 97 out of 169 and a per capita GNI of $US4,002 (US Census Bureau, 2011; UNDP, 2010). The Philippines is located north of the equator 1100 km east of mainland South-east Asia and comprises 7000 islands, fewer than 900 of which are inhabited by the country's people. Major islands include Luzon in the north, Visayas in the centre and Mindanao in the south. The archipelago is within the Pacific "Rim of Fire", indicative of the intense volcanic activity at the margins of the tectonic plates where deposits of epithermal gold, porphyry copper–gold and volcanic-hosted sulfide are found, in addition to arsenic. Consequently, the Philippines is a major producer of chromite, copper, gold, nickel, coal, gypsum, silver and sulfur, most of the ores that are associated with arsenic occurrences. Therefore, the Philippines are at risk for arsenic contamination of surface water and groundwater from at least two sources (volcanism and mining) as well as perhaps geogenic arsenic sources that naturally leach into the environment.

Mining is a known contributor to arsenic contamination of surface water and groundwater in the Philippines, and the leaching process can contaminate the ubiquitous marine environments as well.

Arsenic leaching into surface water and groundwater has been found on Marinduque Island about 160 km south of Manila. A mine operated by Placer Dome partner Marcopper Mining Corporation of the Philippines was closed by the Philippine government after the mining dam failed, polluting the Boac River and coral reefs with 1.4 million tonnes of pulverized mining waste and toxic tailings laden with arsenic, cadmium, lead, manganese, nickel and sulfate. The CMI Mine on Marinduque is near Mogpog and adjacent to the mouth of the Boac River, which drains into the ocean. Obviously, such a site has, just by its sensitive location, a potentially negative environmental impact. However, its infrastructure is clearly critical, because it is a large open pit filled with water and replete with tailings of yellow secondary salts eroding into the ocean, dispersing potential contaminants, including arsenic.

Marinduque has a total population of 217 392 (2000 census), with populations adjacent to mining operations in the towns of Boac (population 48 504) and Mogpog (population 31 330) (National Statistical Coordination Board, 2005). Those living closest to the Placer Dome mine number about 1500, and this is the (conservative) figure that has been used in assessing the number of people who are exposed to arsenic-contaminated groundwater (Plumlee *et al.* 2000).

Acid rock drainage and metal-rich near-neutral rock drainage can adversely affect the environment, especially in streams and groundwater into which the drainage waters flow. Highly acidic water (pH 3.5) is draining from one of the mine dumps, and vegetation has been killed by acid rock drainage. The dump is composed of porphyry deposits that have high geological potential to generate acid rock drainage with arsenic. Iron, copper, lead, zinc, arsenic, cadmium and others in a mineral deposit are released into surface water and groundwater by sulfide weathering. Less acidic waters draining carbonate-rich mineral deposits can contain elevated levels of some metals, such as arsenic. Acid rock drainage and near-neutral rock drainage can form as a result of natural weathering and erosion processes. Mineralized areas (including Marcopper's mine) had acid rock drainage/near-neutral rock drainage prior to mining. Mining greatly accelerates the arsenic flow from mining sites or mineralized areas (Plumlee *et al.* 2000):

> As ARD [acid rock drainage] and NRD [near-neutral rock drainage] flow from mine sites or mineralized areas, they are typically diluted by less acidic waters draining unmineralized rocks. The increased pH (decreased amount of acid) caused by this dilution commonly leads to precipitation of orange to white iron-rich and aluminium-rich mineral particulates in the stream. Arsenic, lead, and copper tend to adsorb onto and precipitate with these iron and aluminium particulates. As the particulates settle to the stream bed, they remove these sorbed metals from the stream waters, thereby improving water quality. However, metals such as zinc and cadmium tend to stay dissolved in waters affected by ARD and NRD because they tend not to sorb onto the iron and aluminium particulates. Other metals such as copper, arsenic, uranium, and selenium can desorb from the particulates back into solution in the stream waters if the pH of the stream increases sufficiently to near-neutral conditions.

Arsenic was measured at nearly 100 μg/l in the seawater at the outflow of Boac River from tailings disposal in Calancan Bay (Plumlee *et al.* 2000) and exceeds by a factor of 10 the USEPA's Recommended Chronic or Acute Water Quality Criteria for salt water, which is less than 10 μg/l. This could pose ecological and health risks in both the Boac and Mogpog rivers if the metals are taken up by marine organisms that live on the ocean bottom in the near-shore marine environment. Arsenic levels have not been tested further offshore, nor have bioassays been conducted on edible marine fauna.

There is also the issue of the redistribution of the tailings in the ocean and along the coast away from the tailings causeway. "Continued erosion of the tailings from and adjacent to the causeway will result in a significant movement of metal bearing, tailings-derived sediments along the Marinduque coastline, and possibly towards the inhabited mainland" (Plumlee *et al.* 2000).

In addition to arsenic leaching into surface water and groundwater associated with mining operations, there is also the release of arsenic from geothermal waters that may be caused by thermal power production.

The occurrence of arsenic contamination of groundwater in south-central Mindanao has been a major issue of contention on the island. In 1992, Dr Nelia Maramba, a toxicologist from the University of the Philippines, identified arsenic contamination of groundwater in six villages in south-central Mindanao.. However, the governor of the area opposed her further involvement in the investigation of the reputed arsenic contamination. For seven years, the government continued to postpone arsenic testing of the Mount Apo area in south-central Mindanao, 32 km west of Davao City. This was explained by Health Undersecretary Antonio Lopez (2003), who stated that the government did "not have the capacity to meet the financial requirements needed by a body of experts and professionals tasked to revalidate the reported increase of arsenic level in Mount Apo area." The provincial government of Cotabato then vowed to raise at least P500 000 for the study.

In 2002, Maramba and her team from the National Poison Control Information Service returned to the scene and took not only samples of soil, water, plants and air from the 10-km area around Mount Apo, but also bioassays (blood, nails, hair and urine) from the residents of the six villages. Presenting results of their study in 2003, Maramba pointed out that the levels of arsenic were elevated, but her team could not directly say if the monitored high levels of arsenic were in fact directly caused by the operations of the state-owned Philippine National Oil Corporation, because arsenic is endemic to the volcanic geology of the area. However, Maramba's team also found high levels of arsenic in the groundwater of the barangay of Ilomavis, a village where Philippine National Oil Corporation geothermal plants are located, which seems to support her hypothesis (Tupaz, 2004).

In August 2005, arsenic levels in the soil samples taken from the Mount Apo area in Kidapawan City and nearby Makilala town (population 54 463) exceeded the risk-based concentration for soil, with the highest record in the Lake Agco relocation site. The Lake Agco site lies within the Mount Apo National Park and is home to 39 barangay communities, which, according to the 1997 census, numbered 25 252 individuals (Mount Apo National Park, 2005; Manar, 2005).

As Jenny Webster (1999) has reported, the Marbel River–Matingao River catchment that drains the Mount Apo geothermal field and the power development project site is not at risk in its upper reaches for arsenic contamination, but it is at risk in its lower reaches through the influent of natural hot springs. Although the rivers are no longer used for drinking-water, residents of the catchment area may still be ingesting arsenic through their consumption of aquatic plants and animals (Webster, 1999).

3.24 RUSSIAN FEDERATION (A, C, M, V)

The Russian Federation, with a population of 138.7 million, has a Human Development Index rank of 65 out of 169 and a per capita GDP of $US 15 258(US Census Bureau, 2011; UNDP, 2010).

Arsenic-enriched coals are found in several geological settings. In the Gorlov basin, as a result of both epigenic hydrothermal fluids and post-Permian intrusions into Permian strata, arsenic-enriched anthracites measure up to 200 mg/kg. In the Palaeogene lignites of the Taldu-Dyurgun deposit in the Altai Mountains, arsenic-enriched coals measure between 32.8 and 201 mg/kg, with an average of 148 mg/kg, and the ash of 32.8 mg arsenic/kg coals measures 113 mg/kg. In the Chulymo-Enissey basin, western Siberia, enriched, weathered arsenic coal fragments are in association with grey-coloured alluvial sands rich in uranium (Arbuzov, 1998; Arbuzov *et al.* 2000; Potseluev *et al.* 2001). It has been shown that in the brown coals of western Siberia's Kansk-Achinsk basin, Krasnoyarsk, where Jurassic brown coal has become cracked and friable, the uranium content has increased, with concentrations ranging from 100 to 1000 mg/kg. In these locations, signs of underground fires were visible, suggesting that burnt organic matter was the source of enriched uranium and, by extension, perhaps arsenic as well (Gavshin & Miroschnichenko, 2000; Baboshkina *et al.* 2003). In the Tuva Federal Republic, enriched-arsenic coals are located in the Ulughem basin, especially in thin seams in the Mezhgey and Eleget deposits, with concentrations up to 84 mg/kg and with an average of 7.8 mg/kg (Arbuzov *et al.* 2001). Tests on the grains and hay fed to cattle in the area indicate that they bioaccumulate elevated concentrations of arsenic from coal emissions from both domestic settings and coal-fired power plants (Chysyma *et al.* 2003).

Ores are continually forming near the discharge centres of geothermal springs enriched with arsenic as well as zinc, copper, antimony and mercury. Precipitation of arsenic sulfide phases has been described in several geothermal systems around the circum-Pacific rim, including Kamchatka, Japan, the Philippines, New Zealand, Argentina and the Pacific coast of the USA (White *et al.* 1963; Karpov & Pavlov, 1982; Spycher & Reed, 1989; Seki, 2000; Welch *et al.* 2000). Detailed arsenic geochemistry provides a basis for understanding arsenic–sulfur–oxygen system reactions (Mironova & Zotov, 1980; Mironova *et al.* 1984, 1990; Webster, 1990; Pokrovski *et al.* 1996; Zakaznova-Iakovleva *et al.* 2000), thermodynamics (Spycher & Reed, 1989; Zotov *et al.* 1994; Pokrovski *et al.* 1996; Shock *et al.* 1997; Nordstrom & Archer, 2002) and theoretical modelling (Spycher & Reed, 1989; Helz *et al.* 1995; Tossel, 1996). While these studies are essential to understanding the parameters of potential arsenic contamination of ground waters in volcanogenic and geothermal areas, further detailed treatment of these issues is beyond the scope of the present work. Despite their natural occurrence, geothermal sources of energy remain largely untapped in the Russian Federation. Fossil fuel power stations currently produce 69% of the total electric power, hydroelectric and nuclear stations produce 20% and 11%, respectively, and geothermal power stations produce only 0.06%.

In 1995, the Moscow-based Union for Chemical Safety began a study to identify the ecological and medical problems created by Soviet-era chemical weapons. One of these, lewisite, is an arsine (AsH_3), the simplest compound of arsenic. It is flammable and highly toxic and used as a chemical weapon because of its lethality. Lewisite is an oily, brown liquid with a distinct geranium-like odour, acting as a skin irritant, blistering agent and lung irritant, and it can be used in combination with mustard gas. Lewisite can easily penetrate ordinary clothing and even rubber. Sufficient absorption can cause systemic poisoning, leading to liver necrosis or death. Lewisite can leach into surface water and groundwater if improperly stored.

Professor Lev Federov, President of the Union of Chemical Safety, wrote a book-length manuscript on the findings, subsequently translated into English as "The Ecological Consequences of Chemical Armament in Russia", but it was never published (Federov, undated). According to Federov's estimate, the former Soviet Union produced 200 000 tonnes of chemical munitions between 1924 and 1987, many of them arsenic based. The chemical weapons, according to Federov, are stored mainly underground at 350 sites across the Russian Federation, with a smaller but more vulnerable marine stockpile of 160 tonnes of early chemical weapons in the Baltic Sea. Another 150 000 tonnes of mustard agents and arsenic-bearing lewisite and possibly the nerve gases tabun and sarin appear to have been dumped into the Barents and Kara Seas between 1945 and 1982 (MacKenzie, 1998). "It is unlikely that anything more than local damage resulted, as chemicals break down in seawater", wrote Deborah MacKenzie, a correspondent for *The New Scientist*. "However, the water is close to freezing temperature, and arsenic is more persistent" (MacKenzie, 1998).

The two studies of Federov and MacKenzie are specific about certain geographical areas where the weapons are stored but are less specific on quantities, specifically of arsenic-bearing chemical agents. The Russian Federation's official stockpile, declared under the terms of the Chemical Weapons Convention, is only 36 000 tonnes, although other estimates suggest that the true amount of chemical weapons currently stored secretly in the former Soviet Union is over 350 000 tonnes.

In the United States Senate, there was general agreement that remediation should be supported through USAID, and funds were forthcoming in 2002 to begin the identification and remediation process. Under Title II of the *Department of Defense Appropriations Act, 2004* (P.L. 108-87, 30 September 2003) dealing with the "Former Soviet Union Threat Reduction" (Martin, 2004) is an appropriation of:

> $450,000,000 to remain available until September 30, 2006 for the republics of the former Soviet Union, including assistance provided by contract or by grants, for facilitating the elimination and the safe and secure transportation and storage of nuclear, chemical and other weapons. This is also for establishing programs to prevent the proliferation of weapons, weapons components, and weapons-related technology. This is also for programs relating to the training and support of defense and military personnel for demilitarization and protection of weapons, weapons components and weapons technology and expertise, and for defense and military contacts.

Kamchatskaya Oblast

Kamchatka, the largest of the Russian Federation's peninsulas, extending about 1200 km meridionally, lies on the tectonic boundary of the Pacific plate subducting beneath the Eurasian plate–hence its extreme volcanism. The two major volcanic ridges, the Sredinny (Median) and the Eastern, converge in the active Eastern Volcanic Plateau, a 700- to 900-m-high terrain with 22 (of the 28) active volcanoes. Porous tephra allows infiltration of precipitation and geothermal runoff into surface water and groundwater as well as into lakes that were once volcanic craters (Ivanov, 2003). The potential for arsenic contamination is high. More than 150 clusters of thermal springs and 500 geothermal features, which are relics of Pleistocene acidic volcanism, are concentrated mostly in the Eastern Volcanic Plateau.

The largest formation is the Uzon-Geyser tectonic–volcanic caldera 15 m × 7.5 km. A geothermal power plant designed to tap into this geothermal resource has a heat flow of 140 000 kcal/s. The Pauzhetka power plant (11 MWe) has been joined by the new Severo-Mutnovka plant (12 MWe) and two other units of 50 MWe on the Severo-Mutnovka hydrothermal system; when combined, they are projected to produce 300 MWe. Effluent from these hydrothermal plants is of concern, since geothermal waters are very high in arsenic. Arsenic in the range 100–5900 μg/l is average for thermal springs and active volcanoes in this region (Ivanov, 2003). The Uzon caldera is now being studied to assess arsenic transport in an active geothermal field, especially the distribution of the aqueous and solid arsenic–sulfur phases, their relative stability and stoichiometry in relation to physical and chemical changes in the precipitating fluids (Cleverley *et al.* 2003).

Kliuchevskoi volcano erupted on 30 September 1994. Kamchatka River is the down slope runoff destination for any arsenic compounds leaching from the lava flows into the surface water riverine environment. The 1994 eruption included massive ejections of gas, vapour and ash, which reached altitudes of 20 000 m. Melting snow triggered mudflows on the north flank of the volcano, which may have threatened agricultural zones and other settlements in the valley to the north.

The Northern subprovince includes 16 thermal spring clusters, including Tymlat, Palana and Rusakov, which are most attractive for geothermal use. In the Median subprovince, 24 hydrothermal springs include Esso and Anavgai springs and two high-temperature systems, Kireuna and Apapel. About one third of all springs are located in the Eastern subprovince. Thermal springs in the Nalycheva River valley provide water for agriculture and swimming pools. The Karym, Semyachik, Uzon and Geyser valleys in this subprovince are located in Kronotskii Reservation, where any industrial or agricultural activity is forbidden. Finally, 55 hydrothermal springs connected partially with the Pauzhetka, Nizhne-Koshelev, Severo-Mutnovka and Bolshe-Bannoe high-temperature systems are found in the Southern subprovince.

Several geothermal springs in Kamchatskaya have high mineralization, including arsenic. Viluchinski volcano is south-west of Petropavlovsk-Kamchatskiy, behind Avacha Bay. The 2175-m volcano is not active, and at its base is a deposit of arsenic and other metalloids that leach into thermal springs with temperatures ranging from 40°C to 60°C. Another thermal spring is Nalychevo, the biggest in Kamchatka, which discharges between the Goryachaya and Zholtaya rivers in a zone of more than 2 km^2. It is situated at the foot of the Kruglaya Mountains, with arsenic deposits spanning more than 50 000 km^2. The springs are weak acid or weak alkaline (pH 5.9–7.8), with an average mineralization of 3.7–12 g/l, which includes iodine and arsenic. Taloviye thermal springs, 6 km from Nalychevo, has arsenic-laced waters (>50 μg/l) that discharge over 1 km^2 with temperatures of 31–38°C. Malkinski thermal hot springs are 5 km east of Malki village, on the left bank of the Kluchoyvka River (the left tributary of the Bistraya River). Its springs are hydrocarbonic–chloride–sulfate–sodium, high in arsenic acids and with a general mineralization up to 0.7 g/l (Kamchatka's Vision, 2004). Verchne-Paratunski thermal springs discharge into the Paratunka River, which descends 2.5 km before emptying into the Karimshino River. Sixteen thermal springs on the slopes of Goryachaya hill descend to the valleys of these two rivers in three parallel streams, finally cascading in a warm waterfall. The temperature of water in the springs is about 39–70°C. The water contains chlorides, sulfates, calcium and sodium, with a general mineralization of about 1000 μg/l; the arsenic concentration in the springs is 30 μg/l (Layla Tour, 2006).

Kurile Islands

Dissolved compounds of arsenic have been studied not only in Kamchatka but also on the Kurile Islands (Ozerova *et al.* 1971), with concentrations in the range 100–5900 μg/l for thermal waters, comparable with those of Iceland (White *et al.* 1963). Two small geothermal power plants are in operation in the Kuriles (Kunashir and Iturup islands), with a capacity of 2.6 and 6 MW, respectively.

Mendeleev andesitic-dacitic strato volcano has two nested calderas 7 and 3 km, respectively, above the Goriachy Pliazh geothermal field. The Baransky (Sashiusu-dake) andesitic-dacitic volcanic complex on Iturup Island (45°5′50″N, 148° 1′27″E) is an eroded Pleistocene volcano capped by a Holocene strato volcano with a young summit lava dome cut by a north-west chain of explosion craters. The only historical and weak eruption occurred in 1951. The south-west geothermal field contains hot springs and geysers that experienced a small hydrothermal explosion in 1992 near an exploratory well in the south-west geothermal field (Smithsonian Institution, undated, a).

An associated geothermal phenomenon is the shallow water submarine hydrothermal release of arsenic that has been observed in the Kurile Islands, in Papua New Guinea and on the island of New Britain (Tarasov *et al.* 1986, 1999; Price & Pilcher, 2005).

Tuva Republic

The Tuva Republic in the Russian Federal Okrug of eastern Siberia, with an area of 1 705 000 km^2 and a population of 309 000, is situated in a 600-m-high mountain basin surrounded by the Sayan and Tannu-Ola ranges capped by Mongun Tayga at 3970 m, the highest mountain in Siberia. Among its 8000 rivers, mostly Yenisei River tributaries, there are numerous springs that leach arsenic-bearing ores, which pose a threat to the safety of the groundwater. A 2001 report detailed the arsenic leaching of the Khovu–Aksy mine tailings into the groundwater (Gaskova *et al.* 2001).

Bryanskaya Oblast

Bryanskaya oblast in the central Russian Federal Okrug, with an area of 349 000 km^2 and a population of 1 480 000, borders Ukraine to the south and five oblasts to the west, north and east. It was drastically affected by caesium fallout from the 1986 Chernobyl disaster and is now under threat from stored chemical weapons, including arsenic-bearing lewisite. There are a number of rivers in the Bryanskaya Oblast that, if contaminated with arsenic as well as other chemical agents, threaten the public health of major cities and towns.

Republic of Chuvashia

The Republic of Chuvashia in the Russian Federal Okrug of Privolzhsky in the Volgo-Vyatsky region has an area of 18 300 km^2 and a population of 1 361 000. The main river, which is dammed into a reservoir, is adjacent to the city of Cheboksary. Potential chemical weapons stockpiles near this watershed pose a threat to the drinking-water of the city (Federov, undated).

Kurganskaya Oblast

Kurganskaya Oblast in the Russian Federal Ural Okrug is bordered by Kazakhstan to the south, Chelyabinskaya Oblast to the west, Sverdlovskaya Oblast to the north and Tyumenskaya Oblast to the east. It has an area of 71 000 km^2 and a population of 1 112 000. There are two rivers, of which the Tobol is the main one, and there are several lakes in the south-east and south-west that are vulnerable to chemical weapons contamination. The most important stockpile is located near the town of Schychje, and the destruction of its chemical weapons, which is under the supervision of the Volgograd Institute, probably will not meet the international deadline of April 2007; as of May 2006, only development work was under way. In addition, local people have voiced their concerns for their health and ecological consequences once the plan for the destruction of the weapons is in full operation (Gyschin, 2006). As of 14 June 2006, the programme was being contemplated, but the deadline was still considered unrealistic (Gyschin, 2006). There is no further information available on the program of remediation as of July 2012.

Novgorodskaya Oblast

Novgorodskaya Oblast is in the Northwestern Russian Federal Okrug. It has an area of 55 300 km^2 and a population of 743 000. The major rivers in the oblast drain into Ilmen Lake, which would be at risk if chemical weapons stockpiles leached into the groundwater.

Penzenskaya Oblast

Penzenskaya Oblast in the Russian Federal Okrug of Privolzhsky, with an area of 432 000 km^2 and a population of 1 562 000, is at risk from contamination of the Surskoye Reservoir near the city of Penza where chemical weapons were manufactured. Drinking-water would be threatened if stockpiles leached into the groundwater.

Samarskaya Oblast

Samarskaya Oblast in the Russian Federal Okrug of Privolzhsky in the region of Povolzhsky covers an area of 53 600 km^2, with a population of 3 312 000. The northern sector of the oblast has a riverine oxbow that has been dammed into two major reservoirs, the Saratovskaya and the Kuibyshevskoye, both of which border the cities of Syzran, Chaaevsk, Pokhvistnevo, Zhigulevsk and Togliatti. The adjacent oblasts of Saratovskaya and Ulyanovskaya would also be at risk if chemical weapons stockpiles in the area leached into the groundwater.

Saratovskaya Oblast

Saratovskaya Oblast in the Russian Federal Okrug of Privolzhsky in the region of Povolzhsky has an area of 100 200 km^2 and a population of 2 739 000. The major source of drinking-water is the Volgogradskoye Reservoir, which bisects the oblast, and therefore its contamination would affect most of the population, especially those who live in Engels, Saratov, Volsk, Kvalynsk and Balakovo. Saratov on the Volga has arsenic contamination of the groundwater (Ivanov, 1995).

Volgogradskaya Oblast

Volgogradskaya Oblast in the Southern Russian Federal Okrug has an area of 113 900 km^2 and a population of 2 704 000. Two major watercourses bisect the oblast, Tsumlanskoye Reservoir and the Volgogradskopye Reservoir, which, if contaminated with leaching chemical weapons, would affect the water supply of the entire oblast.

Republic of Udmurtia

The 42 100-km^2 Republic of Udmurtia in the Russian Federal Okrug of Privolzhsky in the Ural Mountains has a population of 1 639 000 people who depend for most of their drinking-water on dammed sources of the three rivers that are most at risk for arsenic contamination because of Udmurtia's prominent role in the chemical weapons programme: the Vala, the Kilmez and the Cheptsa. The stockpiles in the town of Kambarka in Sverdlovskaya Oblast began remediation in May 2006 as the programme was gearing up to full capacity (Gyschin, 2006).

Republic of Dagestan

The Republic of Dagestan, situated in the North Caucasus Mountains, is the southernmost part of the Russian Federation. The most active hydrothermal resources are used in Krasnodar territory, where the arsenic concentration in the effluent is high but its fate and transport are unknown (Gadzhiev *et al.* 1980). The neogene hydrothermal area in the Republic of Dagestan has 5 known hydrogeothermal zones and 10 prospective zones that are being considered for further development; as is the case with neogenic hydrogeothermal sources, arsenic contamination in effluent from such extraction processes is always a risk. As Kurbanov (2003) indicated: "vapor thermal (150–230°C) rare-metal brines, thermal (70–120°C) weakly mineralized (3–15 g/l) subterranean waters and weakly thermal artesian waters" are found in these zones.

3.25 SAUDI ARABIA (G, P)

Saudi Arabia with a population of 28 million has a human development index of 56 out of 169 and a per capita GNI of US$23,274 (US Census Bureau, 2011; UNDP, 2010).

The Jubail petrochemical complex covers an area of about 170 km^2 on the southern Saudi coast of the Arabian Gulf. About 40 km^2 was landfilled with sand from the nearby desert to construct petrochemical industries. There has been considerable growth in the size of the complex since 1992. This is impacting the shallow aquifers (which are up to about 30 meters in depth). These shallow aquifers are inter-connected with the Arabian Gulf so that contamination of them may adversely impact terrestrial vegetation and the Gulf marine ecosystem. Reports indicate that these shallow aquifers are threatened by leaching of the industrial effluent containing arsenic into the soil and the marine ecosystem. In fact, shrimp from the Jubail area is contaminated with arsenic (Alam & Sadiq, 1993).

Groundwater aquifers in the area around Mecca are also a possible source of arsenic contamination. In a highly publicized 5 May 2011 BBC investigative report, water from the holy Zamzam well was tested for arsenic and found to be positive. While Saudi Arabian authorities deny this allegation, the use of the well by all devout Muslims journeying to Mecca suggests that the report should be taken seriously, since the number of impacted individuals is of the order of 1.7 million for the Mecca urban area which swells to over 3 million during the yearly Dhu al-Hijjah or Hajj. (Lynn, 2011).

3.26 SRI LANKA (A, G)

Sri Lanka, with a population of 21.3 million, has a Human Development Index rank of 91 out of 169 and a per capita GNI of $US 4886 (US Census Bureau, 2011; UNDP, 2010).

Sri Lanka's mineralization is not conducive to arsenic leaching into groundwater except where there are copper and iron deposits, such as in the Saruwile area in the south-west near Colombo, where there is a petroleum distillation site. The country is known for its precious and semiprecious gemstones, such as beryl, chrysoberyl, garnet, ruby, sapphire, spinel, topaz and tourmaline. Graphite is a major industrial mineral, along with coastal mineral sands rich in ilmenite, monazite, rutile and zircon (Kuo, 1996).

Sri Lanka has had floods and landslides, but none of the magnitude of the 2004 tsunami that killed 40 000 people and destroyed nearly 100 000 houses along with thousands of vehicles and infrastructure. "Tsunami" is a Japanese term for a wave created by an earthquake that breaks on shore. The 2004 tsunami waves, which were up to 9 m high and had the momentum and surge of deep water, brought up black sediments that had resulted from Gangetic–Brahmaputra outwash from West Bengal, India, and Bangladesh over the last 10 000 years. These are the same sediments that are responsible for the arsenic crisis in those areas. These Gangetic black sediments have penetrated from tens to hundreds of metres inland (Caldecott & Wickremasinghe, 2005).

The effects of toxic exposure to this deep sea sedimentation that had now been brought to the surface were immediate. In the south-west town of Matara, residents complained of itchiness after contact with wave water. In the south-east town of Hambantota, a slimy water texture was reported (MENR, 2005). An early inference was that marine sediments that contained unfamiliar and possibly toxic chemicals had been washed ashore (Senanayake, 2005). Subsequent analyses have established that contamination by arsenic has occurred in some of the tsunami-affected areas, occasionally at seawater concentrations approaching 10 μg/l. This concentration for seawater is very high, since normal seawater contains only 1 μg/l or less (Caldecott & Wickremasinghe, 2005).

The public health impact of this arsenic contamination is enormous, because the waters have receded and now the land and the littoral areas have become the repository of arsenic contamination. At risk are agricultural crops and aquatic plankton, fish and other animals in the food-chain that will ingest this toxin. Remediation also poses a threat, since accumulated debris concentrates arsenic, which can then leach over time into groundwater. The prognosis is that testing for persistent pollutants in tsunami-affected areas is not only "desirable" (as reported by Caldecott & Wickremasinghe, 2005) but critical to the health of the countries affected by the 2004 tsunami's deep sedimental surge.

Seemingly minor incidences of arsenic poisoning have been reported in the literature. In 1969 and 1972, epidemiologists studying the neuropathy of Mee's lines (white bands across fingernails indicative of long-term exposure to arsenic concentrations above 50 μg/l) found that patients had ingested or been exposed to arsenic. The patients were in two cohort groups, one with long-term ingestion of illicit "liquor" and the other with long-term ingestion of arsenic-contaminated well water (Munasinghe *et al.* 1969; Senanake *et al.* 1972). The former group may have ingested liquor contaminated by high-arsenic ground waters, possibly from the same area of exposure for the latter group, in a zone of presumed iron and copper mineralization that is known to occur on the east coast in Seruwila, Trincomale District. It is also possible that the source of arsenic was pesticides, which have been a known threat in Sri Lanka (Roberts *et al.* 2003). Another source of chronic arsenic poisoning in Sri Lanka is the ubiquitous arsenic-laced opium (Jayasinghe & Misbah, 1983). No further follow-up on these cases could be located in the published literature.

3.27 TAJIKISTAN (C, M)

Tajikistan, with a population of 7.6 million, has a Human Development Index rank of 112 out of 169 and a per capita GNI of $US 2020 (US Census Bureau, 2011; UNDP, 2010).

Coal-related arsenic is found in the Jurassic Nazar-Ailock anthracite deposit near the tin–mercury ores of the Khshert deposit, where hydrothermal fluids are believed to leach from these ores bordering the Tajik basin and penetrate the coal seams (Valiev *et al.* 1993).

Arsenic occurs in association with gold and other minerals in the highly mineralized Zeravshan Valley, in Leninabad Oblast in far north-west Tajikistan, east of Samarkand in Uzbekistan, south of Tashkent in Uzbekistan and near the ancient Silk Road. Zeravshan Valley runs east from Samarkand for some 250 km and was the corridor of ancient travel and commerce. The guardians of the corridor are the 4000-m Turkestan range to the north and the 5550-m Zeravshan range to the south (Chadwick, 1996).

The Taror-Jilau goldfield in the Zeravshan Valley is south-east of Panjakent on the left (south) bank of the Zeravshan River. Here the Jilau open pit and Taror mines are located. The Taror mine has been in production since 1937, when it was opened as an arsenic mine, but after World War II the former Soviet Union opened it as a gold mine. Naturally, the ore was high in arsenic, so leachate from the slag, which was dumped into an unlined tailings dam, was gravity fed to the local Shing and Magyar rivers, tributaries of the Zeravshan. Testing and remediation of the valley and river sediments have not been accomplished, so it is to be expected that arsenic has contaminated much of the surface water and groundwater, since such large volumes of leachate were emitted in the period prior to the current operations.

After 1989, when the Russian engineers left, because the Tajiks were not mining engineers, they reopened the Jilau open pit, which was easier to access, and commenced processing the higher-grade gold ore through crushing, grinding and flotation. However, this ore was high in arsenic, and the tailings from this operation began leaching into the surface water and groundwater. This operation continued until 1996.

To the credit of the present mine operators, the tailings dam has been extensively retrofitted, and a 300-mm-thick clay membrane lined with a polyvinyl chloride membrane below the hydrostatic head has been installed to curtail leaching. In addition, a floating barge installation was constructed for water recycling.

3.28 THAILAND (A, G, M)

Thailand, with a population of 63.7 million, has a Human Development Index rank of 92 out of 169 and a per capita GNI of $US 8,001 (US Census Bureau, 2011; UNDP, 2010).

Arsenicosis in Thailand is related to mineralization and mining activity rather than alluvial and deltaic aquifers, since the area lies within the South-east Asian tin belt. Primary tin–tungsten arsenic mineralization and alluvial placer tin had been mined for over 100 years, with arsenopyrite-rich waste piles and waste from village prospector panning. Waste piles from former bedrock mining contain $\leq$30% arsenic (Williams *et al.* 1996), and alluvial soils contain arsenic at $\leq$0.5% (Fordyce *et al.* 1995).

The affected aquifer is Quaternary alluvial sediment that has been extensively dredged during tin mining operations. The mobilization of arsenic in the groundwater is thought to be due to oxidation of arsenopyrite, exacerbated by the former tin mining activities (Smedley & Kinniburgh, 2002). Arsenic mobilization in groundwater has occurred during post-mining groundwater rebound (Williams, 1997), leading to the contamination of many shallow wells. Arsenic concentrations up to 5000 µg/l have been reported (Smedley & Kinniburgh, 2002). Tin mining has led to arsenic pollution of surface waters, with levels up to 583 µg/l reported (Smedley & Kinniburgh, 2002).

The affected area, in the Ron Phibun District in Nakhon Si Thammarat Province, southern Thailand, lies within the South-east Asian tin belt and is related to mining activity. Ron Phibun arsenic contamination differs from that in many other high-arsenic groundwater provinces in Asia with its shallow aquifer and more oxidizing conditions, distinguishing it from Bangladesh and West Bengal, India. Deeper aquifers of Ron Phibun are more similar to those from other high-arsenic Asian aquifers. The occurrence of arsenic in drinking-water has been attributed to the fact that tube wells were drilled too deep (Pavittranon *et al.* 2003) or that leaching from slag in the mining of tin was a factor (Foy *et al.* 1992/1993). Geologically, the Ronna Mountains are rich in arsenic-bearing minerals, and these leach into the Ronna River, thereby contaminating the drinking-water (Siripitayakunkit *et al.* 2001).

Mining activity in Ron Phibun District was first recognized in 1987 as a source of arsenic poisoning, with 1000 cases diagnosed with arsenic-related skin disorders (Williams, 1997). Fifteen thousand people were exposed to arsenic concentrations above 50 µg/l (Fordyce *et al.* 1995). Arsenic in both surface water and shallow groundwater is thought to be leached by oxidation of arsenopyrite, exacerbated by a rise in groundwater levels (Williams, 1997). The highest arsenic concentrations (≤580 μg/l) were 2–7 km downstream of the mining area (Figure 3.27; Williams *et al.* 1996). Shallow ground waters (<15 m) are from alluvial and colluvial deposits, and deeper ground waters (>15 m) are from an older carbonate aquifer. The shallow aquifer shows the greatest contamination with arsenic, with concentrations up to 5100 μg/l. In the shallow aquifer, 39% of samples collected randomly had arsenic concentrations above 50 μg/l, while in the deeper aquifer, 15% exceeded 50 μg/l (Table 36.1; Williams *et al.* 1996). In another arsenic analysis, 12.7% of shallow wells in the Ron Phibun area were shown to contain levels greater than 50 μg/l (Oshikawa *et al.* 2001).

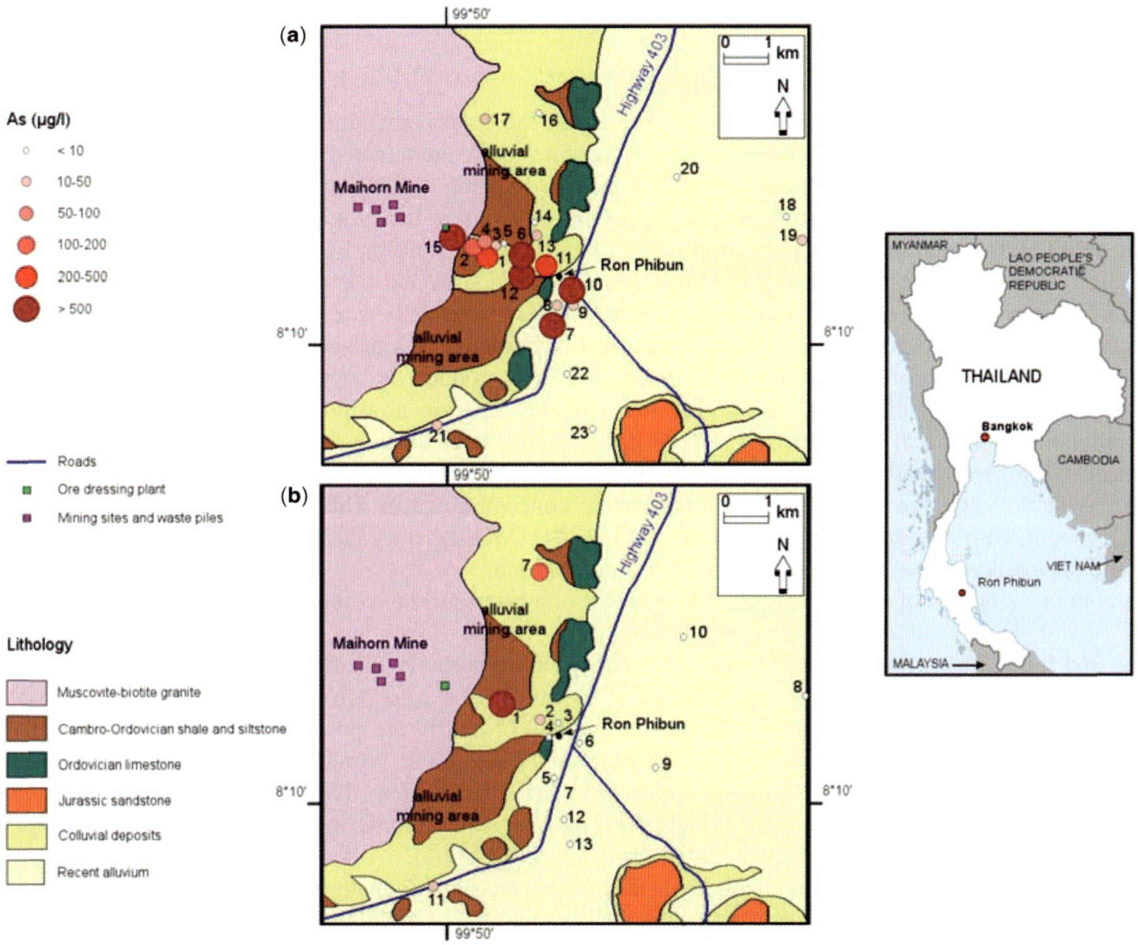

Figure 3.27 Distribution of arsenic in analysed ground waters in the Ron Phibun area, Thailand. Note: The distributions are (a) arsenic in groundwater from shallow tube wells (<15 m depth); (b) arsenic in groundwater from deeper tube wells (>15 m). Numbers refer to samples given in Williams *et al.* (1996) (modified from Williams *et al.* 1996).

Based on the Ron Phibun population (approximately 15 000) and data on the number of affected wells (12.7%), it is estimated that 2000 people are exposed to elevated levels of arsenic, assuming that each well serves a similar number of people (Chongsuvivatwong *et al.* 1997). In a 2000–2002 study of 1051 households, the population at risk was estimated to be 18 000 (Pavittranon *et al.* 2003). This figure has subsequently been updated to 250 000 at risk, with 2000 skin cancers attributed to arsenicosis. The case rate has been determined to be 24.7% of the total population in high-risk areas (Pavittranon *et al.* 2003), with additional projected cases of arsenicosis estimated at 18 600 (Caussy, 2004). In a study of seven villages, only three of which submitted data, arsenic concentrations in drinking-water ranged from 9.2–25.5 to 30.5 μg/l (Chongsuvivatwong *et al.* 1997).

Table 3.30 Frequency distribution of arsenic concentrations in water from the Ron Phibun area, Thailand.

Aquifer	Number of samples (%)			Total number of samples
	As < 10 μg/l	As 10–50 μg/l	As > 50 μg/l	
Surface water	1 (4)	2 (8)	20 (83)	24
Groundwater from shallow aquifer (<15 m)	7 (30)	7 (30)	9 (39)	23
Groundwater from deeper aquifer (>15 m)	9 (69)	2 (15)	2 (15)	13

Source: Williams *et al.* 1996; World Bank, 2005a

Food security and arsenic tainted rice

In countries such as Thailand with concerns for food security, the issues surrounding agricultural produce irrigated with arsenic-tainted water are serious ones. Besides consumption of arsenic contaminated drinking water, contamination of crops irrigated with high arsenic water is a danger throughout the rice-growing regions of Southeast Asia (Achin, 2011). During the lean summer months from January to May a large amount of groundwater is used for irrigation which is contaminated with arsenic resulting from the oxygenated decomposition of arsenopyrite forming Fe^{++} and Fe^{+++} sulfates and H_2SO_4 acid which mobilize arsenic. The mobilized arsenic is then accumulated in soils and taken up by crops like rice. While the exact mechanism of arsenic release is controversial (following oxidation of arsenopyrite, reductive dissolution of Fe hydroxides or anion exchange of arsenic with fertilizer phosphates) its effect is dramatic for it provides another pathway – through the food chain. Arsenic contamination of agricultural soils, crops and farm products is coupled with the arsenic contamination of groundwater problem. In fact the levels of arsenic in Thai paddy rice are now possibly on an increasing trajectory. For instance, Williams *et al.* (2005) reported 0.11 ± 0.01 g/g dry weight of arsenic in Thai rice and in a study a year later Williams *et al.* (2006) showed that the concentration of arsenic in Thai rice was 0.10 g/g dry weight (range 0.06-0.14 g/g dry weight). A survey two years later revealed that arsenic concentrations in Thai paddy rice ranged between 0.01 and 0.39 g/g dry weight. with a mean of 0.14 g/g dry weight (n= 54) (Meharg *et al.* 2009). Clearly, this issue must be followed to monitor the potential increase in arsenic concentrations in Thai rice.

Another complex situation of arsenic contamination is that of a reservoir that could be contaminated by either mining or a high-arsenic coal-burning source: a power generation facility. Arsenic fluxes and arsenic distribution in tailings from a lignite mine and the effluent from a power generation facility (in northern Thailand) were studied by Bashkin & Wongyai (2002). The investigators reported high concentrations of arsenic in the Mae Moh Reservoir, which received acid mine drainage sediment (5213 mg/kg sediment). Geochemical sampling of the fly ash pond also showed increased arsenic levels, which could also have contributed to the elevated levels of arsenic (Bashkin & Wongyai, 2002).

Arsenic-related health effects due to drinking-water were first identified in Thailand in 1987 (Williams *et al.* 1996; Choprapawon & Rodcline, 1997). The main effects reported are skin lesions, including skin cancer (Choprapawon & Rodcline, 1997; Choprapawon & Porapakkham, 2001; Oshikawa *et al.* 2001).

There is likely to be some exposure through mining in the affected area, although no specific reports of health effects from this route of exposure have been located. Arsenic is also contained in a number of traditional medicines (Choprapawon & Rodcline, 1997). According to Oshikawa *et al.* (2001), in 1987, over 1200 subjects who had skin lesions were identified; these lesions may have been attributable to more than one route of exposure. No estimates of the number of people exposed through mining activity or contact with solid mining wastes have been located.

The main mitigation measure instituted in this area is the provision of alternative sources of water. There has been an expansion in the provision of piped water; where this has not yet been achieved, there is extensive use of rainwater (Chongsuvivatwong *et al.* 2000). In 1997, over 93% of the households examined used safe water for drinking, although the percentages using safe water for cooking and washing food and utensils were lower (80.9%, 59.1% and 53.8%, respectively; Chongsuvivatwong *et al.* 2000). These levels, however, may not reflect the situation during the whole of the year as, in a number of cases, households using rainwater sources reported that they reverted to well use during the dry season (Oshikawa, 1997).

3.29 TURKEY (C, G, M, V)

Turkey, with a population of 78.8 million, has a Human Development Index of 83 out of 169 and a per capita GNI of $US 13 359 (US Census Bureau, 2011; UNDP, 2010). Arsenic-enriched coal beds are found where there is potential to mobilize into surface water and ground waters. The Gokler coalfield in western Turkey mainly contains high arsenic (170–3854 mg/kg, average 833 mg/kg) and antimony (8.3–2347 mg/kg, average 134 mg/kg). These deposits are associated with epigenetic stibnite (Sb_2S_3) and iron sulfides (comprising pyrite, marcasite, arsenic and zinc) that formed during post-Miocene hydrothermal mineralizations in microfractures and micropores in the coal seams (Karayigit *et al.* 2002). The lower Eocene coal seam mined in Yeniceltek, Kucukkohne and Ayridam from the Sorgun basin has arsenic-enriched coals associated with many elements and clay minerals, but arsenic, zinc, selenium and antimony are probably concentrated in pyrites and present in low to moderate concentrations (Karayigit *et al.* 2000). In the south Marmor Sea, lignites contain an average of 67 mg arsenic/kg (range 9–346 mg arsenic/kg) and 247 mg arsenic/kg in coal ash. In the Aegean Sea, lignites contain an average of 81 mg arsenic/kg (range 2–686 mg arsenic/kg) and 314 mg arsenic/kg in coal ash. In the western Black Sea, lignites average 109 mg arsenic/kg (range 19–228 mg arsenic/kg) and 493 mg arsenic/kg in coal ash. In the Mediterranean Sea lignites, arsenic averages 197 mg/kg and 488 mg/kg in coal ash (Karayigit *et al.* 2000; Karayigit *et al.* 2001).

Geogenic arsenic occurs in Turkey from arsenic minerals embedded in dense clay matrices that promote the mobilization of arsenic into the groundwater. One instance is in the Hisarcik area that encompasses Igdeköy, Emet and Kütahya villages. This area's aquifers have extremely high arsenic in the groundwater, ranging between 0.07 and 7.754 mg/l. This high arsenic level is caused by the natural dissolution of arsenic minerals, including realgar (AsS) and orpiment (As_2S_3) embedded in colemanite nodules, in a Neogene borate clay zone composed of $Ca_2B_6O_{11}{\cdot}5H_2O$. Groundwater arsenic concentrations vary from spring to spring because of the varying mineralogy and geochemistry of the lithofacies of the Neogene colemanite clay zone of the aquifer that contains the embedded arsenic minerals (Çolak *et al.* 2003).

Mining also contributes to the exposure to arsenic mineralization in Turkey. For instance, in the Tuztafl gold–silver–lead–zinc–copper mineralized zone in the Kazda dome in the Biga Peninsula in western Turkey, arsenic occurs as the most significant geochemical anomaly in the northern half of the Ayale (Ayvack-Altnova-Evciler) area. Granitoids intrude in this matrix where arsenic occurs as both a coherent marker for gold deposits and a potential leachate. Consequently, mining in this zone requires monitoring; even without extraction, some natural geogenic arsenic leaching may occur due to natural weathering processes (Yilmaz, 2007).

Volcanogenic sources of arsenic have been indirectly identified for Izmir residents dependent on the water from the thermal wells in the district of Manisa. The city of Izmir in western Turkey, with a total metropolitan population of 4 million, including a downtown population of 1.95 million, is dependent for its drinking-water on wells in neighbouring Manisa, but the citizens of the city are growing restive about the safety of the water (Molden *et al.* 2001). In fact, Manisa is located in an area of high thermal activity, with hot springs and thermal baths that go back to antiquity. Indirect proof of the elevated arsenic in the groundwater is the fact that breastfeeding mothers have elevated arsenic in their milk (4.219 ± 0.079 µg/l), and cows have even higher levels (4.932 ± 0.38 µg/l) (Ulman *et al.* 1998).

3.30 TURKMENISTAN (A)

Turkmenistan, with a population of 5.0 million, has a Human Development Index rank of 87 out of 169 and a per capita GNI of $US 7052 (US Census Bureau, 2011; UNDP, 2010). Turkmenistan is characterized by an arid, desertified terrain and has the lowest number of water sources in Central Asia. Water supplies come primarily from the human-made Karakum canal, which feeds into the Amu Darya (river), which in turn joins the Aral Sea, as well as numerous wells in the Yaskhan area. These sources are becoming overtaxed, a situation worsening due to population increases.

Overuse of the Amu Darya for irrigation has caused serious depletion of the Aral Sea, reducing its volume by almost 75% and leading to an increasing concentration of geogenic and anthropogenic toxins in the water, including arsenic and various salts. In addition, anthropogenic chemicals containing arsenic have been deposited on soils and surface waters through use of agricultural pesticides, and other anthropogenic chemicals have rendered the groundwater undrinkable. The concentration of arsenic in the Syr Darya has been measured as 1.7 mg/l (Kulmatov, 2004).

Approximately 50% of the drinking-water falls below WHO standards. Conditions are particularly serious in the Balkan district, where rationing is frequent and the occurrence of waterborne diseases is high.

Remediation is complicated by an outdated infrastructure that has caused losses of over half of the water supply through leaky pipes and lack of monitoring. As a result of a 2001 initiative, a wide assortment of activities will be implemented. Geared towards improving, rehabilitating and expanding urban and rural water systems, work will range from upgrading of the entire transmission network, distribution centres and wells to the provision of pipelines and the construction of wastewater pumping stations. Some 147 wells will be drilled, supplementing groundwater sources by an additional 35 000 m^3/day. Since the quality of this water is high, purification will be unnecessary. Additionally, a 40-km main transmission line to the city of Nebitdag will be installed. Another key component entails halting damaging environmental practices, thus helping to ensure the project's ongoing success (Anon., 2001).

3.31 UZBEKISTAN (A, G, M)

Uzbekistan, with a population of 28.1 million, has a Human Development Index rank of 102 out of 169 and a per capita GNI of $US 3085 (US Census Bureau, 2011; UNDP, 2010).

Uzbekistan is bounded by two river systems, the Amu Darya and the Syr Darya, which flow into the Aral Sea from the south-west and north-east, respectively.

Anthropogenic arsenic in Uzbekistan has its sources from discharges of untreated wastewater that flow into the tributaries of the Amu Darya from municipal and industrial sources, because most of the wastewater treatment plants are low capacity, non-functional or unused, a problem that the four governmental and municipal regulatory agencies (the Ministry of Agriculture and Irrigation, the Ministry of Energy, which controls power stations, the Tashkent City and the Municipal Water Supply and Sewage Disposal Enterprises) have been unable to control. The Chirchik River, a heavily arsenic-polluted Amu Darya tributary, is used for drinking-water for the 3.5 million inhabitants of Tashkent, as well as for irrigation of agricultural fields. The primary source of riverine and surficial arsenic pollution is from the production and use of agricultural arsenical pesticides. The production of these pesticides was located near the site of Charvak Reservoir, the sediment bed of which contains arsenic and bismuth concentrations 8 times higher than the maximum permissible levels, with increasing concentrations of zinc, chromium and copper also observed, according to the Centre of Lawyers for Environment Protection (Naumov, 2007). Bioaccumulation has been so severe in the river that riverine fishes and other aquatic life forms have been seriously compromised, according to the head of the State Biological Inspection Office, Alexander Grigoryants (Naumov, 2007), but no testing of aquatic tissue or farm produce, which are additional toxic arsenic pathways, has been undertaken.

Geogenic sources of arsenic are found mainly in the Tien Shan Mountains and the Amantaytau region as arsenides (nickeline, breithauptite, loellingite, safflorite, rammelsbergite, smaltite-chloantite), nickel safflorite, non-hydrated arsenates (rooseveltite, olivenite, adamite, austinite, conichalcite, staszicite, mimetesite, svabite) and hydrated arsenates (legrandite, skorodite, erithrine, annabergite, tyrolite, pharmacosiderite, mixite), all of which have the potential for leaching into groundwater (Badalov *et al.* 1975–1977).

Mining that has exposed these arsenic minerals has also contributed to the leaching of arsenic. In the Amantaytau region of the Kyzylkum Desert, three mining areas are associated with gold mining: Amantaytau Centralny, Amantaytau OP Gold Mine and Uzunbulak. These all have associated arsenopyrite deposits (Mindat.org, 2007b). In addition, arsenic minerals in association with gold and silver–lead ores were intensively mined in the mountain areas of Koksu and Chatkal adjacent to the Brichmulla-Nanai area, which incorporates villages situated in the south-east of the Charvak water reservoir on the western extremity of the Koksu ridge of the western Tien Shan (Embassy of the Republic of Uzbekistan in Israel, 2007).

Furthermore, the overuse of the Amu Darya for irrigation has caused serious depletion of the Aral Sea, reducing its volume by almost 75%. This drawdown has concentrated geogenic and anthropogenic toxins, including riverine and surficial arsenical compounds, such as agricultural pesticides and various mineral salts, in the Aral Sea, rendering the groundwater undrinkable. The concentration of arsenic in the Syr Darya has been measured as 1.7 mg/l (Kulmatov, 2004). Partly, these rivers course through desertified lands where the basic source of water is poor. In this region, the ground waters are most vulnerable to contamination. The rivers also traverse the mountains, where they are fed by rain, snow and glacial runoff, which is less in the autumn, winter and spring than in the summer. In the valleys in this region, the rivers supply much-needed irrigation for a variety of crops, and the drawdown of the rivers and aquifers has led to a concentration of geogenic arsenic in the groundwater. The Amu Darya's capacity is 78 km^3 per year, whereas the Syr Darya's capacity (and that of its tributaries, Chirchik, Ahangaran and Keles) is 36 km^3 per year (Kulmatov, 2004); these capacities have been steadily decreasing since 1990.

3.32 VIET NAM (G,M)

Viet Nam, with a population of 90 million, has a Human Development Index rank of 113 out of 169 and a per capita GNI of $US 2995 (US Census Bureau, 2011; UNDP, 2010).

Studies on the occurrence of arsenic in ground waters in Viet Nam have, to date, concentrated on two areas: north along the Red River (Song Hong) Delta near Hanoi (Figure 3.28) and south in the Mekong River delta (Berg *et al.* 2001). Suspicions of elevated arsenic levels were raised by the geological similarity of this region to the Ganges-Brahmaputra Delta (or Bengal Delta) in South Asia, where arsenic problems are well documented (Smedley & Kinniburgh, 2002). The affected area extends over

11 000 km^2, and levels of arsenic have been found to range between 1 and 3050 µg/l (Berg *et al.* 2001). Elevated arsenic levels were identified in 1998; however, private tube well construction began only in the mid-1990s. By June 2006, there has been growing public awareness of arsenicosis cases, and the prevalence of arsenic-contaminated groundwater has become a major public health issue (VietNamNet Bridge, 2006):

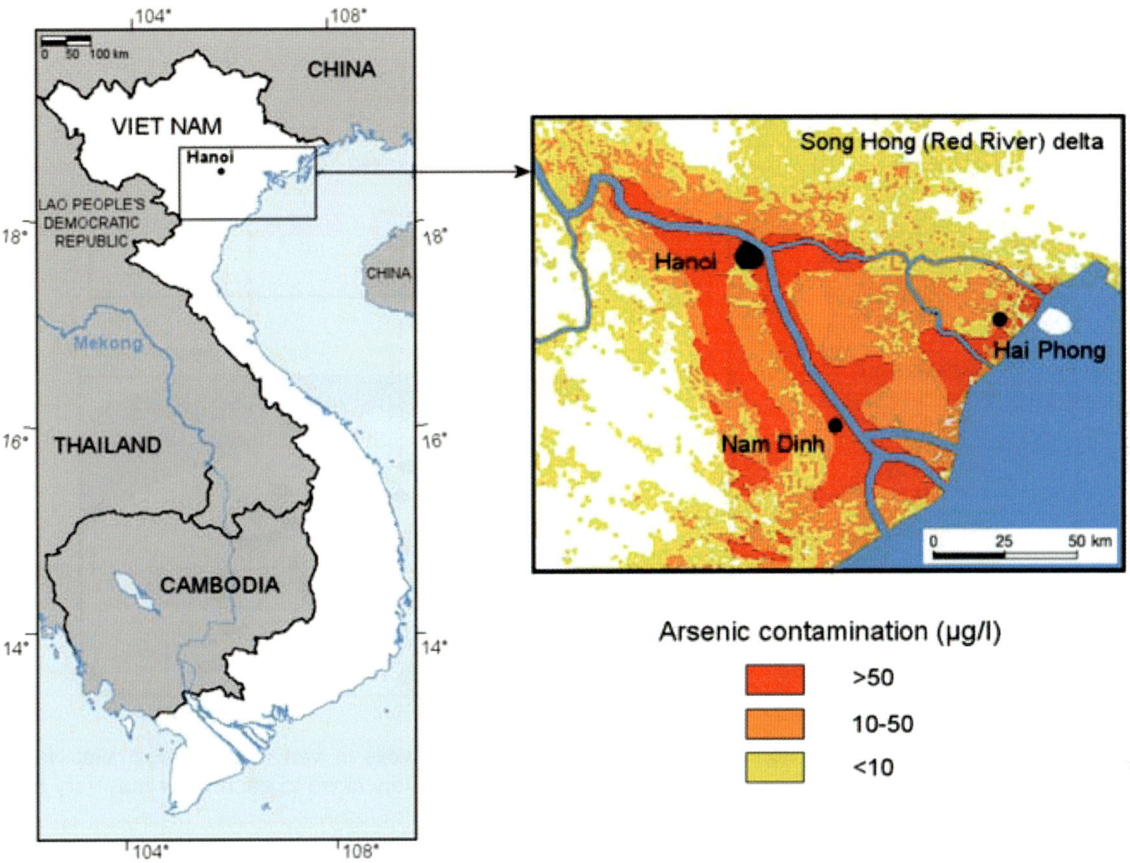

Figure 3.28 Landsat image of Red River (Song Hong) delta, Viet Nam, with five ranges of arsenic contamination indicated. This is a gross interpretation and not intended to indicate the micro level on the ground, where proximate wells may differ in arsenic concentration by as much as 250 µg/l or more (see Figure 3.29) (modified from Treng & Viet, 2005).

Recently the International Company for Chemicals and Water technologies took four samples from Thong Nhat village, Ung Hoa Commune in Ha Tay for tests.

The result was shocking, as each of the four samples contained excessive iron and arsenic, which are 17 to 29 times higher than safe levels. The company has strongly recommended that the water needs treating before use otherwise it would badly affect people's health. However, arsenic pollution is not only a problem in Ha Tay. The survey done by the Institute of Geography to assess water quality in Binh Luc, Ly Nhan, Duy Tien, Kim Bang, Thanh Liem districts in Ha Nam Province reveals that there is too much arsenic in water in these areas. According to UNICEF, the level of arsenic pollution in Vietnam is just as high as in Bangladesh, which has been identified as a country with serious arsenic pollution.

In some areas in Hai Ba Trung District of Hanoi, ground water was found polluted with arsenic in 1996. Scientists said arsenic pollution in Hanoi's water supply is increasing by five to six times every year. Dr Tran Huu Hoan of the Institute of Industrial Chemicals said Vietnam has now been marked on the world map of arsenic pollution. A survey carried out in 12 provinces collected 12,461 water samples from wells. The survey found that millions of Vietnamese people are using ground water polluted with arsenic every day. Using water polluted with arsenic causes cancer, problems with kidneys, lungs and diabetes. UNICEF has warned that arsenic pollution is putting 10 million Vietnamese at great risk. In many cases, even if households use methods to filter the water, drinking this water is still not guaranteed safe.

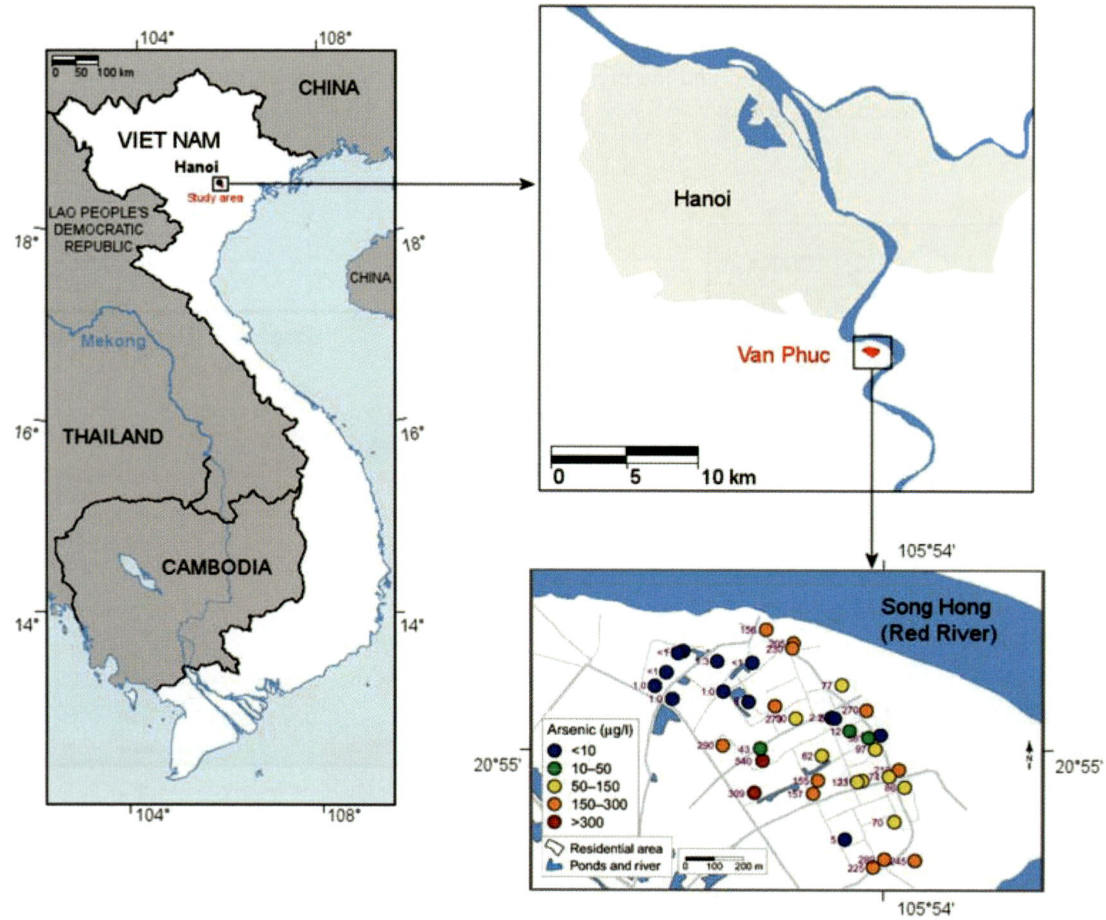

Figure 3.29 Spatial variation of arsenic contamination between proximate wells in Van Phuc Village, Viet Nam, July 2002 (modified from Treng & Viet, 2005). This figure is intended to show that wells very close to each other may vary by as much as 250 µg/l or more. Compare numbers 43 and 540.

Red River (Song Hong) and Delta

The Song Hong is 1175 km long, rises in Yunnan province, south-eastern China, and flows south-east, in deep, narrow gorges, through northern Viet Nam to form a great delta before entering the Gulf of Tonkin. It is one of the most densely populated regions in the world, with about 1160 people/km^2. Sixteen million Mio people live in this region of whom 11 million have no access to public water supplies and therefore depend on other supplies such as private tubewells. The river carries a large quantity of silt rich in iron oxide, which gives it a red colour. It has an irregular flow dependent on the seasonal monsoonal flooding during the June–October high water period; dykes and canals protect the delta from floodwaters. North-west of Hanoi, the river flows onto the coastal plain and receives the Clear and Black rivers, its chief tributaries.

The Red River Song Hong delta is 120 km long and 120 km wide and is the economic centre of northern Viet Nam, whose chief port, Hai Phong, is on the delta's north branch. Its 15 000-km^2 area is underlain by unconsolidated formations and Holocene sediments. It is one of the most developed economic areas of Viet Nam.

Economic development is closely linked to groundwater resources, and extraction of groundwater has caused a lowering of the groundwater table, causing saline intrusion and arsenic contamination in the northern provinces, of which the most affected appear to be (based on very sparse sampling by UNICEF) southern Hanoi, Ha Nam, Ha Tay, Hung Yen, Nam Dinh, Ninh Binh, Binh Dinh, Thai Binh and Hai Duong (Ha, 2004; UNICEF, 2004a,b). Figure 3.30 shows the overall levels of arsenic contamination in the Hanoi area due to the drawdown of groundwater by the concentrated population in the region. Table 3.31 presents arsenic well water test results current as of October 2004 (UNICEF, 2004a).

A round of tests produced in 2005 confirm the extent of arsenic contamination in the three provinces most at risk: Ha Tay, Ha Nam and Hung Yen (Table 3.32).

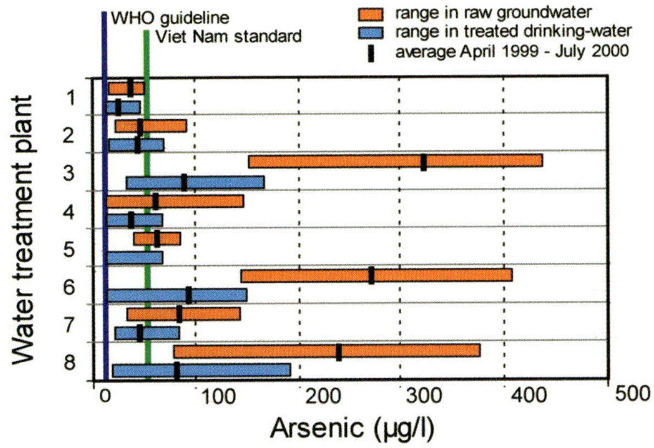

Figure 3.30 Arsenic contamination at water stations in Hanoi, Viet Nam, 1999–2000 (modified from Treng & Viet, 2005).

Table 3.31 Compilation of UNICEF arsenic test results and percentage of wells tested in Viet Nam[a].

	Province	No. wells	Number tested	% tested	No. wells >10 µg/l	No. wells >50 µg/l
1	An Giang	1 453	240	16.52	61	10
2	Binh Phuoc	n.a.	52	n.a.	0	0
3	Cao Bang	n.a.	35	n.a.	2	0
4	Dong Thap	7 780	212	2.73	88	83
5	Gia Lai	n.a.	293	n.a.	9	1
6	Ha Nam	49 000	7 042	14.37	4 517	3 534
7	Hanoi	n.a.	824	n.a.	414	199
8	Ha Tay	180 891	1 368	0.76	638	338
9	Hai Duong	57 938	480	0.83	34	3
10	Ho Chi Minh City	n.a.	240	n.a.	0	0
11	Thua Tien Hue	16 560	322	1.94	17	1
12	Hung Yen	147 933	3 384	2.29	700	310
13	Lam Dong	n.a.	50	n.a.	11	0
14	Long An	2 272	235	10.34	0	0
15	Nam Dinh	42 964	605	1.4	156	104
16	Ninh Binh	n.a.	75	n.a.	26	8
17	Phu Tho	n.a.	150	n.a.	0	0
18	Quang Nam	n.a.	546	n.a.	0	1
19	Quang Ninh	4 960	240	4.8	5	1
20	Quang Tri	n.a.	128	n.a.	14	1
21	Tay Ninh	n.a.	603	n.a.	0	0
22	Thai Binh	136 172	125	0.09	66	1
23	Thai Nguyen	n.a.	240	n.a.	7	2
24	Thanh Hoa	n.a.	347	n.a.	17	17
25	Tien Giang	n.a.	n.a.	n.a.	n.a.	n.a.
26	Vinh Phuc	n.a.	161	n.a.	0	0
	Total	**647 923**	**12 543**[b]	**1.9 average**[c]		

[a] Provinces with <35 samples are not recorded here. Dark shading indicates >1% wells >50 µg As/l; light shading indicates >1% wells >10 µg As/l.
[b] This represents the total only of those provinces whose total number of wells is disclosed by UNICEF. We were unable to establish the total number of wells for those provinces not given by UNICEF.
[c] In other words, less than 2% of all wells have been tested in those provinces where the total number of wells has been disclosed.
Source: UNICEF, 2004a,c

Table 3.32 Arsenic pollution of the surveyed communes in Viet Nam, 2004–2005, under UNICEF support[a].

Province	No. surveyed communes	No. communes where >10% of tube wells have arsenic >10 μg/l	%	No. communes where >10% of tube wells have arsenic >50 μg/l	%	No. communes where >10% of tube wells have arsenic >100 μg/l	%
Ha Tay	207	89	43.0	65	31.4	10	4.8
Ha Nam	111	110	99.1	77	69.4	50	45.0
Hai Duong	166	19	11.4	1	0.6	0	0.0
Hai Phong	75	0	0.0	0	0.0	0	0.0
Hung Yen	269	80	29.7	41	15.2	12	4.5
Nam Dinh	86	17	19.8	6	7.0	0	0.0
Thai Binh	84	6	7.1	4	4.8	0	0.0
Ninh Binh	38	6	15.8	1	2.6	0	0.0
Thai Nguyen	18	1	5.6	0	0.0	0	0.0
Vinh Phuc	12	8	66.7	5	41.7	0	0.0
Phu Tho	72	4	5.6	2	2.8	0	0.0
Bac Giang	20	3	15.0	2	10.0	0	0.0
Long An	14	0	0.0	0	0.0	0	0.0
Dong Thap	10	5	50.0	5	50.0	5	50.0
An Giang	16	7	43.8	1	6.3	0	0.0
HCMC	16	0	0.0	0	0.0	0	0.0
TT Hue	20	0	0.0	0	0.0	0	0.0
Total	**1234**	**355**		**210**		**77**	
%		**28.8**		**17.0**		**6.2**	

[a]Note the elevated arsenic levels in Ha Tay, Ha Nam and Hung Yen (highlighted).
Source: Badloe & Hoa, 2005

Because of the extent of the problem, the phenomenon of proximate wells having orders of magnitude differences in arsenic contamination and the large number of wells that need to be tested, the alternative method of bioassay is attractive. Bioassay of arsenic in hair, nails and urine is an efficient way to collect samples in the field from suspected persons exposed to elevated arsenic and to test these samples using secure laboratory methods to ensure a precise determination so that the correlation with groundwater can be corroborated later (Treng & Viet, 2005).

Arsenic concentrations in drinking-water in the north range from 240 to 430 μg/l (Ninh *et al.* 2004). Of 12 439 tube wells, which are used by 20.48% of the rural population of 16.5 million (or 3.38 million), 30% of those contain arsenic levels above 10 μg/l (affecting about 1 million residents) and 11% have arsenic levels above 50 μg/l (affecting about 371 000). In the provinces of Tu Liem and Gia Lam (west and east of Hanoi), arsenic concentrations in tube wells range from 10 to >50 μg/l. A study completed in 2004 showed that many areas in the Red River (Song Hong) Delta provinces of Hanoi, Ha Nam, Ha Tay, Hung Yen, Nam Dinh, Binh Dinh and Thai Binh, and one in the Mekong delta, Dong Thap Province, have arsenic concentrations higher than both the national and WHO standards set for drinking-water. According to Dr Nguyen Huy Nga, Deputy Director of the Health Ministry's Preventive Health and HIV/AIDS Control Agency, the number of people with arsenic poisoning (with effects on the skin, the nervous system and the cardiovascular system) in Hoa Hau, Vinh Tru communes in Ly Nhan District and Bo De Commune in Luc Binh District, Ha Nam Province, are high after 6 years of exposure to arsenic-contaminated water (Ninh *et al.* 2004; Thanh, 2004).

About 13.5% (12 million) of Viet Nam's 90 million population uses water from wells, which are easily contaminated with arsenic (Thanh, 2004). The magnitude of populations exposed to elevated arsenic in Viet Nam is open to debate (Fredericks, 2004):

> UNICEF shows that there are 4,614 wells already identified as being contaminated. If we assume the rate of contamination in each province is representative of the whole province (not necessarily true) and using the estimates for the total number of wells in each province (where available) then about 100,000 wells are likely to be contaminated above 0.05 mg/L. Assuming all wells are used

> for drinking and cooking purposes, then exposed population is likely to be between 500,000 assuming one family of five per well and 2.5 million assuming five families of five persons per well. The World Bank [2005a] report "Towards a more effective response" quotes UNICEF as estimating the "Population at Risk" to be 10 million.

Further evidence of the arsenic contamination involves drawdown of arsenic tainted water from shallower Holocene to deeper formerly pristine Pleistocene aquifers. Viet Nam Red River Delta inhabitants have been extracting water from Pleistocene aquifers for over 100 years. Recently, it has been discovered that arsenic has migrated vertically, induced by the large-scale pumping. The alarming results of a large-scale survey in 2011 has revealed that approximately 7 million people in the Red River Delta consume groundwater contaminated with arsenic, among other toxic elements, including manganese, selenium and barium (Winkel *et al.* 2011). This catastrophic situation is a cautionary lesson for people in other arsenic affected regions that engage in extensive pumping of uncontaminated aquifers that underlie arsenic contaminated layers. Figures 3.31 and 3.32 below indicate the scope and severity of this dire situation.

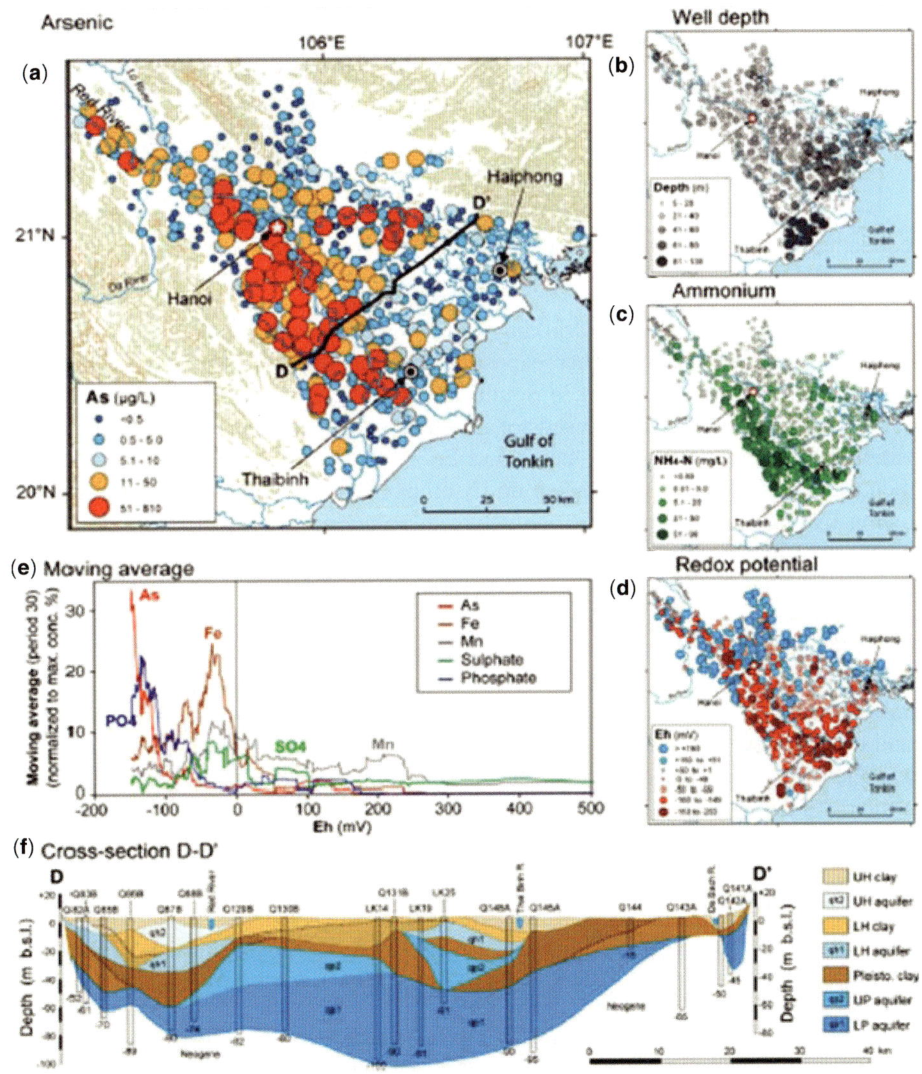

Figure 3.31 Concentrations of As and selected parameters observed in groundwater of the Red River Delta. (Winkel, 2011).

Independently of these new 2011 data with its 7 million estimate of people exposed to arsenic in the Red River Delta, the World Bank (2005b) estimates 10 million "at risk" for the country as a whole. This is interpreted here to mean exposure to arsenic levels above 50 µg/l, which is the country standard, and this high estimate is probably intended to indicate the severity and magnitude of arsenic contamination of wells in Viet Nam so that the misunderstandings that occurred in Bangladesh will not be repeated. However, local, on-the-ground estimates are lower, even though the percentage of tested wells is also small. According to Ninh *et al.* (2004), Viet Nam has 5.8 million people using water with arsenic levels above 10 μg/l. Therefore, the projected number probably falls in the range between 5.8 and 10 million people exposed to arsenic levels above 10 μg/l.

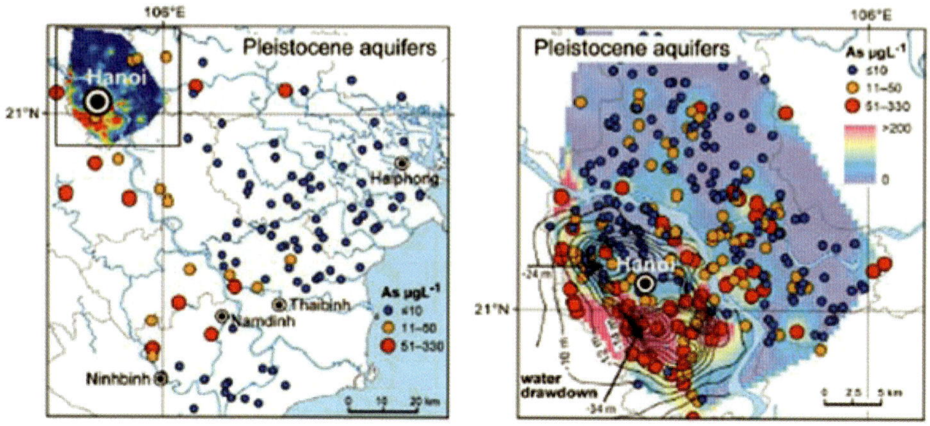

Figure 3.32 Arsenic concentrations of the Hanoi area. (Winkel, 2011).

Besides consumption of arsenic contaminated drinking water, ingestion of arsenic tainted produce, meat or fish, and especially grains, such as the staple, rice, is another exposure pathway of considerable concern. Contamination of crops irrigated with high arsenic water is a known danger in Southeast Asia (Achin, 2011). During the lean summer months from January to May a large amount of groundwater is used for irrigation which is contaminated with arsenic resulting from the oxygenated decomposition of arsenopyrite forming Fe^{++} and Fe^{+++} sulfates and H_2SO_4 acid which mobilize arsenic. The mobilized arsenic is then accumulated in soils and taken up by crops like rice. While the exact mechanism of arsenic release is controversial (following oxidation of arsenopyrite, reductive dissolution of Fe hydroxides or anion exchange of arsenic with fertilizer phosphates) its effect is dramatic for it provides another pathway – through the food chain. Food security coupled with arsenic contamination of agricultural soils and produce is, therefore, in Viet Nam a looming concomitant of arsenic contamination of groundwater. Two recent studies devoted to the effects of phyto-absorption of arsenic by agricultural crops such as Vietnamese paddy rice have found that Vietnamese rice was 0.03–0.47 g/g dry weight (Phuong *et al.* 1999; Williams *et al.* 2005).

To complicate matters further, the seasonal monsoon also induces a decrease in arsenic contamination. For instance, the average arsenic concentrations in the wet season in Thanh Tri south of Hanoi is 60 μg/l, but in the dry season it is 87 μg/l; in Hanoi, the wet season level is 38 μg/l and the dry season level is 54 μg/l; in Gia Lam, the level is 11 μg/l in the wet season and 46 μg/l in the dry season. Accordingly, it is important to note the season in which the data were obtained. Many of the data presented in the literature cited here did not specify the season of collection. Where the season is specified, it is noted.

The following is a geological assessment of Viet Nam from Thanh (2004):

> Arsenic is detected in mining areas, ore outcrops and in areas where industrial wastes contains arsenic. Arsenic occurs naturally in areas mineralized with sulfide minerals and gold, in areas of volcanic activities. The main arsenic liberating mechanism is weathering. The main areas in Vietnam of arsenic concentration are in Pia Oac, Ngan Son, Pac Lang, Chay River, North Tam Dao, Cho Don-Cho Dien, Dao Vien-Chiem Hoa-Na Hang-Tung and Bac Me.
>
> Arsenic also occurs in metamorphic intrusive volcanic formations containing arsenopyrite, pyrite and gold such as in the Hien river, Binh Gia, east of the Chay river, in Nam Xe-Tam Duong, in the downstream part of the Da river. Arsenic concentration in gold ore in the basalts of the Viet-nam formation in Doi Bu (Hoa Binh) is 50-204ppm. In sandstone, siltstone and in the siliceous shale of the Than Sa formation in a mineralized zone of sulfur-gold-quartz in Khau Au-La Hien (Bac Can, Thai Nguyen) the value is 13.2ppm; in the ore it is 1292-1442ppm.
>
> In the sericite shale and clay shale of the Coc Xo formation in the zinc-lead ore area of Cho Don (Bac Can) it is 97.8ppm, and 8.205-261.824ppm in the ore.
>
> According to Dang Van Can, arsenic in water of the ore zone in Lang Vai (Dam Hong) is 0.73mg/l. In Bo Sinh-Moc Chau district, Son La province the arsenic level in springs of the right side of the Ma river is 0.43-1.13 mg/l.

Do Tuyet reports arsenic concentrations in karst water at Tay Bac of 2.8mg/l in Tra commune, 2.67mg/l in Chieng hamlet, 2.29mg/l in Ngan commune and 3.14mg/l in Chieng Bui.

In the Red River Delta plain, the industrial zones discharge waste containing arsenic. In the Viet Tri industrial zone the arsenic concentration in groundwater is higher than the standard limitation (max. value is 0.32mg/l). In Thuong Dinh industrial zone, Hanoi area, arsenic in wastewater is 0.145-0.346mg/l, higher than the standard limitation (Nguyen Van Duc). This wastewater is not treated.

In 1999 with sponsorship of UNICEF, the arsenic level has been studied in 7 provinces of the Red River Delta plain. A total of 1228 samples were taken and analyzed. The results showed that 740 out of 1228 have an arsenic concentration below 0.01mg/l, 1075 samples have an arsenic concentration below 0.05mg/l and 153 samples have an arsenic concentration over 0.05mg/l with a maximum value of 0.6mg/l.

Since 2001, the Northern Hydro-geological Engineering Geological Division has been studying arsenic levels in groundwater in the Red River delta plain. Samples are taken twice per year, stored and preserved in accordance with [International Organization for Standardization] ISO 11969 and Vietnamese standard-TCVN 6626. The results show that out of a total of 34 samples, 17 samples have over 0.001mg/l of arsenic, 5 samples have over 0.05mg/l. The max value is 0.428mg/l (Q58a-Dan Phuong-HaTay). In 2002 there are 18 samples out of 85 samples (21%) having a higher concentration than the standard limitation [>50 μg/l]. 14 of these samples are from the qp aquifer with 0.067mg/l (Q88b-Ha Nam) and 0.406mg/l (Q58a-Ha Tay), the four remaining samples are from the qh aquifer with values of 0.088mg/l (Q85a-Ha Nam) and 0.440mg/l (Q56-Ha Tay).

In 2001–2002, UNICEF Hanoi, the Northern Hydrogeological Engineering Geological Division and the University for Natural Science/National University tested groundwater in Hanoi City and the provinces of Thai Binh, Nam Dinh and Ninh Binh. The results showed that the arsenic contamination was directly proportional to the amount of rainfall and that the levels during the rainy season were generally less than during the dry season. For instance, in Hanoi City, north of the Song Hong and Duong rivers, 8 out of 112 samples in Dong Anh district had an arsenic concentration above 50 μg/l during the dry season. In the Gia Lam area, which is a Quaternary aquifer, 8 samples (40%) during the dry season had an arsenic concentration above 50 μg/l, but in the rainy season, only 2 samples were above this limit. In the Tu Liem area, which is also a Quaternary aquifer, 8 samples (14.5%) during the dry season had arsenic concentrations above 50 μg/l, but in the rainy season, only 1 sample was elevated. In the Thanh Tri area, another Quaternary aquifer, 43 samples (59.7%) had elevated arsenic concentrations in the dry season, but in the wet season only 29 samples (40.3%) were elevated. This scenario was repeated for the Hanoi urban area and the provinces of Nam Dinh, Ninh Binh and Thai Binh (Thanh, 2004).

This scenario leads to the 2-fold mechanism of arsenic contamination of the Song Hong delta. First, iron and methane can combine with arsenic to produce water-soluble arsenic compounds. Unconsolidated sediments (clay, peat and organic matter) in the Song Hong delta act as storage media for arsenic. Arsenic concentrations in soils in four wells in Hanoi (reported by the University for Natural Science/National University and the Northern Hydrogeological Engineering Geological Division) ranged from 6 to 33 mg/g, which correlates with the concentration of iron oxyhydroxide to which arsenic is adsorbed in areas of heavy groundwater pumping. Second, another form of binding arsenic to produce a water-soluble compound is through disintegration of organic matter in peat, in which methane is formed, combining with arsenic to produce methyl-arsenic compounds that dissolve in groundwater. In the Hanoi area, arsenic sources can be clay and peat layers at depths of 0–40 m (Thanh, 2004).

Mekong Delta

The status of arsenic in the Cuu Long delta of the Mekong River in southern Viet Nam indicates that natural arsenic originates in the upper Mekong basin in Cambodia, rather than from the local geology. According to one study (Stanger *et al.* 2005):

> [arsenic] is widespread in soils at typical concentrations of between 8 and 16 ppm (dry weight). Industrial and agricultural arsenic is localized and relatively unimportant compared to the natural alluvial arsenic. Aquifers most typically contain ground waters of no more than 10 μg/L, although scattered anomalous areas of 10 to 30 μg/L are also quite common. The most serious, but possibly ephemeral arsenic anomalies, of up to 600 μg/L, are associated with iron and organic-rich flood-plain sediments subject to very large flood-related fluctuations in water level, resulting in transient arsenopyrite dissolution under oxidizing conditions. In general, however, high-arsenic ground waters result from the competing interaction between sorption and dissolution processes, in which arsenic is only released under reducing and slightly alkaline conditions. High arsenic ground waters are found both in shallow water-tables, and in deeper aquifers of between 100 and 120 m depth. There is no evidence of widespread arsenicosis, but there are serious localised health-hazards, and some risk of low-level arsenic ingestion through indirect pathways, such as through contaminated rice and aquaculture. An almost ubiquitous presence of arsenic in soils, together with the likelihood of greatly increased groundwater extraction in the future, will require continuing caution in water resources development throughout the region.

In the Mekong River Delta in Dong Thap Province, the arsenic concentration in the groundwater was measured at >50 μg/l, affecting Hoa Hau and Vinh Tru communes in Ly Nhan District and Bo De Commune in Luc Binh District. In Ha Nam Province, levels of arsenic are high after 6 years of monitored exposure to arsenic-contaminated water (Thanh, 2004). Holocene, Pleistocene and Pliocene aquifers of the Mekong delta yielded arsenic concentrations of <1–5 µg/l for ground waters from Holocene deposits (9 samples, depth range 4–19 m), <1–32 µg/l for those from Middle and Upper Pleistocene deposits (39 samples, depth range 5–120 m), <1–7 µg/l for groundwater from Lower Pleistocene deposits (12 samples, depth range 113–191 m) and <1–57 µg/l for groundwater from Pliocene deposits (39 samples, depth range 85–330 m). The highest concentrations in the Pliocene aquifer were in the Ben Tre area of the central Mekong delta. Pleistocene and Pliocene sediments are the most exploited aquifers in the region. The Holocene aquifers appear to be largely low yielding and are not heavily used. UNICEF tested Mekong ground waters for arsenic in Viet Nam, but results were negative (Doan, undated).

A Danish research project funded in 2004 by the Council for Development Research under DANIDA (in cooperation with the Technical University of Denmark, DHI Water & Environment, Hanoi University of Mining and Geology, Hanoi University of Science and the Northern Hydrogeological Engineering Geological Division) has investigated the processes leading to arsenic contamination in the Song Hong delta (Tanabe *et al.* 2006). For reasons of hygiene, the Vietnamese want to access groundwater, but there is not enough information about the arsenic problem. So far they have only measured where the arsenic occurs in the water, and they know very little about why it occurs (Hinsby, 2004).

Thanh (2004) outlined some methods to mitigate arsenic in groundwater in Viet Nam:

- Investigation into the degree of arsenic pollution and the scale of distribution of arsenic in the vulnerable areas in the Red River Delta plain and vicinity.
- Assessment of water sources free of arsenic for water supply as part of integrated groundwater management.
- Treatment of waste from industrial, domestic and agricultural activities.
- Study the technology and treatment models for community health.
- Intensive collaboration between national and international organizations to exchange information, experience.

Section 4

European Region

4.1 ALBANIA (A, G, M, V)

Albania, with a population of 3.0 million, has a Human Development Index rank of 64 out of 169 and a per capita GNI of $US 7976 (US Census Bureau, 2011; UNDP, 2010)

An anthropogenic source of arsenic contamination in Albania relates to the former Soviet Union's cache of chemical weapons, which were stored in clandestine bunkers without adequate safeguards to protect the surrounding environment. A recent report by Warrick (2005) describes the process by which these weapons have come to light:

> Near the end of his 40 years in power, Enver Hoxha prepared his tiny country for an invasion he warned was sure to come. The Marxist dictator built 750,000 concrete bunkers in the 1970s and 1980s and imported large quantities of weapons to repel an expected attack by Americans, Soviets, Yugoslavs or perhaps all three at once. But his most prized weapons acquisition was a state secret known only to the Albanian leader and his closest advisers–a secret that only now is coming fully to light. In the mid-1970s, US and Albanian officials now believe, Hoxha arranged the purchase of several hundred canisters of lethal military chemicals to be used in weapons against invading armies. The chemicals included yperite, or sulfur mustard, one of the chemicals used by Iraqi leader Saddam Hussein to slaughter thousands of Kurdish civilians in the 1980s, as well as lewisite and adamsite, which are based on arsenic.
>
> This deadly stockpile was hidden in one of Hoxha's bunkers, then forgotten after Hoxha died in 1985. The communist regime fell in 1991. The current Albanian government's surprise discovery of the canisters, acknowledged to US and [United Nations] officials several months ago, has also led to the disclosure of the country that apparently supplied the chemicals: China. Albanian officials recently allowed a reporter from *The Washington Post* to view the stockpile, a move that comes as there are ongoing efforts by the fledgling democracy to renounce the country's past and bolster its international standing. While the stockpile is small compared with the vast chemical weapons holdings of Russia and the United States, it is worrisome to US officials because of what it represents: one of scores of undocumented or poorly secured weapons caches worldwide that could be exploited by terrorists with deadly effect.

Another anthropogenic occurrence of arsenic relates to the development of commercial chemicals without adequate environmental oversight. The nitrogen fertilizer plant in Fier City polluted the soil in the area with arsenic compounds that subsequently leached into the groundwater. The plant in Fier produced fertilizer from 1967 to 1992. From 1967 to 1971, it used crude oil with a high sulfur content as well as arsenate and arsenite. This produced 850 m^3 of effluent containing arsenic(III) and arsenic(V), which were stored in 1970 in four steel columns set in a 200-m^3 concrete basin with a compromised lining. A water sample taken from the outlet of the basin showed 97 700 μg arsenic/l, and soil samples taken within 50 m contained 830–172 300 mg arsenic/kg, which contaminated the upper groundwater aquifer that drains into the Gjanicës River and is a source of drinking-water downstream (UNEP Balkans, 2000). Thanks to a 1 million euro European Union (EU) cleanup project in March 2004, the site has been remediated as of 2005 (European Union, 2005).

Fier City is not the only instance of such a crisis. An additional 1620 tonnes of chemicals are stored at 12 enterprises, and among them 60.2 tonnes (i.e. cyanides, mercury salts and arsenicals) are of a hazardous nature (Regional Environmental

Center for Central and Eastern Europe, 2000). It is known that arsenic is leaching into the groundwater (Lazo, 2003), but the state of its remediation is not known.

Arsenic contamination related to mining and geogenic sources is due to extensive mineral deposits of chromite, copper, natural gas, nickeliferous iron ore and petroleum, all of which contribute to Albania's GDP (Steblez, 2003a). Specifically, chromite and copper ores are associated with secondary or trace amounts of arsenic minerals, such as Hollingworthite [(Rh, Pt,Pd)AsS], Anduoite [$(Ru,Os)As_2$] and Sperrylite [$PtAs_2$], which can leach into surface water and groundwater through natural weathering. When these are mined, arsenic also becomes more accessible to weathering (Beccaluva *et al.* 1998; Bonadiman *et al.* 2002).

Volcanogenic sources of arsenic are potentially found in numerous thermal mineral springs in Albania with temperatures up to 65.5°C. Three main geothermal areas are Kruje, Ardenica and Peshkopi, with a combined geothermal potential of 39–63GJ/m^2. The Kruje geothermal area is the largest geothermal area in Albania from the Adriatic coast north of Rodoni Cape, south-east to Tirana to the Albanian–Greek border and into Greece's Konica district. Aquifer temperatures range from 145°C to 250°C, and surface water temperatures range from 30°C to 65.5°C, with flow rates of 3.5–15 l/s. These aquifers are of the neogenic Ardenica Formation in the Semani anticline, the northern pericline of the Patos-Verbas carbonatic structure and the neogenic Verbas Formation (Battocletti, 2001):

> The Ardenica geothermal area extends on that part of the peri-Adriatic Depression where the Vlora-Elbasan-Diber transverse passes. Water temperatures from deep wells in this area are 32–38°C; flow rates are 15–18 l/s. The Peshkopi geothermal area is situated in northeast Albania in the Korabi hydrogeologic zone. Two kilometers east of the town of Peshkopi, thermal springs flow out on the Banja river slope, which is composed of flysch deposits. These springs are linked with the disjunctive tectonic zone, in the Ohri-Diber deep fault, peripherically of the Permian-Triassic gypsum diapir, that has penetrated the Eocene flysch which surrounds it in a ring-like pattern. Water temperature is 43.5°C; flow rates are 14–17 l/s. In addition to Kruje, Ardenica, and Peshkopi, numerous thermal springs are found across Albania. Situated primarily in tectonic fractures, thermal springs have temperatures of 21°C to 58°C.

The following list was prepared for the United States Department of Energy in order to assess the feasibility of using these springs as power sources, but they also represent a potential source of arsenic leaching and should be investigated as such (Battocletti, 2001):

- Ardenica – wells drilled, 32–38°C
- Ballsh – wells drilled, 36–38°C
- Bubullima – wells drilled, 48–50°C
- Divjaka – wells drilled
- Frakulla – wells drilled, 33–36°C
- Gagliati – mineral spring, 45–50°C
- Gorisht – wells drilled, 32–38°C
- Ishmi – mineral spring, 64°C
- Kolonja – wells drilled
- Kozan-Elbasan – under study, 57°C
- Kozani – mineral spring, 66°C
- Krane-Sarande – under study, 34°C
- Langareci-Permet – mineral spring, 6–31°C
- Llixha-Elbasan – mineral spring, 58–60°C
- Mamurras-Tirane – under study, 21°C
- Marinze – wells drilled, 40°C
- Peshkopi – mineral spring, 5–106°C
- Sarandaporo-Leskovik – mineral spring, 27°C
- Semani – wells drilled, 35°C
- Shupal-Tirane – under study, 30°C
- Tervoll-Gramsh – under study, 24°C
- Verbasi – wells drilled, 29°C

4.2 AUSTRIA (A, G, M, V)

Austria, with a population of 8.2 million, has a Human Development Index rank of 25 out of 169 and a per capita GDP of $US 37 056 (US Census Bureau, 2011; UNDP, 2010).

Anthropogenic arsenic comes from 9 industrial sites that are responsible for directly discharging 1.70 tonnes of arsenic per year into surface water, 1 site that is responsible for indirectly discharging 0.029 tonne of arsenic per year indirectly into surface water and 11 facilities that are responsible for discharging 1.75 tonnes of arsenic per year into the air (European Pollutant Emission Register, 2004).

Geogenic sources of arsenic are associated with arsenic minerals that are found in many locales in Austria. Most prominent is arsenopyrite, but others include Loellingite ($FeAs_2$) (named after the locality in which it was originally identified), which is an iron arsenate, sometimes found in association with cobalt and nickel. If the iron is replaced by more than 50% nickel, the mineral is Rammelsbergite, and if it is replaced by more than 50% cobalt, the mineral is Safflorite. These arsenic minerals, although rare, have an environmental impact.

Mining sources of arsenic are associated with mined ores of metals that are in association with the arsenic minerals found geogenically, because mining promotes the release of arsenic into surface waters.

Volcanogenic sources of arsenic contamination have been identified in extreme eastern Austria bordering the Great Pannonian Plain in a geological environment identical to that of Hungary (see Section 3.11; Hlavay & Polyak, 1998).

Geogenic, mining and volcanogenic sources of arsenic contamination of surface water and groundwater have been inferentially established by assessing the human intake of arsenic as 27 μg/day (Woidich & Pfannhauser, 1979) and by a study of arsenic concentrations in soil and its distribution and fractionalization (Ahmed, 1999).

4.3 BELGIUM (A, G, M)

Belgium, with a population of 10.4 million, has a Human Development Index rank of 18 out of 169 and a per capita GNI of $US 34 873 (US Census Bureau, 2011; UNDP, 2010).

Anthropogenic arsenic contamination in Belgium is related to 24 industrial direct and indirect arsenic emissions into surface water and groundwater and 21 sources of emissions into air. Total emissions into air from these industrial sources are 2.91 tonnes of arsenic per year; direct and indirect emissions into water are 1.09 and 0.07 tonne per year, respectively (European Pollutant Emission Register, 2004). In a study undertaken to measure only slightly elevated exposures to arsenic in drinking-water and in inhaled air, no significant elevations in mortality were found, except in men who had had previous exposure to arsenic through zinc smelter work or through presumed elevated levels of cigarette smoking (Buchet & Lison, 1998).

Geogenic hidden arsenic has been identified in the Scheldt estuary in Belgium (Andreae & Andreae, 1989). The absence of this hidden species of arsenic during the biologically dormant period of the estuarine and lacustrine year may be attributed to depressed metabolic levels of phytoplankton and algae: with lower excretory levels of methylarsenicals or arsenicals that are excreted but not detected by the traditional hydride generation method, they are hidden from traditional detection methods.

Forty-seven facilities engaged in the metal industry, metal ore roasting or sintering installations and installations for the production of ferrous and non-ferrous metals are responsible for mining-associated arsenic release, emitting 1.8 tonnes of arsenic into the air and 0.6575 tonne of arsenic into water per year (European Pollutant Emission Register, 2004).

While the issue of arsenic pollution in Belgium as a local phenomenon is not alarming, arsenic pollution is also attributable to non-local sources from industrialized regions of Central Europe. The more pernicious species of arsenic, arsenic (III), is elevated in proportion to arsenic (V) in the upper reaches of the Zenne River, which flows through Brussels' industrial zone. It is this river that drains Central Europe's industrialized region via the Scheldt watershed. Reasons for elevated levels of arsenic (III) in the Zenne River are related to a relatively constant level of anthropogenic arsenic effluents, both domestic and industrial, being discharged from cities in Central Europe. These effluents create the anoxic conditions in the upper regions of the estuary that promote release of arsenic (III) into the flowing water column. The combination of relatively constant anthropogenic levels of arsenic emissions and a seasonal fluctuation in the amount of water column flow helps explain the non-linear relationship of arsenic concentrations and salinity. As can be predicted, the farther downstream one is from the head of the Zenne River, the lower the ratio of arsenic (III) to arsenic (V), as the study of Andreae & Andreae (1989) shows.

The form of arsenic that is accumulated by marine life in the Zenne River, and others like it, is of great significance to the upstream and downstream areas. The average arsenic concentration in the sea is about 1.7 μg/l. In oxygenated marine and brackish waters (i.e. downstream), the dominant species is arsenate, arsenic (V), whereas upstream, the dominant species is the more toxic arsenite, arsenic (III). In oxic seawaters, arsenic (III) rarely exceeds 20%. Marine organisms bioaccumulate organoarsenic compounds, and their tissues contain high concentrations of arsenic (1–100 μg/g dry weight), which is mostly organoarsenic in the form of arsenobetaine. Therefore, humans who consume seafood products downstream of anoxic conditions also consume arsenic, but in the form of organoarsenic compounds (Neff, 1997).

4.4 BOSNIA AND HERZEGOVINA (A, G, M, V)

Bosnia and Herzegovina, with a population of 4.6 million, has a Human Development Index rank of 68 out of 169 and a per capita GDP of $US 8222 (US Census Bureau, 2011; UNDP, 2010).

Anthropogenic sources of arsenic contamination, including domestic and industrial waste, have polluted the Danube and Sava rivers (United States Department of Defence Health Clinic Center, 2005).

Geogenic sources of arsenic are related to Bosnia and Herzegovina's mineral resources, which include bauxite, coal, industrial minerals, iron ore, lead and zinc. Natural leaching of arsenopyrite and related arsenic minerals is responsible for the presence of arsenic in surface water and groundwater.

Arsenic from mining occurs from the extraction of these same resources. Mining and processing facilities are exploited in two politically distinct regions with two distinct administrative structures. Although these administrative structures are embraced within a formal national government organization (Steblez, 2003a), there is no environmental coordination between these two political entities.

North-east Bosnia and Herzegovina, including Srebrenica, Maglaj, Zavidovici and Teslić, has Tertiary igneous rocks associated with zinc, lead, arsenic, tin, iron and sulfur, which generate acid mine drainage when mined. There has been a lack of attention to derelict and historic mines as the cause of contamination of surface water and groundwater, even though the Srebrenica lead and zinc mine is still operating at 10% of capacity. In fact, no abandoned mines have been remediated, and many are used as landfill sites or fish farms, such as the open pit at Vares and Mrkonjic – Jajc-Prozor – Konjic, with high concentrations of arsenic, iron, manganese, barium and copper (Midžić & Silajdžić, 2005a).

Effluent from aluminium production operations is discharged into Lake Skadar, which officially has been identified as "too polluted" to provide water for the city of Podgorica. Copper and uranium mining in Bor has led to extensive pollution of the Borska Reka River, a tributary of the Timok River, which is a tributary of the Danube. The pollution results from the discharge of flotation water and releases from tailings ponds. The release of solid wastes from the smelter is estimated at 22 million tonnes per year, including 91 tonnes of uranium and many more tonnes of arsenic. The Borska Reka and the Timok rivers' contaminants include arsenic, copper, lead, uranium and zinc (United States Department of Defense Health Clinic Center, 2005).

Volcanogenic sources of arsenic abound in mineral and hot springs emanating from deep in the Dinaric Alps. Among these are the hot springs at Ilidža near Sarajevo, the chalybeate spring at Kiseljak and a spring impregnated with arsenic at Srebrenica. Two additional mineral thermal springs have been identified at Bosanski Samac, which is used as mineral waters at a temperature of 86°C, and at Kakanj, with a temperature of 54°C (Battocletti, 2001). During Roman times, the Ilidža area was the location of the town *Aquae Sulphurae* ("sulfuric thermal spring"). This was a Roman colony and the main settlement in all of Bosnia and Herzegovina at the time. The source of the river Vrbas is some 90 km to the south of Banja Luka, and the tributary rivers Suturlija, Crkvena and Vrbanja flow into the Vrbas at Banja Luka, the second largest city in Bosnia and Herzegovina. Banja Luka also has a number of thermal springs, which, like those of the other countries bordering the Great Pannonian Plain, are subject to high concentrations of arsenic, which can find its way into the groundwater and even the river systems. The use of these springs as geothermal sources of energy is being seriously investigated (Loy *et al.* 2004).

4.5 BULGARIA (A, C, M)

Bulgaria, with a population of 7.1 million, has a Human Development Index rank of 58 out of 169 and a per capita GDP of $US 11 139 (US Census Bureau, 2011; UNDP, 2010).

Anthropogenic arsenic contamination is believed to be responsible for atmospheric arsenic deposition measured in the moss *Hypnum cupressiforme* (mean, 2.14 mg arsenic/kg; range, 0.3–59 mg arsenic/kg) at Chiprovtzi, a town across the border from Serbia. The copper smelters at Bor in Serbia are thought to be the transboundary source of this atmospheric arsenic (Stamenov *et al.* 2002; Buse *et al.* 2003).

Coal-related arsenic is found in anthracite, bituminous, sub-bituminous and lignite deposits. Carboniferous Svige anthracites have arsenic concentrations ranging from 58 to 88 mg/kg, with concentrations in ash ranging from 88 to 238 mg/kg. In these anthracite occurrences, elevated arsenic appears to be related to epigenetic pyrite (Kortenski *et al.* 1990). Bituminous Cenomanian deposits have coal ash with arsenic concentrations averaging 1302 mg/kg. Bituminous coals are in association with pyrite, marcasite, chalcopyrite and galena. In addition to these sulfidic-related coal deposits, it is thought that some organic arsenate coals are also present (Kortenski *et al.* 1999). Sub-bituminous coals of the Pernik basin have a positive correlation with pyrite (up to 10%) and average 82 ± 13 mg/kg, with an ash content of 339 ± 49 mg arsenic/kg (Kostova *et al.* 1996). Finally, the 20-m-thick Eocene Stanyansky deposit has a lignite seam that is positively correlated with high sulfur and high arsenic. The ash content of this lignite has a mean arsenic concentration of 110 mg/kg (Kortenski *et al.* 1999).

Bulgaria was a significant regional producer of copper until 1991. The country's major copper deposits are in the Srednogorie-Panagjurishte region of the country. Three mines were in operation: the Asarel-Medet and Elatsite surface mines and the Chelopech underground mine. The Chelopech mining operation worked a polymetallic deposit containing arsenic, germanium, gold, iridium, palladium and silver, in addition to copper. Gold is the most important economic element in the Chelopech deposit, estimated at 60 million tonnes, a major portion of Bulgaria's known gold resources. However, the high arsenic content of the ore also presents serious environmental hazards. Owing to the high arsenic content of the Topolnitz tailings dam, the environmental law of 1991 caused the Government of Bulgaria to prohibit the processing of Chelopech ore at MDK, with copper concentrate produced at this facility containing as much as 0.75% arsenic. One result of such mining in Bulgaria was that groundwater became seriously polluted (i.e. >50 μg arsenic/l) (Criaud & Fouillac, 1989). In some areas, metal working factories have so polluted the soil and water with lead, arsenic and cadmium that the land is unusable for agriculture. Wastewater treatment facilities frequently are absent or do not operate properly. As a result, some rivers are considered dead or nearly dead (Steblez, 2004). After 2000, the world market for high-arsenic gold and copper shrank as a result of environmental concerns, so the current mining company, Dundee Precious Metals of Canada, has a target of producing 3 million tonnes of gold and copper per annum using a cyanide extraction process that is designed to capture arsenic as arsenic (V) but has replaced toxic arsenic (III) with also toxic cyanide effluents (Dimitrova, 2007).

4.6 CROATIA (V)

Croatia, with a population of 4.5 million, has a Human Development Index rank of 51 out of 169 and a per capita GNI of $US 16 389 (US Census Bureau, 2011; UNDP, 2010).

Portions of Croatia bordering the southern part of the Great Pannonian Plain have been exposed to elevated concentrations of arsenic in a geological situation identical to that of Hungary (see Section 3.11; Hlavay & Polyak, 1998; Biacs, 2005). This situation needs serious investigation, as the Great Pannonian Plain is a major geological feature of Croatia. The geothermal resources of Croatia provide recreation as well as heating and power generation (Battocletti, 2001):

> Croatia can be divided into two geological regions. In the southeastern part, which belongs to the Dinaride area, Mesozoic carbonate rocks prevail. Towards the northeast, in the area belonging to the Pannonian basin, Quaternary and Tertiary sedimentary rocks predominate, overlying crystalline bedrock and occasional Mesozoic sedimentary rocks. Croatia is endowed with significant geothermal potential in the Pannonian area. The temperature gradient in this region ranges from 0.03 to 0.07°C/m. The terrestrial heat-flow density is also high, ranging from 60 to 100 mega watt/m^2, and occasionally up to 120 mega watt/m^2. INA, in the process of exploring for oil and gas, has discovered, explored, and tested geothermal fields since 1976. More than 50 deep boreholes have been drilled through geothermal aquifers in the Pannonian region. Temperatures are 40–170°C. Some geothermal fields have been defined, but the majority requires further investigation. A small fraction of known geothermal energy potential in Croatia is currently under exploitation. As of 2000, Croatia had 113.9 mega watt of installed capacity. The bulk – 77.24 mega watt – is used for bathing and swimming, followed by 36.66 mega watts for space heating.... The country's direct use potential from tested reservoirs is estimated at 815 mega watt.

The following list of 28 springs was prepared for the United States Department of Energy in order to assess the feasibility of using them as power sources, but they represent a potential source of arsenic leaching and should be investigated as such (Battocletti, 2001):

- Babina Greda – under study, 125°C
- Bizovac – thermal spring, 91–96°C
- Daruvar – thermal spring, 47°C
- Ernestinovo – under study, 80°C
- Ferdinandovac – well(s) or hole(s) drilled
- Hrvatsko Zagorje – thermal spring
- Istra – thermal spring
- Ivanic Grad – thermal spring, 62°C
- Krapinske – thermal spring, 41°C
- Lesce – thermal spring
- Lipik – thermal spring, 60°C
- Livade – thermal spring, 28°C
- Lunjkovec-Kutnjak – under study

- Madarince – under study, 96°C
- Recica – under study, 120°C
- Samobor – thermal spring, 28°C
- Stubicke – thermal spring, 53°C
- Sveta Jana – thermal spring, 26°C
- Sveta Nedjelja – well(s) or hole(s) drilled, 68°C
- Topusko – thermal spring, 62°C
- Tuhelj – thermal spring, 32°C
- Varazdinske Toplice – thermal spring, 58°C
- Velika – thermal spring, 25°C
- Velika Ciglena – under study
- Vuckovec – thermal spring
- Zagreb – thermal spring, 55–82°C
- Zelina – thermal spring, 40°C
- Zlatar – thermal spring, 32°C

4.7 CZECH REPUBLIC (A, C, G, M)

The Czech Republic, with a population of 10.2 million, has a Human Development Index rank of 28 out of 169 and a per capita GNI of $US 22 678 (US Census Bureau, 2011; UNDP, 2010).

Anthropogenic sources of arsenic in the Czech Republic include 9 industrial facilities that emit 2.23 tonnes of arsenic per year directly into surface water and 33 facilities that emit 4.40 tonnes of arsenic per year into the air (European Pollutant Emission Register, 2004). The Czech Republic is also a producer of heavy industrial goods manufactured by the country's tool, machine and chemical industries. Steelmaking, the mining and processing of industrial minerals and the production of construction materials continue to be of domestic and regional importance.

An indirect indication of the arsenic contamination of surface waters is the fact that fish in the River Elbe have high concentrations of arsenic in their flesh. In a recent study along 256 km of the River Elbe, arsenic concentrations were measured in tissues of bream (*Abramis brama*) and perch (*Perca fluviatilis*) caught in various localities from Verdek to Hfiensko. The following is a summary of the findings (Svobodová & Randák, 2005):

> The highest muscle tissue arsenic concentrations were found in Zelãín in bream in 2003 (0.17 ± 0.08 mg·kg^{-1}) and in perch in 2002 (0.084 ± 0.012 mg·kg^{-1}). The highest arsenic concentrations in perch were found in the kidneys, followed by the liver, milt, muscle tissue and eggs. An increase ($P < 0.01$) in arsenic concentrations in 2003 (after floods) compared with 2002 was found in bream muscle tissues in Zelãín and in perch muscle tissues in Podolí; a decrease in arsenic concentrations was demonstrated in perch musculature in Verdek, Nûmãice, Hfiensko ($P < 0.01$) and in bream musculature in Hfiensko ($P < 0.05$). Arsenic contamination in the localities under study does not pose a major environmental or health risk.

However, it should be pointed out that arsenic concentrations vary along the river directly in relation to sources of arsenic release, and therefore the situation requires environmental monitoring.

Arsenic-enriched coal is found in five basins. The Cheb sedimentary basin in the north-western Bohemian massif was formed between the Late Oligocene and the Pliocene (Spicáková *et al.* 2000). It has arsenic concentrations up to 3245 mg/kg, which are correlated positively with sulfur. The Ostrava and Karvina Pennsylvanian bituminous deposits have up to 165 and 110 mg arsenic/kg, respectively. The Sokolov brown coal basin lies within the Karlovy Vary volcanogenic thermal waters, which measure from 25°C to 40°C, with total dissolved solids concentrations between 6000 and 9000 mg/l. Bedrocks consist of Carlsbad Pluton granites and crystalline schists. Organogenic sedimentation formed the Josef coal seam up to 10 m thick with overlying impermeable volcanogenic tuffs and claystones (Hanzlik & Trckova, 1999). The Zitava basin is bounded by the Luzice Mountains, with the Sluknov Hills in the west and the Jizera Mountains with the Frydlant Hills on the east; within its confines are arsenic-enriched brown coal seams (Simon & Haylly, 1984). The modern exploitation of these resources puts stress on the environment. A recent assessment indicates some of the issues involved (Steblez, 2003b):

> As of December 31, 2003, total resources of brown coal amounted to more than 9.501 Gt. Brown coal was used mainly as a fuel in the country's electric power industry; a minor amount was consumed by the chemicals sector. In 2003, major foreign commerce in brown coal centered on exports of about 1.3 Mt; Slovakia (63%) and Hungary (22%) were the major recipients. According to [the Czech Geological Survey's] GEOFOND, Czech standards for coal describe high-volatile lignite as a variety of brown coal that has undergone the least amount of coalification and still has xylitic characteristics (fragments of wood, preserved tree trunks, etc.). Its

dry and ash free calorific value is less than 17 MJ/kg. The boundary between brown coal and high-volatile lignite is not distinct. Lignite, which usually was consumed by the electric-power-generating sector, also was used for heating. The chief deposits occur in the Vienna basin, which extends from Austria to Moravia. Total resources of lignite in the Czech Republic at the end of 2003 amounted to about 1.012 Gt.

Less carbonaceous, softer, high-sulfur, high-arsenic coals used by power plants have arsenic levels as high as 1500 mg/kg (Cmarko, 1963), and arsenic concentrations in flue dust can range from 43 to 110 mg/kg (Zdrazil & Picha, 1966). Fifty-six 10-year-old children living near a power plant burning coal that had a high arsenic content ingested flue dust and had a mean arsenic blood level of 0.0045 mg/kg (range 0.0025–0.0082 mg/kg). These children showed significant hearing losses in both air and bone conduction at a high frequency range, indicating inner ear damage (Bencko & Symon, 1977; Bencko *et al.* 1977).

Geogenic sources of arsenic have been detected from natural weathering of arsenic-rich mineralized granodiorite and arsenic-rich volcano-sedimentary rocks in the vicinity of the Mokrsko gold deposit. Surface water and groundwater were supersaturated with crystalline and hydrous ferric oxides and calcium–iron and potassium (barium)–iron arsenates of arsenic (V), but undersaturated with scorodite, the oldest secondary arsenate mineral in the district. In the stream water, ponds, groundwater and pore water in the district, evidence of high concentrations of arsenite, arsenic(III), the most toxic species, was present (Drahota, 2005):

> In the pore water of soil profiles studied, arsenite may account for up to 91% of total arsenic (anaerobic soil below the water table, where the low redox conditions influence arsenate reduction) in contrast to less than 10% in the aerobic soils. In the stream water, the reduction of arsenate to arsenite is predominantly controlled by microbial activity. The concentration of arsenic in stream water successively decreases from 393 μg/l in the deposit to 10 μg/l in the distance of 1500 m from the deposit. The contents of arsenite in the stream are highly variable (0–95% of total arsenic) depending on the local conditions of sample location. The highest arsenite/arsenate ratio was found in the pond, where arsenite forms 94–99% of total arsenic.

Mining has exposed many arsenic minerals in the Czech Republic. Deposits of gold, silver, copper, tin and tungsten in association with arsenic minerals have been mined since the Middle Ages. Arsenic concentrations in surface water near deposits average 0.5 mg/l, and that not in proximity to mine wastes average about 0.02 mg/l (Grmela & Rapantova, 2005). Provisions in the mining law address issues of compliance with environmental regulations during the exploration and exploitation of mineral deposits and the reclamation of mined-out areas.

4.8 DENMARK (A, C)

Denmark, with a population of 5.5 million, has a Human Development Index rank of 19 out of 169 and a per capita GNI of $US 36 404 (US Census Bureau, 2011; UNDP, 2010).

Nine facilities discharge 2.23 tonnes of arsenic per year into surface water, and 33 facilities discharge 4.40 tonnes per year into the air (European Pollutant Emission Register, 2004). Since precipitation from airborne sources is a possibility, these sources are included here, but no further data are available on the concentrations of arsenic in surface water or groundwater adjacent to or downwind from those sites that discharge arsenic into air.

After World War II, 160 tonnes of chemical weapons, including lewisite, an arsenic-based compound, were dumped in the Baltic Sea (Federov, undated) (as well as the Barents & Kara Seas; see Section 3.23). When leached into the sea, lewisite contaminates fish and other marine biota, which, if not screened for arsenic, may pose a health threat.

Hundreds of chemical weapons were brought to the destruction centre on Denmark's Bornholm Island between 1979 and 2002. The weapons came almost exclusively from the five known dumping areas in the Baltic Sea; therefore, any other sites of chemical weapons disposal containing arsenic-based compounds are not represented and may still pose a threat (HELCOM, 2002).

Lithogenic coal in Denmark usually contains arsenic within the range of 1–4 mg/kg. Testing of a domed peat column revealed an arsenic concentration of 28 mg/kg at a depth of 15 cm. Also present in this sample were mercury and lead. For this reason, it has been speculated that the arsenic peak was attributable to coal-fired plants, and the 15-cm depth would place the date of contamination in the 1950s, a time when the United Kingdom was engaged in massive coal combustion (Shotyk *et al.* 2003).

4.9 ESTONIA (A, G)

Estonia, with a population of 1.3 million, has a Human Development Index rank of 34 out of 169 and a per capita GNI of $US 17 168 (US Census Bureau, 2011; UNDP, 2010).

The Urban Geochemistry of Tallinn Project, supported by the Scientific and Environmental Affairs Division of the North Atlantic Treaty Organization, has analysed 532 topsoil samples from Tallinn, which is the largest industrial centre of Estonia, with a (2007) population of 400 200 (Bityukova *et al.* 2000):

> Intensive local concentrations were determined for As, Cr, Mn, Ni, Pb, S, V and Zn, which are typical for the local pollution of the soils by industrial sources. The levels of As, Cr, Mn and V are more than three times [background levels]. Pb and Zn are more than five times higher in the geochemical anomalies than for background levels. For the Tallinn region two major associations of elements connected with industrial pollutions of the soil are typical: the first association includes Ba, Cr, Mn, Ni and partly Fe and the second one includes As, Pb and Zn.

A study undertaken by Stockholm's Kungliga Tekniska Högskolan is investigating arsenic retention in peat bogs in Estonia, where extremely high concentrations of iron (43% by weight), lead (2.5% by weight) and other metals and metalloids, including arsenic, accumulate. The long-term exposure of the peat bog to metal-rich groundwater is a natural analogue to municipal solid waste landfills. Geochemical and transport models have identified different distributions of metals at different depths with iron: arsenic and manganese are in a lower anaerobic–aerobic stratum, and lead and copper are in a higher stratum, with groundwater moving up from below (Syrovetnik & Neretnieks, 2003).

4.10 FINLAND (A, G, M)

Finland, with a population of 5.3 million, has a Human Development Index rank of 16 out of 169 and a per capita GNI of $US 33 872 (US Census Bureau, 2011; UNDP, 2010).

Anthropogenic sources of arsenic in Finland come from 21 industrial, chemical, metal and mining facilities that discharge 1.75 tonnes of arsenic into the air each year and 9 that discharge 1.70 tonnes per year directly into surface waters (European Pollutant Emission Register, 2004).

Geogenic sources include naturally occurring arsenic minerals, such as arsenopyrite. Elevated levels of arsenic are found in central Lapland and areas of south-west Finland (Kurttio, 1999). Levels up to 1 mg/l in water have been recorded, but typical levels are much lower than this (Kurttio *et al.* 1998; Kurttio, 1999). In central Lapland, the affected area coincides with the greenstone belt, whereas in south-west Finland, the underlying geology comprises volcano-sedimentary schists and gneiss (Kurttio, 1999).

In an edited volume of studies on arsenic occurrences in Finland (Loukola-Ruskeeniemi & Lahermo, 2004), several researchers provided a number of different perspectives on the problem; the following is a summary of the papers taken from that volume:

> **Pasi Eilu** and **Raimo Lahtinen** note that there are many geologically distinct regions in Finland containing higher than average concentrations of arsenic. Arsenic concentrations may be as much as one thousand times greater around ore deposits and occurrences than in the surrounding bedrock. Arsenic values above background are particularly likely in bedrock close to gold occurrences. The most prevalent arsenic mineral in Finland is arsenopyrite, not the arsenic-bearing pyrite that is significant in many other parts of the world. Elevated concentrations of arsenic are found especially in the Tampere–Hämeenlinna region where there also are several gold occurrences. When arsenopyrite is disseminated in bedrock or in quartz veins, it is mainly in unweathered form, and the arsenic does not easily leach out into the groundwater. Occasionally, the arsenopyrite occurs in fracture zones, in locations easily accessible to groundwater, and arsenic may then be released into the water from weathered mineral surfaces arsenic. In his article on arsenic in soil, **Timo Tarvainen** describes how in places more than 4 mg/kg arsenic occurs in glacial till. The chemical composition of till reflects the composition of the bedrock beneath, particularly the more easily weathered rocks: basic rocks rich in dark minerals (amphiboles and pyroxenes), fine-grained sedimentary rocks containing more than average amounts of carbon and sulphur (black shales), and calcareous rocks. Lighter coloured minerals like quartz and feldspars are more abundant in Finnish bedrock, but they weather less easily. Arsenic and heavy metals are more abundant in the clayey soils bordering the Gulf of Finland and Gulf of Bothnia than in more infertile till soils, which are widespread in other parts of the country. About 5,000 to 7,000 years ago, in the Littorina Sea stage of the Baltic Sea, the climate was warmer and the Baltic Sea was larger than it is at present. Sulphur, heavy metals and arsenic were precipitated into clay sediments on the sea bottom. Already then the sediment acted as a sink for potentially harmful elements, removing part of them from the ecosystem. Arsenic [clays emerged as] the sea retracted, the bottom sediments emerged to form expansive flat beds of clay alongside the sea. In areas of Ostrobothnia these marine clays contain elevated sulphur concentrations, which contribute to the acidification of river and stream water. Median arsenic concentrations in different topsoil types in Finnish arable land vary from 0.5 mg/kg in coarser grained varieties to 2.4 mg/kg in topsoil rich in fine-grained material. Arsenic is present in higher concentrations in clayey soil than in sediments consisting mostly of silty material. However, most of the arsenic in soils is not in bio-available form and does not get transported into cereal crops and potatoes or other parts of the ecosystem. Arsenic-bearing drinking water has not caused significant health problems in Finland because the diet is diversified.

According to **Kimmo Virtanen**, the average concentration of arsenic in peat in the mires of Northern Ostrobothnia region in Western Finland is 2.9 mg/kg.... Peat used as a planting medium in horticulture must not contain more than 10 mg/kg arsenic. Peat used as a soil improvement agent, in garden beds, in turn, may contain up to 50 mg/kg....

Jarmo Poikolainen and **Juha Piispanen** estimated the arsenic deposition from air in a study of the amounts of arsenic in moss. Concentrations of arsenic were determined in moss in 274 areas all over Finland in 1995 and 2000. The mean concentration in 1995 was 0.26 mg/kg, while in 2000 it was a little less, 0.19 mg/kg. Concentrations higher than 0.50 mg/kg were measured in 1995 in Southern Finland and around Kokkola, Seinäjoki and Pori in western Finland. But levels were lower in 2000, even in Southern Finland, where almost all concentrations were under 0.50 mg/kg. The deposition of arsenic in Finland is low.

Jari Mäkinen's investigations suggest the average arsenic concentration in bottom sediments of Finnish lakes to be 8 mg/kg. Most of these sediments are clayey gyttja. The portion of organic matter in the sediment varies from 6 to 15%. The highest arsenic concentrations are encountered in fairly large lakes (over 10 km^2). Arsenic concentrations are also elevated in the clay-rich sediments formed 9 000–10 000 years ago, just after the last Ice Age. The water level in ice-dammed glacial lakes formed at the edge of the retreating ice suddenly dropped as the water gouged out new channels to the sea, and an array of smaller lakes was left behind.

Mirja Leivuori and **Henry Vallius** report that the concentration of arsenic is higher in marine sediments than in lake sediments. The average concentration of arsenic in surface sediments in the Gulf of Finland varies between 14 and 16 mg/kg. Concentrations are significantly higher in Bothnian Bay, where the average values range between 109 and 239 mg/kg. Much of the arsenic in Bothnian Bay may have originated in emissions from mining and other industry in Sweden. The arsenic has evidently been transported to the sea from rivers on the Swedish side.

Timo Tarvainen and **Jaakko Mannio** examine the occurrence of arsenic in small streams and lakes. In streams at the head of watercourses the average arsenic concentration in water is 0.36 mg/L, while in small lakes it is 290 µg/L. In organic sediments of the same streams, concentrations range between 0.8 and 15.0 mg/kg. The regional distribution of the arsenic in stream sediment resembles that in till and the stream water, both describing arsenic transported from geological environments.

The survey of arsenic concentrations in groundwater presented by **Birgitta Backman** and **Pertti Lahermo** shows arsenic concentrations in springs and shallow dug wells to be very low in Finland. Greater amounts are found in drilled wells, which, hydrologically, represent older groundwater of longer residence time. Besides the direct water–rock interaction of primary bedrock and groundwater, also secondary precipitates on soil particles and fracture coatings and fillings are significant for the geochemical behaviour of arsenic in groundwater. The Ministry of Social Affairs and Health (STM) has set a limit of 10 µg/L for arsenic in potable groundwater and this value is exceeded only rarely. In the survey "One Thousand Wells" carried out by the Geological Survey of Finland in 1999, which was based on representative samples from throughout the country, only 3% of arsenic concentrations in water from drilled wells exceeded the upper limit.... Arsenic is nevertheless a potential problem, which needs to be taken into consideration in sparsely settled areas where water is obtained from wells drilled into bedrock.

The study carried out by **Risto Juntunen**, **Sirkka Vartiainen** and **Arto Pullinen** of drilled wells in the Pirkanmaa region around the city of Tampere supports the conclusion described above regarding caution in sparsely settled areas. The median and average concentrations of arsenic in Pirkanmaa drilled wells are 1.8 µg/L and 9.7 µg/L, which are ten times the values found in the One Thousand Wells project. Arsenic concentrations in the Pirkanmaa area exceeded the STM limit in the water of 17% of drilled wells. In addition, iron, manganese, fluoride, chloride and nickel concentrations occasionally exceeded the recommended limits. Arsenic concentration in the water was not dependent on the type of rock or rock association. Rather, the water of wells drilled into the same type of rock often contained very different amounts of arsenic. The appearance of arsenic in Pirkanmaa nevertheless shows a clear correlation with the structural features of the bedrock. Knowing the location of fracture zones in the bedrock can thus assist in the recognition and delimiting of the risk areas for arsenic.

Heikki Tanskanen, **Pertti Lahermo** and **Kirsti Loukola-Ruskeeniemi** studied the appearance of arsenic in groundwater in Central Lapland, in areas characterized by elevated arsenic concentrations in bedrock and till. There are also many gold occurrences in the area. The results of the study show that weakly basic groundwater and groundwater containing considerable amounts of dissolved substances in places also contain significant amounts of arsenic. The highest arsenic concentration, 36.2 µg/L was found in a spring on the northeast shore of Lake Soasjärvi, which is located to the northeast of the village of Kiistala. This is the largest concentration of arsenic measured in spring water in Finland.

Hanna Kahelin determined the speciation in which arsenic appears in the Kittilä area, and showed that, in spring water saturated with oxygen, arsenic mostly occurs in oxidation state As^{5+}.

Conclusions: Elevated concentrations of arsenic in the fine fraction of till are most extensive in the Tampere – Hämeenlinna region. Virtually all values there exceed 5 mg/kg. Gold occurrences appear in the same area. A consistent dependency between gold ores and arsenic concentrations in till has not been demonstrated, however. Gold deposits also occur in the area to the north of Tampere, where concentrations of arsenic in till are not elevated. In the Tampere–Hämeenlinna region, weathering of arsenopyrite in fracture zones and fissure surfaces releases the arsenic, which easily finds its way into groundwater. Indeed, there are wells in Pirkanmaa with exceptionally high concentrations of arsenic. Elevated levels of arsenic, and of heavy metals, are also found in the clayey soils near the coasts of the Gulf of Finland and Gulf of Bothnia, and these are reflected in the concentrations in groundwater. Arsenic does not pose as great a risk to human health in Finland as it does in many other countries. The only significant natural path to humans is through drinking water from drilled wells in rural areas. Concentrations of arsenic are nevertheless elevated in bedrock and soil in some areas, and levels in groundwater, surface waters and stream and lake sediments are higher there than in other parts of the country.

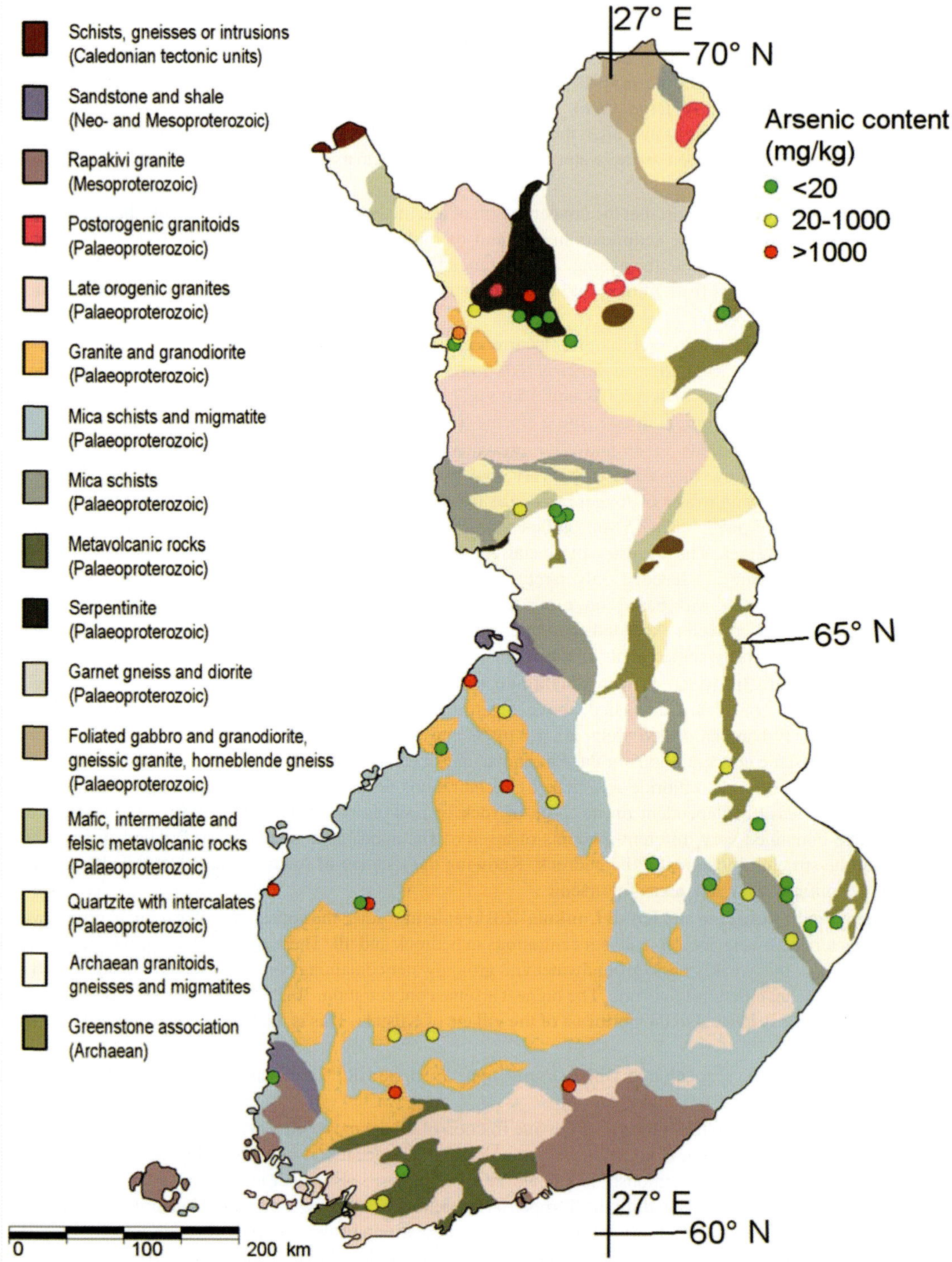

Figure 4.1 Arsenic contamination of groundwater by leaching from bedrock in Finland (modified from Loukola-Ruskeeniemi & Lahermo, 2004).

In addition, a study by Loukola-Ruskeeniemi (2003) showed naturally high arsenic contents in groundwater in areas with abundant gold occurrences. Another study (Pasanen, 2003) showed that arsenic occurs in various pools associated with contaminated soils in eastern Finland and that these have the potential to leach into the groundwater.

The geological occurrences of arsenic are outlined in the bedrock map produced by Loukola-Ruskeeniemi & Lahermo (2004). The map indicates that areas with arsenic contamination of medium and high levels occur throughout the peninsula and may occur in juxtaposition with areas of low concentrations. This indicates that close monitoring and accurate measurements of arsenic must be made regardless of the adjacent rock formations (see Figure 4.1).

The carcinogenic effect of arsenic in drinking-water in Finland was studied in an epidemiological study by Kurttio *et al.* (1999). The aims of the study were to determine whether arsenic exposure from well water is associated with increased risk of bladder and kidney cancers and to assess the concentrations of arsenic in drilled wells used for drinking-water. The subjects were selected from the 12–14% of the Finnish population who lived outside areas with municipal water supply and used private wells in 1967–1980 ($n = 144\,627$):

> The final study population consisted of 61 bladder cancer cases and 49 kidney cancer cases diagnosed between 1981 and 1995, as well as an age- and sex-balanced random sample of 275 subjects (reference cohort). Water samples were obtained from the wells used by the study population at least during 1967–1980. The total arsenic concentrations in the wells of the reference cohort were low (median = 0.1 μg/L; maximum = 64 μg/L), and 1% exceeded 10 μg/L. Arsenic exposure was estimated as arsenic concentration in the well, daily dose, and cumulative dose of arsenic. None of the exposure indicators was statistically significantly associated with the risk of kidney cancer. Bladder cancer tended to be associated with arsenic concentration and daily dose during the third to ninth years prior to the cancer diagnosis; the risk ratios for arsenic concentration categories 0.1–0.5 and 0.5 μg/L relative to the category with < 0.1 μg/L were 1.53 [95% confidence interval (CI), 0.75–3.09] and 2.44 (CI, 1.11–5.37), respectively. In spite of very low exposure levels, we found some evidence of an association between arsenic and bladder cancer risk. More studies are needed to confirm the possible association between arsenic and bladder cancer risk at such low exposure levels.

In other studies in Finland, a statistically significantly elevated risk of bladder cancer was observed for arsenic concentrations above 0.5 μg/l. Exposures earlier than 10 years before cancer diagnoses did not show an association with bladder cancer risk. Hence, relatively recent arsenic exposure appears to be more relevant for bladder cancer risk (Loukola-Ruskeeniemi & Lahermo, 2004). This is in concordance with the hypothesis that arsenic compounds act as promoters and/or co-carcinogens in the late stage of carcinogenesis (Kurttio *et al.* 1999).

4.11 FRANCE (A, G, M, V)

France, with a population of 65 million, has a Human Development Index rank of 14 out of 169 and a per capita GNI of $US 34 341 (US Census Bureau, 2011; UNDP, 2010).

Anthropogenic arsenic in France comes from 43 pharmaceutical, chemical and metal manufacturers along with mining companies that directly or indirectly discharge 5.52 tonnes of arsenic-bearing effluents into air, 33 facilities that emit 4.46 tonnes of arsenic directly into surface waters and 2 that emit 0.0184 tonne of arsenic indirectly into surface waters each year (European Pollutant Emission Register, 2004).

France also has geogenic arsenic in abundance in association with arsenopyrite and gold. In addition, geogenic hidden arsenic has been identified in the Beaulieu and Seine estuaries (Howard *et al.* 1984; Michel *et al.* 1999). The absence of this hidden species of arsenic during the biologically dormant period of the estuarine and lacustrine year may be attributed to depressed metabolic levels of phytoplankton and algae: with lower excretory levels of methylarsenicals or arsenicals that are excreted but not detected by the traditional hydride generation method, they are hidden from traditional detection methods.

France also has elevated concentrations of arsenic in French rice from the Camargue region (0.28 mg/kg) with a range from 0.09 to 0.56 (mg/kg) total arsenic. This data must be correlated with the per capita rice consumption rate for France at 15 g/d as compared to rice intensive economies like Bangladesh at 445 g/d to assess its oncological implications. It is notable that developed countries like Spain have highly skewed high arsenic grain concentrations, while developing countries have normally distributed total grain arsenic concentrations (Meharg *et al.* 2009). This factor may be due to the local application of arsenical pesticides and herbicides.

The tailings of many mines pose an environmental threat, with acid mine drainage containing arsenic. For instance, Carnoules has posed an environmental threat for some time (Casiot *et al.* 2003). More importantly, since the second millennium BC, Salsigne in the south of France has been an important centre for mining of iron, copper, lead, silver, gold and arsenic. Because of the weathering of ancient mine dumps and tailings, the Orbiel River has been contaminated with arsenic (Blachère *et al.* 2005).

The North Mortagene area is characteristic of sites contaminated with arsenic as a result of sulfuric ore processing. To prevent further groundwater contamination in the area of a former zinc refinery, black slag has been used to immobilize the arsenic. This has been shown to be an effective trap for both arsenite and arsenate (Charlet *et al.* 2001).

To biologically remediate toxic acid mine drainage into the groundwater, an experiment was run in which samples of acid mine drainage were collected over a period of 14 months. Heterotrophic and non-obligatory acidophilic bacteria in the samples were cultured on a solid medium (pH 7.0–7.2), and three strains were isolated. The three different strains belong to the genus *Thiomonas*. Culturing in mineral media demonstrated that the isolated strains used thiosulfate as an energy source and oxidized iron in the presence of thiosulfate, but none of the strains oxidized arsenic in the presence of thiosulfate, nor could they use iron or arsenic alone as an energy source. In vitro experiments demonstrated that two of the *Thiomonas* strains were able to oxidize more than 90% of the arsenic (III) present in the acid mine drainage, whereas no abiotic oxidation of arsenic occurred. These results represent the first reported oxidation of arsenic by *Thiomonas* sp. Biologically mediated oxidation and subsequent immobilization of arsenic are of great interest for the remediation of contaminated mine sites (Bruneel *et al.* 2003). A similar experiment was reported by Duquesne *et al.* (2003) using *Acidithiobacillus ferrooxidans* as the biological binding agent.

Arsenic, including arsenic(III), also occurs naturally in French thermal springs in the Massif Central, and their drainage into groundwater and rivers poses threats to the environment as well (Criaud & Fouillac, 1989).

4.12 GERMANY (A, C, G, M)

Germany, with a population of 81.5 million, has a Human Development Index rank of 10 out of 169 and a per capita GNI of $US 35 308 (US Census Bureau, 2011; UNDP, 2010).

Anthropogenic arsenic in Germany comes from 21 pharmaceutical, chemical and metals manufacturing and mining companies that directly discharge 1.66 tonnes of arsenic-bearing effluents per year into surface waters, 14 that indirectly discharge 3.08 tonnes of arsenic-bearing effluents per year into surface waters, and 55 that emit 8.06 tonnes of arsenic effluents per year into the air, according to the European Pollutant Emission Register (2004). Another anthropogenic source of arsenic in Germany was uncovered when Raessler *et al.* (2000) found levels of arsenate up to 147 μg/l in ground waters from Kelheim. Levels were found to be highest during the autumn. The arsenic is believed to have leached from pyrites and iron oxides, which were deposited in the Kelheim region or used as filling materials in road construction and railway embankments.

Arsenic-enriched coals are found in two Mississippian bituminous deposits. The Ebersdorf deposit has arsenic at concentrations ranging from 200 to 400 mg/kg, which is directly correlated with sulfide enrichment. The Ehrenkammer deposit has a maximal coal ash arsenic concentration of 3000 mg/kg (Leutwein & Roesler, 1956).

Arsenic from mining sources in Germany has an ancient history that indicates how long-abandoned mines can be potential sites for environmental contamination. Although Germany's earliest mines date to the Neolithic, they were flint quarries in Bavaria and north of the Alps and did not involve arsenic (Weisgerber, 1993). Arsenic minerals were first exposed to weathering in the Bronze Age, when copper was mined in the Black Forest. Later, the Romans mined copper, lead, silver and beautiful pearly blue-green Smithsonite ($ZnCO_3$), and with these they inadvertently exposed associated arsenic minerals, such as arsenopyrite, Loellingite ($FeAs_2$), Iridarsenite [$(Ir,Ru)As_2$], Anduolite [$(Ru,Os)As_2$] and Omeiite [$(Os,Ru)As_2$]. Mining of these metals and minerals continued through the Middle Ages and the Renaissance, as arsenic minerals in association with ferrous and non-ferrous metals were exposed to weathering and leached into the surface water and groundwater (Clement, 1996; Goldenberg *et al.* 2003). Many of these ancient mines therefore pose environmental risks, because many are not identified or properly remediated. Historical mining, too, has led to heavy metals leaching from tailings as acid mine drainage. The list of such sites is large, as the locations of arsenic minerals in Germany number in the hundreds.

For instance, Freiberg mine water has a low pH and elevated concentrations of heavy metals, including arsenic and aluminium (Baacke, 2000). Also, the northern Palatinate region is characterized by elevated levels of arsenic and antimony in the soil due to the presence of ores. A skarn complex east of Schwarzenberg district, Ergebirge, Saxony, has tailings where arsenic-bearing ores were mined for centuries. An epidemiological study confirmed the existence of (non-hazardous) arsenic uptake by local populations in the area (Gebel *et al.* 1998), with pathways established in both food and water.

Geogenic arsenic is also found in Germany. Northern Bavaria, an area of 8000 km^2, has been found to have elevated levels of arsenic released from the Upper Triassic Keuper Sandstone into the groundwater, and no anthropogenic sources have been detected in the area. A survey of approximately 500 deep water wells revealed that 160 had arsenic at concentrations ranging from 10 to 150 μg/l. The hypothesis for the release of arsenic is that it is related to specific lithofacies of the aquifers and is geogenic in origin (Heinrichs & Udulft, 1999).

Since mining is so ancient in Germany, remediation of high-arsenic tailings that leach into the groundwater is an issue of great current interest. The criticality of remediating arsenic in Germany's drinking-water is highlighted by the 1999 proposal to privatize drinking-water supplies to the public and thereby potentially relax the EU arsenic standard of 10 μg/l. "In Germany, the Czech Republic and France, municipalities guarantee payments to companies if consumption or prices are not sufficient to ensure a profit" (CUPE, 1999). Estimates of relaxing the EU standard of arsenic by 10%, 30% and 100% in untreated Rhine-Westphalian drinking-water would result in 1094, 3281 and 10 938 arsenicosis cases, respectively, according to a recent modelling scenario (Fehr *et al.* 2003). Also in the news is a proposal to remediate the arsenic in the Rhine River. The project will reduce arsenic, zinc, aluminium, iron, manganese, phosphorus, barium and lead emissions, leading to an improvement in the quality of the water in the Rhine (European Report, 2005).

In 2004, when the macrophyte *Lemna gibba* (duckweed) was identified in tailings waters of two abandoned mining sites, Lengenfeld and Neuensalz-Mechelgrun, in Saxony, and it was found that duckweed was very successful in accumulating high concentrations of arsenic, researchers felt that duckweed might be used to biomonitor arsenic transfer (Mkandawire *et al.* 2004; Mkandawire & Dudel, 2005). After many laboratory trials, it was confirmed that duckweed can be used as a preliminary bioindicator for arsenic transfer from substrate to plants and might be used to monitor the transfer of arsenic from lower to higher trophic levels in abandoned mine sites. It may also be used for arsenic phytoremediation of mine tailing waters because of its high accumulation capacity.

4.13 GREECE (A, C, G, M)

Greece, with a population of 10.8 million, has a Human Development Index rank of 22 out of 169 and a per capita GNI of $US 27 580 (US Census Bureau, 2011; UNDP, 2010).

Anthropogenic sources of arsenic in Greece come from three industrial facilities that discharge 1.32 tonnes of arsenic per year into the air and two facilities that discharge 1.82 tonnes per year into surface waters (European Pollutant Emission Register, 2004). This arsenic precipitates out in the soil and surface waters.

Arsenic-enriched coals are found in north-west Greece in Miocene Serra lignite deposits, with concentrations up to 124 mg arsenic/kg. Arsenic-enriched Oligocene lignite deposits are found in the Drama region, with concentrations up to 50 mg/kg. Neogene lignite deposits occur in the Ioannina basin, with concentrations up to 43.8 mg arsenic/kg (possibly originating from syngenetic hydrothermal fluids seeping into coal-forming peat strata) (Filippidis *et al.* 1996; Gentzis, Goodarzi & Foscolos, 1997). In the Domeniko region of central Greece, arsenic-enriched Miocene lignite has up to 150 mg arsenic/kg, but its low ash yield has only 32 mg arsenic/kg (Pentari *et al.* 2004). The island of Crete has Middle Miocene lignites enriched with arsenic, measuring from 14.2 to 47.7 mg/kg, with a maximal ash content of 207 mg arsenic/kg (Gentzis *et al.* 1996).

Geogenic sources of arsenic have been identified in the drinking-water of the northern cities of Aksios and Kalikratia. In 21 representative locations, arsenic concentrations were typically 10–70 μg/l. Aksios groundwater has redox potential values of 87–172 mV and a pH of 7.5–8.2, with arsenic(III) predominating. Manganese concentrations were mostly above 50 μg/l (100–700 μg/l). In Kalikratia, groundwater has redox potential values of 272–352 mV and a pH of 6.7–7.5, and arsenic (V) predominated (Katsoyiannis *et al.* 2007).

Mining has also contributed to arsenic contamination. The Lavrion region of Greece, associated with lead and silver mining activity, has been found to have elevated levels of arsenic in groundwater as a result of the mining activity (Komnitsas *et al.* 1995). The former mining site (nowadays used as a technological and cultural park) has arsenic pollution mainly from slag, sulfur-containing waste, smelting waste and low-grade lead concentrates. Metal and metalloid concentrations were found to be extremely high, posing significant hazards for both human health and the ecosystem. Retention experiments were carried out using different soil/additive ratios, and the performance of untreated fly ash in combination with zeolites was also examined. Results showed a satisfactory retention with respect to copper, selective retention for iron, lead and zinc and rather poor retention for manganese and arsenic (Moutsatsou *et al.* 2003).

Central Macedonia (area 18 811 km^2) is one of 13 peripheries of Greece, situated between the peripheries of East Macedonia and Thrace in the north, West Macedonia and Thessaly to the south and the Republic of Macedonia to the west. With a population of 1 931 870 (2005), the second largest in Greece, and a population density of 103/km^2, the Periphery of Central Macedonia is at risk from arsenic in enriched coal, from geogenic sources and from mining sources. It has extensive arable plains that make it one of the most agriculturally prosperous peripheries, along with Thessaly and Thrace, so these crops might also be susceptible to bioaccumulation of arsenic.

Pleistocene Drama lignites in the Periphery of Central Macedonia are enriched with arsenic in the range of 102–606 mg/kg. In the Serres Prefecture deposit, the arsenic range is 856–1131 mg/kg. In both deposits, arsenic may have been hydrothermally present in the coal-forming peat bogs from mobilized sulfides (arsenopyrite sphalerite and galena) or been supplied in the Eocene-Oligocene volcanism (Papanicolaou *et al.* 2004).

Large deposits of copper, iron, lead, precious metals and zinc, many of which are in association with arsenic minerals, are identified as being a factor in geogenic arsenic contamination. Mineral and metal deposits have also attracted mining.

Mining is an important industry in the Periphery of Central Macedonia, and acid mine drainage containing arsenic and other metalloids and heavy metals has contaminated many of the watersheds. On 17 May 2007, a news release from Central Macedonia (Water for People & Peace, 2007) indicated that:

> Large amounts of Arsenic [see Figure 4.2], Nitrates[, and] Heavy Metals in areas such as [the] Delta of Axios, Chalastra, Sindos and Tringlia for many years now have been poisoning every life form.... 700 Citizens of Macedonia signed a petition in protest to the European Union for immediate action to be taken to solve the problem.

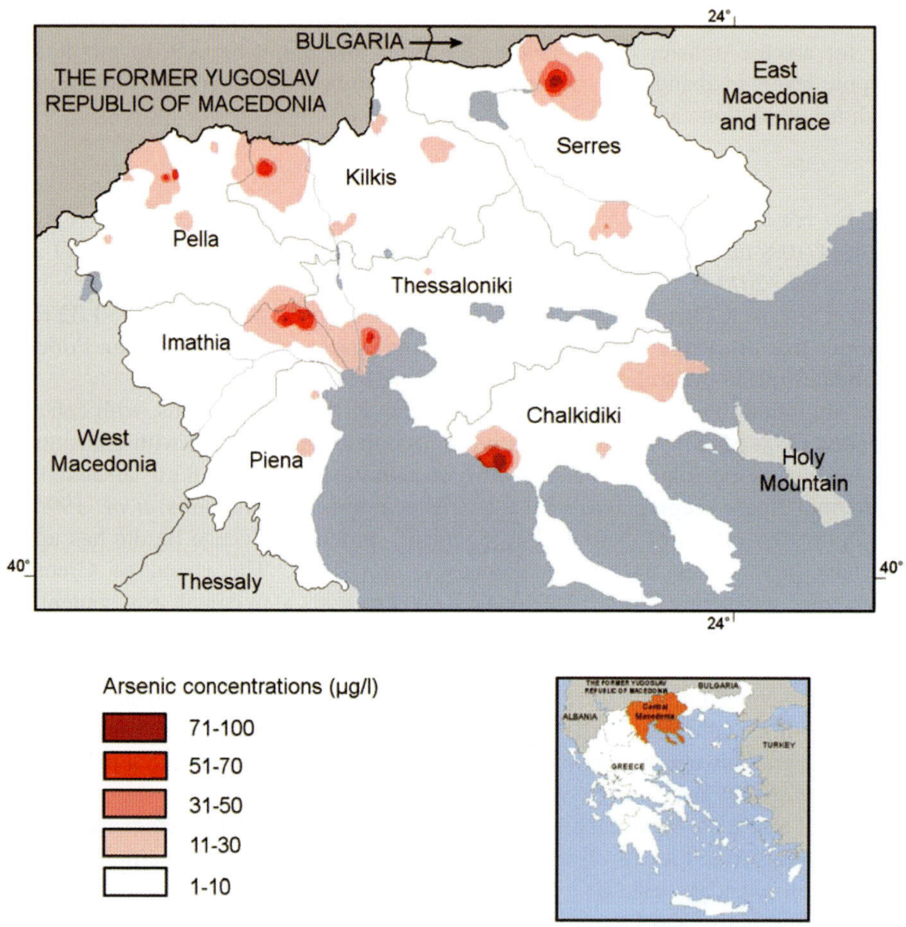

Figure 4.2 Map of the central part of the Periphery of Central Macedonia, Greece, showing arsenic concentrations in groundwater (modified with prefectures from http://www.waterpp.org/images1/Greece_Arsenic_Map.JPG).

In support of the citizens' claim, they also issued a demographic profile of deaths from cancer in both Central Macedonia and all of Greece, showing a higher percentage of deaths from cancer in Central Macedonia compared with the Greek average. A causal connection between elevated arsenic and mortality from cancer was implied but is not yet scientifically proven.

4.14 HUNGARY (A, G, M, V)

Hungary, with a population of 10.0 million, has a Human Development Index rank of 36 out of 169 and a per capita GNI of $US 17 472 (US Census Bureau, 2011; UNDP, 2010).

Anthropogenic sources of arsenic in Hungary come from three chemical plants that discharge 0.169 tonne of arsenic per year into the surface water and one facility that discharges 0.042 tonne per year into the air (European Pollutant Emission Register, 2004).

In Hungary, geogenic arsenic compounds are naturally leached from arsenic-bearing rocks situated on the Great Pannonian (Hungarian) Plain in the presence of elevated concentrations of ammonium, iron, manganese, humic acids, dissolved gases and temperature above 30°C (Polyak & Hlavay, 1999). In a study of urinary arsenic of cohort groups of nearly equal sex distributions from Bacs, Békés, Csongrad and Jasz-Nagykun-Szolnok, there was a significant correlation between arsenic in water and arsenic in urine ($R^2 = 0.46$). High body mass index and cigarette smoking correlated with elevated urinary arsenic (III). At low water concentrations, urinary dimethylarsinic acid was elevated, indicating exposure via food. Maximum concentrations of arsenic in drinking-water were 39 µg/l at Bacs, 31 µg/l at Békés, 40 µg/l at Csongrad and 88 µg/l at Jasz-Nagykun-Szolnok (Lindberg *et al.* 2006). Countries adjacent to this geological situation should be alert to arsenic contamination of groundwater, especially Croatia, Serbia, Romania, Ukraine and Slovakia.

The Great Pannonian Plain in the Danube basin has an underlying, reducing Quaternary alluvial aquifer. In Figure 4.3, its outline can be seen in areas where arsenic in groundwater is in the range of 10 µg/l and higher. A list of those counties at risk is given in Table 4.1, and an estimate of 1.5 million of the population at risk is given by Biacs (2005). Given the population figures for each county and the relative areas of each county affected, this estimate seems very modest, and a rough upward estimate is 2.5 million.

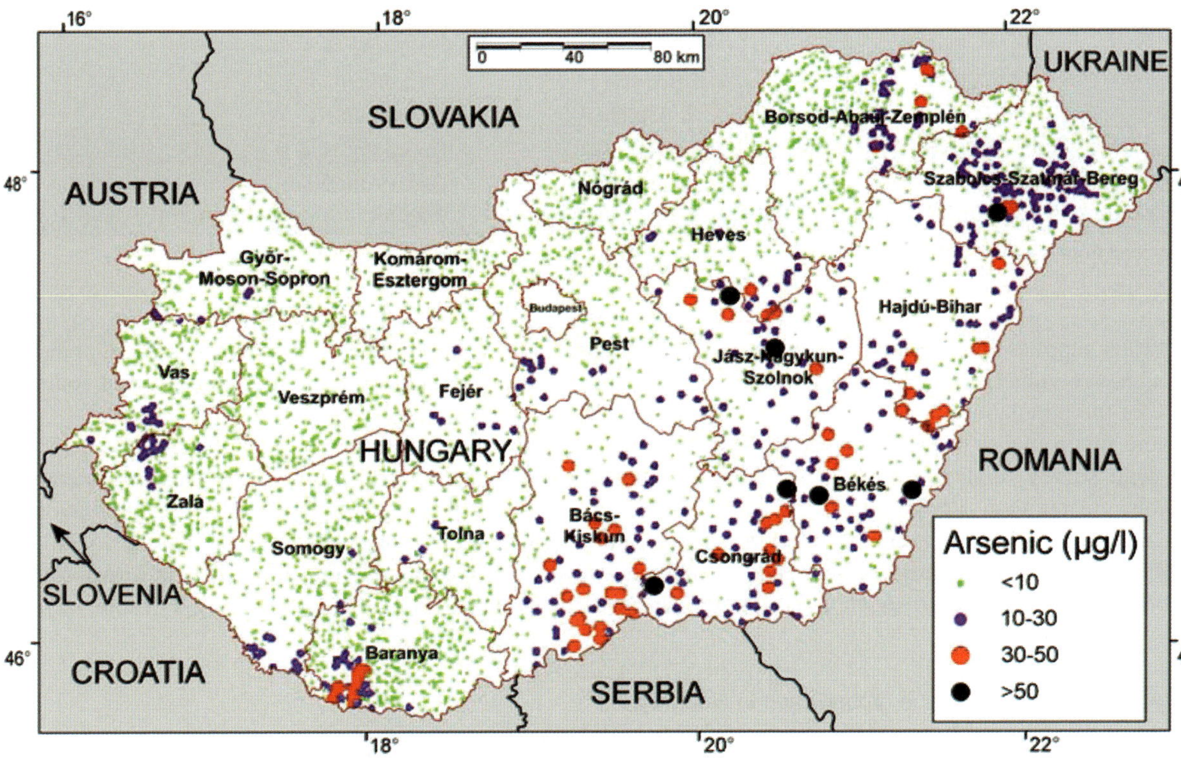

Figure 4.3 Arsenic concentrations in Hungarian drinking-water (modified from Biacs, 2005).

Testing has been undertaken in all counties, and concentrations of arsenic above 50 µg/l have been identified from alluvial sediments associated with the River Danube in the southern part of the Great Pannonian Plain. Data from these tests provide details as to the occurrences of arsenic in groundwater in individual counties. For instance, in Csongrád County, with a population of about 426 000, arsenic in groundwater measured on average 32 µg/l for 85 samples with concentrations up to 150 µg/l (Varsányi *et al.* 1991). Csanády *et al.* (1996) reported levels up to 330 µg/l in a wide-scale survey that indicated that the extent of contamination was county-wide.

Table 4.1 Hungarian counties with arsenic concentrations in groundwater above 10 μg/l, and estimated populations at risk.

County	Population	Population potentially at risk of exposure to >10 μg arsenic/l
Bács-Kiskun	541 584	Throughout county
Baranya	402 260	Area south-west of Pecs
Békés	392 845	Throughout county
Borsod-Abaúj-Zemplén	739 143	Eastern third of county
Csongrád	425 785	Throughout county
Fejér	428 579	Sporadic
Győr-Moson-Sopron	440 138	Very sporadic
Hajdú-Bihar	550 265	Half of county
Heves	323 769	South border with Jász-Nagykun-Szolnok
Jász-Nagykun-Szolnok	413 174	Throughout county
Komárom-Esztergom	315 886	Sporadic
Nógrád	218 218	Not at risk
Pest	1 124 395	Sporadic
Somogy	334 065	Sporadic
Szabolcs-Szatmár-Bereg	583 564	About half of country
Tolna	247 287	Sporadic
Vas	266 342	Sporadic
Veszprém	368 519	Not at risk
Zala	269 705	Sporadic
Estimated total population at potential risk of exposure to >10 μg arsenic/l	**2.5 million**	Based on estimate of percentage of county affected, from this table and Figure 4.3.

Source: Compiled from data in Biacs, 2005

Surface water supplies only 10% of the population, whereas groundwater supplies 90%. Thirty-three per cent of the country's water resources are vulnerable to arsenic poisoning, as they are associated with karstic, ground, bank-filtered sources, which supply 67% of the population.

Wells were first found to be contaminated in 1981, although deep wells had been in use since the end of the 18th century. A countrywide survey, undertaken in 1981, indicated that 400 000 people were exposed to drinking-water containing arsenic levels greater than 50 μg/l (Csanády *et al.* 1996). As a result of mitigation measures, the population exposed to elevated levels of arsenic in drinking-water decreased, and estimates now range between 10 000 and 15 000 (Csanády *et al.* 1999; Smedley & Kinniburgh, 2002; Biacs, 2005). However, the WHO guideline of 10 μg/l is a real challenge for Hungary, because 1.4–2.5 million people are drinking water with arsenic concentrations above 10 μg/l (Table 4.2).

Table 4.2 Arsenic concentrations in groundwater in Hungary.

Arsenic concentration in groundwater (μg/l)	Number of settlements affected	Population affected
50–330	6	10 000–15 000
30–49	64	131 000
10–29	345	1 400 000
Total	**415**	**1 541 000**

Source: Biacs, 2005

Geogenic sources of arsenic in Hungary occur in zones of arsenopyrite, which leaches into surface water and groundwater, such as in the Matra Mountains in north-central Hungary, at Recsk and at Gyongyosoroszi. Recsk is characterized by porphyry copper–polymetallic veins and a copper/gold skarn overlain by an argillic alteration in a matrix of Palaeogene volcanic andesitic rocks on a basement of Triassic limestone. Within the argillic alteration, copper–gold ores occur in association with arsenopyrite. The formation has weathered into an upland valley where there are several small villages.

Mining sources of arsenic come from mine dumps and tailings, especially Lahoca. These leach acid mine drainage into streams and groundwater, exposing villages to high levels of arsenic in their drinking-water. A similar situation occurs at Gyongyosoroszi, where a lead–zinc polymetallic vein on the southern flank of the Matra Mountains contains sphalerite, galena and arsenopyrite. Acid mine drainage containing high concentrations of arsenic is leaching into the surface water and groundwater, which flow through the village to Toka Creek, where intermittent flooding has deposited contaminants, including arsenic, in the village and on the floodplain, which is used as a vineyard (Berger, 1995).

Skin lesions are the main arsenic-related health effect reported in Hungary. It has been suggested, however, that the numbers of stillbirths and spontaneous abortions are significantly higher in areas where arsenic levels exceed 100 μg/l (Börzsönyi *et al.* 1992). These rates were found to decrease following arsenic mitigation measures (Rudnai *et al.* 1996). Reports of other non-occupational arsenic-related health effects have not been identified.

A countrywide project to reduce arsenic levels in drinking-water to below 50 μg/l was conducted over a 15-year period. Several mitigation measures were employed. Primarily, alternative water supplies were sought, or water from wells containing elevated levels of arsenic was mixed with sources containing low levels. In some villages, co-precipitation technology (involving oxidation, coagulation and precipitation processes) was used (Csanády *et al.* 1999). In response to the challenge of meeting the WHO guideline, the University of Veszprem has been developing new arsenic adsorbents (Hlavay & Polyak, 1998; Polyak & Hlavay, 1999). The 2005 solution is to continue to seek technological solutions and to not use the WHO guideline, at least for the present (Biacs, 2005).

4.15 ICELAND (A, V)

Iceland, with a population of 311 000, has a Human Development Index rank of 17 out of 169 and a per capita GNI of $US 22 917 (US Census Bureau, 2011; UNDP, 2010).

Although Iceland's thermal springs are associated with relatively young basalts and in general pose no serious arsenic contamination, arsenic has been measured in the range 50–120 μg/l in thermal waters (White *et al.* 1963). The possibility of these waters leaching into aquifers and being used for drinking-water is not known. However, elevated arsenic is found in surface and thermal waters in the north of the island (Arnórsson, 2003):

> concentrations of As in surface- and up to 90°C ground waters in a tholeiite flood basalt area in N-Iceland lie in the range <0.03–10 μg/kg. With few exceptions surface waters contain <0.5 μg/kg As whereas ground waters generally contain >0.5 μg/kg As. The As content of ground waters increases on the whole with rising temperature. Arsenic is highly mobile in the basalt-water environment of the study area. An insignificant fraction of the As dissolved from the rock is taken up into secondary minerals. Arsenic is less mobile than B but considerably more mobile than Na which has the highest mobility among the major aqueous components. A significant fraction of the As in the basalt occurs in an easily soluble form. The As hosted in the primary minerals is expected to be concentrated in the titano-magnetite. This mineral is stable in contact with both surface- and ground waters and does not, therefore, supply As to the water, explaining the difference in mobility between As and B. Aqueous As concentrations are a reflection of water/rock ratios, i.e. how much rock a given quantity of water has dissolved. This ratio increases with increasing temperature and increasing residence time of the water in contact with the rock. The distribution of As species has been calculated on the assumption of equilibrium at the redox potential retrieved from measurement of aqueous Fe(II) and Fe(III) concentrations. These calculations indicate that pentavalent As is stable in surface waters and in ground waters with an in situ pH of <10 and would occur mostly as $H_2AsO_4^-$ and $HAsO_4^{-2}$. In higher pH ground waters the concentrations of the arsenite species $H_2AsO_3^-$ is significant at equilibrium, up to 65% of the total dissolved As.

In addition, the use of thermal energy has increased, and therefore the effluent from thermal power plants demands constant monitoring. One example of such a plant is reported below (Wetang'ula & Snorrason, 2003):

> An environmental assessment of potential effects of current geothermal wastewater disposal options in the exploitation of the Nesjavellir geothermal field (Iceland) … is made. At Nesjavellir co-generation power plant hot wastewater from the plant is either pumped into shallow drill holes or disposed of in the Nesjavellir stream, which disappears into the lava and finds its way into Lake Thingvallavatn, a rift lake of high conservational value. Measurement of quantities of chemical constituents in effluents and lakeshore outflow sites (springs), and in the biota at outflow sites, and at a control site, Vatnskot, was done. Also an assessment of thermal stress at lake shore line outflow sites was done. Recent increase in the volume of wastewater has caused a 15–17°C rise in water temperature at lake shore line outflow sites. In calm periods during the ice free period and when the lake freezes over in the

winter this causes temporal thermal stress near the outflow sites. Concentrations of SiO_2, As, Al and B are elevated in geothermal effluents but are diluted before the wastewater reaches the lake. There is no detectable rise or accumulation of trace elements in the biota at the geothermally influenced sites. However, taking into account the conservational value of Lake Thingvallavatn, sound wastewater management by deep re-injection and regular monitoring of thermal influence, chemical constituents in effluents and lake outflow sites water should be adopted.

4.16 IRELAND (A, C, M)

Ireland, with a population of 4.7 million, has a Human Development Index rank of 5 out of 169 and a per capita GNI of $US 33 078 (US Census Bureau, 2011; UNDP, 2010).

Anthropogenic arsenic in Ireland comes from one alumina production facility and three electrical power generating plants, which report 0.16 tonne of arsenic per year released into the air (European Pollutant Emission Register, 2004).

Also, Ireland has major stores of lead and zinc, and these have been mined at Tara Mine (County Meath, active), Glenadalough Mines (County Wicklow, abandoned), Galmoy Mine (County Kilkenny, active) and Lisheen Mine (County Tipperary, active). In addition, there are stores of silver in County Tipperary that was once mined at the Silvermine and stores of copper and sulfur in County Wicklow that have been mined at Avoca Mine (Wolkersdorfer & Bowell, 2005). These deposits and their tailings suggest that arsenic is probably present in some; their natural occurrence also suggests the potential for some geogenic release of arsenic as well. In an effort to remediate the acid mine drainage from these mine tailings in Ireland, there has been an aggressive programme of constructing wetlands. The abandoned mine tailings at the monastic site of Glenalough, County Wicklow, for instance, are the site of a constructed wetland for the retention of metals from lead and zinc extraction (Beining & Otte, 1996).

4.17 ITALY (A, C, G, M, V)

Italy, with a population of 61.0 million, has a Human Development Index rank of 23 out of 169 and a per capita GNI of $US 29 619 (US Census Bureau, 2011; UNDP, 2010).

Anthropogenic arsenic in Italy is linked to 22 chemical, coal-burning and thermoelectric utilities that discharge 2.02 tonnes of arsenic per year into the air, 26 facilities that discharge 4.12 tonnes per year directly into surface water and 10 facilities that indirectly discharge 0.778 tonnes per year into surface water (European Pollutant Emission Register, 2004).

Geogenic arsenic in Italy is attributable to highly mineralized areas in Tuscany, Sardinia and Sicily, often with supplemental volcanogenic mechanisms that further aid in the dissolution of arsenic from arsenopyrite and other arsenic-bearing minerals. Geogenic hidden arsenic has also been found in fish in the south Adriatic. Skate (*Raje* spp.), hake (*Merluccius merluccius*) and blue whiting (*Micromesistius poutassou*) contained arsenic at concentrations measuring 14.4–61.5 mg/kg, 8.5–21.5 mg/kg and 5.9–16.1 mg/kg, respectively. Most arsenic was organic, whereas the inorganic arsenic component was only 0.47–3.5% of the total arsenic (Storelli & Marcotrigiano, 2000).

A survey of Italian rice ($n = 38$) found a mean total arsenic concentration of 0.07 to 33 mg/kg. This data must be correlated with the per capita rice consumption rate for Italy at 25 g/d (husked rice) and 17 g/d (polished rice) as compared to rice intensive economies such as Bangladesh at 445 g/d (polished rice) to assess its oncological implications. Another oncological factor is the total inorganic arsenic ($n = 5$) of 0.506 found in Italian rice which compares with Bangladesh 0.719 ($n = 15$). It is also notable that developed countries like Italy have highly skewed high arsenic grain concentrations, while developing countries have normally distributed total grain arsenic concentrations (Meharg *et al.* 2009). This factor may be due to the local application of arsenical pesticides and herbicides.

Mining and volcanogenic sources of arsenic can be closely interrelated. For instance, the mining history of Tuscany from 1000 BC to the 1970s documents the exploitation of tin, copper, lead, zinc, iron and arsenic. Although almost all of the mines are closed because of depletion or economic and environmental conditions, water contamination from acid mine drainage is prevalent around tailings ponds at Gavorrano, Campiano and Fenice Capanne. For instance, the tailings pond of the Campiano mine, which closed in 1995, feeds a spring with a 16 l/s discharge of acid mine drainage with elevated concentrations of sulfate, iron, copper, zinc and arsenic that is mixed with geothermal waters (also possibly with arsenic) with a pH of 4. This pond is leaching into the Merse River, which happens to be an area of particular natural and tourist interest (Garziono, 2005). In southern Tuscany, at the Gabellino mine's flotation tailings, arsenic was measured at 50–388 mg/kg. At the Fontalcinaldo waste rock dump, arsenic was measured at 55–1075 mg/kg (Benvenuti *et al.* 1997). These tailings and dump were a serious hazard, as they were leaching into the groundwater system.

Arsenic pollution of surface water and groundwater from mining is common in Tuscany. For instance (Di Basilio *et al.* 2002):

> A geochemical study was carried out in an abandoned mining district located in southern Tuscany … to evaluate the environmental effects of past mining activities. Mineralogical, chemical and physical investigations of heavy metal-rich tailings and flotation waste were undertaken to estimate their impact on water and soil quality. Soil cores were obtained by a dry rotation drilling technique. Surface waters were sampled at 8 stations along a bordering stream and at 1 from an artificial lake (flooded open pit). Both soil and water samples were analyzed for Hg, As and Pb by ICP-MS [inductively coupled plasma mass spectrometry]. Soil chemical-physical properties (density, porosity, granulometry, permeability, pH, CEC [cation exchange capacity], TOC [total organic carbon] etc.) were also determined. Mineralogical composition of solid samples was examined by X-ray powder diffraction. Soil concentrations were as high as 100 mg kg^{-1} for Hg, 2000 mg kg^{-1} for As and 200 mg kg^{-1} for Pb. These anomalous values are closely related to the great amount of polluted tailings that seems to cause high water contamination downstream of the mine area.

In addition, complex interrelationships can exist between anthropogenic and geogenic sources of arsenic (Marchettini *et al.* 2002):

> The problem of arsenic pollution was studied in the area of an industrial plant which produces sulfuric acid from pyrite combustion. The plant is located in a former wetland, close to the sea, in the county of Scarlino, southwest of Tuscany. Horizontal distribution, shows a regular decrease in the As concentration going far from the plant, but with the presence of hot spot at the distance of 4–6 km from the industry, where the As level were found very high ($As > 600$ mg kg^{-1}). Other irregularities in the vertical distribution of As concentration in the clay-rich soil, were found. This induced some chemists and geologists to propose the hypothesis of Natural origin of arsenic anomalies, due to the fact that the site under study is downstream of a large area that is characterized by large metal mineralizations. On the other hand, several facts suggested that hematite dusts that are residue of the pyrite combustion, and the stock of fine pyrite particles that have been disposed on the soil close to the plant without control, are the major factors responsible for the widespread pollution. Analytical and hydrological data, were collected and a geo-chemical map was produced in order to verify which one of these two hypotheses was correct.

Arsenic contamination is probably attributable to both geogenic and anthropogenic sources.

Volcanogenic arsenic is known from its occurrence on the island of Sardinia at Furtei at the Furtei Volcanic Complex, a major Tertiary-age sub aerial calc-alkaline volcanic centre 5 km in diameter. The volcano-sedimentary sequence is dominated by andesitic lavas and domes, block and ash deposits, interlayered pyroclastics and sedimentary rocks. A central diatreme, measuring 1000 m north–south by 600 m east–west, crosscuts the volcano-sedimentary sequence. Miocene limestone overlies the Furtei Volcanic Complex to the east and west. The earliest phase of gold mineralization is hosted in sheeted microfractures developed in flat-lying sediments and ignimbrite sandwiched between andesite flows. Gold is associated with silica, pyrite and lesser sphalerite (Gold Mines of Sardinia PLC, undated).

One of the most common sulfide minerals bearing arsenic in the Furtei mine is enargite, a copper arsenic sulfide (Cu_3AsS_4). It has surface reactivity properties that impact the environment and copper mining and ore processing. Enargite as a polysulfide is a potential source of mobilizable arsenic and may cause environmental problems upon oxidation. The occurrence of arsenic in the groundwater on Sardinia, although minor, is of concern and is related to both geogenic and mining sources (Fantauzzi *et al.* 2004).

4.18 LATVIA (A)

Latvia, with a Human Development Index of 48 out of 69 and a per capita GNI of $US 12 944 (UNDP, 2010), is bordered by Estonia to the north and by Lithuania to the south and south-west, whilst on the east there is a frontier with the Russian Federation and to the south-east a border with Belarus. Latvia's territory of 64 000 km^2 is home to about 2.2 million people (US Census Bureau, 2011), about 70% of whom are citizens.

Prior to 2000, there had been little thought given to environmental concerns, and this has negatively impacted water resources. In 2004, Latvia joined both the EU and the North Atlantic Treaty Organization, and this has had a positive effect on environmental remediation projects that affect water quality. Latvia's environment has benefited from a shift to service industries after the country regained independence from the former Soviet Union on 21 August 1991. The main environmental priorities are improvement of drinking-water quality and sewage treatment, household and hazardous waste management, and the reduction of air pollution. In 2001, Latvia committed to full enforcement of EU environmental directives by 2010 (World Facts, 2007).

There are approximately 550 landfills in Latvia that had been established under Soviet control in the 1970s and 1980s without any consideration of environmental impacts. The Ministry of Environment Protection and Regional Planning has instituted a strategy of waste disposal as well as legislation to reduce the number of landfills and to engineer new waste disposal sites (Blumberga, 2001).

However, the lack of environmental concerns in the past has created serious remediation problems with regard to drinking-water supplies. In the Liepaja region in western Latvia on the Baltic Sea, 27 ageing landfills had been constructed in the Soviet era without any environmental protection. The disposal technology simply involved filling the landfill with liquid and solid wastes, compacting it with a bulldozer and covering it with a thin layer of soil. Rainwater infiltrated these landfills, increasing the water volume in the pit, which then leached inorganic and organic contaminants into the groundwater (Bedient *et al.* 1999; Blumberga, 2001).

The biggest landfill in Liepaja is at Skede, which covers 7.5 ha, of which 6 ha are covered by waste. It is located 10 km north of Liepaja city and is the repository for 79% of the waste of the entire region. When it was built in 1972, no hydro-isolation barrier was installed, so leaching for 30 years has created serious contamination of groundwater and surface water, which has spread in two plumes, one to Tosmare Lake and the other to the Baltic Sea. The Tosmare Lake basin into which one toxic plume from the landfill flows is 73 km^2, which is the main area affected by the contaminants. The area occupied by the lake is only about 4.05–4.15 km^2 (with an open water surface of only 50 ha because of eutrophication), so this is only a fraction of the affected area. The effect of the contaminants on the Baltic Sea sediments and water is unknown. Soils surrounding the landfill in the direction of the lake also have been polluted with toxic airborne particles. Close to the landfill is a cooperative garden, the produce of which bioaccumulates heavy metals and arsenic. In addition, prior to 2000, Skede had burned raw waste into the atmosphere (Blumberga, 2001).

The level of arsenic contamination in the groundwater is 15.5 μg/l (2001), but this level is expected to reach 200 μg/l in 2070 based on a modelling projection using LandSim (Blumberga, 2001).

The 140 000 inhabitants of the Liepaja region not only have been subjected to these sources of contamination in produce, fish, water and air, but also have experienced malignant tumours that have been increasing at a rate of 7.7% from 1997 to 2006 (Blumberga, 2001; Latvia Health Statistics Department, 2007).

The geological structure of the aquifers at the Skede site has the following strata (Blumberga, 2001):

(1) Quaternary sediments
- ~13 m of technogenic deposits of household/commercial and industrial/demolition waste
- 1–2 m of boggy deposits of peat
- 7–8 m of Eolic deposits of fine sand
- 3–4 m of marine deposits of the Litorina Sea of sand interbedding with silt and mud
- 7–20 m of Glacigene deposits of the Latvian Formation of loam, gravel and pebble

(2) Upper Devonian sediments:
- <5 m of Muri Formation of alternating dolomite and clay
- >15 m of Zagare-Joniski Akmene Formation of alternating dolomite, dolomite mar, siltstone
- Amata Gauja sandstones

(3) Permian limestone

The total water extraction from all these water horizons in 1993 was 62 000–63 000 m^3/day from 750–770 registered borehole wells. The highest water yield, and the source of drinking-water for the region, is from the Upper Devonian and Permian formations. The flow of artesian groundwater is north-west towards the Baltic Sea, and the flow of shallow groundwater is south-east towards Tosmare Lake. In other words, contaminants, including arsenic, flow in two distinct plumes contaminating the Baltic Sea and Tosmare Lake, in addition to contaminating the underlying aquifers upon which the region depends. The main part of the leachate (about 7688 m^3/year or 21.1 m^3/day) from the Skede landfill is discharged into Tosmare Lake (Blumberga, 2001).

4.19 LITHUANIA (A)

Lithuania has a population of 3.5 million, with a Human Development Index of 44 out of 169 and a per capita GNI of $US 14 824 (US Census Bureau, 2011; UNDP, 2010).

Offshore sediments have been analysed for arsenic pursuant to identifying the extent of contamination from chemical weapons munitions dumps. It is known that arsenic occurs as a long-term constituent in many chemical weapons and can be used as an indicator of degradation of released munitions in the marine environment. Arsenic munitions exist in Baltic Sea sediments in the Lithuanian economic zone in the Gotland and Bornholm basins offshore, and arsenic concentrations range from 1.1 to 19.0 mg/kg, with an average of 3.4 mg/kg. Elevated arsenic concentrations in sediments near the weapons dumpsite were low in comparison with those at other sites in the Baltic and North seas, but they still require monitoring (Veiderma, 2005; Garnega *et al.* 2006).

Elevated concentrations of anthropogenic arsenic above 1.4 mg/kg have been detected in *Pleurozium schreberi* mosses around the Akmenės Cementas (cement) factory, in Naujoji Akmenė, a city in Akmenė district municipality near the Lithuanian border, and one of the newest cities in Lithuania. Akmenės Cementas is the largest cement company in the Baltic region (http://www.cementas.lt/old/?obj=changeLng&lng=en) and the only cement company in Lithuania (Buse *et al.* 2003). Because the city is a relatively new industrial centre (the cement factory was built in 1952), it did not have a well-developed infrastructure or environmental monitoring system. Akmenės Cementas produces 711 200 tonnes of cement annually and meets EU standards.

4.20 MACEDONIA (THE FORMER YUGOSLAV REPUBLIC OF) (A, G, M, V)

The former Yugoslav Republic of Macedonia has a population of 2.1 million, with a Human Development Index of 71 out of 169 and a per capita GNI of $US 9487 ((US Census Bureau, 2011; UNDP, 2010). It has a land area of 25 333 km^2, has 24% arable land used for agriculture and vineyards and is bordered by Greece to the south, Bulgaria to the east, Serbia to the north and Albania to the west. Since 2005, it has been a candidate for membership in the EU and the North Atlantic Treaty Organization.

Environmentally, The former Yugoslav Republic of Macedonia is comparable with other Central and Eastern European countries in which centrally planned economies were seen largely in terms of increasing production in the industrial and energy sectors, resulting in overexploitation of natural resources and severe environmental degradation (UNEP, 2000). In an effort to address this problem, the first official environmental report by The Former Yugoslav Republic of Macedonia was issued in 2001 (Republic of Macedonia, 2001).

In addition, a study measuring the bioaccumulation of arsenic in mosses was carried out throughout the country, and the results are shown in Figure 4.4. Mineral waters in Kavadarci (affected by the Kozuf mine) and Kocani are within the range 20–100 μg arsenic/l, whereas normal arsenic concentrations in mineral waters in The former Yugoslav Republic of Macedonia range between 1 and 10 μg/l, so this may contribute to the accumulation of arsenic in mosses at Kavadarci, but Kocani appears unaffected (Barandovski *et al.* 2006). Sepiste appears to be the most seriously at risk, followed by Radovic. The area bordering the Periphery of Central Macedonia in Greece is not well sampled. For the other important heavy metal contaminants and their areas of impact, see Barandovski *et al.* (2006).

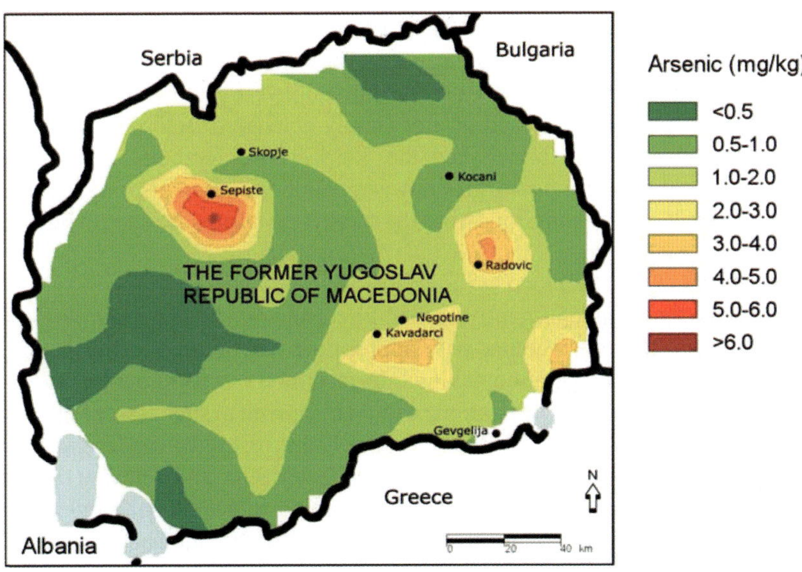

Figure 4.4 Zones with arsenic-contaminated mosses in the Former Yugoslav Republic of Macedonia (modified from Barandovski *et al.* 2006).

The Former Yugoslav Republic of Macedonia is semi-arid, with the Ovce Pole region being the driest area in the Balkans. There are about 250 watersheds larger than 20 km^2, the largest being the Vardar River. The Crni Drim River watershed in the west has more water than the central and eastern areas. Three natural lakes, Ohrid, Prespa and Dojran lakes, are important for their local municipalities.

Water resources vary, with the west having more water. Total water resources are estimated to be 6.63 billion m^3. These are carried in the Vardar basin (4.77 billion m^3), the Strumica basin (0.18 billion m^3) and the Crni Drim basin (1.68 billion m^3), which is the predominant drinking-water source, encompassing 30 springs with a flow greater than 100 l/s. These springs have not been monitored for arsenic, and so they may be a possible source of arsenic contamination, as are similar springs in the neighbouring Periphery of Central Macedonia of Greece.

Groundwater is not effectively monitored for volume and quality. Water use categories are 34% for irrigation, 11% for industry and 10% for domestic consumption, with the remainder unaccounted for. Conflicts arise between the different water user groups, highlighting the need for better water resource planning. In fact, the existing Water Resource Plan, which was prepared using data from 1949–1950 and 1968–1969, needs to be updated with more recent data (Republic of Macedonia, 2001).

Regarding overall water resources, the following observations were presented by UNEP (2000):

> large parts of the country are affected by surface-water and groundwater pollution. The Vardar River, which supplies 75% of the country's total water resources, is heavily polluted by untreated urban and industrial pollution. After passing through FYR [the Former Yuglosav Republic] of Macedonia, the river flows south into Greece, entering the Aegean Sea near Thessaloniki. The principal sources of water pollution are the major cities and more than 130 industrial facilities throughout the country. Only three municipalities, Ohrid, Prespa, and Dojran, have wastewater treatment plants. Ohrid's, however, is the only one functioning to capacity; the collector systems of the other two are not completed. The largest contributors of industrial pollution are the metallurgical, chemical and mining industries.... The surface-water monitoring network, for the Vardar and other rivers, is now being substantially modernized. Groundwater monitoring, however, was stopped in 1981 due to lack of financial resources.

As a result, "Water pollution is evident in rivers and ground waters" (Republic of Macedonia, 2001).

Geogenic arsenic contamination can be identified in four areas adjacent to monitored arsenic contamination hot spots in the Periphery of Central Macedonia in Greece. Groundwaters in the municipalities of Novaci and Prilep adjacent to the Pella Prefecture in Greece contain arsenic concentrations of at least 11–30 µg/l. Gevgelija (see Figure 4.4) and the area around Dojran Lake in Star Dojran municipality adjacent to the arsenic hot spots in Kilkis Prefecture in Greece also have arsenic contamination of at least 11–30 µg/l.

In addition, UNEP (2000) indicated that the following are potential contamination sites for heavy metals, presumably including metalloids such as arsenic. They should be monitored immediately:

(1) *Ferro-alloy plant at Jegunovce ("HEK Jugochrom")*
   - (a) To prevent further contamination, the chromium processing, facilities should be cleaned immediately, and the chromium waste should be stored and disposed of pursuant to appropriate waste handling methods.
   - (b) The chromium remediation program requires urgent revision and acceleration. Additional mapping of the chromium plume in the groundwater should be considered. Additional investigation wells may also be necessary. Chromium should be removed and properly disposed of, not simply converted from Cr(VI) to Cr(III). The program should encompass the area between the factory and the Vardar River and not be limited to implementation of a hydrological barrier. The potential impact of groundwater pollution on Rasce Spring should be thoroughly investigated.
   - (c) Existing proposals to reduce dust emissions from the three largest furnaces should be implemented. The six smaller furnaces require similar attention....

(2) *Lead and zinc smelter, Veles ("MHK Zletovo")*
   - (a) If commitments for the financing of urgently needed filters and other clean technologies cannot be obtained within the next year, then management, together with appropriate Government officials, should review the costs and benefits of the plant's continued operation, taking into account the costs to public health and other relevant factors.
   - (b) In the short term, existing air monitoring stations should be moved to sites closer to the plant that will indicate more quickly when elevated pollution levels are increasing health risks. In the longer term, additional stations that provide continuous data should be installed to provide citizens and relevant authorities, including regional public health and environmental officials, with early warning of acute health risks.
   - (c) Urgent measures should be taken to reduce emissions of sulfur dioxide and dust containing heavy metals.
   - (d) The proposed reconstruction of the wastewater treatment plant should be accelerated. The volume of process water used should be reduced through recirculation in a closed system. Untreated wastewater must not continue to enter the Vardar River.
   - (e) A soil and groundwater monitoring program is required and should encompass the vicinity of the plant itself, the adjacent flood plain, and areas immediately downstream.

(3) *Zinc and lead mine, Probistip ("Rudnici Zletovo")*
   - (a) High priority should be given to constructing a wastewater treatment plant for the concentration plant.
   - (b) There is an urgent need for investigation of wells downstream of the hydrotailings sedimentation basins.

(c) An investigation program should be established for private water wells downstream from the discharge points of the Koritnico and Kiselia Rivers into the Zletovska River. The Kiselia River should be diverted from its channel under the old hydro-tailing area to a position less vulnerable to the influence of leached pollutants.

(d) Measures should be implemented to reduce dust levels from ore crushing.

(4) *Thermal power plant, Bitola ("REK Bitola")*

(a) Several steps are required for tackling the problem of contamination from the dumpsite: divide the area into compartments; separate the compartments, using layers of fly ash and slag covered with coarse material to prevent ash and slag from spreading to the surrounding area; use only one compartment at a time; vegetate inactive compartments to increase rainfall evaporation and reduce leaching of contaminants into the groundwater; and establish a drainage system in the bottom of the dumpsite to handle leachate and percolate.

(b) Technical solutions should be implemented to reduce fly-ash dust and sulfur dioxide emissions from the plant.

(c) Soil and groundwater at the power plant site and in adjacent areas is likely to be contaminated, and polluted groundwater may be entering the river. This requires establishment of an investigation program that includes the wastewater canal, river water and private wells downstream of the power plant.

(d) The flow of untreated wastewater from the plant should be halted by replacing the neutralization basins with proper wastewater treatment facilities and by reconstructing the oil separator to match wastewater volumes.

(e) The feasibility of using excess heat to supply the city of Bitola should be investigated....

(5) *Copper mine, Radovis ("Buchim S.C.")*

(a) Construction of a wastewater treatment that includes removal of heavy metals should be accelerated.

(b) In view of the high risk of groundwater contamination, the plant's monitoring program should be expanded to cover heavy metals such as lead, cadmium and chromium.

(c) The hydro-tailing dam poses a high risk of infiltration by heavy-metal contaminated water into the upper groundwater aquifer downstream, especially during high water periods. Additional monitoring is required to establish the presence and extent of contamination.

(d) Efforts to stabilize tailing dust at the hydro-tailing dam should ensure that emissions affecting the village downwind are eliminated....

(6) *Lojane Mine*

(a) The site should be immediately subject to a comprehensive environmental investigation. Recommendations for short-term risk reduction should be implemented without delay, and longer-term remediation and environmental protection measures fully supported.

An additional source is the Krstov Dol arsenic and antimony mine, at the river from Krstov Dol flowing into Kriva Reka River (Republic of Macedonia, 2001).

The Lojane mine has an open dumpsite for flotation waste, which was created by the over 1 million tonnes of tailings containing arsenic, antimony and various hazardous substances. Soils measured at the dump's edge contained 8.093 mg arsenic/kg, which is more than 50 times the German threshold for arsenic in industrial soil (Kupusovic *et al.* 2002). Lead and zinc mines are located at Zletovo in Probistip, Toranica in Kriva Palanka (with 21 000 inhabitants) and Sasa in Makedonska Kamenica (with 8110 inhabitants) in the north-eastern part of The former Yugoslav Republic of Macedonia, close to Bulgaria. Contamination of water and soil by heavy metals has been observed near the Probistip lead and zinc mine since the area was first mined during the Roman period. Acid mine drainage has polluted the Kiselica and Koritnica rivers, which flow into the Zletovica River, which flows through Bregalnica and Vardar to the Aegean Sea, thus creating transboundary pollution problems (Sarajevo Center for Environmentally Sustainable Development, 2003). Although these issues have been identified, little has been done to prevent pollution or decontaminate the sites. Mine safety has been taken very seriously, but environmental aspects have largely been ignored (Midžić & Silajdžić, 2005b).

Volcanogenic sources of arsenic are potentially found in numerous thermal mineral springs in The former Yugoslav Republic of Macedonia (Battocletti, 2001):

> FYR [The former Yugoslav Republic of] Macedonia is located in the central part of the Balkan Peninsula, in the geothermal zone which runs from Hungary in the north and Italy in the west, and crosses Greece, Turkey, and beyond to the east. Specifically, the country is situated in the southernmost part of the Bosnian-Serbian-Macedonian geothermal area which includes the mountains of the internal Dinarides and parts of the Serbian-Macedonian massif.
>
> The country contains six geotectonic zones: the Cukali-Krasta zone, the West Macedonian zone, the Pelagonian horst anticlinorium, the Vardarian zone, the Serbo-Macedonian massif, and the Kraisthide zone. Geothermal manifestations are mainly connected to the Vardarian zone where the earth's crust is about 32 km.

FYR Macedonia has 18 geothermal fields. More than 50 thermal springs, boreholes, and wells discharge almost 1,400 l/s water with temperatures of 20–79°C. Thermal waters are primarily bicarbonate with equal amounts of Na, Ca and Mg. Dissolved minerals range from 0.5 to 3.7 g/l. All are of meteoric origin.

FYR Macedonia's hot springs have been used since Roman times for bathing. More recently, in 1963, the first geothermal greenhouse in the world was built near Bansko. The 1979–1984 energy crisis stimulated more geothermal exploration. Six large greenhouse heating projects, a rice drying project, a complete hotel heating project, and several smaller heating installations were developed during this period. Some have been abandoned or operate below their design capacity. Development ceased after independence and during the UN [United Nations] and Greek embargoes. The bulk of Macedonia's low-enthalpy geothermal resources are used for greenhouses.

The following list of 32 springs was prepared for the United States Department of Energy in order to assess the feasibility of using these springs as power sources, but they represent a potential source of arsenic leaching and should be investigated as such (Battocletti, 2001):

- Banja Spa – mineral waters, 40–100°C
- Banjishte Spa – mineral waters, 39–52°C
- Bansko – mineral waters, 35–120°C
- Debarska Banja – mineral waters, 39–70°C
- Deribash – preliminary report, 20°C
- Gevgelia – mineral waters, 54–100°C
- Gornitchet – preliminary report, 24–150°C
- Istibanja (Vinica) – mineral waters, 30–100°C
- Katlanovo Spa – mineral waters, 28–115°C
- Kezovica Spa – mineral waters, 30–115°C
- Kochani – mineral waters, 65–90°C
- Kosovrasti Spa – mineral waters, 48°C
- Kratovo-Zletovo – well(s) or hole(s) drilled, 28–49°C
- Kumanovo Spa – mineral waters, 38°C
- Lci – preliminary report, 100–115°C
- Mrezichko – preliminary report, 21–28°C
- Negorska Banja (Negorci) – mineral waters, 32–100°C
- Paltchiste – preliminary report, 23°C
- Podlog – mineral waters, 22–100°C
- Povisica – preliminary report, 80°C
- Proevci – preliminary report, 28–31°C
- Raklesh – mineral waters, 26–30°C
- Sabota voda – well(s) or hole(s) drilled, 21–100°C
- Stip – mineral waters, 32–59°C
- Skopje – preliminary report, 54–115°C
- Smokvica – mineral waters, 34–70°C
- Strnovec – well(s) or hole(s) drilled, 40–50°C
- Toplec – mineral waters, 25–26°C
- Topli dol – preliminary report, 28°C
- Toplik – preliminary report, 22°C
- Trkanie – well(s) or hole(s) drilled, 71°C
- Volkovo – well(s) or hole(s) drilled, 22–90°C

The source of arsenic in Macedonia possibly connected with volcanism lies in the south, adjacent to the Periphery of Central Macedonia in Greece. This area is associated with the Alshar metallogenic ore body, a Carlin type of gold deposit with 15 000 tons of arsenic which exhibits a hydrothermal alteration of host rocks similar to those in Nevada (Volkov *et al.* 2006).

4.21 NETHERLANDS (A, M)

The Netherlands, with a population of 16.7 million, has a Human Development Index rank of 7 out of 169 and a per capita GNI of $US 40 658 (US Census Bureau, 2011; UNDP, 2010).

Anthropogenic arsenic comes from 19 chemical, mining and mineral plants that discharge 0.968 tonne of arsenic per year directly into surface water, 8 facilities that discharge 0.10 tonne per year indirectly into surface water and 3 that emit 0.226 tonne per year into the air (European Pollutant Emission Register, 2004).

The Netherlands Environmental Assessment Agency (2004) reported that:

> Levels of heavy metals in the shallow groundwater (5–15 m deep) under a considerable proportion of the sandy areas exceed the desired quality standard [of 10 μg/l]. This is the result of atmospheric deposition, the use of fertilisers and acidification of the soil. Relatively low levels are found for copper and lead. No clear changes have been observed [from 1990 to 2003].

The percentages of measurements above the desired quality standard for arsenic in The Netherlands are shown in Table 4.3.

Table 4.3 Percentage of measurements exceeding the desired quality standard for arsenic in The Netherlands, 1990–2003.

Substance, by type of soil and land use	Percentage of measurements over desired quality standard (10 μg/l)			
	1990	1995	2000	2003
Sand (agriculture)	4	4	2	4
Sand (nature)	2	0	2	2
Fluvial clay (agriculture)	15	22	18	18
Marine clay (agriculture)	2	5	5	5

(Continued)

In The Netherlands, relatively small amounts of arsenic have been found, but its prevalence has been suspected. This situation has stimulated research into its fate and transport in both the subsoil and drinking-water of The Netherlands. The following research papers studying the occurrence of arsenic in soil and water in The Netherlands were presented at a symposium convened in Utrecht by the International Association of Hydrogeologists – Netherlands Chapter and the Netherlands' Hydrological Society on 29 November 2006 (Van der Valk, 2006).

Dr Tony Appelo presented a scenario (Van der Valk, 2006) of the geological aspects of the fate and transport of arsenic in The Netherlands:

> Pleistocene iron hydroxide was able to take up arsenic into its structure. During the Holocene transgressions sulphur from seawater was taken up to form pyrite (FeS_2) after which the deposits became covered by peat that formed near the outflow of the major river systems into the sea. During the cutting of the peat bogs since the Middle Ages, the pyrite became exposed to air, at which time arsenic was released.

In line with this scenario, Sophie Vermooten provided research (Van der Valk, 2006) to show that:

> groundwater can take up arsenic while flowing from higher Pleistocene deposits in the east of the country towards the western seacoast, after which it might accumulate in the sand dunes that protect the coast (and which are used as major drinking water reservoirs). The outcome of the project would be the drawing up of risk maps and a soil management map.

This then relates to The Netherlands' "national project in which policy-makers are to be advised about the use of soil, sediments and groundwater in areas affected by relatively high arsenic concentration levels in The Netherlands" (Van der Valk, 2006).

Arsenic also has a role in the production of Dutch drinking-water, according to Professor Pieter Stuyfzand (Van der Valk, 2006):

> At a few locations, the arsenic level is above the WHO norm of 10 μg/l, but the situation appears to be stable. While there is no large-scale problem, individual wells have been more closely examined, indicating that at the local level a few problematic cases might arise in the future, due to changes in the chemical composition of the infiltrating water (e.g., gases like CH_4 may give rise to mobilization of arsenic, which might lead to rather high arsenic levels locally [up to 1000 μg/l]).

In a 1985 study, arsenic contamination was found to be high in the Dutch subspecies of the dunlin (*Calidris alpina*), a red-backed sandpiper that breeds in northern or arctic regions and winters in The Netherlands. The study showed that the bird's habitat is the Waddenzee, which is polluted by the Rhine and the Scheldt rivers. The study measured the arsenic contained in the kidneys and the glands used to preen its feathers in both the young and the adults (Goede & De Bruin, 1985). This finding correlates with the finding that in mussel grist, there is generally a high arsenic content (Netherlands

Environmental Assessment Agency, 2004), suggesting that the dunlin that feeds on marine plankton and crustaceans ingests arsenic directly from this food source. Shellfish and crustaceans should be monitored for their content of arsenic and may be hazardous when consumed by humans.

These findings can find some support in clinical data obtained from a recent study of arsenic ingested as part of a typical Dutch daily diet (Vaessen & Van Ooik, 1989):

> 311 duplicate portions of the 24-h diet of Dutch citizens were collected and lyophilized. All samples were assayed for total arsenic and, in addition, the 20 samples which ranked highest in total arsenic were selected for arsenic speciation i.e., the separate determination of organic-bound arsenic and inorganic arsenic.

What is disturbing about these findings is that while they represent a very small population, the relative amounts of inorganic arsenic detected were high: "The highest oral inorganic arsenic intake was 60 μg; the tolerable daily intake (TDI) for a 60-kg individual is 120 μg inorganic arsenic" (Vaessen & Van Ooik, 1989).

4.22 NORWAY (A, M)

Norway, with a population of 4.7 million, has a Human Development Index rank of 1 out of 169 and a per capita GNI of $US 58 810 (US Census Bureau, 2011; UNDP, 2010).

Five smelters discharge 0.823 tonne of arsenic per year directly into surface water, and three emit 0.242 tonne per year into the air (European Pollutant Emission Register, 2004).

Potential arsenic contamination is from the Killingdal area, where arsenic in massive basaltic sulfide deposits measured 1200 mg/kg at Lokken mine and 500 mg/kg at Skorovas mine (Banks *et al.* 1997). Additional testing of ground waters is recommended in areas where mining is associated with arsenopyrite.

4.23 POLAND (A, C, G, M, V)

Poland, with a population of 38.4 million, has a Human Development Index rank of 41 out of 169 and a per capita GNI of $US 17 803 (US Census Bureau, 2011; UNDP, 2010).

Anthropogenic arsenic comes from one manufacturing plant and one mining facility that discharge 0.351 tonne of arsenic per year directly into surface water, one industrial plant and one sanitation facility that discharge 0.023 tonne per year indirectly into surface water and seven facilities (including three power plants) that discharge 1.29 tonnes per year into the air (European Pollutant Emission Register, 2004).

Arsenic-enriched bituminous coals found in the Pennsylvanian Formation in the Lyublin basin averaging 179 mg/kg are probably due to pyrite as the carrier because of the positive correlation of arsenic, sulfur and iron (III) oxide (Cebulak & Rozkowska, 1983).

Natural thermal springs are found in the Polish part of the Sudetes Mountains at Cieplice and Lądek (Dowgiallo, 2002):

> Discharge temperatures of these springs are 44 and 29°C, respectively, and bathing is their traditional use. Drilling performed during the last three decades in the above-mentioned and other localities revealed considerable amounts of thermal waters occurring in tectonically disturbed zones of Sudetic metamorphic rocks and granitoids. The existence of thermal waters may also be expected in numerous areas of the Sudetes, where carbonated waters occur. CO_2 connected with Tertiary and Pleistocene basic magmatism may be an indicator of regional rock heating.

Mining is also a source of arsenic contamination of drinking-water. The Sudetes Mountains contain polymetallic deposits that have been exploited since the Middle Ages. Distinct concentrations of arsenic, mercury, fluoride and chromium in surface water near Zloty Stok suggested that groundwater in the area could also contain elevated metal concentrations. Water samples from 15 locations including Zloty Stream and selected springs have arsenic concentrations ranging from 990 to 26 160 μg/l. Mercury concentrations were locally as high as 11 μg/l. These high arsenic and mercury concentrations exceed EU water quality standards and make Zloty Stream unsuitable as a drinking-water source (Marszalek & Wasik, 2000).

Mining in the vicinity of the former arsenic ore and smelting site Zelezniak in the Kaczawskie Mountains in south-western Poland has contributed to high levels of arsenic in the groundwater (Sculz & Karczawska, 2003).

It is assumed that the contamination of the Odra River sediments with high arsenic concentrations is anthropogenic in origin (Ademiec & Helios-Rybicka, 2002; Helios-Rybicka *et al.* 2003) (see Figure 4.5).

Additionally, Kedzierski (1983) found a high concentration of iron, zinc, tin, copper, lead and arsenic in Polish beers that made use of groundwater sources. The residual arsenic in the human body from these sources (0.41 mg/g, 0.37 mg/g and 0.18 mg/g) has been confirmed by analysis of human bone from individuals in southern and central Poland (Krakow,

Silesia and Lodz, respectively) (Wiechula *et al.* 2003). This is presumably from water in Quaternary alluvial sediments and demands further testing.

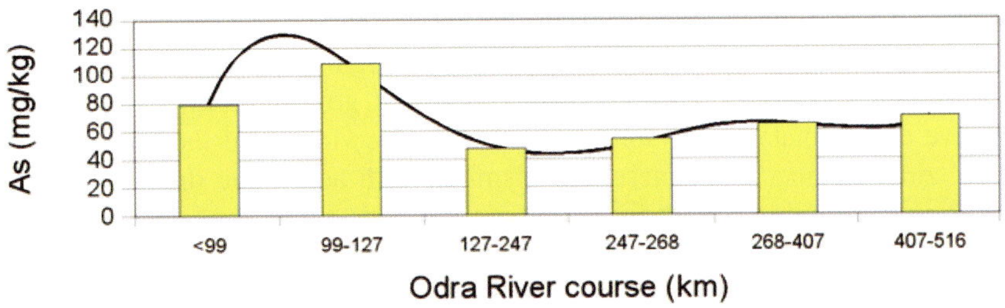

Figure 4.5 Mean arsenic concentrations in suspended matter in the upper and middle Odra River course, Poland (modified from Helios-Rybicka *et al.* 2003).

4.24 PORTUGAL (A, C, G, M, P, V)

Portugal, with a population of 10.8 million, has a Human Development Index rank of 40 out of 169 and a per capita GNI of $US 22 105 (US Census Bureau, 2011; UNDP, 2010).

Anthropogenically produced arsenic contamination is an issue in Portugal, with two waste disposal facilities discharging 0.023 tonne of arsenic per year indirectly into surface water and seven facilities (three power stations and four manufacturing plants) discharging 1.29 tonnes per year into the air (European Pollutant Emission Register, 2004).

High-arsenic coals in Portugal have not been studied systematically. However, they do exist and are a cause of some concern for the contamination of groundwater as well as air. On the basis of estimates in only three coalfields, resources of coal of all ranks in Portugal total at least 76 million tonnes. In 2004, 5285 tonnes of bituminous coal of unknown arsenic content were used in power generation facilities (International Energy Agency, 2007). The presence of high-arsenic coal has been definitely established by an indirect method: arsenic species were bioassayed in lichen transplants and assayed in airborne particulate matter (fine and coarse fractions) from samples collected in 1994–1995 near coal-fired power plants, and it was found that inorganic forms of arsenic (arsenite and arsenate) were present in significant amounts in most of the samples (Farinha *et al.* 2004). Studies of bioaccumulation of arsenic by mosses to concentrations above 1.4 mg/kg have confirmed the contamination of soils and surface waters from coal-fired plants in the vicinities of Lisbon, Oporto and Sines (Buse *et al.* 2003).

Geogenically occurring arsenic (possibly Quaternary arsonium compounds or arsenosugars) is released into marine estuarine ecosystems in the spring in the Targus estuary, where hidden arsenic amounted to 19–25% of the total arsenic in the estuarine water (de Bettencourt *et al.* 1994). This level of increase can bring the level of arsenic exposure above the WHO guideline and therefore must be monitored carefully, as it may affect human consumption of marine organisms.

Mining has played an important role in Portugal's history for the last three millennia. Copper, silver, gold, iron and antimony were mined in Portugal since the Phoenicians about 600 BC. The Romans mined gold in the north, and copper, silver and gold were mined in the sulfide belt. These ancient areas and historically mined areas have become prone to contamination of groundwater by arsenic-containing acid mine drainage because of the geochemical signatures of arsenic-associated ores in these deposits:

(1) tungsten ores: arsenic, copper, molybdenum, lead
(2) tin ores: arsenic, beryllium, (copper)
(3) tungsten – tin ores: arsenic, zinc, cadmium
(4) gold – silver ores: arsenic, zinc, cadmium, (silver), lead
(5) lead – zinc ores: (arsenic), lead, zinc, cadmium, (silver)
(6) sulfide ores: arsenic, zinc, lead, copper, iron, antimony
(7) coal: (arsenic), (iron)

Two metallic ore mines with arsenic minerals are still active, the Neves-Corvo with copper and antimony and the Panasqueira with tungsten and tin. They were started before the implementation of national environmental legislation that requires controls, monitoring, closure procedures and remediation. The Panasqueira mine is located in central Portugal, 300 km north-east of Lisbon on the southern Sierra de Estrela. It was first prospected in 1886, and mining began in 1896 and has continued with two brief hiatuses. The deposit is Precambrian-Cambrian Beira Schist with horizontally stacked hydrothermal quartz bearing wolframite, arsenopyrite, chalcopyrite and cassiterite and is 2500 m long, 400–2200 m wide and at least 500 m deep. Although the average arsenic in copper tailings was nearly 20% in 2001, the Barroca Grande dump was measured at 25% arsenic. The copper plant tailings contain much arsenopyrite, and it is planned to contain the acid mine drainage in a larger tailings pond. The pH of waters released ranged from 3.49 to 3.73. At the water discharge mouth, Boca de Mina, waters exceeded all Portuguese regulatory limits for water quality (Primary Metals Inc., 2007). The Caveira and Lousal mines in the Iberian sulfide province pollute 15 km downstream from the mine, with high concentrations of copper, lead, zinc, arsenic, molybdenum, selenium, cadmium and iron. Castromil is an abandoned gold mine in north-west Portugal, with pyrite and arsenopyrite tailings that have contaminated local soils and a well in the vicinity of the mine. The Penedono gold mine has high levels of copper, lead, zinc, arsenic, manganese and bismuth, with acid mine drainage and surface waters with high contents of copper, zinc, arsenic, cadmium, manganese, aluminium and sulfate, but the groundwater is not polluted. The Argoselo tin–tungsten mine in the north has almost 1 million tonnes of tailings with significant levels of copper, zinc, arsenic, tungsten, iron, cadmium, primary sulfides (arsenopyrite, pyrite, chalcopyrite, sphalerite), arsenates, sulfates and oxides. Stream sediments and soils near the site are contaminated with copper, zinc and arsenic. At the Segura mining area (44 small mines), tin, tungsten, barium and lead were exploited until 1953. Waters associated with mineralized veins have high levels of arsenic, iron and manganese and should not be used for human consumption or agriculture; soils must not be used for agriculture or residences because of high concentrations of tin, bismuth, arsenic and barium, whereas stream sediments show copper, lead, zinc, arsenic, tin, tungsten, bismuth and barium (Cabral Pinto & Silva, 2005). In addition, two mining facilities discharge 0.351 tonne of arsenic per year directly into surface water (European Pollutant Emission Register, 2004).

In a particularly prominent case of present-day anthropogenic arsenic contamination impacting an urban area, a pyrite roasting plant on the south side of the Tejo River has been identified as emitting arsenicals into the surface water. Across the Tejo River is the city of Lisbon, with a population of 564 477, the Lisbon Metropolitan Area, with 2.8 million inhabitants, and the Lisbon Metropolitan Region, with 3.34 million (Soares, 2004). The arsenic emissions from the Barreiro/Seixal industrial complex are in the range of 1000–2000 tonnes per year, and with such a large proximate population, the health impacts are correspondingly high (Andreae *et al.* 1983).

Derelict mining operations continue to have an impact on groundwater and on agricultural soils. A study of controlled population groups discovered increased accumulated water and atmospheric arsenic exposure in the inland north and central areas of the country (Figueira *et al.* 2007). Castromil, the gold-mining area in northern Portugal mentioned above that has been abandoned since 1940, is now a residential area suffering from considerable consequences of poorly regulated mining activities. Leaching from old tailings has impacted local agricultural soils from this region, which have high concentrations of arsenic and lead (Ferreira da Silva *et al.* 2004).

Petroleum-related arsenic releases into air and soils have been confirmed by bioaccumulation studies of mosses with arsenic concentrations above 1.4 mg/kg in the region of Sines, where refineries and petrochemical plants release discharges and emissions (Buse *et al.* 2003).

Volcanogenic arsenic may exist in hot springs that may have geothermal energy potential in the Minho district in the north-west. The 1527-m Larouco Mountains of Minho province form a western-facing escarpment of metamorphic crystalline schists, slates and quartzites and intrusive granitics, cut with fault lines from which hot springs emerge. The metamorphic and granitic matrices do contain some arsenic minerals, but the extent of their leaching into the hot springs is not well known. Furthermore, the geothermal potential has not been developed. Geothermal energy resources exist in São Miguel Island, Azores, and a programme of evaluation and exploration with technical assistance from the USGS was undertaken in the 1980s (Brock & Brock, 1967; Argonne National Laboratory, 1981).

4.25 ROMANIA (A, G, P, V)

Romania, with a population of 21.9 million, has a Human Development Index rank of 50 out of 169 and a per capita GNI of $US 12 844 (US Census Bureau, 2011; UNDP, 2010).Anthropogenic sources of arsenic contamination have been inferred from bioaccumulation studies of mosses. A study of atmospheric arsenic deposition in the moss *Hypnum cupressiforme* was conducted in the province of Moldavia within the catchment of Prut, a border river between Romania, the Republic of Moldova and Ukraine. Using inductively coupled plasma mass spectrometry, a maximum value of arsenic of $>$1.5 mg/kg was determined. The distribution of arsenic is thought to reflect the agricultural use of arsenical pesticides in eastern Romania (Stan *et al.* 2001; Cucu-Man *et al.* 2004).

Geogenic and anthropogenic sources of arsenic may account for a number of health effects attributable to elevated levels (up to 176 μg/l) of arsenic in drinking-water (Polyak & Hlavay, 1999; Gurzau & Gurzau, 2001). These include skin lesions, spontaneous abortions, peripheral neuropathy, bladder cancer and cardiovascular disease. The areas principally affected are the north-west region of Transylvania (where approximately 14 000 people are exposed to drinking-water containing arsenic levels over 50 μg/l) and the Danube basin region (Gurzau & Gurzau, 2001; Smedley & Kinniburgh, 2002). Well water has been known to contain elevated levels of arsenic since the 1940s. In a study of urinary arsenic of cohort groups of nearly equal sex distributions from Bihor and Arad, there was a significant correlation between arsenic in water and arsenic in urine ($R^2 = 0.46$). High body mass index and cigarette smoking correlated with elevated urinary arsenic (III). At low water concentrations, urinary dimethylarsinic acid was elevated, indicating exposure via food. Maximum concentrations of arsenic in drinking-water were 24 μg/l at Bihor and 95 μg/l at Arad (Lindberg *et al.* 2006).

Petroleum-related arsenic emissions are associated with the oil refineries in the south that cause atmospheric pollution across the border in northern Bulgaria (Buse *et al.* 2003). The Romanian National Oil Company runs refineries in the Ploiesti and Campinu regions that produce 60 000 tonnes of acid tars annually, a major agent in generating arsenic contaminants to the air and water supplies (SEC Info, 2001).

Portions of Romania bordering the southern part of the Great Pannonian Plain are potentially exposed to elevated concentrations of arsenic in a volcanogenic situation identical to that of Hungary, with elevated concentrations of ammonium, iron, manganese, humic acids and dissolved gases and temperature above 30°C (see Section 3.11; Hlavay & Polyak, 1998; Biacs, 2005). This situation needs further investigation, as the Great Pannonian Plain is a major geological feature of Romania.

4.26 SERBIA (A, G, M, V)

Serbia, with a population of 7.3 million, a Human Development Index of 60 out of 169 and a per capita GNI of US$10 499, extends across the central part of the Balkan Peninsula and the southern part of the Great Pannonian Plain (US Census Bureau, 2011; UNDP, 2010).

Anthropogenic sources of arsenic in Serbia are related to the industrial release of chemicals into water and soil, which has subsequently contaminated foods (harvested near industrial sites) and fish (both marine and freshwater) with heavy metals and metalloids, including arsenic (United States Army, 2004). After the devastating May 2005 flood in the northern province of Vojvodina, daily monitoring of water quality by the sanitary inspectors and epidemiologists of the Institute for Public Health in Zrenjanin revealed high levels of arsenic in the water. It was thought that the water pipes would have to be phased out after the floodwater subsided. The report of arsenic contamination immediately led to a ban on local tap water, which affected 100 000 people in Vojvodina (Anon., 2004a). Such vulnerability of water supply sources is expected to continue.

The contamination of soils in western Bulgaria at the town of Chiprovtzi is thought to be from emissions from the copper smelter at Bor (Stamenov *et al.* 2002; Buse *et al.* 2003).

The Danube River in Serbia has potential sources of geogenic arsenic from sulfide ores and mines in a large watershed, but the arsenic levels are comparatively low in three zones measured, with averages of 5.5, 5.5 and 0.1 μg/l. However, in riverine sediments, arsenic is enriched up to 100 times in clay, depending on the mineralogical composition and adsorption properties of the sediment. These sediment concentrations may impact the water quality of riparian, shallow, alluvial aquifers along the Danube in Serbia, because they are hydraulically connected to the river and riverine towns and cities, such as Novi Sad, Belgrade, Smederevo, Kostolac, Veliko, Gradiste and Golulac which are actual or potential users of these aquifers. The Sava River, which is the Danube's largest tributary, flowing into it at Belgrade, is used for drinking-water at the nearby Makis well, which has 3.3 μg arsenic/l. The well may also be affected by the known arsenic effluents from the Obrenovac power plant 90 km upstream and the 25 mg/l acid mine drainage from the abandoned mercury mine at Suplja Stena, which contributes to the 100 μg arsenic/l concentration of the Sava 5 km downstream (Dangic, 2007).

Groundwater is also prone to geogenic arsenic contamination in Serbia. Levels of arsenic in groundwater have been reported to range from 0.080 to 120 μg/l (Karanjac, 2004). Aquifers in the Banat region between the Hungarian and Romanian borders in the north and east are accessed by deep wells (150–200 m). These aquifers have an elevated temperature of 22°C and a pH of 8.2, with arsenic concentrations ranging from 12 to 300 μg/l, indicating that the several towns that use this water supply may be at risk (Dangic, 2007).

Mining sources of arsenic in eastern and central Serbia are found in porphyry copper and massive sulfide ores in the Bor metallogenic zone, where very old mining deposits are still extant. For instance, the Early Neolithic Rudna Glava mine, begun about 4500 BC, was exploited until 1960, and archaeological research in Lepenski Vir and Vinca has revealed evidence of Indo-European mining of lead and silver at Crveni Breg mine and mercury at Šuplja Stena mine. Roman mining extracted lead and zinc, which were also exploited from the 14th century, making mining settlements in Serbia some of the largest towns in Europe, such as Novo Brdo, with a population of 28 000 (Jović *et al.* 2002). The Bor metallogenic zone (gold) and the volcanic complex of Lece (hydrothermal lead–zinc–copper) have associated arsenic mineral deposits as well.

Consequently, ancient and historical mines in Serbia represent a regional problem threatening the environment of neighbouring countries and international watercourses. For instance, acid mine drainage from the Bor mine flows into the Bor, Krivelj, Brestovac and Ravna rivers and endangers all communities downstream, including the Timok River bordering with Bulgaria. Acid mine drainage has a high content of copper and arsenic and low pH. Effluent from a sulfuric acid plant (RTB Bor operations) has a low pH and contains copper, arsenic, lead and zinc compounds that are discharged without treatment from a holding tank to the Borska, Timok and Danube rivers at the rate of 300–500 tonnes of sulfuric acid per year, 300–350 tonnes of arsenic per year, 30–100 tonnes of lead per year and 10–35 tonnes of zinc per year (Jović *et al.* 2002). The River Danube is constantly being contaminated, because there are no mining regulations that are being enforced to process wastewaters (Sarajevo Center for Environmentally Sustainable Development, 2003).

Volcanogenic sources of arsenic impact the central Balkan region (see Section 3.11; Hlavay & Polyak, 1998) in a geological situation similar to that of Hungary (Battocletti, 2001):

> Serbia is located in the tectonic center of the Balkan peninsula where three main tectonic-structural systems are developed, all of Alpine origin: the Dinarides in the west, the Serbian-Macedonian Massif (or Rhodopes) in the center, and the Carpatho-Balkanides in the east and southeast. The Pannonian basin in northern Serbia has the most promising high-enthalpy geothermal resources. The location of all Yugoslav oil and gas deposits, the Pannonian region has been well investigated geologically. S. Radavanovic, the "father of Serbian hydrogeology and geothermology" first described the area's geothermal resources in 1897. Thermal spring exploration began between the two World Wars. The first borehole in the Pannonian basin was drilled in 1969. Serbia has 60 convective hydrogeothermal systems: 25 in the Dinarides, 20 in the Carpatho-Balkanides, 5 in the Serbian-Macedonian Massif, and 5 in the Pannonian basin. Conductive hydrogeothermal systems are developed in basins filled with Paleogene and Neogene sedimentary rocks, and are primarily located in the Pannonian basin in Vojvodina in northern Serbia. Thermometric measurements in northern Serbia, down to 5 km, have produced temperatures as high as 240°C. At 3 km, temperatures close to 150°C have been found in the Pannonian basin, the Vardar Zone, and the Serbo-Macedonian Massif. Serbia's geothermal potential or geothermal resource base to a depth of 10 km is estimated at 250×10^{21} J.

Mineral and thermal waters in Serbia have a range of arsenic concentrations that are highest in volcanic massifs (V), metamorphic complexes (M) and hydrogeological basins (H) (Dangic, 2007):

(1) Russanda well, ~1000 m, H, 8.567 mg arsenic/l
(2) Crni Guber spring, V, 4.9 mg arsenic/l
(3) Rudovici well, M, 0.343 mg arsenic/l
(4) Lukovska Banja spring, V, 0.321 mg arsenic/l
(5) Rusanda Lake, H, 0.321 mg arsenic/l

Given these known concentrations of arsenic in Serbia's thermal springs, the following list prepared for the United States Department of Energy in order to assess the use of these springs as power sources is also useful as a list of those sites that represent potential sources of arsenic leaching and should be investigated as such (Battocletti, 2001):

- Arandjelovac – thermal spring, 34–60°C
- Avala – under study, 26°C
- Backi Manastir – under study, 73°C
- Banja "Junakovic"-Apatin – thermal spring, 56°C
- Banja Istok – thermal spring
- Banja Rusanda-Melenci – thermal spring, 92°C
- Banja Topilo – thermal spring, 37°C
- Banja Vuca – thermal spring, 30°C
- Banjska – thermal spring, 54–120°C
- Becej – thermal spring, 24–65°C
- Bioska Banja – thermal spring, 36°C
- Biostanska Banja – thermal spring, 37°C
- Bogatic – under study, 75–90°C
- Bogutovac – thermal spring
- Brestovacka Banja – thermal spring, 30–100°C
- Bujanovacka Banja – thermal spring, 24–70°C
- Bukovicka Banja – thermal spring, 28–34°C
- Cedovo – under study, 26°C

- Cibutkovica – under study, 21°C
- Dag – under study, 30°C
- Debrc – thermal spring, 48–70°C
- Despotovac – under study, 26°C
- Donjobadanjska – thermal spring
- Dublje – well(s) or hole(s) drilled, 44–85°C
- Dvorovi – thermal spring, 50–90°C
- Gamzigradska Banja – thermal spring, 24–80°C
- Gornja Trepca – thermal spring, 24–50°C
- Grocka – under study, 30°C
- Indjija – under study, 60–70°C
- Josanicka Banja – thermal spring, 40–130°C
- Jugovo – under study, 42°C
- Kanjiza – thermal spring, 26–65°C
- Kikinda – thermal spring, 26–51°C
- Klokot Banja – thermal spring, 25–80°C
- Knjazevac – thermal spring
- Koviljaca – thermal spring, 24–40°C
- Kravlje – under study, 32–40°C
- Kula – thermal spring, 25–53°C
- Kupinovo – under study, 54–70°C
- Kursumlijska Banja – thermal spring, 25–140°C
- Ljig – thermal spring, 32°C
- Lukovska Banja – thermal spring, 35–90°C
- Macva – feasibility study, 80–110°C
- Malo Laole – well(s) or hole(s) drilled, 38–40°C
- Mataruge – thermal spring, 24–110°C
- Mataruska Banja – thermal spring, 43–52°C
- Melenci – thermal spring, 33°C
- Metkovic – well(s) or hole(s) drilled, 55–90°C
- Miljakovac – under study, 33–40°C
- Milkovac – under study, 30°C
- Mladenovac – thermal spring, 25–90°C
- Mlakovac – under study, 28–30°C
- Mokrin – thermal spring, 26–51°C
- Nikolicevo – under study, 37°C
- Nikolocevska Banjica – thermal spring, 34°C
- Niska Banja – thermal spring, 25–60°C
- Novopazarska Banja – thermal spring, 52–54°C
- Ovcar Banja – thermal spring, 27–60°C
- Palanacki Kiseljak – thermal spring, 55°C
- Palic – thermal spring, 48°C
- Pazar – thermal spring, 28–120°C
- Pec – thermal spring, 26–48°C
- Pecka Banja – thermal spring, 25–80°C
- Petrovac – thermal spring, 24–46°C
- Pirotska Banjica – thermal spring, 31°C
- Pozarevac – under study, 36°C
- Pribojska Banja – thermal spring, 30–60°C
- Prigrevica – thermal spring, 25–54°C
- Prolom Banja – thermal spring, 24–60°C
- Radalj – under study, 28–60°C
- Rajcinovica Banja – thermal spring, 28–100°C

- Rakovac – under study, 42°C
- Rgoska Banjica – thermal spring, 32°C
- Rgoste – feasibility study, 30–45°C
- Ribarska Banja – thermal spring, 25–110°C
- Sabac – well(s) or hole(s) drilled
- Sarbanovac – under study, 29–100°C
- Selo – thermal spring, 26–43°C
- Selters – thermal spring, 50°C
- Sierinska Banja – thermal spring, 68°C
- Sijarinska Banja – thermal spring, 72–130°C
- Sisevac – under study, 26–40°C
- Slanaca – under study, 37°C
- Smederevska Palanka – thermal spring, 25–75°C
- Soko Banja – thermal spring, 33–42°C
- Srbobran – thermal spring, 24–63°C
- Stari Slankamen – thermal spring
- Sumrakovac – under study, 24–100°C
- Suva Cesma – under study, 22–60°C
- Svackovci – under study, 28°C
- Temerin – thermal spring, 25–41°C
- Vica – under study, 21°C

4.27 SLOVAKIA (A, C, G, M, V)

Slovakia, with a population of 5.5 million, has a Human Development Index rank of 31 out of 169 and a per capita GNI of $US 21 658 (US Census Bureau, 2011; UNDP, 2010).

Anthropogenic arsenic contamination is related to effluents from power generation stations, steel, cement and other manufacturing plants that discharge 31.07 tonnes of arsenic per year into the air and two facilities that discharge 0.099 tonne per year directly into surface water (European Pollutant Emission Register, 2004).

Novaky in the Prievidza district of Slovakia has been found to be contaminated with arsenic as a result of the coal-fired power station using coal rich in arsenic (Thornton *et al.* 2002). A study was initiated in 2003 to assess the arsenic exposure of a population living in the vicinity of a coal-burning power plant in the Prievidza district with high arsenic emissions (Nieuwenhuijsen, 2003):

> 548 spot urine samples were speciated for inorganic As(III), monomethylarsonic acid (MMA), dimethylarsinic acid (DMA), and their sum. The urine samples were collected from the population of a case–control study on non-melanoma skin cancer (NMSC). A total of 411 samples with complete As speciations and sufficient urine quality and without fish consumption were used for statistical analysis. Although current environmental As exposure and urinary As concentrations were low (median As in soil within 5 km distance to the power plant, 41 μg/g; median urinary arsenic 5.8 μg/l), there was a significant but weak association between As in soil and urinary arsenic ($r = 0.21$, $p < 0.01$).
>
> Persons living in the vicinity of the plant had 27% higher arsenic values ($p < 0.01$), based on elevated concentrations of the methylated species. A 32% increase of MMA occurred among subjects who consumed homegrown food ($p < 0.001$). NMSC cases had significantly higher levels of arsenic, DMA, and As(III). The methylation index As(III)/(MMA + DMA) was about 20% lower among cases ($p < 0.05$) and in men ($p < 0.05$) compared with controls and females, respectively.

Measurement of arsenic in mosses from the area has confirmed the deposition of atmospheric arsenic, with a median value of 0.7 mg arsenic/kg (Barandovski *et al.* 2006).

Geogenic and mining sources of arsenic in Slovakia are related to the many deposits of arsenic minerals. Three potential sources of arsenic contamination have the following acid mine drainage and total dissolved solids, as of 1997 (Bajtoš, 2005):

(1) Quartz-sulfidic ore veins in Palaeozoic rocks with calcium–magnesium–bicarbonate acid mine drainage located at Zlatá Idka mine with total dissolved solids (arsenic, iron, manganese) of 1.9–2.2 g/l and Cucma mine with total dissolved solids (arsenic, antimony) of 0.3–1 g/l;

(2) Carbonate–sulfidic ore veins in Palaeozoic rocks with calcium–magnesium–bicarbonate–sulfate acid mine drainage located at Rudnany mine with total dissolved solids of 1.4 mg/l (antimony, arsenic, sulfate), Bindt mine with total

dissolved solids of 0.4–0.8 mg/l (antimony, arsenic, sulfate) and Slovinky mine with total dissolved solids of 1.4–1.9 mg/l (iron, manganese, arsenic, antimony, sulfate);

(3) Sedimentary manganese deposits in Palaeogene sediments with calcium–bicarbonate–carbon dioxide with total dissolved solids of 2.5–4.1 mg/l (iron, manganese, arsenic) and Handlová mine with calcium–magnesium–bicarbonate with total dissolved solids of 0.5–1 mg/l (iron, manganese, arsenic, sulfate).

Slovakia's ancient heritage of mining has been recorded in historical mining maps that are now housed in the Main Chamber–Count Office in Banská Štiavnica. These maps can be of inestimable value for identifying the location of ore deposits and the hydrology of the watersheds within which they are found in order to help identify potential ways in which copper and arsenopyrite are mobilized into the aquifers that now underlie various towns in the region (UNESCO, 2005):

> "Mining maps and plans of the Main Chamber–Count Office in Banská Štiavnica" represents a collection of almost 20 thousand items from the 17th century up to the early years of the 20th century, which are an essential part of the archival finds of this the most significant authority for mining, metallurgy and minting in the Hungarian part of the Austro-Hungarian Empire. The maps and plans display underground mining works as well as above-ground objects in all the mining regions of Slovakia or abroad (Hungary, Transylvania, Banat, Slovenia and many others). Some of the maps display mining works and other technical sights, including a unique system of water reservoirs, in the town Banská Štiavnica and its surroundings, which was inscribed on the UNESCO [United Nations Educational, Scientific and Cultural Organization] World Heritage List in 1993. The collection of maps also contains new patterns for drawing of mining maps and construction plans designed by Gabriel Svaiczer and Anton Péch, the most important officials of the Main Chamber–Count Office in Banská Štiavnica (MCHO).

The central Slovakia mining region was so mineralogically important that it was regarded as the "treasury" and economic centre of the former Hungarian Empire. Its most crucial areas were "Golden" Kremnica, "Silver" Banská Štiavnica and "Copper" Banská Bystrica. Columbus's ships were sheathed with Banská Bystrica copper, which was mined from about 1450 to 1900.

Slovakia's Slovenské Rudohorie Mountains in the West Carpathians have several ancient mineral deposits of gold, silver, copper, iron, manganese, antimony, lead and zinc. Many of these mines are in association with arsenic minerals. About 460 mines have contaminated local surface water and groundwater with a total acid mine drainage of about 1.5 m^3/s, but not all discharges have been registered, so this figure needs to be adjusted upwards. Today, many abandoned mines have two basic types of acid mine drainage, with contaminants in the tens of milligrams per liter. One is acid mine drainage of quartz–antimony deposits with high antimony and arsenic; the other is acid mine drainage from iron, manganese, antimony and arsenic deposits (Bajtoš, 2005).

Arsenic contamination and its potential leaching into the groundwater can come from exposed tailings from these derelict mines. In a study of urinary arsenic of cohort groups of nearly equal sex distributions from Banská Bystrica and Nitra, there was a significant correlation between arsenic in water and arsenic in urine ($R^2 = 0.46$). High body mass index and cigarette smoking correlated with elevated urinary arsenic (III). At low water concentrations, urinary dimethylarsinic acid was elevated, indicating exposure via food. Maximum concentrations of arsenic in drinking-water were 37 μg/l at Banská Bystrica and 39 μg/l at Nitra (Lindberg *et al.* 2006).

High-arsenic coal is also a source of arsenic produced by coal-burning factories and thermal power plants that has contaminated surface soils (Strecansky, 1996):

> The soil has been poisoned by arsenic which has been emitted into the atmosphere by a coal-burning power station. The bulk of the waste produced in the district is produced by the thermal power plants in Zemianske Kostolany and mines in Handlova. The main producer of hazardous waste is the Novaky Chemical Plant. Fifty percent of the plant's waste is landfilled: 179 waste dumps have been reported on the district territory.
>
> The main environmental assets of the district are forested areas, swamps, upper parts of river Nitra and protected, mountainous areas. The upper part of the Vtaèník mountain range is a National Nature Preserve. There are several historical and cultural monuments in the region which attract tourists–Bojnice chateau and Bojnice thermal spa....
>
> A working group retreat was held to select priority environmental problems related to health risks. A number of uncertainties still existed in additional information collected, especially regarding vinylchloride and arsenic. Evidence of vinyl chloride's teratogenic or mutagenic impact on the local population was unclear and contradictory, especially on former workers of the PVC [polyvinyl chloride] production line in Novaky Chemical Works. Arsenic and small sources of air pollution represented another problem because of the ambiguity of available data: arsenic levels have dropped, but its presence in the water and air is still high. Possible impacts on food and crop production were considered by the working group. Concerns were expressed that SOx [sulfur

> oxides] and NOx [nitrogen oxides] from household heating [pose] a high health risk for certain localized areas, but reliable data [are] missing.
> Thus, the working group collected data and information from different monitoring and research institutes, industry and their environmental departments, including information about public opinion of some pollutants and associated health risks, as well as concerns and views of experts representing polluters and enforcement agencies, or civil servants and NGOs [nongovernmental organizations].
>
> Working group members agreed that, based on available information, no clear statement can be made about all sources and the health impacts of different environmental problems. The group selected air pollution as a priority for action, though the problem has been recognized as very broad and uncertain. The concerns of working group members were only partly addressed by the strategy and the substantial amount of additional information collected. Monitoring tendencies in the development of the problem and collecting data, estimating the impact and residuals of arsenic in the soil and water was agreed as the most important tasks for regional institutions.
>
> The main outcomes of the working group discussions were:
>
> (1) An agreement on, and identification of, air pollution as a priority environmental problem. The problem description contained all available information related to ambient and occupational air quality, pollution sources, tendencies, health impacts and the uncertainties of classic pollutants, vinyl chloride and arsenic. In particular, it was agreed that all available research must be collected according to methodology adapted to the realities in Slovakia, and targeting investments to reduce risks. Though the group did not conduct additional monitoring activities, it achieved new knowledge by pooling data and information together.
> (2) Recognition of two important uncertainties and constraints: ambiguity of information for decision making and limited or non-existent financing resources for implementing solutions. Both constraints have been the focus of further discussions.
> (3) Identification of the number of uncertainties related to information problems:
>
> - Limited or contradictory information about the severity of other environmental problems and their impact on health, such as soil contamination and impacts through the food chain;
> - Outdated information which did not reflect the decline in production or some of the clean-up activities recently undertaken;
> - Lack of information about geographical distribution and affected age and social groups, which made it impossible to decide whether ambient air or air in the occupational environment posed a greater risk; or whether certain parts of the district were more affected than others.

Volcanogenic sources of arsenic are located on the northern zone of the Great Pannonian Plain, identical to volcanogenic sources of arsenic in Hungary (see Section 3.11; Hlavay & Polyak, 1998; Biacs, 2005). Based on the arsenic concentrations above 50 μg/l in groundwater in adjacent areas of Hungary, the risk is deemed high. Elevated arsenic concentrations above 1.4 mg/kg measured in mosses have identified areas around thermal power plants and factories in central Spis, the Upper Nitra basin, Ziar basin and Váh valley (Buse *et al.* 2003).

4.28 SLOVENIA (A, V)

Slovenia, with a population of 2.0 million, has a Human Development Index rank of 29 out of 169 and a per capita GNI of $US 25 857 (US Census Bureau, 2011; UNDP, 2010).

Anthropogenic arsenic is emitted by two thermoelectric plants and one manufacturing plant that discharge 0.109 tonne of arsenic per year into the air and one facility that discharges 0.0143 tonne per year directly into surface water (European Pollutant Emission Register, 2004).

Portions of Slovenia bordering the northern part of the Great Pannonian Plain are potentially exposed to elevated concentrations of arsenic in a geological situation identical to that of Hungary (see Section 3.11; Hlavay & Polyak, 1998; Biacs, 2005). Based on adjacent geological sources in Hungary registering arsenic concentrations of 10–30 μg/l, groundwater in Slovenia should be tested for its levels of arsenic contamination. The geological situation of the mineralized thermal springs is critical to their development (Battocletti, 2001):

> With the Julian and Karawanken Alps in the north, and the Dinaric Alps in the south, Slovenia is extremely mountainous. The country's complicated tectonic and stratigraphical setting is reflected in the Earth's thermal field. In the upper few kilometers of the crust, temperatures increase from southwest to northeast. The geothermal conditions in western Slovenia are influenced by the large crustal thickness in the Outer Dinarides and Southern Alps. Thermal springs in this area located southwest of the Pannonian basin, have temperatures of less than 45°C. Northeastern Slovenia is affected by the high enthalpy geothermal system, Termal II, in the Pannonian basin. At depths of more than 2500 m, thermal fluids reach temperatures of 100–200°C. About 6% of the country's total area, or 3200 km^2, is promising for geothermal development.

The following list of springs was prepared for the United States Department of Energy in order to assess the feasibility of using these springs as power sources, but they represent a potential source of arsenic leaching and should be investigated as such (Battocletti, 2001):

- Banovci – thermal spring, 25–68°C
- Bled – thermal spring, 15–22°C
- Catez (Brezice) – thermal spring, 28–64°C
- Celje's Hollow – thermal spring, 20°C
- Cerkno – thermal spring, 28–30°C
- Dabinka – thermal spring
- Dahlenske Toplice – thermal spring
- Dobova – thermal spring, 38–63°C
- Dobrna – thermal spring, 25–36°C
- Dolenjske Toplice – thermal spring, 32–37°C
- Dragonja Vas – well(s) or hole(s) drilled, 19°C
- Gozd Martuljek – well(s) or hole(s) drilled, 9°C
- Koprivica – thermal spring, 131°C
- Kostanjevica – well(s) or hole(s) drilled
- Krsko – thermal spring, 23–64°C
- Lasko – thermal spring, 30–41°C
- Lendava – construction under way
- Ljubljana – well(s) or hole(s) drilled, 21°C
- Ljutomer – under study
- Lucija – well(s) or hole(s) drilled, 28°C
- Marezige – well(s) or hole(s) drilled, 30°C
- Maribor – thermal spring, 21–69°C
- Medija – thermal spring, 21–24°C
- Moravci – thermal spring, 36–66°C
- Moravci-Buckovci – thermal spring, 25–43°C
- Murska Sobota – thermal spring, 25–51°C
- Nova Gorica – thermal spring, 16–28°C
- Okonina – well(s) or hole(s) drilled, 18°C
- Osp – well(s) or hole(s) drilled, 20°C
- Otocec – well(s) or hole(s) drilled, 15°C
- Podcektrek – thermal spring, 15–35°C
- Portoroz – thermal spring, 16–23°C
- Posavje's Faults – thermal spring, 18–23°C
- Ptuj – thermal spring, 29–39°C
- Radenci – thermal spring, 25–40°C
- Rimske Toplice – thermal spring, 32–38°C
- Rogaska Slatina – thermal spring, 30–55°C
- Secovlje – well(s) or hole(s) drilled, 22°C
- Sempeter – well(s) or hole(s) drilled, 32°C
- Slovenj Gradec – well(s) or hole(s) drilled, 38°C
- Smarjeske Toplice – thermal spring, 17–34°C
- Snovik – thermal spring, 18–30°C
- Topolsica – thermal spring, 20–32°C
- Trbovlje – thermal spring, 20–25°C
- Trebavle – thermal spring
- Tuhnije Valley – thermal spring, 22–28°C
- Vaseno – well(s) or hole(s) drilled, 22°C
- Vihnitka – thermal spring
- Vrhnika – thermal spring, 15–22°C

- Zagorje Valley – thermal spring, 25°C
- Zalec – well(s) or hole(s) drilled, 40°C
- Zrece – thermal spring, 22°C

4.29 SPAIN (A, C, G, M, V)

Spain, with a population of 46.8 million, has a Human Development Index rank of 20 out of 169 and a per capita GNI of $US 29 661 (US Census Bureau, 2011; UNDP, 2010).

Anthropogenic arsenic comes from 41 chemical, mining and mineral plants as well as high-arsenic coal-burning and hydro-thermoelectric utilities that discharge arsenic into the air, 12 facilities that discharge 2.93 tonnes of arsenic per year directly into surface water and 3 that discharge 0.043 tonne per year indirectly into surface water (European Pollutant Emission Register, 2004).

Arsenic-enriched bituminous coals are found in the Puertollano basin, with concentrations up to 83 mg arsenic/kg; a positive sulfur correlation indicated that the source is probably pyrite (Alastuey *et al.* 2001).

Geogenic sources of arsenic are associated with the mineralogy of certain areas. Salamanca Province has relatively abundant granitic and metamorphic zones that have been hydrothermally altered; as a consequence, the arsenic concentrations are elevated, which can be a source of contamination to soils and ground waters. Arsenic concentrations in soils, due to low supergene mobility, are only slightly enriched as compared with parent rocks where tailings containing arsenic are high (>1000 mg/kg). Associated groundwater arsenic plumes usually do not exceed 500 m in areal extent, probably due to the low solubility of scorodite ($Fe_3^+AsO_4 \cdot H_2O$) and iron oxyhydroxide adsorption. As a result, ground waters generally have low arsenic concentrations, but some have higher concentrations, over 10 μg/l, perhaps from sulfides in altered granites. Surface waters, however, have high arsenic concentrations in some samples close to mining areas (Garcia-Sanchez & Alvarez-Ayuso, 2003).

A study has identified naturally occurring arsenic leaching into the groundwater of the Madrid or Targus Basin (Lillo *et al.* 2007). In fact, the Madrid Tertiary Detrital Aquifer, one of the largest in Spain and the major supply for Madrid's drinking-water, does have elevated arsenic concentrations, up to 91 μg/l, and these are positively correlated with elevated pH (8.3–9). Twenty per cent of the samples analysed showed arsenic concentrations exceeding the Spanish limit of 50 μg/l (Hernandez-Garcia, 2002).

Two surveys of Spanish rice ($n = 24$) found a mean total arsenic concentration of 0.21 mg/kg and 0.20 mg/kg ($n = 76$). This data must be correlated with the per capita rice consumption rate for Spain at 31 g/d as compared to rice intensive economies like Bangladesh at 445 g/d to assess its oncological implications. It is notable that developed countries like Spain have highly skewed high arsenic grain concentrations, while developing countries have normally distributed total grain arsenic concentrations (Meharg *et al.* 2009). This factor may be due to the local application of arsenical pesticides and herbicides.

Mining in Spain goes back to the Phoenicians and later the Romans. The leaching of acid mine tailings into groundwater has posed a constant threat, and a recent spill from a pyrite mine has confirmed fears about the existence of other such incidents that have gone unnoticed. Aguilar *et al.* (2004) studied the soil pollution after the spill of the Aznalcóllar pyrite mine between 1998 and 2001, analysing arsenic, zinc, cadmium, copper and lead, both total concentrations as well as soluble and bioavailable forms (Aguilar *et al.* 2004):

> The main remediation measures were: clean-up of the tailings and polluted soils, plus application of amendment materials (liming). The results indicate that, after three years, 50–70% of the acidic soils and 25–30% of the basic soils are still highly polluted in total arsenic. The limit of 0.04 mg/kg for water-soluble arsenic is exceeded in 15–20% of all soils. The EDTA [ethylenediaminetetraacetic acid] extractable arsenic (bioavailable) exceeds the limit of 2 mg/kg only in the acidic sectors. After clean-up, the homogenization of the upper 20–25 cm of the soils appears to be the most recommended measure in the reduction of pollution.

4.30 SWEDEN (A, G, M)

Sweden, with a population of 9.1 million, has a Human Development Index rank of 9 out of 169 and a per capita GDP of $US 36 976 (US Census Bureau, 2011; UNDP, 2010).

In Sweden, anthropogenic arsenic comes from 20 chemical, mining and metal plants and pulp factories that discharge 2.27 tonnes of arsenic per year into the air and 25 facilities that discharge 1.29 tonnes per year directly into surface water (European Pollutant Emission Register, 2004).

Geogenically occurring arsenic in drinking-water has been experimentally determined from a survey completed in 2003 in which zones of high arsenic content in tube wells in the central and northern regions were found (see Figure 4.6, left). These areas of arsenic concentration were also found to be hydrologically similar to the concentrations of arsenic found in water from dug wells and from surface water (see Figure 4.6, right).

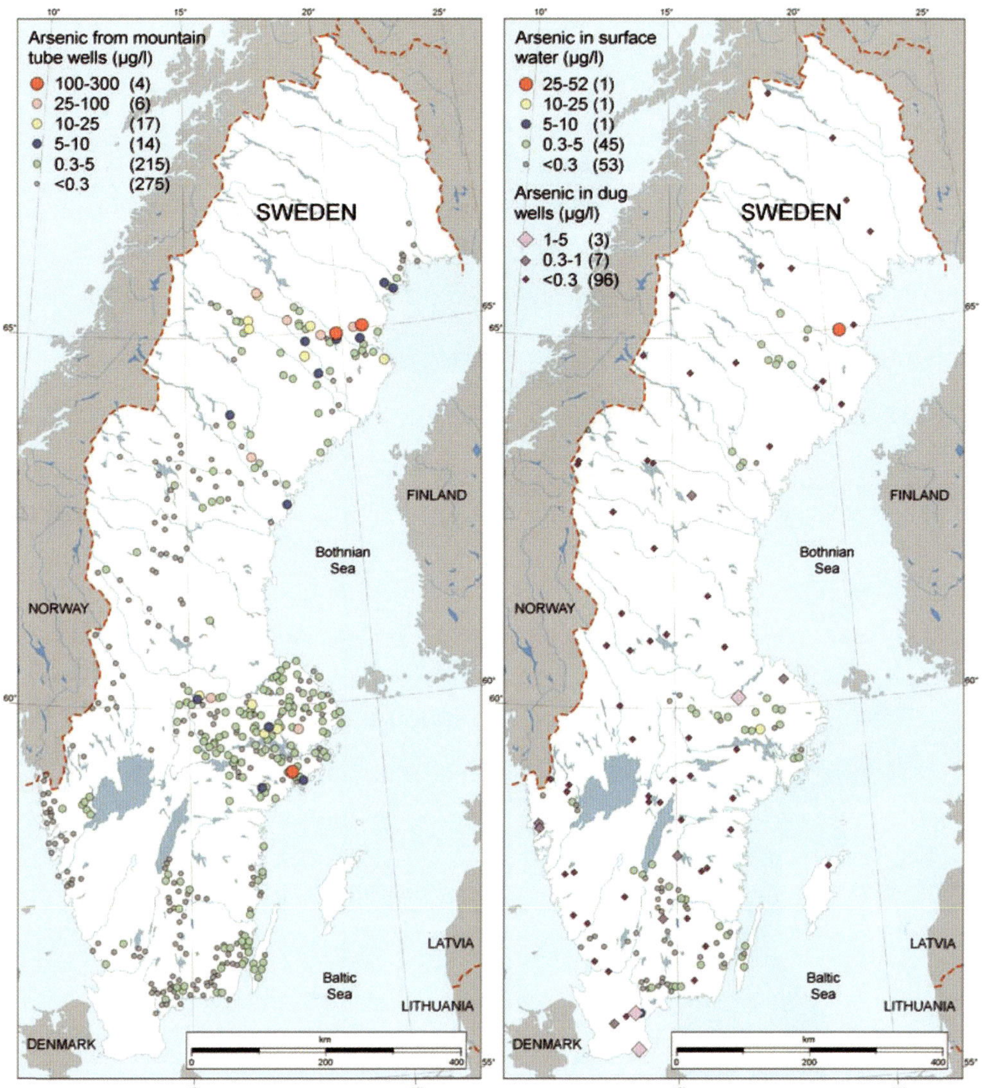

Figure 4.6 Arsenic concentrations from mountain tube wells ($n = 531$) (*left*), surface water and dug wells (*right*) in Sweden (modified from Svergiges geologiska undersoekning, 2005).

These experimental findings correlate with the geological sources of mineralized arsenic (see Figure 4.7). As these areas are of importance to the mining industry, they are also areas where exposed arsenic-bearing rocks can leach into the groundwater and surface water.

The Rönnskär copper smelter case-study is based mainly on investigations by Lars Järup and co-workers (Järup *et al.* 1989; Järup & Pershagen, 1991), as summarized in Berglund *et al.* (2001):

Setting. The Rönnskär copper smelter in the north of Sweden has for a long time been known as "the dirtiest industry in Sweden". It was established in 1928 to process ore from the Boliden mine. The ore was highly complex and was rich in arsenic (up to 20%). The mining company bought two small islands just outside the coast from the local community. The idea was that the winds would carry the emissions from the smelter out to sea, thus minimising the environmental air pollution and the fallout on the surrounding soil.

Exposure. The work environment was heavily polluted with many compounds; it was said that the workers were exposed to "half the periodic system". The main pollutants connected with health problems at the factory were arsenic, sulphur dioxide (sulphuric acid) and lead. Nickel, bismuth and selenium, as well as other contaminants, were also present in some instances. The main concern regarding environmental exposure has been related to arsenic and sulphur dioxide.

Occupational health effects. Since the start of the smelter operation, both acute and chronic health problems have been registered. Early health effects at the smelter were etching injuries, perforation of the nasal septum and chemical bronchitis.

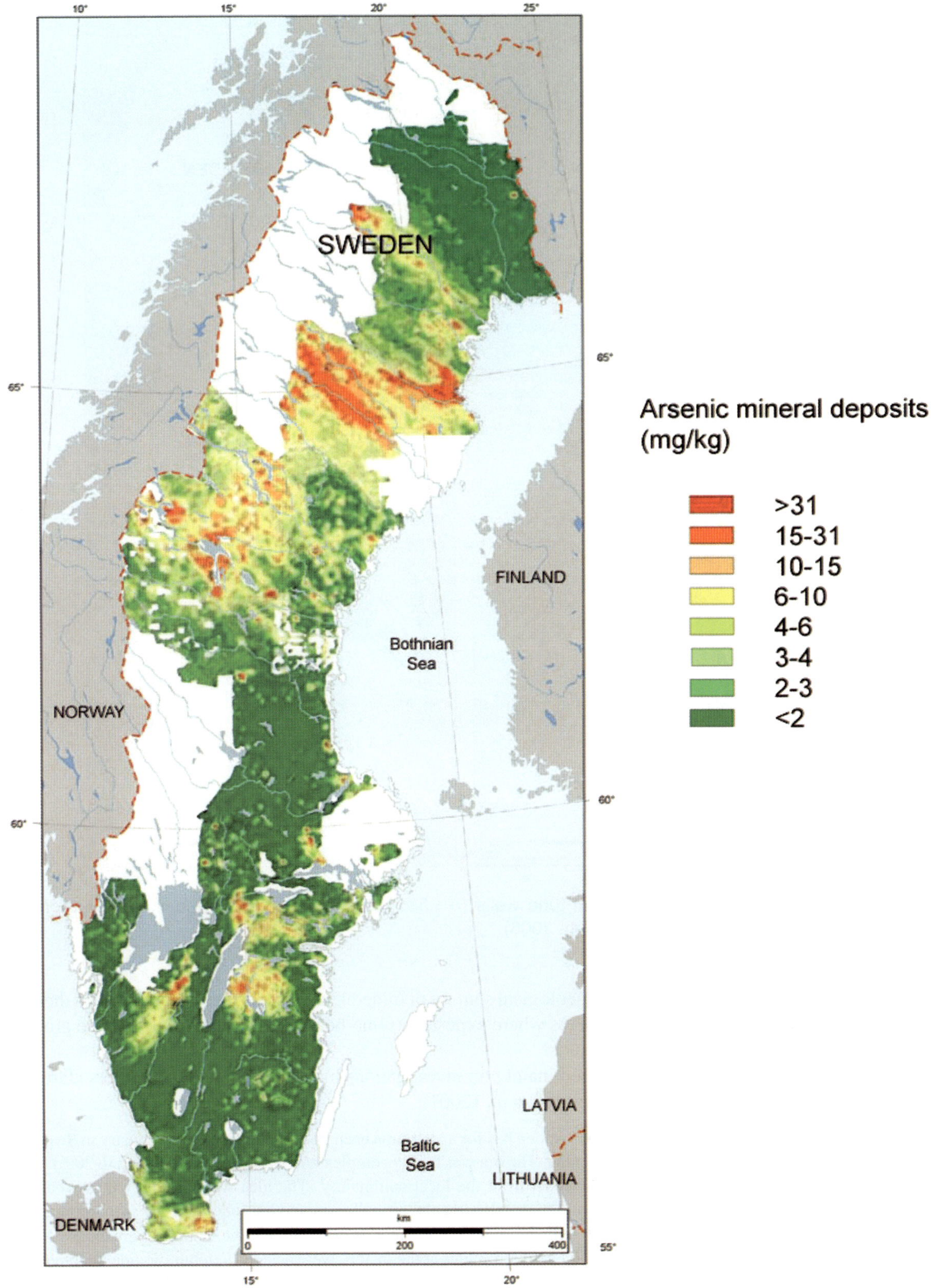

Figure 4.7 Geologically occurring arsenic mineral deposits in Sweden (modified from Svergiges geologiska undersoekning, 2005).

The effects of the smelter on the groundwater situation have yet to be analysed. However, the slag from such an operation strongly suggests some leaching into the groundwater.

In another study, it was found that mine tailings were high in arsenic, which was potentially leaching into the groundwater supply. At Stekenjokk, unoxidized tailings were measured with copper, zinc and arsenic (1178 mg/kg). At Laver, Norrbotten, unoxidized tailings contained arsenic at 68 mg/kg, and oxidized tailings contained 109 mg/kg (Holmstrom *et al.* 1999).

Since all rivers and streams flow to the sea, another aspect of the leaching of arsenic into surface waters is that two zones of high-arsenic marine sediments have been discovered in the Baltic Sea (Figure 4.8). This phenomenon is of critical importance not only for the high levels of arsenic but also because of its potential threat to marine biota, especially molluscs, crustaceans and seafloor-dwelling bony fish that are part of the human marine food supply.

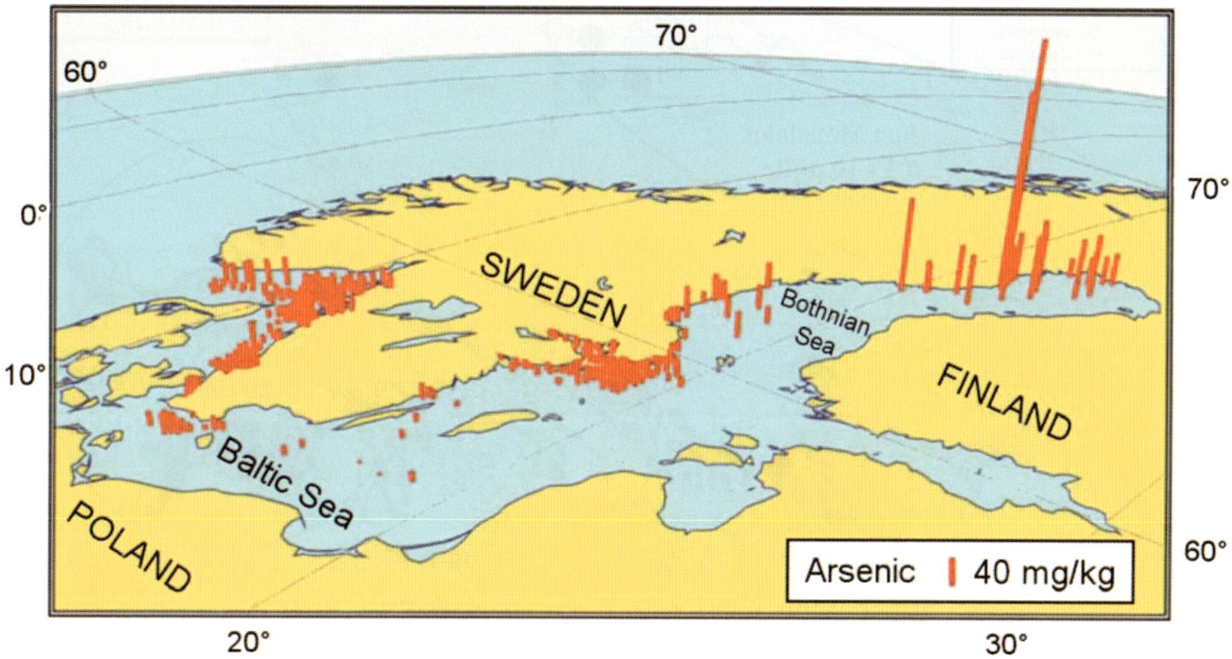

Figure 4.8 High-arsenic marine sediments in the Baltic and Bothnian seas (modified from Svergiges geologiska undersoekning, 2005).

4.31 SWITZERLAND (G, M, V)

Switzerland, with a population of 7.6 million, has a Human Development Index rank of 13 out of 169 and a per capita GNI of $US 39 849 (US Census Bureau, 2011; UNDP, 2010).

Geogenic and/or volcanogenic arsenic occurs in four major areas in Switzerland: in the north-east, where numerous mineral and thermal springs are located; in the Jura, where it occurs with iron, limestone and clays; in the Alps, where it is found in ore deposits and crystalline formations and isolated thermal and mineral springs; and in Lake Greifen, which is situated on the Swiss plateau. In the north-east, arsenic emanates from deep granites and schists in thermal springs at Baden, Zurzach, Schnizach and Bad Saeckingen (130 μg/l), St Moritz, Saxon and Leukerbad. In the lower Engardine Mountains, arsenic concentrations in the disused cold mineral springs of Val Sinestra reach 3000 μg/l. In the Jura, arsenic is found in the Dogger formation brown limestones, the Cretaceous formation limestones (10–20 mg/kg in limestone and 150 mg/kg in soils, mainly at Solothum, Aargau and Neuenberg), the Bohberz formation of red clays with iron nodules (500 mg/kg in bedrock, no direct soil measurements, but dry plant mass measures 500 μg/kg) and the cantons of Waadt, Delémont and Schffhausen. The Swiss Alps have elevated arsenic in crystalline silicates (amphibolites, gneisses and schists) and ores. Some of these deposits are localized, but there are some deposits in the cantons of Wallis Ticino and Graubuenden/Grisons over several hundred square kilometres that contaminate the groundwater in these areas. For instance, in the area around Lugano in Canton

Ticino, a dozen communities with a total population of about 5000 relied on groundwater with elevated arsenic. These communities included Val Isone and Val Colla, with 11–50 μg arsenic/l, Malcantone, with 80 μg arsenic/l, Barbengo-Morcote, with 10–50 μg arsenic/l, and Sottoceneri (Pfeifer & Zobrist, 2001). In Lake Greifen, a small, eutrophic lake in north-eastern Switzerland with a seasonally anoxic hypolimnion, arsenate disappears in the summer in the epilimnion and is replaced by arsenite, whereas arsenate remains the dominant species in the hypolimnion. This links eutrophic conditions with the release of arsenite, in a situation comparable with that found in eutrophic Lake Biwa, Japan (Kuhn & Sigg, 1993).

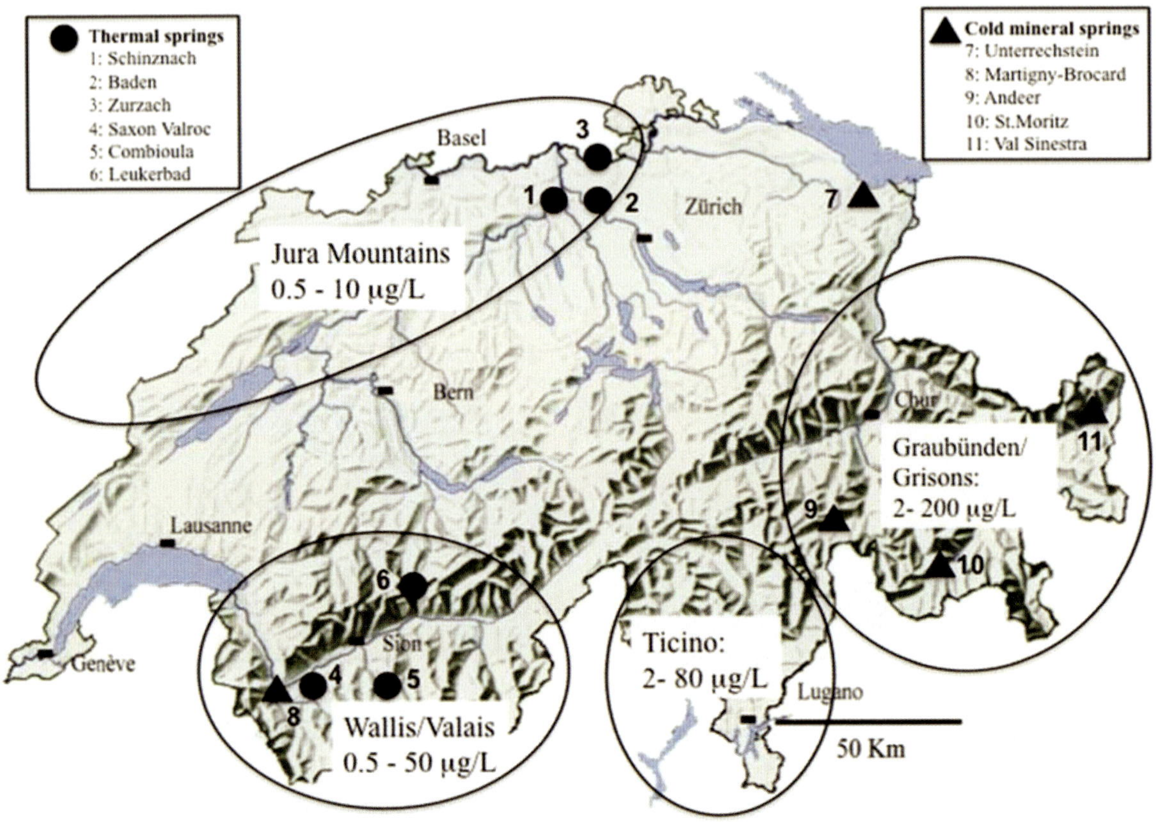

Figure 4.9 Major areas with elevated concentrations of arsenic in Switzerland. In several areas there are thermal or cold mineral springs with As-contents above 10 μg/L. The main spring waters are marked with large circles and the marked ranges refer to drinking waters used up to 2002. (Pfeifer *et al.* 2010)

The regions of elevated arsenic in the northeast mineral springs, the Jura Mountains and the Alps are illustrated in Figure 4.9 and detailed in Table 4.4.

Apart from the major geogenic or volcanogenic contributors of arsenic described above, mining has also contributed to arsenic contamination in Switzerland, especially the surface water and groundwater surrounding several former gold mines (Pfeifer & Dold, 2005). Elevated arsenic levels are present in the Jura where the Bohnerz formation is iron-rich in Dogger limestones and in the Alps where sulfide and silicate deposits are rich in pyrite or haematite.

The official Swiss limit for arsenic in drinking water is 50 μg/l. Data on drinking-water are available for southern Switzerland in Canton Ticino, where a systematic survey of the whole area was undertaken. About a dozen communities have drinking-water containing arsenic at concentrations between 2 and 80 μg/l (Figure 4.9). Data for arsenic in drinking water is also available for water systems in 160 communes of Valais, shown in Table 4.5.

In derelict mines and isolated groundwater drill holes with reducing conditions, arsenic concentrations can approach 300 μg/l. State authorities, having monitored critical springs and wells, have excluded springs with concentrations >10 μg/l. from consumption. Water treatment procedures to reduce arsenic in drinking-water to levels of about 5 μg/l are being implemented.

Table 4.4 The different areas in Switzerland with elevated arsenic in the environment (Pfeifer *et al.* 2010).

AREA Region No. people affected (in 2001)	Origin of As	Typical range of As rocks (r) soils (s) plants (p) waste (w)	Range of As in spring water and surface water
JURA MOUNT. Délémont Weissenstein Probably 0 persons	– Fe-bearing limestones – Fe-rich goethite-bearing clays (Bolus/sidérolitique)	r: 20–800 mg/kg s: 20–150 mg/kg p: 0.1–0.4 mg/kg w: no data	0–5 μg/l
WALLIS/VALAIS Collonges Dorénaz Vernayaz Finhaut Bagnes St.Niklaus Eisten Blatten Oberwald 14 000 persons	– Ore deposits and local veins rich in pyrite and arsenopyrite – A few occurrences of more rare As-rich minerals such as gersdorffite, skutterutite, As-bearing sulfosalts	r: 20–1000 mg/kg (ore: max. 46%) s: 20–1600 mg/kg p: 0.1–3.6 mg/kg (ferns) w: up to 40% (former mines)	0–50 μg/l
TICINO Morcote Malcantone Gambarogno Valcolla Val Isone 4000 persons	– Ore deposits and local veins rich in pyrite, pyrrhotite, arsenopyrite, goethite and scorodite – Glacial deposits rich in Fe-oxyhydroxides, clays and adsorbed As – Silicate rocks with dispersed pyrite, arsenopyrite and or allanite	r: 20–90 000 mg/kg (ore max. 46%) s: 20–2000 mg/kg p: 0.2–10 mg/kg w: up to 40%	0–80 μg/l
GRAUBUENDEN/ GRISONS Upper Engadine Lower Engadine (Val Sinestra) Val Poschiavo 1000 persons	– Silicate rocks with dispersed pyrite, arsenopyrite, hematite, goethite, rarely veins or local ore deposits – Cold CO_2-rich mineral springs	r: 20–500 mg/kg (ore max. 46%) s: no data, except peat: 100–1400 mg/kg p: no data w: no data	0–180 μg/l (Val Sinestra: 3000 μg/L !)

Table 4.5 Low, high and mean values arsenic levels in the water systems of 160 communes of Valalis (Haldimann *et al.* 2005).

Commune	Samples	Low (μg/l)	High (μg/l)	Median (μg/l)	Mean (μg/l)
Agarn	5	0.26	1.91	1.51	1.35
Albinen	2	0.04	0.19	0.12	0.12
Arbaz	10	0.04	0.11	0.07	0.08
Ardon	1	0.13	0.13	0.13	0.13
Ausserberg	10	0.03	3.82	0.16	0.82
Ausserbinn	4	0.01	0.03	0.02	0.02
Ayent	17	0.05	0.30	0.17	0.17

(*Continued*)

Table 4.5 Low, high and mean values arsenic levels in the water systems of 160 communes of Valalis (Haldimann *et al.* 2005) (*Continued*).

Commune	Samples	Low (μg/l)	High (μg/l)	Median (μg/l)	Mean (μg/l)
Ayer	7	0.03	5.34	2.86	3.04
Bagnes	51	0.06	16.76	0.43	2.40
Baltschieder	10	0.48	3.45	1.22	1.61
Beliwaid	25	0.07	2.92	0.30	0.64
Betten	7	0.31	2.47	0.82	1.12
Binn	10	0.03	3.50	0.11	0.68
Birgisch	1	0.36	0.36	0.36	0.36
Bister	1	0.35	0.35	0.35	0.35
Bitsch	9	0.14	2.26	1.40	1.25
Bistten	6	0.07	2.18	0.79	1.00
Bitzingen	10	0.41	4.09	0.51	0.95
Bourg St. Pierre	3	0.24	4.90	1.98	2.37
Bovernier	2	0.32	0.63	0.48	0.48
Bratsch	4	0.04	0.09	0.06	0.06
Brig-Gils	21	0.06	3.80	0.78	1.35
Bürchen	11	0.88	8.87	1.85	2.99
Chaiais	2	0.41	0.63	0.52	0.52
Chamoson	10	0.02	0.11	0.07	0.06
Champery	4	0.05	0.36	0.08	0.15
Chandolin	3	1.12	1.85	1.69	1.55
Charrat	4	0.44	4.82	1.46	2.05
[1]Chermignon	7	0.04	0.07	0.05	0.05
Chippis	1	1.01	1.01	1.01	1.01
Collombey-Muraz	4	0.06	0.41	0.20	0.22
Collonges	3	0.11	18.65	10.69	9.82
Conthey	18	0.03	0.33	0.08	0.10
Dorenaz	1	19.31	19.31	19.31	19.31
Eggerberg	2	1.49	1.79	1.64	1.64
Eischoll	1	7.80	7.80	7.80	7.80
Eisten	10	1.67	24.06	11.40	11.21
Embd	10	0.85	10.26	4.69	4.92
Ergisch	2	2.01	2.10	2.06	2.06
Emen	10	0.04	25.17	0.56	6.09
Erschmatt	2	0.23	0.24	0.24	0.24
Evlonnaz	2	0.31	0.75	0.53	0.53
Evolene	20	0.10	3.71	0.55	0.66
Ferden	3	0.93	5.32	3.95	3.40
Flesch	8	0.10	2.71	0.45	0.92
Fleschertal	8	0.03	2.31	0.23	0.46
Fllet	7	0.02	1.55	0.49	0.60
Flnhaut	4	22.56	27.30	24.72	24.82
Fuly	19	0.09	8.70	0.39	1.19
Gampel	1	6.07	6.07	6.07	6.07
Geschinen	3	0.11	0.39	0.17	0.22

(*Continued*)

Table 4.5 Low, high and mean values arsenic levels in the water systems of 160 communes of Valalis (Haldimann *et al.* 2005) (*Continued*).

Commune	Samples	Low (μg/l)	High (μg/l)	Median (μg/l)	Mean (μg/l)
[2]Gluringen	2	1.57	2.08	1.83	1.83
[2]Goppisberg	7	0.06	1.53	1.08	0.82
Grachen	16	1.61	20.27	5.38	7.93
[2]Grafschaft	2	1.57	2.08	1.83	1.83
[3]Greich	2	1.42	2.25	1.84	1.84
Grengiois	16	0.01	1.30	0.14	0.31
Grimentz	4	0.30	1.13	0.53	0.62
[4]Grimisust	10	0.04	0.11	0.07	0.08
Gröne	1	0.65	0.65	0.65	0.65
Guttet-Feschel	2	0.19	0.45	0.32	0.32
[5]Heremence	10	0.21	29.56	5.11	6.81
Hohtenn	15	0.75	6.01	3.06	2.94
Icogne	7	0.04	0.07	0.05	0.05
Inden	4	0.04	0.10	0.05	0.06
Iserabies	13	0.21	4.60	2.41	2.55
Kippel	6	1.08	6.95	3.76	3.84
Laiden	3	0.51	1.01	0.69	0.74
Lax	12	0.03	1.46	0.25	0.46
Lens	8	0.03	0.09	0.05	0.06
Les Agettes	10	1.18	14.39	6.89	7.48
Leuk	34	0.04	18.47	1.29	1.72
Leukerbad	12	0.03	0.12	0.05	0.06
Leytron	11	0.05	2.67	0.23	0.53
Liddes	3	0.56	4.06	0.85	1.82
Martigny	8	0.06	10.62	7.82	6.58
Martigny-Combe	4	0.69	17.78	8.26	8.75
Martisberg	5	0.05	0.33	0.07	0.14
Mase	1	0.59	0.59	0.59	0.59
Massongex	2	0.42	0.80	0.61	0.61
Mex	1	0.32	0.32	0.32	0.32
Miege	2	0.05	0.07	0.06	0.06
Molles	10	0.04	0.20	0.06	0.08
Montana	12	0.01	0.13	0.03	0.04
Monthey	10	0.04	0.68	0.11	0.15
Mörel	6	0.24	2.37	0.30	0.83
Mühiebach	2	7.53	10.16	8.85	8.85
Mund	11	0.07	1.69	0.50	0.65
Münster	5	0.59	4.72	3.49	2.97
Naters	22	0.04	1.27	0.17	0.36
Nax	1	1.79	1.79	1.79	1.79
Nendaz	17	0.11	6.08	1.14	1.83
Niedergestel	5	0.79	1.24	1.06	1.03
Niederwaid	6	0.15	3.64	0.34	0.89
Oberems	10	1.24	7.87	3.64	3.86

(*Continued*)

Table 4.5 Low, high and mean values arsenic levels in the water systems of 160 communes of Valalis (Haldimann *et al.* 2005) (*Continued*).

Commune	Samples	Low (μg/l)	High (μg/l)	Median (μg/l)	Mean (μg/l)
Obergestein	1	0.38	0.38	0.38	0.38
Oberwaid	12	0.06	101.42	16.87	29.01
Orsieres	10	0.02	1.60	0.21	0.38
Port-Vailais	4	0.18	0.44	0.21	0.26
Randa	7	1.64	10.09	3.39	4.54
Randogne	22	0.01	0.17	0.07	0.08
Raron	2	0.17	0.66	0.42	0.42
Reckingen	6	0.11	2.08	0.25	0.73
Riddes	13	0.05	13.69	0.81	1.80
Ried-Brig	6	0.02	0.23	0.08	0.10
Ried-Mörel	3	0.13	0.18	0.14	0.15
Saas Fee	4	0.13	0.92	0.50	0.51
Saas-Aimagel	5	0.18	1.60	0.34	0.61
Saas-Baien	9	0.61	5.22	1.30	1.65
Saas-Grund	4	0.26	11.21	0.86	3.30
Salion	2	0.33	1.15	0.74	0.74
Saigesch	1	0.09	0.09	0.09	0.09
Salins	7	0.50	9.52	1.76	2.58
Saivan	9	0.63	25.91	14.02	12.33
Saviese	6	0.04	0.14	0.07	0.07
Saxon	8	0.05	0.56	0.18	0.20
Sembrancher	4	0.05	0.62	0.24	0.29
Sierre	5	0.20	0.60	0.31	0.35
Simpion	19	0.10	16.33	0.50	1.51
Sion	36	0.05	79.47	0.08	6.30
St Gingoiph	4	0.09	0.23	0.17	0.17
St Jean	4	0.99	3.25	2.07	2.10
St Leonhard	1	0.48	0.48	0.48	0.48
St Luc	1	1.60	1.60	1.60	1.60
[5]St Martin	22	0.15	29.56	0.79	2.89
St Maurice	1	0.67	0.67	0.67	0.67
St Nickisus	16	0.90	43.77	12.64	15.38
Staiden	9	0.77	4.80	1.82	2.48
Staidenried	19	0.37	17.15	2.58	3.55
Steg	4	1.98	13.89	2.39	5.16
Steinhaus	8	2.57	8.26	5.47	5.45
Tasch	2	0.20	0.62	0.41	0.41
Termen	6	0.01	0.52	0.03	0.13
Törbel	6	4.91	24.72	6.79	9.54
Trient	5	0.26	22.40	0.65	9.12
Troistorrents	14	0.03	0.14	0.05	0.06
Turtmann	15	0.23	1.74	1.03	0.98
Uirichen	6	0.67	20.44	6.87	8.00
Unterbach	5	0.42	22.42	4.38	7.04

(*Continued*)

Table 4.5 Low, high and mean values arsenic levels in the water systems of 160 communes of Valalis (Haldimann *et al.* 2005) (*Continued*).

Commune	Samples	Low (μg/l)	High (μg/l)	Median (μg/l)	Mean (μg/l)
Unterems	5	0.78	2.53	1.25	1.41
Val d'Illez	8	0.01	0.20	0.09	0.09
Varen	6	0.07	0.15	0.08	0.10
Venthone	3	0.07	0.45	0.21	0.24
Vemamiege	1	0.57	0.57	0.57	0.57
Vemayaz	2	17.50	25.17	21.34	21.34
Verossaz	3	0.05	0.17	0.08	0.10
Vetroz	3	0.11	0.57	0.14	0.27
Vex	53	0.11	98.53	2.32	8.22
Veyras	1	0.28	0.28	0.28	0.28
Veysonnaz	11	0.19	12.04	2.23	4.27
Vionnaz	2	0.29	0.33	0.31	0.31
Visp	23	0.07	6.40	1.18	1.31
Visperterminen	33	0.04	3.74	0.51	0.80
Vissoie	1	1.99	1.99	1.99	1.99
Volieges	14	0.06	0.49	0.25	0.26
Vouvry	3	0.13	0.24	0.21	0.19
Wiler	5	1.15	10.39	3.86	4.15
Zeneggen	1	0.21	0.21	0.21	0.21
Zermatt	23	0.03	1.91	0.27	0.46
Zwischbergen	20	0.03	5.73	0.33	0.85

[1]Water supplied from icogne
[2]Water supplied from two supply systems from Reckingen (2 data ponts)
[3]One common supply
[4]Water supplied from Arbaz
[5]One common supply

Canton Ticino's Sottoceneri has over 4,960 inhabitants living in areas with arsenic at concentrations above 10 μg/l (Figure 4.10). Arsenic occurs in locations shown in Figures 4.11 and 4.12 under four types of hydrogeological scenarios:

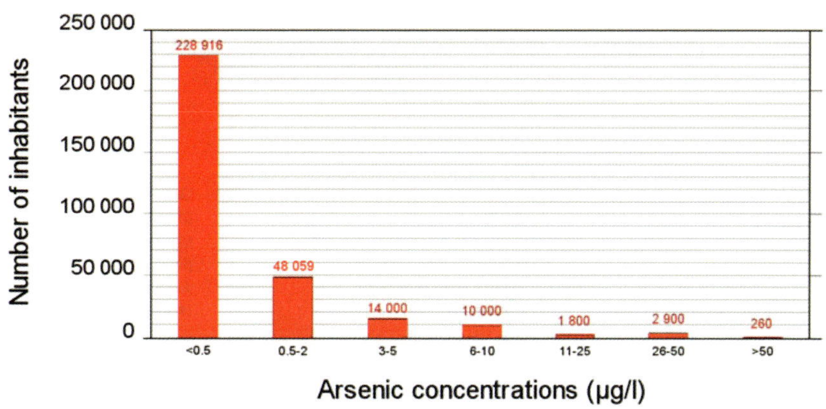

Figure 4.10 Number of people in Canton Ticino, Switzerland, affected by elevated arsenic in drinking-water (modified from Pfeifer *et al.* 2002)

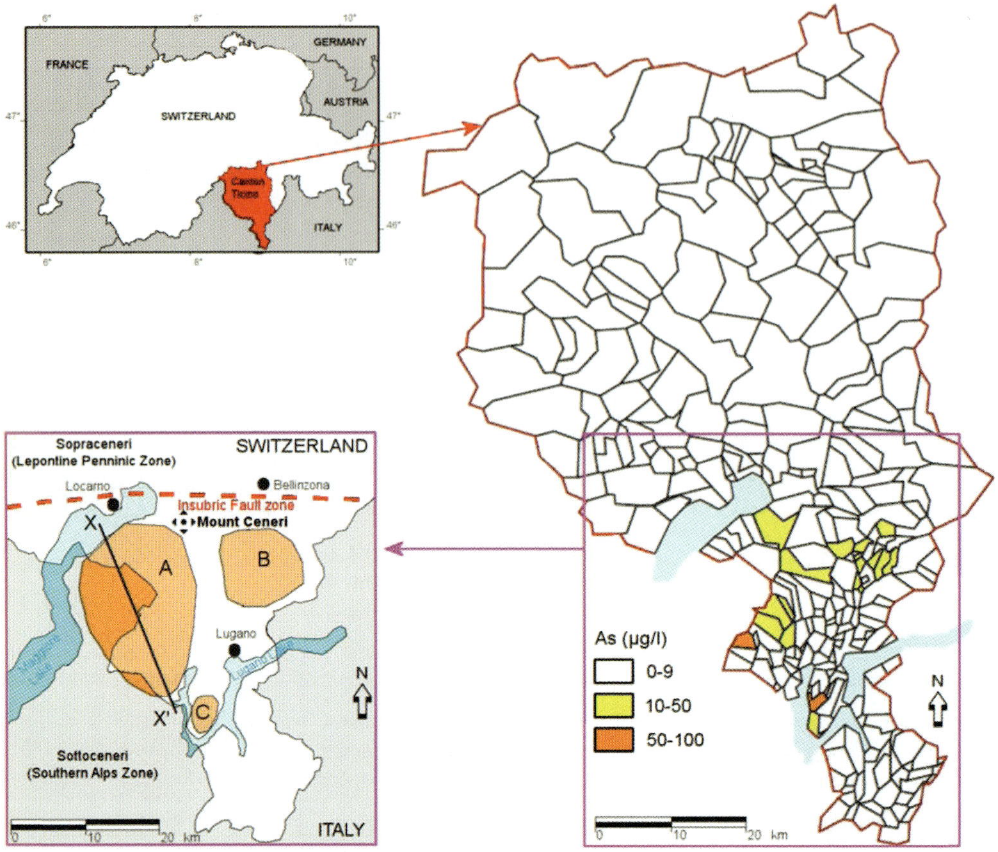

Figure 4.11 Areas in southern Switzerland with elevated natural arsenic in soils and waters. Bottom left: A: Gambarogno-Malcantone-Val Veddasca (Italy), B: Val Isone-Val Colla, C: Barbengo-Figino-Morcote, X-X': cross-section; Bottom right: arsenic concentrations (modified from Pfeifer *et al.* 2002).

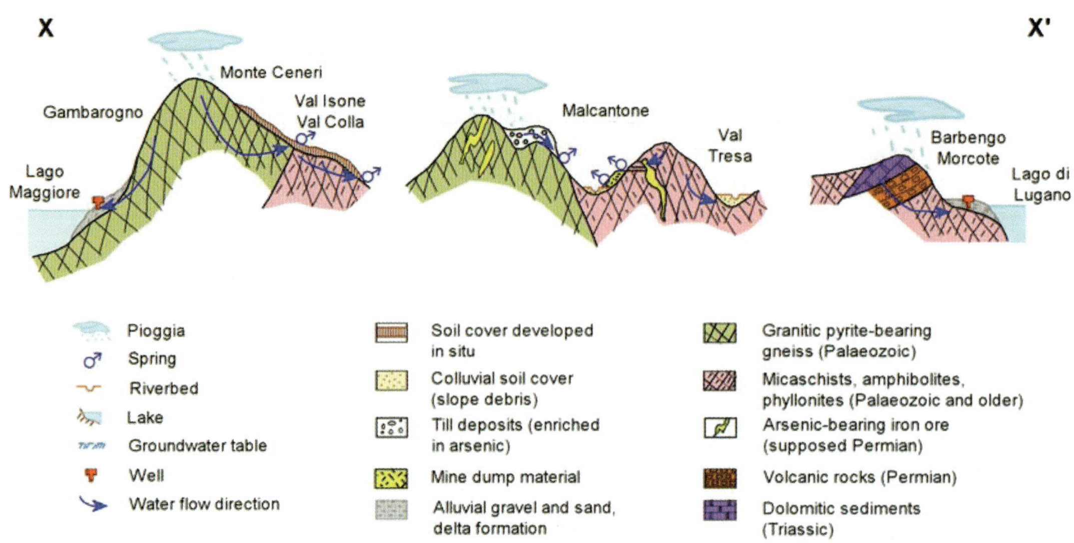

Figure 4.12 Schematic diagram of cross-section X-X′ of Figure 4.11 showing arsenic occurrences in different hydrogeological scenarios (modified from Pfeifer *et al.* 2002).

(1) spring water and local creeks enriched in arsenic (20–50 μg/l), related to a fissured hard rock aquifer with an elevated arsenic content (gneisses, mica schists and phyllonites containing elevated amounts of arsenopyrite and iron oxides);
(2) spring waters rich in arsenic (50–80 μg/l), related to a porous glacial aquifer that is enriched in arsenic, due to the weathering of iron–arsenic deposits;
(3) mine waters very rich in arsenic (100–800 μg/l), issuing from abandoned shafts and/or percolating through mine dump material rich in arsenic; and
(4) reduced iron- and arsenic-rich groundwater (200–300 μg/l for arsenic), slowly circulating or stagnating in alluvial river deposits on the valley grounds, often close to the shore line of lakes and rivers. Here, the exact origin of the arsenic is often unknown (Pfeifer *et al.* 2002).

4.32 UKRAINE (A, C, V)

Ukraine, with a population of 45.1 million, has a Human Development Index rank of 69 out of 169 and a per capita GNI of $US 6535 (US Census Bureau, 2011; UNDP, 2010).

Arsenic-enriched coals with arsenic concentrations in ash up to 1000 mg/kg are found in the Upper Mississippian Buzhanskaya Formation of the L'vov-Volynsk basin (Plaska *et al.* 1971).

Pennsylvanian coal beds in the Donbas region have arsenic concentrations ranging from 0.1 to 121 mg/kg (Lenin mine, central region), 130 mg/kg (Glubokaya mine, Donets-Makeevka region) and 208 mg/kg (Dzerzhinski mine, central region) (Kolker *et al.* 2002). In the Donets coal basin, arsenic enrichment was identified in three roof layers and is thought to positively correlate with high sulfur in some, but not all, beds (Yudovich, 2003). Arsenic concentrations in the Donets Basin coals range from 32.7 mg/kg from anthracites in the eastern basin to 80 mg/kg arsenic from coals in general (Kler, 1988; Kizilstein & Kholodkhov, 1999).

Anthropogenic arsenic has also contaminated the air and soils of the cities of Donets, Makeevka and Yasinovataya, as well as many smaller towns and villages, because they were subjected to the ash from the burning of coke. In the Donets city area, arsenic contamination of soils was measured, with concentrations ranging from 85 to 250 mg/kg and a mean of 63 mg/kg (Yudovich & Ketris, 2005).

The ground waters of these areas have also been contaminated with arsenic, because many of the coal mining pits contained arsenic in concentrations up to 1 kg/tonne. In fact, one of these pits leached a plume of arsenic into groundwater (with concentrations reaching 1200 μg/l) that extended over an area of 15 km by 30 km in the Donets region (Panov *et al.* 1988, 1999).

The extent of the arsenic contamination of the groundwater in the Donets area has been known since the 1980s. Those estimated to be at risk include not only the cities of Donets (1 131 700 population in 2005), Makeevka (389 589 population in 2001) and Yasinovataya (37 000 population in 2001), with a combined population of 1.56 million (Wikipedia, 2007t, u,v), but also small villages associated with coal mines. For instance, Lidievka is adjacent to Lidievka coal mine, one of three high-arsenic coal mines in Kirovsky district, Donets basin, along with Skochynsky and Abakumov. A study of 7000 coal miners from these villages is being undertaken by the University of Illinois at Chicago (USA) to ascertain the underlying causes of acute lung disease among these miners (University of Illinois, 2007). Arsenic concentrations in soils in the Lidievka village area range from 100 to 120 mg/kg, and the soils cover an area of 0.5 km^2 (Panov *et al.* 1988). Surface water and groundwater are also contaminated with arsenic. Sources of arsenic contamination of the biosphere in these and many similar locations include coal waste heaps, coal mines and coke chemical plants. Based on these investigations, the estimated population at risk from elevated arsenic coal contamination in Ukraine is in the neighbourhood of 1.6 million.

Portions of Ukraine bordering the northern part of the Great Pannonian Plain are potentially exposed to elevated concentrations of arsenic in a geological situation identical to that of Hungary (see Section 3.11; Hlavay & Polyak, 1998; Biacs, 2005). As the levels of arsenic in adjacent areas of Hungary are in the range of 10–30 μg/l in the groundwater, tests should be initiated in Ukraine to determine the level of risk.

4.33 UNITED KINGDOM OF GREAT BRITAIN AND NORTHERN IRELAND (A, C, G, M, P)

The United Kingdom, with a population of 62.7 million, has a Human Development Index rank of 26 out of 169 and a per capita GNI of $US 35 087 (US Census Bureau, 2011; UNDP, 2010).

Anthropogenic sources of arsenic have been linked with industrial riverine discharges. For instance, arsenic has been discharged into the lower Humber estuary in large quantities (Grant, 1999), with concentrations measuring 0.54–4.07 μg/l (Neal *et al.* 1996). In the Caron River, arsenic concentrations measured 1–80 μg/l (water) and 1–3800 μg/l (particulate) (Johnson & Thornton, 1987).

Additional anthropogenic sources have been linked with the emissions from 25 power generation plants (including nuclear), metal manufacturing plants and brewing facilities that discharge 0.959 tonne of arsenic per year into the air, 48 facilities that discharge 3.24 tonnes per year directly into surface water and 11 facilities that discharge 0.254 tonne per year indirectly into surface water (European Pollutant Emission Register, 2004). The problem of ash from high-arsenic coal is an old one. Anthropogenic high-arsenic coal-burning industrial furnaces bioaccumulated arsenic in forage that cattle ingested and induced arsenic-related diseases in the cattle (Bayet & Slosse, 1919). This was an early study, and the epidemiological impact on humans is not known. In Leeds, high-arsenic coals produced fine ash containing 30–230 mg arsenic/kg that penetrated city buildings and was also undoubtedly inhaled by the inhabitants (Dunn & Bloxham, 1933). Again, the epidemiological effects are not recorded.

Arsenic-enriched coals are found in several locations in the United Kingdom. In south Yorkshire, vitrinite bituminous coals from the Beeston seam contain up to 12.4 mg arsenic/kg, with an ash concentration of 1879 mg arsenic/kg (Yudovich & Ketris, 2005). Newcastle low-ash coals contain as much as 8000 mg arsenic/kg (Goldschmidt & Peters, 1934). East Midland coals contain arsenic in the range from 19 to 13 000 mg/kg (White *et al.* 1989). South Wales coal beds have elevated arsenic associated with hydrothermal fluids, local mineralization and metamorphism that mobilizes arsenic to penetrate ground waters (Gayer *et al.* 1999).

Geogenic sources of arsenic are linked to highly mineralized areas, such as Cornwall, where arsenopyrite and other arsenic minerals are found in association with copper and tin. In addition, another unusual geogenic pathway of arsenic is through mutton from sheep that have grazed on kelp from naturally occurring arsenic-tainted littoral sediments. Thirteen different arsenic species (arsenite, arsenate, dimethylarsinic acid, monomethylarsonic acid, trimethylarsine oxide, tetramethylarsonium ion, arsenobetaine, arsenocholine, dimethylarsinoyl ethanol and four common dimethylarsinoylribosides, or arsenosugars) were identified in seaweed extracts and in the urine and blood serum of seaweed-eating North Ronaldsay sheep from northern Scotland. The sheep daily eat 2–4 kg of seaweed, which is washed ashore on the most northern Island of Orkney. The following technical data (Feldmann *et al.* 2000) provide an assessment of the biochemistry of the problem:

> The sheep have a body burden of approximately 45–90 mg arsenic daily. Since the metabolism of arsenic species varies with the arsenite and arsenate being the most toxic, and organoarsenic compounds such as arsenobetaine the least toxic compounds, the determination of the arsenic species in the diet and their body fluids are important. The major arsenic species in their diet are arsenoribosides. The major metabolite excreted into urine and blood is DMAA [dimethylarsinic acid] (95–4.1%) with minor amounts of MMAA [monomethylarsonic acid], riboside X, TMA [tetramethylarsonium ion] and an unidentified species. The occurrence of MMAA is assumed to be a precursor of the exposure to inorganic arsenic, since demethylation of dimethylated or trimethylated organoarsenic compounds is not known (max. MMAA concentration 259 μg/l). The concentrations in the urine (3179–2667 μg/l) and blood (44–19 μg/kg) are at least two orders of magnitude higher than the level of arsenic in the urine of the control sheep or literature levels of blood for the unexposed sheep. The tissue samples (liver: 292–99 μg/kg, kidney: 565–193 μg/kg, muscle: 680–224 μg/kg) and wool samples (10 470–5690 μg/kg) show elevated levels which are also 100 times higher than the levels for the unexposed sheep.

It is unknown whether these elevated arsenic levels in the littoral sediments affect the molluscs that are gathered for food in the area; the extent of this problem in other littoral areas is also unknown. However, elevated arsenic levels have been identified in the liver, kidney and muscle of kestrels (*Falco tinnunculus*), sparrowhawks (*Accipiter nisus*) and barn owls (*Tyto alba*) from south-west England, an area with geogenic and mining-associated arsenic releases (Erry, 1999).

In a related instance, geogenic hidden arsenic has been identified in the Southampton, Tamar and Severn estuaries of southern England (Howard *et al.* 1988; Howard & Comber, 1989). The absence of hidden species of arsenic during the biologically dormant period of the estuarine and lacustrine year may be attributed to depressed metabolic levels of phytoplankton and algae: with lower excretory levels of methylarsenicals or arsenicals that are excreted but not detected by the traditional hydride generation method, they are hidden from traditional detection methods. In the Severn Estuary, for instance, the top 10 highest individual arsenic concentrations ranged from 5 to 71 μg/l, with an average of 20.9 μg/l (Marine Biological Association, 2003).

Geogenic arsenic can be elevated near rock-forming minerals close to the redox zones in semi-contained aquifers, such as in the Chalk and Sherwood Sandstone formations and in the Wilmslow Sandstone Formation. Sherwood Sandstone occurs as a broad belt between Nottingham and south Yorkshire, enhancing the Sherwood Forest landscape. The deposit thickens from the south (100 m) to the north (300 m) and serves as an important aquifer for a wide area (Nottingham County Council, 2003). Similarly, the Permo-Triassic Red Bed sediments of the Cheshire basin and the Wilmslow Sandstone Formation are mineralized with copper and lead, with minor zinc, silver, copper, vanadium, nickel, arsenic, antimony, manganese and barite along basin faults. The Wilmslow Sandstone Formation is cemented by authigenic quartz enclosing iron sulfides, which predate sulfidic cements of tennantite, bornite and sulfur arsenides (Naylor *et al.* 1989). In a test borehole at Abbey Arms Wood, Delamere, Cheshire, total arsenic concentrations ranged from 5 to 15 mg/kg, with an average of 8 mg/kg.

Arsenic concentrations were not positively correlated with depth, but pore water centrifuged from the core was positively correlated with depth, ranging from 8 μg arsenic/l at 10 m to 30 μg arsenic/l at 150 m. Lack of discrete mineralization zones suggests that arsenic desorbed from iron oxides in rock-forming minerals within sandstones and that desorption may be correlated with a slightly higher pH (up to 8.0) at 150 m (Kinniburgh *et al.* 2006).

Mining sources of arsenic are found in the south-west of England. The history of the mining in the region is centuries old, and the extent of contamination is extensive. Since the late 18th century, intensive exploitation of metalliferous ore deposits and geogenic leaching from mineralized parent rocks have resulted in a zone of arsenic-contaminated soils, leachate and tailings. The scale of the dispersion of these arsenic contaminants by both anthropogenic and geogenic processes encompasses at least 722 km^2, with concentrations above 110 μg/g, or more than twice the maximum that might be expected in uncontaminated soil (Mitchell & Barre, 1995). In a report dealing with arsenic cycling in the environment based on its release during mining operations, it was cited that arsenopyrite in granitic copper–tin veins such as those in Cornwall in south-west England leach into soils with bioaccessible concentrations ranging from 3 to 2700 mg/kg (Bowell & Parshley, 2004). Another study confirmed this finding by comparing the bioaccessibility of arsenic in soils in the vicinity of an abandoned arsenic mine and background arsenic concentrations in agricultural soils, also in south-west England (Palumbo, 2003).

Petroleum sources that contribute to arsenic release are associated with hydrocarbon effluents from petroleum fractionation terminals that create conditions that promote the release of arsenic. While these terminals are registered with the European Pollutant Emission Register, ongoing monitoring of emission targets is imperative.

Section 5

Region of Australia and Oceania

5.1 AUSTRALIA (A, C, G, M)

Australia, with a population of 21.8 million, has a Human Development Index rank of 2 out of 169 and a per capita GNI of $US 38 692 (US Census Bureau, 2011; UNDP, 2010).

Anthropogenic arsenic related to arsenic chemicals used in sheep wool processing has been found at Lake Yangebup, south of Perth, Western Australia, which is part of the Beeliar wetland chain and has high arsenic concentrations in both its water and sediments. The arsenic contamination appears as a by-product of the Jandakot wool scouring operation located on its shores. The lake sustains high nutrient loads and is classed as hypertrophic with respect to both nitrogen and phosphorus. Despite the relatively well-defined source of arsenic, sediment and water analyses of the lake indicate that the total concentration is spatially homogeneous throughout the lake, which suggests that some process is responsible for the remobilization of arsenic. The resuspension of arsenic in these sediments at Lake Yangebup is cause for concern (Ling & Oldham, 2002). The question is whether there are other wool processing locations where arsenic chemicals were used. Other lakes in the Perth metropolitan area (e.g. Lake Monger) also exhibit high arsenic concentrations (Oldham, 2003).

In the Perth suburb of Stirling, arsenic has been found in borehole wells at 1700 μg/l (Dortch, 2005) and households have been instructed to stop using borehole well water. The arsenic is believed to be related to the excavation for new housing in development areas that disturbs underlying peat beds and releases arsenic from its sorbed state. In this context, 3200 households or 12 800 people are exposed in Stirling (Government of Western Australia, 2005). Additionally, anthropogenic arsenic has been emitted from industrial wastes and runoff from agricultural land (Australian Water Resources Council, 1982).

Arsenic-enriched coals are found in the Gunnedah Basin, New South Wales, which is about 15 000 km^2 and forms the central part of the Sydney-Gunnedah-Bowen basin system of Australia's eastern margin. It comprises rocks of Permian and Triassic age and is unconformably overlain by the Jurassic-Cretaceous strata of the Surat Basin. The upper strata of Permian coals are dominated by quartz, illite and other epigenetic detrital minerals, whereas kaolinite is abundant in the lower strata. Infills of calcite and dolomite (associated with strontium and barium) occur in upper and lower fractures. Pyrite is abundant in strata where coals formed under marine influence, and these coals are associated with high concentrations of arsenic and minor concentrations of thallium. Chlorine is inversely correlated with pyrite concentrations and appears to be associated with organic matter, as are germanium and gallium (Ward *et al.* 1999). With increased mining of the basin, as is planned, exposure to arsenic-enriched coals should be monitored, as the basin's aquifer of 965 445 ha is the major groundwater resource for the area, with a recharge estimate of 304 406 million litres per year (New South Wales Department of Natural Resources, 2006).

Geogenic and mining arsenic sources have been found in surface soil and groundwater, resulting from both natural hydrogeological conditions and the presence of contamination resulting from gold mining operations (Department of Manufacturing and Industry Development, 1991). For instance, in Western Australia, in the soil adjacent to the silver, gold and copper tailings of the Nevoria Mine, arsenic levels in soil are 500 mg/kg; in the soil at the gold, silver and zinc tailings of the Marvel Loch Mine, the level of arsenic in soil is 2875 mg/kg. These levels can be compared with those in South

Australian coastal waters, which measure 0.06–13 μg/l (Mueller, 1997). In addition, levels of arsenic up to 4 mg/l have been reported in some streams in Australia, attributed to arsenical pesticides and mining activity (Naidu, 2000).

Arsenic-rich waste was also observed to leach from a sawmill in Western Australia (Southwell, 2003a,b):

> FOIA [Freedom of Information Act] documents from the State Government revealed the presence of cancer-causing arsenic at the Pemberton sawmill site, showing that the Department of Environmental Protection was alerted about the pesticide waste dump in 1989, but did little to stop the pollution or warn the public until late 2002. Documents obtained by The West Australian using FOIA reveal that arsenic contamination had been detected in brook sediments and soil on the river flats near the toxic sludge pits by early 1990. Government officials agreed to share the cost of cleaning the 56,000 total cubic meters of contaminated material with the landowners.

In some rural areas of the town of Victoria, water supplies that are based on groundwater extraction have been found to have elevated arsenic concentrations (Department of Manufacturing and Industry Development, 1991). The presence of elevated arsenic in the coastal sandy aquifers of the Yarrahapinni-Stuarts Point groundwater systems (about 1480 ha in extent) in the Richmond River coastal plain on the north coast of New South Wales 500 km north of Sydney is a regional phenomenon in coastal aquifer systems in eastern Australia (Smith *et al.* 2006). Dissolved arsenic in the Stuarts Point coastal sand aquifer has been reported at more than 350 μg/l, 50 times the Australian drinking-water limit of 7 μg/l. Since anthropogenic contamination has been eliminated as a cause, the source of Stuarts Point arsenic contamination has been attributed to the weathering of arsenic-rich stibnite deposits on Yarrahapinni Mountain and mine tailings in the upper Macleay River catchment (O'Shea, 2005).

Identification of arsenic in groundwater associated with coastal barrier sand dune environments is rare, since coastal sand dunes are usually clean and reliable sources of water. This is usually due to a system of coastal sandy embayments. The New South Wales coast, oriented north-northeast–south-southwest, is characterized by three types of coastal embayments. The first type of coastal embayment is a Pleistocene inner barrier beach ridge and quartzose sand dunes impregnated with humic colloids forming a hardpan up to 10 m thick that completely fills the many curved sandy bays of the north coast of New South Wales, now partly eroded by streams and wave action. The second type of coastal embayment is a system of dual barriers (Pleistocene and Holocene) separated by a beach ridge barrier system. The third type of coastal embayment is a Holocene beach ridge barrier. At Stuarts Point, marine transgression has formed two Holocene elongate sand barriers separated by shallow water and subject to strong tidal and river currents. The downward slope of bedrock provides a gradient from Yarrahapinni Mountain through unconsolidated, highly conductive Pleistocene barrier sands to the sea.

Mobilization of arsenic in a sand dune coastal environment can occur in four ways (Smith *et al.* 2006). First, oxidation of arsenical pyrite is increased when the water table is lowered, thereby inducing oxygen to enter the soil, producing sulfuric acid. It has been found that in the Richmond River coastal plain, the mean pyrite concentration is 2%, which produces soil pH in the range 3–1.8. A flood mitigation programme exacerbated this mobilization by drawing down the water table, so the wetland could be reclaimed for dry agriculture. The second way in which mobilization of arsenic can occur in a coastal environment is by its release from oxhydroxides, where bicarbonate from Pleistocene molluscan shells is the carbon contributor (Smedley *et al.* 1998, 2002; Nickson *et al.* 2000). A third route to mobilization occurs in the presence of carbon originating from non-molluscan organic matter, which may or may not be highly dissolved (Xu *et al.* 1991; Kalbitz & Wennrich, 1998; Harvey *et al.* 2002; Smedley *et al.* 2002). A fourth route to arsenic mobilization in a coastal environment can occur when clays in a sandstone matrix are present in an aquifer with high pH and alkaline conditions (Schlottman & Breit, 1992).

The Yarrahapinni-Stuarts Point groundwater systems flow in a gradient from Yarrahapinni Mountain through unconsolidated, highly conductive Pleistocene barrier sands and estuarine and marine Holocene clays (Smith *et al.* 2006). The scenario that led to the mobilization of arsenic was probably through the first mobilization route outlined above, the oxidation of arsenical pyrite, facilitated by the construction of drains and tidal floodgates to reclaim salt marsh for agriculture. When the amount of sulfuric acid exceeds the neutralizing capacity of the soil, acidification ensues (Dent, 1986; Smith *et al.* 2006). O'Shea (2005) proposed the following scenario for arsenic release into the Stuarts Point geomorphological system:

> The heterogeneity of the aquifer sediments causes redox stratification to occur, which in turn governs arsenic mobility in the groundwater. The dominant geochemical processes governing arsenic mobility are reductive dissolution of iron oxides and the formation of arsenian pyrite under sulfate reducing conditions. Withdrawal of the aquifer's groundwater promotes seawater intrusion into the fresh water aquifer and coincides with increased concentrations of arsenic in groundwater. Average dissolved arsenic concentrations in the bulk of the aquifer generally do not exceed 15 μg/l. However, dissolved arsenic begins to increase with increasing depth, suggesting an additional source of arsenic to the aquifer. Dissolved arsenic concentrations increase to more than 80 μg/l in ground waters affected by seawater intrusion.

However, the second and third routes of arsenic mobilization may have played a role as well. Further investigations of coastal sandy aquifers are needed to determine the extent of elevated arsenic, especially where local populations rely on groundwater for human consumption.

5.2 GUAM (A)*

Guam, an island of 541 km^2 with a population of 183 000 (US Census Bureau, 2011) is located in the Northern Mariana Islands and has been an unincorporated territory of the USA since 1898. Its major industry is tourism in Tumon Bay.

Guam's main water source is from a limestone aquifer in the northern part of the island. An estimated 80% of the total water source is derived from this aquifer, and the remaining water comes from the surface on the southern part of the island. There are more than 100 wells that serve the population, and most of these (about 80) serve about 80% of the population (or 129 000) and are located on the northern part of the island. Some of these wells are connected through conduits that flow out through springs along Tumon Bay.

Studies in 2001 conducted by the Guam Environmental Protection Agency along Tumon Bay spring outflows showed unusually high levels of arsenic, ranging from 16 to 100 μg/l, with an average of 37 μg/l from nine stations. These levels were clearly high and warranted further investigation. The high levels of arsenic in the spring water pose a serious threat to tourists who rely heavily on Tumon Bay's beaches and water for recreation. All the springs on Tumon Bay discharge directly into the bay (Vuki, 2004).

However, subsequent speciation studies of arsenic in the same springs indicated low levels of arsenic (Table 5.1). The discrepancy between the two sets of data was complicated by the fact that one set of data was analysed by field test kits and the other by laboratory tests.

Table 5.1 Inorganic arsenic in Tumon Bay samples, Guam.

		Arsenic concentration (μg/l)	
Sample	**Spring location**	**7 July 2003**	**29 July 2003**
TB1	Hilton on-site spring	0.6	0.2
TB2	Hilton onshore spring (duplicate)	0.4	0.2
TB3	Marriott onshore spring	0.4	0.2
TB4	Reef onshore spring	0.6	0.3
TB5	Westin onshore spring	0.6	0.3
TB6	Outrigger offshore spring	0.4	0.3
TB7	Hyatt offshore spring	0.4	0.3
TB8	Wet Willie's onshore spring	0.4	0.2
TB9	Gun Beach onshore spring (background)	0.4	0.4
TB10	Gun Beach rock ledge onshore spring	0.4	0.4
TB11 (experimental blank)	Blank	9.8	0.2
TB2SP(10)	Hilton onshore spring (spiked)	5.8	n.a.
TB9SP(5)	Gun Beach onshore spring	n.a.	n.a.
Std. 5	Calibration standard	n.a.	4.9
Std. 10	Calibration standard	n.a.	9.1

n.a., not available
Source: Vuki, 2004

A project was developed to establish which data were valid and also to extend the investigation into sediments and biological materials to establish the possible sinks and the mobilization conditions for arsenic species. This study began in March 2004 and ended in February 2005. The objectives were (Vuki, 2004):

- To investigate the levels of arsenic in Tumon Bay, the connecting freshwater wells on Guam and other potential sites during the wet and dry seasons;
- To conduct speciation studies of arsenic to ascertain the levels of the different forms of arsenic, both organic and inorganic;
- To conduct a speciation study on the sediment and biota samples;

*Unlike most of the other entries in this Sourcebook, Guam is not a country, but a territory of the USA.

- To conduct an interlaboratory validation exercise;
- To correlate the levels of arsenic to the likely sources and sinks.

An update (Vuki *et al.* 2007) has shown that:

> total arsenic concentrations in the spring water samples ranged from <0.3–1.2 micrograms/L. Inorganic arsenate, As(V), appears to be the dominant species in the spring water samples tested. The concentrations are much lower than previously reported, probably due to a much more rigorous methodological approach and requires further investigations on the status of As contamination in groundwater on the island.

It has been hypothesized that a source of arsenic that is at present unknown has leached into the aquifer under the Won Pat International Airport and that this aquifer serves 80 wells on the north part of the island, which service 129 000 residents (Vuki, 2004).

5.3 NEW ZEALAND (A, C, G, M, V)

New Zealand, with a population of 4.3 million, has a Human Development Index rank of 3 out of 169 and a per capita GNI of $US 25 348 (US Census Bureau, 2011; UNDP, 2010). Arsenic occurs naturally in the areas of New Zealand where arsenopyrite is found. This is generally in regions of volcanic hydrothermal veins where gold and mercury are found (Figure 5.1). Gold mining has exposed many of these veins, and so the possibility of arsenic being released into the groundwater in these regions must be considered.

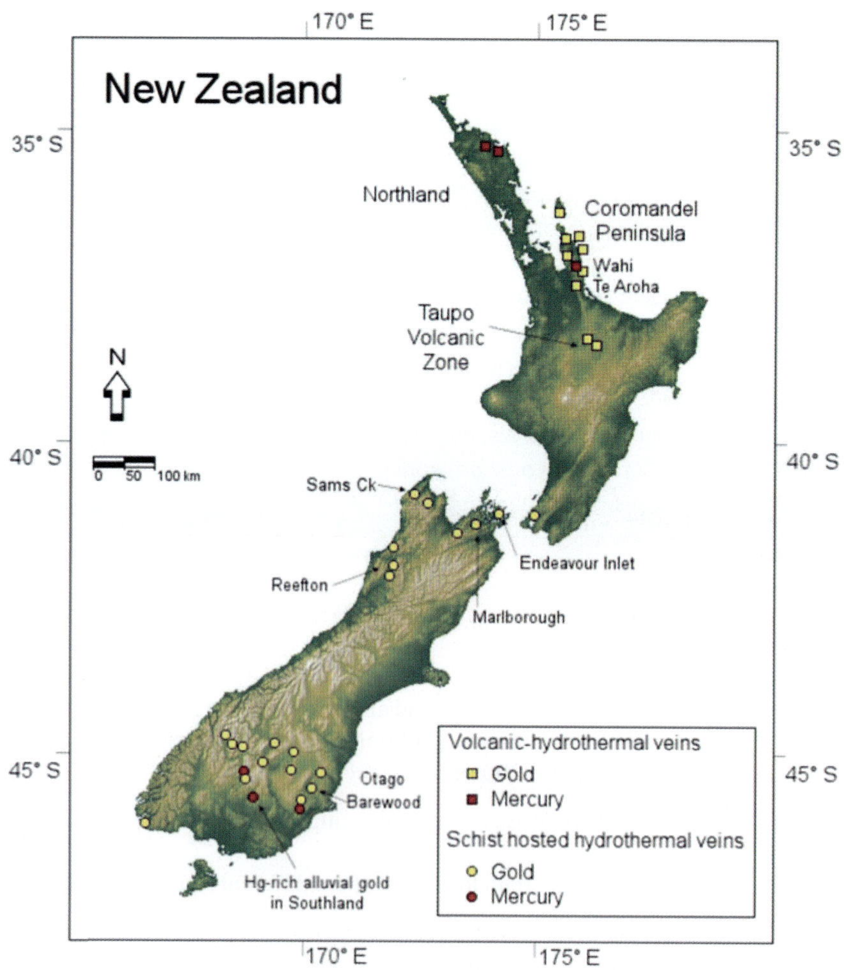

Figure 5.1 Volcanogenic and geogenic sources of arsenic in New Zealand (modified from Otago University, 2012).

For instance, natural weathering of arsenopyrite in gold-bearing veins in east Otago has been accelerated by mining. Acid mine drainage has pH values of 6–7, and this promotes arsenic mobilization, with arsenic concentrations up to 4000 μg/l, compared with background levels of 100 μg/l. At Barewood, century-old tailings have arsenic concentrations of up to 33 000 mg/kg as scorodite ($Fe_3^+AsO_4{\cdot}H_2O$), and dissolved arsenic (from arsenopyrite) produces concentrations of 750 μg/l at a pH of about 5.5. In the Macraes mining area, arsenic concentrations in groundwater are in excess of 200 000 μg/l at pH 10. Dissolved arsenic in these environments is partially remediated both by being precipitated as scorodite and by being adsorbed by (predominantly) fine-grained phyllosilicates, which can bind from 50 to 100 mg/kg (Craw *et al.* 2000).

Basement rock (bedrock) is exposed on ridges and mountains with valleys, basins and coastal plains filled with sediments. The basement rock of New Zealand is greywacke (ancient ocean sediments) and schist (greywacke metamorphosed with heat and pressure). South Island is dominated by greywacke and schist. Exceptions are granites on the west coast of South Island, which are not important for arsenic mobility. Locally thick sediment covers some basement rock, especially on North Island, and some are coal-bearing, which has some environmental significance. Young volcanic rocks were erupted onto basement rock on North Island, especially Taupo Volcanic Zone, Coromandel Peninsula and Northland, and it is here where the arsenic-bearing pyrites are found (Table 5.2).

Table 5.2 Arsenic occurring in common New Zealand mineral deposits.

Deposit type	Metallic minerals	Metals discharged	pH
Schist-hydrothermal (mesothermal)	pyrite, FeS_2 stibnite, Sb_2S_3 **arsenopyrite, FeAsS** gold, Au scheelite, $CaWO_4$ cinnabar (rare), HgS	**As**, Sb	6–8
Volcanic-hydrothermal (epithermal)	**a) gold-related** pyrite, FeS_2 sphalerite, ZnS galena, PbS **arsenopyrite, FeAsS** chalcopyrite, $CuFeS_2$	**As**, Cd, Cu, Pb, Zn, Fe, Mn, Al	1–7
	b) mercury-related pyrite, FeS_2 **marcasite, FeS_2 (+ trace As, Hg)** cinnabar, HgS stibnite, Sb_2S_3	**As**, Hg, Sb, Fe, Mn, Al	1–7
Coal	**pyrite (+ trace As)**	**As**, Ni, Fe, Zn, Cu, B	3–7
Alluvial gold	gold, Au gold–mercury amalgam, Au-Hg **pyrite and/or marcasite**	**As**, Hg, Ni, Fe, Zn, Cu	3–8

Source: Otago University, 2012

In two cases associated with natural leaching of arsenic into the environment, the levels were high. The first was Waikato River near Hamilton, in which the arsenic level was measured as 32.1 μg/l (McLaren & Kim, 1995; Robinson *et al.* 1995). The second was Ohakuri Lake, which contained arsenic at a concentration of 38.3–441 μg/l (Johnson & Thornton, 1987).

Volcanogenic release of arsenic can occur in thermal springs related to hydrothermal deposits in areas where there is still active volcanogenic activity.

Arsenic mobility is enhanced by mining when rocks are exposed to weathering agents or smelting practices, and the discharge may be channelled at higher concentrations in selected zones. Although natural erosion can have similar effects on metal mobility, mining accelerates both the scale and time horizon of arsenic discharge. Monitoring of tailings is essential to ensure that trace metal levels do not leach out, especially during periods of low rainfall when downstream dilution is lessened.

Gold is the main metallic resource that has been mined in New Zealand for over 150 years. Deposits of other metals do occur, but nearly all these are geological curiosities only and have had no economic significance. New Zealand has had a favourable geological setting for gold, which is accompanied by other metals that were discarded by early miners. Two different gold deposit types occur. Both of these are hydrothermal deposits, formed when hot water moved through cracks in the rocks before the rocks were deeply eroded. Arsenic concentrations in New Zealand mining areas are compared in Table 5.3 with the Australia and New Zealand Environment Conservation Council (ANZECC) guidelines for agricultural water and aquatic ecosystems and Ministry of Health drinking-water limits.

Table 5.3 Arsenic concentrations in New Zealand (NZ) mining areas compared with ANZECC guidelines for agricultural water and aquatic ecosystems and Ministry of Health drinking-water limits.

Mine waters			Water (µg/l)					Solids (mg/kg)		
			NZ drinking-water limits	Agriculture, irrigation guidelines		Aquatic ecosystems		Mine rocks		Aquatic ecosystem sediment guidelines
Lower	Upper	Mine type		Lower	Short-term	99% protection	80% protection	Lower	Upper	
100	500 000	Schist-hydro	10	10	20	1	14	100	20 000 (2%)	20

Source: Otago University, 2012

In addition to the geological occurrence of arsenic in New Zealand, arsenic has been anthropogenically introduced. One case (UPI, 2004) involved the building of 1155 homes on a former orchard contaminated with dangerous levels of DDT, arsenic, lead and copper from dumped horticultural sprays.

5.4 PAPUA NEW GUINEA (M)

Papua New Guinea with a population of 7 million has a human development index of 153 out of 169 and a per capita GNI of US$2,271 (US Census Bureau, 2011; UNDP, 2010).

EnviroGulf Consulting (EGC), an independent environmental consulting group with experience in arsenic contamination, has undertaken several arsenic evaluation projects in Papua New Guinea. EGC was instrumental in the design and implementation of sampling programs to assess the scalp hair arsenic concentrations of people inhabiting the Fly River Delta, Papua New Guinea and arsenic concentrations in dietary seafood. EGC also investigated the arsenic exposure levels associated with the mining activities along the Strickland River and Lake Murray, Papua New Guinea, and indicated the impact on public health due to exposure by miners to arsenic either as dust or from the leaching of mine tailings. EGC also evaluated the arsenic contamination of the biota associated with the following mining operations: *Ok Tedi Gold Mine, PNG (1982–1985)* – three-year assessment of river biological effects of mill tailing and mine waste disposal; *Lihir Gold Project, PNG* – assessment of the marine biological effects of at-depth marine tailing disposal and near-surface dumping of soft oxidised waste, on the basis of aeration-dilution and bioassay tests, field dump trials and predictions of waste transport and dispersion (with NSR/Kennecott-Niugini Mining Joint Venture); *Misima Gold Project Environmental Plan (1985–86)* – assessment of the marine biological effects of at-depth marine tailing disposal and near-surface dumping of soft oxidised waste, on the basis of aeration-dilution and bioassay tests, field dump trials and predictions of waste transport and dispersion (with NSR for Placer (PNG) Pty. Limited) (EnviroGulf Consulting, 2011).

References

Abernathy J. R. (1983). Role of arsenical chemicals in agriculture. In: Arsenic: Industrial, Biomedical, Environmental Perspectives, W. H. Lederer and R. J. Fensterheim (eds), Van Nostrand Reinhold, New York, NY, USA, pp. 57–60.

Academy of Science of Georgia (1998). Pollution Forecast of Black Sea Coastal Zone and Western Georgia's Rivers of Black Sea Basin. Academy of Science of Georgia, Hydro–meteorological Institute, Tbilisi, Georgia.

Acharyya S. K. (2002). Arsenic contamination in ground water affecting major parts of southern West Bengal and part of western Chhattisgarh: source and mobilization process. *Current Science*, **82**, 740–744.

Acharyya S. K. and Shah B. A. (2004). Risk of arsenic contamination in groundwater affecting the Ganga Alluvial Plain, India [correspondence]. *Environmental Health Perspectives*, **112**(1), A19–A20.

Acharyya S. K., Chakraborty P., Lahiri S., Raymahashay B. C., Guha Saumyen and Bhowmik Amitava (1999). Arsenic poisoning in the Ganges delta. *Nature*, **401**, 545.

Acharyya S. K., Lahiri S., Raymahashay B. C. and Bhowmik A. (2000). Arsenic toxicity in groundwater in parts of Bengal Basin in India and Bangladesh: role of quaternary stratigraphy and holocene sea-level fluctuation. *Environmental Geology*, **39**, 1127–1137.

Achebe C. C. and Epstein P. R. (2004). Oil: prize or curse? The Nigerian Village Square, 4 July. http://www.energybulletin.net/node/1067 (accessed 7 July 2012).

Achin K. (2011). Arsenic Expert: Millions at risk. Voice of America, September 29. http://www.voanews.com/content/arsenic-expert-millions-of-indians-at-risk-130807418/168045.html (accessed 9 July 2012).

ActionAid (2007). Gold Rush: The Impact of Gold Mining on Poor People in Obuasi in Ghana. ActionAid, Johannesburg, South Africa. http://www.actionaid.org.uk/doc_lib/gold_rush.pdf.

Adaikpoh E. O., Nwajei G. E. and Ogala J. E. (2005). Heavy metals concentrations in coal and sediments from River Ekulu in Enugu, Coal City of Nigeria. *Journal of Applied Science and Environmental Management*, **9**(3), 5–8.

Ademiec E. and Helios-Rybicka E. (2002). Distribution of pollutants in the Odra River system. Part V. Assessment of total and mobile heavy metals content in the suspended matter and sediments of the Odra River system and recommendations for river chemical monitoring. *Polish Journal of Environmental Studies*, **11**(6), 675–688.

Adriano D. (2001). Trace Elements in Terrestrial Environments: Biogeochemistry, Bioavailability and Risks of Metals, 2nd edn. Springer, New York, NY, USA.

Aelion C. M. and Davis H. T. (2006). Use of a general toxicity test to predict heavy metal concentrations in residential soils. *Chemosphere*, **67**(5), 1043–1049.

Aguilar J., Dorronsoro C., Fernandez E., Fernandez A., Garcia I., Martin F. and Simon M. (2004). Soil pollution by a pyrite mine spill in Spain: evolution in time. *Environmental Pollution*, **132**(3), 395–401.

Ahmad T. (2004). Arsenic an emerging issue: experiences from Pakistan. In: People-Centred Approaches to Water and Environmental Sanitation. 30th Water, Engineering and Development Centre (WEDC), International Conference, Vientiane, Lao People's Democratic Republic. WEDC Conference No. 30.

Ahmed K. M., Huq S. M. and Naidu R. (2006). Extent and severity of arsenic poisoning in Bangladesh. In: Managing Arsenic in the Environment: From Soil to Human Health, R. Naidu, E. Smith, G. Owens, P. Bhattacharya and P. Nadebaum (eds), CSIRO Publishing, Collinwood, Victoria, Australia.

Ahmed M. (1999). Environmental Impact of Arsenic in Austria and Bangladesh: Its Distribution and Fractionation in Relation o Soil Properties. PhD thesis, University of Agriculture, Institute of Soil Research, Vienna, Austria, p. 190.

Akabzaa T. (2000). Vital statistics Ghana's gold mining industry. *Drillbits & Tailings*, **5**(19).

Akabzaa T. (2002). Boom and Dislocation: The Environmental and Social Impacts of Mining in the Wassa, West District of Ghana. Third World Network, Accra, Ghana.

Akoijam A.B. (1997). Distribution of fluoride in groundwater and pollution of shallow aquifers in parts of Imphal valley, Manipur, India. *Chem Environ. Res.*, **6**(3&4), 30.

Akter Kazi F., Owens G., Davey D. E. and Naidu R. (2005). Arsenic speciation and toxicity in biological systems. *Reviews of Environmental Contamination and Toxicology*, **184**, 97–149.

Alaerts G. J., Khouri N. and Kabir B. (2001). Strategies to Mitigate Arsenic Contamination of Water Supply (Chapter 8). Revision 20 April 2001. Prepared for the World Health Organization, Geneva, Switzerland. http://www.who.int/water_sanitation_health/dwq/arsenicun8.pdf (accessed 5 July 2012).

Allafrica. (2011). Namibia: Tsumeb smelter awaits final report on arsenic allegations. Allafrica.com, 23 November 2011. http://allafrica.com/stories/201111250940.html (accessed 5 July 2012).

Alam I. and Sadiq M. (1993). Investigation of the Size of Industrial Wastes in the Industrial City of Jubail, Final Report. King Abdulaziz City for Science and Technology (KACST) Project AR-09-059. KACST, Riyadh, Saudi Arabia.

Alastuey A., Jiménez A., Plana F., Querol X. and Suárez-Ruiz. (2001). Geochemistry, mineralogy and technological properties of the main Stephanian (Carboniferous) coal seams from the Puertollano Basin, Spain. *International Journal of Coal Geology*, **45**, 247–265.

Alfaro S. O. (2006). Sustainable Development and Local Responses in Rural El Salvador. Paper prepared for delivery at the Annual Conference of the Canadian Association of Latin America and Caribbean Studies: From Local to Global in Latin America and the Caribbean: Where Have We Come from and Where Are We Headed, 28–30 September 2006, Rosza Centre, University of Calgary, Calgary, Alberta, Canada.

Alfaro-Valero C. (2005). Geochemistry of hot spring at Cundinamarca Department, Colombia. In: Proceedings of the World Geothermal Congress. International Geothermal Association IGA, Antalya, Turkey, 24–29 April 2005. p. 1–8.

Allen G. T., Nash T. J. and Janes D. E. (1995). Contaminants evaluation of Marais des Cygnes National Wildlife Refuge in Kansas and Missouri, USA. *Environmental Management*, **19**(3), 393–404.

Alpers C. N., Blowes D. W., Nordstrom D. K. and Jambo J. L. (1994). Secondary minerals and acid mine-water chemistry. In: The Environmental Geochemistry of Sulfide Mine-Wastes. (Short Course Handbook 22), J. L. Jambor and D. W. Blowes (eds), Mineralogical Association of Canada, Waterloo, Ontario, Canada, pp. 247–438.

Amasa S. K. (1975). Arsenic pollution at Obuasi goldmine, town and surrounding countryside. *Environmental Health Perspectives*, **12**, 131–135.

Amat P. D. (2000). Influencia de la contaminación por metales pesados en la Bahia de Manzanillo. MSc thesis, Instituto Superior Minero Metalurgico de Moa (I.S.M.M), Manzanillo, Cuba.

Amat P. D., Pierra A. and Casals I. (2002). Estudio de la contaminación por metales pesados en sedimentos y ostiones de la Bahía de Manzanillo, Cuba. *Rev Soc Quím México*, **2002**(46), 357–361.

American Rivers. (2010). Upper Delaware River, Threat: Natural Gas Extraction. American Rivers, Pennsylvania, New York. See: http://www.americanrivers.org/assets/pdfs/mer-2010/upperdelaware_factsheet_2010.pdf (accessed 5 July 2012).

Amonoo-Neizer E. H. and Amekor E. M. K. (1993). Determination of total arsenic in environmental samples from Kumasi and Obuasi, Ghana. *Environmental Health Perspectives*, **101**(1), 46–49.

An D., Li D., Liang Y. and Jing Z. (2007). Unventilated indoor coal-fired stoves in Guizhou Province, China: reduction of arsenic exposure through behavior changes resulting from mitigation and health education in populations with arsenicosis. *Environmental Health Perspectives*, **115**(4), 659–662.

Anawar H. M. Akai J., Mostofa K. M., Safiullah S. and Tareq S. M. (2002). Arsenic poisoning in groundwater: health risk and geochemical sources in Bangladesh. *Environment International*, **27**(7), 597–604.

Andreae M. O. and Andreae T. W. (1989). Dissolved arsenic species in the Scheldt estuary and watershed, Belgium. *Estuarine, Coastal and Shelf Science*, **29**(5), 421–433.

Andreae M. O., Byrd J. T. and Froelich P. N. (1983). Arsenic, antimony, germanium, and tin in the Tejo estuary, Portugal: modelling a polluted estuary. *Environmental Science and Technology*, **17**, 731–737.

Andrews G. (2007). Arsenic Contamination in Wells Raises Concerns [Media Release]. News and Communications Services, Oregon State University, Corvallis, OR, USA. 16 May 2007. See: http://oregonstate.edu/dept/ncs/newsarch/2007/Apr07/arsenic.html

Animal Health Institute (1996). Feed Additive Compendium. The Miller Publishing Company, Minnetonka, Minnesota, USA.

Anon. (1986). Raw materials report. The mining industry of Zambia. *Zambia*, **4**(4), 14–27.

Anon. (1999). Living with deadly water: Tasik Biru's squatters. http://www.rengah.c2o.org/assets/xml/de0057t.xml (accessed 7 July 2012).

Anon. (2001). Turkmenistan. *OPEC Fund Newsletter*, September–December, 10(3), 11.

Anon. (2002). BOBS sets drinking water quality standard. Daily News Online. http://www.gov.bw/cgi-bin/news.cgi?d=20020529.

Anon. (2004a). Arsenic fears lead to ban on water in northern Serbian town. AP Worldstream, 20 January. See: http://www.highbeam.com/doc/1P1-89581644.html (accessed 5 July 2012).

Anon. (2004b). Arsenic contamination in groundwater and drinking water quality surveillance in Lao PDR. In: Interregional Conference on Water Quality, 23–26 November 2004, Taiyuan, China.

Anon. (2007). Canadian Greed and Fear Breed in Alberta's Oil Sands. http://1loneranger.wordpress.com/2007/03/06/canadian-greed-and-fear-breed-in-albertas-oil-sands/ (accessed 5 July 2012).

Anon. (2009). Q+A: Environmental fears over US shale gas drilling. Reuters, 23 December 2009. http://www.reuters.com/article/2009/12/23/us-naturalgas-shale-qa-idUSTRE5BM1XN20091223 (accessed 31 March 2012).

Anon. (2011). Namibia: Tsumeb Smelter Awaits Final Report On Arsenic Allegations. 23 November 2011. http://allafrica.com/stories/201111250940.html (accessed 5 July, 2012).

ANZECC/ARMCANZ. (2000). Australian and New Zealand guidelines for fresh and marine water quality. Australia and New Zealand Environment and Conservation Council and Agriculture and Resource Management Council of Australia and New Zealand, Canberra, Australia. http://www.environment.gov.au/water/publications/quality/nwqms-guidelines-4-vol1.html (accessed 5 July 2012).

Aowal A. F. S. A. (1981). Design of an iron eliminator for hand tube wells. *Journal of the Indian Water Works Association*, **13**(1), 65.

Appelo C. A. J. and Postma D. (1993). Geochemistry, Groundwater and Pollution. A. A. Balkema, Rotterdam, Netherlands.

Appelo C. A. J., Van Der Weiden M. J. J., Tournassat C. and Charlet L. (2002). Surface complexation of ferrous iron and carbonate on ferrihydrite and the mobilization of arsenic. *Environmental Science and Technology*, **36**, 3096–3103.

Aquastat. (1995). AQUASTAT Database. Food and Agriculture Organization of the United Nations, Rome, Italy. http://www.fao.org/nr/water/aquastat/main/index.stm (accessed 5 July 2012) (Database covers 1960–2012).

Arbuzov S. I. (1998). Rare earth elements and scandium in coals of the Kuznetsk Basin. *Fuel and Energy Abstracts*, **39**(3), 168.

Arbuzov S. I., Rikhvanov A. Y. and Ershov V. V. (2001). Trace element-potential of the Siberian coals. Geology, prospection and exploration on the Siberian mineral resources. *Proceedings of the Tomsk Polytechnic University*, **301**(1), 130–147 (in Russian).

Arbuzov S. I., Ershov V. V., Potseluev A. A. and Rikhvanov L. P. (2000). *Rare Elements in Coals of the Kuznetsk Basin*. Russian Federation, Kemerovo (in Russian).

Arehart G. B., Chryssoulis S. L. and Kesler S. E. (1993). Gold and arsenic in iron sulfides from sediment hosted disseminated gold deposits: implications for depositional processes. *Economic Geology*, **88**(1), 171–185.

Argonne National Laboratory (1981). Portugal and United States Cooperative Energy Assessment. Volume 3. Reference Reports, Part 1. United States Department of Energy, Argonne National Laboratory, Argonne, IL, USA. http://www.osti.gov/bridge/product.biblio.jsp?osti_id=6693344 (accessed 5 July 2012).

Armienta M. A., Rodriguez C. R., Ongley L. K., Brust H., Morales F., Aguayo A., Cruz O. and Ceniceros N. (2007). Origin and fate of arsenic in a historic mining area of Mexico. In: Arsenic in Soil and Groundwater Environment, P. Brattacharya, A. B. Mukherjee, J. Bundschuh, R. Zevenhoven and R. H. Loeppert (eds), Elsevier, Oxford, UK.

Arizona Department of Health Services. (1999). Public health assessment: ASARCO Hayden smelter site (aka ASARCO Incorporated Hayden plant, Hayden, Gila County, Arizona). http://www.atsdr.cdc.gov/hac/pha/pha.asp?docid=905&pg=0 (acessed 5 July 2012).

Arkansas Pollution Control and Ecology Commission. (2002). Petition to Initiate Rulemaking: In the matter of: Regulation No. 30 (Hazardous Substance Remedial Action Trust Fund Priority List) Addition; Ouachita-Nevada Wood Treater, General Dynamics Corporation; and On-Call Contract Assessment Sites (Docket No. 05-002-R) Submitted by Mike Bates, Chief, Hazardous Waste Division, Arkansas Department of Environmental Quality. http://www.adeq.state.ar.us/regs/drafts/reg30_draft_docket_05-002-R/reg30_draft_docket_05-002-R_petition.pdf (accessed 5 July 2012).

Arkansas Water Resources Center. (2006). Annual technical report FY 2005. US Geological Survey, Reston, VA, USA. http://water.usgs.gov/wrri/AnnualReports/2005/FY2005_AR_Annual_Report.pdf (accessed 5 July 2012).

Armienta A., Rodriguez R., Morton O., Cruz O., Ceniceros N., Aguayo A. and Brust H. (2000). Health risk and sources of arsenic in the potable water of a mining area. In: Interdisciplinary Perspectives on Drinking Water Risk Assessment and Management, E. G. Reichard, F. S. Hauchman and F. Ana Maria Sancha (eds), Proceedings of the Santiago, Chile, Symposium, September 1998. International Association of Hydrological Sciences Publication No. 260. IAHS, Velp, Netherlands, pp. 9–16. http://cat.inist.fr/?aModele=afficheN&cpsidt=1176888 (accessed 9 July 2012).

Armienta M. A., Rodriguez C. R., Ongley L. K., Brust H., Morales F., Aguayo A., Cruz O. and Ceniceros N. (2007). Origin fate of arsenic in a historic mining area of Mexico. In: Arsenic in Soil and Groundwater Environment: Biogeochemical Interactions, Health Effects and Remediation, P. Battacharya, A. B. Mukherjee, J. Bundschuh, R. Zevenhoven and R. H. Loeppert (eds), Elsevier, Oxford, UK.

Arnórsson S. (2003). Arsenic in surface- and up to 90°C ground waters in a basalt area, N-Iceland: processes controlling its mobility. *Applied Geochemistry*, **18**(9), 1297–1312.

Ashton P., Love D., Mahachi H. and Dirks P. (2001). An overview of the Impact of Mining and Mineral Processing Operations on Water Quality in the Zambezi, Limpopo and Olifants Catchments in Southern Africa. Contract report to the Mining, Minerals and Sustainable Development (Southern Africa) Project CSIR Environmentek, Pretoria, South Africa, and Geology Department, University of Zimbabwe, Harare, Zimbabwe. (Report No. ENV-P-C 2001-042). http://pubs.iied.org/pdfs/G00599.pdf (accessed 5 July 2012).

Associated Press. (2006). New Arsenic Standards Costly in Bucks County. Associated Press, 25 November 2006. http://abclocal.go.com/wpvi/story?section=local&id=4795240&ft=lg) (accessed 5 July 2012).

ATSDR (Agency for Toxic Substances and Disease Registry). (1988). Public Health Evaluation of Crystal Chemical Company on Westpark Drive in Houston, Texas. ATSDR, Atlanta, GA, USA. http://www.atsdr.cdc.gov/hac/pha/pha.asp?docid=77&pg=0 (accessed 5 July 2012).

ATSDR (Agency for Toxic Substances and Disease Registry). (1998). Armour Road Site Public Health Assessment. ATSDR, Atlanta, GA, USA. http://www.atsdr.cdc.gov/hac/pha/pha.asp?docid=852&pg=0 (accessed 5 July 2012).

ATSDR (Agency for Toxic Substances and Disease Registry). (2000). Public Health Evaluation of Arsenic Contaminated Private Drinking Water Wells in Northeastern Connecticut: Woodstock, East Hampton and Colchester Counties. ATSDR, Atlanta, GA, USA. http://www.atsdr.cdc.gov/HAC/PHA/HCPHA.asp?State=%27CT%27 (accessed 5 July 2012).

Australian Government. (2007). Managing Acid and Metalliferous Drainage. Leading Practice Sustainable Development Program for the Mining Industry. Australian Government, Department of Industry, Tourism and Resources, Canberra. http://www.dmp.wa.gov.au/documents/file_MAMD20070227104556.pdf (accessed 5 July 2012).

Australian Water Resources Council. (1982). Groundwater Resources of Australia. Australian Government Publishing Service, Canberra.

Avert. (2009). HIV & AIDS in Botswana. http://www.avert.org/aids-botswana.htm (accessed 5 July 2012).

Ayerza A. (1918). Arsenicismo regional endémico (keratodérmia y melanodérmia combinados). *Boletim da Academia Nacional de Medicina*, **1**, 11–41.

Ayotte J. D., Nielsen M. G., Robinson G. R. and Moore R. B. (1999). Relation of arsenic, iron, and manganese in ground water to aquifer type, bedrock lithogeochemistry, and land use in the New England coastal basins. United States Geological Survey (USGS) Water-Resources Investigations Report 99-4162. USGS, United States Department of the Interior, Pembroke, NH, USA. http://pubs.usgs.gov/wri/wri994162/ (accessed 5 July 2010).

Ayotte J. D., Montgomery D. L., Flanagan S. M. and Robinson K. W. (2003). Arsenic in ground water in eastern New England: occurrence, controls, and human health implications. *Environmental Science and Technology*, **37**(10), 2075–2083. http://www.ncbi.nlm.nih.gov/pubmed/12785510 (accessed 9 July 2012).

Ayuso R. A. and Robinson M. S. (2001). Tracing Pb isotopic compositions of common arsenical pesticides in a coastal Maine watershed containing arsenic-enriched groundwater. In: Annual International Conference on Soils, Sediments and Water. University of Massachusetts, Amherst, MA, USA.

Azcue J. M. and Nraigu J. O. (1994). Arsenic: historical perspectives. In: Arsenic in the Environment. Part I: Cycling and Characterization, J. O. Nraigu (ed.), John Wiley & Sons, New York NY, USA, pp. 1–16.

Azzie B. (2007). Coal Mine Waters in South Africa: Their Geochemistry, Quality and Classification. Unpublished PhD thesis. University of Capetown, Capetown, South Africa.

Baacke D. (2000). Geochemisches Verhalten umweltrelevanter Elemente in stillgelegten Polysulfiderzgruben am Beispiel der Grube "Himmelfahrt" in Freiberg/Sachsen. Dissertation, TU Bergakademie Freiberg - Fakultät für Geowissenschaften, Geotechnik und Bergbau.

Baboshkina S. V., Puzanov A. V. and Mal'gin M. A. (2003). Arsenic in Altai chestnut soils. *Geography and Natural Resources*, **2**, 73–78. (in Russian).

Badalov S. T. (1975–1977). Mineraly Uzbekistana. 4 vols. Fan, Tashkent, Institut geologii i geofiziki im. M. Kh Abdullaeva, Uzbekistan.

Badloe C. and Hoa N. Q. (2005). Update of UNICEF-Supported Arsenic Activities in Vietnam. United Nations Children's Fund, Hanoi, Viet Nam.

Bailey K. V. (1991). Zambia: review of national IDD control programme. Summary of report to World Health Organization/International Council for Control of Iodine Deficiency Disorders. *IDD Newsletter*, **7**(3). http://iccidd.server295.com/media/IDD%20Newsletter/1991-2006/idd891.htm#Zambia (accessed 5 July 2012).

Baioumy H. M. (2005). Preliminary data on cadmium and arsenic geochemistry for some phosphorites in Egypt. *Journal of African Earth Sciences*, **41**(3), 266–274.

Bajtoš P. (2005). Mine water issues in Slovakia. In: Contemporary Reviews of Mine Water Studies in Europe, C. Wolkersdorfer and R. Bowell (eds), Springer Verlag, Berlin, Germany (Mine Water and the Environment, 24: Supplementary Material; http://www.imwa.info/docs/10.1007_s10230-005-0081-3.pdf (accessed 5 July 2012).

Bakken B. M., Hochella M. F., Marshall A. F. and Turner A. M. (1989). High-resolution microscopy of gold in unoxidized ore from the Carlin Mine, Nevada. *Economic Geology*, **84**(1), 171–179.

Bakuradze T. (1995). Environmental Information Systems in Georgia. Report prepared by United Nations Environment Programme/GRID-Arendal. http://enrin.grida.no/htmls/georgia/age_p1.htm (accessed 5 July 2012).

Ball J. W. (1998). Chemical Analyses of Hot Springs, Pools, Geysers, and Surface Waters from Yellowstone National Park, Wyoming, and Vicinity, 1974–1975. United States Geological Survey, United States Department of the Interior, Boulder, CO, USA, (USGS Open-File Report 98-182. http://www.brr.cr.usgs.gov/projects/GWC_chemtherm/pubs/ofr98-182.pdf (accessed 5 July 2012).

Banks D., Younger P. L., Arnesen R.-T., Iversen E. R. and Banks S. B. (1997). Mine-water chemistry: the good, the bad and the ugly. *Environmental Geology*, **32**, 157–174.

Baraboshkina T. A. and Ogorodnikova E. N. (1999). Ecologic-geochemical topics of an ash disposal impact on the environment and some ways of ash utility. *Energetik*, **10**, 18–20. (in Russian) [cited by Yudovich & Ketris, 2005].

Barandovski L., Cekova M. V., Frontasyeva S. S., Pavlov T., Stafilov E., Steinnes V. and Urumov M. (2006). Air pollution studies in Macedonia using the moss biomonitoring technique, NAA, AAS and GIS technology. Intended for submission to Environmental Monitoring and Assessment. http://www1.jinr.ru/Preprints/2006/160(E18-2006-160).pdf).

Barbiero L., Furquim S. A. C., Vallès V., Furian S., Sakamoto A., Rezende Filho A. T. and Fort M. (2007). Natural arsenic in groundwater and alkaline lakes at the upper Paraguay basin, Pantanal, Brazil. In: Arsenic in Soil and Groundwater Environment: Biogeochemical Interactions, Health Effects and Remediation, P. Battacharya, A. B. Mukherjee, J. Bundschuh, R. Zevenhoven and R. H. Loeppert (eds), Elsevier, Oxford, UK, pp. 101–126. Abstract with access to article: http://goo.gl/3CP77 (accessed 9 July 2012).

Barnes V. C. (1981). Coal mining activities in South Africa. *Mining Congress Journal*, **67**(4), 53–57.

Bashkin V. N. and Wongyai K. (2002). Environmental fluxes of arsenic from lignite mining and power generation in northern Thailand. *Environmental Geology*, **41**(8), 883.

Basu N. and Hu H. (2010). Toxic Metals and Indigenous Peoples Near the Marlin Mine in Western Guatemala: Potential Exposures and Impacts on Health. Physicians for Human Rights, Cambridge, MA.

Bates M. N., Smith A. H. and Cantor K. P. (1995). Case-control study of bladder cancer and arsenic in drinking water. *American Journal of Epidemiology*, **141**, 523–530.

Battocletti L. (2001). Geothermal Resources in the Balkans. Report No. INEEL/EXT-2001-677. Prepared for Idaho National Engineering & Environmental Laboratory (INEEL) and the United States Department of Energy. http://www.bl-a.com/ECB/PDFFiles/GeoRes Balkans.pdf (accessed 5 July 2012).

Bayet A. and Slosse A. (1919). L'intoxication arsénicale dans les industries de la houille et ses dérivés (intoxication houillère arsénicale). *Comptes Rendus de l'Académie des Sciences*, **168**, 704–706.

Beccaluva L., Coltorti M., Ferrini V., Saccani E., Siena F. and Zeda O. (1998). Petrological modelling of Albanian ophiolites with particular regard to the Bulqiza chromite ore deposits. *Periodico di Mineralogia*, **67**, 7–23.

Beck J. N. *et al.* (1990). Heavy metal content of sediments in the Calcasieu River/Lake Complex, Louisiana. *Hydrobiologia*, **192**(2–3), 149–165.

Bedient P. B., Rifai H. S. and Newell C. J. (1999). Groundwater Contamination: Transport and Remediation. Second edition. Prentice–Hall, Englewood Cliffs, NJ, USA.

Beining B. and Otte M. L. (1996). Retention of metals from an abandoned lead-zinc mine by a wetland at Glendalough, Co. Wicklow, Ireland. *Biology of the Environment*, **96**, 117–126.

Bejarano Sifuentes G. and Nordberg E. (2003). Mobilisation of Arsenic in the Rio Dulce Alluvial Cone, Santiago del Estero Province, Argentina. Master's thesis. Kungliga Tekniska Högskolan, Department of Land and Water Resources Engineering, Stockholm, Sweden, p. 40. TRITA-LWR-EX-03-06; http://www.lwr.kth.se/publikationer/PDF_Files/LWR_EX_03_6.PDF (accessed 5 July 2012).

Bencko V. and Symon K. (1977). Health aspects of burning coal with a high arsenic content. *Environmental Research*, **13**, 378–385.

Bencko V., Symon K., Chládek V. and Pihrt J. (1977). Health aspects of burning coal with a high arsenic content. II. Hearing changes in exposed children. *Environmental Research*, **13**, 386–395.

Benin A. L., Sargent J. D., Dalton M. and Roda S. (1999). High concentrations of heavy metals in neighborhoods near ore smelters in northern Mexico. *Environmental Health Perspectives*, **107**(4), 279–284.

Benson N. R. (1976). Retardation of apple tree growth by soil arsenic residues from old insecticidal treatments. *Journal of the American Society of Horticultural Science*, **101**(3), 251–253.

Bentellis-Mosbah A. and Rached O. (2006). Trace elements soil contamination and plants uptake of wadi banks situated in the vicinity of an old Sb-mining area in Algeria. *Chinese Journal of Geochemistry*, **25**(Suppl.), 32.

Benvenuti B., Mascaro I., Corsini F., Lattanzi P. and Parrini P. (1997). Mine waste dumps and heavy metal pollution in abandoned mining district of Boccheggiano (Southern Tuscany, Italy). *Environmental Geology*, **30**, 238–243.

Berg M., Tran H. C., Nguyen T. C., Pham H. V., Schertenleib R. and Giger W. (2001). Arsenic contamination of ground water and drinking water in Vietnam: a human health threat. *Environmental Science and Technology*, **35**(13), 2621–2626.

Berg M., Stengel C., Pham T. K., Pham H. V., Sampson M. L., Leng M., Samreth S. and Fredericks D. (2006). Magnitude of arsenic pollution in the Mekong and Red River Deltas, Cambodia, and Vietnam. *Science of the Total Environment*, **372**, 413–442.

Berger B. R. (1995). Joint USGS Project with Hungary and Slovakia on Mineral Resources and Environmental Risks. States Geological Survey, United States Department of the Interior, United Denver, CO, USA. http://mine-drainage.usgs.gov/archive/berger1.html (accessed 5 July 2012).

Berger B. R. and Bethke P. M. (1984). Geology and geochemistry of epithermal systems. *Reviews in Economic Geology*, **2**, 169–202.

Berger V. and Theodore T. G. (2005). Implications of Stratabound Carlin-Type Gold Deposits in Paleozoic Rocks of North-Central Nevada: Slide Presentation. United States Department of the Interior, United States Geological Survey, Reno, NV, USA (Open-File Report 2005-1156, Version 1.0; http://pubs.usgs.gov/of/2005/1156/ (accessed 5 July 2012).

Berglund M., Elinder C. G. and Järup L. (2001). Human Exposure Assessment: An Introduction. World Health Organization, Geneva, Switzerland (WHO/SDE/OEH/01.3).

BGS (British Geological Survey) & Water Aid. (2001). Groundwater Quality: Pakistan. BGS, Natural Environment Research Council, Keyworth, Nottingham, UK.

BGS (British Geological Survey). (2011). Groundwater Studies for Arsenic Contamination in Ghana. BGS, Keyworth, Nottingham, UK. http://www.bgs.ac.uk/research/groundwater/health/arsenic/Ghana.html (accessed 10 July 2012).

BGS (British Geological Survey) & DPHE (Department of Public Health Engineering). (2001). Arsenic contamination of groundwater in Bangladesh. In: British Geological Survey Technical Report WC/00/19, D. G. Kinniburgh, P. L. Smedley (eds), British Geological Survey, Keyworth, Nottingham, UK.

Bhattacharya P., Chatterjee D. and Jacks G. (1997). Occurrence of As-contaminated groundwater in alluvial aquifers from the delta plains, eastern India: options for safe drinking water supply. *Water Resource Development*, **13**, 76–92.

Bhattacharya P., Claesson M., Fagerberg J., Bundschuh J., Storniolo Adel R., Martin R. A., Thir J. M. and Sracek O. (2005). Natural arsenic in the groundwater of the alluvial aquifers of Santiago del Estero Province, Argentina. In: Natural Arsenic in Groundwater: Occurrences,

Remediation and Management, J. Bundschuh, P. Bhattacharya and D. Chandrashekharam (eds), Taylor & Francis Group, London, UK. pp. 57–66.

Bhattacharya P., Claesson M., Bundschuh Jochen, Sracek O., Fagerberg J., Jacks G., Martin R. A., Storniolo Adel R. and Thir J. M. (2006a). Distribution and mobility of arsenic in the Río Dulce alluvial aquifers in Santiago del Estero Province, Argentina. *Science of the Total Environment*, **358**, 97–120.

Bhattacharya P., Claesson M. and Fagerberg J. (2006b). Arsenic in Groundwater of Santiago del Estero Province, Northwestern Argentina. http://goo.gl/oeK0V (accessed 9 July 2012).

Battacharya P., Mukherjee A. B., Bundschuh J., Zevenhoven R. and Loeppert R. H. (eds). (2007). Arsenic in Soil and Groundwater Environment: Biogeochemical Interactions, Health Effects and Remediation. Elsevier, Oxford, UK.

Biacs P. (2005). Clear water-food safety, drinking water for Hungary [extrapolated from map]. East-West Agrarian Forum, 21 January 2005, Berlin, Germany. Budapest, Hungary, Hungarian Food Safety Office. http://lhu.emu.ee/downloads/Welfood/WP3T2L2.pdf (accessed 5 July 2012).

Bickford C. G. C. and Kenyon C. (1987). Coalfield geology of eastern Vancouver Island, Canada. Paper 1988-1. British Columbia Ministry of Energy, Mines and Petroleum Resources, Geological Fieldwork Victoria, British Columbia. http://www.empr.gov.bc.ca/Mining/Geoscience/PublicationsCatalogue/Fieldwork/Documents/1987/441-450-bickford.pdf (accessed 9 July 2012).

Biderman J. (2003). Groundwater and drought management in the Southern African Development Community (SADC) in Africa. http://www.devex.com/en/projects/63777/secure?mem=ua&src=tender (accessed 5 July 2012)

Bigot P. B. (2004). A contribution to the study of areas with natural arsenic contamination in Nicaragua. In: Book of Abstracts for the Third Regional Conference on the Evaluation and Management of Drinking Water Sources Contaminated with Arsenic, Universidad de Chile, Division de Recursos Hidricos y Medio Ambiente (eds), Universidad de Chile, Santiago, Chile, 8-11 November 2004. Abstract: http://phys4.harvard.edu/~wilson/arsenic/conferences/arsenic_project_conferences.html (accessed 9 July 2012).

Birkle P. and Merkel B. (2002). Mineralogical-chemical composition and environmental risk potential of pond sediments at the geothermal field of Los Azufres, Mexico. *Environmental Geology*, **42**(1), 115.

Biswas B., Dhar R. K., Samanta G., Mandal B. K., Chakraborti D., Faruk I., Islam K. S., Chowdhury M. M., Chowdhury M. and Islam A. (1998). Detailed study report of Samta, one of the arsenic affected villages of Jessore District, Bangladesh. *Current Science*, **74**(2), 134–145.

Bityukova L., Shogenova A. and Birke M. (2000). Urban geochemistry: a study of element distributions in the soils of Tallinn (Estonia). *Environmental Geochemistry and Health*, **22**(2), 173–193.

Blachère A., Morel S. and Lefort D. (2005). France. In: Contemporary Reviews of Mine Water Studies in Europe, C. Wolkersdorfer and R. Bsowell (eds), Springer Verlag, Berlin, Germany (Mine Water and the Environment, 24: Supplementary Material; http://www.imwa.info/docs/10.1007_s10230-005-0081-3.pdf).

Blacksmith Institute. (2011). Mailuu-Suu Legacy Uranium Dumps. Blacksmith Institute, New York, N.Y. http://www.blacksmithinstitute.org/projects/display/129 (accessed 5 July, 2012).

Bland D. and Scholle P. A. (2007). Uranium: is the next boom beginning? New Mexico Earth Matters, Winter. http://geoinfo.nmt.edu/publications/periodicals/earthmatters/7/EMv7n1.pdf (accessed 6 July 2012).

Blumberga M. (2001). Risk Assessment of the Skede Landfill in Liepaja, Latvia. Stockholm University, Stockholm, Sweden. http://www.lwr.kth.se/Publikationer/PDF_Files/AMOV_EX_2001_4.pdf (accessed 6 July 2012).

Bonadiman C., Roncni F. and Sacerdoti M. (2002). Determinazone del rapporto Fe 3+/Fe 2+ in alcune cromiti ad alto contenuto in cromo delle ofioliti dell' Albania. *Plinius*, **28**, 67–69.

Boockhoff L. A., Baron D., Horton R., Negrini R. and Parker J. (2005). Arsenic sediments and groundwater from a well in the southern San Joaquin valley, California [poster]. In: Arsenic Occurrence and Fate in Hydrogeologic Systems. Annual Meeting of the Geological Society of America, Salt Lake City, Utah, USA (Session 164.T25). Abstract: https://gsa.confex.com/gsa/2005AM/final program/abstract_96001.htm (accessed 9 July 2012).

Borba R. P., Figueiredo B. R., Rawlins B. and Matschullat Jorge. (2000). Arsenic in water and sediment in the Iron Quadrangle, Minas Gerais state, Brasil. *Revista Brasileira de Geociências*, **20**, 554–557.

Borgoño J. M., Vicent P., Venturino H. and Infante A. (1977). Arsenic in the drinking water of the city of Antofagasta: epidemiological and clinical study before and after the installation of a treatment plant. *Environmental Health Perspectives*, **19**, 103–105.

Borregaard N., Lewin P., Gonzalez P. and Newbold J. (2000). Confronting sustainability in the mining sector: What role for a sustainability fund? Draft working paper prepared by Centre for Environmental Investigation and Planning (CIPMA), United Nations Environment Programme, Chile. http://www.unep.ch/etu/etp/events/upcming/chile_rep.pdf (accessed 6 July 2012).

Börzsönyi M., Bereczky A., Rudnai P., Csanady M. and Horvath A. (1992). Epidemiological studies on human subjects exposed to arsenic in drinking water in southeast Hungary. *Archives of Toxicology*, **66**, 77–78.

Bosnia & Herzegovina. (2007). Heart-shaped land. http://www.bhtourism.ba/eng/search.wbsp (accessed 6 July 2012).

Bostick B., Chiew H. and Sampson M. L. (2009). Response to comment on "effect of groundwater iron and phosphate on efficacy of arsenic removal by iron-amended biosand filters." *Env. Sci. & Tech.*, **43**(22), 8696–8697.

Botswana Bureau of Standards. (2004). Botswana Standards. BOS 93:2004 Water quality, waste water, physical, microbiological and chemical requirements specifications (Annex C) ICS: 13.060.20. http://www.bobstandards.bw/web_en/cat/items%5Cd468.html (accessed 6 July 2012).

Botswana, National Council on Population and Development. (2010). *Revised National Population Policy: Improving the Quality of Life.* Botswana National Council on Population and Development, Ministry of Finance and Development Planning, Gabarone.

Bottomley D. J. (1984). Origins of some arseniferous groundwaters in Nova Scotia and New Brunswick, Canada. *Journal of Hydrology*, **69**, 223–257.

Bounakhla M. (2011). PGAA metals analysis in tailings in Zaida abandoned mine, high Moulouya, Morocco. *Journal of Radioanalytical and Nuclear Chemistry*, **291**(1), 129–135.

Bowell R. and Parshley J. (2004). Arsenic cycling in the mining environment. In: *SRK Consulting, US EPA Workshop on Managing Arsenic Risks to the Environment: Characterization of Waste, Chemistry, and Treatment and Disposal Proceedings and Summary Report.* May 1–3, 2001. SRK, Denver Colorado. pp. 10–11. http://goo.gl/p029I (accessed 6 July 2012).

Boyle D. R., Turner R. J. W. and Hall G. E. M. (1998). Anomalous arsenic concentrations in groundwaters of an island community, Bowen Island, British Columbia. *Environmental Geochemistry and Health*, **20**, 199–212.

Boyle R. W. (1979). The geochemistry of gold and its deposits. *Bulletin of the Geological Survey Canada*, **280**, 1–584.

Bragg L. J., Oman J. K., Tewalt S. J., Oman C. J., Rega N. H., Washington P. M. and Finkelman R. B. (1997). United States Geological Survey (USGS) Coal Quality (COALQUAL) Database: Version 2.0. USGS Open-file Report 97-134, CDROM. USGS National Center, Reston, Virginia, USA. http://energy.er.usgs.gov/products/databases/CoalQual/index.htm (accessed 9 July 2012).

Brammer H. (2009). Mitigation of arsenic contamination in irrigated paddy soils in south and southeast Asia. *Environment International*, **35**, 856–863. Abstract: http://www.ncbi.nlm.nih.gov/pubmed/19394085 (accessed 6 July 2012).

Brazil. (2010). Ministerio da Saúde, Decreto Nº 7.217, de 21 de Junho de 2010. Regulamenta a Lei no 11.445, de 5 de janeiro de 2007, que estabelece diretrizes nacionais para o saneamento básico, e dá outras providências. http://www.planalto.gov.br/ccivil_03/_ato2007-2010/2010/Decreto/D7217.htm (accessed 7 July 2012).

Bright D. A., Dodd M. and Reimer K. J. (1996). Arsenic in subArctic lakes influenced by gold mine effluent: the occurrence of organoarsenicals and "hidden" arsenic. *Science of the Total Environment*, **180**(2), 165–182.

Brink A. B. A., Van Schalkwyk A., Partridge T. C., Midgley D. C., Ball J. M. and Geldenhuis S. J. J. (1990). The changing impact of urbanization and mining on the geological environment. *South African Journal of Science*, **86**(7–10), 434–440.

Britt I. and Jones J. (2009). Design and Appraisal of Hydraulic Fractures. Society of Petroleum Engineers, Allen, Texas.

British Geological Survey (BGS). (2001). Groundwater Quality: Zambia. BGS, Natural Environment Research Council. http://www.wateraid.org/documents/plugin_documents/zambiagroundwater.pdf (accessed 6 July 2012).

British Geological Survey (BGS). (2002a). Groundwater Quality: Burkina Faso. BGS, Natural Environment Research Council. http://www.wateraid.org/documents/plugin_documents/gw_burkinafaso.pdf (accessed 6 July 2012).

British Geological Survey (BGS). (2002b). Groundwater Quality: Mali. BGS, Natural Environment Research Council. http://www.wateraid.org/documents/plugin_documents/maligroundwater.pdf (accessed 6 July 2012).

British Geological Survey (BGS). (2011c). Groundwater studies for arsenic contamination in Ghana. http://www.bgs.ac.uk/arsenic/ghana/home. html (accessed 18 December 2011).

Britton L. J., Anderson C. L., Goolsby D. A. and Van Haveren B. P. (eds). (1989). Summary of the USGS and US Bureau of Land Management National Coal Hydrology Program, 1974–84. United States Geological Survey (USGS) Professional Paper 1464. United States Department of the Interior, USGS National Center, Reston, Virginia, USA.

Britwum A., Jonah K. and Tay F. D. (2001). Structural Adjustment Participatory Review Initiative (SAPRI): Ghana country report. Tripartite National Steering Committee, SAPRI Network (SAPRIN), Washington, DC, USA. http://www.saprin.org/ghana/research/gha_country_rpt.pdf (accessed 6 July 2012).

Brock T. D. and Brock M. L. (1967). The hot springs of the Furnas Valley, Azores. *Internationale Revue der gesamten Hydrobiologie und Hydrographie*, **52**(4), 545–558.

Brooks W. E. (2011). USGS Minerals Information: Arsenic Statistics and Information. http://minerals.usgs.gov/minerals/pubs/commodity/arsenic/mcs-2011-arsen.pdf (accessed 6 July 2012).

Brown B. L., Slaughter A. D. and Schreiber M. E. (2005). Controls on roxarsone transport in agricultural watersheds. *Applied Geochemistry*, **20**(1), 123–133.

Brown V. J. (2004). Reconstructing the environment in Iraq. *Environmental Health Perspectives*, **112**(8), A464.

Brumbaugh W. G., Ingersoll C. G., Kemble N. E., May T. W. and Zajicek J. L. (1994). Chemical characterization of sediments and pore water from the upper Clark Fork River and Milltown Reservoir, Montana. *Environmental Toxicology and Chemistry*, **13**(12), 1971–1983.

Bruneel O., Personné J. C., Casiot C., Leblanc M., Elbaz-Poulichet F., Mahler B. J., Le Flèche A. and Grimont P. A. (2003). Mediation of arsenic oxidation by Thiomonas sp. in acid-mine drainage (Carnoules, France). *Journal of Applied Microbiology*, **95**(3), 492–499.

Brysk A. (2000). From Tribal Village to Global Village: Indian Rights and International Relations in Latin America. Stanford University Press, Palo Alto, CA, USA.

Buchet J. P. and Lison D. (1998). Mortality by cancer in groups of the Belgian population with a moderately increased intake of arsenic. *International Archives of Occupational and Environmental Health*, **71**(2), 125–130.

Bukharbayeva B. (2005). Kazakhstan set to open pipeline to China. AP Online, 14 December. http://www.chinadaily.com.cn/english/doc/2005-12/15/content_503505.htm (accessed 9 July 2012).

Bundschuh J. (2007). Geogenic arsenic in sedimentary aquifers of southwestern Sébaco Valley, Nicaragua. (Paper No. 192-8). Annual Meeting of the Geological Society of America, 28-31 October 2007, Denver, CO, USA. http://gsa.confex.com/gsa/2007AM/finalprogram/abstract_125801.htm (accessed 6 July 2012).

Bundschuh J., Farias B., Martin R., Storniolo A., Bhattacharya P., Cortes J., Bonorino G. and Albouy R. (2004a). Groundwater arsenic in the Chaco-Pampean Plain, Argentina: case study from Robles County, Santiago del Estero Province. *Applied Geochemistry*, **19**(2), 231–243.

Bundschuh J., Bhattacharya P., Claesson M., Sracek O., Fagerberg J., Storniolo A., Martin R. and Thir J. (2004b). Geogenic arsenic in shallow aquifers of Río Dulce Alluvial Cone: Santiago del Estero Province NW-Argentina. In: Proceedings Evaluation and Management of Drinking Water Resources Contaminated with Arsenic. Int. Congress, Nov. 2004, Universidad de Chile, Santiago de Chile, Chile, 8–11 November 2004.

Bundschuh J., Litter M., Ciminell V. S. T, Morgad M. E., Cornej L., Hoyo S. G., Hoiki J., Alarcon-Herrer M. T., Armient M. A. and Bhattacharya P. (2010). Emerging mitigation needs and sustainable options for solving the arsenic problems of rural and isolated urban areas in Latin America – a critical analysis. *Water Research*, **44**, 5828–5845.

Bundschuh J., Litter M., Parvez F., Roman-Ross G., Nicolli H. B., Jean J. S., Liu C. W., Lopez D., Armienta M. A., Builherme L. R. G., Cuevas A. G., Cornejo L., Cumbal L. and Toujaguez R. (2011a). One century of arsenic exposure in Latin America. A review of history and occurrence from 14 countries. *Science of the Total Environment.*

Bundschuh J., Nath B., Bhattacharya P., Liu C. W., Armienta M. A., Lopez M. V. M., Lopez D. L., Jean J. S., Cornejo L., Macedo L. F. L. and Filho A. T. (2011b). Arsenic in the human food chain: the Latin American perspective. *Science of the Total Environment.*

Bureau of Geological Consultancy. (1993). Golden Opportunities in Liptako. Bureau of Geological Consultancy, Brussels, Belgium.

Buse A., Norris D., Marmens H., Bueker P., Ashenden T. and Mills G. (2003). Heavy metals in European mosses: 2000–2001 survey. UN Economic Commission for Europe, International Cooperative Programme on Effects of Air Pollution on Natural Vegetation and Crops. http://192.171.153.213/products/publications/documents/HeavyMetalsinEuropeanMosses.pdf (accessed 9 July 2012).

Buschmann J., Berg M., Stengel C. and Sampson M. (2007). Arsenic and manganese contamination of drinking water resources in Cambodia: coincidence of risk areas with low relief topography. *Environmental Science and Technology*, **41**(7), 2146–2011.

Buschmann J., Berg M., Stengel C., Winkel L., Sampson M. L., Trang P. T. and Viet P. H. (2008). Contamination of drinking water resources in the Mekong delta floodplains: Arsenic and other trace metals pose serious health risks to population. *Environment International*, **34**(6), 756–764.

Business news agency. (2009). China arsenic pollution explosion. Business news agency, 1 May 2009 (in Chinese). http://china.toocle.com (accessed 30 March 2012).

Business Wire. (2004). Indusmin Receives Five-Year Extension of Licenses in Armenia. Vancouver, British Columbia, Canada. Business Wire, 6 June 2004. http://findarticles.com/p/articles/mi_m0EIN/is_2004_June_6/ai_n6055847 (accessed 6 July 2012).

Cabral Pinto M. M. S. and Silva M. M. V. G. (2005). Portugal. In: Contemporary Reviews of Mine Water Studies in Europe, C. Wolkersdorfer and R. Bowell (eds), Springer Verlag, Berlin, Germany. (Mine Water and the Environment 24: Supplementary Material; http://www.imwa.info/docs/10.1007_s10230-005-0081-3.pdf.

Cáceres L., Gruttner E. and Contreras R. (1992). Water recycling in arid regions: Chilean case. *Ambio*, **21**, 138–144.

Caldecott J. and Wickremasinghe W. R. M. S. (2005). Sri Lanka: post tsunami environmental assessment. United Nations Environment Programme and Sri Lanka Ministry of Environment and Natural Resources. http://postconflict.unep.ch/publications/dmb_srilanka.pdf (accessed 6 July 2012).

Caldéron J., Ortiz-Pérez D., Yáñez L. and Díaz-Barriga F. (2003). Human exposure to metals: pathways of exposure, biomarkers of effect, and host factors. *Ecotoxicology and Environmental Safety*, **56**(1), 93–103.

Cambodia, Kingdom of, National Arsenic Database (2005). WellMap: The online well database of the Kingdom of Cambodia. http://www.cambodiawellmap.com (accessed 9 July 2012).

Canova J. L. (1999). Elements in South Carolina inferred background soil and stream sediment samples. *South Carolina Geology*, **41**, 11.

Canyellas C. (2004). Stability analysis of arsenic containing sludges produced in the treatment of surface waters in rural areas of Peru. In: Universidad de Chile, Division de Recursos Hidricos y Medio Ambiente (eds), Book of abstracts for the Third Regional Conference on the Evaluation and Management of Drinking Water Sources Contaminated with Arsenic. Universidad de Chile, Santiago, Chile, 8–11 November 2004. Abstract: http://phys4.harvard.edu/~wilson/arsenic/conferences/arsenic_project_conferences.html (accessed 9 July 2012).

Cardoso S., Grajeda C., Argueta S. and Garrido S. E. (2010). Experiencia satisfactoria para la remoción de arsénico en Condado Naranjo, Mixco, Guatemala. In: Tecnologías Económicas Para el Abatimiento de arsénico en aguas, M. I. Litter, A. M. Sancha and A. M. Ingallinella (eds), Editorial Programa Iberoamericano de Ciencia yTecnologia para el Desarrollo, Buenos Aires, Argentina, pp. 211–222.

Carpenter R., Peterson M. L. and Jahnke R. A. (1978). Sources, sinks, and cycling of arsenic in the Puget Sound region. In: Estuarine Interactions, M. L. Wiley (ed.), Academic Press, New York, NY, USA, pp. 459–480.

Carrillo A. and Drever J. I. (1998). Adsorption of arsenic by natural aquifer material in the San Antonio El Triunfo mining area, Baja California, Mexico. *Environmental Geology*, **35**, 251–257.

Carrillo-Chavez A., Drever J. I. and Martinez M. (2000). Arsenic content and groundwater geochemistry of the San Antonio El Triunfo, Carrizal and Los Planes aquifers in southernmost Baja California, Mexico. *Environmental Geology*, **39**, 1295–1303.

Carter J. M., Sando S. K. and Hayes T. (1997). Source, occurrence, and extent of arsenic in the Grass Mountain area of the Rosebud Indian Reservation, South Dakota. USGS Water-Resources Investigations Report 97-4286. United States Geological Survey, Department of the Interior.

Casiot C., Morin G., Juillot F., Bruneel O., Personné J. C., Leblanc M., Duquesne K., Bonnefoy V. and Elbaz-Poulichet F. (2003). Bacterial immobilization and oxidation of arsenic in acid mine drainage (Carnoules Creek, France). *Water Research*, **37**(12), 29–36.

Castellanos M. E., Sosa L., Moreira A., Maya H., Perez S., León A. R. and Gomez B. M. (2005). Concentración de arsenic en macroalgas de la Bahía de Cienfuegos. Rev Investig March 2005, 21–26. Revista de Investigación Marina (RIM), Cuba. http://www.cim.uh.cu/rim/pdf/2005/1/2005-21.PDF (accessed 9 July 2012).

Castro de Esparza MLC. (2000). The Problem of Arsenic in Drinking Water in Latin America. Pan American Center for Sanitary Engineering and Environmental Sciences (CEPIS)/Pan American Health Organization (PAHO)-World Health Organization (WHO) Regional Adviser, Lima, Peru.

Castro de Esparza MLC. (2005). Estudeio para Mejoramiento de la Calidad del Agua de Pozos en Zonas Rurales de Puno. Pan American Center for Sanitary Engineering and Environmental Sciences (OPS/CEPIS), Lima, Peru.

Castro de Esparza MLC. (2009). The presence of arsenic in drinking water in Latin America and its effect on public health. In: Natural Arsenic in Groundwater of Latin America, J. Bundschuh, M. A. Armienta, P. Birkle, P. Bhattacharya, J. Marschullat, A. B. Mukherjee (eds), Leiden, The Netherlands: CRC Press/Balkema Publisher, pp. 17–29.

Catskill Mountainkeeper. (2009). Marcellus Shale: The Marcellus Shale - America's next super giant. Catskill Mountainkeeper website. http://www.catskillmountainkeeper.org/our-programs/fracking/marcellus-shale/ (accessed 6 July 2012).

Caussy D. (2004). Epidemiological Methods for Mitigating Health Impacts of Arsenic in South East Asia Region of the World Health Organization. World Health Organization, New Delhi, India. Abstract: http://www.bvsde.paho. org/bvsacd/ arsenico/Arsenic2004/theme1/paper1.2.pdf (accessed 9 July 2012).

CBCL Ltd, Conestoga-Rovers & Associates. (1999). Phase 1 site assessment, Muggah Creek Watershed, Sydney, Nova Scotia. Final Report No. 98628B. Supply and Services Canada, Sydney, Nova Scotia; CBCL Limited, Sydney, NS; Conestoga-Rovers & Associates, Waterloo, Ont. http://www. tarpondscleanup. ca/reports /CBCL-10.pdf (accessed 6 July 2012).

CBC News. (2007). Alberta oilsands rush threatening environment. CBS News, 22 August 2007. http://www.cbc.ca/news/canada/calgary/story/2006/08/01/oil-sands.html (accessed 7 July 2012).

Cebrián M. E., Albores A., Aguilar M. and Blakely E. (1983). Chronic arsenic poisoning in the north of Mexico. *Human Toxicology*, **2**, 121–133. http://www.ncbi.nlm.nih.gov/pubmed/6840787 (accessed 9 July 2012).

Cebrián M. E., Albores A., García-Vargas G. and Del Razo L. M. (1994). Chronic arsenic poisoning in humans. The case of México. In: Arsenic in the Environment. Part II: Human Health and Ecosystem Effects, J. O. Nriagu (ed.), John Wiley & Sons, New York, NY, USA, pp. 93–107.

Cebulak S. and Rozkowska A. (1983). Geochemical correlation of coal seams of the Central Region of Lubezk coal basin. *Kwartalnik Geologiczny*, **27**(1), 25–40 (in Polish).

Central Pollution Control Board (CPCB). (2001). Arsenic Contamination in Groundwater and its Control. CPCB, Ministry of Environment and Forests, Government of India, Delhi, India.

Center for Entrepreneurship in International Health and Development & University of California, School of Public Health. (2005). Partnership for Clean Indoor Air: Household Energy, Indoor Air Pollution and Health, Overview of Experiences and Lessons in China. University of California, School of Public Health, Berkeley, CA, USA. http://www.pciaonline.org/files/China_Household_Energy_and_Health_Overview.pdf (accessed 6 July 2012).

CEO (Caucasus Environment Outlook). (2002a). Regional trends overview. Caucasus Environment Outlook, Tblisi, Georgia. http://www.eionet.europa.eu/seris/alp/alpine_1 (accessed 6 July 2012).

CEO (Caucasus Environment Outlook). (2002b). State of the Caucasus environment and policy measures: a retrospective from 1972 to 2002. Caucasus Environment Outlook, Tblisi, Georgia. http://www.docstoc.com/docs/80122648/Caucasus-Environment-Outlook-_CEO_-2002 (accessed 6 July 2012).

Chadha D. K. and Sinha Ray S. P. (1999). High Incidence of Arsenic in Ground Water in West Bengal. pnMinistry of Water Resources, Central Ground Water Board, Faridabad, India.

Chadwick J. (1996). Nelson's gold on the Great Silk Road: Nelson Gold Corp.–Tajikistan's Zeravshan Valley. Mining Magazine, 1 April. http://www.highbeam.com/doc/1G1-18301740.html (accessed 6 July 2012).

Chakraborti D. K. (2006). Worsening arsenic contamination in Bihar: another Bengal delta calamity in the making? Harvard Chronic Arsenic Poisoning Website, 6 March. 2006. http://phys4.harvard.edu/~wilson/arsenic/references/Bihar.pdf (accessed 6 July 2012).

Chakraborti D. Biswas B. K., Chowdhury T. R, Basu G. K., Mandal B. K., Chowdhury U. K., Mukherjee S. C., Gupta J. P., Chowdhury S. R. and Rathore K. C. (1999). Arsenic groundwater contamination and sufferings of people in Rajnandgaon district, Madhya Pradesh, India. *Current Science*, **77**(4), 502–504.

Chakraborti D., Mukherjee S. C., Pati S., Sengupta M. K., Rahman M. M., Chowdhury U. K., Lodh D., Chanda C. R. and Chakraborti A. K. & Basu G. K. (2003). Arsenic groundwater contamination in Middle Ganga Plain, Bihar, India: a future danger? *Environmental Health Perspectives*, **111**, 1194–1201.

Chakraborti D., Rahman M. M., Chowdhury U. K., Paul K., Sengupta M., Lodh D., Basu G. K., Chanda C. R., Saha K. C. and Mukherjee S. C. (2002b). Present groundwater arsenic contamination status in West Bengal, India. 5th International Conference on Arsenic Exposure and Health Effects. San Diego, CA, USA, 14–18 July. Abstract: http://phys4.harvard.edu/~wilson/arsenic/countries/SDAbstracts SPEAKERS.pdf (accessed 9 July 2012).

Chakraborti D., Rahman M. M., Paul K., Chowdhury U. K., Sengupta M. K., Lodh D., Chanda C. R., Saha K. C. and Mukherjee S. C. (2002a). Arsenic calamity in the Indian subcontinent: What lessons have been learned? *Talanta*, **58**, 3–22. Cited in: IARC WHO 2004 84 p.65 and p. 69. http://monographs.iarc.fr/ENG/Monographs/vol84/index.php (accessed 6 July 2012).

Chakraborti D., Sengupta M. K., Rahman M. M., Ahamed S., Chowdhury U. K., Hossain M. A., Mukherjee S. C., Pati S., Saha K. C., Dutta R. N. and Quamruzzaman Q. (2004). Groundwater arsenic contamination and its health effects in the Ganga-Meghna-Brahmaputra plain. *Journal of Environmental Monitoring*, **6**(6), 74N–83N.

Chakraverty S. (2005). Personal communication. *Dec*. **30**, 2005.

Charlet L., Ansari A. A. and Musso M. L. (2001). Risk of arsenic transfer to a semi-confined aquifer and the effect of water level fluctuation in North Mortagene France at a former industrial site. *Science of the Total Environment*, **277**, 133–147.

Chatterjee A., Das D., Mandal B. K., Roy Chowdhury T., Samanta G. and Chakraborti D. (1995). Arsenic in groundwater in six districts of West Bengal, India: the biggest arsenic calamity in the world. Part 1. Arsenic species in drinking water and urine of the affected people. *Analyst*, **120**(3), 643–650.

Chea S., Mao S., Uy D., Ngai T. and Mahin T. (2008). Kanchan Arsenic Filter: Evaluation of Applicability to Cambodia. Phase I Technical Report, September 2008. http://web.mit.edu/watsan/tech_hwts_chemical_k anchanarsenicfilter.html (accessed 6 July, 2012).

Chen C.-J., Chen C. W., Wu M. M. and Kuo T. L. (1992). Cancer potential in liver, lung, bladder and kidney due to ingested inorganic arsenic in drinking water. *British Journal of Cancer*, **66**, 888–892.

Chen C.-J., Chuang Y. C., You S. L., Lin T. M. and Wu H. Y. (1986). A retrospective study on malignant neoplasms of bladder, lung and liver in blackfoot disease endemic area in Taiwan. *British Journal of Cancer*, **53**(3), 399–405.

Chen C.-J., Kuo T. L. and Wu M. (1988). Arsenic and cancers. *Lancet*, **1**, 414–415.

Chen C. and Li G. (1990). Dexing copper mine. *Mining Magazine*, **162**(4), 287–288.

Chen M., Ma Lena Q., Hoogeweg C. G. and Harris W. G. (2001). Arsenic background concentrations in Florida, USA surface soils: determination and interpretation. *Environmental Forensics*, **2**(2), 17–126.

Cheng Z., Buckley B. M., Katz B., Wright W., Bailey R., Smith K. T. and Geen van A. (2005). Arsenic uptake by trees at a highly contaminated site. https://goo.gl/vibZF (accessed 9 July 2012).

Cherry J. A., Morel F. M. M., Rouse J. V., Schnoor J. L. and Wolman M. G. (1986a). Hydrogeochemistry of sulfide and arsenic-rich tailings and alluvium along Whitewood Creek, South Dakota (Part 1 of 3 parts). *Mineral & Energy Resources*, **29**(4), 12.

Cherry J. A., Morel F. M. M., Rouse J. V., Schnoor J. L. and Wolman M. G. (1986b). Hydrogeochemistry of sulfide and arsenic-rich tailings and alluvium along Whitewood Creek, South Dakota (Part 2 of 3 parts). *Mineral & Energy Resources*, **29**(5), 15.

Cherry J. A., Morel F. M. M., Rouse J. V., Schnoor J. L. and Wolman M. G. (1986c). Hydrogeochemistry of sulfide and arsenic-rich tailings and alluvium along Whitewood Creek, South Dakota (Part 3 of 3 parts). *Mineral & Energy Resources*, **29**(6), 15.

China Daily. (2004). Magic plant joins war on pollution. China Daily, 7 April 2004. http://www.chinadaily.com.cn/english/doc/2004-04/07/content_321270.htm (accessed 6 July 2012).

China Daily. (2007). Daqing oilfield to turn green. China Daily, 10 June 2007. http://www.gasandoil.com/news/2007/07/cns72856 (accessed 6 July 2012).

Chiou H. Y., Hsueh Y. M., Liaw K. F., Horng S. F., Chiang M. H., Pu Y. S., Lin J. S., Huang C. H. and Chen C. J. (1995). Incidence of internal cancers and ingested inorganic arsenic: a seven year follow-up study in Taiwan. *Cancer Research*, **55**, 1296–1300.

Chongsuvivatwong V., Ritsmithchai S. and Duerawee D. (1997). Preliminary report: follow-up of water use among Ron Phibun villagers, Nakhon Si Thammarat. Proceedings of Toxic Metal Study of Pak Panang and Pattani River Basin, PSU, Hat Yai, Songkhla, 17 October. http://medinfo.psu.ac.th/AnnualResearch/1998/cviral.htm (accessed 6 July 2012).

Chongsuvivatwong V., Lim A., Dueravee M., Geater A., Ritsamitchai S. and Oshikawa S. (2000). Follow up of water use in a tin mining area affected with arsenic. *Southeast Asian Journal of Tropical Medicine and Public Health*, **31**(4), 769–774.

Choprapawon C. and Rodcline A. (1997). Chronic arsenic poisoning in Ronpibool Nakhon Sri Thammarat, the southern province of Thailand. In: Arsenic Exposure and Health Effects, C. O. Abernathy, R. L. Calderon and W. R. Chappell (eds), Chapman & Hall, London, UK, pp. 69–77.

Choprapawon C. and Porapakkham Y. (2001). Occurrence of cancer in arsenic contaminated area, Ronpibool district, Nakorn Srithmmarat Province, Thailand. In: Arsenic Exposure and Health Effects IV, W. R. Chappell, C. O. Abernathy and R. L. Calderon (eds), Elsevier, Amsterdam, Netherlands, pp. 201–206.

Chowdhury U. K., Biswas B. K. and Chowdhury T. R. (2000). Groundwater arsenic contamination in Bangladesh and West Bengal, India. *Environment Health Perspectives*, **108**, 393–397.

Chowdhury T. R., Mandal B. K., Samanta G., Basu G. K., Chowdhury P. P., Chanda Chitta R., Karan N. K., Lodh D., Dahr R. K., Das D., Saha K. C. and Chakraborti D. (1997). Arsenic in groundwater in six districts of West Bengal, India: the biggest arsenic calamity in the world: the status report up to August 1995. In: Arsenic Exposure and Health Effects, C. O. Abernathy, R. L. Calderon and W. R. Chappell (eds). Chapman & Hall, London, UK.

Christen K. (2001). The arsenic threat. *Environmental Science and Technology*, **35**, 286A–291A.

Christensen O. D. (1993). Carlin trend geologic overview. In: Gold Deposits of the Carlin Trend, Nevada. Society of Economic Geologists, O. D. Christensen (ed.), Society of Economic Geologists Guidebook Series 18, Littleton, CO, USA, pp. 3–26.

Chung E.-H., Lee J. S., Chon H. T. and Sager M. (2005). Environmental contamination and bioaccessibility of arsenic and metals around the Dongjeong Au-Ag-Cu mine, Korea. *Geochemistry*, **5**, 69–74.

Chysma R. B., Petukhov V. L., Kuzmina E. E., Barinov E. Y., Dukhanov Yu. A. and Korotkova G. H. (2003). The content of heavy metals in feeds of the Tyva Republic. *Journal de Physique IV (Proceedings)*, **107**, 297–299.

CIA (Central Intelligence Agency). (2009a). The World Factbook: Botswana. https://www.cia.gov/library/publications/the-world-factbook/geos/bc.html (accessed 6 July 2012).

CIA (Central Intelligence Agency). (2009b). The World Factbook: Mongolia. https://www.cia.gov/library/publications/the-world-fact book/geos/mg.html (accessed 5 July 2012).

CIA (Central Intelligence Agency). (2012). World Fact Book: Indonesia. https://www.cia.g ov/library/publications/the-world-factbook/geos/id.html (accessed 6 July 2012).

Claesson M. and Fagerberg J. (2003). Arsenic in Ground Water of Santiago del Estero: Sources, Mobility Patterns and Remediation with Natural Materials. Master's thesis. Department of Land and Water Resources Engineering, Kungliga Tekniska Högskolan, Stockholm, Sweden, p. 59 (TRITA-LWR-EX-03-05).

Clark P. J., Zingaro R. A., Irgolic K. J. and McGinley A. N. (1980). Arsenic and selenium in Texas lignit. *International Journal of Environmental and Analytical Chemistry*, **7**, 295–314.

Clasen T. F. (2006). Household Water Treatment for the Prevention of Diarrhoeal Disease. PhD thesis. University of London, School of Tropical Medicine and Hygiene, London, UK.

Clement M. (1996). Tausend Jahre Metallerzbergbau in Mitteleuropa. Ein Beitrag zu seinem Ende, dargestellt am Blei-Zink-Silber-Erzbergwerk Rammelsberg. Verlag Glückauf, Essen, Germany.

Cleverley J. S., Benning L. G. and Mountain B. W. (2003). Reaction path modelling in the As-S system: a case study for geothermal arsenic transport. *Applied Geochemistry*, **18**, 1325–1345.

Cline C. S., Rigger J. T. and Raymond J. R. (1985). Ground-Water Monitoring at the Hanford Site, January–December 1984. United States Department of Energy, Pacific Northwest Laboratory, Richland, WA, USA, (PNL-5408).

Cmarko V. (1963). Hygienic problems of arsenic exhalations of ENO plant. *Ceskoslovenska Hygiena*, **8**, 359–362 (in Slovak, with English summary).

Coakley G. J. (1995). The Mineral Industry of Ghana. United States Department of the Interior, United States Geological Survey, Minerals Information. http://minerals.usgs.gov/minerals/pubs/country/1995/9213095.pdf (accessed 6 July 2012).

Coalition Press. (2000). Environmentalists call on South Africa to reject imports of hazardous waste. Press release. http://www.ban.org/ban_news/enviros_call.html (accessed 6 July 2012).

Cohen S. (2002). Possible role of dimethylarsinous acid in dimethylarsinic acid-induced urothelial toxicity and regeneration in the rat. *Chemical Research and Toxicology*, **15**(9), 1150–1157.

Coleman S. L. and Bragg L. J. (1990). Distribution and mode of occurrence of arsenic in coal. In: Recent Advances in Coal Geochemistry, L. L. Chyi and C.-L. Chou (eds). Geological Society of America, Boulder, CO, USA, pp. 13–26 (Special Paper 248).

Çolak M., Gemici Ü. and Tarcan G. (2003). The effects of colemanite deposits on the arsenic concentrations of soil and ground water in Igdeköy-Emet, Kütahya, Turkey. *Water Air and Soil Pollution*, **149**(1–4), 127–143.

Colorado Department of Public Health and Environment. (2003). Analysis of diagnosed vs. expected cancer cases in residents of the Vasquez Boulevard/I-70 Superfund site study area. Citizen's summary, November. Colorado Department of Public Health and Environment, Colorado.

Companhia de Tecnologia de Saneamento Ambienta. (1991). Avaliacäo da qualidade ambiental do Rio Ribeira do Iguape – consideracoes preliminares. Relatorio Técnico, Säo Paulo, Brazil.

CONAMA (Comision Nacional del Medio Ambiente). (2000). Política Ambiental Para el desarrollo Sustentable. CONAMA, Antofagasta, Chile.

COREMI (2004). Anuario Estadístico de la Mineria Mexicana 2003. Consejo de Recursos Minerales, Pachuca. Cited in Armienta *et al.* (2007).

Corey G., Tomasini R. and Pagura J. (2005). Estudio epidemiológico de la exposición al arsenico a través del consumo de agua. Provincia de Santa Fe, República Argentina. http://www.sertox.com.ar/img/item_full/CARATULA,%20PROLOGO,%20INDICE%20E%20 INTRODUCCION.pdf (accessed 6 July 2012).

Cortes S. A., Pino P., Atalah E., Silva C. and Jara M. (2004). Exposure to environmental arsenic in children of Antofagasta, II Region, Chile. Session 2.12. In: Universidad de Chile, Third International Conference on the Evaluation and Management of Drinking Water Sources Contaminated with Arsenic. Universidad de Chile, Santiago, Chile, 8–11 November 2004. http://www.Bvsde.paho.org/bvsacd/arsenico/Arsenic2004/index.htm (accessed 9 July 2012).

Cortés Muñoz J., Rivera Ma. L., Bedolla L., Piña M. and Martín A. (1999). Evaluación de filtros intradomiciliarios, puesta en marcha de dos plantas potabilizadoras en Zimapán, Hidalgo y efectos a la salud asociados con la exposición a arsénico. Informe final. Instituto Mexicano de Tecnología del Agua (TC-9910), Morelos, Mexico. Abstract: http://www.medicontacto.org/medicontacto/content/el-agua-en-m% C3%A9xico-lo-que-debemos-saber-tercera-parte (accessed 9 July 2012).

Craw D., Chappell D., Reay A. and Walls D. (2000). Mobilisation and attenuation of arsenic around gold mines, east Otago, New Zealand. *New Zealand Journal of Geology & Geophysics*, **43**, 373–383.

Criaud A. and Fouillac C. (1989). The distribution of arsenic (III) and arsenic (V) in geothermal waters: examples from the Massif Central of France, the Island of Dominica in the Leeward Islands of the Caribbean, the Valles Caldera of New Mexico, USA, and southwest Bulgaria. *Chemical Geology*, **76**, 259–269.

Csanády M., Kárpáti Z. and Csalagovits I. (1996). Arsenic in drinking water in Hungary: data on occurrence and remediation. In: International seminar on arsenic, health effects, remediation and treatment costs. Santiago, Chile, 8–10 October 1996.

Csanády M., Karpati Z. and Csalagovits I. (1999). Arsenic in Drinking Water. National Institute of Environmental Health, Budapest, Hungary.

Cucu-Man S., Mocanu R., Culicov O., Steinnes E. and Frontasyeva M. (2004). Atmospheric deposition of metals in Romania studied by biomonitoring using the epiphytic moss Hypnum cupressiforme. *Journal of Environmental Analytical Chemistry*, **84**, 845–854.

Cunningham C. G. (1988). The relationship between some disseminated gold deposits, the western edge of the Precambrian Craton, and paleothermal anomalies in Nevada. In: Bulk Mineable Precious Metal Deposits of the Western United States: Symposium Proceedings. R. W. Shafer, J. J. Cooper and P. G. Vikre (eds), Geological Society of Nevada, Reno, NV, USA, pp. 35–48.

CUPE (Canadian Union of Public Employees). (1999). Profiting from Water: Turning a Life Source into a Commodity – Annual Report on Privatization: 1999. CUPE, Canada. www.portaec.net (accessed 6 July 2012).

DACAAR (Danish Committee to Aid Afghan Refugees). (2005). Annual report 2005. DACAAR, Kabul, Afghanistan. http://www.nzdl.org/gsdlmod?. (accessed 6 July 2012).

Dangic A. (2007). Arsenic in surface and groundwater in central parts of the Balkan peninsula (SE Europe). In: Arsenic in Soil and Groundwater Environment: Biogeochemical Interactions, Health Effects and Remediation, P. Battacharya, A. B. Mukherjee, J. Bundschuh, R. Zevenhoven and R. H. Loeppert (eds), Elsevier, Oxford, UK, pp. 127–156.

Danielson L. (2006). Architecture for Change: an Account of the Mining, Minerals and Sustainable Development Project. Global Public Policy Institute, Berlin, Germany.

Das D., Chatterjee A., Mandal B. K., Samanta G., Chanda B. and Chakraborti D. (1995). Arsenic in ground water in six districts of West Bengal, India: the biggest arsenic calamity in the world. Part 2. Arsenic concentration in drinking water, hair, nails, urine, skin scale and liver tissue (biopsy) of the affected people. *Analyst*, **120**(3), 917–924.

Das D. K. (2007). Effects of arsenic-contaminated irrigation water, zinc and organic matter on the mobilization of arsenic in soils in relation to rice (Oryza sativa L.). In: Arsenic in Soil and Groundwater Environment: Biogeochemical Interactions, Health Effects and Remediation, P. Battacharya, A. B. Mukherjee, J. Bundschuh, R. Zevenhoven and R. H. Loeppert (eds), Elsevier, Oxford, UK, pp. 339–362.

Davenport J. R. and Peryea F. J. (1991). Phosphate fertilizers influence leaching of lead and arsenic in a soil contaminated with lead arsenate. *Water, Air, and Soil Pollution*, **57–58**, 101–110.

Davis A., de Curnou P. and Eary L. E. (1997). Discriminating between sources of arsenic in the sediments of a tidal waterway, Tacoma, Washington. *Environmental Science and Technology*, **31**, 1985–1991.

Davis C. J. (2001). An overview of the UNICEF supported arsenic mitigation programme in Bangladesh. In: Arsenic Exposure and Health Effects IV, W. R. Chappell, C. O. Abernathy and R. L. Calderon (eds), Elsevier, Amsterdam, Netherlands, pp. 381–385.

De Bettencourt A. M., Florencio M. H., Duarte M. F. N., Gomes M. L. R. and Vilas Boas L. F. C. (1994). Refractory methylated arsenic compounds in estuarine water: tracing back elusive species. *Applied Organometallic Chemistry*, **8**(1), 43–56.

De Jong S. J. and Kikietta A. (1980). Une particuliarité heureusement bien localisée; la presence d'arsenic en concentration toxique dans un village près de Mogtedo (Haute Volta), Ouagadougou, Burkina Faso. Bulletin du Comité Interafricain d'Etudes Hydrauliques (CIEH) No. 42-43. CIEH, Ouagadougou, Burkina Faso.

Del Razo L. M., Arellano M. A. and Cebrián M. E. (1990). The oxidation states of arsenic in well-water from a chronic arsenicism area of northern Mexico. *Environmental Pollution*, **64**, 143–153.

Del Razo L. M., Hernandez J. L., Garcia-Vargas G., Ostiosky-Wegman P., Cortinas De Nava C. and Cebrian E. (1994). Urinary excretion of arsenic species in a human population chronically exposed to arsenic via drinking water. A pilot study. In: Arsenic Exposure and Health, W. R. Chappell, C. O. Abernathy and C. R. Cothern (eds), Science and Technology Letters, Northwood, UK, pp. 91–100.

Del Razo L. M., García-Montalvo Eliud A. and Valenzuela Olga L. (2003). Chapter 10: Arsenic exposure alters purine metabolism in rats, mice, and humans. In: Arsenic Exposure and Health Effects V: Proceedings of the Fifth International Conference on Arsenic Exposure and Health Effects, W. R. Chappell, C. O. Abernathy, R. L. Calderon and D. J. Thomas (eds), July 14-18, 2002, Elsevier, San Diego, New York, CA, USA, pp. 135–145.

Del Razo L. M., Garcia-Vargas G. G., Garcia-Salcedo J., Sanmiguel M. F., Rivera M., Hernandez M. C. and Cebrian M. E. (2002). Arsenic levels in cooked food and assessment of adult dietary intake of arsenic in the Region Lagunera, Mexico. *Food and Chemical Toxicology*, **40**(10), 1423–1431.

Demongeot M. (2006). Oil rises for fourth day; Nigeria, Iran support. Reuters. http://royaldutchshellplc.com/2006/03/03/the-new-york-times-oil-rises-for-fourth-day-nigeria-iran-support/ (accessed 6 July 2012).

De-Moraes-Flores E.-M., Saidelles A. P. F., Barin J. S., Mortari S. R. and Martins A. F. (2001). Hair sample decomposition using polypropylene vials for determination of arsenic by hydride generation atomic absorption spectrometry. *Journal of Analytical Atomic Spectrometry*, **16**, 1419–1423.

Dent D. (1986). Acid Sulphate Soils: A Baseline for Research and Development. Publication No. 39. International Institute for Land Reclamation and Improvement (ILRI), Wageningen, The Netherlands.

Department of Manufacturing and Industry Development. (1991). Arsenic in the Environment. Stage 1. Department of Manufacturing and Industry Development, Victoria, Australia.

Deschamps E., Ciminelli V. S. T., Lange F. S., Matschullat J., Raue B. and Schmidt H. (2002). Soil and sediment geochemistry of the Iron Quadrangle, Brazil: The case of arsenic. *Journal of Soils & Sediments*, **2**, 216–222.

DFID (Department of International Development). (1997). A Guide to the Regional Environmental Action Plan for the Antofagasta Region of Chile. DFID, Environmental Resources Management, Santiago, Chile.

Dahr R., Kumar Biswas B. K., Samanta G., Mandal B. K., Chakraborti D., Roy S., Jafar A., Islam A., Ara G., Kabir S., Khan A. W., Ahmed S. A. and Hadi S. A. (1997). Groundwater arsenic calamity in Bangladesh. *Current Science*, **73**(1), 48–59.

Díaz-Barriga F., Santos M. A., Mejía J. J., Batres L., Yáñez L., Carrizales L., Vera E., Del Razo L. M. and Cebrian M. E. (1992). Health effects in children exposed to arsenic. The San Luis Potosí case. In: International Seminar Proceedings: Arsenic in the Environment and its Incidence on Health, Universidad de Chile, Santiago, Chile, pp. 41–49.

Di Basilio M. (2002). Chemical and geological characterization of a mine site in central Italy. In: Proceedings of the 17th World Congress of Soil Science. Bangkok, Thailand, 14–21 August, pp. 1–11 (Symposium No. 25, Paper No. 798).

Diebe N. H. (2004). Hydrochemistry of Arsenic in Shallow Aquifers in Cambodia. Unpublished MSc thesis, Department of Earth Sciences (now SEAES), University of Manchester, p. 120.

Diehl S. F. (2002). Mineralogic residence and sequence of emplacement of arsenic and other trace elements in coals of the Warrior Basin, Alabama. In: Proceedings of the 19th International Pittsburgh Coal Conference, 23–27 September, Pittsburgh, PA, USA (CD ROM, Section 16, Trace Metal Contamination).

Diehl S. F., Goldhaber M. B. and Hatch J. (2004). Modes of occurrence of mercury and other trace elements in coal of the Warrior Basin, northwestern Alabama. *International Journal of Coal Geology*, **59**, 193–208.

Dilday R. H., Yan Wengui, Slaton N., Gibbons J. and Moldenhauer K. (2001). Straighthead of rice as influenced by arsenic and nitrogen. In: B. R. Wells Rice Research Studies, R. J. Norman and C. A. Beyrouty (eds), University of Arkansas Agricultural Experiment Station, Fayetteville, AK, USA, pp. 201–214 (Research Series 476). Abstract: http://www.ars.usda.gov/research/publications/publications.htm?seq_no_115=121104 (accessed 9 July 2012).

Dimitrova M. (2007). Gold or arsenic and cyanides: a tricky dilemma for Chelopech. http://www.bnr.bg/print.htm?id=%7B186347D8-BEFD-43FA-9966-BFED6A2A7AB1%7D&lng=en

Doan N. T. (undated). Assessment of Arsenic Pollution in Groundwater of Mekong Delta. Division of Hydrogeology and Engineering Geology for the South of Vietnam & Arcadis, Haskoning, Viet Nam.

Domato M. M., Neely K. W., Hoffman B. and Benner S. (2005). Geochemical processes and mechanisms of arsenic in southwestern Idaho groundwater. *Geological Society of America, Abstracts with Programs*, **37**(7), 376–376 (Paper No. 164-8).

Dong A., Li D., Zhou Y., Liu J. and Waalkes M. P. (2006). Current Status and Prevention Strategy for Coal-arsenic Poisoning in Guizhou. *China J Health Popul Nutr.*, **24**(3), 273–276.

Dong A., Li D., Liang Y. and Jing Z. (2007). Unventilated indoor coal-fired stoves in Guizhou Province, China: Reduction of arsenic exposure through behavior changes resulting from mitigation and health education in populations with arsenicosis. *Environ Health Perspect.*, **115** (4), 659–662.

Donglou X. (2004). The prevention and control of endemic arsenicosis in China. In: Interregional conference on water quality. 23–26 November 2004, Taiyuan, China. UNU, ESCAP, UNICEF, WHO, Washington, DC.

Dortch E. (2005). Arsenic find prompts call to stop using bore water. *Australasian Business Intelligence*, 8 November. http://www.highbeam.com/doc/1G1-138457476.html (accessed 6 July 2012).

Dou X., Zhang Y., Yang M., Pei Y., Huang X. and Kato S. (2006). Occurrence of arsenic in groundwater in the suburb of Beijing and its removal using an Fe-Ce bimetal oxide adsorbent. Source: Min Yang (corresponding author); e-mail: yangmin@mail.rcees.ac.cn.

Dowgiallo J. (2002). The Sudetic geothermal region of Poland. *Geothermics*, **31**(3), 343–359.

DPHE/BGS/DFID (Bangladesh Department of Public Health Engineering/British Geological Survey/UK Department for International Development). (2000). Groundwater Studies of Arsenic Contamination in Bangladesh. Final report. DPHE/BGS/DFID, Dhaka and London.

DPHE/BGS/MML (Department of Public Health Engineering/British Geological Survey/Mott MacDonald Ltd). (1999). Groundwater Studies of Arsenic Contamination in Bangladesh. Phase I: Rapid Investigation Phase. Technical report to United Kingdom Department for International Development, Bangladesh. DPHE/BGS/MML, Dhaka and London.

Drahota P. (2005). Subproject: Biogeochemistry of Arsenic in the Mokrsko Gold Deposit. Migration into Soil, Water and Atmosphere as the Result of Bedrock Weathering. In: Institute of Geology Annual Report 2005. Academy of Sciences of the Czech Republic, Praha, Czech Republic. http://library.gli.cas.cz/type.php?slnm=dtt&w9=1&r7=1 (accessed 6 July 2012).

Dunbar N. W., Chapin C. E. and Brandvold L. A. (2002). Arsenic in New Mexico's drinking water. *New Mexico Earth Matters*, **2**(2), 1–4.

Dunn J. Y. and Bloxham H. C. (1933). The occurrence of lead, copper, zinc and arsenic compounds in atmospheric dusts, and the sources of these impurities. *Journal of the Society of Chemical Industry*, **52**, 189–192.

Duquesne K., Lebrun S., Casiot C., Bruneel O., Personné J.-C., Leblanc M., Elbaz-Poulichet F., Morin G. and Bonnefoy V. (2003). Immobilization of arsenite and ferric iron by *Acidithiobacillus ferro-oxidans* and its relevance to acid mine drainage. *Applied and Environmental Microbiology*, **69**(10), 6165–6173.

Dzhaoshvili V. S. (1968). Naselenie Gruzii: Ekonomiko-Geograficheskoe Issledovanie. Metsniereba, Tbilisi, Georgia.

Earth Rights Institute. (2003). Niger Delta Fund Initiative. http://www.earthrights.net/nigeria/news/definition.html. (accessed 6 July 2012).

Eccles L. A. (1976). Sources of Arsenic in Streams Tributary to Lake Crowley, California. United States Geological Survey (USGS) Water-Resources Investigations Report 76-36. United States Department of the Interior, USGS, Reston, Virginia.

Edelstein D. (1997). Arsenic, Annual Review: 1996. United States Department of the Interior, United States Geological Survey (USGS), Mineral Industry Surveys. USGS National Center, Reston, Virginia.

Edelstein D. (1998). Arsenic, Annual Review: 1997. United States Department of the Interior, United States Geological Survey (USGS), Mineral Industry Surveys. USGS National Center, Reston, Virginia.

Einarsson S., Vides R. A. and Cuellar G. (1975). Disposal of geothermal waste water by reinjection. In: Second United Nations Symposium on the Development and Use of Geothermal Resources. Vol. 2. United States Government Printing Office, Washington, DC. pp. 1349–1363.

El-Hajjar D. (2004). Arsenic Testing at the Nebraska Public Health Laboratory. Conservation and Survey Division, University of Nebraska-Lincoln, Lincoln, NE, USA. http://www.nphl.org/documents/AresenicTesting.pdf (accessed 6 July 2012).

Embassy of the Republic of Uzbekistan in Israel. (2007). Tourism in Brichmulla-Nanai area. http://www.parus87.narod.ru/Articles/Brichmulla-nanay.htm (accessed 6 July 2012).

Engel R. and Smith A. (1994). Arsenic in drinking water and mortality from vascular disease: An ecologic analysis in 30 countries. *Archives of Environmental Health*, **49**, 418–427.

Engel R. R., Hopenhayn-Rich C., Receveur Olivier and Smith Allan H. (1994). Vascular effects of chronic arsenic exposure: a review. *Epidemiologic Reviews*, **16**(2), 184–209.

Enkhtsetseg S. (2004). Issue of arsenic in Mongolia. In: Interregional conference on water quality. 23-26 November 2004, Taiyuan, China. Washington, D.C.: UNU, ESCAP, UNICEF; WHO: Inter-Regional Conference on Water. http://health.groups.yahoo.com/group/FluoridePoisoning/message/2843 (accessed 6 July 2012).

Enomenga M. (1996). Maxus Oil: destruction and violence in the Ecuadorian Oriente. *Huaorani Leader*, March. http://forests.org/archived_site/today/recent/1996/maxecua.htm (accessed 6 July 2012).

EnviroGulf Consulting. (2011). http://www.envirogulf.com/arsenic.html (accessed 18 December 2011).

Environmental Integrity Project. (2005). Arsenic Contamination found in Pennsylvania Fly Ash Spill. Environmental Integrity Project, Washington, DC, 2 February. http://www.environmentalintegrity.org/pdf/publications/flyashfact.jan05.pdf (accessed 6 July 2012).

Environment Canada. (2001). Threats to Sources of Drinking Water and Aquatic Ecosystem Health in Canada. Burlington, Ontario, Canada. National Water Research Institute (NWRI) Scientific Assessment Report Series No. 1. Ontario, Environment Canada.

Erickson M. L. (2005). Arsenic in Upper Midwest Ground Water: Occurrence and Geochemical Mobilization Mechanisms. PhD thesis. University of Minnesota, Minneapolis, MN, USA.

Erickson M. L. and Barnes R. J. (2005a). Arsenic in Upper Midwest drinking water. In: Proceedings of the 2005 National Ground Water Association Naturally Occurring Contaminants Conference: Arsenic, Radium, Radon, and Uranium. National Ground Water Association, Charleston, SC, USA, pp. 135–149.

Erickson M. L. and Barnes R. J. (2005b). Well characteristics influencing arsenic concentrations in ground water. *Water Research*, **39**, 4029–4039.

Erry B. V. (1999). Arsenic residues in predatory birds from an area of Britain with naturally and anthropogenically elevated arsenic levels. *Environmental Pollution*, **106**(1), 91–95.

ESCAP (Economic and Social Commission for Asia and the Pacific). (1993). Atlas of Mineral Resources of the ESCAP Region Vol. 10: Cambodia: Explanatory Brochure. ESCAP, United Nations, New York, p. 87.

Esch M. (2009). NYC says Catskill Gas Drilling Risks are too Great. Associated Press, 23 December 2009. Associated Press, New York, NY.

Esch M. (2008). Estimated gas yield from Marcellus shale goes up. International Business Times, 4 November 2008. http://articles.dailyamerican.com/2008-11-03/news/26315526_1_marcellus-shale-natural-gas-cubic-feet (accessed 6 July 2012).

Eshete D. W. and Chyl L. L. (2003). Source identification and natural attenuation of arsenic contaminated groundwater in northeastern Ohio. *Geological Society of America, Abstracts with Programs*, **35**(6), 565. Paper No. 232-8; https://gsa.confex.com/gsa/2003AM/finalprogram/abstract_63548.htm (accessed 6 July 2012).

Espinoza C. M. and Donnell M. K. (2011). Evaluation of the Kanchan™ Arsenic Filter Under Various Water Quality Conditions of the Nawalparasi District, Nepal. MIT, Cambridge, MA. http://dspace.mit.edu/ (accessed 6 July 2012).

Espinoza M. (2005). Distribution of arsenic pollution in the groundwater of the south west basin of Sebaco valley, Matagalpa, Nicaragua. Paper presented at the Congreso Caragua, Hydrogeology and Water Resources Management: Working together for the future, 9–11 March. Congreso Caragua, Managua, Nicaragua.

Espinoza-Ojeda E. (2003). Evaluacion de riesgo a la salud por exposición a arsénico en agua potable en los poblados de la costa agrícola de H. Caborca. *MS thesis*. Universidad de Sonora, Sonora, Mexico.

Eurasian Natural Resources Corporation. (2004). Eurasian Energy Corporation. http://66.129.79.18/Operations/DisplayPage.asp?PageTitle=EEC).

European Pollutant Emission Register. (2004). EPER database. European Commission. http://ec.europa.eu/environment/ippc/eper/index.htm (accessed 6 July 2012).

European Report. (2005). State aid/environment: Rhine pollution project subsidy approved for Germany (Solvay Soda). *Goliath Industry & Business News*. http://www.highbeam.com/doc/1G1-133174371.html (accessed 6 July 2012).

European Union. (2005). The elimination of the arsenic stock in Fier. http://goo.gl/JKKHn (accessed 6 July 2012).

Evans P. O. (1997). Destruction of Abandoned Chemical Weapons in China. Paper 13. Bonn International Center for Conversion, Bonn, Germany.

Ezzati M. (2005). Iran's advantage: the country holds ten percent of the world's proved oil reserves. *Energy*, 22 March. http://www.amazon.com/Irans-advantage-country-percent-reserves/dp/B000CPMV50 (accessed 6 July 2012).

Falling Rain Genomics Inc. (2007). Global Gazetteer Version 2.1. http://www.fallingrain.com/world/ (accessed 6 July 2012).

Fan P. (2000). Arsenic toxins flowing on 'Gold and Silver villages'. Yangtze Evening Daily [cited in Z. M. Xie, Y. M. Zhang and R. Naidu (2006)].

Fantauzzi M., Atzei D., Elsener B., Lattanzi P. and Rossi A. (2004). Enargite surface reactivity in an acidic mine effluents environment an electrochemical and XPS investigation. http://eapclu.iap.tuwien.ac.at/www/ecasia05/allpdf/1403.pdf (accessed 6 July 2012).

FAO/WHO (Food and Agriculture Organization of the United Nations/World Health Organization). (2010). *Joint FAO/WHO Expert Committee on Food Additives: Summary and Conclusions 72nd Meeting*. Rome 16-25 Feb. 2010. FAO/WHO, Rome/Geneva.

Faria S., Atzei D., Elsener B., Lattanzi P. and Rossi A. (2004). Evaluacion toxicologica alimentaria. Determination de arsenico total e inorganico en carne de llama y su relacion con arsenico total presente in aguas de bebida en la provincia de Jujuy, Argentina. In: Third regional conference on the evaluation and management of drinking water sources contaminated with arsenic. Universidad de Chile, Santiago, Chile, 8–11 November 2004. Abstract: http://eapclu.iap.tuwien.ac.at/www/ecasia05/allpdf/1403.pdf (accessed 9 July 2012).

Fariña S. (2009). Geology and Hydrology. Vol. 2. (Asuncion: SAG-PY [Sistema Acuifero Guarani, Paraguay]). http://www.sag-py.org/ Informes Tecnicos (accessed 11 Nov. 2011).

Farinha M. M., Šlejkovec Z., van Elteren J. T., Wolterbeek H. T. and Freitas M. C. (2004). Arsenic speciation in lichens and in coarse and fine airborne particulate matter by HPLC-UV-HG-AFS. *Journal of Atmospheric Chemistry*, **49**(1–3), 343–353.

Federal Ministry of Industry. (2004). The Directory of Industries in Nigeria. Federal Republic of Nigeria. http://www.fmind.gov.ng/detail.php?recordID=7&num=23.

Fedorov L. A. (2010). Chemical armament: a country's water against its own people. *Social Sciences and Humanities - Middle East Studies Online Journal* (MESOJ), **1**(2).

Fehr R., Merkel O., Lacombe M. and Wolf U. (2003). Towards health impact assessment of drinking-water privatization: The example of waterborne carcinogens in North Rhine-Westphalia (Germany). *Bulletin of the World Health Organization*, **81**, 408–414. http://www.who.int/bulletin/volumes/81/6/fehr.pdf (accessed 9 July 2012).

Feinglass E. J. (1973). Arsenic intoxication from well water in the United States. *New England Journal of Medicine*, **288**, 828–830.

Feldman P. R. and Rosenboom J.-W. (2001). Final Activity Report: Drinking Water Quality Assessment in Cambodia. Unpublished final report to the Cambodian Ministry of Rural Development and the Ministry of Industry, Mines and Energy (with the support of the World Health Organization), Cambodia.

Feldman P. R., Rosenboom J.-W., Saray M., Navuth P., Samnang C. and Iddings S. (2006). Assessment of the chemical quality of drinking water in Cambodia. *Journal of Water and Health*, **5–1**, 101–116.

Feldmann J., John K. and Pengprecha P. (2000). Arsenic metabolism in seaweed-eating sheep from northern Scotland. *Fresenius' Journal of Analytical Chemistry*, **368**(1), 116–121.

Fenelon J. M., Bayless E. R. and Watson L. R. (1995). Groundwater Quality in Northeastern St. Joseph County, Indiana. United States Geological Survey (USGS) Water-Resources Investigations Report 95-4092. United States Department of the Interior, USGS, Reston, VA, USA.

Feng L. (2006). Preventive Policy Against Hazardous Effects of Drinking Arsenic-Polluted Well Water in Shanxi, China. Shanxi Heath Bureau. http://www.cepis.ops-oms.org/bvsacd/arsenico/arseniccont/preventive.pdf (accessed 12 February 2006).

Fengthong T., Dethoudom S. and Keosavanh O. (2002). Drinking water quality in the Lao People's Democratic Republic. In: Seminar on the environmental and public health risks due to contamination of soils, crops, surface and groundwater from urban, industrial and natural sources in South East Asia, Hanoi, Viet Nam, December. United Nations Economic and Social Commission for Asia and the Pacific, Bangkok, Thailand.

FEPA (Federal Environmental Protection Agency). (1989). Our National Environmental Goals. (Special Publication No. 3). FEPA, Lagos, Nigeria.

FEPA (Federal Environmental Protection Agency). (1991a). Guidelines and Standards for Environmental Pollution Control in Nigeria. FEPA, Lagos, Nigeria.

FEPA (Federal Environmental Protection Agency). (1991b). S.I.8 National Environmental Protection (Effluent Limitation) Regulations of 1991. FEPA, Lagos, Nigeria.

FEPA (Federal Environmental Protection Agency). (1991c). S.I.9 National Environmental Protection (Pollution Abatement in Industries and Facilities Generating Wastes) Regulation of 1991. FEPA, Lagos, Nigeria.

Ferreccio C., González P. C., Milosavjlevic S. V., Marshall G. G. and Sancha A. M. (1998). Lung cancer and arsenic exposure in drinking water: a case-control study in northern Chile. *Cad Saude Publica*, **14**(Suppl. 3), 193–198.

Ferreccio C., González C., Milosavjlevic V., Marshall G., Sancha A. M. and Smith A. H. (2000). Lung cancer and arsenic concentrations in drinking water in Chile. *Epidemiology*, **11**, 673–679.

Ferreccio C., Marshall G., Bates M. and Selvin S. (2004). Arsenic in drinking water: ten-year study of the excess cancer mortality in Region II, northern Chile. In: Third regional conference on the evaluation and management of drinking water sources contaminated with arsenic. Universidad de Chile, Santiago, Chile, 8–11 November 2004.

Ferreira da Silva E., Zhang C., Pinto L. S., Patinha C. and Reis P. (2004). Hazard assessment on arsenic and lead in soils of Castromil gold mining area, Portugal. *Applied Geochemistry*, **19**(6), 887–898.

FGN (Federal Government of Nigeria). (1988a). Federal Environmental Protection Agency Decree No. 58, 30 December 1988. FGN, Government Press, Lagos, Nigeria.

FGN (Federal Government of Nigeria). (1988b). Harmful Wastes Decree No. 42 of 30 November 1988. FGN, Government Press, Lagos.

FGN (Federal Government of Nigeria). (1993). Water Resources Decree No. 101 of August 1993. FGN, Government Press, Lagos.

Fhilippidis A., Georgakopoulos A., Kassoli-Fournaraki A., Misaelides P., Yiakkoupis P. and Broussoulis J. (1996). Trace element content in composted samples of three lignite seams from the central part of the Drama lignite deposit, Macedonia, Greece. *International Journal of Coal Geology*, **29**, 219–234.

Figueira R., Sérgio C., Lopes J. L. and Sousa A. J. (2007). Detection of exposition risk to arsenic in Portugal assessed by air deposition in biomonitors and water contamination. *International Journal of Hygiene and Environmental Health*, **210**(3–4), 393–397.

Figueiredo B. R., Borba R. P. and Angélica R. S. (2005). Arsenic occurrence in Brazil and human exposure. Presented at Workshop on medical geology: Metals, health and the environment, Rio de Janeiro, Brazil, June 2005. http://www.cprm.gov.br/publique/media/Palestra04.pdf

Figueiredo B. R., Borba R. P. and Angélica R. S. (2007). Arsenic occurrence in Brazil and human exposure. *Environmental Geochemistry and Health*, **29**(2), 109–118.

Figueroa M. E. and Montes F. A. (undated). Removal of Arsenic by Chemical Coagulation. Department of Sanitary Works (Part 1). Los Andes Department, Salta, Argentina (in Spanish) [cited by Esparza, 2000].

Finkelman R. B. (2000). Health Impacts of Coal Combustion. United States Geological Survey Fact Sheet FS-094-00. United States Department of the Interior, USGS, Reston, VA.

Finkelman R. B., Belkin H. E. and Zheng B. (1999). Health impacts of domestic coal use in China. *Proceedings of the National Academy of Sciences of the United States of America*, **96**, 3427–3431.

Fitzpatrick F., Arnold T. L. and Colman J. A. (1998). Surface-Water Quality Assessment of the Upper Illinois River Basin in Illinois, Indiana, and Wisconsin: Spatial Distribution of Geochemicals in the Fine Fraction of Streambed Sediment. United States Geological Survey (USGS) Water-Resources Investigations Report 98-4109. United States Department of the Interior, USGS, Reston, VA.

Fleet M. and Mumin H. (1997). Goldbearing arsenian pyrite and marcasite and arsenopyrite from Carlin Trend gold deposits and laboratory studies. *American Mineralogist*, **82**, 182–187.

Flores R. M., Stricker G. D. and Nichols D. J. (2006). Coal palynology: clues to depositional environments controlling quality and chemistry of minable coals in the Powder River basin. *Geological Society of America, Abstracts with Programs*, **38**(7), 204.

Florida Department of Health. (2004). *Investigation of Cancer Incidence Community Adjacent to Coronet Industries, Plant City, Florida*. Florida Department of Health, Division of Environmental Health, Office of Environmental Public Health and Medicine http://www.myfloridaeh.com/medicine/SUPERFUND/pdf/Coronet/Coronet-Cancer-Rpt.pdf (accessed 6 July 2012).

Flowers L. (2004). New Mexico Environmental and Health Effects Tracking Program: Arsenic in Drinking Water Project. http://www.sandia.gov/water/Arsenic2004/themesession/flowers_pres.pdf (accessed 6 July 2012).

Flynn H. C., McMahon V., Diaz G. C., Demergasso C. S., Corbisier P., Meharg A. A. and Paton G. I. (2002). Assessment of bioavailable arsenic and copper in soils and sediments from the Antofagasta region of northern Chile. *Science of the Total Environment*, **286**(1–3), 51–59.

Focazio M. J., Welch A. H., Watkins S. A., Helsel D. R. and Horn M. A. (1999). A retrospective analysis on the occurrence of arsenic in ground-water resources of the United States and limitations in drinking-water-supply characterizations. United States Department of the Interior, United States Geological Survey, 21 pp. (USGS Water-Resources Investigations Report 99-4279. http://www.pubs.usgs.gov/wri/wri994279/pdf/wri994279.pdf (accessed 6 July 2012).

Folkes D. J., Helgen S. O. and Litle R. A. (2001). Impacts of historical arsenic pesticide use on residential soils in Denver, Colorado. In: Arsenic Exposure and Health Effects IV, W. R. Chappell, C. O. Abernathy and R. L. Calderon (eds), Elsevier, Amsterdam, Netherlands, pp. 97–113.

Folkes D. J., Kuehster T. E. and Litle R. A. (2001). Contributions of pesticide use to urban background concentrations of arsenic in Denver, Colorado, USA. *Environmental Forensics*, **2**, 127–139.

Ford C. J., Byrd J. T., Grebmeier J. M., Harris R. A., Moore R. C., Madix S. E., Newman K. A. and Rash C. D. (1995). Final Project Report on Arsenic Biogeochemistry in the Clinch River and Watts Bar Reservoir. Vol. 1. Main text. Energy Systems Environmental Restoration Program and Clinch River Environmental Restoration Program. Prepared by Environmental Sciences Division, Oak Ridge National Laboratory, Oak Ridge, TN, USA, for United States Department of Energy, Office of Environmental Management (ESD Publication 4185; ORNL/ER-206/V1. http://www.ornl.gov/~webworks/cpr/rpt/78555.pdf (accessed 9 July 2012).

Ford R. G. (2004). Final project report (2 September 2004): Natural attenuation study, groundwater, surface water, soil and sediment investigation, Industri-Plex Superfund site, Woburn, Massachusetts. United States Environmental Protection Agency, Washington, DC. EPA 600/X-04/019; http://www.epa.gov/ne/superfund/sites/industriplex/230912.pdf.

Ford R. (2006). Natural Attenuation of Arsenic in an Urban, Industrialized Watershed. USEPA/ORD/NRMRL/GWERD, Washington, DC. http://www.docstoc.com/docs/7859370/Natural-Attenuation-of-Arsenic-in-an-Urban-Industrialized-Watershed (accessed 6 July 2012).

Fordyce F. M., Williams T. M., Paijitprapapon A. and Charoenchaisri P. (1995). Hydrogeochemistry of Arsenic in an Area of Chronic Mining-Related Arsenism, Ron Phibun District, Nakhon Si Thammarat Province, Thailand: Preliminary Results. British Geological Survey, Keyworth, Nottingham, UK. http://nora.nerc.ac.uk/8397/ (accessed 9 July 2012).

Forrester Group. (2003). Site Management: Westside Business Park, Kansas City, Missouri, Brownfield Redevelopment of 22-Acre Abandoned Railyard. The Forrester Group, Springfield, MO, USA.

Foster S. S. D., Lewis W. J. and Farr J. L. (2004). The pollution hazard to village water supplies in eastern Botswana. *Proceedings of the Institution of Civil Engineers*, **69**(2), 281–293.

Foust R. D. Jr., Mohapatra P., Compton-O'Brien A.-M. and Reifel J. (2004). Groundwater arsenic in the Verde Valley in central Arizona, USA. *Applied Geochemistry*, **19**(2), 251–255.

Foy H. M., Tarmapai S., Eamchan P. and Metdilogkul O. (1992/1993). Chronic arsenic poisoning from well water in a mining area in Thailand. *Asia-Pacific Journal of Public Health*, **6**(3), 150–152.

Francesconi K. S. and Edmonds J. S. (1998). Arsenic species in marine samples. *Croatica Chemica Acta*, **71**(2), 342–359.

Franklin R. E., Duis L., Smith B. R., Brown R. and Toler J. E. (2003). Elemental concentrations in soils of South Carolina. *Soil Science*, **168**(4), 280–291.

Fredericks D. (2004). Promotion of Household Level Mitigation for Arsenic and Fluoride Removal: Summary of Mission to Cambodia and Viet Nam, (APW Report). World Health Organization, Washington, DC. http://www.wpro.who.int/environmental_health/documents/apw_arsenic_flouride_removal/en/index.html (accessed 6 July 2012).

Fredericks D. (2005). Research Needs for Household Level Treatment to Remove Arsenic and Fluoride in Drinking Water in SE Asia. Prepared for World Health Organization (WHO). WHO, Geneva, Switzerland. http://www.who.int/household_water/resources/Fredericks.pdf (accessed 6 July 2012).

Freyssinet P., Lecomte P. and Edimo A. (1989). Dispersion of gold and base metals in the Mborguéné lateritic profile, east Cameroun. *Journal of Geochemical Exploration*, **32**(1–3), 99–116.

Friends of the Earth International. (2004). Annual Report 2004. Friends of the Earth. http://www.foei.org/en/resources/publications/annual-report/archive/ar2004.pdf/view?searchterm=annual%20report%202004 (accessed 6 July 2012).

Friske P. W. B., McCurdy M. W. and Day S. J. A. (1998). National Geochemical Reconnaissance: Ontario Compilation – Arsenic in Lake Sediment. (Map, geochemical, arsenic distribution, 1:500 000). Geological Survey of Canada, Ontario, Open File 3379d, 1998.

Fujii R. and Swain W. C. (1995). Areal Distribution of Selected Trace Elements, Salinity, and Major Ions in Shallow Ground Water, Tulare Basin, Southern San Joaquin Valley, California. United States Geological Survey (USGS) Water-Resources Investigations Report 95-4048. United States Department of the Interior, USGS, Reston, VA.

Gabarino J. R., Bednar A. J., Garbarino J. R., Needham R., Staver K. W. and Wershaw R. L. (2003). Environmental fate of roxarsone in poultry litter. I. Degradation of roxarsone during composting. *Environmental Science and Technology*, **37**, 1509–1514.

Gadzhiev A. G. (1980). The Problems of Geothermal Energy in Dagestan. Nedra Publishing House, Moscow, 208.

Gaeta Tassinari C. C., Diaz Pinzon F. and Buena Ventura J. (2008). Age and sources of gold mineralization in the Marmato mining district, NW Colombia: a Miocene-Pliocene epizonal gold deposit. *Ore Geology Reviews*, **33**, 505–518.

Galbraith H., LeJeune K. and Lipton J. (1995). Metal and arsenic impacts to soils, vegetation communities and wildlife habitat in southwestern Montana uplands contaminated by smelter emissions: I. Field evaluation. *Environmental Toxicology and Chemistry*, **14**, 1895–1903.

Garai R., Chakraborty A. K. and Dey S. B. (1984). Chronic arsenic poisoning from tubewell water. *Journal of the Indian Medical Association*, **82**, 34–35.

Garcia M. E. (2006). Control mechanisms of seasonal variation of dissolved arsenic and heavy metal concentrations in surface waters of Lake Poopo basin, Bolivia. Geological Society of America, Abstracts with Programs, **38**(7), 320.

Garcia M. E., Sracek O., Fernández D. S. and Hidalgo M. del V. (2007). Factors affecting arsenic concentration in groundwaters from Northwestern Chaco-Pampean Plain, Argentina. *Environmental Geology*, **52**(7), 1261–1275.

Garcia-Guinea J. and Harffy M. (1998). Bolivian mining pollution: past, present, and future. *Ambio*, **27**, 251.

Garcia-Sanchez A. and Alvarez-Ayuso E. (2003). Arsenic in soils and waters and its relation to geology and mining activities (Salamanca Province, Spain). *Journal of Geochemical Exploration*, **80**(1), 69–79.

Garlipp C. R., Bottini P. V., Capitan E. M. de, Pinho M. C., Panzan A. D. N., Sakuma A. M. A. and Paoliello M. B. (2003). Urinary protein excretion profile: a contribution for subclinical renal damage identification among environmental heavy metals exposure in southeast Brazil. *Journal de Physique IV (Proceedings)*, **107**, 513.

Garnega G., Wyse E., Azemard S., Stankevicius A. and de Mora S. (2006). Arsenic in sediments from the southeastern Baltic Sea. *Environmental Pollution*, **144**(3), 855–861.

Garrido Hoyos S. E., Avilés Flores M., Rivera Huerta M. L. and Nájera Flores M. C. (2007). Diagnóstico de la presencia de arsénico en agua de pozo, Mixco, Guatemala. Final report TC-0711.3. Instituto Mexicano de Tecnología del Agua, Jiutepec, Mexico.

Garziono C. A. (2005). Italy: some examples of mine water problems in Tuscany. In: Contemporary Reviews of Mine Water Studies in Europe, C. Wolkersdorfer and R. Bowell (eds), Springer Verlag, Berlin, Germany, (Mine Water and the Environment, 24: Supplementary Material).

Gaskovao L., Bessonova E. P. and Bortnikova S. B. (2001). Leaching experiments on trace element release from the arsenic-bearing tailings of Khovu-Aksy (Tuva Republic, Russia). *Applied Geochemistry*, **18**(9), 1361–1371.

Gavshin V. M. and Miroschnichenko L. V. (2000). Uranium concentration in altered brown coals located under burnt rocks from Kansk-Achinsk basin, west Siberia. *Geostandards Newsletter, the Journal of Geostandards and Geoanalysis*, **24**(2), 241–246 (abstract available at http://www.crpg.cnrs-nancy.fr/Geostandards/GGR_abs.v24/msvm107_abstr. html).

Gayer R. A., Rose M. and Dehmer J. (1999). Impact of sulphur and trace element geochemistry on the utilization of marine-influenced coal: case study from South Wales Variscan foreland basin. *International Journal of Coal Geology*, **40**, 151–174.

Gbadebo A. M. (2005). Occurrence and fate of arsenic in the hydrogeological systems of Nigeria. Geological Society of America, Abstracts with Programs, **37**(7), 376.

Gebel T. W., Suchenwirth R. H. R., Bolten C. and Dunkelberg H. H. (1998). Human biomonitoring of arsenic and antimony in case of an elevated geogenic exposure. *Environmental Health Perspectives*, **106**(1), 33–39.

Gennett J. A. and Raymond A. L. (1986). Changes in floral composition with depositional environment in Texas Eocene Manning Formation lignites. *American Association of Petroleum Geologists Bulletin*, **70**(9), 1181–1182.

Gentzis T. and Goodarzi F. (1997). Trace element geochemistry of brackish water coals in the central Alberta Plains, Canada. *Energy Sources*, **19**, 493–505.

Gentzis T., Goodarzi F. and Foscolos A. E. (1997). Geochemistry and mineralogy of Greek lignites from the Ioannina basin. *Energy Sources*, **19**, 111–128.

Gentzis T., Goodarzi F., Koukouzas C. N. and Foscolos A. E. (1996). Petrology, mineralogy and geochemistry of lignites from Crete, Greece. *International Journal of Coal Geology*, **30**, 131–150.

Geological Survey of Iran (1976). Seismotectonic Atlas of Iran. Islamic Republic of Iran, Geological Survey of Iran, Tehran.

Gideon R. (1991). The potential impact of industrial wastes on water resources in Amman-Zarqa Basin. Proceedings of the Second Environmental Pollution Symposium, 1990, Friederich Ebert Stiftung Goethe-Institut, Amman Water Research and Study Center. University of Jordan, Amman.

Gillespie D. M., Goodarzi F., Koukouzas C. N. and Foscolos A. E. (1989). Macrofauna Distributions and Sediment Analyses from the Brunswick, Georgia Ocean Dredged Material Disposal Site and Environs. (Technical Report TRS-89/1). University of Georgia, Marine Extension Service, Savannah, GA, USA. Abstract and text: http://www.sciencedirect.com/science/article/pii/016651629 5000348 (accessed 9 July 2012).

Ginarte Z. (1999). Estudio de la contaminación por metales pesados en el entorno de la fábrica de acumuladores de Manzanillo. MSc Thesis, Instituto Superior Minero Metalurgico de Moa (I.S.M.M), Manzanillo, Cuba.

Glass R. L. (1996). Ground-water Conditions and Quality in the Western Part of Kenai Peninsula, Southcentral Alaska. United States Geological Survey (USGS) Open-File Report 96–466. United States Department of the Interior, USGS, Reston, VA.

Glass G. L. (2003). Tacoma Smelter Plume Site. Credible Evidence Report: The ASARCO Tacoma Smelter and Regional Contamination in Puget Sound. Final report, September 2003. Prepared for Tacoma-Pierce County Health Department and Washington State Department of Ecology. www. ecy.wa.gov/ programs/tcp/ sites_brochure/tacoma_smelter/Sources/Credible_Evidence/ web%20pieces/X_cred%20entire%20report.pdf (accessed 6 July 2012).

Gleason S. V. (2010). Environmental Health Hazards for the Presence of Arsenic and Fluoride in the Aquifers of Mexico. National Water Commission. Wastewater Management and Water Quality, 549 Avenue San Bernabe, San Jeronimo Lidice, CP 10200, Mexico DF, Unpublished paper.

Goddard K. E. (1988). Applied Research Studies of the Cheyenne River System, South Dakota: Description and Collation of Data, 1985 and 1986 Water Years. United States Geological Survey (USGS) Open-File Report 88–484. United States Department of the Interior, USGS, Reston, VA.

Goddard K. E. (1989). Composition, Distribution, and Hydrologic Effects of Contaminated Sediments from the Discharge of Gold Mine Wastes to the Whitewood Creek at Lead and Deadwood, South Dakota. United States Geological Survey (USGS Water-Resources Investigations Report 87-4051). United States Department of the Interior, USGS, Reston, VA.

Goede A. A. and De Bruin M. (1985). Arsenic in the dunlin (*Calidris alpina*) from the Dutch Waddenzee. *Bulletin of Environmental Contamination and Toxicology*, **34**(5), 617–622.

Goff F. and Shevenell L. (1987). Travertine deposits of Soda Dam, New Mexico and their implications for the age and evolution of the Valles caldera hydrothermal system. *Geological Society of America Bulletin*, **99**(2), 292–302.

Goff F., Grisby C. O., Shevenell L. A. and Gutierrez J. W. (1986b). Geochemistry at Honduras Geothermal Sites. *Los Alamos Science*, **14** (Fall), 90–93.

Goff F., Shevenell L., Truesdell A. H., Janik C. J., Grigsby C. O. and Paredes R. (1986a). Hydrogeochemistry and prelimninary reservoirmodel of the Platanares geothermal system, Honduras, Central America. Los Alamos National Laboratory Report LA-UR-86-1404/1986. Los Alamos National Laboratory, Los Alamos.

Ģolak M., Gemici Ü. and Tarcan G. (2003). The effects of colemanite deposits on the arsenic concentrations of soil and ground water in Igdeköy-Emet, Kütahya, Turkey. *Water, Air, and Soil Pollution*, **149**(1–4), 127–143.

Goldenberg G., Maass A., Steffens G. and Steuer H. (2003). Hematite mining during the linear ceramics culture in the area of the Black Forest, southwest Germany. In: Deutsches Bergbau-Museum, Man and Mining, T. Stöllner, G. Körlin, G. Steffens and J. Ciemy (eds), Bochum, Germany, pp. 179–186.

Goldhaber M., Irwin E. R., Hatch J. R. and Pashin J. C. (2001). Potential Environmental Consequences of Localized Arsenic Enrichment in Appalachian Basin Coals. United States Environmental Protection Agency, Washington, DC, USA. http://nepis.epa.gov/Exe/ZyPURL.cgi?Dockey=P100AFIX.txt (accessed 6 July 2012).

Gold Mines of Sardinia PLC (2003). Furtei: Geology and Exploration. Gold Mines of Sardinia PLC, Surrey, UK.

Goldschmidt V. M. and Peters C. (1934). Zur Geochemie des Arsens. In: Nachrichten von der Gesellschaft der Wissenschaften zu Göttingen, Mathematisch-Physikalische Klasse, Fachgruppe VI. Gesellschaft der Wissenschaften zu Göttingen, Germany, pp. 11–12.

Gomez A. (2004). Chronic arsenicosis in El Zapote, Nicaragua, 1994–2002. In: Third Regional Conference on the Evaluation and Management of Drinking Water Sources Contaminated with Arsenic. Universidad de Chile, Santiago, Chile, 8–11 November 2004.

Goodarzi F. and Swaine D. J. (1993). Chalcophile elements in western Canadian coals. *Advances in Organic Petrology and Geochemistry*, **24** (1–4), 281–292.

Goodarzi F. and Van der Flier-Keller E. (1989). Organic petrology and geochemistry of intermontane coals from British Columbia: the Blackburn opencast mine near Tulameen, British Columbia, Canada. *Chemical Geology*, **75**, 227–247.

Goodbred S. L. and Kuehl S. A. (2000). The significance of large sediment supply, active tectonism, and eustasy on margin sequence development: late Quaternary stratigraphy and evolution of the Ganges-Brahamaputra delta. *Sedimentary Geology*, **133**, 227–248.

Gosselin D. C., Klawer L. and Noe A. (2004). Arsenic in Nebraska's Groundwater and Public Water Supplies. Nebraska Natural Resource Surveys, University of Nebraska-Lincoln, School of Natural Resources, Conservation and Survey Division, Lincoln, NE, USA. Earth Science Notes No. 7; http://www.nlc.state.ne.us/epubs/U2375/H016.0007-2004.pdf.

Gosselin D. C., Klawer L. and Noe A. (2004). Arsenic in Nebraska's groundwater and public water supplies. *Geological Society of America*, Abstracts with Programs, **36**(5), 356.

Gotkowitz M. B. (2005). Almost everywhere: naturally occurring arsenic in Wisconsin's aquifers. *Geological Society of America*, Abstracts with Programs, **3**(5), 69.

Government of Canada (1996). Northern River Basins Study Final Report. Alberta Environment, Northern River Basins Study Board, Edmonton, Alberta, Canada. http://www3.gov.ab.ca/env/water/nrbs/toc.html).

Government of Western Australia (2005). Media statement, 8 November. http://72.14.205.104/search?q=cache:0CLearnyHEEJ:portal.environment.wa.gov.au/pls/portal/docs/PAGE/DOE_A DMIN/MEDIA_REPOSITORY/TAB108318/TAB108358/STIRLING%2520BORE%2520WATER%2520TESTING%2520FINAL%252008.11.05.PDF+australia+arsenic+%22Stirling%22&hl=en&ct=clnk&cd=1&gl=us (accessed 6 July 2012).

Granada J. (2002). Potabilización de agua mediante abatimiento de arsénico. Aguas de Antofagasta, Chile.

Grant A. (1999). UK Report for Period November 1998 to July 1999. University of East Anglia, International Geological Correlation Programme Project 396. http://www2.env.uea.ac.uk/gmmc/igcp/capetown/nat99_2.html (accessed 6 July 2012).

Grantham D. A. and Jones J. F. (1977). Arsenic contamination of water wells in Nova Scotia. *Journal of the American Water Works Association*, **69**, 653–657.

Grmela A. and Rapantova N. (2005). Mine water issues in the Czech Republic. In: Contemporary Reviews of Mine Water Studies in Europe, C. Wolkersdorfer and R. Bowell (eds), Springer Verlag, Berlin, Germany (Mine Water and the Environment, 24: Supplementary Material).

Groen J., Schuchmann J. B. and Geirnaert W. (1988). The occurrence of high nitrate concentration in groundwater in villages in northwestern Burkina Faso. *Journal of African Earth Sciences*, **7**, 999–1009.

Grosz A. E., Grossman J. N., Garrett R., Friske P., Smith D. B., Darnley A. G. and Vowinkel E. (2004). A preliminary geochemical map for arsenic in surficial materials of Canada and the United States. *Applied Geochemistry*, **19**(2), 257–260.

Grosz A. E., Hathaway J. C. and Escowitz E. C. (1986). Placer deposits of heavy minerals in Atlantic continental shelf sediments. In: Proceedings of the Eighteenth Annual Offshore Technology Conference, Houston, Texas, pp. 387–394.

Guérèquiz A. R., Mañay N., Goso Aguilar C. and Bundschuh J. (2006). Evaluación de riesgo ambiental por presencia de arsénico en el sector oeste del acuífero Raigón, Departamento de San José, Uruguay. Presentación del proyecto. In: Libro de actas del taller de distribución del arsénico en Iberoamérica, M. Litter (ed.), Red Temática 406RT0282 IBEROARSEN, Centro Atómico Constituyentes, Buenos Aires, Argentina.

Guérèquiz A.R., Mañay N., Goso C., Fernandez J. and Garcia M. (2008). Environmental risk assessment of arsenic in the Raigon Aquifer, Uruguay: Research Advancements. PA23. Abstract volume II International Congress: Arsenic in the Environment: From Nature to Humans, May 21–23, Valencia, Spain.

Guérèquiz A.R., Mañay N., Goso-Aguilar C., Fernández-Turiel J.L. and García-Valles M. (2009). Environmental risk assessment of arsenic in the Raigón aquifer, Uruguay. *Biologist (Lima)*, **7**(1–2) Special Issue, C0130.

Gueye D. (2002). Small-scale Mining in Burkina Faso 2001. International Institute for Environment and Development and World Business Council for Sustainable Development. Mining, Minerals and Sustainable Development Report No. 73.

Gulens J., Champ D. R. and Jackson R. E. (1979). Influence of redox environments on the mobility of arsenic in ground water. In: Chemical Modeling in Aqueous Systems, E. A. Jenne (ed.), American Chemical Society, Washington, DC, USA.

Gunson A. J. (2004). Mercury and Artisanal Small Scale Gold Miners in China. Master's of Applied Science thesis, University of British Columbia, Vancouver, British Columbia, Canada. http://www.pdfport.com/view/680345-mercury-and-artisanal-small-scale-gold-miners-in-china.html (accessed 6 July 2012).

Guo X., Fujino Y., Kaneko S.i, Wu K., Xia Y. and Yoshimura T. (2001). Arsenic contamination of groundwater and prevalence of arsenical dermatitis in the Hetao Plain area, Inner Mongolia, China. *Molecular and Cellular Biochemistry*, **222**(1–2), 137–140.

Gurdak J. J. and Thyne G. (2005). Sorption controlled arsenic mobility in groundwater of the Zimapán Valley, Mexico. *Geological Society of America*, Abstracts with Programs, **37**(7), 376. Paper No. 164–6.

Gurzau E. S. and Gurzau A. E. (2001). Arsenic in drinking water from groundwater in Transylvania, Romania: an overview. In: Arsenic Exposure and Health Effects IV, W. R. Chappel, C. O. Abernathy and R. L. Calderon (eds), Elsevier, Amsterdam, Netherlands, pp. 181–184.

Guynup S. (2012). The Fracking Industry Buys Congress. *Environment News Service*, 16 February 2012. http://www.ens-newswire.com/ens/feb2012/2012–02-16-02.html (accessed March 31, 2012).

Gyschin A. (2006). Good bye, chemical weapons. Defense and Security/Oblastnaya Gazeta, 14 June. http://www.partnershipforglobalsecurity.org/PrinterFriendly.asp?Doc=614200642857PM.html#1I (accessed 6 July 2012).

Ha L. (2004). Red River Delta despoiled by arsenic pollution. VietNamNet Bridge. http://www.asiafinest.com/forum/index.php?showtopic=24589 (accessed 7 July 2012).

Hails J. R. and Hoyt J. H. (1968). Barrier development on submerged coasts; problems of sea-level changes from a study of the Atlantic Coastal Plain of Georgia and parts of the East Australian coast. *Zeitschrift für Geomorphologie*, **7**, 24–55.

Hamamé M. (2002). Regional Risk Assessment in Northern Chile. Chalmers University of Technology, Environmental Systems Analysis, Goteborg, Sweden. http://www.esa.chalmers.se/Publications/PDF-files/Thesis/esa20021.pdf (accessed 7 July 2012).

Hammarlund L. and Piñones J. (2009). Arsenic in Geothermal Waters of Costa Rica: A Minor Field Study. MSc Thesis. Department of Land and Water Resources Engineering, KTH Royal Institute of Technology, Stockholm, Sweden, ISSN 1651-064X. http://www.2.lwr.kth.se/Publikationer/PDF_Files/LWR_EX_09_02.pdf (accessed 27 July 2012).

Hammermeister D. P. and Walmsley S. J. (1985). Hydrologic Data for Leviathan Mine and Vicinity, Alpine County, California, 1981–83. United States Geological Survey (USGS) Open-File Report 85–160. United States Department of the Interior, USGS, Reston, VA.

Hanzlik J. and Trckova J. (1999). Consequences of Open-Pit Coal Mining on the Geohydraulical System, Case Study. Institute of Rock Structure and Mechanics, Praha, Czech Republic. http://www.iahr.org/membersonly/grazproceedings99/doc/000/000/018.htm (accessed 7 July 2012).

Haque R., Ghosh N., Dey B. K., Santra A. and Chakraborti D. (1999). Arsenic groundwater contamination and sufferings of people in Rajnandgaon district, Madhya Pradesh, India. *Current Science*, **77**(4), 502–504.

Harrington J. M. Middaugh J. P., Morse D. L. and Housworth J. (1978). A survey of a population exposed to high concentrations of arsenic in well water in Fairbanks, Alaska. *American Journal of Epidemiology*, **108**(5), 377–385.

Harris R. and Duncan D. (2005). The Development of Niger's First Gold Mine. Etruscan Resources Inc., Canada.

Hartwig T., Owor M., Muwanga A., Zachmann D. and Pohl W. (2005). Lake George as a sink for contaminants derived from the Kilembe Copper Mining Area, Western Uganda. *Mine Water and Environment*, **24**, 114–124, IMWA Springer-Verlag.

Harvey C. K., Swartz C. H., Badruzzaman A. B., Keon-Blute N., Yu W., Ali M. A., Jay J., Beckie R., Niedan V., Brabander D., Oates P. M., Ashfaque K. N., Islam S., Hemond H. F. and Ahmed M. F. (2002). Arsenic mobility and groundwater extraction in Bangladesh. *Science*, **298**, 1602–1606.

Hasegawa H. (1996). Seasonal changes in methylarsenic distribution in Tosa Bay and Uranouchi Inlet. *Applied Organometallic Chemistry*, **10**(3), 233–240.

Haselow J. S., Aiken S. C. and Rehm Bernd W. (2003). In-situ co-precipitation of Metals. Contaminated Soil Sediment and Water, (January/February). Association for Environmental Health and Sciences, Amherst, Massachusetts, USA. http://scholarworks.umass.edu/cgi/viewcontent.cgi?article=1001&context=soils_conf_abstracts (accessed 7 July 2012).

Hatch J. R. (1987). Elementary geochemistry. In: Geological Investigations of the Vermillion Creek Coal Bed in the Eocene Niland Tongue of the Wasach Formation, Sweetwater County, H. W. Roehler and L. Martin (eds), Wyoming, United States Geological Survey (USGS) Professional Paper 1314–G7. United States Department of the Interior, USGS, Reston, Virginia, USA. pp. 121–131.

Hatt R. (2007). A little bit of trouble: mercury & arsenic in coal. As published in World Coal. http://www.coalcombustion.com/PDF%20Files/A%20Little%20Bit%20of%20Trouble03.pdf (accessed 7 July 2012).

Hawley J. E. (1955). Germanium content of some Nova Scotian coals. *Economic Geology*, **50**, 517–532.

Heakin A. (1997). USGS Survey Activities Related to American Indians and Alaska Natives Fiscal Year 1997: Water Supply and Water-quality Assessment of the Pine Ridge Indian Reservation. United States Department of the Interior, United States Geological Survey pubs.usgs.gov/amerind/amerind.pdf (accessed 7 July 2012).

Heinrichs G. and Udluft P. (1999). Natural arsenic in Triassic rocks: a source of drinking-water contamination in Bavaria, Germany. *Hydrogeology Journal*, **7**, 468–476.

HELCOM (2002). Chemical warfare agents. In: HELCOM Manual on Co-Operation in Response to Marine Pollution within the Framework of the Convention on the Protection of the Marine Environment of the Baltic Sea area (Helsinki Convention). Vol. 2. December. http://www.helcom.fi/stc/files/ResponseManual/ResponseManualVol_2.pdf (accessed 7 July 2012).

Helios-Rybicka E., Adamiec E. and Skwarczek M. (2003). Assessment of the odra river contamination with heavy metals. Proceedings, German- Chinese-Polish Environmental Engineering. http://baztech.icm.edu.pl/baztech/cgi-bin/btgetdoc.cgi?BPOB-0007-0027 (accessed 7 July 2012).

Helmer R. and Hespanhol I. (1997). Water Pollution Control: A Guide to the Use of Water Quality Management Principles. E & FN Spon, London, UK (on behalf of United Nations Environment Programme, Water Supply & Sanitation Collaborative Council and the World Health Organization). http://www.who.int/water_sanitation_health/resources/watpolcontrol/en/index.html (accessed 7 July 2012).

Helz G. R., Tossell J. A., Charnock J. M., Pattrick R. A. D., Vaughan D. J. and Garner C. D. (1995). Oligomerization in As(III) sulfide solutions: theoretical constraints and spectroscopic evidence. *Geochimica et Cosmochimica Acta*, **59**, 4591–4604.

Hering J. G. and Chiu V. Q. (2000). Arsenic occurrence and speciation in municipal ground-water-based supply system. *Journal of Environmental Engineering*, **126**(5), 471–474.

Hernandez-Garcia M. E. (2002). Naturally occurring arsenic in groundwaters of the Madrid tertiary detrital aquifer (Spain). 2002 Denver Annual Meeting of the Geological Society of America: Paper No. 84–1. https://gsa.confex.com/gsa/2002AM/finalprogram/abstract_40496.htm (accessed 7 July 2012).

Herrera M. T. A., Montenegro I. F., Navar P. R., Dominguez I. R. M. and Vazquez R. T. (2001). Arsenic content of the drinking water sources for the Guadiana Valley, Mexico. *Ingenieria Hidraulica en Mexico*, **16**, 63–70.

Hickmott D. D. and Baldridge W. S. (1995). Application of PIXE microanalysis to macerals and sulfides from the lower Kittang coal of western Pennsylvania. *Economic Geology*, **90**, 246–254.

Hilson G. M. (2004). Structural adjustment in Ghana: assessing the impacts of mining-sector reform. *Africa Today*, December. http://www.muse.jhu.edu/login?auth=0&type=summary&url=/journals/africa_today/v051/51.2hilson.html (accessed 7 July 2012).

Hindmarsh J. T., McLetchie O. R., Hefferman L. P. M., Hayne O. A., Ellenberger H. A., McCurdy R. F. and Thiebaux H. J. (1977). Electromyographic abnormalities in chronic environmental arsenication. *Journal of Analytical Toxicology*, **1**, 270–276.

Hinkle S. R. and Polette D. J. (1999). Arsenic in ground water of the Willamette Basin, Oregon. United States Geological Survey (USGS) Water-Resources Investigations Report 98–4205. United States Department of the Interior, USGS, Reston, VA. oregon.usgs.gov/pubs_dir/Online/Pdf/98-4205.pdf (accessed 7 July 2012).

Hinsby K. (2004). Arsenic Pollution of the Groundwater: New Research Project in Vietnam. Geological Survey of Denmark and Greenland. www.geus.dk/News (accessed 7 July 2012).

Hlavay J. and Polyak K. (1998). Advanced Arsenic Removal Processes for Drinking Water Purification. University of Veszprem, Veszprem, Hungary. http://phys4.harvard.edu/~wilson/arsenic/conferences/Unidentified_conference/Tuesday/Abstracts8.html (accessed 7 July 2012).

Hollerman J. T. (1992). In Arkansas, which comes first, the chicken or the environment? *Tulane Environmental Law Journal*, **6**(1).

Holm T. R. (1995). Ground-Water Quality in the Mahomet Aquifer, McLean, Logan, and Tazewell Counties. Illinois State Water Survey, Champaign, IL, USA (Contract Report 579).

Holm T. R. and Curtiss C. D. III (1988). Arsenic Contamination in East-Central Illinois Ground Waters. Report ILENR/RE-WR-88/16). Champaign, IL, USA, Illinois State Water Survey.

Holmes C. W., Rodrigues F. Shinn and Marot E. (2000). Be-7/Pb-210 Content of Saharan Dust Measured in the Azores Sampled During Mid-Winter and Spring, 2000. American Geophysical Union Meeting, San Francisco, CA, USA. Abstract: http://130.203.133.150/showciting;jsessionid=2711AE769AC46DF61CDC107DFDCF603F?cid=13479436 (accessed 9 July 2012).

Holmes C. W. and Shinn E. A. (2001). Atmospheric transport and deposition of arsenic and potential link to Florida groundwater. Presented at American Geophysical Union, Spring Meeting 2001, Abstract No. H61C-06; https://gsa.confex.com/gsa/2002AM/finalprogram/abstract_38066.htm (accessed 7 July 2012).

Holmes C. W., Shinn E. A. and Rodrigues F. (2001). The atmospheric transport and deposition of arsenic and other metals in southeastern United States. http://wwwbrr.cr.usgs.gov/projects/GWC_chemtherm/FinalAbsPDF/holmes.pdf (accessed 7 July 2012).

Holmstrom H., Ljungberg J. and Ohlander B. (1999). Role of carbonates in mitigation of metal release from mining waste: evidence from humidity cells tests. *Environmental Geology*, **37**, 267–280.

Hoover J. H. (2004). Toxic metals in groundwater on the Navajo Reservation, National Science Foundation CAREER Award, 2004–09, Franz M. Geiger Northwestern University Department of Chemistry. http://www.asdlib.org/directoryUploads/38_41facultydir.pdf (accessed 7 July 2012).

Hopenhayn-Rich C., Biggs M. L., Fuchs A., Bergoglio R., Tello E. E., Nicolli H. and Smith A. H. (1996). Bladder cancer mortality associated with arsenic in drinking water in Argentina. *Epidemiology*, **7**, 117–124.

Hopenhayn-Rich C., Biggs M. L. and Smith A. H. (1998). Lung and kidney cancer mortality associated with arsenic in drinking water in Cordoba, Argentina. *International Journal of Epidemiology*, **27**, 561–569.

Hopenhayn-Rich C., Browning S. R., Hertz-Picciotto I., Ferreccio C., Peralta C. and Gibb H. (2000). Chronic arsenic exposure and risk of infant mortality in two areas of Chile. *Environmental Health Perspectives*, **108**, 667–673.

Hoque B. A., Islam S. and Rahaman M. (2001). Arsenic mitigation and water supply in Bangladesh. In: Arsenic Exposure and Health Effects IV, W. R. Chappell, C. O. Abernathy and R. L. Calderon (eds), Elsevier, Amsterdam, Netherlands, pp. 407–412.

Horiguchi S., Teramoto K., Kurono T. and Ninomiya K. (1978). The arsenic, copper, lead, manganese and zinc contents of daily foods and beverages in Japan and the estimate of their daily intake. *Osaka City Medical Journal*, **24**, 131.

Hossain M. (2009). The Impact of Shallow Tubewells and Boro Rice on Food Security in Bangladesh. The International Food Policy Research Institute. http://www.ifpri.org/sites/default/files/publications/ifpridp00917.pdf (accessed 7 July 2012).

Hossain M. A., Sengupta M. K., Ahamed S., Rahman M. M., Mondal D., Lodh D., Das B., Nayak B., Roy B. K., Mukherjee A. and Chakraborti D. (2005). Ineffectiveness and poor reliability of arsenic removal plants in West Bengal, India. *Environmental Science and Technology*, **39**(11), 4300–4306.

Howard A. G. and Apte S. C. (1989). Seasonal control of arsenic speciation in an estuarine ecosystem. *Applied Organometallic Chemistry*, **3**, 499–509.

Howard A. G., Arbabzavar M. H. and Apte S. (1984). The behavior of dissolved arsenic in the estuary of the River Beaulieu. *Marine Chemistry*, **11**, 493–498.

Howard A. G., Apte S. C., Comber S. D. W. and Morris R. J. (1988). Biogeochemical control of the summer distribution and speciation of arsenic in the Tamar estuary. *Estuarine, Coastal and Shelf Science*, **27**, 427–443.

Howard A. G. and Comber S. D. W. (1989). The discovery of hidden species in coastal waters. *Applied Organometallic Chemistry*, **3**, 509–514.

Howe A., Fung L. H., Lalor G., Rattray R. and Vutchkov M. (2005). Elemental composition of Jamaican foods: a survey of five food crop categories. *Environmental Geochemistry and Health*, **27**, 19–30.

Hower J. C., Calder J. H., Eble C. F., Scott A. C., Richardson J. D. and Blanchard L. J. (2000). Metalliferous coals of the Westphalian A Joggins Formation, Cumberland Basin, Nova Scotia, Canada: petrology, geochemistry and palynology. *International Journal of Coal Geology*, **42**, 185–206.

Hower J. C., Taulbee D. N., Rimmer S. M. and Morrell L. G. (1994). Petrographic and geochemical anatomy of lithotypes from the Blue Gem coal bed, southeastern Kentucky. *Energy Fuels*, **8**, 719–728.

Hsu K.-H., Froines J. R. and Chen C.-J. (1997). Studies of arsenic ingestion from drinking water from northeastern Taiwan: chemical speciation and urinary metabolics. In: Arsenic Exposure and Health Effects, C. O., R. L. Calderon and W. R. Chappell (eds), Abernathy, Chapman Hall, London, UK. pp. 190–209.

Htut T. (2004). Arsenic mitigation initiatives in Myanmar: lessons from the work in progress. In: Interregional Conference on Water Quality. Taiyuan, China, 23–26 November 2004.

Hu X., Murao S. and Huang X. (1996). Genetic model of gold-antimony-tungsten mineralization: evidence from geology and mineralization of the Wuxi deposit, Hunan, China. *Bulletin of the Geological Survey of Japan*, **47**(11), 577–597.

Huang R.-Q., Gao S.-F., Wang W. L., Staunton S. and Wang G. (2006). Soil arsenic availability and the transfer of soil arsenic to crops in suburban areas in Fujian Province, southeast China. *Science of the Total Environment*, **368**(2–3), 531–541.

Huang Y. Z., Qian X. C., Wang G. Q., Xiao B. Y., Ren D. D., Feng Z. Y., Wu J. Y., Xu R. J. and Zhang F. E. (1985). Endemic chronic arsenism in Xinjiang. *Chinese Medical Journal*, **98**(3), 219–222.

Huang Z. and Xie H. (2006). In the southwest high arsenic coal and gold genesis research on origin between high-As coal and gold deposit in southwestern area of Guizhou Province. *Journal of Hunan University of Science & Technology*, **21**(1), 77–81 (in Chinese).

Hubbard T. E., Miller T. R., Hower J. C., Ferm J. C. and Helfrich C. T. (2002). The upper Hance coal bed in southeastern Kentucky: Palynologic, geochemical and petrographic evidence for environmental succession. *International Journal of Coal Geology*, **49**, 177–194.

Huddle J. (2009). Pennsylvania lawsuit says drilling polluted water. Reuters, 9 November 2009. http://www.reuters.com/article/2009/11/09/us-fracking-suit-idUSTRE5A80PP20091109 (accessed 31 March 2012).

Huggins F. E., Huffman G. P., Kolker A., Mroczkowski S. J., Palmer C. A. and Finkelman R. B. (2002). Combined application of XAFS spectroscopy and sequential leaching for determination of arsenic speciation in coal. *Energy and Fuels*, **16**(5).

Hutton L. G. and Lewis W. J. (1980). Nitrate pollution of groundwater in Botswana. In: Water and Waste Engineering in Africa: Proceedings of the 6th Water, Engineering and Development Centre (WEDC) Conference, 24–28 March 1980, S. Ball and J. Pickford (eds), Ahmadu Bello University, Zaria, Nigeria.

Huttrer G. W. (2001). The status of world geothermal power generation 1995–2000. *Geothermics*, **30**, 1–27.

IAEA (International Atomic Energy Agency). (2005). Upgrading Nuclear Analysis Techniques for Air Pollution Monitoring. CRO 8/008. IAEA, Vienna, Austria. http://tc.iaea.org/tcweb/projectinfo/projectinfo_body.asp (accessed 7 July 2012).

Infomine.com. (2012). Countrymine: Mining intelligence by country: Mining in Indonesia. http://www.infomine.com/countries/indonesia.asp (accessed 25 March 2012).

Imai N., Ando A. and Takeda E. (1984). Minor elements in Japanese coal (II). *Bulletin of the Geological Survey of Japan*, **35**, 287–314.

IMERCSA. (2004). State of the Environment, Zambezi Basin. India Musokotwane Environment Resource Centre for Southern Africa, Zambezi. http://www.sardc.net/imercsa/Background.htm (accessed 7 July 2012).

International Mining (im-mining.com). (2010). The global technical magazine–written for miners by miners. http://www.im-mining.com/2010/04/07/arsenic-management-in-sulphide-concentrates/#more-2175 (accessed 25 March 2012).

India Institute of Hygiene and Public Health. (2006). Groundwater arsenic contamination in West Bengal – India: 20 years study. http://www.soesju.org/arsenic/wb.htm (accessed 7 July 2012).

Institute of Public Health. (2003). Prevalence of Arsenicosis due to Ingestion of Arsenic Through Drinking Water: An Epidemiological Survey from Seven Districts of Punjab. Government of Punjab, Institute of Public Health, ctPunjab, India.

International Development Research Center. (1994). Discontinuous aquifers (Niger). http://idris.idrc.ca/app/Search?request=directAccess&projectNumber=871018&language=en (accessed 7 July 2012).

International Energy Agency. (2010). Coal and peat in Portugal 2009. http://www.iea.org/stats/coaldata.asp?COUNTRY_CODE=PT (accessed 7 July 2012).

International Rice Research Institute. (2006). Rice: IRRI High Yield Producing Countries in River Plain Facing Arsenic Contamination. International Rice Research Institute, Los Baños, Philippines, 22 September. http://www.sos-arsenic.net/english/contamin/river_plain.html (accessed 7 July 2012).

Iqbal S. Z. (2001). Arsenic Contamination in Pakistan. United Nations Economic and Social Commission for Asia and the Pacific (UN-ESCAP) report. Expert Group Meeting on Geology and Health, Bangkok, Thailand.

IranMania. (2006). Iran [sic] biggest gold mine inaugurated. Iranmania. http://www.middleeaststar.com/index.php?sid/8023881/scat/1b76a2b4cf7810bd (accessed 7 July 2012).

Ishihara S., Wang P., Kajiwara Y. and Watanabe Y. (2003). Origin of sulfur in some magmatic-hydrothermal ore deposits of South China. *Bulletin of the Geological Survey of Japan*, **54**(3/4), 161–169.

Ishizaki M. (1979). Concentrations of arsenic in agricultural and marine products and the arsenic intake of people in Ibaraki prefecture, Ibaraki-ken Eisi. *Kenkyusho Nempo*, **17**, 49–54.

ISIR. (1996). National Standard of Drinking Water. Institute of Standard and Industrial Research, Tehran, Islamic Republic of Iran.

Ivanov A. (1995). Russia: Environment: the Legacy of Chemical Weapons Manufacture. Inter Press Service English News Wire, 14 August.

Ivanov A. (2003). The far east. In: The Physical Geography of Northern Eurasia, M. Shahgedanova (ed.), Oxford University Press, Oxford, UK.

IWACO. (1975). Dutch Consultancy Report on Burkina Faso. IWACO Ltd, Consultants for Water and Environment, a Division of Royal Haskoning, Rotterdam, Netherlands.

Jackson B. P. and Miller W. P. (1999). Soluble arsenic and selenium species in flyash/organic waste-amended soils using ion-chromatography-inductively coupled plasma mass spectrometry. *Environmental Science and Technology*, **33**(2), 270–275.

Jacobs A. (2000). Lawsuits threatened over arsenic in New Jersey water. *The New York Times*, **9** March, p. B5.

Jakariya M. (2000). The Use of Alternative Safe Water Options to Mitigate the Arsenic Problem in Bangladesh: A Community Perspective. MSc thesis, University of Cambridge, Department of Geography, Cambridge, UK.

Janik C. J., Truesdell A. H., Goff F., Shevenell L., Stallard M. L. and Trujillo P. E. (1991). A geochemical model of the Platanares geothermal system, Honduras. *Journal of Volcanology and Goethermal Research*, **45**, 125–46.

Japan Times (2003). Redress unlikely for residents poisoned by WWII arms. *The Japan Times*, 10 October. http://www.nautilus.org/archives/napsnet/dr/0310/OCT21-03.html#item28.

Järup L. and Pershagen G. (1991). Arsenic exposure, smoking, and lung cancer in smelter workers: a case-control study. *American Journal of Epidemiology*, **134**, 545–551 [cited in Berglund M., Elinder C. G. and Järup L. (2001)].

Järup L., Pershagen G. and Wall S. (1989). Cumulative arsenic exposure and lung cancer in smelter workers: a dose-response study. *American Journal of Industrial Medicine*, **15**, 31–41.

Jayasinghe K. S. A. and Misbah S. A. (1983). Contaminated opium as a source of chronic arsenic poisoning in Sri Lanka. *Ceylon Medical Journal*, **8**, 22–27.

Jdid E. A., Blazy P., Kamoun S., Guedria A., Marouf B. and Kitane S. (1998). Environmental impact of mining activity on the pollution of the Medjerda River, north-west Tunisia. *Bulletin of Engineering Geology and the Environment*, **57**(3), 273–280.

Jin Y. (2006). Current status and prevention strategy for coal arsenic poisoning in China. Chinese Centre for Disease Control and Prevention, Institute of Environmental Health and Related Product Safety. *J Health Popul Nutr.*, 24(3): 273–276.

Jin Y., Liang C., He G. and Cao J. (2003). Study on distribution of endemic arsenism in China. *Wei Sheng Yan Jiu*, **32**(6), 519–540 (in Chinese).

Jishnu L. (2004). The Water Holocaust. Business World, 6 December. Manila, Philippines.

Johnson P. (2004). Unit 516: Inhuman chemical warfare. War Without End Forum, 20 August. http://www.warwithoutend.co.uk/the-americas/2004/08/05/the-holocausts-of-wwii-with-u-s-and-allied-implications-page-2.php (accessed 7 July 2012).

Johnson C. A. and Thornton I. (1987). Hydrological and chemical factors controlling the concentrations of Fe, Cu, Zn, and As in ariver system contaminated by acid mine drainage. *Water Resources*, **21**, 359–365.

Johnston R., Heijnen H. and Wurzel P. (2001). Arsenic in Drinking Water. Final Draft. World Health Organization, Geneva, Switzerland.

Jones J. (2006). Native groups join to save water supply. 7 September. http://water-is-life.blogspot.com/2006/09/native-groups-join-to-save-water.html (accessed 7 July 2012).

Jonnalagadda S. B. and Nenzou G. (1996). Studies on arsenic rich mine dumps. Effect on the river water. *Journal of Environmental Science and Health, Part A*, **31**, 2547–2555.

Jonnalagadda S. B. and Nenzou G. (1997). Studies on arsenic rich mine dumps II. Effect on the heavy metal uptake by vegetation. *Journal of Environmental Science and Health, Part A*, **32**(2), 455–464.

Jovic D., Nikolic V., Vukadinovic I. and Grzetic I. (2002). State of mineral resources, mining industry and their impact to the environment in the Federal Republic of Yugoslavia. Paper presented at the Workshop on Geo- and Mining Hazards: what they are and how to cope with them, November 18–26, Hannover, Germany. Bundesanstalt für Geowissenschaften und Rohstoffe, Hanover, Germany.

Kalbitz K. and Wennrich R. (1998). Mobilization of heavy metals and arsenic in polluted wetland soils and its dependence on dissolved organic matter. *Science of the Total Environment*, **209**(1), 27–39.

Kamchatka's Vision. (2004). Kamchatka's thermal hot springs. http://www.kamchatka.org.ru/thermal.html.

Kansas Geological Survey. (1997). Elements found in high concentration in Kansas coal. *Current Research in Earth Sciences, Bulletin*, **240**(4).

Karanjac J. (2004). Water Resources Management. University of the West Indies, Mona Campus Kingston, Jamaica, West Indies.

Karanjac J. (2005). Vulnerability of ground water in the karst of Jamaica. Paper prepared for the conference: Water resources & environmental problems in Karst, 14–19 September, Beograd & Kotor, Serbia & Montenegro.

Karayigit A. I. (2001). Trace elements in the Lower Pliocene fossiliferous Kangil lignites, Silvas, Turkey. *International Journal of Coal Geology*, **47**, 73–89.

Karayigit A. I., Spears D. A. and Booth C. A. (2000). Distribution of environmental sensitive trace elements in the Eocene Sorgun coals, Turkey. *International Journal of Coal Geology*, **42**(4), 297–314.

Karayigit A. I., Spears D. A. and Booth C. A. (2002). Antimony and arsenic anomalies in the coal seams from the Gokler coalfield, Gediz, Turkey. *International Journal of Coal Geology*, **44**(1), 1–17.

Karcher S., Caceres L., Jekel M. and Contreras R. (1999). Arsenic removal from water supplies in northern Chile using ferric chloride coagulation. *Journal of the Chartered Institute of Water and Environmental Management*, **13**, 164–169.

Karczawska A., Krajewski J. and Bogda A. (2003). Arsenic in the environment in the vicinity of former arsenic ore mining and smelting site Zelezniak in the Kaczawskie Mountains, SW Poland. In: Proceedings of the 7th international conference on the biogeochemistry of trace elements, G. R. Gobran and N. Lepp (eds), Vol. **2**. Uppsala, pp. 80–81.

Karim M. (2000). Arsenic in groundwater and health problems in Bangladesh. *Water Research*, **34**(1), 304–310.

Karner F. R., Schobert H. H., Falcone S. K., Benson S. A. (1986). Elemental distribution and association with inorganic and organic components in North Dakota lignites. In: Mineral Matter in Ash and Coal. American Chemical Society (ACS) Symposium Series, K. S. Vorres (ed.), **301**, ACS, Washington, DC, USA, pp. 70–89.

Karpov G. A. and Pavlov A. L. (1982). Zoning of mineral deposits in the discharge areas of recent hydrotherms. In: Hydrothermal Mineral Forming Solutions in the Areas of Active Volcanism, S. I. Naboko (ed.), Oxonian Press, New Delhi, India, pp. 233–237.

Katsoyiannis I. A., Hug S. J., Ammann A., Zikoudi A. and Hatziliontos C. (2007). Arsenic speciation and uranium concentrations in drinking water supply wells in northern Greece: correlations with redox indicative parameters and implications for groundwater treatment. *Science of the Total Environment*, **383**(1–3), 128–140.

Kedzierski B. (1983). Concentration of iron, zinc, tin, copper, lead and arsenic in Polish beers. *Roczniki Panstwowego ZakLadu Higieny*, **34**(5–6), 475–480 (in Polish).

Kelly W. R. and Wilson S. D. (2000). Deep Well Sampling in the Central Mahomet Valley Aquifer: Preliminary Results. Illinois State Water Survey, Groundwater Division, 26 September (internal report), Champaign, IL, USA.

Ketris M. P. and Yudovich Y. E. (2002). Lithogenes i geochimiya osadochnykh formatisiy Timano-Ural'skogo regiona [Calculation procedure of the coal Clarkes]. Komi Nauchnyj tsentra, Ural'skoe Rossii Academiae Nauk. Translation In: Proceedings, Institute of Geology, Komi Scientific Center, Ural Division of the Russian Academy of Sciences, 111.

KHS Environmental Management Group Ltd. (2004). Arsenic trioxide management project description–progress report. Elec. File - S:\5620 Giant Mine\5620-8 Public Registry\5620-8-C-WaterLic\C6; Original report: File C6. Listed in http://reviewboard.ca/upload/project_document/EA0809-001_A_Partial_Annotative_list_of_the_Giant_Mine_Remediation_Plan_Process_1328900870.pdf (accessed 7 July 2012).

Kim O. (2004). The Caspian Sea: an uncertain future. EnvironmentWatch, UN Chronicle, 1 September.

Kim M.-J., Ahn K.-H. and Jung Y. (2002). Distribution of inorganic arsenic species in mine tailings of abandoned mines from Korea. *Chemosphere*, **49**(4), 307–312.

Kim M.-J., Ahn K.-H., Jung Y., Lee S. and Lim B.-R. (2003). Arsenic, cadmium, chromium, copper, lead, and zinc contamination in mine tailings and nearby streams of three abandoned mines from Korea. *Bulletin of Environmental Contamination and Toxicology*, **70**, 5.

King J. W. (1956). High-grade uraniferous lignites in Harding County, South Dakota. In: Proceedings of the international conference on peaceful uses of atomic energy. Geneva, Switzerland, pp. 473–483.

King T. V. V. (ed.) (1995). Environmental Considerations of Active and Abandoned Mine Lands: Lessons from Summitville, Colorado. United States Geological Survey (USGS) Bulletin 2220. United States Department of the Interior, USGS, Reston, Virginia, USA.

Kinniburgh D. G., Newell A. J., Davies J., Smedley P. L., Milodowski A. E., Ingram J. A. and Merrin P. D. (2006). The arsenic concentration in groundwater from the Abbey Arms Wood observation borehole, Delamere, Cheshire, UK. In: Fluid Flow and Solute Movement in Sandstones: The Onshore UK Permo-Triassic Red Bed Sequence, R. D. Barker and J. H. Tellam (eds), Geological Society of London, London, UK. pp. 265–284.

Kizilstein L. Y. and Kholodkhov Y. I. (1999). Ecologically hazardous elements in coals of the Donets basin. *International Journal of Coal Geology*, **40**, 189–197.

Klepal D. (2005). Crosby Turnpike cleanup nowhere. Cincinatti.com, 6 April. http://news.enquirer.com/apps/pbcs.dll/article?AID=/20050406/NEWS01/504060397/1056/news01 (accessed 2005).

Kler V. R. (1988). Minor element concentrations in the coals and coal-bearing formations. In: Metallogeny and Geochemistry of Coal-Bearing and Oil Shale-Bearing Strata of the USSR: the Regularities of the Element Concentrations and Methods of their Study, V. R. Kler, V. F. Nenakhova, F. Ya. Saprykin, M. Ya. Shpirt, L. I. Rokhlin, A. F. Kulachkova and R. I. Iovchev (eds), Nauka (Publishers), Moscow, pp. 67–142 (in Russian).

KMPO (Kurdistan Management and Planning Organization). (2000). Annual report. KMPO, Information and Statistics Adjutantship, Erbil (Hewlêr).

Kneebone P. E. and Hering J. G. (2000). The behavior of arsenic and other redox-sensitive elements in Crowley Lake, CA, a reservoir in the Los Angeles aqueduct system. *Environmental Science and Technology*, **34**, 4307–4312.

Kneebone P. E., O'Day P. A., Jones N. and Hering J. G. (2002). Deposition and fate of arsenic in iron- and arsenic-enriched reservoir sediments. *Environmental Science and Technology*, **36**, 381–386.

Knoll M. A., Modi A. L. and Candlish J. R. (2005). Occurrence and distribution of arsenic in the hydrogeologic system of southern Cumberland plateau of Tennessee: a preliminary report. *Geological Society of America, Abstracts with Programs*, **37**(7), 376 (Paper No. 164–9).

Ko I., Ahn J. S., Park Y. S. and Kim K.-W. (2003). Arsenic contamination of soils and sediments from tailings in the vicinity of Myungbong Au mine, Korea. *Chemical Speciation and Bioavailability*, **15**(3), 67.

Kolker A., Cannon W. F., Westjohn D. B. and Woodruff L. G. (1998). Arsenic-rich pyrite in the Mississippian Marshall Sandstone: source of anomalous arsenic in southeastern Michigan ground water. *Geological Society of America, Abstracts with Programs*, **30**(7), A–59.

Kolker A., Panov B. S., Landa E. R., Panov Y. B., Korchemagin V. A. and Conko K. M. (2002). Trace metal geochemistry and environmental implications of selected Donbas coals and associated mine water in the vicinity of Donetsk, Ukraine. In: Proceedings of the 19th International Pittsburgh Coal Conference, 23–27 September, Pittsburgh, PA, USA (CD ROM, Section 16, Trace Metal Contamination).

Komnitsas K., Xenidis A. and Adam K. (1995). Oxidation of pyrite and arsenopyrite in sulphidic soils in Lavrion. *Minerals Engineering*, **12**, 1443–1454 [cited in Smedley & Kinniburgh, 2002].

Koren I., Todd M. C., Kaufman Y. J. and Rudich R. W. (2006). The Bodélé depression: a single spot on the Sahara that provides most of the mineral dust to the Amazon forest. *Environmental Research Letters*, **1**, 1–4.

Korte N. (1991). Naturally occurring arsenic in groundwaters of the Midwestern United States. *Environmental Geology and Water Science*, **18**(2), 137–141.

Kortenski J. I., Ilieva I. E. and Popova G. (1990). Trace elements in coal seams of the Svoge anthracite basin. *Geologica Bavarica*, **20**, 63–77.

Kortenski J. I., Popov K. and Kostova I. (1999). An occurrence of chalcophile elements in the VI coal bed, Kachulka Mine, Balkan basin. *Review of the Bulgarian Geological Society*, **60**(1–3), 83–91.

Kortenski J., Popov K. and Kupenov A. (1999). Mineralogical and geochemical peculiarities of the Stanyanski basin lignites and their comparison with other basins of Sofia Province. *Annual of the University of Mining and Geology "St. I. Rilski"*, **42**(1), 59–64.

Kostova I., Petrov O. and Kortenski J. (1996). Mineralogy, geochemistry and pyrite content of Bulgarian subbituminous coals, Pernik basin. In: Coalbed Methane and Coalbed Geology. Geological Society Special Publications No. 109, R. Gayer and I. Harris (eds), Geological Society, Sofia, Bulgaria, pp. 301–314.

Kowalczyk K. (2004). Slow death by arsenic. 28 November. http://www.boloji.com/wfs3/wfs315.htm (accessed 7 July 2012).

Kreiss K., Zack M. M., Landrigan P. J., Feldman R. G., Niles C. A., Chirico-Post J., Sax D. S., Boyd M. H. and Cox D. H. (1983). Neurologic evaluation of a population exposed to arsenic in Alaskan well water. *Archives of Environmental Health*, **38**(2), 116–121.

Kribek B., Mihaljevič M., Sracek O., Knésl I., Ettler V. and Nyambe I. (2011). The extent of Arsenic and of metal uptake by aboveground tissues of Pteris vittata and Cyperus involucratus growing in copper- and cobalt-rich tailings of the Zambian copperbelt. *Archives of Environmental Contamination and Toxicology*, **61**(2), 228–242.

Kuhn A. and Sigg L. (1993). Arsenic cycling in eutrophic Lake Greifen, Switzerland: influence of seasonal redox processes. *Limnology and Oceanography*, **38**(5), 1052–1059.

Kulmatov R. (2004). Modern problems of natural environment change: water and land resources in Uzbekistan. In: First International Workshop on the Human Dimensions of Climate and Environmental Change in Central Asia, May 21–22, 2004, Grand Valley State University, Grand Rapids, Michigan, USA. Grand Valley State University, Grand Rapids, MI.

Kumar F. and Siddiqui M. S. (1989). Food and feeding habits of the carp Labeo calbasu Ham. in north Indian waters. *Acta Ichthyologica et Piscatoria*, **19**(1), 33–48.

Kunli L. (2005). Arsenic content and distribution pattern in Chinese coals. *Toxicological and Environmental Chemistry*, **87**(4), 427–438.

Kuo T.-L. (1968). Arsenic content of endemic well water in endemic area of arsenic poisoning. *Reports of the Institute of Pathology, National Taiwan University*, **20**, 7–13.

Kuo C. S. (1996). *The Mineral Industry of Sri Lanka*. United States Department of the Interior, United States Geological Survey. http://minerals.er.usgs.gov/minerals/pubs/country/1996/9329096.pdf

Kupusovic T. (2002). Environmental regulation of mine waters in the European Union, Macedonia. *European Commission Fifth Framework Programme*. http://cordis.europa.eu/search/index.cfm?fuseaction=proj.document&PJ_RCN=4947727 (accessed 7 July 2012).

Kurbanov M. (2003). Geothermal power charge resources of Dagestan, their status and prospect of their complex development. In: Abstracts of the 2003 International Geothermal Workshop. http://www.jardhitafelag.is/media/PDF/Session05.pdf (accessed 7 July 2012).

Kurttio P. (1999). Arsenic and fluoride in well waters: exposure and health effects. National Public Health Institute, Department of Environmental Medicine, Kuopio, Finland.

Kurttio P., Komulainen H., Hakala E., Kahelin H. and Pekkanen J. (1998). Urinary excretion of arsenic species after exposure to arsenic present in drinkingwater. *Archives of Environmental Contamination and Toxicology*, **34**, 297–305.

Kurttio P., Pukkala E., Kahelin H., Auvinen A. and Pekkanen J. (1999). Arsenic concentrations in well water and risk of bladder and kidney cancer in Finland. *Environmental Health Perspectives*, **107**(9), 705–710.

Kuwabara J. S., Chang C. C. Y. and Pasilis S. P. (2003). Effects of benthic flora on arsenic transport in Whitewood Creek, South Dakota. In: Toxic Substances in Surface Waters and Sediments: a Study to Assess the Effects of Arsenic-Contaminated Alluvial Sediment in Whitewood Creek, South Dakota, J. S. Kuwabara and C. C. Fuller (eds.), United States Geological Survey (USGS) Professional Paper 1681. United States Department of the Interior, USGS, Reston, VA. See: pubs.usgs.gov/pp/pp1681/ (accessed 7 July 2012).

Kyodo World News Service. (2003). Ibaraki detects prewar poison gas. Kyodo World News Service, 15 May. Noted in USA CHPPM Health Information Operations (HIO) Update, 18 April 2003, p. 12; http://www.scribd.com/doc/1739390/US-Army-HIOWEEKLY UPDATE041803 (accessed 7 July 2012).

Kyoung-Woong K. and Ilwon K. (2003). Remediation of As-Contaminated Soils in the Vicinity of Myungbong Au Mine, Korea. Kwangju Institute of Science and Technology, Department of Environmental Science and Engineering, Soil Environment Laboratory.

Lacayo M. L., Cruz A., Calero S., Lacayo J. and Fomsgaard I. (1992). Total arsenic in water, fish, and sediments from Lake Xolotlan, Managua, Nicaragua. *Bulletin of Environmental Contamination and Toxicology*, **49**(3), 463–470.

Lado L., Polya D. A., Winkel L., Berg M. and Hegan A. (2008). Modelling arsenic hazard in groundwater in Cambodia: a geostatistical approach using ancillary data. *Applied Geochemistry*, **23**, 3010–3018.

Lai M. S., Hsueh Y. M., Chen C. J., Shyu M. P., Chen S. Y., Kuo T. L., Wu M. M. and Tai T. Y. (1994). Ingested inorganic arsenic and prevalence of diabetes mellitus. *American Journal of Epidemiology*, **139**(5), 484–492.

Lalor G., Rattray R., Simpson P. and Vutchkov M. (1999). Geochemistry of an arsenic anomaly in St. *Elizabeth, Jamaica. Environmental Chemistry and Health*, **21**(1), 3–11.

Lambert T. W. and Lane S. (2004). Lead, arsenic, and polycyclic aromatic hydrocarbons in soil and house dust in the communities surrounding the Sydney, Nova Scotia, tar ponds. *Environmental Health Perspectives*, **112**, 35–41.

Latvia Health Statistics and Medical Technology State Agency (2007). Statistical overview on health and health care, Latvia 2006. http://www.vsmtva.gov.lv (accessed 5 July 2012).

Layla Tour (2006). Kamchatka sights. http://trip-to-moscow.com/kamchatka/sightseeing.htm#m27; expired URL, 5 July 2012).

Lazo P., Cullaj A., Arapi A. and Deda T. (2007). Arsenic in soil environments in Albania. *Trace Metals and other Contaminants in the Environment*, **9**, 237–256.

Lederer W. H. and Fensterheim R. J. (eds). (1983). Arsenic: Industrial, Biomedical, Environmental Perspectives. Van Nostrand Reinhold, New York, NY, USA.

Lee I. C. (1976). Study on the concentration of some trace elements in the diet in Korea. *Journal of the Korean Nuclear Society*, **8**, 195–201.

Lee J.-S. and Chon H.-T. (2003). Exposure assessment of heavy metals on abandoned metal mine areas by ingestion of soil, crop plant and groundwater. *Journal de Physique IV (Proceedings)*, **107**, 757.

Lemay T. G. (2003). Arsenic Concentrations in Quaternary Drift and Quaternary-Tertiary Buried Channel Aquifers in the Athabasca Oil Sands (In Situ) Area, Alberta. Publication GEO 2002–04. Alberta Geological Society. http://www.ags.gov.ab.ca/publications/abstracts/GEO_2002_04.html (accessed 5 July 2012).

Lerda D. E. and Prosperi C. H. (1996). Water mutagenicity and toxicology in Rio Tercero, Cordoba, Argentina. *Water Research*, **30**, 819–824.

Leutwein F. and Roesler H. J. (1956). Geochemische Untersuchungen an palaeolithischen und mesozoischen Kohlen Mittel- und Ostdeuschlands. Akademie Verlag, Berlin, GDR.

Lewis D. R., Southwick J. W., Ouellet-Hellstrom R., Rench J. and Calderon R.L. (1999). Drinking water arsenic in Utah: a cohort mortality study. *Environmental Health Perspectives*, **107**, 359–365.

Li D. and An D. (2005). Epidemiology of arsenicosis caused by coal-burning in Guizhou Province. *Chinese Medicine*, **6**, 50–53.

Li Z. (2006). SEPA slams local governments over pollution. People's Daily, 14 September. http://journal.probeinternational.org/2006/09/15/sepa-slams-local-governments-over-pollution/ (accessed 5 July 2012).

Li Z. and Peters S. G. (1998). Comparative Geology and Geochemistry of Sedimentary-Rock-Hosted (Carlin-Type) Gold Deposits in the People's Republic of China and in Nevada, USA. United States Geological Survey (USGS) Open-File Report 98–466. United States Department of the Interior, USGS, Reston, VA. http://pubs.usgs.gov/of/1998/of98–466/ (accessed 5 July 2012).

Liebscher K. and Smith H. (1968). Essential and nonessential trace elements. *Archives of Environmental Health*, **17**, 881–890.

Lillo J., Aracil E. and Iglesias J. A. (2007). Geochemical identification of the sources of natural arsenic in groundwater of the Madrid Basin (Spain). Presented at the 23rd International Applied Geochemistry Symposium, Oviedo (Asturias, North-Western Spain). In: Abstracts 23rd International Applied Geochemistry Symposium, Oviedo. June 2007.

Lin N. F. (1991). Biogeochemistry of arsenic and its toxicity. In: *Medical and Environmental Geochemistry*. Jilin Science and Technology Press, Changchun, China, pp. 186–210.

Lin Y. H., Guo M. X. and Gan W. M. (1997). Mercury pollution from small gold mines in China. *Water, Air, and Soil Pollution*, **97**, 209–221.

Lin M. C. and Lin S. Y. (2005). Risk assessment of arsenic exposure caused by consumption of cultured milkfish (Chanos chanos) from the arsenic-contaminated area in southwestern Taiwan. *Bulletin of Environmental Contamination and Toxicology*, **83**, 125–129.

Lin H. T., Wong S. S. and Li G. C. (2004). Heavy metal content of rice and Shellfish in Taiwan. *Journal of Food and Drug Analysis*, **12**, 167–174.

Lindberg A.-L., Goessler W., Gurzau E., Koppova K., Rudnai P., Kumar R. and Fletcher T. (2006). Arsenic exposure in Hungary, Romania and Slovakia. *Journal of Environmental Monitoring*, **8**, 203–208.

Linge K. L. and Oldham C. E. (2002). Arsenic remobilization in a shallow lake: the role of sediment resuspension. *Journal of Environmental Quality*, **32**, 822–828.

Lipfert G. and Reeve A. (2004). Characterization of Three Water Types in a Fractured Schist, Highly Fractured Watershed in Maine. University of Maine, Orono, ME, USA. http://clu-in.org/products/siteprof/2004fracrockconf/cdr_pdfs/indexed/group1/638.pdf (accessed 5 July 2012).

Liu J., Zheng B., Aposhian H. V., Zhou Y., Chen M. L., Zhang A. and Waalkes M. P. (2002). Chronic arsenic poisoning from burning high-arsenic-containing coal in Guizhou. *Environmental Health Perspectives*, **110**(2), 119–120.

Living on Earth (2007). Rice and arsenic? High arsenic levels in rice may be caused by pollution from cotton farming. *Living on Earth*, 23 March. http://www.organicconsumers.org/articles/article_4969.cfm (accessed 5 July 2012).

Lo M.-C., Hsen Y.-C. and Lin K.-K. (1977). Second Report on the Investigation of Arsenic Content of Underground Water in Taiwan. Taiwan Provincial Institute of Environmental Sanitation, Taichung, Taiwan, China.

Loebenstein J. R. (1994). The Materials Flow of Arsenic in the United States. Information Circular 9382. United States Department of the Interior, Washington, DC.

Loiselle M. C., Marvinney R. G. and Smith A. E. (2002). Arsenic in ground water wells in Maine. In: Arsenic in New England: Proceedings of a Conference, May 29–31, 2001, Manchester, New Hampshire. New Hampshire Consortium on Arsenic, Dartmouth, NH.

Lopez B. V. (2003). Arsenic Mount Apo. INQ7.net, 22 October. http://www.accessmylibrary.com/coms2/summary_0286-24762799_ITM (accessed 5 July 2012).

López D.L., Bundschuh J., Birkle P., Armienta M. A., Cumbalg L., Sracek O. *et al.* (2012). Arsenic in geothermal waters of volcanic-magmatic systems of Latin America. *Science of the Total Environment*, **429**, 57–75. doi:10.1016/j.scitotenv.2011.08.043

Louisiana-El Salvador Partners of the Americas (2006). Chronology of Louisiana-El Salvador Partners of the Americas Projects, Years 2002–2006. (Available from Gene L. Jefferies., President jeff8350@bellsouth.net).

Loukola-Ruskeeniemi K. (2003). Naturally high arsenic contents in groundwater in areas with abundant gold occurrences, Finland [poster]. In: Symposium 1 (SYM 1): Arsenic in Soil and Groundwater Environment: Biogeochemical Interactions. Final poster papers. 15–19 June. www.gtk.fi/ (accessed 5 July 2012).

Loukola-Ruskeeniemi K. and Lahermo P. (2004). Arsenic in Finland: distribution, environmental impacts and risks. *Geological Survey of Finland*, **13**, 18.

Loy D., Fütterer H., Jüttermann P. and Reiche D. (2004). Energy-policy Framework Conditions for Electricity Markets and Renewable Energies 21 Country Analyses June 2004. Deutsche Gesellschaft für Technische Zusammenarbeit (GTZ) GmbH, Division Environment and Infrastructure, Eschborn Germany. http://www2.gtz.de/dokumente/bib/04-0110.pdf (accessed 6 July 2012).

Lu X. and Zhang X. (2005). Environmental geochemistry study of arsenic in Western Hunan mining area, P.R. China. *Environmental Geochemistry and Health*, **27**(4), 313–320.

Luo F. J., Luo Z. D. and Ma L. (1995). A study on the relationship between drinking water with high arsenic content and incidence of malignant tumours in Heihe village, western part of Huhehot, Inner Mongolia. *Zhonghua Liu Xing Bing Xue Za Zhi*, **16**(5), 289–291 (in Chinese).

Luo K., Zhang X., Chen C. and Lu Y. (2004). Estimate of Arsenic Emission Amount from the Coal Power stations in China. http://d.wanfangdata.com.cn/Periodical_kxtb-e200420013.aspx.

Luo Z.-D., Zhang Y. and Ma L. (1993). Epidemiological survey of chronic arsenic poisoning at Tie Mengeng and Zhi Jiliang villages in Inner Mongolia. *Chinese Public Health*, **9**(8), 347–348.

Luo Z.-D., Zhang Y. M. and Ma L. (undated). Chronic arsenicism and cancer in Inner Mongolia. Consequences of well-water levels greater than 50 μg/L. http://www.physics.harvard.edu/~wilson/arsenic/countries/inner_mongolia/inner_mongolia_paper1.html (accessed 7 July 2012).

Luo Z. D., Zhang Y. M., Ma L., Zhang G. Y. and He X. (1997). Chronic arsenicism and cancer in Inner Mongolia: consequences of well-water arsenic levels greater than 50 μg/l. In: Arsenic Exposure and Health Effects. C. O. Abernathy, R. L. Calderon and W. R. Chappell (eds) , Chapman & Hall, New York, NY, USA, pp. 55–68.

Luong T. V., Guifan S., Liying W. and Dianjun S. (2004). Endemic chronic arsenic poisoning: China. In: *People-Centred Approaches to Water and Environmental Sanitation*. Proceedings of the 30th Water, Engineering and Development Centre (WEDC) International Conference, Vientiane, Lao People's Democratic Republic, pp. 586–589.

Lyday T. Q. (2003). The Mineral Industry of Afghanistan. United States Department of the Interior, United States Geological Survey. http://minerals.er.usgs.gov/minerals/pubs/country/2003/afmyb03.pdf).

Lynn G. (2011). Contaminated "Zam Zam" holy water from Mecca sold in the UK. BBC Online, 5 May 2011. http://www.bbc.co.uk/news/uk-england-london-13267205#story_continues_1

Lyons M. (2004). A case study in multinational corporate accountability: Ecuador's indigenous peoples struggle for redress. *Denver Journal of International Law and Policy*, **32**, 701.

Ma L. Q., Komar K. M., Tu C., Zhang W., Cai Y. and Kennelley E. D. (2001). A fern that hyperaccumulates arsenic. *Nature (London)*, **409**, 579.

MacDonald & Partners. (1990). Hydrogeological Map of Zambia. 1:1 500 000 Scale. Philip Print Ltd, London, UK.

MacKenzie D. (1998). Plumbing the depths of the Cold War. *New Scientist*, **14**(2121), 20, 14 February.

MacRae J. D. (2005). Microbes and arsenic contamination of groundwater in Maine: is there a link? Annual International Conference on Soils, Sediments and Water. University of Massachusetts, Amherst, MA, USA. http://scholarworks.umass.edu/soilsproceedings/vol11/iss1/8/ (accessed 5 July 2012).

Madeic H., Cepero E., Mozziconacci D., Rivero S. and Alvarez J. A. (1999). The removal of arsenic. In: Proceedings, ISWA XXII World Congress and Exhibition, Buenos Aires, Argentina. International Solid Waste Association, Vienna, Austria.

Madhavan N. and Subramanian V. (2000). Sulphide mining as a source of arsenic in the environment. *Current Science*, **78**, 702–709.

Mafa B. and Vogel H. (2004). Effects of mining and urban expansion on groundwater quality in francistown, Botswana. In: Water Resources of Arid Areas: Proceedings of the International Conference on Water Resources of Arid and Semi-Arid Regions of Africa, T. R. Chaoka, D. Stephenson and E. M. Shemang (eds), Garborone, Botswana, 3–6 August 2004. Taylor & Francis, London, pp. 133–141.

Mahmood S. N., Naeem S., Siddiqui I. and Khan F. R. (1998). Studies on physico-chemical nature of ground water of Korangd Landhi (Karachi). *Journal of the Chemical Society of Pakistan*, **19**, 42–48.

Makeyevka (Makiivka). (2012). Official site of Makeyevka City Council, Executive Committee and City Head. http://www.makeyevka.dn.ua/eng/ (accessed 5 July 2012).

Makhoukh M., Sbaa M., Berrahou A. and Vanclooster M. (2011). Assessing water and sediment quality of the Moulouya River (East-Morocco). *European Journal of Scientific Research*, **61**(4), 508–523.

Malaysian Tin Bulletin. (2004). Opening up more mines, quarries. *Malaysian Tin Bulletin*, **15**(8), 2.

Malaysian Tin Bulletin. (2005). Production by mining methods. *Malaysian Tin Bulletin*, **16**(2), 8.

Manar M. C. (2005). Kidapawan City water free from arsenic poisoning, says LGU. MindaNews, 29 August. http://www.afrim.org.ph/Archives/2005/MindaNews/August/30/Kidapawan%20City%20water%20free%20from%20arsenic%20poisoning,%20says%20LGU.htm (accessed 5 July 2012).

Manful G. A. and Verloo M. (1989). Environmental effects of arsenic contamination around a gold smelter in Ghana. In: Proceedings, International Conference on Heavy Metals in the Environment. WHO, Geneva, **2**, 329–333.

Mango H. (2005). Arsenic contamination in groundwater in southwestern Vermont. *Geological Society of America, Abstracts with Programs*, **37**(1), 80. http://gsa.confex.com/gsa/2005NE/finalprogram/abstract_83266.htm).

Marchettini N., Donati A., Pulselli F., Bastianoni S. and Tiezzi E. (2002). High soil as level in the area of a plant for pyrite processing: natural anomaly vs pollution hypothesis [poster]. 17th World Congress of Soil Science, Bangkok, Thailand, 14–21 August (Symposium No. 54, Paper No. 2268). citeseerx.ist.psu.edu/viewdoc/download?doi=10.1.1.10.rep (accessed 9 July 2012).

Marin A. R., Masscheleyn P. H. and Patrick W. H., Jr (1992). The influence of chemical form and concentration of arsenic on rice growth and tissue arsenic concentration. *Plant and Soil*, **139**, 175–183.

Marine Biological Association. (2003). Characterisation of European Marine Sites. The Severn Estuary: Possible Special Area of Conservation, Special Protection Area. Marine Biological Association Occasional Publication No. 13. Marine Biological Association, Plymouth, Devon, UK. http://www.mba.ac.uk/NMBL/publications/charpub/pdf/Severn_Estuary.pdf (accessed 5 July 2012).

Mariner R. H. and Willey L. M. (1976). Geochemistry of thermal waters in Long Valley, Mono County, California. *Journal of Geophysical Research*, **81**(B5), 792–800.

Marshall M. (2006). Arsenic poisoning, well water: Burkina Faso. International Society for Infectious Diseases, ProMed-mail post, 30 December, Archive Number: 20061230.3651. http://www.promedmail.org/ (accessed 5 July 2012).

Marszalek H. and Wasik M. (2000). Influence of arsenic-bearing gold deposits on water quality in Zloty Stok mining area (SW Poland). *Environmental Geology*, **39**(8), 888–892.

Martin A. (1999). Hidrogeologia de la provincia de Santiago del Estero. Ediciones del Rectorado, Universidad Nacional de Tucumán, Tucumán, Argentina.

Martin K. (2004). Fiscal year 2004 security assistance legislation and funding allocations. *DISAM Journal of International Security Assistance Management*, **26**(3), 1–60. http://www.disam.dsca.mil/pubs/Archives/Journal%2026-3.pdf (accessed 5 July 2012).

Mased V. G. and Masie M. (2006). Spatial distribution of arsenic in groundwater, a GIS and geostatistics approach: Gomoti Valley, Lower Okavango Delta, Botswana. Presented at the 2nd International Conference on Groundwater for Sustainable Development, Jawaharlal Nehru University (INU), New Delhi, India, 1–4 February (contact: masedi_vg@yahoo.co.uk).

Mastalerz M., Hower J. C., Drobniak A., Mardon S. M. and Lis G. (2004). From in-situ coal to fly-ash: a study of coal mines and power plants from Indiana. *International Journal of Coal Geology*, **59**, 171–192.

Mate K. (1998). Boom in Ghana's golden enclave. Major new investments boost output, but environmental concerns are growing. *Africa Recovery*, **11**(3).

Matschullat J., Borba R. P., Deschamps E., Figueiredo B. R., Gabrio T. and Schwenk M. (2000). Human and environmental contamination in Iron Quadrangle, Brazil. *Applied Geochemistry*, **15**, 181–190.

Matschullat J. and Deschamps E. (2004). Arsenic hazards to humans, plants, and animals from gold mining. *Reviews of Environmental Contamination and Toxicology*, **180**, 133–165.

Maurer D. K., Plume R.W., Thomas J. M. and Johnson A. K. (1996). Water Resources and Effects of Changes in Ground-Water use Along the Carlin Trend, North-Central Nevada. United States Geological Survey (USGS) Water-Resources Investigations Report 96-4134. United States Department of the Interior, USGS, Reston, VA. http://pubs.usgs.gov/wri/1996/4134/report.pdf (accessed 9 July 2012).

Mazumder D. N. G. (2001). Arsenic and liver disease. *Journal of the Indian Medical Association*, **99**(6), 314–315, 318–320.

Mazumder D. N. G. (2005). Mitigation activity forarsenicosis. DNGM Research Foundation (PowerPoint presentation; available from dngm@apexmail.com or guhamazumder@yahoo.com).

Mazumder D. N. G., Chakraborty A. K., Ghose A., Gupta J. D., Chakraborty D. P. and Dey S. B. (1988). Chronic arsenic toxicity from drinking tubewell water in rural Bengal. *Bulletin of the World Health Organziation*, **66**(4), 499–506.

Mazumder D. N. G., Das Gupta J., Chakraborty A. K., Chatterjee A., Das D. and Chakraborty D. (1992). Environmental pollution and chronic arsenicosis in South Calcutta. *Bulletin of the World Health Organization*, **70**(4), 481–485. http://whqlibdoc.who.int/bulletin/1992/Vol70-No4/bulletin_1992_70%284%29_481-485.pdfaccessed 9 July 2012).

Mazumder D. N. G., Das Gupta J. and Santra A. (1997). Non-cancer effects of chronic arsenicosis with special reference to liver damage. In: Arsenic Exposure and Health Effects, C. O. Abernathy, R. L. Calderon and W. R. Chappell (eds), Chapman & Hall, London, UK.

Mazumder D. N. G., Haque R., Ghosh N., De Binay K., Santra A., Chakraborty D. and Smith A. (1998). Arsenic levels in drinking water and the prevalence of skin lesions in West Bengal, India. *International Journal of Epidemiology*, **27**(5), 871–877.

Mazumder D. N. G., Haque R., Ghosh N., De Binay K., Santra A., Chakraborty D. and Smith A. (2000). Arsenic in drinking water and the prevalence of respiratory effects in West Bengal, India. *International Journal of Epidemiology*, **29**(6), 1047–1052.

MBendi (2004). Coal Mining in South Africa: Overview. MBendi Information Services Ltd, Cape Town, South Africa. http://www.mbendi.co.za/indy/ming/coal/af/sa/p0005.htm (accessed 5 July 2012).

MBendi (2007). Armenia: An Overview. MBendi Information Services Ltd, Cape Town, South Africa. http://www.mbendi.com/land/as/am/p0005.htm (accessed 5 July 2012).

McArthur J. M., Ravenscroft P., Safiulla S. and Thirlwall M. F. (2001). Arsenic in groundwater: testing pollution mechanisms for sedimentary aquifers in Bangladesh. *Water Resources Research*, **37**, 109–117.

McGoff J. (2007). State Advances Arsenic Remediation at Orange Elementary School. Indiana Finance Authority, Indianapolis, IN, USA. http://www.in.gov/ifa/files/031407OrangeElemArsenic.pdf (accessed 5 July 2012).

McLaren S. J. and Kim N. D. (1995). Evidence for a seasonal fluctuation of arsenic in New Zealand's longest river and the effect of treatment on concentrations in drinking water. *Environmental Pollution*, **90**(1), 67–73.

McVickar L. (2002). Pownal Tannery Superfund site, Pownal, Vermont. Final Remedial Investigation. Vol. 1. Prepared by Metcalf & Eddy, Inc., Wakefield, MA, USA (EPA Contract No. 68-W6-0042; EPA Work Assignment No. 122-RICO-01N9; http://www.epa.gov/ne/superfund/sites/pownal/34939.pdf (accessed 5 July 2012).

Mead M. N. (2005). Arsenic: in search of an antidote to a global poison. *Environmental Health Perspectives*, **113**(6), A378–A387.

Medilinks. (2006). Burkina Faso: Privatisation Threatens Small-Scale Miners. Medilinks, Friday 13 October 2006. Medilinks, New York, NY. http://www.irinnews.org/report.asp?ReportID=55952 (accessed 5 July 2012).

Meharg A. A. (2007). Elevated Arsenic Levels Found in Rice Grown in the South Central States of the USA. European Virtual Institute for Speciation Analysis (EPA Contract No. 68-W6-0042; EPA Work Assignment No. 122-RICO–01N9. http://www.speciation.net/Public/News/2007/03/07/2699.html (accessed 5 July 2012).

Meharg A. A., Williams P. N., Adomako E., Lawgali Y. Y., Deacon C. and Villada A. (2009). Geographical variation in total and inorganic arsenic content of polished (white) rice. *Environmental Science and Technology*, **43**, 1612–1617.

MEND NigerDelta.com 2012. http://mendnigerdelta.com/ (to access current news).

Meng X., Korfiatis G. P., Christodoulatos C. and Bang S. (2001). Treatment of arsenic in Bangladesh well water using a household co-precipitation and filtration system. *Water Research*, **35**(12), 2805–2810.

MENR (Ministry of Environment and Natural Resources). (2005). Rapid Assessment of Tsunami Damage to Human Settlements and Associated Contamination Hazards in the Coastal Zone of Sri Lanka. MENR, Battaramulla, Sri Lanka.

Mensah G. G. (1976). Water Quality and Pollution Survey of Inland and Coastal Water of Ghana. Council of Scientific and Industrial Research, Water Resources Research Unit, Accra, Ghana.

Meranger J. C. and Subramanian K. S. (1984). Arsenic in Nova Scotian groundwater. *Science of the Total Environment*, **39**, 49–55.

Mexican Government. (2000a). Modificación a la Norma Oficial Mexicana NOM-127-SSA1-1994, Salud Ambiental. Agua para uso y consumo humano. Límites permisibles de calidad y tratamiento a que debe someterse el agua para su potabilización. Diario Oficial de la Federación, 22 November 2000, pp. 1–5.

Mexican Government. (2000b). Translation of Official Mexican Standard NOM-018-STPS 2000, Appendices C and D. http://www.ilpi.com/msds/other/mexico/nom18appcd.pdf).

Meza M. M., Kopplin M. J., Burgess J. L. and Gandolfi A. J. (2004). Arsenic drinking water exposure and urinary excretion among adults in the Yaqui Valley, Sonora, Mexico. *Environmental Research*, **96**, 119–126.

Michel P., Chiffoleau J. F., Averty B., Auger D. and Chartier E. (1999). High resolution profiles for arsenic in the Seine estuary. Seasonal variations and net fluxes to the English Channel. *Continental Shelf Research*, **19**, 2041–2061.

Midžic S. and Silajdžic I. (2005a). Bosnia and Herzegovina. In: Contemporary Reviews of Mine Water Studies in Europe, C. Wolkersdorfer and R. Bowell (eds), Springer Verlag, Berlin, Germany, pp. 68–70 (Mine Water and the Environment, 24). http://www.springerlink.com.libproxy.mit.edu/content/ghv047g061803222/fulltext.pdf (accessed 5 July 2012).

Midžic S. and Silajdžic I . (2005b). Macedonia. In: Contemporary Reviews of Mine Water Studies in Europe, C. Wolkersdorfer and R. Bowell (eds), Springer Verlag, Berlin, Germany (Mine Water and the Environment, 24).

Mindat.org (2007a). Matsuo mine, Kijoh-cho, Miyazaki Prefecture, Kyushu Island, Japan. http://www.mindat.org/loc-69718.html (accessed 5 July 2012).

Mindat.org (2007b). Uzbekistan. http://www.mindat.org/loc-47392.html (accessed 5 July 2012)

Mindat.org (2007c). Locality regional search: localities within 'China', Guizhou Province. http://www.mindat.org/lsearch.php? tloc=China (accessed 5 July 2012).

Mindat.org (2007d). BaterMines (Batra Mines), Bater-Martanesh District, Mat District, DiberCounty. http: //www.mindat.org/loc-8580.html (accessed 5 July 2012).

Mindat.org (2007e). Bulqiza Complex (Bulqiza Massif), Bater-Martanesh District, Mat District, Diber County. http://www.mindat.org/loc-30761.html (accessed 5 July 2012).

Mining.com (2011). http://www.mining.com/2011/12/30/outotec-offers-new-partial-roasting-process-to-purify-contaminated-copper-and-gold-concentrates/ (accessed 25 March 2012).

Mining Review Africa (2003). Construction of Niger's first gold mine underway. *Mining Review Africa*, **34**(16). http://www.miningreview.com/magazine_archives (accessed 5 July 2012).

Ministry of Environment of Georgia & UNDP (2003). National assessment report for sustainable development (NARDS): 2001–2002. Draft. Ministry of Environment of Georgia & UNDP.

Minnesota Department of Health (2001a). The Minnesota Arsenic Study (MARS). Minnesota Department of Health, Division of Environmental Health, with support from the United States Agency for Toxic Substances and Disease Registry. http://www.health.state.mn.us/divs/eh/hazardous/topics/arsenicstudy.pdf

Minnesota Department of Health (2001b). Well management: arsenic in well water. http://www.health.state.mn.us/divs/eh/wells/waterquality/arsenic.html

Minnesota Pollution Control Agency (1998). Arsenic in Minnesota's ground water. Ground Water, November 2. http://www.pca.state.mn.us/water/groundwater/pubs/arsenic.pdf

Mirecki J. (2004). ASR Water Quality Trends in Southern Florida. Aquifer Storage Recovery IV Forum, Tampa, FL, USA. American Ground Water Trust, Aquifer Storage Recovery Programs, Tampa, Florida.

Mironova G. D. and Zotov A. V. (1980). Solubility studies of the stability of As(III) sulfide complexes. *Geochemistry International*, **17**, 46–54.

Mironova G. D., Zotov A. V. and Gulko N. I. (1984). Determination of the solubility of orpiment in acid solutions at 25–150 °C. *Geochemistry International*, **21**, 53–59.

Mironova G. D., Zotov A. V. and Gulko N. I. (1990). The solubility of orpiment in sulphide solutions at 25–150T and the stability of arsenic sulphide complexes. *Geochemistry International*, **27**, 61–73.

Mississippi Department of Environmental Quality (2002). Risk evaluation Procedures for voluntary cleanup and redevelopment of Brownfield sites. http://.deq.state.ms.us/mdeq.nsf/pdf/GARD_brownfieldrisk/$File/Proced.pdf?openelement. (accessed 5 July 2012).

Mitchell P. and Barre D. (1995). The nature and significance of public exposure to arsenic: a review of its relevance to south-west England. *Environmental Geochemistry and Health*, **17**(2), 57–82.

Mizuta N., Mizuta M., Ito F., Ito T., Uchida H., Watanabe Y., Akama H., Murakami T., Hayashi F., Nakamura K., Yamaguchi T., Mitzuia W. and Oishi Matsamura H. (1956). An outbreak of acute arsenic poisoning caused by arsenic contaminated soysauce (shoyu): a clinical report of 220 cases. *Bulletin of the Yamaguchi Medical School*, **4**(2–3), 132–149.

Mkandawire M. and Dudel E. G. (2005). Accumulation of arsenic in Lemna gibba L. (duckweed) in tailing waters of two abandoned uranium mining sites in Saxony, Germany. *Science of the Total Environment*, **236**, 81–89.

Mkandawire M., Taubert B. and Dudel E. G. (2004). Capacity of Lemna gibba L. (duckweed) for uranium and arsenic phytoremediation in mine tailing waters. *International Journal of Phytoremediation*, **6**(4), 347–362.

MMSD (Mining, Minerals and Sustainable Development) (2001). Impacts of Mining and Minerals Processing on the Biophysical Environment in Southern Africa: Final Report. International Institute for Environment and Development, MMSD Project. http://www.docin.com//p-99685076.html (accessed 5 July 2012).

Mobbs P. M. (2010). The Mineral Industry of Nigeria. United States Department of the Interior, United States Geological Survey, Reston, VA. http://minerals.usgs.gov/minerals/pubs/country/2010/myb3-2010-ni.pdf (accessed 5 July 2012)

Mobbs P. M. (2010). The Mineral Industry of Zimbabwe. United States Department of the Interior, United States Geological Survey, Reston, VA.

Mok W. M., Riley J. A. and Wai C. M. (1988). Arsenic speciation and quality of groundwater in a lead-zinc mine, Idaho. *Water Research*, **22**(6), 769–774.

Molden D. J., Keller L. and Sakthivadivel R. (2001). Hydronomic zones for developing basin water conservation strategies. (Research Report 56). International Water Management Institute, Colombo, Sri Lanka.

Moore J. N. (1994). Contaminant mobilization resulting from redox pumping in a metal-contaminated river-reservoir system. In: Environmental Chemistry of Lakes and Reservoirs, L. A. Baker (ed.), American Chemical Society, Washington, DC, pp. 451–471.

Mora M. A., Mora M. A., Papoulias D., Nava I. and Buckler D. A. (2001). A comparative assessment of contaminants in fish from four resacas of the Texas, USA-Tamaulipas, Mexico border region. *Environment International*, **27**(1), 15–20.

Morgan B. (2006). Arrests after China river polluted by arsenic compound. Agence France-Presse, 12 September. http://www.terradaily.com/reports/Arrests_After_China_River_Polluted_By_Arsenic_Compound_999.html (accessed 5 July 2012)

Morrison J. L. (1969). Distribution of arsenic from poultry litter in broiler chickens, soil, and crops. *Journal of Agronomy and Food Chemistry*, **17**, 1288–1290.

Mortefolio M. and Ridenour J. A. (2006). Derivation of site-specific arsenic background in soil: a case study. In: Annual International Conference on Soils, Sediments and Water. University of Massachusetts, Amherst, MA, USA. http://scholarworks.umass.edu/cgi/viewcontent.cgi?article=1034&context=soilsproceedings (accessed 5 July 2012)

Mosaferi M., Yunesian M., Mesdaghinia A., Nadim A., Nasseri A. and Mahvi A. H. (2005a). Arsenic Occurrence in Drinking Water of I.R. of Iran: The Case of Kurdistan Province. Tehran University of Medical Sciences, School of Public Health, Department of Environmental Health Engineering, Tehran, Iran. www.bvsde.paho.org/bvsacd/arsenico/arsenic/Mosaferi.pdf (accessed 5 July 2012)

Mosaferi M., Yunesian M., Mesdaghinia A., Nadim A., Nasseri A. and Mahvi A. H. (2005b). Correlation between arsenic concentration in drinking water and human hair. *Iranian Journal of Environmental Health, Science and Engineering*, **2**(1), 13–21. http://www.bvsde.paho.org/bvsacd/arsenico/arsenic/Mosaferi.pdf (accessed 9 July 2012).

Motzer W. E. and Will R. K. (2007). Arsenic in California's mother lode belt. *Geological Society of America*, Abstracts with Programs, **39**(6), 519 (Paper No. 192–12).

Mount Apo National Park (2005). Mount Apo National Park (MANP): park management plan. Manila. p. 74.

Moutsatsou A., Gregou M., Liokalou M., Matsas D. and Protonotarios V. (2003). Mobilization of metals and metalloids from contaminated soil and readsorption on synthesized zeolites and zeolite fly-ash mixtures. In: Proceedings of the 8th International Conference on

Environmental Science and Technology. Vol. A. Lemnos Island, Greece, 8–10 September, pp. 642–649. www.gnest.org/cest8/8cest_papers/abstracts./p33_Moutsatsou.pdf (accessed 5 July 2012).

Movement for Health of La Oroya (2006). Peruvian congressman denounces Doe Run's 'subversive attitude'. 13 May. www.manosperu.org/leer.php/106 (accessed 5 July 2012).

Mozziconacci D., Madiec H. and Cepero E. (1998). Arsenic Reduction by Coagulation on a Filter. (Review No. 40). Ingeniería Sanitaria y Ambiental, Buenos Aires, Argentina (in Spanish).

Mudur G. (2000). News: half of Bangladesh population at risk of arsenic poisoning. *British Medical Journal*, **320**, 822.

Mueller G. A. (1997). The Nevoria gold skarn deposit in Archean iron-formation, Southern Cross greenstone belt, Western Australia; I, Tectonic setting, petrography, and classification. *Economic Geology*, **92**(2), 181–209.

Mukhopadhyay D. and Sanyal S. K. (2005). Transport of aqueous arsenic across and down some surface soils of West Bengal, India [poster]. 17th World Congress of Soil Science, Bangkok, Thailand, 14–21 August (Symposium No. 01; Paper No. 469).

Munasinghe D. R., Rajasuriya K. and Thenabadu P. N. (1969). Polyneuropathy following acute arsenic poisoning. *Ceylon Medical Journal*, **14**(2), 85–89.

Murcott S. (2001). A comprehensive review of low-cost, tubewell water treatment technologies for arsenic removal. In: Arsenic Exposure and Health Effects IV, W. R. Chappell, C. O. Abernathy and R. L. Calderon (eds), Elsevier Science Ltd, Amsterdam, Netherlands, pp. 419–429.

Murdur G. (1996). Arsenic poisons 220 000 in India. *British Medical Journal*, **313**, 9.

Mutandwa B. and Mutandwa E. (2007). An appraisal of environmental management systems in small scale mines of the Midlands Province of Zimbabwe: the case of Indarama Mine. *Journal of Sustainable Development in Africa*, **9**(1).

Naicker K., Cukrowska E. and McCarthy T. S. (2003). Acid mine drainage arising from gold mining activity in Johannesburg, South Africa and environs. *Environmental Pollution*, **122**(1), 29–40.

Naidu R. (2000). Arsenic contamination of rural groundwater supplies in Bangladesh, India and Australia: implications for soil quality, animal and human health. In: Arsenic in sedimentary aquifers. 31st International Geological Conference, Rio de Janeiro, Brazil.

Naidu R., Smith E., Owens G. and Bhattacharya P. (2006). Managing Arsenic in the Environment: From Soil to Human Health. CSIRO Publishing, Melbourne, Australia.

Nakadaira H., Nakamura K., Mutoh K., Yamamoto M. and Katoh K. (2000). Arsenic residues in well water 36 y endemic arsenic poisoning. Arch Environ Health. 55(5): 364.

Nakao M. (1960). A study of the arsenic content in daily food consumption in Japan. *Osaka City Medical Journal*, **9**, 541–571.

Nakayama S. M. M., Ikenaka Y., Hamada K., Muzandu K., Choongo K., Teraoka H., Mizuno N. and Ishizuka M. (2010). Metal and metalloid contamination in roadside soil and wild rats around a Pb-Zn mine in Kabwe, Zambia. *Environmental Pollution*, **159**(1), 175–181.

Naranjo-Pulido A., Romero-Schmidt H., Mendez-Rodriguez L., Acosta-Vargas B. and Ortega-Rubio A. (2002). Soil arsenic contamination in the Cape Region, B.C.S., Mexico. *Journal of Environmental Biology*, **23**(4), 347–352. http://www.bashanfoundation.org/ortega/ortegasoilarsenic.pdf (accessed 9 July 2012).

National Coalition for Alternatives to Pesticides (2007). Arsenic and old waste. minerals.usgs.gov/minerals/mflow/d00–0195/ (accessed 6 July 2012).

National Statistical Coordination Board (NSCB) (2005). Marinduque. *NSCB Fact Sheet*, April 8.

Naumov S. (2007) Nobody's river, everyone's headache. *Asia Water Wire*. www.cawater-info.net/rivertwin/documents/pdf/naumov_e.pdf (accessed 6 July 2012).

Naylor H., Turner P. and Vaughan D. J. (1989). Genetic studies of red bed mineralization in the Triassic of the Cheshire Basin, northwest England. *Journal of the Geological Society*, **146**(4), 685–699.

Neal C. (1996). Trace element concentrations in the major rivers entering the Humber estuary, NE England. *Journal of Hydrology*, **182**, 37–64.

Neff J. M. (1997). Ecotoxicology of arsenic in the environment: review. *Environmental Toxicology and Chemistry*, **16**(5), 917–927.

Nelson J. R., Batchelder G. L., Radville M. E. and Albert S. A. (2006). Using multiple lines of evidence to demonstrate that elevated arsenic groundwater concentrations are naturally occurring. In: Annual International Conference on Soils, Sediments and Water. University of Massachusetts, Amherst, MA, USA. http://scholarworks.umass.edu/soilsproceedings/vol12/iss1/7/ (accessed 6 July 2012).

Nemmers S., Ulery A. and Greenlee J. (2006). Characterization of arsenic in the soils of the Mimbres watershed: utilizing remote sensing and GIS techniques. In: New Mexico Water Research Symposium, 15 August 2006 (Poster Session E-2).

Netherlands Environmental Assessment Agency (2004). Metals and Arsenic in Shallow Groundwater in the Netherlands, 1990–2003. Environmental Data Compendium. www.mnp.nl/mnc/i-en-0272.html (accessed 6 July 2012).

News24.com (2004). Drinking water 'health threat'. News24.com. http www.news24.com/./News/Drinking-water-health-threat-20041019 (accessed 6 July 2012).

News-Dominica.com (2000). Finland funding local geothermal study. *The Chronicle*, 9 June. http://www.news-dominica.com/5yearsago/index.cfm?ID=1412; archive no longer available on-line).

New South Wales Department of Natural Resources (2006). Gunnedah Basin Groundwater Source: Report Card. Porous rock groundwater sources: September 2006. (Contact: hcr@cma.nsw.gov.au).

Ngai T. N. K. (2007). Personal communication. Arsenic in Nepal.

NGO (National Geographical Organization) (2001). The Gazetteer of Townships in the I.R. of Iran: Sanandaj Township. NGO, Tehran, Islamic Republic of Iran.

NHMRC (National Health and Medical Research Council) (2004). National Water Quality Management Strategy: Australian Drinking Water Guidelines. Australian Government, NHMRC, Canberra, Australia. http://www.nhmrc.gov.au/_files_nhmrc/publications/attachments/adwg_11_06.pdf?q=publications/synopses/_files/adwg_11_06.pdf (accessed 6 July 2012).

Nickson R. T., McArther J., Burgess W., Ahmed K. M., Ravenscroft P. and Rahman M. (1998). Arsenic poisoning of Bangladesh groundwater [letter]. *Nature*, **395**, 338.

Nickson R. T., McArthur J. M., Ravenscroft P., Burgess W. G. and Ahmed K. M. (2000). Mechanism of arsenic release to groundwater, Bangladesh and West Bengal. *Applied Geochemistry*, **15**, 403–413.

Nickson R. T., McArthur J. M., Shrestha B., Kyaw-Myint T. O. and Lowry D. (2004). Arsenic and other drinking water quality issues, Muzaffargarh District, Pakistan. *Applied Geochemistry*, **20**(1), 55–68.

Nicolli H. B. and Merino M. H. (2002). High contents of F, As and Se in groundwater of the Carcarañá river basin, Argentine Pampean Plain. (Contact via M.P. Etcheverría, e-mail: postmaster@aga.inv.org.ar).

Nicolli H. B. Suriano J. M., Gomez Peral M. A., Ferpozzi L. H. and Balean O. M. (1989). Groundwater contamination with arsenic and other trace-elements in an area of the Pampa, province of Córdoba, Argentina. *Environmental Geology*, **14**, 3–16.

Nieuwenhuijsen M. J. (2003). Association between arsenic exposure from a coal-burning power plant and urinary arsenic concentrations in Prievidza District, Slovakia. *Environmental Health Perspectives*, **111**(7), 889–89.

NigeriaBusinessInfo.com (2005). The new face of Nigeria's oil industry. http://www.chinadialogue.net/article/show/single/en/548-The-new-face-of-Nigeria-s-oil-industry (accessed 6 July 2012).

Nimick D. A. (1994). Arsenic transport in surface and ground water in the Madison and upper Missouri river valleys, Montana. In: Proceedings of the American Geophysical Union Fall Meeting. American Geophysical Union, 247.

Nimick D. A. (1998). Arsenic hydrogeochemistry in an irrigated river valley: A reevaluation. *Ground Water*, **36**(5), 743–753.

Nimick D. A., Moore J. N., Dalby C. E. and Savka M. W. (1998). The fate of arsenic in the Madison and Missouri rivers, Montana and Wyoming. *Water Resources Research*, **34**(11), 3051–3067.

Ninh N. D. (2004). Arsenic mitigation initiatives in Vietnam. In: Interregional conference on water quality. 23–26 November 2004, Taiyuan, China. Inter-Regional Conference on Water: UNU, ESCAP, UNICEF & WHO, Washington, DC. http://health.groups.yahoo.com/group/FluoridePoisoning/message/2843 (accessed 6 July 2012).

Nishikawa Y. (2000). The Caucasus mineral industry. *Mining Magazine*, 1 March. http://www.highbeam.com/doc/1G1–60371035.html (accessed 6 July 2012).

Niu S., Cao S. and Shen E. (1997). The status of arsenic poisoning in China. In: Arsenic Exposure and Health Effects, C. O. Abernathy, R. L. Calderon and W. R. Chappell (eds), Chapman & Hall, London, UK. pp. 78–83

NNASC (Nepal National Arsenic Steering Committee) (2001). Nepal's Interim Arsenic Policy and Preparation Report. NNASC, Kathmandu, Nepal.

Noguchi K. and Nakagawa R. (1969). Arsenic and arsenic-lead sediments from Tamagawa Hot Springs, Akita Prefecture. *Proceedings of the Japan Academy*, **45**, 45–50.

Nordstrom D. K. and Archer D. G. (2002). Arsenic thermodynamic data and environmental geochemistry. In: Arsenic in Groundwater: Geochemistry and Occurrence, A. H. Welch and K. G. Stollenwerk (eds), Kluwer Academic Publishers, Boston, MA, USA.

Norman D. I., Miller G. P., Branvold L., Thomas T., Appiah H., Ayamsegna J. and Nartey R.(undated). Arsenic in Ghana, West Africa, Ground Waters. *United States Department of the Interior*, United States Geological Survey.

Norrgren L., Pettersson U., Orn S. and Bergqvist P. (2000). Environmental monitoring of the Kafue River, located in the Copperbelt, Zambia. *Archives of Environmental Contamination and Toxicology*, **38**, 334–341.

Norvell J. L. S. (1995). Distribution of, Sources of, and Processes of Mobilizing Arsenic, Chromium, Selenium, and Uranium in the Central Oklahoma Aquifer. MS thesis. Colorado School of Mines, Golden, CO, USA.

Nottingham County Council (2003). Sherwood Sandstone: Chapter 7. http://www.nottinghamshire.gov.uk/large-static/mineralslocalplan/depo-chapter-7.pdf.

NRDC (Natural Resources Defense Council) (1997–1999). Arsenic in Selected Bottled Waters. NRDC, New York, NY, USA (www.nrdc.org/water/drinking/bw/figure5.html (accessed 9 July 2012).

NRDC (Natural Resources Defense Council) (2000). Arsenic and Old Laws A Scientific and Public Health Analysis of Arsenic Occurrence in Drinking Water, Its Health Effects, and EPA's Outdated Arsenic Tap Water Standard. NRDC, Washington, DC, USA. http://www.nrdc.org/water/drinking/arsenic/appa.asp (accessed 6 July 2012).

Nriagu J. O. (1983). Arsenic enrichment in lakes near the smelters at Sudbury, Ontario. *Geochimica et Cosmochimica Acta*, **47**, 1523–1526.

Offshore Technology.com (2010). Jubilee fields project. *Offshore-technology.com*, March 2010. http://www.offshore-technology.com/projects/jubilee-field/ (accessed 5 December 2011).

Ogola J. S., Mitullah W. V. and Omulo M. A. (2002). Impact of gold mining on the environment and human health: A case study in the Migori Gold Belt, Kenya. *Environmental Geochemistry and Health*, **24**(2), 141–157.

O'Grady B., Galant M., O'Brien W., VanDerLoop C. and Grodin M. (2003). The arsenic nobody wanted: Now what? A panel discussion on arsenic in communities. Annual International Conference on Soils, Sediments and Water. University of Massachusetts, Amherst, MA, USA. www.umasssoils.com/abstracts2003/Monday/monworkshops.htm (accessed 6 July 2012).

Oklahoma Department of Environmental Quality (2007). *Arsenic and the Blackwell Zinc Smelter site in Blackwell, Oklahoma*. Land Fact Sheet, June. www.deq.state.ok.us/factsheets/land/arsenicbwz.pdf (accessed 6 July 2012).

Oklahoma Office of the Secretary of the Environment (2000). *Governor Frank Keating's Tar Creek Superfund Task Force*. Final Report. Oklahoma City, OK, USA, Office of the Secretary of the Environment, 1 October. http://digitalprairie.ok.gov/cdm/singleitem/collection/stgovpub/id/15026/rec/4 (accessed 6 July 2012).

Oldham C. (2003). Relationships between Sediment Mineralogy and the Release of Contaminants under varying Physical Conditions in Shallow Wetlands. MS thesis. University of Western Australia, Perth, Australia.

Ongley L. K., Armienta A. and Mango H. (2003). Concentrations of heavy metals in soil, Zimapan, México. *Journal de Physique IV (Proceedings)*, **107**, 983.

Ori L. V., Amacher M. C. and Sedberry J. E. Jr. (1993). Survey of the total arsenic content in soils in Louisiana. *Communications in Soil Science and Plant Analysis*, **24**(17), 2321–2332.

O'Shea B. (2005). Arsenic distribution in a coastal aquifer affected by seawater intrusion. *Geological Society of America, Abstracts with Programs*, **37**(7), 170. (Paper No. 69–6).

Oshikawa S. (1997). Re-examination of a Cohort of Subjects with Arsenical Skin Lesions Ten years after Inception. Master's thesis. Prince of Songkla University, Thailand.

Oshikawa S., Geater A. and Chongsuvivatwong V. (2001). Arsenic in Ronphibun, Thailand: past, present and future. International Workshop on Arsenic Pollution of Drinking Water in South Asia and China, Tokyo, 10 March 2001.

Otago University. (2012). Metals in the New Zealand Environment. Department of Geology, Otago University, Dunedin, New Zealand. http://www.otago.ac.nz/geology/research/environmental_geology/metals/arsenic.html (accessed 23 June 2012).

Owen-Joyce S. J. (1984). Hydrology of a Stream-Aquifer System in the Camp Verde Area, Yavapai County, Arizona. Arizona Department of Water Resources Bulletin 3. Geological Survey, United States Department of the Interior, Tucson, Arizona.

Owen-Joyce S. J. and Bell C. K. (1983). Appraisal of Water Resources in the Upper Verde River Area, Yavapai and Coconino Counties, Arizona. Arizona Department of Water Resources Bulletin 2. Arizona Department of Water Resources, Tuscon, Arizona. http://www.azwater.gov/azdwr/Hydrology/Library/Bulletins.htm (accessed 23 June 2012).

Owor M., Hartwig T, Muwanga A., Zachmann D. and Pohl W. (2007). Impact of tailings from the Kilembe copper mining district on Lake George, Uganda. *Environmental Geology*, **51**(6), 1065–1075.

Ozerova N. A., Naboko S. I. and Vinogradov V. I. (1971). Sulphides of mercury, antimony and arsenic, forming from the active thermal spring of Kamchatka and Kuril islands. In: Proceedings of the 7th General Meeting of the International Mineralogical Association. Institute of Geological Ore Deposits, Moscow, USSR, pp. 164–170 (Paper No. 2).

PAHO/ECO. (1993). Environmental Epidemiology: A Project for Latin America and the Caribbean. Pan American Centre for Human Ecology and Health. 525 23rd Street, NW Washington, DC.

Pakistan Council of Research in Water Resources (2003a). Arsenic Contamination in Groundwater of Southern Punjab. Ministry of Science and Technology, Islamabad.

Pakistan Council of Research in Water Resources (2003b). Innovative Low Cost Arsenic Removal Technologies for Developing Countries. Pakistan Ministry of Science and Technology, Islamabad.

Pakistan Council of Research in Water Resources (2004). Arsenic Contamination in Groundwater of Central Sindh. Pakistan Ministry of Science and Technology, Islamabad.

Palache C., Berman H. and Frondel C. (1951). (Dana's) System of Mineralogy, 7th edn. **Vol. II**. John Wiley & Sons, New York, NY, USA.

Palmer C. A. and Lyons P. C. (1996). Selected elements from major minerals from bituminous coal as determined by INAA: implications for removing environmentally sensitive elements from coal. *International Journal of Coal Geology*, **32**, 151–166.

Palumbo B. (2003). A comparison of arsenic bioaccessibility from soils in the vicinity of an abandoned arsenic mine and background agricultural soils in SW England. In: *Symposium 1 (SYM 1): Arsenic in soil and groundwater environment: Biogeochemical interactions*. Final poster papers. 15–19 June.

Panov B. S., Dudik A. M., Shevchenko O. A. and Matlak E. S. (1999). On pollution of the biosphere in industrial areas: the example of the Donets coal basin. *International Journal of Coal Geology*, **40**(2–3), 199–210.

Panov B. S., Shevchenko O. A., Dudik A. M. and Selyakov S. Y. (1988). On geo-ecology of the Donetsk basin. *Izvestiya Vysshikh Uchebnykh Zavedeni, Geologika Razved*, **5**, 138–145. (in Russian)

Papanicolaou C., Kotis T., Foscolos A. and Goodarzi F. (2004). Coals of Greece: a review of properties, uses and future perspectives. *International Journal of Coal Geology*, **58**, 147–169.

Parisio S., Keimowitz A. R., Simpson H. J., Lent A. and Blackman V. (2005). Arsenic-rich iron floc deposits in seeps downgradient of solid waste landfills. In: Annual International Conference on Soils, Sediments and Water. University of Massachusetts, Amherst, MA, USA. http://www.highbeam.com/doc/1P3-1124894361.html (accessed 6 July 2012).

Parsons M. (1999). Jacksonville, Florida, Ocean Dredged Material Disposal Site (ODMDS) benthic surveys 1995 and 1998. United States Environmental Protection Agency, Science and Ecosystem Support Division, Ecological and Assessment Branch, Athens, GA, USA. digitalcorpora.org/corp/nps/files/govdocs1/020/020630.pdf (accessed 6 July 2012).

Partida E. G., Rodriguez V. T. and Birkle P. (1995). Plio-Pleistocene Volcanic History of the Ahuachapán Geothermal System, El Salvador: the Concepcion de Ataco Caldera. Institute de Investigaciones Ekctricas (HE), Departamento Geotermia [cited in Prevost, 2004], Cuernavaca, Morelos, Mexico.

Pasanen T. (2003). Arsenic in various pools in contaminated soils in eastern Finland. In: Symposium 1 (SYM 1): Arsenic in soil and groundwater environment: Biogeochemical interactions. Final poster papers. 15–19 June.

Pavittranon S., Sripaoraya K., Ramchuen S., Kachamatch S. and Puttaprug W. (2003). Laboratory case identification of arsenic in Ronpibul Village, Thailand (2000–2002). *Journal of Environmental Science and Health, Part A*, **38**(1), 213–221.

Pazirandeh A., Barati A. H. and Ghannadi M. (1998). Determination of arsenic in hair using neutron activation. *Applied Radiation and Isotopes*, **49**(7), 753–759.

Peard R. J. (2006). Living with arsenic in Hawai'i: case study of a former sugar cane plantation. In: Annual International Conference on Soils, Sediments and Water. University of Massachusetts, Amherst, MA, USA.

Peng J. T., Hu R., Lin Y. and Zhao J. (2002). Sm-Nd isotope dating of hydrothermal calcites from the Xikuangshan antimony deposit, Central Hunan. *Chinese Science Bulletin*, **47**(13), 1134–1137.

Pentari D., Foscolos A. E. and Perdikatis V. (2004). Trace element contents in the Domeniko lignite deposit, Elassona basin, central Greece. *International Journal of Coal Geology*, **58**, 261–268.

Peplow M. (2005). US rice may carry an arsenic burden: legacy of cotton pesticides might be poisoning crops. *Nature News*. www.nature.com/news/2005/050801/full/050801-5.html (accessed 6 July 2012).

Perlez J. (2004). Jakarta report raises tension over mine. *The New York Times*, 9 November.

Perlez J. and Rusli M. (2004). Spurred by illness, Indonesians lash out at US mining giant. *The New York Times*, 8 September 2004. http://www.nytimes.com/2004/09/08/international/asia/08indo.html (accessed 6 July 2012).

Perry D. D., Cahill T. A. and Eldred R. A. (1997). Long-range transport of North African dust to the eastern United States. *Journal of Geophysical Research*, **102**, 225–228.

Peters L. P. (1977). Summary of the Occurrence of Arsenic and Selected Heavy Metals in Rural Domestic Water Supplies in the York and Charlotte Counties of the Province of New Brunswick, Canada. New Brunswick Department of the Environment, Fredericton, New Brunswick, Canada, p. 27 (Publication No. T7709).

Peters L. P. (1996). Heavy Metals Toxicity, Sampling and Migration in Ground Water and the Integration of Existing Heavy Metal Geochemical Surveys for Environmental Health Purposes in New Brunswick. University of New Brunswick, Fredericton, New Brunswick, Canada, Ground Water Studies Group, p. 264 (unpublished report).

Peters L. P. (1999). A New Brunswick Case Study Investigating Fusion of Geochemical Data for Environmental Health Purposes. University of New Brunswick, Fredericton, New Brunswick, Canada, Unpublished article available from the author: 197 University Avenue, Fredericton, New Brunswick, Canada E3B 4H6, (506) 454–1726, e-mail: lpeters@nbnet.nb.ca.

Peters S. (2007). Arsenic in Pennsylvania Groundwater. The Lehigh Earth Observatory, Bethlehem, PA.

Peters S. C., Blum J. D., Klaue B. and Karagas M. R. (1999). Arsenic occurrence in New Hampshire drinking water. *Environmental Science and Technology*, **33**, 1328–1333. (pubs.acs.org/doi/abs/10.1021/es980999e (accessed 6 July 2012).

Pettersson U. T., Ingri J. and Andersson P. S. (2000). Hydrogeochemical processes in the Kafue River upstream from the Copperbelt mining area, Zambia. *Aquatic Geochemistry*, **6**, 385–411.

Pettry D. E. and Switzer R. E. (2001). Arsenic Concentrations in Selected Soils. Mississippi State University, Division of Agriculture, Forestry and Veterinary Medicine (MAFES Bulletin 1104. http://msucares.com/pubs/bulletins/b1104.pdf).

Pfeifer H.-R., Beatrizotti G., Berthoud J., Lavanch J.-C., De Rossa M., Reymond D., Righetti G., Schlegel C., Schmit V. and Temgoua E. (2002). Natural arsenic-contamination of surface and ground waters in southern Switzerland (Ticino). *Bulletin of Applied Geology*, **7**(1), 81–103.

Pfeifer H.R. and Dold B. (2005). Switzerland. In: Contemporary Reviews of Mine Water Studies in EuropeC. Wolkersdorfer and R. Bowell (eds), Springer Verlag, Berlin, Germany. (Mine Water and the Environment, 24: Supplementary Material).

Pfeifer H.-R. and Zobrist J. (2001). Arsenic in drinking water: also a problem in Switzerland? *EAWAG [Swiss Federal Institute of Aquatic Science and Technology] News*, **53**, 15–17. http://www.eawag.ch/publications/eawagnews/www_en53/en53e_screen/en53e_pfeifer_s.pdf (accessed 6 July 2012).

Philip D. (1979). Africanization, Nationalization and Inequality: Mining Labor and the Copperbelt in Zambian Development. Cambridge University Press, London, UK.

Phillips I. R. (1999). Copper, lead, cadmium and zinc sorption by waterlogged and air-dry soil. *Journal of Soil Contamination*, **8**, 343–364.

Phuong T. D., Chuong P. V., Khiem D. T. and Kokot S. Elemental content of Vietnamese rice Part 1 Sampling, analysis and comparison with previous studies. *Analyst*, **124**, 553–560.

Pinedo M. and Zigarán A. (1998). Hydroarsenicism in Córdoba province, Argentina; updating the risk map and incidence. In: Twenty-sixth Inter–American Congress of Sanitary and Environmental Engineering, Lima, Peru (in Spanish).

Pino P., Atalah E., Amigo H., Silva C., Caceres D., Jara M. and Cumsille F. (2004). Exposure to arsenic in three Chilean cities. In: Third Regional Conference on the Evaluation and Management of Drinking Water Sources Contaminated with Arsenic. Universidad de Chile, Santiago, Chile, 8–11 November 2004.

Piperingos G. N. (1966). Origin of Elements Associated with Uranium in the Cave Hills area Harding County, South Dakota. United States Geological Survey (USGS) Professional Paper 476–B. USGS, United States Department of the Interior, Reston, VA, USA.

Planer-Friedrich B., Armienta M. A. and Merkel B. J. (2001). Origin of arsenic in the groundwater of the Rioverde basin, Mexico. *Environmental Geology*, **40**(10), 1290–1298.

Planet Ark (2002). Workers clean up arsenic after toxic spill in Jingxiu County. http://www.planetark.org/envpicstory.cfm/newsid/ 19085 (accessed 8 May 2007).

Plaska Y. P. (1971). On the trace element contents in sink-float fractions, L'vov-Volyn basin. *Geologii i Geochimika Goryuch. Kopalin*, **61**, 141–196. (in Russian) [cited by Yudovich & Ketris, 2005].

Plumlee G. S., Morton R. A., Boyle T., Medlin J. H. and Centeno J. A. (2000). An Overview of Mining-Related Environmental and Human Health Issues, Marinduque Island, Philippines: Observations from a Joint US Geological Survey–Armed Forces Institute of Pathology Reconnaissance Field Evaluation, 12–19 May 2000. United States Department of the Interior, United States Geological Survey pubs.usgs.gov/of/2000/ofr-00–0397/ofr-00–0397.pdf (accessed 6 July 2012).

Pohn H. A. (2001). Lateral Ramps in the Folded Appalachians and in Overthrust Belts Worldwide: A Fundamental Element of Thrust-Belt Architecture. USGS Bulletin 2163. United States Department of the Interior, United States Geological Survey. pubs.usgs.gov/bul/b2163/ (accessed 6 July 2012).

Pokrovski G., Gout R., Schott J., Zotov A. and Harrichoury J.-C. (1996). Thermodynamic properties and stoichiometry of As(III) hydroxide complexes at hydrothermal conditions. *Geochimica et Cosmochimica Acta*, **60**, 737–749.

Polizzotto M. L., Kocar B. D., Sampson M., Ouch K., Ung M., Oum R., Benner S. G. and Fendorf S. (2005). Arsenic Contamination in the Flood Plain Aquifer of the Mekong River, Cambodia. http://www.rdic.org/PolizzottoGSA2005new.pdf (accessed 6 July 2012).

PollutionWatch. (2004). Water Releases of Arsenic in 2003. Environmental Defence and Canadian Environmental Law Association, Toronto, Ontario.

Polya D. A., Gault A. G., Diebe N., Feldmann P., Rosenboom J. W., Gilligan E., Fredericks D., Milton A. H., Sampson M., Rowland H. A. L., Lythgoe P. R., Jones J. C., Middleton C. and Cooke D. A. (2005). Arsenic Hazard in Shallow Cambodian Groundwaters [Poster]. The Engineering and Physical Sciences Research Council of Manchester University Conference. Manchester University, Engineering and Physical Sciences Research Council, Manchester, UK.

Polya D. A., Gault A. G., Diebe N., Feldmann P., Rosenboom J. W., Gilligan E., Fredericks D., Milton A. H., Sampson M., Rowland H. A. L., Lythgoe P. R., Jones J. C., Middleton C. and Cooke D. A. (2005). Arsenic hazard in shallow Cambodian groundwaters. *Mineralogical Magazine*, **69**, 807–823.

Polyak K. and Hlavay J. (1999). Arsenic Problem in Hungary: Development of a Low Waste Technology. University of Veszprem, Veszprem, Hungary.

Porter S. D., White K. D. and Clark J. R. (1995). Water-Quality Assessment of the Kentucky River Basin, Kentucky: Distribution of Metals and Other Trace Elements in Sediment and Water, 1987–90. United States Geological Survey (USGS) Water-Resources Investigations Report 94–4134. United States Department of the Interior, USGS, Reston, VA, p. 184.

Potseluev A. A., Rikhvanov L. P., Arbuzov Lyapunov P. I., Rubinov I. M., Bazhenov M. I. and Zadorin L. I. (2001). Trace elements in deposits of Jurassic paleovalleys of the Chulym-Yenisei depression (Malinovskoe uranium deposit). *Geologia Geofizika*, **42**(6), 891–899. (in Russian) Abstract: http://library.iem.ac.ru/geo-geop/6-4201.html (accessed 10 July 2012).

PRC (Peoples Republic of China) National Cooperation Group. (2003). Study on distribution of endemic arsenism in China [Taiwan not included]. Institute for Environment Hygiene and Health Related Product Safety, Chinese Center for Disease Control and Prevention, Beijing 100050, China. http://www.ncbi.nlm.nih.gov/pubmed/14963897 (accessed 6 July 2012).

Preul H. C. (1997). Case Study XII-Kingdom of Jordan. In: Water Pollution Control: A Guide to the Use of Water Quality Management Principles, H. Richard and H. Ivanildo (eds), E & F Spon, London, UK, for the United Nations Environment Programme, the Water Supply & Sanitation Collaborative Council and the World Health Organization. http://www.who.int/water_sanitation_health/resourcesquality/wpccasestudy12.pdf (accessed 31 March 2012).

Prevost J. K.-J. (2004). The Geothermal Energy Industry of El Salvador. Harvard University, Cambridge, MA, USA, John F. Kennedy School of Government. http://web.mit.edu/10.391J/www/proceedings/Geothermal_Prevost2004.pdf (accessed 6 July 2012).

Price M. (ed.) (2000). Cooperation in the European Mountains. 2: The Caucasus. (Environmental Research Series 13). World Conservation Union (IUCN), Gland, Switzerland.

Price R. E. and Pilcher T. (2005). Distribution, speciation and bioavailability of arsenic in a shallow-water submarine hydrothermal system, Tutum Bay, Ambitle Island, Papua New Guinea. *Chemical Geology*, **224**, 122–135.

Prieto G. R. and Gonzalez M. L. (1998). Diagnosis of environmental problems related to vein gold mining in Colombia. Environmental Geochemistry in the Tropics. *Lect Notes Earth Sci*, **72**, 185–191.

Primary Metals Inc. (2007) Primary Metals Announces Increased NI 43-101 Resource Estimates for Panasqueira Tungsten Mine [Portugal]. News Release: 29 March 2007. Primary Metals Inc., Vancouver, BC. http://www.24hgold.com/english/news-company-gold-silver-increased-ni-43-101-resource-estimates-for-panasqueira-tungsten-mine.aspx?articleid=130249 (accessed 6 July 2012).

Priyadarshi N. (2004). Distribution of arsenic in permian coals of North Karanpura Coalfield, Jharkhand. *Journal of the Geological Society of India*, **63**(5), 533–36.

Prospero J. M. (1999). Long-term measurements of the transported African mineral dust to the southeastern United States: impact on regional air quality. *Journal of Geophysical Research*, **104**, 15917–15927.

PR Web Press Release Newswire. (2007). Doe Run Peru Reports Significant Drop in La Oroya Main Stack Emissions. PR Web Press Release Newswire, 19 April. http://www.prnewswire.com/news-releases/doe-run-peru-reports-significant-drop-in-la-oroya-main-stack-emissions-58604732.html (accessed 6 July 2012).

Putrohari R. D. (2005). [Geologi UGM] Menghirup debu atau Minum racun. Free Lists, 28 January (No longer hosted, 9 July 2012, at http://www.freelists.org/archives/geologiugm/01-2005/msg00263.html

Quamruzzaman Q. and Rahman M. (2002). Only painting tubewells red or green, does not help arsenicosis patients. In: 5th International Conference on Arsenic Exposure and Health Effects. San Diego, CA, USA, 14–18 July 2002.

Quintanilla J. (1992). Evaluation of arsenic in bodies of superficial water of the South Lipez of Bolivia (south-west). In: International Seminar Proceedings: Arsenic in the Environment and its Incidence on Health. Universidad de Chile, Santiago, Chile, pp. 109–121.

Raessler M., Michalke B., Schulte-Hostede V. and Kettrup A. (2000). Long-term monitoring of arsenic and selenium species in contaminated groundwaters by HPLC and HG-AAS. *Science of the Total Environment*, **258**, 171–181.

Rahman M. H. (2002). Water resources management and water-related diseases: Bangladesh experience. In: International Conference on environmental threats to the health of children: hazards and vulnerability. Bangkok, Thailand, March.

Rahman M., Tondel M., Ahmad A. and Axelson O. (1998). Diabetes mellitus associated with arsenic exposure in Bangladesh. *American Journal of Epidemiology*, **148**(2), 198–203.

Rahman M., Tondel M., Ahmad S. A., Chowdhury I. A., Faruquee M. H. and Axelson O. (1999a). Hypertension and arsenic exposure in Bangladesh. *Hypertension*, **33**(1), 74–78.

Rahman M., Tondel M., Chowdhury I. A. and Axelson O. (1999b). Relations between exposure to arsenic, skin lesions and glucosuria. *Occupational and Environmental Medicine*, **56**(4), 277–281.

Rahman M. M., Chowdhury U. K., Mukherjee S. C., Mondal B. K., Paul K., Lodh D., Biswas B. K., Chanda C. R., Basu G. K., Saha K. C., Roy S., Das R., Palit S. K., Quamruzzaman Q. and Chakraborti D. (2001). Chronic arsenic toxicity in Bangladesh and West Bengal, India: a review and commentary. *Journal of Toxicology and Clinical Toxicology*, **39**(7), 683–700.

Ramaswami A., Tawachsupa S. and Isleyen M. (2001). Batch-mixed iron treatment of high arsenic waters. *Water Research*, **35**(18), 4474–4479.

Ratnaike R. (2006). Arsenic in health and disease. In: Managing Arsenic in the Environment: From Soil to Human Health, R. Naidu, E. Smith, G. Owens, P. Bhattacharya and P. Nadebaum (eds), CSIRO Publishing, Collinwood, Victoria, Australia.

Ravencroft P. (2007). Predicting the Global Extent of Arsenic Pollution of Groundwater and its Potential Impact on Human Health. UNICEF, New York.

Ravenscroft P., Brammer H. and Richards K. (2009). Arsenic Pollution: A Global Synthesis. John Wiley Sons Ltd., West Sussex, UK.

Regional Environmental Center for Central and Eastern Europe. (2000). Strategic and Environmental Analysis of Albania, Bosnia and Herzegovina, Kosovo and Macedonia. Regional Environmental Center for Central and Eastern Europe. Office, Tirana, Albania. http://goo.gl/Se1jj (accessed 6 July 2012).

Reid J. C. and Carpenter R. H. (1993). A Geochemical Atlas of North Carolina. North Carolina Geological Survey, Raleigh, NC, USA. www.geology.enr.state.nc.us/NUREgeochem/geochem2.htm (accessed 6 July 2012).

Reid J. C. and Wooton R. (2002). Arsenic in Groundwater Study. North Carolina Geological Survey, Raleigh, NC, USA. http://www.geology.enr.state.nc.us/As%20drill%20core/Asingroundwater.htm (accessed 6 July 2012).

Reid J. C., Haven W. T., Eudy D. D., Milosh R. M. and Stafford E. G. (2007). Arsenic groundwater investigations: Eeastern slate belt (ESB), Nash and Halifax counties, North Carolina. Geological Society of America, Abstracts with Programs, **39**(2), 76.

Republic of Azerbaijan (2004). Country Study on Biodiversity of Azerbaijan Republic. Baku, Azerbaijan. aoa.ew.eea.europa.eu/virtual-library-viewer/answer_6109382793 (accessed 6 July 2012).

Republic of Macedonia (2002). State of Environment Report. http://www.johannesburgsummit.org/html/prep_process/national_ reports/508_macedonia_natl_assess.pdf (accessed 6 July 2012).

Resources Development International-Cambodia (RDIC). (2012). Groundwater Arsenic in Cambodia. http://www.rdic.org/groundwater-arsenic-in-cambodia.php (accessed 25 March 2012) and http://www.rdic.org/dwqi-ground-water-summary.php (accessed 25 March 2012).

Reuters News Service. (2005). US-run smelter in Peru harming children: Scientists. Reuters News Service, 7 December. deaconesshospital.adam.com/content.aspx?productId=16&pid. (accessed 6 July 2012).

Richards J. P., Wilkinson D. and Ullrich T. (2006). Geology of the Sari Gunay epithermal gold deposit, northwest Iran. *Economic Geology*, **101** (8), 1455–1496.

Rivara M. I., Cebrián M., Corey G., Hernández M. and Romieu I. (1997). Cancer risk in an arsenic-contaminated area of Chile. *Toxicology and Industrial Health*, **13**, 321–338.

Rivera M. L. and Pena M. (2004). A Pilot Study of Water by Adsorption in Natural Zeolite. Adsorption in Presence of Iron and Manganese. Instituto Mexicano de Tecnología del Agua. http://www.bvsde.paho.org/bvsacd/arsenico/Arsenic2004/theme3/paper3.14.pdf (accessed 6 July 2012).

Roberts W. L., Campbell T. J. and Rapp G. R. (1990). Encyclopedia of Minerals. 2nd edn., Van Nostrand Reinhold Co., New York, NY, USA.

Roberts D. M., Karunarathna A., Buckley N. A., Manuweera G. and Sheriff M. H. (2003). Influence of pesticide regulation on acute poisoning deaths in Sri Lanka. *Bulletin of the World Health Organization*, **81**(11), 789–798.

Roberts K., Stearns B. and Francis R. L. (1985). Investigation of Arsenic in Southeastern North Dakota Ground Water. A Superfund Remedial Investigation Report. North Dakota State Department of Health.

Robertson F. N. (1989). Arsenic in groundwater under oxidizing conditions, south-west United States. *Environmental Geochemistry and Health*, **11**, 171–185.

Robinson R. E. (2003). The Amanzi concept for acid mine drainage. Science in Africa, 2003 (May). www.scienceinafrica.co.za/2003/may/amd.htm (accessed 6 July 2012).

Robinson B., Outred H., Brooks R. and Kirkman J. (1995). The distribution and fate of arsenic in the Waikato River System, North Island, New Zealand. *Chemical Speciation and Bioavailability*, **7**, 89–96.

Rodríguez R., Ramos J. A. and Armienta A. (2004). Groundwater arsenic variations: the role of local geology and rainfall. *Applied Geochemistry*, **19**(2), 245–250.

Roller J. W., Schreiber M., Tadanier C., Widdowson M. and Johnson J. A. (2003). Arsenic release due to dissimilatory reduction of iron oxides in petroleum-contaminated aquifers. *Geological Society of America*, Abstracts with Programs, **35**(6), 575.

Romero L., Alonso H., Campano P., Fanfani L., Cidu R., Dadea C., Keegan T., Thornton I. and Farago M. (2003). Arsenic enrichment in waters and sediments of the Rio Loa (Second Region, Chile). *Applied Geochemistry*, **18**(9), 1399–1416.

Romero L., Alonso H., Espinoza S., Campano P., Fanfani L., Cidu R., Lorrai M., Lopez L. and Ferreyra R. E. (2001). Arsenic and boron contamination in the Loa basin, Northern Chile. In: *Proceedings of the 10th International Symposium on Water-Rock Interaction*, R. Cidu (ed.), Balkema A. A., Lisse, Netherlands, pp. 1127–1130 (WRI–10).

Romero F. M., Prol-Ledesma R. M., Canet C., Núñez Alvares L. and Pérez-Vázquez R. (2010). Acid drainage at the inactive Santa Lucia mine, western Cuba: natural attenuation of arsenic, barium and lead, and geochemical behavior of rare earth elements. *ApplGeochem*, **25**, 716–727.

Rosas I., Belmont R., Armienta A. and Baez A. (1999). Arsenic concentrations in water, soil, milk and forage in Comarca Lagunera, Mexico. *Water, Air, and Soil Pollution*, **112**(1–2), 133–149.

Rosenboom J. W. (2004). Not Just Red or Green: An Analysis of Arsenic Data from 15 Upazilas in Bangladesh. Government of the People''s Republic of Bangladesh, Ministry of Local Government, Rural Development and Co-operatives, Department of Public Health Engineering, Arsenic Policy Support Unit&; Department for International Development (UK), and UNICEF Bangladesh.

Rosi-Marshall E. J., Meyer L., Neumann K. and Lyons B. (2001). Defining away metal contamination in Georgia streams. In: Proceedings of the 2001 Georgia Water Resources Conference, University of Georgia, Athens, GA, USA. http://www.gwri.gatech.edu/uploads/proceedings/2001/Rosi-MarshallE-01.pdf (accessed 6 July 2012).

Rudnai P., Csanady S. E., Kiss V. and Mucsi I. (1996). Association between arsenic level of drinking water and reproductive outcomes in Hungary. In: International seminar on arsenic, health effects, remediation methods and treatment costs. Santiago, Chile, 8–10 October 1996.

Ruppert L. F., Minkin J. A., McGee J. J. and Cecil C. B. (1992). An unusual occurrence of arsenic-bearing pyrite in the upper Freeport coal bed, west-central Pennsylvania. *Energy Fuels*, **6**, 120–125.

Rutherford D. W., Bednar A. J., Garbarino J. R., Needham R., Staver K. W. and Wershaw R. L. (2003). Environmental fate of roxarsone in poultry litter. Part II. Mobility of arsenic in soils amended with poultry litter. *Environmental Science and Technology*, **37**, 1515–1520.

Ryker S. J. (2001). Mapping arsenic in groundwater. *Geotimes*, **46**(11), 34–36. http://water.usgs.gov/nawqa/trace/pubs/geo_v46n11/fig1.pdf

Saad A. and Hassanien M. A. (2001). Assessment of arsenic level in the hair of the nonoccupational Egyptian population: pilot study. *Environment International*, **27**, 471–478.

Sadiq M. and Alam I. (1996). Arsenic Chemistry in a groundwater aquifer from the eastern Province of Saudi Arabia. *Water, Air, and Soil Pollution*, **89**(1–2), pp. 67–76.

Sadr K. (2001). Water markets and pricing in Iran. In: Water Management in Islam, N. I. Faruqui, A. K. Biswas and M. Bino (eds.), UN University Press, Tokyo, Japan, pp. 103–114.

SAG-PY (2009). Geology and Hydrology. 2. http://www.sag-py.org/Informes Tecnicos. (accessed 11 November 2011).

Saha K. C. (1984). Melanokeratosis from arsenic contaminated tubewell water. *Indian Journal of Dermatology*, **29**, 37–46.

Saha K. C. (2002). Saha's grading of arsenicosis progression and patient treatment. In: Proceedings of the 5th international conference on arsenic exposure and health effects. San Diego, CA, USA, pp. 14–18 July 2002.

Saha K. C. and Chakraborti D. (2001). Seventeen years experience of arsenicosis in West Bengal, India. In: Arsenic Exposure and Health Effects IV, W. R. Chappell, C. O. Abernathy and R. L. Calderon (eds.), Amsterdam, Netherlands, pp. 387–395.

Sakanoue M. (1960a). As, Ge and Se in hot springs of the Misasa. *Journal of the Chemical Society of Japan, Pure Chemistry Section*, **81**, 242.

Sakanoue M. (1960b). Geochemical studies on the radioactive sediments: IV. Uranium and germanium in the carbonaceous material found in the 'Misasa-group' sedimentary beds of Tertiary age. *Journal of the Chemical Society of Japan, Pure Chemistry Section*, **81**, 1520–1523.

Sakulpitakphon T. (2004). Tracking mercury from the mine to the power plant: chemistry of the Manchester coal bed, Clay County, Kentucky. *International Journal of Coal Geology*, **57**, 127–141.

Saltori R. (2004). Arsenic contamination in Afghanistan: preliminary findings. In: Interregional conference on water quality. 23–26 November 2004, Taiyuan, China. UNU, ESCAP, UNICEF; WHO, Inter–Regional Conference on Water, Washington, DC. http://health.groups.yahoo.com/group/FluoridePoisoning/message/2843 (accessed 6 July 2012).

Sancha A. M. and Castro M. L. (2001). Arsenic in Latin America: occurrence, exposure, health effects and remediation. In: Arsenic Exposure and Health Effects IV, W. R. Chappell, C. O. Abernathy and R. L. Calderon (eds.), Elsevier, Amsterdam, Netherlands, pp. 87–96.

Sancha A. M. and Frenz P. (2000). Estimate of the current exposure of the urban population of northern Chile to arsenic. In: Proceedings of the Interdisciplinary perspectives on drinking water risk assessment and management, Santiago, Chile, Symposium, September 1998. International Association of Hydrological Sciences (IAHS) Publication No. 260. IAHS, Oxford, UK.

Sancha A. M., Ryan R. and Perez O. (2000). The removal of arsenic from drinking water and associated costs: the Chilean case. In: Proceedings of the Interdisciplinary perspectives on drinking water risk assessment and management, Santiago, Chile, Symposium, September 1998. International Association of Hydrological Sciences (IAHS). IAHS, Oxford, UK. pp. 17–25.

Sandy G. (2005). The GS web site autobiography of Emil Sandi: Index. http://www.tundria.com/Esandi/esandi-index-e.shtml (accessed 6 July 2012).

Sarajevo Center for Environmentally Sustainable Development (2003). Regional Action Plan from Mine Water Pollution Control. Sarajevo, Bosnia and Herzegovina.

Sarver S. (2004). The Pines, Indiana: arsenic, boron, lead, manganese in ground water. Post-Tribune (Indiana), 15 February. http://www.mindfully.org/Water/2004/Pines-As-B-Pb-Mn15feb04.htm.

Saunders J. A., Lee M.-K., Uddin A., Mohammed S., Wilkin R. T., Fayek M. and Korte N. E. (2005). Natural arsenic contamination of Holocene alluvial aquifers by linked tectonic, weathering, and microbial processes. *Geochemistry, Geophysics, Geosystems*, **6**. http://www.agu.org/ pubs/crossref/2005/2004GC000803.shtml).

Scanlon B. (2006). Arsenic in Groundwater. University of Texas at Austin, Bureau of Economic Geology, Austin, TX, USA.

Schoolmeester W. L. and White D. R. (1980). Arsenic poisoning. *Southern Medical Journal*, **73**, 198–208.

Schlottman J. L. (2005). Naturally Occurring Arsenic in the Central Oklahoma Aquifer. United States Department of the Interior, United States Geological Survey, Oklahoma City, OK, USA. http://wwwbrr.cr.usgs.gov/projects/GWC_chemtherm/FinalAbsPDF/schlottmann. pdf (accessed 6 July 2012).

Schlottman J. L. and Breit G. N. (1992). Mobilization of As and U in the Central Oklahoma aquifer, USA. In: Proceedings of the 7th International Symposium on Water-Rock Interaction, Y. F. Kharaka and A. S. Maest (eds.), Park City, Utah, USA, 13–18 July 1992. Balkema, Rotterdam, Netherlands, pp. 835–838.

Schmidt A. R. and Blanchard S. F. (1997). Surface-Water-Quality Assessment of the Upper Illinois River Basin in Illinois, Indiana, and Wisconsin: Results of Investigations through April 1992. United States Department of the Interior, United States Geological Survey (USGS) Water-Resources Investigations Report 96–4223. USGS, Reston, VA. il.water.usgs.gov/pubs/wrir95_4045.pdf (accessed 6 July 2012).

Schoof R. A., Yost L. J., Crecelius E., Irgolic K., Goessler W. and Guo H. R. Dietary arsenic intake in Taiwanese districts with elevated arsenic in drinking water. *Human and Ecological Risk Assessment*, **4**, 117–135.

SCUK (2001). Water Quality Survey in 327 Villages and Wards Located in Kyonpyaw, Hinthada, Laymyathna and Thabaung townships in Ayeyarwaddy Division, October 2000 – February 2001. Save the Children (UK) Myanmar WatSan Programme.

Sculz A. and Karczawska A. (2003). Arsenic in the environment in the vicinity of former arsenic ore mining and smelting site Zelezniak in the Kaczawskie Mountains, SW Poland. In: Symposium 1 (SYM 1): Arsenic in soil and groundwater environment: Biogeochemical interactions. Final poster papers. 15–19 June. See Karczawska *et al.* (2003).

Seal R. R., Haffner D. P., Meier A. L. and Caldwell J. Preliminary Report of Ground and Surface Water Geochemistry in the Vicinity of the Unmined Bald Mountain Massive Sulfide Deposit. United States Geological Survey (USGS) Open-File Report, United States Department of the Interior, USGS, Reston, VA [cited in Welch A. H., Watkins S. A., Helsel D. R. and Focazio M. J. (2000). http://water.usgs.gov/nawqa/trace/pubs/gw_v38n4/ (accessed 6 July 2012).

SEC Info (2001). Ceva International, Inc., 10–12G/A, On 2/2/01. http://www.secinfo.com/d1z443.49.htm#m03).

Seki Y. (2000). Hydrothermal alteration of the Sunagohara Formation in the Okuaizu geothermal area, Japan; alteration of lacustrine sediments formed by the present geothermal activity. *Chishitsu Chosajo Geppo (Bulletin of the Geological Survey of Japan)*, **51**, 329–367.

Selander L. and Svan P. (2007). Occurence and Distribution of Heavy Metals in Lake Poopo, Bolivia. Masters Thesis. Lund University.

Senanake N., De Silva W. A. S. and Salgado M. S. L. (1972). Arsenical polyneuropathy: a clinical study. *Ceylon Medical Journal*, **17**, 195–203.

Senanayake R. (2005). Lessons of the tsunami: beware the toxic sludge on the shoreline. *The Island*, 28 January, 8.

Sengör C. (1984). The Cimmeride Orogenic System and the Tectonics of Eurasia. Geological Society of America, Boulder, CO, USA.

Sengupta A. (1999). Workshop on groundwater pollution and protection, Science City, Calcutta. Calcutta, Science City, Central GroundwaterBoard.

Senior L. A. and Sloto R. A. (2006). Arsenic, Boron, and Fluoride Concentrations in Ground Water in and Near Diabase Instrusions, Newark Basin, Southeastern Pennsylvania. United States Geological Survey in cooperation with the United States Environmental Protection Agency (Scientific Investigations Report 2006–5261; http://pubs.usgs.gov/sir/2006/5261/pdf/sir2006–5261.pdf (accessed 6 July 2012).

Senza D. (1994). Russia and Armenia Sign Agreement on Various Joint Geological Ventures. *The Armenian Reporter*, 16 July. http://www.highbeam.com/doc/1P1–2266226.html (accessed 6 July 2012).

Servicio Agricola y Ganadero (2000). Informe contaminacion Rio Loa II region: Antofagasta periodo Marzo 1997 – Febrero 2000. Ministerio de Agricultura, Chile.

Sheets C. J. and Kozar M. D. (2000). Ground-Water Quality in the Appalachian Plateaus, Kanawha River basin, West Virginia. United States Geological Survey (USGS) Water-Resources Investigations Report 99–4269. United States Department of the Interior, USGS, Reston, VA. pubs.usgs.gov/wri/wri99–4269/ (accessed 6 July 2012).

Shepard H. H. (1951). The Chemistry and Action of Insecticides. 1st edn. McGraw–Hill, New York, NY, USA.

Shi F. (2004). Arsenic in groundwater in Huhhot alluvial basin in Inner Mongolia, People's Republic of China, TRITA-LWR. Master's thesis. KTH Land and Water Resources Engineering. phys4.harvard.edu/~wilson/arsenic/references/Huhot.pdf (accessed 6 July 2012).

Shields J. (1991). Ambient air arsenic levels along the Texas-Mexico border. *Journal of the Air and Waste Management Association*, **41**(6), 827–831.

Shinn E. A., Prospero J. M., Betzer P., Hayes M. L., Garrison V. and Barber R. T. (2000). African dust on the demise of Caribbean coral reefs. *Geophysical Research Letters*, **27**, 3029–3032.

Shock E. L., Sassani D. C., Willis M. and Sverjensky D. A. (1997). Inorganic species in geological fluids: correlations among standard molal thermodynamic properties of aqueous ions and hydroxide complexes. *Geochimica et Cosmochimica Acta*, **61**, 907–950.

Shotyk W. Goodsite M. E., Roos-Barraclough F., Frei R., Heinemeier J., Asmund G., Lohse C. and Hansen T. S. (2003). Anthropogenic contributions to atmospheric Hg, Pb, and As accumulation recorded by peat cores from southern Greenland and Denmark dated using the 'bomb pulse curve'. *Geochimica et Cosmochimica Acta*, **67**(21), 3991–4011.

Shrestha R. R., Maskey A. and Khadka P. K. (2002). Groundwater arsenic contamination in Nepal: a new challenge for water supply sector. In: Proceedings of the 5th international conference on arsenic exposure and health effects. San Diego, CA, USA, 14–18 July 2002.

Shrestha B. R., Whitney J. W. and Shrestha K. B. (eds) (2004). The State of Arsenic in Nepal: 2003. National Arsenic Steering Committee and Environment and Public Health Organization, Kathmandu, Nepal.

Siddiqi A. (2003). Presence and removal of arsenic from rural water supplies in Canada. In: Aquatic Arsenic Toxicity and Treatment, T. Murphy and J. Guo (eds), Backhuys Publishers, Leiden, Netherlands.

Silva H., (2011). Heavy metals in Jamaican soils and their possible impact on health. Poster session, Jamaica Insitute of Environmental Professionals, 5th conference on the environment: balancing national development and environmental protection, Programme, Jamaica Pegasus Hotel, Kingston, June 6–8. http://jiep.org/drupal/sites/default/files/JIEP%20CONFERENCE%20PROGAMME%202011_0.pdf (accessed 6 July 2012).

Silver S. and Phung L. T. (2005). Genes and enzymes involved in bacterial oxidation and reduction of inoganic arsenic. *Applied and Environmental Microbiology*, **71**, 599–608.

Simo J. A., Freiberg P. G. and Freiburg K. S. (1996). Geologic Constraints on Arsenic in Groundwater with Applications to Groundwater Modeling. Groundwater Research Report WRC GRR 96–01. University of Wisconsin – Madison, Water Resources Center, Madison, WI, USA.

Simon L. and Haylly J. (1984). Rychlé stanoveni arsenu v hnédém uhli. *Acta Montanistica Slovaca, Košice*, **68**, 253–263.

Singh I. B. (2001). Proxy records of neotectonics, climate changes and anthropogenic activity in the Late Quaternary of Ganga plain. In: Proceedings of the national symposium on role of earth sciences in integrated development and societal issues. 2–4 November 2001, Geological Survey of India, Lucknow, India. Calcutta, India, pp. xxxiii–I (Special Publication **65**, Vol. 1).

Singh A. K. (2004). Arsenic contamination in groundwater of north eastern India. In: Proceedings of national seminar on hydrology with focal theme on 'water quality'. Held at National Institute of Hydrology, Roorkee, India, 22–23 November 2004.

Siripitayakunkit U., Lue S. and Chprapawon C. (2001). Possible effects of arsenic on visual perception and visual motor integration of children in Thailand. In: Arsenic Exposure and Health Effects IV, W. R. Chappell, C. O. Abernathy and R. L. Calderon (eds.), Elsevier, Amsterdam, Netherlands, pp. 165–172.

Sky-Peck H. H. (1990). Distribution of trace elements in human hair. *Clinical Physiological Biochemistry*, **8**, 70–80.

Slotnick M. J., Meliker J. and Nriagu J. (2003). Natural sources of arsenic in southeastern Michigan groundwater. *Journal de Physique IV (Proceedings)*, **107**(2), 1247–1250.

Smedley P. (2002). Arsenic in groundwater: Eastern Asia. In: Arsenic in Groundwater: Occurrence and Geochemistry, A. H. Welch and K. G. Stollenwerk (eds.), Kluwer Academic Publishers, Boston, MA, USA.

Smedley P. L. and Kinniburgh D. G. (2002). A review of the source, behaviour and distribution of arsenic in natural waters. *Applied Geochemistry*, **17**, 517–568.

Smedley P. L., Edmunds W. M. and Pelid-Ba K. B. (1996). Mobility of arsenic in groundwater in the Obuasi area of Ghana. In: Environmental Geochemistry and Health, J. D. Appleton, R. Fuge and G. J. H. McCall (eds.), (Geological Society Special Publication 113). Geological Society, London, UK. pp. 163–181.

Smedley P. L., Nicolli H. B., Barros A. J. and Tullio J. O. (1998). Origin and mobility of arsenic in groundwater from the Pampean Plain, Argentina. In: Proceedings of the 9th international symposium on water-rock interaction, G. B. Arehart and J. R. Hulston (eds.), Balkema, Rotterdam, Netherlands, pp. 275–278.

Smedley P. L., Nicolli H. B., Barros A. J. and Tullio J. O. (2002). Hydrogeochemistry of arsenic and other inorganic constituents in groundwaters from La Pampa, Argentina. *Applied Geochemistry*, **17**, 259–284.

Smedley P., Nicolli H. and Macdonald D. (2000). Hydrogeochemistry of arsenic and other problem constituents in groundwaters from La Pampa. *Argentina. Journal of Conference Abstracts*, **5**(2), 936.

Smith A. H., Hopenhayn-Rich C., Bates M. N., Goeden H. M., Hertz-Picciotto I., Duggan H. M., Wood R., Kosnett M. J. and Smith M. T. (1992). Cancer risks from arsenic in drinking water. *Environmental Health Perspectives*, **97**, 259–267.

Smith A. (1998). Marked increase in bladder and lung cancer mortality in a region of Northern Chile due to arsenic in drinking water. *American Journal of Epidemiology*, **147**, 660–669.

Smith A. H. Arroyo A. R., Mazumder D. N., Kosnett M. J., Hernandez A. L., Beeris M., Smith M. M. and Moore L. E. (2000). Arsenic-induced skin lesions among Atacameño people in northern Chile despite good nutrition and centuries of exposure. *Environmental Health Perspectives*, **108**, 617–620.

Smith A. H., Lingas E. O. and Rahman M. (2000). Contamination of drinking water by arsenic in Bangladesh: a public health emergency. *Bulletin WHO*, **78**, 1093–1103.

Smith J. V. S., Jankowski J., Sammut J. (2006). Natural occurrences of inorganic arsenic in the Australian coastal groundwater environment: implications for water quality in Australian coastal communities. In: Managing arsenic in the Environment: From soil to human health, R. Naidu, E. Smith, G. Owens, P. Bhattacharya, P. Nadebaum (eds.), CSIRO Publishing, Collinwood, Victoria, Australia, pp. 129–153.

Smithsonian Institution (undated, a). *Global Volcanism Program: Baransky*. Smithsonian Institution, National Museum of Natural History, Department of Mineral Sciences, Washington, DC, USA. http://www.volcano.si.edu/world/volcano.cfm?vnum=0900–08= (accessed 6 July 2012).

Smithsonian Institution (undated, b). *Global Volcanism Program: Biu Plateau*. Smithsonian Institution, National Museum of Natural History, Department of Mineral Sciences, Washington, DC, USA. http://www.volcano.si.edu/world/volcano.cfm?vnum=0204–05- (accessed 6 July 2012).

Soares N. P. (2004). Uma populacao que se urbaniza. Instituto Geográfico Portugués, Lisbon, Portugal. http://www.igeo.pt/atlas/ (accessed 6 July 2012).

SOES (School of Environmental Studies) (2006). Groundwater Arsenic Contamination in West Bengal, India (20 years Study). SOES, Jadavpur University, West Bengal. www.soesju.org/arsenic/wb.htm (accessed 6 July 2012).

Sohrin Y., Tateishi T., Matsui M., Kawashima M. and Hasegawa H. (1997). Arsenic biogeochemistry affected by eutrophication in Lake Biwa, Japan. *Environmental Science and Technology*, **31**(10), 2712–2720.

Sonderegger J. L. and Ohguchi T. (1988). Irrigation related arsenic contamination of a thin, alluvial aquifer, Madison River Valley, Montana, USA. *Environmental Geology and Water Sciences*, **11**(2), 153–161.

Sonderegger J. L., Sholes B. R. and Ohguchi T. (1989). Arsenic Contamination of Aquifers Caused by Irrigation with Diluted Geothermal Water in the Lower Madison Valley. Montana, Open-file Report. Montana Bureau of Mines and Geology, Montana, USA.

South Africa (2000). *How Energy Generation Causes Environmental Change in South Africa*. South African Government. http://www.docstoc.com/docs/100451802/Energy-profile-of-South-Africa (accessed 6 July 2012).

South Carolina Department of Health and Environmental Control (2003). 2002 Environmental Surveillance and Oversight Program (ESOP) Data Report. http://www.scdhec.gov/environment/envserv/docs/esop_datarpt_02.pdf (accessed 6 July 2012).

Southwell M. (2003a). DEP knew of leaking toxic waste for decade. *The West Australian*, 8 February. www.highbeam.com/doc/1G1–97352302.html (accessed 6 July 2012).

Southwell M. (2003b). Sawmill clean-up cost secret. *The West Australian*, 15 March. www.highbeam.com/doc/1G1–98816851.html (accessed 6 July 2012).

Southwick J. W., Western A. E., Beck M. M., Whitley T., Isaacs R. and Petajan J. (1983). An epidemiological study of arsenic in drinking water in Millard County, Utah. In: Arsenic, Industrial, Biomedical, Environmental Perspectives, W. H. Leaderer and J. F. Robert (eds.), Van Nostrand Reinhold (VNR) Environmental Engineering Series. VNR, New York, NY.

Spicáková L., Ulicny D. and Koudelková G. (2000). Tectonosedimentary evolution of the Cheb basin (north-west Bohemia, Czech Republic) between Late Oligocene and Pliocene: a preliminary note. *Studia Geophysica et Geodaetica*, **44**(4), 556–580.

Spycher N. F. and Reed M. H. (1989). Evolution of the broadlands-type epithermal ore fluid along alternative P-T paths: implications for the transport and deposition of base, precious, and volatile metals. *Economic Geology*, **84**, 328–359.

Staff Writers (2006). Too many Chinese chemical plants said to be built near rivers. *Agence France-Presse*, 11 September. http://www.spacedaily.com/reports/Too_Many_Chinese_Chemical_Plants_Said_Built_Near_Rivers_999.html (accessed 6 July 2012).

Stamenov J., Iovchev M., Vachev B., Gueleva E., Yurukova L., Ganeva A., Mitrikov M., Antonov A., Srentz A., Varbanov Z., Batov I. V., Damov K., Marinova E., Frontasyeva M. V., Pavlov S. S. and Strelkova L. P. (2002). New Results From Air Pollution Studies in Bulgaria (Moss Survey 2000–2001). *Submitted to Journal of Balkan Ecology* (E14–2002–204; http://cdsweb.cern.ch/record/621645 (accessed 6 July 2012).

Stan O., Lucaciu A., Frontasyeva M. V. and Steinnes E. (2001). New results from air pollution studies in Romania. In: Man-Made Radionuclides and Heavy Metals in the Environment, (NATO Science Series IV, Earth and Environmental Sciences, Vol. 5), M. V. Frontasyeva, P. Vater and V. P. Perelygin (eds.), Kluwer Academic Publishers, Dordrecht, Netherlands, pp. 179–190

Stanger G. (2005). A palaeo-hydrogeological model for arsenic contamination in southern and south-east Asia. *Environmental Geochemistry and Health*, **27**(4), 359–368.

Stanger G., Truong T. V., Ngoc K. S., Luyen T. V. and Thanh T. T. (2005). Arsenic in groundwaters of the Lower Mekong. *Environmental Geochemistry and Health*, **27**(4), 341–357.

State Committee on Ecology and Control of Natural Resources Utilization (1998). Annual report. Baku, Azerbaijan.

Stauffer R. E. and Thompson J. M. (1984). Arsenic and antimony in geothermal waters of Yellowstone National Park, Wyoming, USA. *Geochimica et Cosmochimica Acta*, **48**(11), 2547–2561.

Steblez W. G. (2003a). The Mineral Industries of the Southern Balkans: Albania, Bosnia and Herzegovina, Croatia, Macedonia, Serbia and Montenegro, and Slovenia. United States Department of the Interior, United States Geological Survey, Reston, VA. http://minerals.usgs.gov/minerals/pubs/country/2003/albkhrmkmyb03.pdf (accessed 6 July 2012).

Steblez W. G. (2003b). The Mineral Industries of Central Europe: Czech Republic, Hungary, Poland, and Slovakia. United States Department of the Interior, United States Geological Survey, Reston, VA. http://minerals.usgs.gov/minerals/pubs/country/2003/ezhupllomy b03.pdf (accessed 6 July 2012).

Steblez W. (2004). *The Mineral Industry of Bulgaria*. United States Department of the Interior, United States Geological Survey. http://minerals.usgs.gov/minerals/pubs/country/1996/9408096.pdf (accessed 6 July 2012).

Stinchfield M., Moffitt C. M. and Foster S. (2005). Negotiating achievable arsenic soil cleanup standards in the context of high natural arsenic background. In: Annual International Conference on Soils, Sediments and Water. University of Massachusetts, Amherst, MA, USA. http://goo.gl/eRpzO (accessed 6 July 2012).

Stohrer G. (1991). Arsenic: opportunity for risk assessment. *Archives of Toxicology*, **65**, 525–531.

Stolz J. F. and Oremland R. S. (1999). Bacterial respiration of arsenic and selenium. *FEMS Microbiology Reviews*, **23**, 615–627.

Storelli M. M. and Marcotrigiano G. O. (2000). Organic and inorganic arsenic and lead in fish from the South Adriatic Sea, Italy. *Food Additives and Contaminants*, **17**(9), 763–768.

Straskraba V. and Moran R. E. (1990). Environmental occurrence and impacts of arsenic at gold mining sites in the western United States. *Mine Water and the Environment*, **9**(1–4), 181–191.

Stratamodel Inc (2006). Gold potential of the Liptako Belt Southwest Niger, West Africa www.stratamodel.com/Niger_rpt.htm (accessed 6 July 2012).

Strecansky B. (1996). Slovak Republic, upper Nitra Region. In: Developing Local and Regional Environmental Action Plans: case studies of Bulgaria, the Czech Republic, Hungary, Poland and the Slovak Republic, T. Hak (ed.), Regional Environmental Center for Central and Eastern Europe, Budapest, Hungary.

Stricker G. D. and Ellis M. S. (1999). Coal Quality and Geochemistry, Powder River Basin, Wyoming and Montana. United States Geological Survey (USGS) Professional Paper 1625-A. United States Department of the Interior, USGS, Reston, VA. (pubs.usgs.gov/pp/p1625a/Chapters/PQ.pdf (accessed 6 July 2012).

Subtelny O. (2000). Ukraine: A History. University of Toronto Press, Canada.

Sun D. and Yang Z. (2004). Progress on the investigation and prevention of endemic arsenicosis in PRC. In: Interregional conference on water quality, 23–26 November 2004, Taiyuan, China.

Sun G. (2004a). Arsenic contamination and arsenicosis in China. *Toxicology and Applied Pharmacology*, **198**(3), 268–271.

Sun G. (2004b). Arsenicosis patient identification and toxicity transfer from mother to child. In: Interregional Conference on water quality. 23–26 November 2004, Taiyuan, China. Inter–Regional Conference on Water, UNU, ESCAP, UNICEF, WHO, Washington, DC. http://health.groups.yahoo.com/group/FluoridePoisoning/message/2843 (accessed 6 July 2012).

Sun G. F. (2002). The distribution of arsenic concentration in water of China and the relation to patients. In: Proceedings of the 5th international conference on arsenic exposure and health effects. San Diego, CA, USA, 14–18 July 2002.

Sun G. F., Liu S., Li B., Li X., Sun X., Guo X. Y., Qian C. and Pi J. B. (2001a). Current situation of endemic arsenicosis in China. *Environmental Science*, **8**(5), 425–434.

Sun G. F., Pi J., Li B., Guo X., Yamauchi H. and Yoshida T. (2001b). Progresses on researches of endemic arsenism in China: population at risk, intervention actions and related scientific issues. In: Arsenic Exposure and Health Effects IV, W. R. Chappell, C. O. Abernathy and R. L. Calderon (eds.), Elsevier, Amsterdam, Netherlands, pp. 79–85.

Sushella A. K. (2001). A Treatise on Fluorosis. Fluorosis Research and Rural Development Foundation, New Delhi, India, . 15.

Svergiges geologiska undersoekning (2005). Mineralmarknaden, tema: arsenik. www.sgu.se/sgu/sv/./perpubl_2005–4.htm (accessed 6 July 2012).

Svobodová Z. and Randák T. (2005). Arsenic content in tissues of fish from the River Elbe. *Acta Veterinaria*, **74**, 419–425.

Swaine D. J. (1990). Trace Elements in Coal. Butterworth, London, UK.

Syrovetnik K. and Neretnieks I. (2002). Use of peat bogs as analogues for the development and validation of the long-term models for MSW landfills. Tekniska högskolan i Stockholm. Institutionen för kemiteknik, Stockholm.

Tan Y. J. (1994). Geology of Carlin-Type Gold deposits in the Dian-Qian-Gui area. In: Chinese Carlin-Type Gold Deposits, Y. J. Tan, J. Wang and D. S. Liu (eds), University of Nanjing Press, Nanjing, China, pp. 142–159 (in Chinese).

Tanabe S., Saito Y., Vu Q. L., Hanebuth T. I. I., Ngo Q. L. and Kitamura A. (2006). Holocene evolutaion of the Song Hong (Red River) delta system, northern Vietnam. *Sed. Geol.* **187**, 26–61.

Tao C. (1990). Ore-controlling factors and exploration guides of superfinegrained disseminated gold deposits, southwestern Guizhou. *Geology and Prospecting*, **26**(8), 9–15 (in Chinese).

Tarasov V. G., Gebruk A. V., Shulkin V. M., Kamenev G. M., Fadeev V. I., Kosmynin V. N., Malakhov V. V., Starynin D. A. and Obzhirov A. I. (1999). Effect of shallow-water hydrothermal venting on the biota of Matupi Harbour (Rabaul Caldera, New Britain Island, Papua New Guinea). *Continental Shelf Research*, **19**, 79–116.

Tarasov V. G., Propp M. V., Propp L. N., Kamenev G. M. and Blinov S. V. (1986). Shallow-water hydrothermal vents and a unique ecosystem of Kraternaya Caldera. *Kurile Islands Biologiya Morya (Marine Biology)*, **2**, 72–74 (in Russian).

Tariq J., Ashraf M., Jaffar M. and Afza M. (1996). Pollution status of the Indus River, Pakistan, through heavy metal and macronutrient contents of fish, sediment and water. *Water Research*, **30**(6), 1337–1344.

Teaf C., Covert D. J., Coleman R. M., Petrovich M., Murali R. S. and Yarina V. (2006). Risk and background evaluation for arsenic in soil at a planned residential development. In: Annual International Conference on Soils, Sediments and Water. University of Massachusetts, Amherst, MA, USA. http://goo.gl/HfS79 (accessed 6 July 2012).

Tello E. E. (1986). Arsenicismos hidricos: Que es el hidroarseniciso cronico regional endemico Argentino (Hacrea)? *Archivos Argentinos de Dermatologia*, **36**(4), 197–214.

Thanh T. N. (2004). Arsenic pollution in groundwater in the Red River Delta. http://publications.iwmi.org/pdf/H033503.pdf (accessed 6 July 2012)

Thomas M. A. (2003). Arsenic in Midwestern Glacial Deposits: Occurrence and Relation to Selected Hydrogeologic and Geochemical Factors. United States Geological Survey (USGS) Water-Resources Investigations Report 03–4228. United States Department of the Interior, USGS, Reston, VA. http://pubs.usgs.gov/wri/wri034228/pdf/wri034228.pdf (accessed 6 July 2012).

Thomas K. W., Pellizzari E. D. and Berry M. R. (1999). Population-based dietary intakes and tap water concentratioins for selected elements in the EPA region V. National Human Exposure Assessment Survey (NHEXAS). *Journal of Exposure Analysis and Environmental Epidemiology*, **09**, 402–413.

Thomas M. A., Schumann T. L. and Pletsch B. A. (2005). Arsenic in Ground Water in Selected Parts of Southwestern Ohio, 2002–03. United States Geological Survey (USGS) Scientific Investigations Report 2005–5138. United States Department of the Interior, USGS, Reston, VA. http://pubs.usgs.gov/sir/2005/5138/ (accessed 6 July 2012).

Thompson T. S., Le M. D., Kasick A. R. and Macaulay T. J. (1999). Arsenic in well water supplies in Saskatchewan. *Bulletin of Environmental Contamination and Toxicology*, **63**(4), 478–483.

Thornton I. and Farago M. (1997). The geochemistry of arsenic. In: Arsenic Exposure and Health Effects, C. O. Abernathy, R. L. Calderon and W. R. Chappell (eds), Chapman & Hall, London, UK. pp. 1–16.

Thornton I., Farago M. E., Keegan T., Nieuwenhuijsen M. J., Colvile R. N., Pesch B., Ranft U., Miskovic P. and Jakubis P. (2002). Environmental impacts, exposure assessment and health effects related to arsenic emissions from a coal-fired power plant in central Slovakia: the EXPASCAN Study. In: Fifth International Conference on arsenic exposure and health effects. San Diego, CA, USA, 14–18 July 2002. Abstract: http://www.cabdirect.org/abstracts/20043164105.html;jsessionid=12C9DD1DEEE326A766D6069004534281 (accessed 10 July 2012).

Tian D., Ma H., Feng Z., Xia Y., Le X. C. and Ni Z. (2001). Analyses of micronuclei in exfoliated epithelial cells from individuals chronically exposed to arsenic via drinking water in Inner Mongolia, China. *Journal of Toxicology and Environmental Health A*, **64**(6), 473–484.

Tondel M., Rahman M., Magnuson A., Chowdhury I. A., Faruquee M. H. and Ahmad S. A. (1999). The relationship of arsenic levels in drinking water and the prevalence rate of skin lesions in Bangladesh. *Environmental Health Perspectives*, **107**(9), 727–729.

Tossel J. A. (1996). Theoretical studies on As and Sb sulfide molecules. In: M. D. Dyar, C. McCammon and M. W. Schaefer (eds), Mineral Spectroscopy: A Tribute to Roger G. Burns, (Geochemical Society Special Publication). Geochemical Society, Houston, TX, USA, pp. 9–21.

Toujague R. T., Leonarte A., Reyes Verdecia B. L. and Miravet R. M. (2003). Arsénico y metales pesados en aguas del área Delita, Isla de la Juventud. *Cienc Tierra Espacio Cuba*, **4**, 5–8.

Travel Document Systems (2007). Niger, Africa. http://www.traveldocs.com/ne/economy.htm

Trejo-Vazquez R., Bonilla-Petriciolet A. and Lariz-Duron P. G. (2002). Possible correlation of arsenic and fluoride content in the drinking water of the city of Aguascalientes, Mexico. *Afinidad*, **59**, 184–190.

Treng K. H. and Viet P. H. (2005). Arsenic Contamination in Vietnam, Some Results by CETASD (1999–2005). Research Centre for Environmental Technology and Sustainable Development and Vietnam National University, Hanoi, Viet Nam (PowerPoint presentation).

Tsai S. M., Wang T. N. and Ko Y. C. (1999). Mortality for certain diseases in areas with high levels of arsenic in drinking water. *Archives of Environmental Health*, **54**, 186–193.

Tse P.-K. (2005a). The Mineral Industry of Malaysia. United States Department of the Interior, United States Geological Survey, Reston, VA. http://minerals.usgs.gov/minerals/pubs/country/2010/myb3-2010-my.pdf (accessed 2012).

Tse P.-K. (2005b). The Mineral Industry of Mongolia. United States Department of the Interior, United States Geological Survey, Reston, VA. http://minerals.usgs.gov/minerals/pubs/country/2003/mgmyb03.pdf (accessed 6 July 2012).

Tseng W. P., Chu H. M., How S. W., Fong J. M., Lin C. S. and Yen S. (1968). Prevalence of skin cancer in an endemic area of chronic arsenism in Taiwan. *Journal of the National Cancer Institute*, **40**, 453–463.

Tseng W.-P. (1977). Effects and dose-response relationships of skin cancer and blackfoot disease with arsenic. *Environmental Health Perspectives*, **19**, 109–119.

Tseng C. H., Chong C. K., Chen C. J. and Tai T. Y. (1996). Dose-response relationship between peripheral vascular disease and ingested inorganic arsenic among residents in blackfoot disease endemic villages in Taiwan. *Atherosclerosis*, **120**(1–2), 125–133.

Tucker K. (2002). Spraying Orchard, Hood River. Oregon Historical Society. http://www.ohs.org/education/oregonhistory/historical_records/dspDocument.cfm?doc_ID=000704FB-EFA0-1E89-891B80B0527200A7 (accessed 6 July 2012).

Tupaz J. M. (2004). Guv Seeks Removal of Expert in Mt. Apo Arsenic Probe. Sun Star, Davao, Philippines, 24 February. http://www.afrim.org.ph/Archives/2004/Sunstar/February/23/Guv%20seeks%20removal%20of%20expert%20in%20Mt.%20Apo%20arsenic%20probe.txt (accessed 6 July 2012).

Turner J. P., Lee M. K., Uddin A., Jean J. and Saunders J. (2005). Natural arsenic contamination of alluvial aquifers by linked geological and microbial processes [poster]. In: Annual Meeting of the Geological Society of America, Salt Lake City, Utah, USA, 16–19 October 2005 (Paper No. 164-3). www.geosociety.org/meetings/2005/ (accessed 6 July 2012).

Tuttle M., Goldhaber M., Ruppert L. and Hower J. (2001). Arsenic in coal and stream sediments from the Appalachian Basin, Kentucky. www.brr.cr.usgs.gov/projects/GWC_chemtherm/FinalAbsPDF/tuttle.pdf (accessed 6 July 2012).

Tvarlchrelidze A. (1997). Georgia's Solid Mineral Resources. Centre for Strategic Investigation, Tbilisi, Georgia.

Ulman C., Gezer S., Anal Ö., Töre I. R. and Kirca Ü. (1998). Arsenic in human and cow's milk: a reflection of environmental pollution. *Water, Air, and Soil Pollution*, **101**(1–4), 411–416.

UN (United Nations) (1983). Ground Water in West Africa. UN Natural Resources/Water Series No. 18. UN, New York, NY, pp. 53–65.

UN (United Nations) (1989). Ground Water in Eastern, Central and Southern Africa. UN Natural Resources/Water Series No. 19. UN, New York, NY, pp. 301–308.
UNDP (United Nations Development Programme) (2004). Afghanistan National Human Development Report 2004. Security with a Human Face: Challenges and Responsibilities. UNDP Afghanistan, Kabul, Islamic Republic of Afghanistan.
UNDP (United Nations Development Programme) (2005a). Botswana Human Development Report 2005: Harnessing Science and Technology for Human Development. UNDP, Gaborone, Botswana. http://www.facebook.com/note.php?note_id=475776792450994 (accessed 6 July 2012).
UNDP (United Nations Development Programme) (2005b). Reducing Trans-Boundary Degradation of the Kura-Aras/Araks River Basin Project. UNDP.
UNDP (United Nations Development Programme) (2005c). Human Development Report, Serbia. UNDP. http://hdr.undp.org/en/reports/nationalreports/europethecis/serbia/name,3304,en.html (accessed 6 July 2012).
UNDP (United Nations Development Programme) (2006). Human Development Report, Beyond Scarcity : Power Poverty and the Global Water Crisis, United Nations: Development Programme. http://waterwiki.net/index.php/HDR_2006_-_Beyond_Scarcity:_Power,_Poverty_and_the_Global_Water_Crisis (accessed 6 July 2012).
UNDP (United Nations Development Programme) (2006a). Country Fact Sheets. UNDP. http://hdr.undp.org/en/countries/ (accessed 6 July 2012).
UNDP (United Nations Development Programme) (2006b). Draft Country Programme Document for Zimbabwe, 2007–2009. UNDP. http://web.undp.org/africa/programmedocs/Zimbabwe-&per;20CPD&per;202007-2009&per;20&per;28English&per;29.pdf (accessed 6 July 2012).
UNDP (United Nations Development Programme) (2007). Azerbaijan. UNDP. http://europeandcis.undp.org/WaterWiki/index.php/Azerbaijan (accessed 6 July 2012).
UNDP (United Nations Development Programme) (2010). Human Development Report. The Real Wealth of Nations: Pathways to Human Development. 20th Anniversary Issue. hdr.undp.org/en/reports/global/hdr2010/ (accessed 6 July 2012).
UNDP (United Nations Development Programme) (2011). Human Development Reports: Human Development Index 2011 Rankings. http://hdr.undp.org/en/statistics/ (accessed 6 July 2012).
UNEP (United Nations Environment Programme) (2000). Post-Conflict Environmental Assessment: FYROM. UNEP. http://enrin.grida.no/htmls/macedon/reports/postcon/eng/report.htm (accessed 6 July 2012).
UNEP (United Nations Environment Programme) Balkans (2000). Post-Conflict Environmental Assessment: Albania. UNEP, Geneva.
UNESCO (United Nations Educational, Scientific and Cultural Organization) (2005). Slovakia: Mining maps and plans of the Main Chamber: Count Office in Banská Štiavnica. UNESCO, New York, NY. http://www.unesco-ci.org/photos/showgallery.php/cat/547 (accessed 6 July 2012).
UNICEF (United Nations Children's Fund) (2004a). Arsenic Contamination in Vietnam: A Situation Overview & the Mitigation Measures Required. UNICEF, Hanoi, Viet Nam.
UNICEF (United Nations Children's Fund) (2004b). Arsenic Mitigation in Vietnam. Presentation at the workshop on arsenic mitigation, 4 June 2003. UNICEF, Hanoi, Viet Nam.
UNICEF (United Nations Children's Fund) (2004c). Report on assessment of the use of water filter provided by UNICEF in the flood prone area of Am Giang and Dong Thap provinces. UNICEF, Hanoi, Viet Nam.
UNICEF (United Nations Children's Fund) (2007). Predicting the Global Extent of Arsenic Pollution of Groundwater and its Potential Impact on Human Health. New York. Report Prepared by Peter Ravenscroft, Consultant Hydrogeologist, Cambridge, UK.
UNICEF (United Nations Children's Fund) (2009). *Situation Analysis of Children and Women in Cambodia*. www.unicef.org/cambodia/12681_15682.html (accessed 6 July 2012).
UNICEF (United Nations Children's Fund) Kolkata (2004). Laboratory based water quality surveillance for arsenic mitigation in West Bengal. In: Interregional Conference on water quality, 23–26 November. UNICEF, Kolkata, India.
UNICEF (United Nations Children's Fund) & WHO (World Health Organization) (2002). Roundtable on Arsenic and Water Quality, 8–9 March 2002. UNICEF and WHO, New York, NY.
UNICEF (United Nations Children's Fund) & WHO (World Health Organization) (2012). Progress on sanitation and drinking water: 2012 update. UNICEF & WHO, New York. http://www.who.int/water_sanitation_health/monitoring/en/ (accessed 5 July 2012).
United States Army (2004). A Soldier's Guide to Staying Healthy in the Balkans. United States Army Public Health Command, Maryland, USA. http://deploymenthealthlibrary.fhp.osd.mil/Product/RetrieveFile?prodId=98 (accessed 24 June 2012).
United States Census Bureau (2000). Factfinder. http://factfinder.census.gov (accessed 6 July 2012).
United States Census Bureau (2006a). New Hampshire, state and county quick facts. http://quickfacts.census.gov/qfd/states/33000.html (accessed 6 July 2012).
United States Census Bureau (2006b). New Mexico, state and county quick facts. http://quickfacts.census.gov/qfd/states/35000.html (accessed 6 July 2012).
United States Census Bureau (2006c). Geographic area: San Ysidro Village, New Mexico. http://censtats.census.gov/data/NM/ 1603571020.pdf (accessed 6 July 2012).
United States Census Bureau (2006d). State and county quick facts: Albuquerque (city) New Mexico, population 2003 estimate. http://quickfacts.census.gov/qfd/states/35/3502000.html (accessed 6 July 2012).

United States Census Bureau (2006e). Oklahoma–place–Data set 2000. http://factfinder.census.gov/servlet/GCTTable?_bm=y&-geo_id=04000US40&-_box_head_nbr=GCT-PH1&-ds_name=DEC_2000_SF1_U&-format=ST-7 (accessed 6 July 2012).

United States Census Bureau (2011). http://www.census.gov/population/international/data/idb/rank.php (accessed 6 July 2012).

United States Department of Defense, Deployment Health Clinic Center (2005). Environmental exposures: Water contamination. Walter Reed Army Medical Center, Washington, DC.

United States Department of Health and Human Services (1998). Health Consultation, Stephenson-Bennett Mine Site. Organ, Dona Ana County, New Mexico [Agency for Toxic Substances and Disease Registry. Division of Health Assessment and Consultation.Exposure Investigation and Consultation Branch, Atlanta, Ga). http://www.atsdr.cdc.gov/hac/pha/pha.asp?docid=1205&pg=0 (accessed 6 July 2012).

Universidad de Chile (2000). Informe pais: estado del medio ambiente en Chile: 1999. LOM ediciones, Santiago, Chile.

University of Illinois (2007). Ukraine: Occupational Lung Disease in Ukrainian Coal Miners. University of Illinois at Chicago, School of Public Health, Chicago, IL, USA. www.uic.edu/sph/glakes/global/global_miners.htm (accessed 6 July 2012).

UPI (United Press International) (2004). New Zealand Homes on Polluted Land 1,155. United Press International, 28 October, Auckland, New Zealand.

Urbina I. (2011). Regulation lax as gas wells' tainted water hits rivers. *NY Times*, February 27, A1. www.nytimes.com/2011/02/27/us/ 27gas.html?pagewanted=all (accessed 6 July 2012).

USDA (US Department of Agriculture) (1997a). Agricultural Chemical Usage 1996 Field Crops Summary. USDA (National Agricultural Statistics Service Publication), Washington, DC.

USDA (US Department of Agriculture) (1997b). Census of Agriculture. USDA, Washington, DC, USA.

USEPA (United States Environmental Protection Agency) (2000a). Arsenic Occurrence in Public Drinking Water Supplies. Office of Ground and Drinking Water, USEPA, Washington, DC, USA (EPA 815-0-00-001).

USEPA (United States Environmental Protection Agency) (2000b). National Primary Drinking Water Regulations; arsenic and clarifications to compliance and new source contaminants monitoring; proposed rule. Federal Register, **60**(121), 38888–38983.

USEPA (United States Environmental Protection Agency) (2001a). Arsenic in Drinking Water. USEPA. http://www.epa.gov/safewater/arsenic.html (accessed 6 July 2012).

USEPA (United States Environmental Protection Agency) (2001b). Preliminary Results Indicate Isolated Areas of Elevated Lead and Arsenic Contamination in Newhall Street Neighborhood. USEPA New England Press Release, 3 May. http://goo.gl/PzFQI (accessed 6 July 2012).

USEPA (United States Environmental Protection Agency) (2006). Sampling Activities Set to Resume in Spring: South Minneapolis Neighborhood Soil Contamination Site, Minneapolis, Minnesota. USEPA. http://www.epa.gov/Region5/sites/cmcheartland/pdfs/fs_sampling_200604.pdf (accessed 6 July 2012).

USEPA (United States Environmental Protection Agency) (2007a). Monsanto Corp. (Augusta Plant). USEPA. http://www.epa.gov/region4/waste/npl/nplga/monsanga.htm (accessed 6 July 2012).

USEPA (United States Environmental Protection Agency) (2007b). Woolfolk Chemical Works, Inc. USEPA. http://www.epa.gov/Region4/waste/npl/nplga/wolfokga.htm (accessed 6 July 2012).

USEPA (United States Environmental Protection Agency) (2007c). The Arsenic Trioxide Site. USEPA, Denver, CO, USA. http://www.epa.gov/region8/superfund/nd/arsenic/index.html (accessed 6 July 2012).

USEPA (United States Environmental Protection Agency) (2007d). Vasquez Boulevard & I-70. USEPA. http://www.epa.gov/region08/superfund/co/vbi70/ (accessed 6 July 2012).

USEPA (United States Environmental Protection Agency) (2009). TRI Explorer Database: Arsenic. USEPA, Washington, DC, USA. http://www.epa.gov/tri/ (accessed 6 July 2012).

USGS (United States Geological Survey) (1997). Programs in New Jersey. United States Department of the Interior, USGS, Reston, VA. http://pubs.usgs.gov/fs/FS-030-96/ (accessed 6 July 2012).

USGS (United States Geological Survey) (1998). China and US Geological Survey: Working Together on Environmental Issues. United States Department of the Interior, USGS, Reston, VA. www.usgs.gov/newsroom/article.asp?ID=1101 (accessed 6 July 2012).

USGS (United States Geological Survey) (1999). US Geological Survey's Mineral Resources Program Activities in the Upper Midwest. United States Department of the Interior, USGS, Reston, VA. http://pubs.usgs.gov/info/mwinfo/mwmins.pdf (accessed 6 July 2012).

USGS (United States Geological Survey) (2000). Organic Compounds and Trace Elements in the Pocomoke River and Tributaries, Maryland. United States Department of the Interior, USGS, Reston, VA. http://pubs.usgs.gov/of/1999/ofr-99-057/ (accessed 6 July 2012).

USGS (United States Geological Survey) (2001). Groundwater Contamination by Heavy Metals: Tar Creek, Oklahoma. United States Department of the Interior, USGS, Reston, VA. Toxic Substances Hydrology Program. http://toxics.usgs.gov/bib/bib-tar.html (accessed 6 July 2012).

USGS (United States Geological Survey) (2003). Arsenic Concentrations in Private Bedrock Wells in Southeastern New Hampshire. USGS Fact Sheet 051-03. United States Department of the Interior, USGS, Reston, VA. http://pubs.usgs.gov/fs/fs-051-03/ (accessed 6 July 2012).

USGS (United States Geological Survey), Afghanistan Ministry of Mines and Industry (2006). Assessment of Undiscovered Technically Recoverable Conventional Petroleum Resources of Northern Afghanistan. United States Department of the Interior, USGS, Reston, VA. http://pubs.usgs.gov/of/2006/1253/ (accessed 6 July 2012).

USGS (United States Geological Survey) (2012). Mineral commodity summaries: Arsenic. United States Department of the Interior, USGS, Reston, VA. pp. 20–21. http://minerals.usgs.gov/minerals/pubs/commodity/arsenic/ (accessed July 5, 2012).

Uy, Davin (2009). Total of tube-well tested in Cambodia. http://verification-unit.org/documents/end_project_workshop/Water&per;20Davin&per;20Cambodia.pdf (accessed 6 July 2012).

Uy, Davin (2011). Arsenic Situation and Mitigation in Cambodia. http://verification-unit.org/documents/end_project_workshop/Water&per;20Davin&per;20Cambodia.pdf (accessed 25 March 2012).

Vaessen H. A. M. G. and Van Ooik A. (1989). Speciation of arsenic in Dutch total diets: methodology and results. *Zeitschrift für Lebensmitteluntersuchung und -Forschung A*, **189**(3), 232–235.

Vaisman A. G. and Lacer L. D. (2003). Estimated heavy metal emissions to the atmosphere due to projected changes in the Brazilian energy generation matrix. *Regional Environmental Change*, **3**(4), 140–145.

Valentine J. L., Kang H. K. and Spivey G. (1979). Arsenic levels in human blood, urine, and hair in response to exposure via drinking water. *Environmental Research*, **20**, 24–32.

Valiev Y. Y., Gofen G. I. and Pachadzhanov D. N. (1993). Trace elements in the Jurassic anthracites of the Nazar-Ailok deposit, Central Tajikistan. *Geokhimia*, **2**, 243–251.

Van der Valk M. R. (2006). Arsenic in groundwater: a world problem. Symposium organized by the International Association of Hydrogeologists (Netherlands Chapter) and the Netherlands. Hydrological Society, Utrecht, Netherlands, 29 November 2006. www.iah.org/downloads/occpub/arsenic_gw.pdf (accessed 6 July 2012).

Vandiver J. (2004). Arsenic: chicken feed effects questioned. Daily Times (Maryland), 4 January. http://www.mindfully.org/Farm/2004/Arsenic-Chicken-Roxarsone4jan04.htm (accessed 6 July 2012).

Varsányi I., Fodré Z. and Bartha A. (1991). Arsenic in drinking water and mortality in the southern Great Plain, Hungary. *Environmental Geochemistry and Health*, **13**, 14–22.

Veiderma M. (2005). Natural gas in the Baltic Sea region. Paper delivered to the Assembly of the Baltic States on 26 November. www.akadeemia.ee/_repository/File/Veiderma_ettekanneINGL.pdf (accessed 6 July 2012).

Verrengia J. B. (2003). U.N.: Iraq facing huge problems of pollution, environmental destruction. AP Worldstream, 25 May. http://news.google.com/newspapers?nid=2026&dat=20030425&id=Qc4qAAAAIBAJ&sjid=NNEFAAAAIBAJ&pg=2939,3081887 (accessed 6 July 2012).

VietNamNet Bridge (2006). 10mil people exposed to arsenic in water. VietNamNet, 22 June. http://www.searchqu.com/web?src=404&appid=102&systemid=406&q=http%3A%2F%2Fwww.idswater.com%2Fwater%2Fus%2F1%2F2012%2Fpress%2F2%2F187%2Farsenic_water (accessed 6 July 2012).

Vogt R. D., Eilertsen O. and Arnoldussen A. (2011). Preliminary environmental audit antimony and mercury mines and background environmental conditions in the Talas, Osh and Bakten oblast of Kyrgyz Republic. http://foLk.uio.no/rvogt/CV/Reports/ Kyrgyz.pdf (accessed 18 December 2011).

Volkov A. V., Serafimovski T., Kochneva N. T., Tomson I. N. and Tasev G. (2006). The Alshar Epithermal Au-As-Sb-Tl Deposit, Southern Macedonia. *Geology of Ore Deposits*, **48**(3), 175–192.

Vroblesky D. A. (2001). *Mapping a Discharge Zone of Arsenic-Contaminated Ground Water using Diffusion Samplers*. United States Department of the Interior, United States Geological Survey, Colombia, SC, USA. H42A-15; http://adsabs.harvard.edu/abs/2001AGUSM.H42A15V (accessed 6 July 2012).

Vuki M. (2004). Arsenic Speciation Study in Guam Waters. (Report 2003GU26B). University of Guam, Water and Environmental Research Institute of the Western Pacific, Mangilao, Guam. http://water.usgs.gov/wrri/02-03grants_new/prog-compl-reports/2003GU26B.pdf (accessed 6 July 2012).

Vuki M., Limtiaco J., Aube T., Emmanuel J. and Denton G. (2007). Arsenic speciation study in some spring waters of Guam, Western Pacific Ocean. *Science of the Total Environment*, **379**, 175–179.

WALHI. (2004a). Buyat Bay is polluted and a risk to the community: highlights of the joint investigation of Buyat Bay. Compiled by Indonesia Forum for Environment (WALHI) – Friends of the Earth Indonesia, Indonesian Mining Advocacy Network (JATAM), and Indonesian Center for Environmental Law (ICEL), based on Joint Investigation Technical Team report, 9 November 2004. www.minesandcommunities.org/article.php?a=5859 (accessed 6 July 2012).

WALHI (2004b). Critique of the 'Buyat Bay' CSIRO Environmental Monitoring study commissioned by PT Newmont Minahasa Raya, 31 October 2004. Indonesia Forum for Environment (WALHI) – Friends of the Earth Indonesia. www.minesandcommunities.org/article.php?a=5814&l=1 (accessed 6 July 2012).

Wang G. (1984). Arsenic poisoning from drinking water in Xianjiang. *Chinese Journal of Preventative Medicine*, **18**, 105–107.

Wang Z. G., He H., Yan Y. and Wu C. (1999). Arsenic exposure of residents in areas near Shimen arsenic mine. *Journal of Hygiene Research*, **28**(1), 12–14 (in Chinese).

Wang G., Xiao B. and Huang Y. (1995). Epidemological studies on endemic fluorosis and arsenism in Xinjiang. *Zhonghua YuFang Yi Xue ZaZhi*, **29**(1), 30–33 (in Chinese).

Wang G. Q., Huang Y. Z., Xiao B. Y., Qian X. C., Yao H., Hu Y., Gu Y. L., Zhang C. and Liu K. T. (1997). Toxicity from water containing arsenic and fluoride in Xinjiang. Research report. *Fluoride*, **30**(2), 81–84.

Wang L. and Huang J. (1994). Chronic arsenism from drinking water in some areas of Xinjiang, China. In: Arsenic in the Environment, Part II: Human Health and Ecosystem Effects, J. O. Nriagu (ed.), John Wiley, New York, NY, USA, pp. 159–172.

Wang X., Zhu L. and Wang J. (1992). The Forming and Distribution of Coalfields in China. Science Publishing House, Beijing, China.

Wang M., Zheng B., Wang B., Li S., Wu D. and Hu J. (2006). Arsenic concentrations in Chinese coals. *Science of the Total Environment*, **357**, 96–102.

Ward C. R., Spears C. A., Booth I. A., Staton L. and Gurba L. W. (1999). Mineral matter and trace elements in coals of the Gunnedah Basin, New South Wales, Australia. *International Journal of Coal Geology*, **40**(4), 281–308.

Warner K. L. (2001). Arsenic in glacial drift aquifers and the implication for drinking water: lower Illinois River Basin. *Ground Water*, **39**, 433–442.

Warrick J. (2005). Albania's chemical cache raises fears about others; long-forgotten arms had little or no security. The Washington Post, 10 January. http://www.washingtonpost.com/wp-dyn/articles/A61698-2005Jan9.html (accessed 6 July 2012).

Wasserman G. A., Liu X., Parvez F., Ahsan H., Factor-Litvak P., van Geen A., Slavkovich V., LoIacono N. J., Cheng Z., Hussain I., Momotaj H. and Graziano J. H. (2004). Water arsenic exposure and children's intellectual function in Araihazar, Bangladesh. *Environmental Health Perspectives*, **112**(13), 1329–1333.

Water for People and Peace. (2007). Problem of Drinking Water in Greece: S.O.S. from Central Macedonia. Water for People and Peace, Dimosthenous, Greece. http://www.waterpp.org/.

Weber E. F. (1986). A Stochastic Model and Risk Analysis of Arsenic, well Depth, and well yield in the Fairbanks Area, Alaska. MS thesis. University of Alaska, Fairbanks, AK, USA.

Weber D. (2007). Arsenic-tainted site bought for school: seminole officials will have contaminated soil removed. A Sanford-area high school is planned. Orlando Sentinel, 14 April. http://www.highbeam.com/doc/1G1-162047229.html (accessed 6 July 2012).

Webster J. G. (1990). The solubility of As2S3 and speciation of As in dilute and sulphide-bearing fluids at 25 and 90 °C. *Geochimica et Cosmochimica Acta*, **54**, 1009–1017.

Webster J. G. (1999). The source of arsenic (and other elements) in the Marbel-Matingao river catchment, Mindanao, Philippines. *Geothermics*, **28**(1), 95–111.

Weeger W., Lièvremont D., Perret M., Lagarde F., Hubert J. C., Leroy M. and Lett M. C. (1999). Oxidation of arsenite to arsenate by a bacterium isolated from an aquatic environment. *Biometals*, **12**, 141–149. Abstract: http://www.ncbi.nlm.nih.gov/pubmed/10406083 (accessed 10 July 2012).

Weisgerber G. (1993). Quarzit, Feuerstein, Hornstein, Jaspis, Ocker – mineralische Rohstoffe der Steinzeit. In: Alter Bergbau in Deutschland, H. Steuer and U. Zimmermann (eds), Nikol Verlagsgesellschaft, Stuttgart, Germany, pp. 24–34.

Weitz and Luxenberg. (2007). Cedar Lake, Indiana, to get help for arsenic in water. http://www.weitzlux.com/Arsenic/Cedar%20Lake/in_945349.html (accessed 6 July 2012).

Welch A. H. and Lico M. S. (1998). Factors controlling As and U in shallow ground water, southern Carson Desert, Nevada. *Applied Geochemistry*, **13**, 521–539.

Welch A. H., Helsel D. R., Focazio M. J. and Watkins S. A. (1999). Arsenic in ground water supplies of the United States. In: Arsenic Exposure and Health Effects, W. R. Chappell, C. O. Abernathy and R. L. Calderon (eds), Elsevier, Amsterdam, Netherlands, pp. 9–17.

Welch A. H., Lico M. S. and Hughes J. L. (2000). Arsenic in ground water of the western United States. *Ground Water*, **26**(3), 333–347.

Welch A. H., Watkins S. A., Helsel D. R. and Focazio M. J. (2000). Arsenic in ground water of the United States: occurrence and geochemistry. *Ground Water*, **38**(4), 589–604.

Wells B. R. and Gilmour J. T. (1977). Sterility in rice cultivars as influenced by MSMA rate and water management. *Agronomy Journal*, **69**, 451–454.

Wells J. D. and Mullens T. E. (1973). Goldbearing arsenian pyrite determined by microprobe analysis, Cortez and Carlin gold mines, Nevada. *Economic Geology*, **68**, 187–201.

Westerhoff P., Esparza-Soto M., Mata P. C., Parry W. T. and Johnson W. P. (2005). Drinking Water Quality in the US-Mexico Border Region. Final report to Southwest Center for Environmental Research and Policy (Project No. W-030–19).

Westjohn D. B., Kolker A., Cannon W. F. and Sibley D. F. (1998). Arsenic in ground water in the 'Thumb Area' of Michigan. The Mississippian Marshall Sandstone revisited. In: Michigan: Its geology and geologic resources, 5th symposium. East Lansing, MI, USA, pp. 24–25 (abstract).

Wetang'ula G. and Snorrason S. (2003). Ecological risk assessment of Nesjavellir co-generation plant wastewater disposal on Lake Thingvellavatn, SW Iceland. In: Proceedings of the International Geothermal Conference. Reyjavik, Iceland, September.

White R. N., Smith J. V. and Spears D. A. (1989). Analysis of iron sulfides from UK coal by synchrotron radiation X-ray fluorescence. *Fuel*, **68**, 1480–1486.

White N. C. (2006). Gold-bearing mesothermal veins from the Gubong Mine, Cheongyang gold district, Republic of Korea: fluid inclusion and stable isotope studies. *Economic Geology*, **101**(4), 883–901.

White D. E., Hem J. D. and Waring G. A. (1963). Chemical composition of sub-surface waters. In: Data of Geochemistry, M. Fleischer (ed.), 6th edn. United States Geological Survey (USGS) Professional Paper 440-F. United States Department of the Interior, USGS National Center, Reston, VA, USA.

WHO (World Health Organization). (2000). Arsenic. In: *Air quality guidelines for Europe*, WHO (ed.), 2nd edn. WHO Regional Office for Europe, Copenhagen, pp. 125–128.

WHO (World Health Organization) (2004a). Iraq humanitarian Assistance Report. WHO, Geneva, Switzerland, 5 April.

WHO (World Health Organization). (2004b). Safety Evaluation of Certain Food Additives and Contaminants. WHO Food Additives Series No. 52. WHO, Geneva, Switzerland.

WHO (World Health Organization). (2005). Mongolia, Environmental Health Country Profile. WHO, Geneva, Switzerland.

WHO (World Health Organization). (2008). Guidelines for Drinking-Water Quality, 3rd edn. incorporating 1st and 2nd addenda. Vol. 1. Recommendations. WHO, Geneva, Switzerland. pp. 306–308b.

WHO (World Health Organization). (2011a). Guidelines for Drinking Water Quality, 4th edn, WHO, Geneva.

WHO (World Health Organization). (2011b). Arsenic in Drinking Water: Background Documents for the Development of the Guidelines for Drinking Water Quality. 4th edn., WHO, Geneva. http://www.who.int/water_sanitation_health/dwq/chemicals/arsenic.pdf (accessed 24 June 2012).

WHO (World Health Organization), UNICEF (United Nations Children's Fund) (2006). Meeting the MDG Drinking Water and Sanitation Target: The Urban and Rural Challenge of the Decade. WHO and UNICEF, Geneva.

Wiechula D., Jurkiewicz A. and Loska K. (2003). Arsenic content in the femur head of the residents of southern and central Poland. *Biological Trace Element Research*, **92**(1), 17–26.

Wiertz J. and Rozas I. (1996). Technological alternatives for the reduction of arsenic emission in Chilean copper smelters. In: Proceedings of the III International Conference on Clean Technology. M. A. Sanchez, F. Vergard and S. H. Castro (eds), Department of Metallurigical Engineering, University de Concepcion. http://www2.udec.cl/~msanchez/index_arq/book10.html (accessed 24 June 2012)

Wikipedia. (2007b). Chongqing. In: Wikipedia. http://en.wikipedia.org/w/index.php?title=Chongqing&oldid=131857447 (accessed 6 July 2012).

Wikipedia. (2007c). Gansu. In: Wikipedia. http://en.wikipedia.org/w/index.php? (accessed 6 July 2012).

Wikipedia. (2007d). Guizhou. In: Wikipedia. http://en.wikipedia.org/w/index.php? (accessed 6 July 2012).

Wikipedia. (2007e). Hebei. In: Wikipedia. http://en.wikipedia.org/w/index.php?title=Hebei&oldid=131305845 (accessed 6 July 2012).

Wikipedia. (2007f). Heilongjiang. In: Wikipedia. http://en.wikipedia.org/w/index.php?title=Heilongjiang& oldid=129414786 (accessed 6 July 2012).

Wikipedia. (2007g). Jiangsu. In: Wikipedia. http://en.wikipedia.org/w/index.php?title=Jiangsu&oldid=131984514 (accessed 6 July 2012).

Wikipedia. (2007h). Jiangxi. In: Wikipedia. http://en.wikipedia.org/w/index.php?title=Jiangxi&oldid=129446429 (accessed 6 July 2012).

Wikipedia. (2007i). Jilin. In: Wikipedia. http://en.wikipedia.org/w/index.php?title=Jilin&oldid=129956442 (accessed 6 July 2012).

Wikipedia. (2007j). Liaoning. In: Wikipedia. http://en.wikipedia.org/w/index.php?title=Liaoning&oldid=129682136 (accessed 6 July 2012).

Wikipedia. (2007k). Ningxia. In: Wikipedia. http://en.wikipedia.org/w/index.php?title=Ningxia&oldid=131619036 (accessed 6 July 2012).

Wikipedia. (2007l). Shaanxi. In: Wikipedia. http://en.wikipedia.org/w/index.php?title=Shaanxi&oldid=132223466 (accessed 6 July 2012).

Wikipedia. (2007m). Shandong. In: Wikipedia. http://en.wikipedia.org/w/index.php?title=Shandong &oldid=119471213 (accessed 6 July 2012).

Wikipedia. (2007n). Shanxi. In: Wikipedia. http://en.wikipedia.org/w/index.php?title=Shanxi&oldid=129745310 (accessed 6 July 2012).

Wikipedia. (2007o). Sichuan. In: Wikipedia. http://en.wikipedia.org/w/index.php?title=Sichuan&oldid=130697974 (accessed 6 July 2012).

Wikipedia. (2007p). Taiwan. In: Wikipedia. http://en.wikipedia.org/w/index.php?title=Taiwan&oldid=132032184 (accessed 6 July 2012).

Wilcox R. (1997). Concentrations of trace elements and other constituents in the Rio Grande and in fish tissue from the vicinity of Albuquerque, New Mexico, 1994–1996. (Abstract from Open–File Report 97–667). United States Department of the Interior, United States Geological Survey. http://nm.water.usgs.gov/publications/abstracts/ofr97–667.html (accessed 6 July 2012).

Wilkie J. A. and Hering J. G. (1998). Rapid oxidation of geothermal arsenic(III) in streamwaters of the eastern Sierra Nevada. *Environmental Science and Technology*, **32**, 657–662.

Williams M. (1997). Mining-related arsenic hazards: Thailand case study. Technical Report WC/97/49. British Geological Survey, Keyworth, Nottingham, UK.

Williams T. M. and Breward N. (1995). Environmental impact of gold and complex sulphide mining (with particular reference to arsenic contamination): project summary report, Report No. WC/95/002. British Geological Survey, Keyworth, Nottingham, UK.

Williams M., Fordyce F., Paijitprapapon A. and Charoenchaisri P. (1996). Arsenic contamination in surface drainage and groundwater in part of the southeast Asian tin belt, Nakhon Si Thammarat Province, southern Thailand. *Environmental Geology*, **27**, 16–33.

Williams P. N., Islam M. R., Adomako E. E., Raab A., Hossain S. A. and Zhu Y. G. (2006). Increase in rice grain arsenic for regions of Bangladesh irrigating paddies with elevated arsenic in groundwaters. *Environmental Science and Technology*, **40**, 4903–4908.

Williams P. N., Price A. H., Raab A., Hossain S. A., Feldmann J. and Meharg A. A. (2005). Variation in arsenic speciation and concentration in paddy rice related to dietary exposure. *Environmental Science and Technology*, **39**, 5531–5540.

Wilson W. (2004). Letter to Senators Allard and Campbell and Representative DeGette, October 8. http://latimes.image2.trb.com/lanews/media/acrobat/2004-10/14647025.pdf (accessed 6 July 2012).

Wilson W. E. (2007). The Shimen mine: Jiepaiyu, Shimen County, Hunan Province, China. *The Mineralogical Record*, **38**(1), 43–53.

Wilson F. H. and Hawkins D. B. (1978). Arsenic in streams, stream sediments, and ground water, Fairbanks area, Alaska. *Environmental Geology*, **2**(4), 195–202.

Wilson S. D., Kelly Walton R., Roadcap George S. and Holm Thomas R. (2001). Distribution of Arsenic in the Mahomet Aquifer of Central Illinois. Poster presented at the Geological Society of America Annual Meeting, Boston, MA, USA, 5–8 November. http://www.isws.illinois.edu/gws/docs/GSA2001.pdf (accessed 6 July 2012).

Winkel L. H. E., Berg M., Stengel C. and Rosenberg T. (2008). Hydrogeological survey accessing arsenic and other groundwater contaminants in the lowlands of Sumatra, Indonesia. *Applied Geochemistry*, **23**(11), 3019–3028.

Winkel L. H. E., Trangb P. T. K., Vi Mai Lanb V. M., Stengel C., Aminia M., Hac N. T., Vietb P. H. and Berg M. (2011). Arsenic pollution of groundwater in Vietnam exacerbated by deep aquifer exploitation for more than a century. *PNAS*, **108**(4). http://www.pnas.org/content/108/4/1246 (accessed 31 March 2012).

Woidich H. and Pfannhauser W. (1979). Schwermetallbelastung der nahrung in Osterreich: Arsen-Genalte in Lenensmitteln und Monatliche Aufnahme. *Deutsch Lebensmittel-Rundschau: Zeitschrift für Lebensmittelkunde und Lebensmittelrecht*, **75**, 190–195.

Wolkersdorfer C. and Bowell R. (eds) (2005). *Contemporary Reviews of Mine Water Studies in Europe*. Springer Verlag, Berlin, Germany (Mine Water and the Environment, 24: Supplementary Material).

Wong C. K., Renshaw C. E., Feng X. and Sturup S. (2002). New Hampshire apple orchards as a source of arsenic contamination. Presented at the American Geophysical Union (AGU), Spring Meeting 2002 (Abstract No. H42A-05). AGU, Washington, DC.

Woo N. C. and Choi M. J. (2001). Arsenic and metal contamination of water resources from mining wastes in Korea. *Environmental Geology*, **40**(3), 305–311.

World Bank. (1990). Towards the Development of an Environmental Action Plan for Nigeria. World Bank, Washington, DC, USA (Report No. 9002-UNI).

World Bank. (2005a). Towards a More Effective Response: Arsenic Contamination of Groundwater in South and East Asian Countries. Vol. **II**. World Bank, Washington, DC, USA (Technical Report 31303).

World Bank. (2005b). On a Proposed Credit in the Amount of SDR 31.4 Million (US$ 45.87 Million Equivalent) to the Socialist Republic of Vietnam for a Red River Delta Water Supply and Sanitation Project in Support of the First Phase of a Rural Water Supply and Sanitation Program. World Bank, Washington, DC, USA, 15 August (Project Appraisal Document).

World Bank, Pakistan Ministry of Environment. (2005). Strategic country environmental assessment. National stakeholder conference document, 9 May. http://siteresources.worldbank.org/SOUTHASIAEXT/Resources/Publications/448813-1188777211460/pakceavolume1.pdf (accessed 6 July 2012).

World Facts. (2007). Facts about Latvia. http://worldfacts.us/Latvia.htm (accessed 6 July 2012).

World Gazetteer. (2005). Kordestān: agglomérations. http://www.gazetteer.de/wg.php?x=1133288393&men=gcis&lng=en&gln= xx&des=gamelan&dat=200&geo=-106&srt=pnan&col=aohdqcfbeimg&pt=a&va=x&geo=-1920 (accessed 6 July 2012).

World Gazeteer. (2012). Enugu. http://www.world-gazetteer.com/wg.php?men=gpro&des=gamelan&geo=346806750; (accessed 6 July 2012).

World News Network. (2007). Limpopo population. World News Nework. http://population.wn.com/ (accessed 6 July 2012).

World Rainforest Movement. (2007). Ecuador: the Hoaorani people of the Amazonia, self-isolation and forced contact. http://www.wrm.org.uy/bulletin/87/Ecuador.html (accessed 6 July 2012).

World Resources Institute. (2006). Country Profile: Botswana. Population, Health and Human Well-being. World Resources Institute, Washington, DC, USA.

WRUD (Water Resources Utilization Department). (1999). Ministry of Agriculture and Irrigation Conducted Arsenic Testing in Mid-1999. Ministry of Agriculture and Irrigation, WRUD, Myanmar.

Wu J. C. (2005). The Mineral Industry of the Republic of Korea. United States Department of the Interior, United States Geological Survey. http://minerals.er.usgs.gov/minerals/pubs/country/2004/ksmyb04.pdf.

Wu M.-M., Li K. T., Hwang Y. H. and Chen C.-J. (1989). Dose-response relation between arsenic concentration in well water and mortality from cancers and vascular diseases. *American Journal of Epidemiology*, **130**, 1123–1132. Abstract: http://aje.oxfordjournals.org/content/130/6/1123 (accessed 10 July 2012).

Wyatt C. J., Fimbres C., Romo L., Méndez R. O. and Grijalva M. (1998). Incidence of heavy metal contamination in water supplies in northern Mexico. *Environmental Research*, **76**, 114–119.

Xia Y. J. and Liu J. (2004). An overview on chronic arsenism via drinking water in PR China. *Toxicology*, **198**(1), 25–29.

Xie Z. M., Zhang Y. M. and Naidu R. (2006). Extent and severity of arsenic poisoning in China. In: Managing arsenic in the Environment: From soil to human health, R. Naidu, E. Smith, G. Owens, P. Bhattacharya and P. Nadebaum (eds), CSIRO Publishing, Collinwood, Victoria, Australia, pp. 541–552.

Xiong J. B., Wang W., Fan H. and Wang G. (2007). Arsenite-oxidizing Agrobacterium and arsenic resistant microorganisms of a Chinese coal mine ecosystem. *Sciencepaper Online*. http://wenku.baidu.com/view/953e9cf8fab069dc50220157.html (accessed 6 July 2012).

Xu H., Allard B. and Grimvall A. (1991). Effects of acidification and natural organic materials on the mobility of arsenic in the environment. *Water, Air, and Soil Pollution*, **57**, 269–278.

Xu H. N. and Xu J. L. (1996). Formation and distribution of arsenic abnormality in China. *Soils*, **28**, 80–84.

Xu Z., Yu D., Jing L. and Xu X. (2000). Air pollution and daily mortality in Shenyang, China. *Archives of Environmental Health*, **55**(2), 115–120.

Yamaguchi M. (2003). Water supplies in southern Iraq could be undrinkable. *Environmental News Network*, 17 April 2003. http://www.uswaternews.com/archives/arcglobal/3watsup5.html (accessed 6 July 2012).

Yan Y.-H., Chen H.-F., Li C.-H., Hua Z.-L. and Xing F.-W. (2004). The ferns in arsenic-rich areas in Shimen County of Hunan Province: vegetation restoration in polluted environment in arsenic-rich areas. *Journal of Jishou University (Natural Science Edition)*. Abstract: http://www.cnki.net/kcms/detail/detail.aspx?dbcode=cjfq&dbname=cjfq2004&filename=jsdn200403006&uid=&p= (accessed 10 July 2012).

Yang R., Chang Y. and Sun X. (1998). Resource evaluation and development prospect of heavy oil and tar sands in northwestern margin of Junggar *Basin. No. 1998/200*. http://www.oildrop.org/Info/Centre/Lib/7thConf/19980200.pdf.

Yao Z. (1990). Tectonic paleogeographic control of sediment-reformed gold deposits in China. *Bulletin of the Nanjing Institute of Geology and Mineral Resources, Chinese Academy of Geological Sciences*, **11**(2), 87–93 (in Chinese).

Yarling M. (1992). Anomalous Concentrations of Arsenic in the Groundwater at Wakarusa, Indiana: A by-Product of Chemical Weathering of Shales. Indiana Department of Environmental Management, Indianapolis, Indiana, USA.

Yasuní Rainforest Campaign (2005). Huaorani Elect New Anti-Oil Officer to ONHAE, the Huaorani Government. Yasuní Rainforest Campaign News, 10 October 2005. http://saveamericasforests.org/Yasuni/News/Articles/10-10-05%20Huaorani%20Elect%20Anti-Oil%20Government.htm (accessed 6 July 2012).

Yasuní Rainforest Campaign News (2007). President Palacio declares Intangible Zone in Yasuni. *Yasuní Rainforest Campaign News*, 9 January 2007. http://saveamericasforests.org/Yasuni/News/Articles/2007/1-9-07%20YRC-ZonaIntangibleNewsStory.htm (accessed 6 July 2012).

Yasynuvata city administration (2012). Yasynuvata city administration website. http://ya.dn.ua/.

Yilmaz H. (2007). Stream sediment geochemical exploration for gold in the Kazdag Dome in the Biga peninsula, Western Turkey. *Turkish Journal of Earth Sciences*, **16**, 33–55. http://mistug.tubitak.gov.tr/bdyim/abs.php?dergi=yer&rak=0612–4 (accessed 6 July 2012).

Yoshike Y. (1993). Variation in chemical composition of the Tamagawa Hot Springs, Akita Prefecture (1912–1991). *Journal of the Balneological Society of Japan*, **43**(2), 6–67.

Yount J., Breit G., Sopheap S., Moniphea L., Mengieng U., Kagna O., Sampson M. and Roseboom J. W. (2005). Geologic controls on groundwater arsenic distribution in the upper Mekong floodplain, Kien Svay region of Cambodia [poster]. *Geological Society of America, Abstracts with Programs*, **37**(7), 376 (Paper No. 164–4; https://gsa.confex.com/gsa/2005AM/finalprogram/abstract_94468.htm (accessed 6 July 2012).

Yu M., Qin X., Wang S., Zhang L., Li X, Zuo Z., Zhang X. and Wang L. (2009). Child skeletal fluorosis from indoor burning of coal in southwestern China. *Journal of Environmental Public Health*. http://www.ncbi.nlm.nih.gov/pubmed/20041010 (accessed 6 July 2012).

Yu G. Q., Sun D. J. and Zheng Y. (2007). Health effects of exposure to natural arsenic in groundwater and coal in China: an overview of occurrence. *Environmental Health Perspectives*, **115**(4), 636–642.

Yuan H. (2006). China's Hunan shuts 32 plants after toxic river spill (Update 1). Bloomberg.com, 14 September. http://www.bloomberg.com/apps/news?pid=newsarchive&sid=avR5fDJ2Pbm4&refer=asia (accessed 6 July 2012).

Yudovich Y. E. (2003). Notes on the marginal enrichment of germanium in coal beds. *International Journal of Coal Geology*, **56**, 223–232.

Yudovich Y. E. and Ketris M. P. (2005). Arsenic in coal: a review. *International Journal of Coal Geology*, **61**, 141–196.

Yudovich Y. E., Ketris M. P. and Merts A. V. (1985). Trace Elements in Fossil Coals. [Redkiye elementy v iskopayemykh ugliakh]. Nauka Science Publication House, Leningrad, USSR (in Russian).

Zakaznova-Iakovleva V. P., Seward T. M. and Suleimenov O. M. (2000). Spectrophotometric determination of the first ionization constant of H_3AsO_3 from 25 to 300°C. In: Steam, Water and Hydrothermal Systems: Physics and Chemistry Meeting the Needs of Industry. Proceedings of the 13th international conference on the properties of water and steam, P. R. Tremaine, P. G. Hill, D. E. Irish and P. V. Balakrishnan (eds), NRC Research Press, Ottawa, Ontario, Canada, National Research Council Canada, pp. 694–695.

Zaldivar R. (1974). Arsenic contamination of drinking water and foodstuffs causing endemic chronic poisoning. *Beiträge zur Pathologie*, **151**, 384–400.

Zaldivar R. (1977). Ecological investigations on arsenic dietary intake and endemic chronic poisoning in man: dose response curve. *Zentralblatt für Bakteriologie, Parasitenkunde, Infektionskrankheiten und Hygiene. Erste Abteilung Originale. Reihe B: Hygiene, Präventive Medizin*, **164**, 481–484.

Zdrazil J. and Picha F. (1966). The occurrence of the carcinogenetic compounds 3,4-benzpyrene and arsenic in the soil. *Neoplasma*, **13**, 49–55.

Zeng R., Zhuang X., Koukouzas N. and Xu W. (2005). Characterization of trace elements in sulphur-rich Late Permian coals in the Heshan coal field, Guangxi, south China. *International Journal of Coal Geology*, **61**, 87–95.

Zhang H. H. (2006). Spatial distribution and vertical variation of arsenic in Guangdong soil profiles, China. *Environmental Pollution*, **144**(2), 492–499.

Zhang L. and Chen C. (1997). Geographical distribution and exposure population of drinking water with high concentration of arsenic in China. *Wei Sheng Yan Jiu*, **26**(5), 310–313 (in Chinese).

Zheng B., Ding Z., Huang R., Zhu J., Yu X., Wang A., Zhou D., Mao D. and Su H. (1999). Issues of health and disease relating to coal use in southwestern China. *International Journal of Coal Geology*, **40**, 119–132.

Zhang A., Feng H., Yang G., Pan X., Jiang X., Huang X., Dong X., Yang D., Xie Y. and Peng L. (2007). Unventilated indoor coal-fired stoves in [Jiaole township] Guizhou Province, China: cellular and genetic damage in villagers exposed to arsenic in food and air. *Environmental Health Perspectives*, **115**, 653–658.

Zhang J., Ren D., Zheng C., Zeng R., Chou C.-L. and Liu J. (2002). Trace Element Abundances in Major Minerals of Late Permian Coals from Southwestern Guizhou Province, China. *International Journal of Coal Geology*, **53**(1), 55–64.

Zhou D. X. (1993). Investigation of chronic arsenic poisoning caused by high arsenic coal pollution. *Chinese Journal of Preventive Medicine*, **27**(3), 147–150 (in Chinese). Abstract: http://www.ncbi.nlm.nih.gov/pubmed/8243176 (accessed 10 July 2012).

Zhou D. X., Zhou Y. S., Zhou C., Jin D. X., Peng J. H., Lu X. Z., Zhu S. L., Liu D. N., Lu X. Z. and Zheng B. S. (1994). Correlation of total arsenic intake and incidence of arsenic poisoning in coal type endemic area. *Chinese Journal of Epidemiology*, **13**, 215–218 (in Chinese).

Zhou Y. S., Zhou D. X. and Zhu S. L. (1994). Investigation of arsenic content in food dried by normal coal and high arsenic-containing coal. *Chinese Public Health*, **10**, 77–79 (in Chinese).

Zhu Y. (1998). Spatial effects of 'informal urbanization' in China: the case of Fujian Province. *Asia-Pacific Population Journal*, **13**(1).

Zotov A. L., Kudrin A. V., Levin K. A., Shikina N. D. and Varyash L. N. (1994). Experimental studies of the solubility and complexing of selected ore elements (Au, Ag, Cu, Mo, As, Sb, Hg) in aqueous solutions. In: Fluids in the Crust, K. I. Shmulovich, B. W. D. Yardley and G. G. Gonchar (eds), Chapman Hall, London, UK.

Zulkifli S. Z., Mohamat-Yusuff F., Arai T., Ismail A. and Miyazaki N. (2010a). An assessment of selected trace elements in intertidal surface sediments collected from the Peninsular Malaysia. *Environmental Monitoring Assessment*, **169**(1–4), 457–472.

Zulkifli S. Z., Ismail A., Mohamat-Yusuff F., Arai T. and Miyazaki N. (2010b). Johor strait as a hotspot for trace elements contamination in peninsular Malaysia. *Bulletin of Environmental Contamination and Toxicology*, **84**(5), 568–573.